PRINCIPLES OF ELECTRON OPTICS

Volume 3

WAVE OPTICS

P. W. HAWKES
CNRS Laboratory of Electron Optics
Toulouse, France

and

E. KASPER
Institut für Angewandte Physik
Universität Tübingen, Germany

ACADEMIC PRESS
Harcourt Brace & Company, Publishers
LONDON   SAN DIEGO   NEW YORK
BOSTON   SYDNEY   TOKYO   TORONTO

This book is printed on acid-free paper

ACADEMIC PRESS LIMITED
24–28 Oval Road
London NW1 7DX

United States Edition published by
ACADEMIC PRESS INC.
San Diego, CA 92101

**A catalogue record for this book is available from
the British Library**

ISBN 0-12-333343-1

In memory of my son, Sebastian Hawkes

Printed in Great Britain by St Edmundsbury Press Limited,
Bury St Edmunds, Suffolk.

Principles of
Electron Optics:
Wave Optics

Contents of Volume 3
Wave Optics

Contents

PART XII – ELECTRON INTERFERENCE AND ELECTRON HOLOGRAPHY

PART XIII – THEORY OF IMAGE FORMATION

PART XVI – COHERENCE, BRIGHTNESS AND SPECTRAL FUNCTIONS

Contents of Volume 1
Basic Geometrical Optics

Contents of Volume 2
Applied Geometrical Optics

Preface

The last attempt to cover systematically the whole of electron optics was made by the late Walter Glaser, whose *Grundlagen der Elektronenoptik* appeared in 1952; although a revised abridgement was published in the *Handbuch der Physik* four years later, we cannot but recognize that those volumes are closer to the birth of the subject, if we place this around 1930, than to the present day.

The difference between Glaser's work and our own is much greater in the present volume than in the two volumes on geometrical optics, for whole branches of the subject have come into being since 1956. The representation of the image-forming process by transfer functions has yielded a much deeper understanding of the notion of resolution. The development of highly coherent light and electron sources has made holography possible and the invention of the electron biprism has rendered it practical. The widespread availability of large fast computers and the gradual introduction of microscope–computer links, as well as the peculiarities of the electron image-forming process, have generated considerable interest in digital image processing. Lastly, a theoretical advance in optical coherence theory clarifies some obscure points in the related electron optical theory.

Only a very small fraction of the present text finds a counterpart in Glaser's work, therefore, and Parts XII–XVI are entirely new. Part XI, on the basic wave-mechanical formalism, follows the original work of Walter Glaser and Peter Schiske quite closely, however, for there is little new to report there apart from the investigations based on the Dirac equation.

Like Volumes 1 and 2, this work is intended to be both a textbook and a source-book. The fundamentals of the topics covered are presented in detail but the reader who wishes to go more deeply into a particular subject will need to examine the original articles, review articles and more specialized textbooks. This is particularly true of themes that are closer to electron microscopy than to electron optics and of image processing. For the former, we have included two chapters on beam–specimen interactions but these are intended merely to initiate the reader into the beginnings of this vast subject. For the latter, and despite the length of Part XV, we have had to be selective and refer frequently to other texts in this field. Furthermore, image processing is a field in rapid growth and the reader will need to complement our account of the fundamentals with the contents of conference proceedings and the journals that specialize in these themes.

The earlier work on wave optics is all brought together in Glaser's two great texts of 1952 and 1956. Of the other earlier works on electron optics, only those of Picht (1939, 1957, 1963) and de Broglie (1950) devote much space to the subject. The subjects of Parts XII–XVI are all more recent and no single volume attempts to cover them all. Specialized texts do of course deal with their particular topics in much more detail than we can and we refer to—and lean on—these in the relevant chapters. Only Part XIV has generated a classic textbook, *Electron Microscopy of Thin Crystals* by Hirsch *et al.* (1965), the record of courses given at a summer-school in 1963. Once again, therefore, we can say that, although "standing on the shoulders of giants", the present volume differs from other books on the themes discussed here in that developments of the past thirty years are set out in detail and in a uniform presentation.

For whom is this work intended? A knowledge of physics and mathematics to first degree level is assumed, though many reminders and brief recapitulations are included. It would be a suitable background text for a post-graduate or final year course in electron optics or electron image formation or electron image processing and some of the material has indeed been taught for some years in the Universities of Tübingen and Toulouse. Its real purpose, however, is to provide a self-contained, detailed and above all modern account of wave-mechanical electron optics, with image formation, interference, holography and coherence as the principal examples of application. This is complemented by a survey of the ways in which digital image processing can be made to participate in image interpretation and microscope control. The basic equations are given, the applications are discussed at length and ample guidance to the related literature is provided.

Composition of a volume such as this puts us in debt to a host of colleagues: many have permitted us to reproduce their work and have often provided illustrations; the librarians of our institutes and the Librarian and Staff of the Cambridge Scientific Periodicals Library have been unflagging in their pursuit of recondite and elusive early papers; Mrs Ströer has uncomplainingly word-processed hundreds of pages of mathematical and technical prose, aided by Mrs Lannes and Mrs Davoust; most of the artwork has also been produced by Mrs Ströer, using computer software rather than the drawing pen; the remainder has been prepared by Miss Quessette and Mr Caminade; the references have been typed by Mrs Bret and her assistants; Mr Aussoleil and Mrs Altibelli have provided computer expertise; Academic Press has generously supported the production costs. To all of these we are extremely grateful. We also thank the many authors and publishers who have been good enough to allow us to reproduce published drawings. The details are as follows.

AKADEMIE VERLAG: Fig. 63.9 (T. Leuthner *et al.*) from *Physica Status Solidi*.

AMERICAN PHYSICAL SOCIETY: Fig. 72.2 (J.-i. Matsuda *et al.*) from *Review of Scientific Instruments*.

ELSEVIER, NORTH-HOLLAND PUBLISHING CO.: Figs 63.6 (D. Joy *et al.*) and 66.15 (K. Ishizuka) from *Ultramicroscopy*.

JAPANESE JOURNAL OF APPLIED PHYSICS: Fig. 54.3 (H. Tomita *et al.*).

JAPANESE SOCIETY OF ELECTRON MICROSCOPY: Fig. 63.7 (A. Tonomura) from *Journal of Electron Microscopy*.

PHYSICAL SOCIETY OF JAPAN: Fig. 54.1 (A. Tonomura and H. Watanabe) from *Nihon Butsuri Gakkai-shi*; Figs 62.11, 63.1 and 63.8 (A. Tonomura *et al.*) from *Proceedings of the Second International Symposium on the Foundations of Quantum Mechanics in the Light of New Technology*.

PRENTICE–HALL: Fig. 76.10 from *Syntactic Pattern Recognition* by K.S. Fu.

ROYAL MICROSCOPICAL SOCIETY: Figs 71.6 (R.E. Burge *et al.*) and 77.3 (W.O. Saxton *et al.*) from *Journal of Microscopy*.

SCANNING MICROSCOPY INTERNATIONAL: Fig. 75.6 (W.O. Saxton) from Supplement 6 to *Scanning Microscopy*; Fig. 76.4 (C.E. Fiori *et al.*) from *Scanning Electron Microscopy*.

SOCIÉTÉ FRANÇAISE DE MICROSCOPIE ELECTRONIQUE: Fig. 62.5 (G. Möllenstedt) from *Journal de Microscopie*.

SPRINGER–VERLAG: Fig. 54.2 (G. Möllenstedt and H. Wahl) from *Die Naturwissenschaften*; Figs 60.4 (F. Lenz), 60.9 (F. Lenz and E. Krimmel), 61.5 (G. Möllenstedt and H. Düker) and 62.9 (W. Bayh) from *Zeitschrift für Physik*.

TAYLOR & FRANCIS: Fig. 66.11 (W.B. Krakow *et al.*) from *Philosophical Magazine*; Figs. 78.4–7 (J.T. Foley and E. Wolf) from *Journal of Modern Optics*.

WILEY–LISS: Fig. 75.8 (F. Zemlin *et al.*) from *Journal of Electron Microscopy Techniques*.

WISSENSCHAFTLICHE VERLAGSGESELLSCHAFT: Fig. 77.2 (W.O. Saxton) from *Optik*.

> *For while the tired waves, vainly breaking,*
> *Seem here no painful inch to gain,*
> *Far back through creeks and inlets making*
> *Comes, silent, flooding in, the main.*

54

Introduction

54.1 Organization of the subject

The behaviour of beams of free electrons, released from a source and propagating through a vacuum region in some device, is of interest in many diverse fields of instrumentation and technology. The study of such beams forms the subject of electron optics, which divides naturally into geometrical optics, where effects due to wavelength are neglected, and wave optics, where these effects are considered. Volumes 1 and 2 were devoted to geometrical optics. This final volume is concerned with wave optics. A knowledge of this branch of the subject is essential in microscopy, to understand the propagation of electrons from the source to the specimen, through the latter and from it to the image plane of the instrument. It is also needed to explain all interference phenomena, notably holography, and in formal coherence theory.

The various branches of the subject have reached different degrees of sophistication. The laws that govern wave propagation are closely analogous to those already familiar in light microscopy, provided that electron spin is neglected, and can be regarded as well-established. Some of the applications are, conversely, in rapid evolution and new developments are to be anticipated. We have therefore concentrated on the principles, which should remain largely unaffected by the passage of time. We have not, of course, neglected their practical exploitation, in holography for example, or in image formation and processing, and the lifetime of the corresponding sections will no doubt prove limited.

We shall not repeat here the general remarks on the classification of electron optical studies to be found in the Preface to Volumes 1 and 2. Our theme in this third volume is the study of electron propagation through static electric or magnetic fields, including those inside specimens, based on Schrödinger's equation. This is complemented by a Part on digital image processing which, if not in the main line of electron optics, is an inevitable preoccupation of anyone concerned with electron microscope imaging.

The wave theory of electron optics is founded on the Dirac equation but, in practice, it is almost always permissible to replace this by the relativistic form of Schrödinger's equation, for spin is negligible except in a few very specialized situations. The book therefore opens (Part XI) with an account of the relevant material from *quantum mechanics*. A paraxial form of Schrödinger's equation is derived in Chapter 58, which enables us to understand elementary image formation in terms of the wavefunction. In the last two chapters (59 and 60), the laws of diffraction and interference are studied.

In the remainder of the book, the laws of propagation established in Part XI are applied to a variety of different situations, directly in the case of Parts XII, XIII, XIV and XVI, indirectly in the case of Part XV on image processing. *Interference effects* are the subject of Part XII, which is divided into *interferometry* and *holography*. The distinction between the two is not sharp but, in interferometry, we are not concerned with fine diffraction effects in the specimen, whereas in holography it is precisely these effects that render the technique valuable. Holography is an example of a topic of which our account cannot be definitive, for the techniques are just beginning to emerge from the confines of a few research laboratories and, even there, have by no means attained their full potential. We therefore present only the fundamentals of the procedures that at present seem the most promising, but the situation may well change.

The next Part, which fills a substantial fraction of the book, is devoted to *image formation*, in the transmission electron microscope and, more briefly, in the scanning transmission electron microscope (STEM). Here, the relation between the intensity at the image and the wavefunction at the specimen is explored in great detail and the linear theory that is applicable to a certain class of specimen is presented at length. The effects of source-size and energy spread are examined, as are less conventional imaging modes, using tilted or hollow-cone illumination in particular. The chapter on the STEM concentrates on the differences between this instrument and the conventional microscope, notably, the possibility of controlling the detector response, either by configuring the detector surface or by recording the two-dimensional signal generated by each specimen-element as though it were an image and combining the intensity values of this image in any way that seems helpful. We also draw attention to the information that can be obtained about crystalline specimens when the area illuminated coherently by the probe is appreciably smaller than the unit cell.

Part XIV is a brief reminder of the ways in which the propagation of the electron wavefunction through the *specimen* is analysed. Superficial though this presentation inevitably is, for the subject is not central to the theme of the book, we felt that some account of this material was indispensable, for without it certain notions introduced elsewhere, the specimen transparency for example, would remain mysterious. The theory is presented separately for *amorphous* and *crystalline* specimens, for the collective effects in the latter require us to analyse them in terms of concepts totally inappropriate to amorphous materials. We do insist, however, that this Part can do no more than bridge the gap between our detailed presentation of imagery and other specialized texts on the microscopy of specimens of a particular kind.

In Part XV, we turn to *digital image processing*. We cover, even if unevenly, the whole field of image processing and make no apology for including this in a book on electron optics for much of the material presented has been a major preoccupation of microscopists over the years: the work on the phase problem and that on three-dimensional reconstruction are obvious examples and the current studies that aim to use all the information from every object-element in the STEM, so that the image becomes four-dimensional, provide an even more persuasive justification. At a humbler level, image enhancement has been practised in scanning electron microscopy since the earliest days of the instrument. Image processing divides naturally into four large sections: *acquisition, sampling, quantization and coding; enhancement; restoration;* and *image analysis.* We have adopted these divisions, adding to them a chapter on *instrument control* and on the measurement of microscope *operating parameters*, during image acquisition in particular. We have also included a short introduction to *image algebra* for, although this subject is too young to have had much impact on electron image processing as yet, we anticipate that some familiarity with it will be required to read the image processing literature of the future. In this Part, we describe many of the procedures that are used to improve images in some way or render them more informative; in particular, we devote considerable space to the ideas of mathematical morphology, which are already important in scanning electron microscopy, to the work on the phase problem and to three-dimensional reconstruction.

The book concludes with a Part of a rather theoretical nature devoted to *coherence* and in particular, to the relation between coherence and *radiometry* (Chapter 78). A short chapter is also devoted to instrumental aspects of coherence, notably the effect of partial coherence on image formation in terms of the transmission cross-coefficient. The discussion of the various brightness functions in Chapter 78 is inspired by the work of Wolf and his school, who were concerned with light sources. The translation to electron sources is, however, immediate, since the latter are quasi-monochromatic and only the spatial partial coherence raises problems: the contributions from different wavelengths can safely be added 'incoherently'. Nevertheless, some questions still remain without a fully satisfactory answer; in particular, we have preferred to describe the work of Agarwal *et al.*

(1987) out of context in Section 78.10.1, for although it provides a transparent gateway between light and electron optics, some further elucidation is required before we dare pass through.

Most aspects of wave electron optics have thus been covered here, some more thoroughly than others. The emphasis throughout is on physical principles and on their theoretical formulation while technical details of microscopes or ancillary equipment are kept to a minimum. Even more than in Volumes 1 and 2, the inclusion of such details would not only have rendered the book unreasonably large but would also have shortened its useful lifetime, for there are few branches of the subject that are not in rapid development, especially in the fields of holography, high-resolution imagery, STEM imagery and image processing, which fill so many of its pages.

Again as in the earlier volumes, we have adopted a compromise towards the voluminous literature of the subject, though the decisions were hard to take for no single attempt to treat the subjects of this book has hitherto been made. The coverage of the literature is therefore quite full for all Parts except that on interactions in thin specimens, where we list only the seminal papers and refer to the specialized textbooks for more information; there are many excellent titles available here. There are a few deliberate omissions in other Parts too. In the sections in which the Aharonov–Bohm effect is examined, we have preferred to refer to the book by Peshkin and Tonomura (1989) than to list all the contributors to the lively but ultimately bootless controversy that raged around this phenomenon. Even in the chapter on holography, the listing of references is far from exhaustive for several very complete review articles have been devoted to this subject.

We conclude this introductory chapter with a brief historical account of the various themes of this volume. Even more than in Volumes 1 and 2, one is struck by the way in which the human mind invents tools that permit or encourage it to conceive original ideas: the deficiencies of the microscope objective lens, which Scherzer showed by theory to be intrinsic, inspired Gabor to invent holography, long before it could work. The field-emission gun and the laser rendered experimental tests possible but physical limitations made digital rather than optical reconstruction essential—if the computing power needed for digital image processing had not been available, holography would again have been abandoned. We could trace a similar evolution, original thinking alternating with experimental progress, in many fields: in the history of the STEM from the use of a field-emission gun to the exploitation of Rutherford scattering and of the far-field diffraction pattern generated by each object-element, for example, or in image simulation or again in the phase problem. Electron optics is an art, not a science.

54.2 History

The most important step in wave electron optics was taken before geometrical optics came into being, for the notions of electron frequency and subsequently wavelength were introduced by Louis de Broglie in a series of notes to the Académie des Sciences de Paris in 1923, which preceded the full account in his thesis of 1924, published in 1925. When Ernst Ruska learnt of de Broglie's ideas several years later, in 1932, his first reaction was discouragement: "I have a lively memory even today [1979] of the first discussion between Knoll and myself about this new kind of wave, for I was, at the time, extremely disappointed that once more a wave phenomenon would limit the resolution." This feeling was fortunately short-lived: "I was immediately heartened, though, directly I had satisfied myself with the aid of the de Broglie equation, that the waves must be around five orders of magnitude shorter in wavelength than light waves." (Ruska, 1979, 1980).

The pre-war years are characterized by a flurry of activity in electron diffraction but only isolated publications appeared on electron image formation in which the wavelength was not neglected. The main stages in the understanding of electron diffraction are well-known, notably the classic papers of Clinton J. Davisson and Lester H. Germer (1927), George Paget Thomson and Alexander Reid (1927), Thomson (1927, 1928), Seichi Kikuchi (1928) and Nishikawa and Kikuchi (1928); their ideas were taken up in many laboratories and this period can be pleasurably explored with the aid of the many historical essays and reminiscences collected by Goodman (1981). The books of Pinsker (1949, 1953) and Vainshtein (1956, 1964) are useful for tracing the early Russian studies. A paper by Elsasser, published as early as 1925, foreshadowed later developments and although it has been thought that "no notice seems to have been taken of the Elsasser letter, and that in particular it had no influence on the work of either Davisson or G.P. Thomson" (Blackman, 1978), it is included in the historical section of the long account of particle diffraction by Frisch and Stern (1933). The papers by Walter Kossel and Gottfried Möllenstedt (1938, 1939, 1942; Möllenstedt, 1941; Kossel, 1941) on convergent-beam electron diffraction, which inspired the work of Carolina MacGillavry (1940a,b) on the underlying theory, are particularly relevant for modern electron microscopy (Bristol, 1984; Tanaka and Terauchi, 1985; Tanaka et al., 1988). Möllenstedt's recollections of that period may be read in his paper of 1989.

In electron image formation, the high points are the realization that electron image contrast is due not to absorption but to scattering, the recognition that the haloes around specimens are Fresnel fringes and the appreciation that a diffraction pattern is formed in the back-focal plane of the microscope (for a parallel beam incident on the specimen).

There is some uncertainty in the literature as to who first became aware that contrast is due to scattering. Ladislaus Marton believed that "the first clear statement ascribing the contrast in the image to differences in the scattering properties of the object was contained in [his] paper of 8 May 1934..." (Marton, 1968) and claimed that a statement in a slightly earlier paper of Ruska (1934) showed that the latter had not appreciated this point. Ruska pointed out, however, in a private communication to C. Süsskind (cited by the latter, 1985) that he had in fact given an explicit description of what we now call scattering contrast:

> Wegen der an sich kleinen Absorption sind Absorptionsunterschiede praktisch von geringem Einfluss auf die Bildkontraste. Dagegen ist bei relativ zur Strahlspannung sehr dünnen Objekten trotz praktisch konstanter durchsetzender Strahlstromdichte die Bildhelligkeit deswegen stark verschieden, weil die einzelnen Objektstellen mit entsprechend ihrer Massendicke verschiedener Apertur (Intensitätsverteilung auf die Streuwinkel) strahlen, so daß bei genügend kleinen Spulenöffnungen verschieden große Ströme von den Objektpunkten auszugehen scheinen bzw. in das Bild gelangen. Man könnte im Grenzfall von Absorptions- und Diffusionsbildern sprechen, doch überwiegt in ihrer Wirkung meist die Diffusionserscheinung. Wegen der großen Bedeutung dieser Erscheinung für das Zustandekommen der Bildkontraste hängt der Kontrastreichtum der Bilder so wesentlich von der Apertur der einfallenden Objektstrahlung (Brennweite der Kondensorspule) ab.

Moreover, a careful reading of Marton's paper of 1934 (Marton, 1934a) and in particular of his slightly later paper (1934b) does not clearly support his claim. In the earlier paper, the calculation of scattering angle as a function of thickness is made in order to establish the limiting thickness of the metal support film, the purpose of which was to cool the biological specimen to protect it from damage. The notion of scattering contrast is not mentioned, even indirectly. The later paper states unequivocally: "Nous pouvons donc former les images électroniques de chaque objet qui émet des électrons ou faire traverser des objets par des électrons et ainsi rendre visible les objets par absorption." Two years later, however (Marton, 1936), scattering contrast *is* described: "En traversant l'objet, les électrons sont dispersés dans toutes les directions. A la formation de l'image ne contribueront que ceux qui sont dispersés dans l'angle solide délimité par le diaphragme de l'objectif... Les détails de l'image seront perceptibles, si les quantités d'électrons dispersés dans le diaphragme et provenant de deux points voisins de l'objet sont différentes." Nevertheless, it is agreed that Marton's role in understanding the origin of microscope contrast was a central one: "The main source of contrast was *scattering*, and the first to recognize this fact clearly was Bill Marton", wrote Gabor in 1968 and "Ruska does acknowledge that immediately following his brief reference to these questions [quoted above], Marton pursued them more fully" (Süsskind, 1985). See in particular Marton and Schiff (1941).

Another very relevant event in 1936 was the publication by Hans Boersch of two major papers in which the formation of a diffraction pattern in the back focal plane of the objective was demonstrated and a form of selected-area diffraction was adumbrated. From then on, thanks to the work of Ruska, Marton and Boersch, there could be no doubt that electron image contrast is the result of scattering within the specimen and interception of some of the scattered electrons by the objective aperture. Energy was of course deposited in the specimen by virtue of inelastic scattering, especially in those early years before the development of techniques and instruments for preparing very thin specimens.

The haloes seen around the edges of specimens were interpreted as Fresnel fringes a few years later by Boersch (1940, 1943) and by Hillier (1940).

For our present purposes, the 1940s are the decade in which Walter Glaser, first with Peter Schiske and later with Günther Braun, developed the theory of image formation on the basis of the Schrödinger equation. The result of this work was a detailed account of paraxial electron optics in wave-optical terms and the beginnings of a study of the effect of aberrations and of the source brightness function. Although Glaser failed to take the vital step, the Fourier transformation of his equation representing the convolutional relation between object and image wavefunctions, his wave-optical analysis occupies a central position in the contrast-transfer theory of image formation. This work may be traced through the work of Glaser (1943, 1949, 1950a,b) and of Glaser and Schiske (1953) and Glaser and Braun (1954, 1955).

The problem of microscope resolution was studied at several levels. Aberration correction and improvements in specimen preparation both played an important part, of course, and the former has been examined at length in Volume 2. The possibility of high, even atomic, resolution in the electron microscope was studied by James Hillier (1941) and Leonard I. Schiff (1942) and in the immediate post-war years, Hans Boersch published a series of studies on the possibility of imaging atoms (Boersch, 1946, 1947a,b, 1948). In 1949, Otto Scherzer analysed "The theoretical resolution limit of the electron microscope". This important paper foreshadows the contrast-transfer theory that was introduced twenty years later for the role of the wave aberration is clearly recognized – we meet the formula $\gamma = s^2\Theta^4 - \tau s\Theta^2$, identical apart from notation with our equation (65.30b), for the first time in the electron optical literature though it was of course known in light optics. The optimum value of defocus that we now call Scherzer focus appears there, in the form $2.5(\lambda C_A/2\pi)^{1/2}$, where $C_A \equiv C_s$ and we note that $2.5/(2\pi)^{1/2} \approx 1$. The resolution of individual atoms is again discussed.

One way of correcting the unfortunate effects of spherical aberration was not examined in Chapter 41 however: holography. The in-line form of this was introduced in 1948 by Dennis Gabor in an attempt to remedy the undesirable consequences of this aberration. The early attempt of Michael E. Haine and Tom Mulvey (1952) and later Tadatosi Hibi (1956, recollected in Hibi, 1985) to put this idea into practice fell victim to the relatively large emissive area of the thermionic sources and their equally large energy spread, which impaired the quality of the holograms recorded. But even if highly coherent field-emission sources had been available in 1950, the fact that the laser still lay in the future would have vitiated the reconstruction step. Although holography had to await technological developments and the invention of the maser and its successor, the laser, simpler kinds of electron interference did not. In 1952, Marton had attempted to use a crystal as a beam splitter, capable of producing two beams with a fixed phase difference from a single source (Marton, 1952; Marton *et al.*, 1953); the scattering process in the crystal rendered this way of creating two coherent beams inefficient but, a few years later, Gottfried Möllenstedt and Heiner Düker (1955, 1956, see recollections by Möllenstedt, 1991) split an electron beam with an 'electron biprism', a fine thread held at a positive potential relative to its surroundings. Thereupon, electron interferometry became a subject in its own right. Many references to this work and to the interferometric studies of Charles Fert, Jean Faget and Monique Fagot are to be found in Part XII.

During this same period, a calculation of the elastic and inelastic scattering parameters based on a simple but adequate model of the atomic potential was made by Friedrich Lenz (1953, 1954); his formulae were heavily used at a time when computing power was modest and even today they are regularly employed, when extreme accuracy is not required, to study the relative magnitudes of different cross-sections for example.

A series of experimental studies of image formation with crystalline specimens stimulated theoretical work that is the basis of our understanding of these materials. Among the experimental papers, we single out those of Robert D. Heidenreich (1949, 1951); of James Menter (1956) who obtained the first micrographs of edge dislocations in his work on platinum phthalocyanine using a newly acquired Siemens Elmiskop I; of Hatsujiro

Hashimoto (1954); and of Peter Hirsch *et al.* (1956), in which dislocation glide was reported (see Hirsch, 1980, 1986). Soon after, all the basic
theory of the kinematical theory of diffraction contrast of dislocations in
crystalline specimens was developed (Hirsch *et al.*, 1960) and the dynamical
theory appeared immediately after (Howie and Whelan, 1960, 1961, 1962).
Related studies appeared in the same years (Hashimoto *et al.*, 1960, 1961,
1962).

Meanwhile, an observation that was to have a long-lasting effect on image simulation was made by Alex Moodie in Australia: a chance sighting of
a series of images of wire-mesh covering a window, illuminated by a distant
street-light, led him and John Cowley to develop a theory of 'Fourier images', which subsequently proved to have been noticed by H. Fox Talbot in
1836 and discussed long after by Lord Rayleigh (1881), Weisel (1910) and
Wolfke (1913).* The work on Fourier images may be traced through the
papers of Cowley and Moodie (1957a,b, 1958, 1959, 1960); see also Sanders
and Goodman (1981) and Cowley (1981) as well as Cowley's *Diffraction
Physics* (1975, 1981). From these ideas, the multi-slice method of image
simulation emerged (Cowley and Moodie, 1957a,b); it was successfully put
into practice when the necessary computing power became available (Lynch
and O'Keefe, 1972; O'Keefe, 1973) and improved by Ishizuka and Uyeda
(1977) who exploited the mapping of convolution products into direct products by the Fourier transform.

Further attempts to speed up the calculation were made by Dirk van
Dyck, who introduced a real-space procedure in 1980, subsequently perfected by him in collaboration with Wim Coene (van Dyck and Coene,
1984; Coene and van Dyck, 1984).

The next major development concerns Part XIII on image formation
but first we must return to the 1940s when Fourier optics was born. In
1940, Pierre-Michel Duffieux published two papers on the harmonic analysis of optical images, and a third followed in 1942. These appear to have
passed unnoticed and in France, Duffieux's own presentation of his ideas
was found incomprehensible (Duffieux, 1970). He was urged to write them
out clearly and the result was *L'Intégrale de Fourier et ses Applications à
l'Optique* (1946), produced privately for Duffieux by a Rennes printer. For
some years, his work was little known but in 1959 it was fully described in
Born and Wolf's *Principles of Optics* and in 1960 in *Diffraction, Structure
des Images* by Maréchal and Françon. Soon after, Karl-Joseph Hanszen and
colleagues introduced the idea of characterizing the transfer of information
to the image from the object when the latter scatters weakly by means of

---

* Of these papers, only that of Wolfke is listed by Czapski and Eppenstein (1924), in
the context of Abbe's theory of microscope resolution. The paper by Wolfke is the last of
a series of papers on the imaging of gratings.

a linear transfer theory (Hanszen and Morgenstern, 1965; Hanszen, 1966). The experiments of Friedrich Thon (1966a,b) confirmed the correctness of the theory and drew attention to it vividly, after which it was extended to many other practical situations, tilted and hollow-cone illumination, for example. The effects of source-size and energy spread were likewise explored in great detail, notably by Hanszen and Ludwig Trepte (1971) and by Joachim Frank (1973), later in collaboration with Richard Wade (Wade and Frank, 1977). The arrival of the scanning transmission electron microscope (Crewe et al., 1968) was soon followed by the full study of its image-forming mechanism by Elmar Zeitler and Michael Thomson (1970) and in the language of transfer theory by Harald Rose (1974). The possibility of using the detector response as a free parameter was recognized by Dekkers and de Lang (1974) and the repercussions of this observation are with us still.

The idea that the microscope is characterized by a transfer function gave rise to several suggestions for altering the transfer characteristics of the instrument. An early example is the zone plate suggested by Walter Hoppe in 1961. All these ideas may be traced back to the work of André Maréchal and Paul Croce (1953, see Croce, 1956) on filtering in the light microscope.

The years around 1970 witnessed developments of the highest importance in image processing: the first attempts to exploit the sequential nature of the image-forming process in the scanning electron microscope with a view to improving the image in various ways; the first three-dimensional reconstruction from transmission images; and the iterative solution of the phase problem.

The scanning microscope was first made available commercially in 1965, when the Cambridge Instrument Co. Stereoscan was launched. The first attempts to alter the image contrast electronically were made soon after (MacDonald, 1968, 1969; White et al., 1968) and although specially designed circuits were used for the purpose at that time, these efforts heralded the digital image processing of today.

It was in 1968 that the first three-dimensional reconstructions were performed by David de Rosier and Aaron Klug. Although this first reconstruction leaned heavily on the known symmetry of the specimen (tail of the bacteriophage T4), the general reconstruction was set out in full and even the possibility of reconstruction from a "field of particles" was described. We quote from their paper for comparison with the Hoppe quotation below:

The electron microscope image represents a projection of the three dimensional density distribution in the object at all levels perpendicular to the direction of view. According to a theorem familiar to crystallographers, the Fourier coefficients calculated from a projection of a three dimensional density distribution form a section through the three dimensional set of Fourier co-efficients corresponding to that distribution. By collecting many different projections of a structure in the form of electron microscope images, it should therefore be possible to collect, section by section, the full set of Fourier coefficients required to describe that structure... The number of projections needed to fill Fourier space roughly uniformly depends on the size of the particle and on the resolution... If more than one projection of an object is needed to resolve its structure, such projections can be obtained in two ways. The most obvious one is to systematically tilt and photograph a single particle in the electron microscope. If all the necessary projections are collected regardless of particle symmetry, no assumptions are needed to calculate the three dimensional structure... Alternatively, different images from a field of particles are, in principle, projections of the same structure... The exact orientation of each particle in relation to the direction of view must be determined in order to relate correctly the Fourier space section obtained from it to those obtained from other particles.

Their method was soon extended and improved (Crowther, 1971; Crowther *et al.*, 1970a,b). A paper by Hoppe *et al.* that also appeared in 1968 likewise contained a clear description of the principle of the method:

Die Dichteverteilung in einem Gitter lässt sich mathematisch in ihre Fourier-Komponenten zerlegen. Diese Daten können wieder in einem dreidimensionalen "Gitter" (dem reziproken Gitter) geometrisch übersichtlich geordnet werden. Der Fourier-Zerlegung jeder Projektion (also jedes elektronenmikroskopischen Bildes eines Kristallgitters in entsprechender Orientierung) entspricht eine durch den Ursprung gehende Gitterebene in diesem reziproken Gitter. Andererseits lässt sich jedes Gitter aus Bündeln von Gitterebenen aufbauen. Die für das dreidimensionale Bild erforderlichen Daten kann man also erhalten, wenn man elektronenmikroskopische Aufnahmen von Kristallgittern in verschiedenster Orientierung herstellt, diese mathematisch zerlegt, die Fourier-Komponenten zum dreidimensionalen reziproken Gitter ordnet und schliesslich die dreidimensionale Dichteverteilung durch eine Fourier-Synthese berechnet.

We quote it at length here for the late Walter Hoppe always cited it together with the paper of de Rosier and Klug and felt that other authors did not always do justice to it.* It does not, however, contain any actual reconstruction.

Before leaving these first three-dimensional reconstructions, we must also mention the isolated attempt by Roger G. Hart to combine several views of a specimen taken at different angles (Hart, 1968). His original aim was not to obtain information about the three-dimensional structure of the object (a dilute suspension of tobacco mosaic virus and colloidal gold particles, which served as fiduciary marks, sprayed onto a support film and air-dried) but to enhance the contrast. Nevertheless, he did realise that three-dimensional reconstruction should be possible:

> ... if the tilt angle were increased from its present 20 deg to 45 deg, the depth discrimination, for the finest details observable, would be comparable to the lateral resolution and would be limited by only the quality and number of the original micrographs.
>
> Thus the polytropic montage seems to offer a means of determining the three-dimensional structures of low-contrast biological specimens at a resolution of 3 Å .... I have not yet reached this point but preliminary efforts have produced images of tobacco mosaic virus comparable in fineness of detail to those obtained by shadowing... Still to be determined is the extent to which the fine details appearing in the montage represent real structures of the virus rather than residual noise that may have survived this attempt at its elimination.

In the following years, three-dimensional reconstruction progressed from being a difficult exercise to which only a very small number of laboratories could aspire to the major activity that it has become today. The subject gradually separated into two parts, electron crystallography and electron tomography. An important step in the latter was the introduction by Marin van Heel and Joachim Frank (1980, 1981; Frank and van Heel, 1980) of a statistical technique known as correspondence analysis. This enables us to classify images of poor visual quality into groups, each corresponding to a

---

* In a circular letter dated October 1986, which accompanied his retrospective account of 1983, Hoppe wrote: "Similarly disconcerting is also the citation of our first works on 3-dimensional electron microscopy. Almost nowhere is mentioned that [Hoppe *et al.*, 1968] is not only a parallel paper to [de Rosier and Klug, 1968], but theoretically and experimentally showed the way to 3-dimensional analysis of native structures... Apparently somewhat better known is that we carried out the first true 3-dimensional reconstruction." (The last remark refers to Hoppe *et al.*, 1974, 1976.)

particular view through the particle being studied, an important part of the pre-processing stage.

The 'phase problem' was not new when the first successful iterative solution was proposed in 1972 by Ralph Gerchberg and W. Owen Saxton (1972, 1973). The difficulty is easily stated: how can we obtain the modulus and phase of a complex signal when only the intensity (that is, the square of the modulus) can be recorded? It had arisen in coherence theory, in X-ray crystallography and in optics, to cite only closely related fields, but little progress had been made in the search for a solution, though ways of circumventing it had been devised by the crystallographers. The original feature of the problem in electron microscope imagery, where we should like to know the complex wavefunction emerging from the specimen, is that both diffraction patterns and images can be recorded. After a first attempt to obtain a direct, non-iterative solution, which ran into the same difficulties as earlier efforts (some of which are listed in Chapter 74), Gerchberg and Saxton (1972, 1973) devised an iterative solution, constrained by the measured moduli in the image and diffraction pattern of the same specimen area. This generated a vast activity that goes well beyond the electron microscope community, from which other algorithms (associated in particular with James Fienup) and detailed studies of the uniqueness of the solutions in one and more dimensions emerged.

In this connection, we should also mention here the extension by Peter Schiske of the Wiener filter to complex, weakly scattering objects, for which there is a linear relation between complex object transparency and (real) image contrast. His first publication (1968) neglected noise but, in a later paper (1973), noise is included and the analogy with the simple Wiener filter is exact.

While digital electron image processing was coming into being, an old proposal for remedying some of the defects of the electron microscope image was at last successfully tested: holography. With the development of first pointed-filament sources and later field-emission guns and the introduction of the electron biprism, the conditions necessary for electron hologram formation could be met. The laser was by this time in widespread use and so the conditions for reconstruction could likewise be satisfied. Finally, Leith and Upatnieks (1962, 1963) had devised an 'off-axis' method of separating the two images that are superimposed in the original in-line procedure of Gabor and the time was therefore ripe to resume attempts to put the technique into practice. Before mentioning the early landmarks, however, we must just draw attention to a semantic difficulty: in its primitive form, the hologram is formed by interference between the part of the beam that traverses the specimen without being scattered by the atoms that make up the object—this is the reference beam—and those electrons that have been

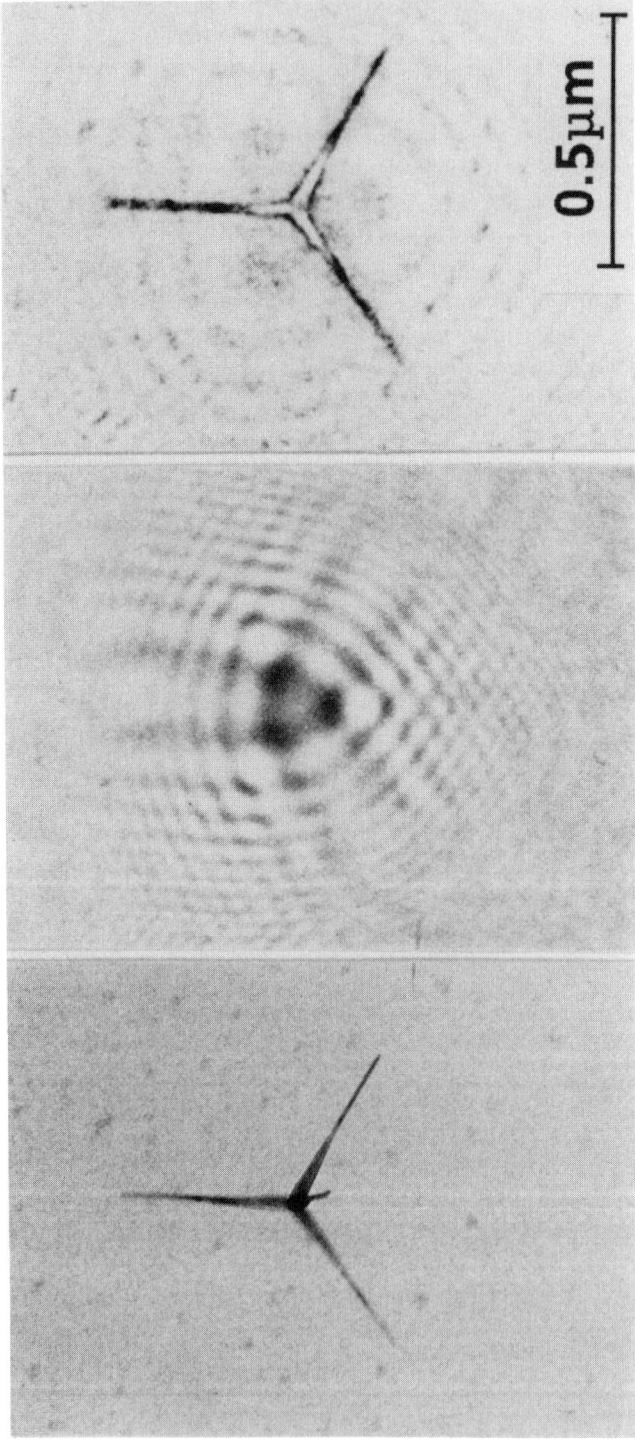

Fig. 54.1: In-line Fraunhofer hologram (centre) and optical reconstruction (right), obtained by Tonomura and Watanabe in 1968. In fact, the Fraunhofer condition is not satisfied for the whole specimen, a zinc oxide crystal (left), but is satisfied for the needles since their diameter is small.

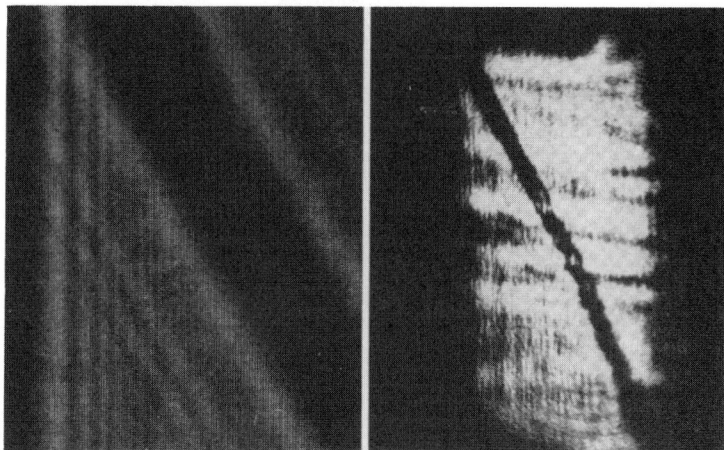

Fig. 54.2: Off-axis Fresnel hologram of a metallized quartz fibre (left) and optical reconstruction (right), obtained by Möllenstedt and Wahl in 1968. The virtual sources are lines, not points, and the fibre is oblique to these lines.

scattered. The hologram is thus essentially the same as the bright-field image. There is hence a large body of literature, especially dating from the 1970s, in which no distinction is made between holography and bright-field imagery and inverse filtering of the image of weakly scattering specimens is described as holography. The interplay between these different points of view is examined with great care by Hanszen in his various surveys (1971, 1973 and 1982).

The earliest example of Fraunhofer in-line electron holography was published by Tonomura and Watanabe (1968) and their remarkable reconstruction is reproduced here (Fig. 54.1); see also Tonomura *et al.* (1968) and Watanabe and Tonomura (1969). Several further attempts to use this approach were made, notably by Gallion *et al.* (1975), Troyon *et al.* (1976), Bonnet *et al.* (1978) and Bonhomme and Beorchia (1980) and by Munch (1975), who used a field-emission gun.

The first off-axis hologram was made with the aid of a biprism and reconstructed using a He-Ne laser by Gottfried Möllenstedt and Herbert Wahl (1968), closely followed by Akira Tonomura (1969) using a crystal to split the beam into two coherent beams. The Möllenstedt and Wahl experiment only partially fulfilled the conditions for holography as the two virtual sources were lines rather than points. The first off-axis hologram using point sources was made by Hiroshi Tomita *et al.* (1970a,b, 1972) and we reproduce in Figs. 54.2–3 the reconstructions of Möllenstedt and Wahl

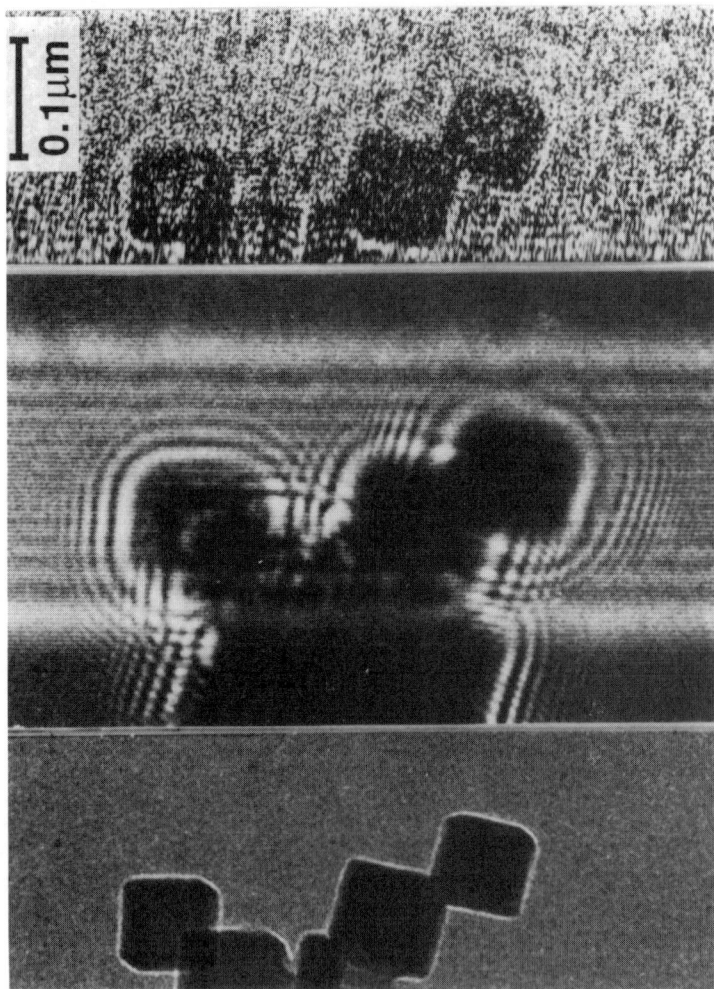

Fig. 54.3: Off-axis Fresnel hologram (b) and optical reconstruction (c) of magnesium oxide crystals, the image of which is shown in (a), obtained by Tomita, Matsuda and Komoda in 1970. This is the earliest example of off-axis electron holography with point virtual sources.

and Tomita *et al.* Wahl then went on to study off-axis image plane holography much more fully (1974, 1975). In the following years, holography was explored in depth by Akira Tonomura and colleagues in Japan, by Hannes Lichte and colleagues in Tübingen and by Karl-Joseph Hanszen, Georg Ade and Rolf Lauer in Braunschweig, with more isolated work elsewhere.

One side-effect of the experimental developments that contributed to the progress of holography was the incontrovertible demonstration by Tonomura *et al.* (1986) that the Aharonov–Bohm effect really exists. This interference phenomenon, in fact noticed by Ehrenberg and Siday in 1949 and rediscovered by Aharonov and Bohm in 1959, who gave credit to Ehrenberg and Siday as soon as they became aware of the earlier work of their colleagues, has a considerable literature. The effect is a shift of the interference fringes formed when two beams that have passed on either side of a local magnetic field are made to overlap; even if the magnetic field is negligibly small in the regions traversed by the electrons, the fringe structure changes if the magnetic field is altered. This is due to the fact that the vector potential A is different along the two paths even if $B = \text{curl } A$ effectively vanishes. A fierce controversy over the reality of the effect raged for many years, the Italian group around P. Bocchieri in Pavia and A. Loinger in Milan manifesting the most obdurate incredulity; the experimental results of Tonomura *et al.* that put the reality of the phenomenon beyond further doubt are reproduced as Fig. 62.11.

The last part of the book is concerned with coherence and largely with the relation between source coherence and the radiometric quantities, a subject that has been explored in depth for light sources during the past two decades but has been neglected for electron emitters. The account here therefore leans heavily on the literature of light optics. The fact that traditional radiometry is essentially applicable only to incoherent sources and that some new definition of brightness was needed was first pointed out by Adriaan Walther in 1968 but the principal developments are due to Emil Wolf and a series of colleagues, E.W. Marchand and W.H. Carter in particular. We cite especially the papers by Marchand and Wolf (1972a,b, 1974) upon which most subsequent developments repose and the recognition of the importance of the quasihomogeneous source by Carter and Wolf (1977).

Part XI

Wave Mechanics

55

The Schrödinger Equation

55.1 Introduction

Electrons, like all other kinds of elementary particles, have a dual character: They behave like *corpuscles* or like *waves*, depending on the particular experimental conditions. This double nature of electrons can be completely understood only within the frame of the general quantum theory and a rigorous treatment would therefore require the whole subject of electron optics to be cast into quantum theoretical form. Such a treatment would, however, be extremely complicated and unnecessarily detailed for almost all practical problems. We thus seek reasonable simplifications.

Wave mechanics is a part of the general quantum theory, in which the *wave nature* of the radiation is of paramount interest. Concepts typical of wave physics, such as frequency, wavelength, diffraction and interference, and experimental situations in which these concepts are useful are studied. Like geometrical electron optics, which is based on classical mechanics, this is again an incomplete description. The full theory, the unification of particle and wave physics by means of the so-called 'second quantization', is, however, practically never needed in electron optical practice.

Electrons have spin and hence satisfy not Schrödinger's but Dirac's equation. In electron optics, effects due to spin are usually negligible with a few exceptions, such as very low voltage scanning electron microscopy. A brief account of the transition from the Dirac formalism to the Schrödinger equation is to be found in Chapter 56.

55.2 Formulation of Schrödinger's equation

The material presented in the remainder of this chapter is dealt with extensively in all the numerous textbooks on quantum mechanics and is quite familiar; the following account is therefore very concise.

We start from the *Hamilton equation* for the motion of a single electron in a stationary electromagnetic field:

$$H(\boldsymbol{r}, \boldsymbol{p}) = \frac{1}{2m_0}\{\boldsymbol{p} + e\boldsymbol{A}(\boldsymbol{r})\}^2 - e\Phi(\boldsymbol{r}) = E = \text{const} \qquad (55.1)$$

in which we have retained the notation used in Vol. 1: m_0 denotes the rest mass of the electron, e the absolute elementary charge, Φ the electrostatic potential and $\boldsymbol{A}$ the vector potential, while $\boldsymbol{p}$ is the *canonical* momentum.

In wave mechanics, the momentum $\boldsymbol{p}$ and the total energy E are replaced by *operators*, which act on a *wavefunction* $\Psi(\boldsymbol{r}, t)$. These operators are chosen as

$$\boldsymbol{p} \to -i\hbar\nabla \equiv -i\hbar\ \text{grad} \qquad (55.2a)$$

$$E \to \ i\hbar\frac{\partial}{\partial t} \qquad (55.2b)$$

$\hbar = h/2\pi = 1.05 \times 10^{-34}\,\text{J s}$ being Dirac's constant. Replacing $\boldsymbol{p}$ and E in (55.1) by these operators and applying the resulting operator equation to a wavefunction Ψ, we arrive at Schrödinger's equation

$$\frac{1}{2m_0}\{-i\hbar\nabla + e\boldsymbol{A}(\boldsymbol{r})\}^2\Psi(\boldsymbol{r}, t) - e\Phi(\boldsymbol{r})\Psi(\boldsymbol{r}, t) = i\hbar\frac{\partial\Psi(\boldsymbol{r}, t)}{\partial t} \qquad (55.3)$$

This is a *complex* linear partial differential equation (PDE) of second order in the space coordinates and of first order in time. The complex nature of (55.3) implies that the solution Ψ cannot have an immediate physical meaning; the calculation of observable quantities is the topic of the next section.

In very many practical applications it is sufficient to consider *time-independent* solutions. These are obtained by seeking a solution in the separated form

$$\Psi(\boldsymbol{r}, t) = \psi(\boldsymbol{r})e^{-i\omega t} \qquad (55.4)$$

Relating the oscillation frequency ω to the total energy E by Einstein's relation

$$E = \hbar\omega \qquad (55.5)$$

and cancelling out a common factor $e^{-i\omega t}$, we arrive at the *time-independent* Schrödinger equation

$$\mathsf{H}\psi(\boldsymbol{r}) := \left\{\frac{1}{2m_0}(-i\hbar\nabla + e\boldsymbol{A})^2 - e\Phi\right\}\psi(\boldsymbol{r}) = E\psi(\boldsymbol{r}) \qquad (55.6)$$

The operator H, defined to be the expression in braces, is called the *Hamilton operator*. This PDE and its relativistic generalization, derived in the next chapter, are the starting point of most subsequent calculations.

A still simpler form is obtained for the motion of electrons in purely electrostatic fields. We may then assume that $A(r) \equiv 0$ and can hence cast (55.6) into the form

$$\nabla^2 \psi(r) + \frac{2m_0}{\hbar^2} \{E + e\Phi(r)\} \psi(r) = 0 \qquad (55.7)$$

This has the form of a *Helmholtz equation*

$$\nabla^2 \psi(r) + k^2(r)\psi(r) = 0 \qquad (55.8)$$

with a wavenumber depending on position:

$$k(r) := \hbar^{-1} \sqrt{2m_0\{E + e\Phi(r)\}} \qquad (55.9)$$

In nonrelativistic classical mechanics, the conservation of energy takes the form

$$\frac{1}{2m_0} g^2 - e\Phi(r) = E$$

and the *kinetic* momentum (2.14) is thus given by

$$g = \sqrt{2m_0\{E + e\Phi(r)\}} \qquad (55.10)$$

Comparing this with (55.9), we immediately obtain de Broglie's relation

$$g = \hbar k \qquad (55.11)$$

The validity of this relation will be studied in later chapters.

55.3 The continuity equation

An analogue of the well-known continuity equation of classical electrodynamics can be derived in wave mechanics. To obtain this, we multiply both sides of (55.3) by the complex conjugate function Ψ^*, giving

$$\Psi^* \frac{1}{2m_0} (-i\hbar\nabla + e\boldsymbol{A})^2 \Psi - e\Phi\Psi^*\Psi = i\hbar\, \Psi^* \frac{\partial\Psi}{\partial t}$$

The complex conjugate of the whole equation is

$$\Psi \frac{1}{2m_0} (+i\hbar\nabla + e\boldsymbol{A})^2 \Psi^* - e\Phi\Psi^*\Psi = -i\hbar\Psi \frac{\partial\Psi^*}{\partial t}$$

Subtraction of the second equation from the first yields an equation from which the terms in Φ have cancelled out and the right-hand side is the time-derivative of a product:

$$\frac{1}{2m_0}\left\{\Psi^*(-i\hbar\nabla + e\boldsymbol{A})^2\Psi - \Psi(i\hbar\nabla + e\boldsymbol{A})^2\Psi^*\right\} = i\hbar\frac{\partial}{\partial t}(\Psi^*\Psi) \quad (55.12)$$

The evaluation of the left-hand side proceeds as follows: expansion of the first quadratic term gives

$$\Psi^*(-i\hbar\nabla + e\boldsymbol{A})^2\Psi$$
$$= \Psi^*(-\hbar^2\nabla^2\Psi + e^2A^2\Psi - 2ie\hbar\boldsymbol{A}\cdot\operatorname{grad}\Psi - ie\hbar\Psi\operatorname{div}\boldsymbol{A})$$

with an analogous result for its complex conjugate*. The terms in e^2A^2 cancel and we can rewrite (55.12) in the form

$$\frac{\partial}{\partial t}|\Psi|^2 + \frac{\hbar}{2m_0 i}(\Psi^*\nabla^2\Psi - \Psi\nabla^2\Psi^*)$$
$$+ \frac{e}{m_0}\left\{\boldsymbol{A}\cdot(\Psi^*\nabla\Psi + \Psi\nabla\Psi^*) + |\Psi|^2\operatorname{div}\boldsymbol{A}\right\} = 0$$

The terms involving spatial derivatives can be expressed as the divergence of a vector and we thus arrive at a continuity equation (cf. 47.6)

$$\frac{\partial}{\partial t}\varrho(\boldsymbol{r}, t) + \operatorname{div}\boldsymbol{j}(\boldsymbol{r}, t) = 0 \quad (55.13)$$

with the scalar density function

$$\varrho(\boldsymbol{r}, t) := |\Psi(\boldsymbol{r}, t)|^2 \quad (55.14)$$

and the current density vector

$$\boldsymbol{j}(\boldsymbol{r}, t) := \frac{\hbar}{2m_0 i}\left\{\Psi^*\operatorname{grad}\Psi - \Psi\operatorname{grad}\Psi^*\right\} + \frac{e\varrho}{m_0}\boldsymbol{A} \quad (55.15)$$

Both density functions are always real and hence have a physical significance. The function $\varrho(\boldsymbol{r}, t)$ is the *particle density*, which here means the probability of finding a single electron at the given point in space; $\boldsymbol{j}(\boldsymbol{r}, t)$

---

  * The factor 2 appears when the commutation relation is employed during the expansion.

is then the corresponding particle current density. If the wavefunction can be normalized to unity

$$\int_{\mathbb{R}^3} \varrho(\boldsymbol{r},t)\, d^3r \equiv \int_{\mathbb{R}^3} |\Psi|^2\, d^3r = 1 \tag{55.16}$$

the continuity equation (55.13) ensures the conservation of each individual particle in space. Creation and annihilation processes cannot occur.

Some other density functions of practical interest are the electric space charge density

$$\varrho_{el}(\boldsymbol{r},t) = -e\varrho(\boldsymbol{r},t) \tag{55.17}$$

and the electric current density vector

$$\boldsymbol{j}_{el}(\boldsymbol{r},t) = -e\boldsymbol{j}(\boldsymbol{r},t) \tag{55.18}$$

We recall that in the physics of electron guns (Part IX), the opposite sign convention for $\boldsymbol{j}_{el}$ is more usual.

An important difference between electron optics and the physics of atoms and molecules is that *unbounded* solutions of (55.3) are of greatest interest in charged particle optics, which implies that the normalization (55.16) is often formally not possible; physically, wavefunctions are always truncated, by apertures for example, and are hence square-integrable.

55.4 The gauge transformation

The wave-optical concepts introduced in the context of Schrödinger's equation have *no* physical meaning in the sense that they are not observable quantities. Only the density functions ϱ and $\boldsymbol{j}$ (and consequently ϱ_{el} and $\boldsymbol{j}_{el}$) can be considered as observable in principle. To show this, we subject all the quantities appearing in the theory to a gauge transformation.

Since (55.3) is a *homogeneous* linear PDE, it is immediately clear that its solution may contain an arbitrary complex factor. To keep the density functions ϱ and $\boldsymbol{j}$ invariant, this factor must have unit norm.

This is, however, not the only possible degree of freedom; within the frame of time-independent potentials, the following gauge transformations are allowed:

$$\begin{aligned}
\boldsymbol{A}(\boldsymbol{r}) &= \boldsymbol{A}'(\boldsymbol{r}) + \operatorname{grad} F(\boldsymbol{r}) & (a) \\
\Phi(\boldsymbol{r}) &= \Phi'(\boldsymbol{r}) + \Phi_0 & (b) \\
E &= E' - e\Phi_0 & (c) \\
\omega &= \omega' - e\Phi_0/\hbar & (d) \\
\Psi(\boldsymbol{r},t) &= \Psi'(\boldsymbol{r},t)\, e^{i\varphi(\boldsymbol{r},t)} & (e) \\
\varphi(\boldsymbol{r},t) &= \alpha + e\{\Phi_0 t - F(\boldsymbol{r})\}/\hbar & (f)
\end{aligned} \tag{55.19}$$

The real constants α and Φ_0 and the scalar function $F(\boldsymbol{r})$ can be chosen at will. The relation between E and Φ must be such that the classical kinetic energy

$$T = E + e\Phi(\boldsymbol{r}) = E' + e\Phi'(\boldsymbol{r}) \tag{55.20}$$

remains invariant; the *kinetic* momentum g (55.10) is hence invariant but not the canonical momentum $\boldsymbol{p}$.

From (55.19a) and (55.19f), a minor calculation shows that

$$(-i\hbar\nabla + e\boldsymbol{A})\Psi = e^{i\varphi}(-i\hbar\nabla + e\boldsymbol{A}')\Psi' \tag{55.21a}$$

and consequently

$$(-i\hbar\nabla + e\boldsymbol{A})^2\,\Psi = e^{i\varphi}(-i\hbar\nabla + e\boldsymbol{A}')^2\,\Psi' \tag{55.21b}$$

The time-dependent term in (55.19f) is chosen so that

$$e\Phi\Psi + i\hbar\frac{\partial\Psi}{\partial t} = \left\{ e\Phi'\Psi' + i\hbar\frac{\partial\Psi'}{\partial t} \right\}e^{i\varphi} \tag{55.22}$$

On introducing all this into (55.3), we notice that a common factor $e^{i\varphi}$ appears throughout and can therefore be cancelled. This means that the Schrödinger equation takes the same form for Ψ' as for Ψ: it is *form-invariant*.

It is obvious that the phase factor $e^{i\varphi}$ cancels out from (55.14), leaving ϱ invariant: $\varrho' = \varrho$. After a minor calculation, we find that $\boldsymbol{j}' = \boldsymbol{j}$ is also an invariant. For *unbounded* states these (and ϱ_{el} and $\boldsymbol{j}_{el}$) are the essential observables.

56

The Relativistic Wave Equation

The Schrödinger equation, dealt with in the previous chapter, is valid only for *nonrelativistic* electron motion. In practice, this means a restriction to kinetic energies not exceeding about 10 keV. In very many cases, the electron energy reaches the relativistic domain and in high-voltage electron microscopy (HVEM), it may even exceed the threshold $2m_0c^2 \approx 1$ MeV. It is therefore obviously necessary to consider *relativistic* wave equations.

56.1 The Dirac equation

The relativistically correct form of the wave equation is Dirac's equation. This can be written in various ways, which show its Lorentz-covariance. A fairly simple form is

$$c\boldsymbol{\alpha} \cdot (-i\hbar\nabla + e\boldsymbol{A})\hat{\Psi} + \alpha_4 m_0 c^2 \hat{\Psi} - e\Phi\hat{\Psi} = i\hbar \frac{\partial \hat{\Psi}}{\partial t} \qquad (56.1)$$

Here $\hat{\Psi}(\boldsymbol{r}, t)$ is a four-element *spinor*, that is, a column vector the four elements of which are complex wavefunctions $\Psi_j(\boldsymbol{r}, t)$, $j = 1 \ldots 4$. The symbol $\boldsymbol{\alpha} = (\alpha_1, \alpha_2, \alpha_3)$ denotes a vector operator; its components α_j ($j = 1, 2, 3$) and also α_4 are Hermitian 4×4 matrices satisfying

$$\alpha_j \alpha_k + \alpha_k \alpha_j = 2\delta_{jk} \qquad (j, k = 1 \ldots 4) \qquad (56.2)$$

A simpler explicit representation of these Clifford matrices is given by

$$\alpha_1 = \begin{pmatrix} 0 & 0 & 0 & 1 \\ 0 & 0 & 1 & 0 \\ 0 & 1 & 0 & 0 \\ 1 & 0 & 0 & 0 \end{pmatrix} \qquad \alpha_2 = \begin{pmatrix} 0 & 0 & 0 & -i \\ 0 & 0 & i & 0 \\ 0 & -i & 0 & 0 \\ i & 0 & 0 & 0 \end{pmatrix}$$

$$\alpha_3 = \begin{pmatrix} 0 & 0 & 1 & 0 \\ 0 & 0 & 0 & -1 \\ 1 & 0 & 0 & 0 \\ 0 & -1 & 0 & 0 \end{pmatrix} \qquad \alpha_4 = \begin{pmatrix} 1 & 0 & 0 & 0 \\ 0 & 1 & 0 & 0 \\ 0 & 0 & -1 & 0 \\ 0 & 0 & 0 & -1 \end{pmatrix}$$

They are simultaneously Hermitian and unitary, have unit determinant and vanishing trace.

The Dirac equation satisfies all the requirements of relativistic wave mechanics: it is Lorentz-covariant, gauge covariant and describes correctly all spin interactions. Moreover, it provides correct expressions for the particle and current densities.

Below the threshold $2m_0c^2$, creation and annihilation processes cannot take place and even in HVEM they are quite unimportant. Hence Dirac's equation describes the propagation of electrons in practically all electron optical systems to a very high degree of accuracy. Nevertheless it is practically never used in electron optics because the calculations would become exceedingly and unwarrantably complicated.

56.2 The scalar wave equation

It is rather usual in electron optical calculation to start with Schrödinger's equation and then apply some relativistic corrections to the results obtained. This is certainly justified if the corrections are chosen in such a way that good agreement with experimental results is achieved. From the theoretical standpoint, however, such a procedure is unsatisfactory and can be avoided, as we shall soon see.

Simplification of Dirac's equation that reduces it to a scalar wave equation is only possible if we are willing to sacrifice some essential aspects of the relativistic theory.

(i) A wavefunction having only *one* component $\Psi(\boldsymbol{r}, t)$ is certainly acceptable only if we ignore completely the electron spin. In fact, spin effects are unimportant in electron optics except in some very specialized situations (see Reimer, 1985, Section 7.3.5) so that this simplification is usually justified.

(ii) There is no need for a Lorentz-covariant formulation. In practically all situations, the description in the *laboratory frame* is perfectly adequate. We may hence sacrifice the possibility of performing Lorentz transformations.

(iii) As in geometrical electron optics, it is convenient to choose a fixed gauge for the electrostatic potential: the cathode surface is assigned the potential $\Phi = 0$. The quantity $e\Phi(\boldsymbol{r})$ is then a direct measure of the energy acquired during acceleration in the electrostatic field.

With these assumptions, the formulation of a scalar relativistic wave equation is possible. The latter can be directly derived from Dirac's equation (Kasper, 1973) but the following argument, hitherto unpublished, is much simpler.

We start from the familiar classical Hamiltonian

$$H(\boldsymbol{r}, \boldsymbol{p}) = c\sqrt{(\boldsymbol{p} + e\boldsymbol{A})^2 + m_0^2 c^2} - e\Phi(\boldsymbol{r}) - m_0 c^2 \qquad (56.3)$$

This is obtained from (4.28) by setting $Q = -e$. A consequence of the assumption (iii), which requires (ii), is that the numerical value of H is very small. At the cathode surface $\Phi = 0$, the value of H is the *kinetic emission energy*, which is usually less then 1 eV. Since H represents the total energy, which is conserved, it must *always* remain very small, and so $|H| \ll m_0 c^2$.

In order to obtain a form that can be replaced by an operator, the square root term must be eliminated. Solving (56.3) for this term and then squaring the whole result, we obtain, still exactly,

$$2H(m_0 + \frac{e}{c^2}\Phi) + \frac{H^2}{c^2} = (\boldsymbol{p} + e\boldsymbol{A})^2 - 2em_0\Phi - \frac{e^2\Phi^2}{c^2}$$

It is now advantageous to introduce a *mass function*

$$m(\boldsymbol{r}) := m_0 + \frac{e}{c^2}\Phi(\boldsymbol{r}) \qquad (56.4)$$

which is the classical value of the relativistic electron mass (cf. 2.3 and 2.10). The second term represents the mass equivalent of the acceleration energy according to Einstein's energy–mass relation. Using (56.4) and ignoring the extremely small term H^2/c^2, we can cast the Hamiltonian into the form

$$H(\boldsymbol{r}, \boldsymbol{p}) = \frac{1}{2m(\boldsymbol{r})}(\boldsymbol{p} + e\boldsymbol{A})^2 - \frac{1}{2m}\left(2m_0 e\Phi + \frac{e^2\Phi^2}{c^2}\right)$$

Next, we notice that it is advantageous to introduce the relativistic acceleration potential (2.18)

$$\hat{\Phi}(\boldsymbol{r}) := \Phi(\boldsymbol{r})\left\{1 + \frac{e}{2m_0 c^2}\Phi(\boldsymbol{r})\right\} \qquad (56.5)$$

(see also Section 2.3). We then obtain the convenient result

$$H(\boldsymbol{r}, \boldsymbol{p}) = \frac{1}{2m(\boldsymbol{r})}\{(\boldsymbol{p} + e\boldsymbol{A})^2 - 2m_0 e\hat{\Phi}(\boldsymbol{r})\}$$

This is still not quite suitable for replacement by an operator, since m appears unsymmetrically. This minor deficiency is removed by writing

$$H(\boldsymbol{r}, \boldsymbol{p}) = \frac{1}{2\sqrt{m}}(\boldsymbol{p} + e\boldsymbol{A})^2 \frac{1}{\sqrt{m}} - \frac{m_0 e}{m}\hat{\Phi} \qquad (56.6)$$

We are now in a position to define the corresponding Hamilton operator:

$$\mathsf{H} = \frac{1}{2\sqrt{m}}(-i\hbar\nabla + e\boldsymbol{A})^2 \frac{1}{\sqrt{m}} - \frac{m_0 e\hat{\Phi}}{m} \qquad (56.7)$$

and finally obtain the wave equation

$$\mathsf{H}\chi = \frac{1}{2\sqrt{m}}(-i\hbar\nabla + e\boldsymbol{A})^2 \left(\frac{\chi}{\sqrt{m}}\right) - \frac{m_0 e\hat{\Phi}}{m}\chi = i\hbar\frac{\partial\chi}{\partial t} \qquad (56.8)$$

Obviously this simplifies to Schrödinger's equation when $|e\Phi| \ll m_0 c^2$, for which $m \to m_0$. The wavefunction $\chi(\boldsymbol{r},t)$, being a solution of (56.8), leads to the correct particle density $\varrho = |\chi|^2$. The factor $m^{-\frac{1}{2}}$, however, is inconvenient in practical calculations. To remove it, we define a new wavefunction,

$$\Psi(\boldsymbol{r},t) := \sqrt{\frac{m_0}{m(\boldsymbol{r})}}\chi(\boldsymbol{r},t) \qquad (56.9)$$

which satisfies the modified wave equation

$$\frac{1}{2m_0}(-i\hbar\nabla + e\boldsymbol{A})^2\Psi - e\hat{\Phi}\Psi = i\hbar\frac{m(\boldsymbol{r})}{m_0}\frac{\partial\Psi}{\partial t} \qquad (56.10)$$

This is more convenient in practice.

56.3 Properties of the relativistic wave equation

In order to demonstrate that (56.10) is a reasonable approximation, we now study its consequences. First of all, we introduce the product separation (55.4) with (55.5), giving

$$\frac{1}{2m_0}(-i\hbar\nabla + e\boldsymbol{A})^2\psi(\boldsymbol{r}) - e\hat{\Phi}(\boldsymbol{r})\psi(\boldsymbol{r}) = \frac{Em(\boldsymbol{r})}{m_0}\psi(\boldsymbol{r}) \qquad (56.11)$$

In the absence of vector potentials ($\boldsymbol{A} \equiv 0$), this can be cast into the form of a Helmholtz equation (55.8) with the wavenumber

$$k(\boldsymbol{r}) = \frac{1}{\hbar}\sqrt{2m_0 e\hat{\Phi}(\boldsymbol{r}) + 2Em(\boldsymbol{r})} \qquad (56.12)$$

where $|E| \ll m_0 c^2$ is necessary for reasons of consistency, since $E^2/c^2 = H^2/c^2$ was neglected in the derivation of (56.10). Within this limit, the square-root factor is still the relativistically correct expression for the kinetic momentum $g(\boldsymbol{r})$. According to the laws of relativistic kinematics (Section 2.3), we have exactly

$$g^2(\boldsymbol{r}) = 2m_0(E + e\Phi(\boldsymbol{r}))\left[1 + \frac{1}{2m_0c^2}\{E + e\Phi(\boldsymbol{r})\}\right]$$

$$= 2m_0e\Phi^{\frac{1}{2}}\left(1 + \frac{e}{2m_0c^2}\Phi\right) + 2E\left(m_0 + \frac{e}{c^2}\Phi\right) + \frac{E^2}{c^2}$$

$$= 2m_0e\hat{\Phi}(\boldsymbol{r}) + 2Em(\boldsymbol{r}) + \frac{E^2}{c^2}$$

The only error in (56.12) is indeed the omission of the very small and quite unimportant term E^2/c^2. Hence de Broglie's relation (55.11) is still valid *relativistically*.

A further consequence is the existence of a continuity equation (55.13). This can be derived just as in Section 55.3. The expressions for ϱ and $\boldsymbol{j}$ are now slightly different:

$$\varrho(\boldsymbol{r},t) = |\chi(\boldsymbol{r},t)|^2 = \frac{m(\boldsymbol{r})}{m_0}|\Psi(\boldsymbol{r},t)|^2 \tag{56.13}$$

$$j(\boldsymbol{r},t) = \frac{\hbar}{2mi}(\chi^*\nabla\chi - \chi\nabla\chi^*) + \frac{e}{m}\boldsymbol{A}|\chi|^2$$

$$= \frac{\hbar}{2m_0i}(\Psi^*\nabla\Psi - \Psi\nabla\Psi^*) + \frac{e}{\sqrt{m_0}}\boldsymbol{A}|\Psi|^2 \tag{56.14}$$

The mass function $m(\boldsymbol{r})$ cannot be eliminated completely, whether we use χ or Ψ.

When we come to consider the effect of the gauge transformation, a modification is necessary. Since we have ascribed the potential $\Phi = 0$ to the cathode surface, which is the natural gauge in electron optics, it makes little sense to depart from this choice. We therefore set $\Phi_0 = 0$ in (55.19), whereupon all the kinematic functions become invariants. The gauge transformation then reduces to

$$\boldsymbol{A}(\boldsymbol{r}) = \boldsymbol{A}'(\boldsymbol{r}) + \operatorname{grad} F(\boldsymbol{r}) \tag{56.15a}$$

$$\Psi(\boldsymbol{r},t) = \Psi'(\boldsymbol{r},t)e^{i(\alpha - eF/\hbar)} \tag{56.15b}$$

It is easily verified that (56.10) is form-invariant with respect to this transform and that ϱ and $\boldsymbol{j}$ are again invariants.

A special form of the wave equation, which is frequently needed in the physics of electron scattering and electron diffraction, arises from (55.8) taken together with (56.12). The electrostatic potential then consists of a large and constant acceleration term U and a small atomic contribution

$V(\mathbf{r})$. A constant energy shift can always be incorporated in the constant U and we may hence set $E = 0$ in (56.12) without loss of generality.

In the physics of electron diffraction it is often convenient to use an alternative definition of the wavenumber, which we write k:

$$\mathit{k} = \frac{k}{2\pi} = \frac{1}{\lambda} = \frac{g}{h} \tag{56.16}$$

λ denoting the wavelength, g the kinetic momentum and h Planck's constant. Equation (56.16) is again the de Broglie relation.

The atomic potential $V(\mathbf{r})$ has a short spatial range and is quite often very weak, which means that $|V(\mathbf{r})| \ll U$. In these conditions, it is advantageous to define the quantities in (56.16) for the *asymptotic* domain, where they become constants. Setting thus $E = 0$ and $\hat{\phi} = \hat{U} = U(1 + eU/2m_0c^2) = U(1 + \epsilon U)$ in (56.12), we obtain the *asymptotic* wavenumber

$$k_\infty = \frac{2\pi}{\lambda_\infty} = \frac{1}{\hbar}\sqrt{2m_0e\hat{U}} \tag{56.17}$$

and the *asymptotic* mass

$$m_\infty = m_0 + \frac{e}{c^2}U = \text{const} \tag{56.18}$$

In the wave equation itself, the term in $V(\mathbf{r})$ must be retained but, because $|V| \ll U$, a linearization is allowed. With (56.17) and (56.18), we then obtain

$$\nabla^2\psi(\mathbf{r}) + \{k_\infty^2 + 2em_\infty\hbar^{-2}V(\mathbf{r})\}\psi(\mathbf{r}) = 0 \tag{56.19}$$

Later we shall drop the subscript ∞, whenever this does not cause confusion.

The wavelengths encountered in electron microscopy are several orders of magnitude smaller than those of visible light. From (56.17), we see that

$$\lambda_\infty/\text{nm} \approx \frac{1.2}{\hat{U}^{\frac{1}{2}}/V^{\frac{1}{2}}} \tag{56.20}$$

so that at 100 kV, $\lambda_\infty = 3.7$ pm and at 500 kV, $\lambda_\infty = 1.4$ pm. (Lists of values are to be found at the end of Grivet, 1965, 1972, Table 3.)

56.4 Rigorous approach

The foregoing reasoning is adequate for almost all practical purposes but its purpose is to justify using the relativistic form of the Schrödinger equation rather than the full Dirac equation. Apart from some preliminiary studies by Rubinowicz (1934, 1957, 1963, 1965, 1966), Durand (1953) and Phan Van Loc (1953, 1954, 1955, 1958a,b, 1960), no attempts to work with the Dirac equation directly and hence to check the correctness of using the relativistic Schrödinger equation were made until the 1980s, when first Ferwerda et al. (1986) and subsequently Jagannathan (Jagannathan et al.,1989; Jagannathan, 1990) reopened the question and investigated it with great thoroughness. Ferwerda et al. present a fully relativistic version of the theory of microscope image formation, which essentially vindicates the use of the Klein–Gordon equation; Jagannathan does not even use the latter but derives the focusing theory for electron lenses, and in particular for magnetic and electrostatic round lenses and quadrupole lenses, from the Dirac equation. We cannot reproduce this work here and the interested reader is referred to the papers listed above for full and largely self-contained accounts.

57

The Eikonal Approximation

Exact solutions of the Schrödinger equation or the relativistic wave equation are known only in very few exceptional cases, which are far from fitting the experimental conditions. A systematic method of finding an approximate solution in realistic conditions is therefore required.

One such procedure is the eikonal method, closely associated in general quantum mechanics with the Wentzel–Kramers–Brillouin (WKB) method. Its main purpose is to show that classical mechanics is an approximation to the more correct quantum mechanical formulation. Moreover, it also offers a way of establishing approximate wavefunctions which may then be used in the theory of electron diffraction.

57.1 The product separation

The solution of (56.10) is a complex-valued function $\Psi(\boldsymbol{r}, t)$. Any such function can certainly be represented in terms of its modulus or 'amplitude' $a = |\Psi|$ and its phase. We make use of (55.4) with (55.5) and come thus to the product separation*

$$\Psi(\boldsymbol{r}, t) = a(\boldsymbol{r}) \exp\left[\frac{\mathrm{i}}{\hbar}\{S(\boldsymbol{r}) - Et\}\right] \tag{57.1}$$

with real functions $a(\boldsymbol{r})$ and $S(\boldsymbol{r})$. After introducing this into (56.10), carrying out the necessary differentiations and cancelling a common exponential factor, we obtain

$$-\hbar^2 \nabla^2 a + a(\nabla S)^2 - \mathrm{i}\hbar(a\nabla^2 S + 2\nabla a \cdot \nabla S) - 2\mathrm{i}e\boldsymbol{A} \cdot (\hbar\nabla a + \mathrm{i}a\nabla S)$$
$$+ a(e^2 A^2 - \mathrm{i}e\hbar\nabla \cdot \boldsymbol{A} - 2m_0 e\hat{\Phi} - 2mE) = 0$$

---

* Two sign conventions for wave propagation are in use in the literature of optics and electron optics, corresponding to (57.1) ('quantum mechanical convention') and its complex conjugate. We are in agreement with Glaser (1952, 1956) and Born and Wolf (1959, 1980) but not with (most of) Spence (1988). For a detailed examination of these conventions, see Saxton et al. (1983), whose table of sign conventions is reproduced with corrections by Spence (1988, p. 156).

This complex equation can be satisfied only if its real and imaginary parts vanish separately. This gives

real part:

$$a\{(\nabla S)^2 + 2e\mathbf{A} \cdot \nabla S + e^2 A^2 - 2m_0 e\hat{\Phi} - 2mE\} - \hbar^2 \nabla^2 a = 0 \qquad (57.2)$$

imaginary part:

$$\hbar\{a\nabla^2 S + 2\nabla a \cdot \nabla S + 2e\mathbf{A} \cdot \nabla a + ea\nabla \cdot \mathbf{A}\} = 0 \qquad (57.3)$$

The imaginary part can be cast into a more convenient form by multiplication by $a/\hbar$. Recalling that $\nabla(a^2) = 2a\nabla a$ and using the product differentiation rules of vector analysis, we may write (57.3) as a *divergence equation*:

$$\mathrm{div}\{a^2(\mathrm{grad}\,S + e\mathbf{A})\} = 0 \qquad (57.4)$$

The real part (57.2) can be cast into the more concise form

$$(\mathrm{grad}\,S + e\mathbf{A})^2 = 2m_0 e\hat{\Phi} + 2mE + \hbar^2 \frac{\nabla^2 a}{a}$$

The first two terms on the right-hand side are just the square of the classical *kinetic* momentum, $g^2(\mathbf{r})$ and the real part therefore becomes

$$(\mathrm{grad}\,S + e\mathbf{A})^2 = g^2 + \hbar^2 \frac{\nabla^2 a}{a} \qquad (57.5)$$

These two differential equations (57.4) and (57.5) hold *exactly*, but they are not simpler to solve than (56.10), as they are coupled and not even linear. Some simplifications are therefore necessary to come to a practical solution.

57.2 The essential approximation

The coupling between (57.4) and (57.5) is essentially due to the last term in (57.5). This is the only term containing $\hbar$. An important simplification will be achieved if this term can be neglected. This requires that $|a^{-1}\nabla^2 a| \ll g^2/\hbar^2 = k^2$ (56.16). Since $4\pi^2 \approx 40 \gg 1$, a sufficient limit is

$$\left|\frac{\nabla^2 a}{a}\right| < \left(\frac{k}{2\pi}\right)^2 = \frac{1}{\lambda^2} \qquad (57.6)$$

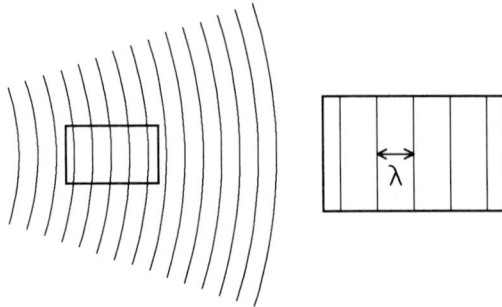

Fig. 57.1: Illustration of the eikonal approximation. Macroscopically, the wavefronts are curved and the amplitude varies in space. The radius of curvature, however, is much larger than the local wavelength. In every sufficiently small domain, shown in the enlarged rectangle, the wavefronts are nearly planar and the amplitude is practically constant.

This is the eikonal approximation in its proper sense. When (57.6) is satisfied, (57.5) becomes independent of the amplitude function $a(\boldsymbol{r})$ and takes the form

$$\{\operatorname{grad} S(\boldsymbol{r}) + e\boldsymbol{A}(\boldsymbol{r})\}^2 = g^2(\boldsymbol{r}) \tag{57.7}$$

This is the reduced *Hamilton–Jacobi equation* derived in Chapter 5. Equation (57.7) is identical with (5.10) and (5.11) for $Q = -e$ and the nominal value $E = 0$, if S is identified* with $\overline{S}$. The solution of the partial differential equation (57.7) gives the *classical electron trajectories* in a time-independent representation. The surfaces $S = \text{const}$, considered in Hamiltonian optics, are now seen to be *surfaces of constant wave phase* and thus take on a wave-optical meaning. We have thus justified some fundamentals of geometrical electron optics in the sense that they emerge from the more rigorous wave mechanics.

In the general case, it is difficult to specify concrete limits for the validity of (57.6). It is trivially obvious that (57.6) is satisfied for every plane or spherical wave. Moreover, (57.6) is justified if the amplitude $a(\boldsymbol{r})$ varies very little in domains that are a few wavelengths in geometrical extent. The true wave can then always be approximated locally by a plane wave

$$\psi(\boldsymbol{r}) = a_0 \exp\{i\boldsymbol{k}_0 \cdot (\boldsymbol{r} - \boldsymbol{r}_0)\}$$

the constants to be chosen appropriately. This approximation is sketched in Fig. 57.1.

---

* In Chapter 5, we used the notation $\overline{S}$ so as to be able to write $S := \overline{S}/(2m_0 e)^{1/2}$ in later chapters. This is not convenient here and S will henceforward denote the function defined in (4.32) and (4.34).

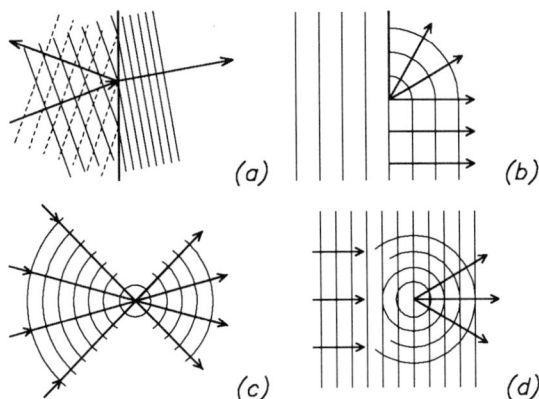

Fig. 57.2: Cases in which the eikonal approximation breaks down: (a) reflection and refraction at surfaces, (b) diffraction at sharp edges, (c) high intensity in foci, (d) scattering in atomic fields.

The eikonal approximation breaks down in the following cases:

(i) in the vicinity of interfaces: here the incident wave splits abruptly into a reflected and a refracted wave;

(ii) in the vicinity of sharp edges: these cause diffraction;

(iii) in the vicinity of foci of every kind: here the amplitude becomes very large in a very small domain;

(iv) in all atomic fields: these vary considerably over distances of the order of an ångström (100 pm) and cause scattering. (We recall that 100 pm is about 30 wavelengths at 100 kV.)

These situations are sketched in Figs. 57.2a–d. In view of all these exceptional cases, it is obvious that the eikonal approximation is valid only for motion in *macroscopic* fields, far from their singularities. Even then it may break down, since a focus with singular amplitude can be produced even in field-free space.

All this does not mean that the eikonal approximation is useless. Reflection, refraction and diffraction can be treated, not in terms of a *single* amplitude–phase solution, but by *superposition* of several solutions. The eikonal method then supplies the necessary tools for calculating the elementary waves.

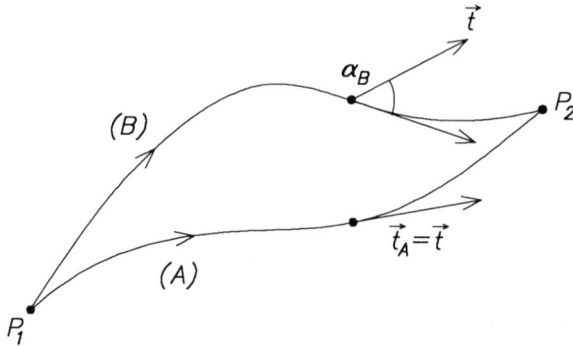

Fig. 57.3: Two paths of integration (A) and (B) joining the points P_1 and P_2. Line (A) is an extremal curve.

57.3 The variational principle

In Part I, we regarded Hamilton's variational principle as fundamental and derived from it all the classical equations of motion including the Hamilton–Jacobi equation; now, however, it is only an approximation. The Hamilton principle, though, is *not* a direct consequence of the wave equation. In this situation the next natural step is to establish a variational principle from the Hamilton–Jacobi equation, which will prove to be the electron optical form of the Euler–Maupertuis principle of least action.

In the following derivations we assume that a finite domain can be found, in which the solution $S(\boldsymbol{r})$ of (57.7) is unique. Equation (57.7) prescribes only the *length* of the vector $(\operatorname{grad} S + e\boldsymbol{A})$. We can convert (57.7) to a vector equation by writing

$$\operatorname{grad} S(\boldsymbol{r}) + e\boldsymbol{A}(\boldsymbol{r}) = \boldsymbol{g}(\boldsymbol{r}) =: g(\boldsymbol{r})\boldsymbol{t}(\boldsymbol{r}) \tag{57.8}$$

with a unit vector $\boldsymbol{t}(\boldsymbol{r})$. This is initially not known but must exist and can be determined *after* the solution of (57.7) has been found. Later this vector $\boldsymbol{t}$ will be identified with the tangent vector of the classical electron trajectory. In any case, there must exist lines (A) along which the local tangent $\boldsymbol{t}_A(s)$ always coincides with the vector $\boldsymbol{t}$ of (57.8). We now study the situation sketched in Fig. 57.3. Curve (A) has the above property and runs from a fixed starting point P_1 to a fixed terminal point P_2. Line (B) is an arbitrary smooth curve joining the same points.

We now evaluate the eikonal function $S(P_1, P_2)$ between these points and satisfying $S(P_1, P_1) = 0$. The assumed uniqueness means that

$$S(P_1, P_2) = \overset{(A)\ P_2}{\underset{P_1}{\int}} \operatorname{grad} S \cdot d\boldsymbol{r} = \overset{(B)\ P_2}{\underset{P_1}{\int}} \operatorname{grad} S \cdot d\boldsymbol{r}$$

Substituting for $\operatorname{grad} S$ from (57.8) and using $d\boldsymbol{r} = \boldsymbol{t}_A \, ds$ on (A) and $d\boldsymbol{r} = \boldsymbol{t}_B \, ds$ on (B), we obtain

$$S = \overset{(A)\ P_2}{\underset{P_1}{\int}} (g\boldsymbol{t} \cdot \boldsymbol{t}_A \, ds - e\boldsymbol{A} \cdot d\boldsymbol{r}) = \overset{(B)\ P_2}{\underset{P_1}{\int}} (g\boldsymbol{t} \cdot \boldsymbol{t}_B \, ds - e\boldsymbol{A} \cdot d\boldsymbol{r})$$

From the initial assumption, we have $\boldsymbol{t} = \boldsymbol{t}_A$, $\boldsymbol{t} \cdot \boldsymbol{t}_A = 1$ on line (A), whereas $\boldsymbol{t} \cdot \boldsymbol{t}_B = \cos \alpha_B \leq 1$ on line (B). We may therefore conclude that

$$S = \overset{(A)\ P_2}{\underset{P_1}{\int}} (g \, ds - e\boldsymbol{A} \cdot d\boldsymbol{r}) = \overset{(B)\ P_2}{\underset{P_1}{\int}} (g \cos \alpha_B \, ds - e\boldsymbol{A} \cdot d\boldsymbol{r})$$

$$\leq \overset{(B)\ P_2}{\underset{P_1}{\int}} (g \, ds - e\boldsymbol{A} \cdot d\boldsymbol{r})$$

Since this is true for any test curve (B) joining the same endpoints P_1 and P_2, we find:

$$S(P_1, P_2) = \overset{(A)\ P_2}{\underset{P_1}{\int}} (g \, ds - e\boldsymbol{A} \cdot d\boldsymbol{r}) = \text{extr.} \tag{57.9}$$

which is a form of the Euler–Maupertuis principle of least action. The curve (A) along which the vector $\boldsymbol{t}(s)$ satisfies (57.8) is an extremal curve. Introducing the vector $\boldsymbol{g} = g\boldsymbol{t}$ from (57.8) and noting that $\boldsymbol{p} = \boldsymbol{g} - e\boldsymbol{A}$ is the canonical momentum, we find

$$S(P_1, P_2) = \int_{P_1}^{P_2} \boldsymbol{p} \cdot d\boldsymbol{r} = \text{extr.} \tag{57.10}$$

in agreement with (4.33). Equation (57.10) is an expression of Fermat's principle.

It is easy to derive the time-independent ray equation directly from (57.8). We rewrite this as

$$\operatorname{grad} S = g\boldsymbol{t} - e\boldsymbol{A}$$

and form the curl of both sides, giving

$$\text{curl}(g\boldsymbol{t}) - e \ \text{curl}\,\boldsymbol{A} = \text{curl}(g\boldsymbol{t}) - e\boldsymbol{B} = 0$$

$\boldsymbol{B}$ being the magnetic field strength. Forming the vector product with $\boldsymbol{t}$, we find

$$\boldsymbol{t} \times \text{curl}(g\boldsymbol{t}) + e\boldsymbol{B} \times \boldsymbol{t} = 0$$

Expansion of the curl of a product gives

$$\text{curl}(g\boldsymbol{t}) = -\boldsymbol{t} \times \text{grad}\,g + g \ \text{curl}\,\boldsymbol{t}$$

and so

$$\boldsymbol{t} \times \text{curl}(g\boldsymbol{t}) = \text{grad}\,g - (\boldsymbol{t} \cdot \text{grad}\,g)\boldsymbol{t} + g\boldsymbol{t} \times \text{curl}\,\boldsymbol{t}$$

The rules of vector analysis tells us that

$$\boldsymbol{t} \times \text{curl}\,\boldsymbol{t} = \frac{1}{2} \text{grad}(\boldsymbol{t} \cdot \boldsymbol{t}) - \boldsymbol{t} \cdot \nabla \boldsymbol{t} = -\boldsymbol{t} \cdot \nabla \boldsymbol{t}$$

since $\boldsymbol{t}^2 = 1$. Putting all this together, we obtain

$$\text{grad}\,g - (\boldsymbol{t} \cdot \text{grad}\,g)\boldsymbol{t} - g\boldsymbol{t} \cdot \nabla \boldsymbol{t} = \text{grad}\,g - \boldsymbol{t} \cdot \nabla(g\boldsymbol{t}) = -e\boldsymbol{B} \times \boldsymbol{t}$$

Next we observe that $\boldsymbol{t} = d\boldsymbol{r}/ds$ for the tangent to the curve and that $\boldsymbol{t} \cdot \nabla = d/ds$ is the operator of differentiation with respect to the arc-length s. Thus we finally arrive at

$$\frac{d}{ds}\left(g(\boldsymbol{r}) \frac{d\boldsymbol{r}}{ds} \right) = \text{grad}\,g(\boldsymbol{r}) + e\boldsymbol{B}(\boldsymbol{r}) \times \frac{d\boldsymbol{r}}{ds} \qquad (57.11)$$

which agrees with (3.1) for $Q = -e$.

The derivation of the fundamental laws of geometrical electron optics as an *approximate* consequence of wave mechanics is now complete. All this is based on the approximation (57.6), which allowed us to replace (57.5) by (57.7).

57.4 The calculation of eikonal functions

The eikonal functions play an important role as elementary waves in the theory of diffraction, as we have already mentioned. We now enquire how they can be determined in practice.

In general this is a very difficult task. In order to determine the point eikonal $S(P_1, P_2)$ between two points P_1 and P_2, we have first to establish the classical ray joining these two points, which requires numerical solution of (57.11). Since we are given two endpoints rather than a starting point and an initial direction, this is mathematically a two-point *boundary-value problem* for trajectories. As this may have more than one solution, the point eikonal $S(P_1, P_2)$ can have several values.

Once the trajectory (or the set of trajectories) had been determined, the eikonal itself would be obtained by numerical evaluation of the integral (57.9). Since electron wavelengths are so short, this would require very high precision of about 14 to 15 significant digits. With modern computers this is feasible if the accuracy is increased up to the machine-dependent limit. This is undeniably a very hard but not an unrealistic task. Since the entire procedure cannot be carried out for *all* point-pairs (P_1, P_2) in practice, it would have to be performed for a suitable pattern of discrete points. The values for intermediate points would then have to be determined by means of suitable interpolation techniques.

It is not always necessary to start from a fixed point P_1. We may equally well prescribe a surface $S(\boldsymbol{r}) = S_0 = \text{const}$ and start the ray trace from this. For simplicity we shall assume here that $\boldsymbol{A}(\boldsymbol{r}) \equiv 0$, at least in the domain of interest. The rays are then orthogonal to surfaces of constant S, as shown in Fig. 57.4. Such a procedure is advantageous whenever the propagation of wavefronts is to be studied only over a short distance. Certain simplifications that considerably facilitate the calculation are then sometimes permissible. This will become more clear when we deal with applications in the physics of scattering.

A third way of determining eikonal functions involves series expansions. These will be dealt with in the context of electron diffraction and aberration theory.

We conclude this section with a brief remark on the gauge transformation. The eikonal S is *not* gauge-invariant. If we introduce (56.15b) into (57.1), we find

$$S(\boldsymbol{r}) = S'(\boldsymbol{r}) - eF(\boldsymbol{r}) + \hbar\alpha \qquad (57.12)$$

which means that not even eikonal *differences* between two points, from which α cancels out, are gauge-invariant! However, the *kinetic* momentum

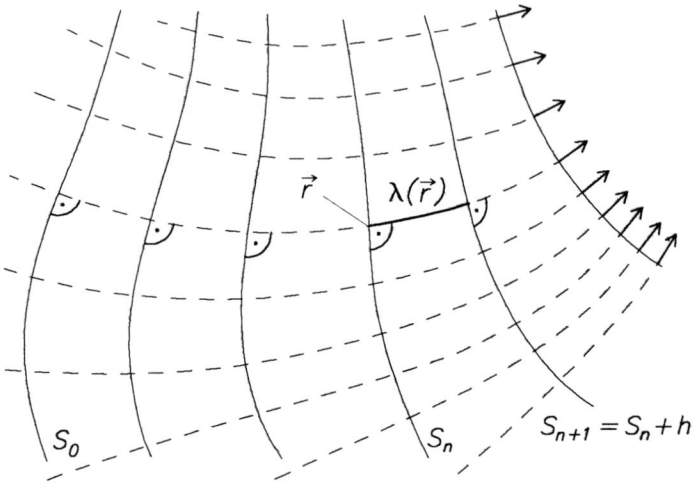

Fig. 57.4: Propagation of wave fronts starting from a surface $S(r) = S_0 = $ const. The local distance between neighbouring surfaces is approximately the wavelength $\lambda(r)$, if the eikonal difference is Planck's constant $\hbar$. The broken lines indicate the classical orthogonal trajectories. The sketched orthogonality holds only for $A(r) \equiv 0$.

is gauge-invariant thanks to (56.15a):

$$g(r) = \operatorname{grad} S(r) + eA(r) = \operatorname{grad} S'(r) + eA'(r) \tag{57.13}$$

as it must be.

57.5 The calculation of wave amplitudes

We shall now discuss the interpretation and solution of the divergence relation (57.4). From (57.8) we can write this in the form

$$\operatorname{div}\{a^2(r)g(r)\} = 0 \tag{57.14}$$

and (57.13) shows that this is a gauge-invariant relation. The physical meaning of (57.14) is that particle current is conserved. To see this, we differentiate (57.1) and multiply by Ψ^*, giving

$$\Psi^*\nabla\Psi = a(\nabla a + \frac{ia}{\hbar}\nabla S)$$

Introducing this and its complex conjugate into the second form of (57.14), we see that ∇a cancels out and find

$$j(r) = \frac{1}{m_0}a^2(\nabla S + eA) = \frac{a^2}{m_0}g \tag{57.15}$$

Hence (57.14) can be rewritten as div $j = 0$ and is evidently the stationary case of the continuity equation (55.13).

Introducing (57.1) into (56.13), we find

$$\varrho(\boldsymbol{r}) = \frac{m(\boldsymbol{r})}{m_0} \, a^2(\boldsymbol{r}) \qquad (57.16)$$

Since $\boldsymbol{v}(\boldsymbol{r}) = \boldsymbol{g}(\boldsymbol{r})/m(\boldsymbol{r})$ is the classical particle velocity, we can cast (57.15) into the familiar classical form

$$\boldsymbol{j}(\boldsymbol{r}) = \varrho(\boldsymbol{r})\boldsymbol{v}(\boldsymbol{r}) \qquad (57.17)$$

We come now to the solution of (57.14). This can be obtained by formally integrating and then using Gauss's integral theorem:

$$\oint a^2(\boldsymbol{r})\boldsymbol{g}(\boldsymbol{r}) \, d\boldsymbol{\sigma} = 0 \qquad (57.18)$$

$d\boldsymbol{\sigma}$ denotes the surface-element vector. We choose now a section cut from a current tube, as shown in Fig. 57.5a. The mantle surfaces are formed by rays and do not contribute to the integral (57.18), since $\boldsymbol{g} \perp d\boldsymbol{\sigma}$ on them. The two cross-sections $\Delta\sigma_1$ and $\Delta\sigma_2$ are orthogonal to the mantle and so small that the midpoint rule of integration may be applied. We find then that the quantity

$$a_1^2 g_1 \Delta\sigma_1 = a_2^2 g_2 \Delta\sigma_2 = a^2(s)g(s)\Delta\sigma(s) \qquad (57.19)$$

is constant along the tube, the latter expression giving this quantity at an arbitrary cross-section specified in position by the arc-length s along the tube.

The amplitude becomes singular if the rays start from a common point P_1 as shown in Fig. 57.5b. We then choose two concentric spheres of radii R_1 and R_2. If these are so small that the curvature of the rays can be ignored and the absolute momentum g is a constant, we find

$$4\pi R_1^2 a_1^2 = 4\pi R_2^2 a_2^2 = 4\pi R^2 a^2 = \text{const}$$

hence

$$a(R) = \text{const}/R \qquad \text{for} \quad R \to 0 \qquad (57.20)$$

Fortunately the eikonal approximation (57.6) does not break down close to such a singularity, since for $R > 0$ the expression in (57.20) is a solution of the Laplace equation in spherical coordinates. This is the basis for the calculation of a *Green's function*, which will be needed in the next chapter.

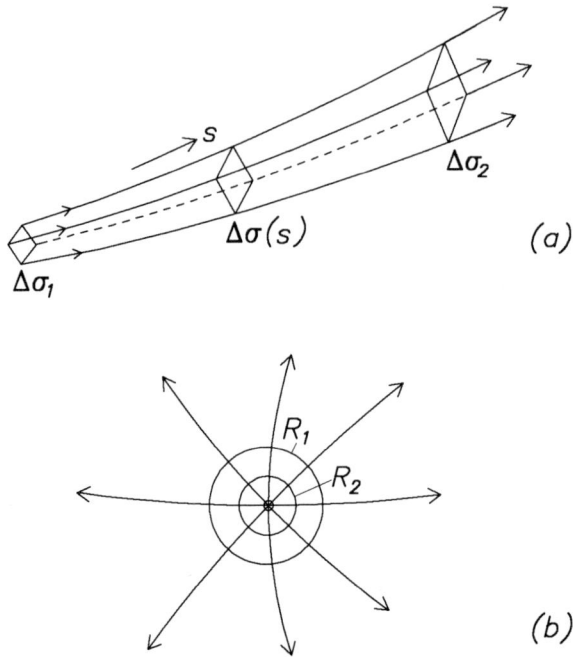

Fig. 57.5: Domains of integration for the continuity law: (a) section of a current tube with endfaces $\Delta\sigma_1$, $\Delta\sigma_2$ and a cross-section $\Delta\sigma(s)$ at an arbitrary position; (b) spherical shell with radii R_1 and R_2 round a point source.

If the rays do not intersect exactly in a common point, the reasoning based on Laplace's equation breaks down. This occurs in all real sources or foci, which must therefore be treated carefully in a more rigorous way with a more accurate solution of (56.10).

Another consequence of the continuity equation is that the amplitude–phase solutions cannot in general be normalized. This is immediately obvious for the purely spherical wave, for which $a = \text{const}/R$ holds for all $R > 0$. Amplitude–phase solutions with a *continuous energy spectrum* must therefore be superposed to build up a wave-packet. It is therefore necessary to use the time-dependent form (56.10) of the wave equation with an energy constant E that does not vanish identically.

58

Paraxial Wave Optics

58.1 The paraxial Schrödinger equation

The relativistically corrected Schrödinger equation (56.10) and its approximate solution in the form of an amplitude–phase product (57.1) are valid whenever (57.6) is satisfied. So far as image formation in electron microscopes is concerned, this constraint is still unnecessarily weak. In order to keep the electron optical aberrations sufficiently small, the beam must be confined to a very narrow domain in the vicinity of the optic axis, and this limitation should be incorporated in the search for an approximate solution of (56.10).

The wave-optical calculations pass essentially through the same stages as the geometrical calculations. In a first step the spatial domain considered is so close to the optic axis that lens aberrations can be ignored; this is the *paraxial domain*. The corresponding solution describes *perfect* image formation in wave-optical terms. When the corresponding wavefunctions have been established, the second step is to use them in a perturbation calculus and thereby include the effect of the lens aberrations. This corresponds to the geometrical perturbation theory.

There are two ways of obtaining an approximate solution of Schrödinger's equation. The first, which will be followed extensively in later chapters, involves finding a general solution and then confining it to the appropriate domain. The alternative is to *simplify* the wave equation before attempting to solve it. The fact that the solutions are applicable only in the paraxial region is then self-evident. This approach was followed by Glaser and Schiske (1953) and is presented in detail by Glaser (1952, 1956). We now give a brief account of this approach, including relativistic correction terms.

We set out from the wave equation (56.10), which it is convenient to reorganize in the form

$$\nabla^2 \Psi + \frac{2ie}{\hbar} \boldsymbol{A} \cdot \nabla \Psi + \frac{1}{\hbar^2}(2m_0 e\hat{\Phi} - e^2 \boldsymbol{A}^2)\Psi = -\frac{2im}{\hbar}\frac{\partial \Psi}{\partial t} \qquad (58.1)$$

The potentials $\Phi(\boldsymbol{r})$ and $\boldsymbol{A}(\boldsymbol{r})$ are now assumed to be rotationally symmetric and A_z to vanish (see Section 7.3). On the optic axis we then have

$$\frac{\partial^2 \Psi_0(z,t)}{\partial z^2} + \frac{2m_0 e}{\hbar^2} \hat{\phi}(z)\Psi_0 = \frac{2m(z)}{i\hbar}\frac{\partial \Psi_0}{\partial t} \tag{58.2}$$

where $\phi(z)$ is as usual the axial potential and we recall that

$$\hat{\phi}(z) := \phi(1 + \epsilon\phi) \quad , \quad m(z) = m_0(1 + 2\epsilon\phi) = \gamma m_0$$
$$\epsilon = \frac{e}{2m_0 c^2} \quad , \quad \gamma = 1 + 2\epsilon\phi$$

(see 2.16 and 2.21).

For monoenergetic electrons, the time-dependence can be represented by the familiar exponential factor, $\exp(-iEt/\hbar)$. In accordance with the conventions of Section 56.3, the very small energy E can be set to zero for electrons having negligible energy at the cathode surface; $E = e\Delta\phi$ is then the chromatic energy shift if emission spectra are taken into account.

After substituting the harmonic time dependence in (58.2), we find

$$\frac{\partial^2 \Psi_0}{\partial z^2} + \frac{g^2(z)}{\hbar^2}\Psi_0 = 0 \tag{58.4}$$

with the axial momentum

$$g(z) = \{2m_0 e\hat{\phi}(z) + 2me\Delta\phi\}^{\frac{1}{2}} \tag{58.5}$$

Apart from quadratic terms in $\Delta\phi$, which can be ignored, this is in agreement with

$$g^2(z) = 2m_0 e(\phi + \Delta\phi)\{1 + \epsilon(\phi + \Delta\phi)\}$$

which again confirms the consistency of the approximations made in the previous chapter. A highly accurate solution of (58.4) is

$$\Psi_0(z,t) = \frac{C}{\sqrt{g(z)}} \exp\left(\frac{i}{\hbar}\int g\,dz - \frac{ie\Delta\phi}{\hbar}t\right) \tag{58.6}$$

The necessary approximation,

$$\left|\frac{3g'^2}{4g^2} - \frac{g''}{2g}\right| \ll \frac{g^2}{\hbar^2} = \frac{4\pi^2}{\lambda^2}$$

is entirely justified in practice: the left-hand side is of the order of 1 mm^{-2} while the right-hand side approaches 1 fm$^{-2} = 10^{24}$ mm^{-2}. Even the less stringent condition

$$\left|\frac{g'}{g}\right| \ll \frac{2\pi}{\lambda} = \frac{g}{\hbar} \tag{58.7}$$

is perfectly justified.

We now seek a solution of (58.1) of the form

$$\Psi(\mathbf{r}, t) = \Psi_0(z, t)\psi_p(\mathbf{r}) \tag{58.8}$$

The suffix p for 'paraxial' is a reminder that—together with the time dependence—the rapidly varying axial phase factor has been removed. We shall later drop this suffix, whenever this does not cause confusion.

Differentiation of (58.8) using (58.6) gives

$$\frac{\partial \Psi}{\partial t} = -\frac{ie}{\hbar}\Delta\phi\psi_p\Psi_0$$

$$\nabla\Psi = (\nabla\psi_p + \frac{i}{\hbar}g\psi_p\mathbf{u}_z)\Psi_0(z, t)$$

$$\nabla^2\Psi = (\nabla^2\psi_p + \frac{2i}{\hbar}g\frac{\partial\psi_p}{\partial z} - \frac{g^2}{\hbar^2}\psi_p)\Psi_0(z, t)$$

$\mathbf{u}_z$ denoting the unit vector in the axial direction. Introducing this into (58.1) and taking (58.2–6) into consideration, a common factor $\Psi_0(z, t)$ cancels, leaving

$$\nabla^2\psi_p + \frac{2i}{\hbar}g\frac{\partial\psi_p}{\partial z} + \frac{2ie}{\hbar}\mathbf{A}\cdot\nabla\psi_p + \frac{1}{\hbar^2}\{2m_0e\hat{\Phi}(\mathbf{r}) + 2me\Delta\phi - p^2 - e^2A^2\}\psi_p = 0 \tag{58.9}$$

The quantity $\hat{\Phi}(\mathbf{r})$ can be replaced by its radial series expansion. In the paraxial approximation this is truncated after the term in r^2:

$$\Phi(z, r) = \phi(z) - \frac{r^2}{4}\phi''(z) + O(r^4)$$

and consequently

$$\hat{\Phi} = \Phi(1 + \epsilon\Phi) = \hat{\phi}(z) - \gamma\frac{r^2}{4}\phi'' + O(r^4)$$

From (58.5) and $m = \gamma m_0$, we find

$$2m_0e\hat{\Phi} + 2m_0e\Delta\phi - g^2 = -\frac{1}{2}m(z)er^2\phi''(z) \tag{58.10}$$

which vanishes everywhere on the optic axis. In the paraxial approximation, we may also write

$$A^2(z, r) = \frac{1}{4}r^2B^2(z) + O(r^4)$$

$B(z)$ being as usual the axial magnetic field strength. Introducing all this into (58.9), we find:

$$\nabla^2\psi_p + \frac{2i}{\hbar}\left\{g\frac{\partial\psi_p}{\partial z} + e\boldsymbol{A}(\boldsymbol{r})\cdot\nabla\psi_p(\boldsymbol{r})\right\} - \frac{r^2}{\hbar^2}\left\{\frac{em}{2}\phi''(z) + \frac{e^2B^2}{4}\right\}\psi_p(\boldsymbol{r}) = 0$$
(58.11)

The term involving the vector potential $\boldsymbol{A}(\boldsymbol{r})$ is inconvenient in practical calculations but can be removed by transforming the wave equation from the laboratory frame to the *rotating frame* (15.7), as we did systematically in the geometrical treatment of magnetic lenses.

We again adopt the notation introduced in Chapter 15: the cartesian coordinates in the laboratory frame are X, Y, z and the coordinates of the rotating system x, y, z. The latter are non-cartesian and defined by

$$x + iy := (X + iY)\exp\{-i\theta(z)\}$$
(58.12)

A consequence of this definition is that we have to consider the derivatives

$$x'(z) = \theta'(z)y(z) \quad , \quad y'(z) = -\theta'(z)x(z)$$
(58.13)

obtained from (58.12) if we hold X and Y constant. This implies that we have to distinguish carefully between partial and total derivatives with respect to z, as will shortly become obvious.

We now write the paraxial wavefunction as

$$\psi_p(X, Y, z) =: \psi(x, y, z)$$
(58.14)

It is easy to verify that

$$\psi_{p|XX} + \psi_{p|YY} = \psi_{|xx} + \psi_{|yy}$$
(58.15a)
$$X\psi_{p|Y} - Y\psi_{p|X} = x\psi_{|y} - y\psi_{|x}$$
(58.15b)

are scalar invariants; the subscripts behind the vertical bar denote partial derivatives with respect to the corresponding coordinates. The differentiation is straightforward, since z is kept constant.

In the paraxial approximation, the vector potential is given by (7.43–44)

$$A_X = -\frac{1}{2}YB(z) \quad , \quad A_Y = \frac{1}{2}XB(z) \quad , \quad A_z \equiv 0$$

and the corresponding term in (58.11) hence becomes

$$\boldsymbol{A}\cdot\nabla\psi_p \equiv A_X\psi_{p|X} + A_Y\psi_{p|Y}$$
$$= \frac{1}{2}B(X\psi_{p|Y} - Y\psi_{p|X}) = \frac{1}{2}B(z)(x\psi_{|y} - y\psi_{|x})$$
(58.16)

in which (58.15b) has been used.

In partial differentiations with respect to z we have to remember that

$$\psi_p(X, Y, z) = \psi(x(z), y(z), z)$$

and that X and Y are to be kept constant. With (58.13), we find

$$\psi_{p|z} = \psi_{|z} + \psi_{|x}x' + \psi_{|y}y' = \psi_{|z} - \theta'(z)(x\psi_{|y} - y\psi_{|x}) \qquad (58.17)$$

Introducing this and (58.16) into (58.11), we see that the mixed term can be eliminated by choosing

$$\theta'(z) = \frac{eB(z)}{2g(z)} \qquad (58.18)$$

This is seen to be in agreement with (15.9) if we recall (58.5); the term in $\Delta\phi$ describes the effect on $g(z)$ of chromatic aberrations, which can easily be included here.

The derivative $\psi_{p|zz}$ transforms to a more complicated expression if we repeat the operations of (58.17); it is, however, not really necessary to perform this calculation. The whole expression can be ignored since it is much smaller than the derivatives $\psi_{|xx}$ and $\psi_{|yy}$. This can be easily understood if we consider the special case of field-free motion ($\phi = $ const, $B \equiv 0$). We then have *eigen*equations

$$\psi_{|xx} = -\hbar^2 g_x^2 \psi \quad , \quad \psi_{|yy} = -\hbar^2 g_y^2 \psi$$

but as a consequence of (58.8):

$$\psi_{|zz} = -\hbar^2(g_z - g)^2 \psi$$

From $g^2 = g_x^2 + g_y^2 + g_z^2$ and $g_x^2 + g_y^2 = \alpha^2 g^2$ with a typically very small slope ($\alpha \approx 10^{-2}$), we can conclude that

$$(g - g_z)^2 \approx \frac{1}{4}g^2\alpha^4 \ll g_x^2 + g_y^2$$

can indeed be ignored in the paraxial approximation. This together with (58.14–18) reduces (58.11) to the simpler wave equation

$$\psi_{|xx} + \psi_{|yy} + \frac{2\mathrm{i}}{\hbar}g(z)\psi - \frac{r^2}{\hbar^2}\left\{\frac{em(z)}{2}\phi''(z) + \frac{e^2 B^2(z)}{4}\right\}\psi = 0$$

This in turn can be cast into a more convenient form by introduction of the lens function (15.13)

$$F(z) := g^{-2}(z)\left\{\frac{e}{2}m(z)\phi''(z) + \frac{e^2}{4}B^2(z)\right\}$$

$$= \frac{2(1 + 2\epsilon\phi)\phi'' + eB^2/m_0}{8\phi(1 + \epsilon\phi + 2\epsilon\Delta\phi)} = \frac{\gamma\phi'' + \eta^2 B^2}{4\hat{\phi}_c} \tag{58.19}$$

with $\hat{\phi}_c = \phi(1 + \epsilon\phi + 2\epsilon\Delta\phi)$. Finally we obtain the paraxial Schrödinger equation

$$\psi_{|xx} + \psi_{|yy} + \frac{2i}{\hbar}g(z)\psi_{|z} - \frac{g^2 r^2}{\hbar^2}F(z)\psi(x,y,z) = 0 \tag{58.20}$$

This differs from Glaser's expression (Glaser, 1952, eq. 159.31 or 1956, eq. 45.10) in the following respects:

(i) Relativistic correction terms are included.

(ii) Chromatic effects are included.

(iii) A term involving the derivative $g'(z)$ has been removed by the inclusion of the square-root factor (58.6): $\psi_{\text{Glaser}} \equiv \psi g^{-\frac{1}{2}}$.

The physical meaning of the different terms in (58.20) becomes more clear when we rewrite it thus:

$$-\frac{\hbar^2}{2m}(\psi_{|xx} + \psi_{|yy}) + \frac{g^2 F}{2m}(x^2 + y^2)\psi = \frac{i\hbar g}{m}\psi_{|z} \equiv i\hbar v\psi_{|z} \tag{58.21}$$

The first term represents the kinetic energy of the *transverse* motion, the second a purely quadratic *focusing potential*. The right-hand side of the wave equation shows that the role of the time t has been taken over by the axial coordinate z: the transformation $dz = v\,dt$ with the velocity $v = g/m$ is self-explanatory.

A continuity equation can be derived by a calculation quite analogous to that of Section 55.3. We multiply (58.20) by $\hbar\psi^*/2ig$ and add the result to its complex conjugate. The terms containing the factor F cancel out and we obtain

$$\frac{\hbar}{2ig}(\psi^*\psi_{|xx} - \psi\psi_{|xx}^* + \psi^*\psi_{|yy} - \psi\psi_{|yy}^*) + \psi^*\psi_{|z} + \psi\psi_{|z}^* = 0$$

This can be cast into the form of a two-dimensional continuity equation:

$$\frac{\partial}{\partial x}J_x + \frac{\partial}{\partial y}J_y + \frac{\partial}{\partial z}(\psi^*\psi) = 0 \tag{58.22}$$

with a flux density vector given by

$$J_x = \frac{\hbar}{g(z)}\Im(\psi^*\psi_{|x}), \qquad J_y = \frac{\hbar}{g(z)}\Im(\psi^*\psi_{|y}) \qquad (58.23)$$

From (58.22) the conservation law of intensity

$$I = \iint |\psi(x,y,z)|^2 \, dx\, dy = \text{const} \qquad (58.24)$$

can be derived, provided that ψ is square-integrable in a natural way, that is, without imposing some cut-off that would violate (58.20). Generally, this is possible only if $\psi(x,y,z)$, or more correctly $\Psi(\boldsymbol{r},t)$, represents a *wave packet* with a finite transverse spectrum.

58.2 Particular solution of the paraxial Schrödinger equation

The solution is obtained in much the same way as in Section 57.1; the only novel aspect is the reduction of dimensions, as the Laplacian in (58.20) is only two-dimensional.

We seek a solution of the form

$$\psi(x,y,z) = C(z)\exp\left\{\frac{i}{\hbar}\overline{S}(x,y,z)\right\} \qquad (58.25)$$

and shall see that this is possible in closed form and with no further approximations. Substitution of (58.25) into (58.20) results in a complex differential equation, which is equivalent to two coupled real ones. Separation into real and imaginary parts gives.

$$\overline{S}_{|x}^2 + \overline{S}_{|y}^2 + 2g\overline{S}_{|z} + g^2 r^2 F = 0 \qquad (58.26a)$$
$$(\overline{S}_{|xx} + \overline{S}_{|yy})C + 2gC' = 0 \qquad (58.26b)$$

We notice that $C(z)$ does not appear in (58.26a), which means that this equation is, in fact, *uncoupled* and that we should try to solve it first.

In the paraxial approximation, the eikonal $\overline{S}(x,y,z)$ must be a quadratic form in x and y and, to be consistent with the assumption of rotational symmetry, the terms in x^2 and y^2 can appear only in the form $x^2 + y^2$; hence

$$\overline{S}(x,y,z) = \frac{1}{2}Q(z)(x^2 + y^2) + \alpha(z)x + \beta(z)y + \gamma(z) \qquad (58.27)$$

with coefficient functions to be determined from (58.26a). Since this must be valid for *all* values of x and y, we obtain the four conditions

$$gQ' + Q^2 + g^2 F = 0 \qquad (a)$$
$$g\alpha' + Q\alpha = 0 \quad , \quad g\beta' + Q\beta = 0 \qquad (b) \quad (58.28)$$
$$g\gamma' + (\alpha^2 + \beta^2)/2 = 0 \qquad (c)$$

Equation (58.28a) is a Riccati equation and can be transformed into a homogeneous, *linear* differential equation; indeed, if we write

$$Q(z) =: g(z)h'_p(z)/h_p(z)$$

in (58.28a) and multiply the result by h_p/g^2, we arrive at

$$\frac{1}{g(z)}\frac{d}{dz}\{g(z)h'_p(z)\} + F(z)h_p(z) = 0 \qquad (58.29)$$

This is the familiar *paraxial ray equation**, as can be seen by comparison with (15.13); we have to bear in mind here that $g \propto \hat{\phi}^{\frac{1}{2}}$. The general solution of the paraxial ray equation requires a second, linearly independent partial solution, which we shall denote here by $g_p(z)$. The Wronskian

$$w = g(g_p h'_p - g'_p h_p)$$

is then a constant and must not vanish.

If we eliminate $Q(z)$ from (58.28b), we obtain the double equation

$$\frac{\alpha'}{\alpha} = \frac{\beta'}{\beta} = -\frac{h'_p}{h_p}$$

which is readily integrated to give

$$\alpha = -\frac{a}{h_p(z)} \quad , \quad \beta = -\frac{b}{h_p(z)}$$

with arbitrary constants of integration, a and b. Equation (58.28c) now becomes

$$\gamma' = -\frac{(a^2 + b^2)}{2gh_p^2}$$

---

* We use $g_p(z)$ and $h_p(z)$ for the solutions of the paraxial equation as the boundary conditions associated with these solutions (15.56) will prove convenient here. The suffix p is added to distinguish these paraxial solutions from the kinetic momentum and Planck's constant.

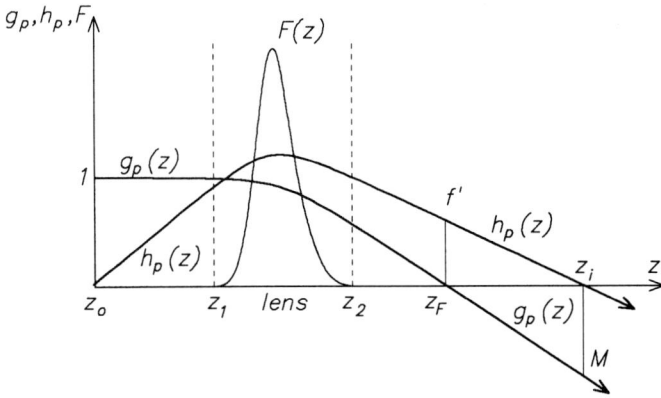

Fig. 58.1: The pair of basic solutions $g_p(z)$ and $h_p(z)$ of the paraxial ray equation with standard initial conditions adapted to the object plane $z = z_o$. The image plane is defined by $h_p(z_i) = 0$, and $g_p(z_i) = M < 0$ is then the magnification. The back focal plane is $z = z_F$ with $g_p(z_F) = 0$. In the most frequent case $g(z) = \text{const}$, the value $f' := h_p(z_F)$ is simply the focal length $f = -1/g_p'(z_F)$. More generally we have $f = f' g_i/g_o$.

and this too can be integrated in closed form, as follows: we divide the Wronskian by gh_p^2, giving

$$\frac{w}{gh_p^2} = \frac{g_p h_p'}{h_p^2} - \frac{g_p'}{h_p} = -\frac{d}{dz}\left(\frac{g_p}{h_p}\right)$$

and so

$$\gamma' = -\frac{a^2 + b^2}{2w}\frac{d}{dz}\left(\frac{g_p}{h_p}\right) \quad , \quad \gamma = \frac{a^2 + b^2}{2w}\frac{g_p(z)}{h_p(z)} + S_0$$

in which S_0 is a constant.

Subsituting all these expressions into (58.27), we find

$$\overline{S} = \frac{1}{h_p(z)}\left\{\frac{g(z)h_p'(z)}{2}(x^2 + y^2) - ax - by + \frac{a^2 + b^2}{2w}g_p(z)\right\} + S_0 \quad (58.30)$$

In order to apply this formula to the theory of image formation, it is advantageous to choose the functions $h_p(z)$ and $g_p(z)$ and the constants of integration in such a way that $\overline{S}(x, y, z)$ vanishes if the reference point (x, y, z) approaches an object point (x_o, y_o, z_o). This is easy with the standard initial conditions (Fig. 58.1)

$$g_p(z_o) = 1 \quad , \quad g_p'(z_o) = 0 \quad , \quad h_p(z_o) = 0 \quad , \quad h_p'(z_o) = 1 \quad (58.31a)$$

We then find

$$w \equiv g(z)\left\{ g_p(z)h'_p(z) - h_p(z)g'_p(z) \right\} = g(z_o) =: g_o \qquad (58.31b)$$

and

$$a = g_o x_o \quad , \quad b = g_o y_o \quad , \quad S_0 = 0$$

Hence

$$\overline{S} = \frac{g_o}{2h_p(z)}\left\{ \frac{g(z)}{g_o}h'_p(z)(x^2 + y^2) - 2(x_o x + y_o y) + (x_o^2 + y_o^2)g_p(z) \right\} \qquad (58.32)$$

This formula is *exactly* valid within the paraxial approximation.

For very small values of $|z - z_o|$ we can set $g(z) \to g_o$, $g_p(z) \to 1$, $h_p(z) \to z - z_o$, giving

$$\overline{S} \approx \frac{g_o}{2(z - z_o)}\{(x - x_o)^2 + (y - y_o)^2)\} \qquad (58.33)$$

This formula demonstrates an important difference between the wave-optical theory and the geometrical approximation. In the latter we have to choose $x(z)$ and $y(z)$ as rays passing through the point (x_o, y_o, z_o) and hence $x - x_o = x'_o(z - z_o)$, $y - y_o = y'_o(z - z_o)$. We then obtain the well-behaved result

$$\overline{S} = \frac{1}{2}g_o(z - z_o)(x_o'^2 + y_o'^2)$$

In the wave-optical theory, however, the coordinate pairs (x, y) and (x_o, y_o) are independent and $\overline{S}$ may hence become *singular* as $z - z_o$ vanishes. This singularity does not really appear in the final results because integrations remove it, as will become clear in later chapters.

It still remains to determine the amplitude $C(z)$ from (58.26b). The Laplacian of $\overline{S}$ from (58.30) is found to be

$$\overline{S}_{|xx} + \overline{S}_{|yy} = 2gh'_p/h_p$$

On introducing this into (58.26b), we see that the factor $2g$ cancels out and so

$$C'/C = -h'_p/h_p$$

This is readily integrated to give

$$C(z) = C_0/h_p(z) \qquad (58.34)$$

with an arbitrary constant of integration C_0. This amplitude also becomes singular for $z \to z_o$, but this singularity too will be removed by integration.

Introducing (58.34), (58.25) and (58.32) into (58.23), we obtain the density functions

$$\overline{\varrho}(z) := = |\psi|^2 = |C_0|^2/h_p^2(z) \qquad (58.35a)$$
$$J_x = \overline{\varrho}(h_p'x - g_o x_o/g)/h_p \qquad (58.35b)$$
$$J_y = \overline{\varrho}(h_p'y - g_o y_o/g)/h_p \qquad (58.35c)$$

It is easy to verify explicitly that the continuity equation (58.22) is satisfied. Here we have an obvious example of the fact that (58.22) does not always permit us to derive (58.24): the wavefunction is not square-integrable since $\overline{\varrho}$ is independent of x and y. The density functions take a slightly different form if we include the factor Ψ_0 of (58.6) and (58.8) and are then in agreement with Glaser's formulae.

58.3 Paraxial image formation

From (58.32) and (58.33), it is obvious that the partial solution (58.34) becomes singular as $z \to z_o$ and is thus not at all satisfactory. This difficulty can be easily circumvented by forming a suitable *wave packet*. We now regard the object coordinates x_0 and y_o as free parameters over which we may integrate. The wave packet is then constructed by forming a weighted linear superposition of solutions of (58.20) of the form

$$\psi(x,y,z) = \iint C(x_o, y_o) \exp\left\{ \frac{i}{\hbar} \overline{S}(x,y,z; x_o, y_o, z_o) \right\} \frac{dx_o dy_o}{h_p(z)} \qquad (58.36)$$

The amplitude function $C(x_o, y_o)$ can be chosen at will provided that it is sufficiently well-behaved for the integration. Glaser and Schiske (1953) were able to show that the function $C(x_o, y_o)$ is proportional to the wavefunction ψ itself in the object plane and exploited this to explain the process of image formation within the frame of wave mechanics. We briefly outline this theory and then discuss its inevitable shortcomings.

With the aid of (58.31b), we can cast the eikonal $\overline{S}$ (58.32) into the more convenient form

$$\overline{S} = \frac{gg_p'}{2g_p}(x^2 + y^2) + \frac{g_o g_p}{2h_p}\left\{ (x_o - x/g_p)^2 + (y_o - y/g_p)^2 \right\} \qquad (58.37)$$

The terms that do not depend on x_o and y_o are taken outside the integral, giving

$$\psi(x, y, z) = \frac{1}{h_p} \exp\left\{\frac{i g g_p'}{2\hbar g_p}(x^2 + y^2)\right\}$$

$$\times \iint C(x_o, y_o) \exp\left[\frac{i g_o g_p \kappa}{2\hbar h_p}\{(x_o - x/g_p)^2 + (y_o - y/g_p)^2\}\right] dx_o\, dy_o$$

$$(58.38)$$

We have to assume here that $g_p \neq 0$. The factor κ that we have included will be useful later; here it is equal to unity.

If we now examine the behaviour of this integral as $z \rightarrow z_0$, we see that the factor $h_p = z - z_o$ in the denominator causes huge but always imaginary exponents, which means that the whole exponential function is an extremely rapidly oscillating factor. The contributions to the integral cancel, unless $x = x_o/g_p$ and $y = y_o/g_p$. The approximation

$$C(x_o, y_o) \rightarrow C(x/g_p, y/g_p)$$

is therefore justified and this factor can then be taken outside the integral. The remaining integral is easily evaluated by introducing polar coordinates

$$x_o = x/g_p + r_o \cos\varphi \quad , \quad y_o = y/g_p + r_o \sin\varphi$$

and then writing $r_o^2/2 =: s$, $r_o\, dr_o = ds$. The integrand is independent of the azimuth φ and we find then, with a truncation radius R_0,

$$\iint \exp\left[i\frac{\Lambda}{2}\left\{(x_o - x/g_p)^2 + (y_o - y/g_p)^2\right\}\right] dx_o\, dy_o$$

$$= \int_0^{R_0} \int_0^{2\pi} \exp(i\Lambda r_o^2/2) r_o\, dr_o d\varphi$$

$$= 2\pi \int_0^{R_0^2/2} \exp(i\Lambda s)\, ds = \frac{2\pi i}{\Lambda}\{1 - \exp(i\Lambda R_0^2/2)\}$$

$$\Lambda := \frac{g_o g_p \kappa}{\hbar h_p}$$

The surviving exponential term depends very sensitively on R_0 if Λ is real. This difficulty can be removed by the following trick: we allow the constant κ to become complex, $\kappa = 1 + i\delta$, where δ is a very small but positive quantity. We now can proceed to the limit $R_0 \rightarrow \infty$, whereupon the exponential term vanishes. The remainder $2\pi i/\Lambda$ does not depend on R_0 and we can now set $\delta \rightarrow 0$.

This procedure is called a *limitation*; it is mathematically correct and physically justified, since we know that for $R_0 \rightarrow \infty$ the paraxial approximation has no meaning; the oscillations that have been eliminated are thus unphysical artefacts.

The quantity $h_p(z)$, which vanishes for $z = z_o$, now cancels in (58.38) and we obtain the well-behaved result

$$\psi(x, y, z) = \frac{2\pi i\hbar}{g_o g_p} C\left(\frac{x}{g_p}, \frac{y}{g_p}\right) \exp\left\{\frac{i g g'_p}{2\hbar g_p}(x^2 + y^2)\right\} \tag{58.39}$$

This formula can be used in various ways. First, we evaluate it in the *object plane*, $g_p = 1$, $g'_p = 0$, $g = g_o$ and obtain immediately

$$C(x_o, y_o) = \psi(x_o, y_o, z_o)\frac{g_o}{2\pi i\hbar} = -\frac{i}{\lambda_o}\psi(x_o, y_o, z_o) \tag{58.40}$$

in which $\lambda_o = 2\pi\hbar/g_o$ is the de Broglie wavelength.

There is, however, a second case of interest: the only condition for the validity of (58.39) was $h_p(z) \rightarrow 0$ and this is true also for the *image plane* $z = z_i$. There we have $g'_p = g'_{pi}$, $g = g_i$ and the magnification $g_{pi} = M$. In the rotating frame the relations between the lateral coordinates take the simple form $x_i = M x_o$, $y_i = M y_o$. From (58.39) we now obtain the simple formula

$$\psi(x_i, y_i, z_i) = \frac{2\pi i\hbar}{M g_o}C(x_o, y_o)\exp\left\{\frac{i g_i g'_{pi}}{2\hbar M}(x_i^2 + y_i^2)\right\}$$

$$= \frac{i\lambda_o}{M}C(x_o, y_o)\exp\left\{\frac{i\pi g'_{pi}}{\lambda_i M}(x_i^2 + y_i^2)\right\}$$

The quadratic phase factor is irrelevant, since only intensities can be measured. In fact it represents the spherical wave surface centred on the image point. After elimination of $C(x_o, y_o)$ by means of (58.40), we obtain the very simple relation:

$$|\psi(x_i, y_i, z_i)|^2 = M^{-2}|\psi(x_i/M, y_i/M, z_o)|^2 \tag{58.41}$$

This expression demonstrates clearly that the intensity distribution at the object plane is reproduced exactly at the image plane with magnification M. The factor M^{-2} guarantees that the total intensity is conserved in the sense that (58.24) is satisfied. We can hence conclude that the paraxial approximation is *self-consistent* in the sense that, after the truncation of the radial series expansions for the potentials beyond the quadratic terms and the omission of $\psi_{|zz}$, no further simplifications are necessary.

Returning to (58.36), we can use (58.40) to obtain an expression for the wavefunction in an arbitrary plane in terms of that in the object plane:

$$\psi(x, y, z)$$

$$= \frac{g_o}{2\pi i \hbar h_p(z)} \iint \psi(x_o, y_o, z_o) \exp\left\{ \frac{i}{\hbar} \overline{S}(x, y, z; x_o, y_o, z_o) \right\} dx_o \, dy_o$$

$$= \frac{1}{i\lambda_o h_p(z)} \iint \psi(x_o, y_o, z_o)$$

$$\times \exp\left[i\pi \left\{ \frac{h'_p}{\lambda h_p}(x^2 + y^2) - \frac{2}{\lambda_o h_p}(xx_o + yy_o) + \frac{g_p}{\lambda_o h_p}(x_o^2 + y_o^2) \right\} \right] dx_o \, dy_o$$

$$= \frac{e^{i\gamma}}{i\lambda_o h_p(z)} \iint \psi(x_o, y_o, z_o)$$

$$\times \exp\left[i\pi \left\{ -\frac{2}{\lambda_o h_p}(xx_o + yy_o) + \frac{g_p}{\lambda_o h_p}(x_o^2 + y_o^2) \right\} \right] dx_o \, dy_o$$

$$\tag{58.42}$$

with $\gamma := (\pi h'_p/\lambda h_p)(x^2 + y^2)$. This is the general *law of propagation* of the wavefunction in integral form. It enables us in principle to calculate the wavefunction anywhere in space provided that we know it in a particular plane. This is a consequence of the fact that the wave equation (58.20) is of first order in z; the omission of $\psi_{|zz}$ is hence a very powerful simplification.

As another example of the usefulness of the law of propagation, we consider the back focal plane $z = z_F$ in which $g_p(z)$ vanishes: $g_p(z_F) = 0$ (Fig. 58.1); we write $g(z_F) = g_f$ and $g_f = 2\pi\hbar/\lambda_f$. For simplicity, we assume that the space between the focal plane and image plane is field-free so that $g_f h_p(z_F)/g_o = \lambda_o h_p(z_F)/\lambda_f$ is equal to the focal length $f = -1/g'_p(z_F)$. From (58.42), we have

$$\psi(x, y, z_F) = \frac{g_f}{2\pi i \hbar f} \exp\left\{ \frac{ig_f h'_p(z_F)}{2\hbar f}(x^2 + y^2) \right\}$$

$$\times \iint \psi(x_o, y_o, z_o) \exp\left\{ -\frac{ig_f}{\hbar f}(x_o x + y_o y) \right\} dx_o \, dy_o$$

$$= \frac{1}{i\lambda_f f} \exp\left\{ \frac{i\pi h'_p(z_F)}{\lambda_f f}(x^2 + y^2) \right\}$$

$$\times \iint \psi(x_o, y_o, z_o) \exp\left\{ -\frac{2\pi i}{\lambda_f f}(x_o x + y_o y) \right\} dx_o \, dy_o$$

The quadratic phase factor $\exp i\gamma$ in front of the integral is of no immediate importance since it disappears when we form the intensity distribution. The integral has the form of a two-dimensional Fourier transform and can be cast into a convenient form by introducing the variables

$$q_x := x/\lambda_f f \quad , \quad q_y := y/\lambda_f f \qquad (58.43)$$

which are the components of the *spatial frequency.* Thus

$$\psi(x, y, z_F) = \frac{e^{i\gamma}}{i\lambda_f f} \iint \psi(x_o, y_o, z_o) \exp\{-2\pi i(x_o q_x + y_o q_y)\} \, dx_o \, dy_o$$

The inverse Fourier transform of any two-dimensional function $U(x_o, y_o)$ is defined by

$$\tilde{U}(q_x, q_y) := \mathcal{F}^-(U) \equiv \int\limits_{-\infty}^{\infty} U(x_o, y_o) \exp\{-2\pi i(q_x x_o + q_y y_o)\} \, dx_o \, dy_o$$

$$(58.44)$$

and so

$$\psi(x, y, z_F) = -\frac{i e^{i\gamma}}{\lambda_f f} \mathcal{F}^-(\psi(z = z_o)) \qquad (58.45)$$

Apart from unimportant factors, therefore, the wavefunction in the back focal plane is equal to the inverse Fourier transform of the object wave.

We state without proof a minor generalization of this result. The quadratic phase factor is absent from the wavefunction at the object plane, which is equivalent to illumination by a plane wave and hence to an electron source effectively at infinity. Let us now place the source at a finite distance from the object, and replace the g-ray by another solution of the paraxial equation , $\bar{g}(z)$, that again intersects the object plane at unit height, $\bar{g}(z_o) = 1$ but intersects the axis in the source plane. A quadratic phase factor now does appear at the object plane and the Fourier transform of the object wave is formed in the plane conjugate to the source plane. Only in the special case of illumination by a plane wave or infinitely distant source does this plane coincide with the back focal plane.

The intensity corresponding to (58.45) is proportional to $\tilde{\psi}\tilde{\psi}^*$, where $\tilde{\psi}$ denotes $\mathcal{F}^-(\psi(z = z_o))$; this is the Fourier transform of the autocorrelation function of the object wavefunction.

Concluding remarks

The paraxial approximation of the wave theory has enabled us to derive some important laws, notably (58.41), which states that a sharp image should be obtained on the basis of this theory. Although this is in full agreement with the corresponding classical corpuscular theory, it is certainly *wrong*: we have completely ignored the inevitable *lens aberrations* and all *diffraction* at beam-confining apertures. The latter could be incorporated into the paraxial theory, as Glaser and Schiske (1953) did, but this is still not entirely satisfactory.

An attempt to extend the paraxial Schrödinger equation (58.20) further, to include lens aberrations, reveals a severe weakness of this theory (Glaser, 1953, 1954; Glaser and Braun, 1954, 1955). It is no longer permissible to ignore the term $\psi_{|zz}$, so that we have no simple propagation law like (58.42). The quantity $\boldsymbol{A} \cdot \nabla \psi_p$ cannot be completely eliminated since we have nonlinear terms in the vector potential. Consequently, mixed second-order derivatives of ψ appear in a complicated manner. The theory then loses all its attraction, and we shall therefore not pursue it further.

In the subsequent chapters we shall deal with diffraction processes in a fairly general manner, before we again start to specialize.

59
The General Theory of Electron Diffraction and Interference

In this chapter we shall develop the fundamentals of electron diffraction and interference. Practical examples are collected in Chapter 60 in order to avoid too many distractions from the general theme by technical details.

Although electron optical diffraction phenomena closely resemble those of light optics, the underlying theory is much more complicated than Kirchhoff's approach. This is a consequence of the fact that an electron diffraction pattern is separated from the object generating it by an electromagnetic field. We shall nevertheless start with the derivation of Kirchhoff's formula, in order to avoid too much abstraction at the beginning.

Electron interference patterns differ from the corresponding light-optical phenomena in two important respects: polarization effects are absent and magnetic flux affects the location of the interference fringes. This so-called 'Aharonov–Bohm effect' will be dealt with in Section 59.6.

59.1 Kirchhoff's general diffraction formula

Kirchhoff's formula is an integral that provides an *approximate* expression for diffraction phenomena *in field-free space*. The propagation of electrons in field-free domains is described by a solution of the wave equation (55.8)

$$\nabla^2 \psi(\boldsymbol{r}) + k^2 \psi(\boldsymbol{r}) = 0 \tag{59.1}$$

Here we have again used the conventional wavenumber

$$k = \frac{g}{\hbar} = \frac{2\pi}{\lambda} \tag{59.2}$$

g being the constant kinetic momentum. A simple solution of (59.1) in spherical coordinates (R, ϑ, φ) is evidently

$$\hat{G}(R) = \frac{1}{4\pi R}\, e^{ikR} \tag{59.3}$$

for $R > 0$. This can be generalized to a function of *two* positions $\boldsymbol{r}$ and $\boldsymbol{r}'$:

$$G(\boldsymbol{r}, \boldsymbol{r}') = \frac{1}{4\pi|\boldsymbol{r} - \boldsymbol{r}'|} e^{ik|\boldsymbol{r}-\boldsymbol{r}'|} \qquad (\boldsymbol{r} \neq \boldsymbol{r}') \tag{59.4}$$

which obviously satisfies (59.1) with respect to *both* positions, r and r', since it exhibits the symmetry property

$$G(r, r') = G(r', r) = \hat{G}(|r - r'|) \tag{59.5}$$

The function G becomes *singular* for $r \to r'$, when $R \to 0$. We shall see that this leads to a partial differential equation for G containing Dirac's δ-function. To show this, we keep the singularity at position r' fixed and consider a sphere S with centre at r' and radius $R = |r - r'|$ in the r-space. Integration of the normal derivative $n \cdot \nabla G = d\hat{G}/dR$ over the surface ∂S gives

$$\oint_{\partial S} n \cdot \nabla G \, d^2 r = (-1 + ikR) e^{ikR}$$

As $R \to 0$, this tends to a finite limit:

$$\lim_{R \to 0} \oint_{\partial S} n \cdot \nabla G \, d^2 r = -1 \tag{59.6}$$

The theory of distributions shows that Gauss's integral theorem can be generalized to include singular functions. It now tells us that

$$\oint_{\partial S} n \cdot \nabla G \, d^2 r = \int_S \nabla^2 G \, d^3 r$$

and consequently

$$\lim_{R \to 0} \int_S (\nabla^2 G + k^2 G) \, d^3 r = -1 \tag{59.7}$$

The additional term vanishes linearly with R and therefore does not alter the limit. *Dirac's function* has the property

$$\int f(r) \delta(r - r') \, d^3 r = f(r') \tag{59.8}$$

so that if we choose $f = -1$, we see that the integrand of (59.7) is $-\delta(r - r')$. Hence

$$\nabla^2 G(r, r') + k^2 G(r, r') = -\delta(r, r') \tag{59.9a}$$

and, owing to the symmetry (59.5), G also satisfies

$$\nabla'^2 G(r, r') + k^2 G(r, r') = -\delta(r, r') \tag{59.9b}$$

A function satisfying these inhomogeneous partial differential equations is called a *Green's function*; in the present context it is the 'free' Green's function.

It is now fairly simple to derive an integral equation for the solution of (59.1). We begin by writing $u = \psi(\boldsymbol{r})$ and $v = G(\boldsymbol{r}, \boldsymbol{r}')$ with fixed $\boldsymbol{r}'$ in Green's integral theorem

$$\int_D (v\nabla^2 u - u\nabla^2 v)\,d^3r = \oint_{\partial D} \boldsymbol{n}\cdot(v\nabla u - u\nabla v)\,d^2r$$

Here D denotes a closed domain (containing the point $\boldsymbol{r}$) and ∂D its surface. On the left-hand side, the terms in k^2, arising from (59.1) and (59.9a) cancel out and only $\psi(\boldsymbol{r})\delta(\boldsymbol{r} - \boldsymbol{r}')$ remains in the integrand. Using (59.8) with $f = \psi$, integration gives $\psi(\boldsymbol{r}')$ and hence

$$\psi(\boldsymbol{r}') = \oint_{\partial D} \boldsymbol{n}\cdot\{G(\boldsymbol{r}, \boldsymbol{r}')\nabla\psi(\boldsymbol{r}) - \psi(\boldsymbol{r})\nabla G(\boldsymbol{r}, \boldsymbol{r}')\}\,d^2r \qquad (59.10)$$

This is Kirchhoff's formula in its most general form. It is exact and, in principle at least, it allows values of the function $\psi(\boldsymbol{r})$ to be calculated inside the domain from a knowledge of the boundary values of ψ and $\boldsymbol{n}\cdot\nabla\psi$ on ∂D.

Unfortunately this integral formula is of no practical use, since the boundary values of ψ and $\boldsymbol{n}\cdot\nabla\psi$ cannot be chosen independently and consistent sets of them of any physical interest are unknown (cf. Sommerfeld, 1954, Section 34). Thus major simplifications are necessary to generate a useful formula.

59.2 Necessary simplifications

We now suppose that the electron wave falls on a large opaque screen with small openings, through which the wave can propagate from the exterior into the domain D of diffraction (Fig. 59.1). Complete opacity means that we can assume that ψ and $\boldsymbol{n}\cdot\nabla\psi$ vanish on the *inner* side of the corresponding parts S. The hemisphere H can be chosen so large that it does not contribute to the diffraction surface integral. This is justified, since for $R \to \infty$ the Green's function and the diffracted wavefunction become asymptotically proportional, so that the integrand of (59.10) vanishes at least as R^{-3}.

The openings A are very small in comparison with the distance of the observation point $\boldsymbol{r}'$ from the screen. On the other hand their diameters are

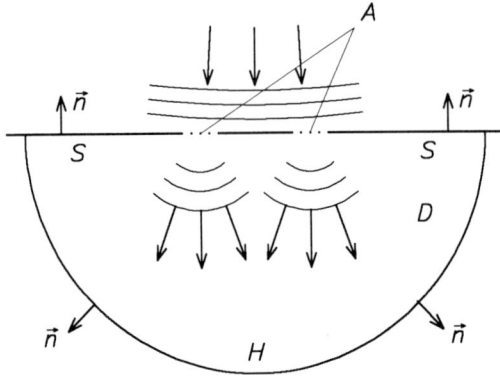

Fig. 59.1: A typical diffraction configuration consisting of a screen with openings A and opaque parts S. The domain D is closed by a large hemisphere H. A wave propagates from outside through the openings A into the interior of D.

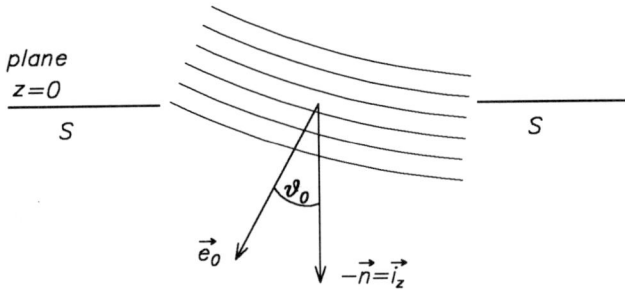

Fig. 59.2: The undiffracted wave ψ_0 in the vicinity of an opening and its mean wave-normal e_0.

much larger than the wavelength. In the plane of the screen, the incident wave is then very little affected by the edges of the apertures. This influence does not extend beyond about one wavelength. It is hence reasonable to assume that the boundary values in the apertures A are equal to the *unperturbed* incident wave $\psi = \psi_0(\mathbf{r})$. The integral equation (59.10) then reduces to the well-defined integral

$$\psi(\mathbf{r}') = i_z \cdot \iint\limits_A \{-G(\mathbf{r}, \mathbf{r}')\nabla\psi_0(\mathbf{r}) + \psi_0(\mathbf{r})\nabla G(\mathbf{r}, \mathbf{r}')\} \, dx \, dy \qquad (59.11)$$

the screen now being located in the coordinate plane $z = 0$ (Fig. 59.2).

Although this integral contains only well-known functions, it is still too complicated for exact evaluation. The assumption that the point of observation is far from the screen implies that $|\mathbf{r} - \mathbf{r}'| \gg \lambda$. This distance is also much larger than the geometrical size of the apertures, which are located near the origin of the coordinate system; this implies $|\mathbf{r} - \mathbf{r}'| \gg r$ in the area of integration. These very strong assumptions are often satisfied in practice.

The following approximations for the Green's function and its normal derivative are then justified:

$$G(\mathbf{r}, \mathbf{r}') = \frac{1}{4\pi R'} e^{ik|\mathbf{r} - \mathbf{r}'|} \qquad (59.12a)$$

$$\mathbf{i}_z \cdot \nabla G = -ik \cos \vartheta' G(\mathbf{r}, \mathbf{r}') \qquad (59.12b)$$

with

$$R' = |\mathbf{r}'| = (x'^2 + y'^2 + z'^2)^{\frac{1}{2}} \qquad (59.12c)$$

and

$$\cos \vartheta' = \frac{z'}{R'} \qquad (59.12d)$$

Moreover we make the approximation

$$\nabla \psi_0(\mathbf{r}) = ik\mathbf{e}_0 \psi_0(\mathbf{r}) \qquad (a)$$
$$\mathbf{i}_z \cdot \nabla \psi_0 = ik \cos \vartheta_0 \psi_0(\mathbf{r}) \qquad (b) \qquad (59.13)$$

and consider the unit vector $\mathbf{e}_0$ (see Fig. 59.2) to be so slowly varying that it may be replaced by its mean value. This implies that the curvature of the wavefronts may be neglected in the slowly varying amplitude factor (but not in the rapidly varying phase factor). Bringing all this together, Kirchhoff's formula (59.11) simplies further to

$$\psi(\mathbf{r}') = \frac{k(\cos \vartheta_0 + \cos \vartheta')}{4\pi i R'} \iint_A \psi_0(\mathbf{r}) e^{ik|\mathbf{r} - \mathbf{r}'|} \, dx \, dy \qquad (59.14)$$

It is the rapidly varying exponential factor in (59.14) that creates difficulties. These can be circumvented, as we shall show in the next section.

59.3 Fresnel and Fraunhofer diffraction

The assumption that we have made concerning the magnitude of $|\mathbf{r} - \mathbf{r}'|$ suggests that we should expand the exponential term in (59.14) as a power series. Using (59.12c and d) and recalling that $z = 0$, we have

$$|\mathbf{r} - \mathbf{r}'| = R' - \frac{1}{R'}(xx' + yy') + \frac{1}{2R'}(x^2 + y^2) - \frac{1}{2R'^3}(xx' + yy')^2 + \dots \quad (59.15)$$

In almost all experimental situations, the diffraction angles are small, which means that $|\vartheta_0| \ll 1$, $|\vartheta'| \ll 1$ and hence that $x'^2 + y'^2 \ll z'^2$. Consequently

$$R' = z' + \frac{x'^2 + y'^2}{2z'} + O\left(\frac{1}{z'^3}\right) \qquad (59.16a)$$

We may then drop the last term in (59.15) and can rewrite this expansion more concisely as

$$|\mathbf{r} - \mathbf{r}'| = z' + \frac{(x - x')^2 + (y - y')^2}{2z'} + O\left(\frac{1}{z'^3}\right) \qquad (59.16b)$$

Substituting all this into (59.14) yields Fresnel's diffraction formula

$$\psi(\mathbf{r}') = \frac{1}{i\lambda z'} e^{ikz'} \iint_A \psi_0(x, y) e^{i\varphi} \, dx \, dy \qquad (59.17a)$$

where

$$\varphi = \frac{k}{2z'} \left\{ (x - x')^2 + (y - y')^2 \right\} \qquad (59.17b)$$

Important examples of the application of this diffraction formula will be found in Chapter 60.

A still further simplification is obtained if the diffracting object is so small that even the second-order terms in x and y can be ignored. It is then not necessary to assume that x' and y' are also small, although this will quite often be the case. The result takes a very convenient form, if we introduce the notion of *spatial frequency* (58.43)

$$\begin{aligned} q_x &:= \frac{kx'}{R} = \frac{2\pi}{\lambda} \sin \alpha_x \\ q_y &:= \frac{ky'}{R} = \frac{2\pi}{\lambda} \sin \alpha_y \end{aligned} \qquad (59.18)$$

$\alpha_x := x'/R'$ and $\alpha_y := y'/R'$ being the diffraction angles (Fig. 59.3). With $\cos \vartheta \approx \cos \vartheta'$, we find the *Fraunhofer formula*

$$\psi(\mathbf{r}') = \frac{1}{i\lambda R'} e^{ikR'} \cos \vartheta' \iint_A \psi_0(x, y) \exp\{-i(xq_x + yq_y)\} \, dx \, dy \qquad (59.19)$$

The essential part of this diffraction formula is now a two-dimensional *Fourier integral*. Since the mathematical techniques for the calculation of

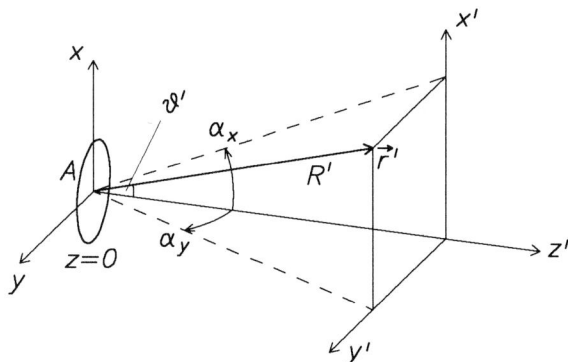

Fig. 59.3: Coordinate system and diffraction angles in Fraunhofer's formula; the diffracting aperture lies in the plane $z = 0$ and close to the origin; the vector $\boldsymbol{r}'$ specifies the observation point. The angles satisfy the relation $\sin^2 \vartheta' = \sin^2 \alpha_x + \sin^2 \alpha_y$.

Fourier integrals are highly advanced, it is fairly easy to evaluate (59.19). This is one reason why many optical techniques rely on Fraunhofer diffraction, as will become more evident in later chapters.

A straightforward generalization of this analysis extends the reasoning to the case of spherical incident waves. This shows that the Fraunhofer diffraction pattern is formed in the plane conjugate to the centre of the spherical incident wavefront, a result of considerable practical importance. In a microscope, for example, this diffraction pattern is formed in the plane conjugate to the crossover of the gun, which may not coincide exactly with the back focal plane of the objective. The presence of the quadratic phase factors in the propagation laws of Chapter 58 is another manifestation of this. For details of this generalization, see for example Section 8.3.3 of Born and Wolf (1980).

59.4 Electron diffraction in the presence of electromagnetic fields

The theory of diffraction outlined above is mathematically identical with the corresponding theory in light optics and fairly simple. We have recapitulated it to familiarize the reader with the concepts and approximations of the diffraction theory. In reality, this field-free approach is unsatisfactory since the diffracted electron waves are strongly influenced by the fields in electron lenses and other electron optical elements. It is therefore necessary to generalize the theory for these cases. The first, fully developed account of this topic appeared in the 1952 treatise of Glaser, who had been pondering over the wave theory of electron optics since 1943 (Glaser 1943; Glaser 1950a, which is based on a course given in Prague in 1942/1943; 1950b,c; 1951b). The detailed theory was developed in collaboration with Schiske (Glaser and Schiske, 1953), as Glaser makes clear (note [236] of Glaser, 1952; Glaser, 1951b, p. 111). The present account proceeds along similar lines to theirs.

The starting point is the time-independent Schrödinger equation (56.11) and the aim is to establish an approximate solution in the form of a diffraction formula. It is convenient to use the expression (56.12) for the wavenumber k even in the presence of a magnetic vector potential, and we can then rewrite (56.11) more concisely:

$$\mathsf{D}\psi := \left[\{\nabla + \frac{ie}{\hbar}\boldsymbol{A}(\boldsymbol{r})\}^2 + k^2(\boldsymbol{r}) \right]\psi(\boldsymbol{r}) = 0 \qquad (59.20)$$

This is evidently a generalization of the Helmholtz equation (59.1) in which the vector potential is present and the wavenumber is position-dependent.

A new aspect is that the operator D, as defined by (59.20), is *not self-adjoint*. The adjoint operator is identical with the complex conjugate:

$$\mathsf{D}^\dagger = \{\nabla - \frac{ie}{\hbar}\boldsymbol{A}(\boldsymbol{r})\}^2 + k^2(\boldsymbol{r}) \qquad (59.21)$$

Green's integral theorem is now not directly applicable but must be replaced by a more general integral formula containing the vector potential. In order to find this, we have to recast the alternating bilinear form $v\mathsf{D}u - u\mathsf{D}^\dagger v$ as a divergence expression which can then be integrated.

Evaluation of the squares of the operators appearing in D and $\mathsf{D}^\dagger$ gives

$$v\mathsf{D}u - u\mathsf{D}^\dagger v = v\nabla^2 u - u\nabla^2 v + \frac{2ie}{\hbar}\{\boldsymbol{A} \cdot (v\nabla u - u\nabla v) + uv\nabla \cdot \boldsymbol{A}\}$$

This can be rewritten as

$$v\mathsf{D}u - u\mathsf{D}^\dagger v = \mathrm{div}(v\ \mathrm{grad}\,u - u\ \mathrm{grad}\,v + \frac{2ie}{\hbar}\boldsymbol{A}uv)$$

Gauss's integral theorem then gives

$$\int_D (v\mathsf{D}u - u\mathsf{D}^\dagger v)\,d^3r = \oint_{\partial D} \boldsymbol{n} \cdot (v\nabla u - u\nabla v + \frac{2ie}{\hbar}\boldsymbol{A}uv)\,d^2r \qquad (59.22)$$

We now identify u with the required solution ψ of (59.20) and v with the corresponding Green's function $G(\boldsymbol{r}, \boldsymbol{r}')$; the latter has to satisfy the *adjoint* PDE:

$$\mathsf{D}^\dagger G(\boldsymbol{r}, \boldsymbol{r}') = -\delta(\boldsymbol{r} - \boldsymbol{r}') \qquad (59.23)$$

Introducing these choices into (59.22), we find immediately

$$\psi(\boldsymbol{r'}) = \oint_{\partial D} \boldsymbol{n} \cdot \{G(\boldsymbol{r}, \boldsymbol{r'})\nabla\psi(\boldsymbol{r}) - \psi(\boldsymbol{r})\nabla G(\boldsymbol{r}, \boldsymbol{r'})\} \, d^2 r$$

$$+ \frac{2ie}{\hbar} \oint G(\boldsymbol{r}, \boldsymbol{r'})\psi(\boldsymbol{r})\boldsymbol{n}(\boldsymbol{r}) \cdot \boldsymbol{A}(\boldsymbol{r}) \, d^2 r \qquad (59.24)$$

which is evidently a generalization of (59.10).

The next task is thus to find the appropriate Green's function, which will meet with only partial success. We consider the reference point $\boldsymbol{r'}$ as the singularity of G and look for a solution of the form

$$G(\boldsymbol{r}, \boldsymbol{r'}) = a(\boldsymbol{r}, \boldsymbol{r'})e^{iS(\boldsymbol{r}, \boldsymbol{r'})/\hbar} \qquad (59.25)$$

$\boldsymbol{r}$ being the variable to which the operator $\mathsf{D}^\dagger$ refers. The function $S(\boldsymbol{r}, \boldsymbol{r'})$ is the point eikonal from the integration point $\boldsymbol{r}$ to the reference position $\boldsymbol{r'}$. A solution of (59.20) of the form (59.25) would require an eikonal S with $\boldsymbol{r}$ as the terminal position. This is consistent with (59.25), since the eikonal is unaltered if the two points are exchanged and $\boldsymbol{A}(\boldsymbol{r})$ changes its sign, as required by (59.21):

$$S(\boldsymbol{r}, \boldsymbol{r'})_{(+\boldsymbol{A})} = S(\boldsymbol{r'}, \boldsymbol{r})_{(-\boldsymbol{A})} \qquad (59.26)$$

Equation (59.25) hence has the appropriate form.

The singularity of the Green's function must have the same strength as in the field-free case, since the fields are unimportant in the vicinity of the singularity:

$$\lim_{|\boldsymbol{r}-\boldsymbol{r'}|\to 0} \{4\pi|\boldsymbol{r} - \boldsymbol{r'}|a(\boldsymbol{r}, \boldsymbol{r'})\} = 1 \qquad (59.27)$$

The amplitude has to satisfy a continuity equation corresponding to (59.12), in which the vector potential term has a negative sign. Since $\boldsymbol{r}$ is now the starting position in $S(\boldsymbol{r}, \boldsymbol{r'})$, ∇S changes sign and the kinematic momentum is hence given by

$$\boldsymbol{g}(\boldsymbol{r}) = -\left(\frac{\partial S}{\partial \boldsymbol{r}} + e\boldsymbol{A}(\boldsymbol{r})\right) \qquad (59.28)$$

Thus (57.14) is again satisfied. To obtain a closed formula for the amplitude, we apply (57.19) to the configuration shown in Fig. 59.4. The quantities with label 1 refer to the element of area $d\sigma$ in the plane $z = \text{const}$, while the quantities with label 2 refer to the element of solid angle subtended at

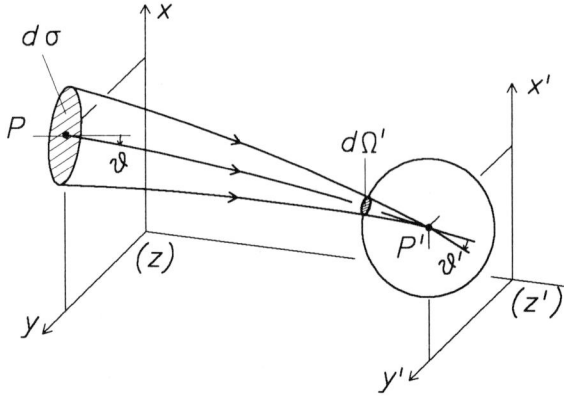

Fig. 59.4: Definition of geometrical parameters: The rays starting in the area $d\sigma$ with centre P intersect at point P'; $d\Omega'$ is the element of solid angle subtended by these rays.

P'. The surface normal at P is inclined at an angle ϑ to the beam axis, and so

$$a_1^2 g_1 \Delta\sigma_1 \rightarrow a^2(\boldsymbol{r}, \boldsymbol{r}') g(\boldsymbol{r}) \Delta\sigma \cos\vartheta$$

From (59.27), and considering the radius $R = 1$:

$$a_2^2 g_2 \Delta\sigma_2 \rightarrow \frac{1}{(4\pi)^2} g(\boldsymbol{r}') \Delta\Omega'$$

Equating these two expressions, we find

$$a^2(\boldsymbol{r}, \boldsymbol{r}') = \frac{g(\boldsymbol{r}')}{16\pi^2 g(\boldsymbol{r}) \cos\vartheta} \frac{\Delta\Omega'}{\Delta\sigma} \tag{59.29}$$

In the limit $\Delta\Omega' \rightarrow 0$, $\Delta\sigma \rightarrow 0$, the expression $\Delta\Omega'/\Delta\sigma$ becomes a Jacobian determinant, which can be expressed in terms of derivatives of S. The ray tangent at $\boldsymbol{r}'$ has the components

$$\sin\alpha_k = \frac{g_k(\boldsymbol{r}')}{g(\boldsymbol{r}')} = -\frac{1}{g(\boldsymbol{r}')}\left\{\frac{\partial S}{\partial x'_k} + eA_k(\boldsymbol{r}')\right\} \quad (k = 1, 2, 3)$$

with $\vartheta' = \pi/2 - \alpha_3$. Evaluation of dg_x and dg_y with respect to $\boldsymbol{r}'$, while retaining x and y as parameters gives

$$d\Omega' = \frac{dg_x dg_y}{g^2(\boldsymbol{r}')\cos\vartheta'} = \frac{1}{g^2(\boldsymbol{r}')\cos\vartheta'} \frac{\partial(g_x, g_y)}{\partial(x, y)} dx\, dy$$

$$= \frac{1}{g^2(\boldsymbol{r}')\cos\vartheta'} \begin{vmatrix} \partial^2 S/\partial x \partial x' & \partial^2 S/\partial y \partial x' \\ \partial^2 S/\partial x \partial y' & \partial^2 S/\partial y \partial y' \end{vmatrix} d\sigma$$

(The vector potential terms do not depend on x and y.) Introducing this into (59.29), we obtain the final result

$$a(\boldsymbol{r}, \boldsymbol{r}') = \{16\pi^2 g(\boldsymbol{r})g(\boldsymbol{r}') \cos \vartheta \cos \vartheta'\}^{-\frac{1}{2}} \begin{vmatrix} \partial^2 S/\partial x \partial x' & \partial^2 S/\partial y \partial x' \\ \partial^2 S/\partial x \partial y' & \partial^2 S/\partial y \partial y' \end{vmatrix}^{\frac{1}{2}}$$

(59.30)

Obviously this expression is *symmetric* with respect to the two positions:

$$a(\boldsymbol{r}, \boldsymbol{r}') = a(\boldsymbol{r}', \boldsymbol{r})$$

(59.31)

The general theory now passes through essentially the same stages as the field-free case. We again make the various simplifications and arrive at a formula containing the undiffracted wave ψ_0. This differs from (59.11) in that an additional term involving the vector potential is now present:

$$
\begin{aligned}
\psi(\boldsymbol{r}') = \boldsymbol{i}_z \iint\limits_{A} & \{-G(\boldsymbol{r}, \boldsymbol{r}')\nabla\psi_0(\boldsymbol{r}) + \psi_0(\boldsymbol{r})\nabla G(\boldsymbol{r}, \boldsymbol{r}')\} \, d^2r \\
& - \frac{2ie}{\hbar} \iint\limits_{A} G(\boldsymbol{r}, \boldsymbol{r}')\psi_0(\boldsymbol{r})A_z(\boldsymbol{r}) \, d^2r
\end{aligned}
$$

(59.32)

Simplifying assumptions concerning the local gradients can now be made. For $\nabla\psi_0$ we have

$$
\begin{aligned}
\{\hbar\nabla + ie\boldsymbol{A}(\boldsymbol{r})\}\psi_0(\boldsymbol{r}) &= ig(\boldsymbol{r})e_0(\boldsymbol{r})\psi_0(\boldsymbol{r}) \\
\boldsymbol{i}_z \cdot (\hbar\nabla + ie\boldsymbol{A})\psi_0 &= ig(\boldsymbol{r})\cos\vartheta_0\psi_0(\boldsymbol{r})
\end{aligned}
$$

(59.33)

which is a generalization of (59.13). The angle ϑ_0 now denotes the angle of *incidence*, which may be different from the angle ϑ in Fig. 59.4. The corresponding derivative of the Green's function can be approximated by

$$\boldsymbol{i}_z \cdot (\hbar\nabla - ie\boldsymbol{A})G(\boldsymbol{r}, \boldsymbol{r}') = -ig(\boldsymbol{r})\cos\vartheta G(\boldsymbol{r}, \boldsymbol{r}')$$

(59.34)

since the contribution from ∇a can again be ignored in the far zone. Introducing all this into (59.32), we obtain the still fairly accurate diffraction formula

$$\psi(\boldsymbol{r}') = -\frac{i}{\hbar} \iint\limits_{A} (\cos\vartheta_0 + \cos\vartheta)g(\boldsymbol{r})G(\boldsymbol{r}, \boldsymbol{r}')\psi_0(\boldsymbol{r}) \, dx \, dy$$

(59.35)

In practice, all direction cosines can be replaced by unity. The diffracting object is so small that the amplitude $a(\boldsymbol{r}, \boldsymbol{r}')$ can be replaced by its *axial* value and the kinetic momentum is then given by

$$g(z) = \{2m_0 e\hat{\phi}(z)\}^{\frac{1}{2}} \quad , \quad g(z') = \{2m_0 e\hat{\phi}(z')\}^{\frac{1}{2}}$$

The Jacobian appearing in (59.30) is replaced by its axial value

$$J(z, z') := \begin{vmatrix} \partial^2 S/\partial x \partial x' & \partial^2 S/\partial y \partial x' \\ \partial^2 S/\partial x \partial y' & \partial^2 S/\partial y \partial y' \end{vmatrix} \tag{59.36}$$

For a diffracting object in the plane $z = 0$, we finally arrive at the diffraction formula

$$\psi(\boldsymbol{r}') = -ib(z') \iint_A \psi_0(x, y) e^{iS(\boldsymbol{r}, \boldsymbol{r}')/\hbar}\, dx\, dy \tag{59.37a}$$

with

$$b(z') = \frac{1}{\hbar} \left\{ J(0, z') \frac{\hat{\phi}(0)}{\hat{\phi}(z')} \right\}^{\frac{1}{2}} \tag{59.37b}$$

In the case of field-free diffraction we have

$$S(\boldsymbol{r}, \boldsymbol{r}') = g|\boldsymbol{r} - \boldsymbol{r}'|$$

for which the Jacobian has the exact form

$$J(\boldsymbol{r}, \boldsymbol{r}') = \frac{g^2(z - z')^2}{|\boldsymbol{r} - \boldsymbol{r}'|^4}$$

Making the small-angle approximation and recalling that $g = \hbar k$ (59.2), we recover (59.14) with $R' = |z'|$: (59.37a,b) essentially contain (59.14) as a special case.

In Glaser's book (1952), the assumption $G(\boldsymbol{r}, \boldsymbol{r}') = 0$ for $z = 0$ was made to eliminate the term involving $\nabla \psi$ from (59.24). The corresponding Green's function then becomes more complicated by virtue of the mirror operations which then become necessary. We have preferred to circumvent this difficulty by making the reasonable and simple assumption (59.33), which leads practically to the same result.

Although we have succeeded in casting the solution of the diffraction problem into the fairly compact form of equations (59.37), we are still not in a position to carry out the necessary integrations. As we have seen in Section 59.3, even Fresnel diffraction in field-free space is fairly complicated. We cannot expect to obtain simple explicit results in closed form. The next steps are therefore *series expansions* of S in the exponential term. Since S is the *characteristic function* (apart from a factor $\sqrt{2m_0 e}$, omitted in Parts III and IV of Vol. 1), we can invoke the vast mass of material on such series expansions derived in geometrical electron optics.

Even this will be of no immediate practical help, since the integrals that arise are still far too complicated to be evaluated in closed form.

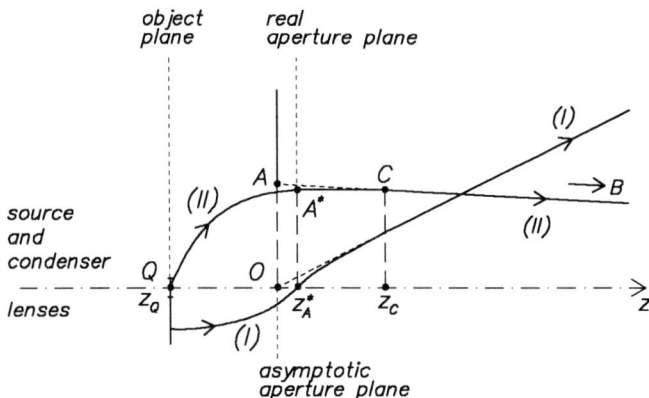

Fig. 59.5: Classical paraxial electron trajectories I and II with their exit asymptotes (broken lines). The physical aperture must be located in the plane $z = z_A^*$, while the asymptotic aperture lies in the plane through the intersection point O.

Efficient numerical integration requires entirely new concepts, which will be developed in later chapters.

59.5 Asymptotic diffraction formulae

Quite often, the beam-confining aperture in an electron microscope is located in the weak fringe-field domain behind the magnetic objective lens. Not only are the electron waves irradiating the aperture from the specimen side influenced by the field, but so too are the diffracted waves. The surfaces of equal phase, and by implication the rays, are weakly curved until they become indistinguishable from their exit asymptotes. This situation is sketched in Fig. 59.5.

A correct treatment of this diffraction problem would be extremely complicated and has never, as far as we know, been attempted. Glaser's contribution, which is presented in considerable detail in his textbook (1952), deals only with *paraxial* imaging and therefore leaves the difficult part of the problem untouched, valuable though his analysis is.

Clearly some intermediate approach is required for it is certainly not permissible to ignore the field beyond the aperture completely, though this is often done for simplicity. The approximate reasoning that follows seems a good compromise between total neglect of the fringing field beyond the aperture and the extreme complication of a rigorous analysis.

We consider a very small specimen close to the optic axis, as is the case in high resolution electron microscopy. Its axial point is Q in Fig. 59.5. The family of rays (I) pass through the specimen without any interaction; the slope in the plane $z = z_Q$ is determined by the condenser system; the axial intersection at z_A^* specifies the plane of the real (physical) aperture. The asymptotic aperture, considered later, is located in the plane through the intersection point O of the corresponding *exit* asymptote; without loss of generality we can choose this point as the origin of the coordinate system.

Ray (II) in Fig. 59.5 is scattered at the axial point Q in the specimen; it intersects the real aperture plane in A^*, the corresponding asymptotic intersection point is A. We assume now that for $z \geq z_C$ all real rays are practically identical with their asymptotes. The whole configuration can easily be calculated by numerical ray tracing, which does not provide any problems.

Let us now consider the eikonal S_{II} from the starting point Q to the terminal point B; this is

$$S_{II} = S_{QA^*} + S_{A^*C} + S_{CB}$$

with

$$S_{CB} = g\,|\boldsymbol{r}_B - \boldsymbol{r}_C| \quad,\quad g = (2m_0 e \hat{U})^{\frac{1}{2}}$$

being the asymptotic momentum. This is not yet a favourable representation; we therefore add and subtract again the optical path from A to C, thus obtaining

$$S_{II} = S_{QA^*} + S_{A^*C} - g\,|\boldsymbol{r}_C - \boldsymbol{r}_A| + g\,|\boldsymbol{r}_B - \boldsymbol{r}_A|$$

The first three terms together converge rapidly to a unique function

$$S_a(x_a, y_a) := S_{QA^*} + S_{A^*C} - g\,|\boldsymbol{r}_C - \boldsymbol{r}_A|$$

x_a, y_a being the cartesian coordinates of the point A in the plane $z = 0$; the subscript a indicates the *asymptotic* character of this function. If we form the line integral over the canonical momentum $\boldsymbol{p}$, as in (57.9), (57.10) with terms of *higher order*, but along the *paraxial* ray, we admit the same errors as in the standard aberration theory (see Part IV); these errors are acceptably small.

Another quite useful representation of the eikonal S_a is obtained in the following way. Here it is not even necessary for the source Q to be situated on the optic axis. In the case of a *perfect* lens all rays emerging from the arbitrary point Q in the object plane $z = z_Q < 0$ would intersect exactly in the conjugate image point with position vector $\boldsymbol{r}_i^{(0)}$, the eikonal along these rays would have the same value $S_{QI}^{(0)}$ and the asymptotic eikonal in the aperture plane would consequently be $S_a^{(0)} = S_{QI}^{(0)} - g\,|r_i^{(0)} - r_A|$.

For a real lens with its inevitable aberrations, this is not correct and we have instead:

$$S_a(r_Q, r_A) = S_{QI}^{(0)} - g\left(|r_i^{(0)} - r_A| + W(r_Q, r_A)\right) \tag{59.38}$$

where the deviation term is called the *wave aberration*. This quantity has the dimension and meaning of a length, and is the additional optical path caused by aberrations. The function $W(r_Q, r_A)$ can be evaluated by means of power series expansions. Apart from the scale and the notation, these are the same as in Part IV. Further examples will be given in Part XIII.

We are now in a position to derive a fairly accurate integral expression for the wavefunction $\psi_a(x_a, y_a)$. We identify the 'object' in (59.37a) with the source plane $z = z_Q$ so that $\psi_0(x, y) \to \psi_Q(x_Q, y_Q)$. The recording plane is now our aperture plane, hence $r' \to r_A = (x_a, y_a, 0)$. The eikonal S is identified with the S_a of (59.38), in which the additive constant $S_{QI}^{(0)}$ is irrelevant and ignored. Apart from a complex amplitude factor c_a, we obtain, using $k = g/\hbar$,

$$\psi_a(x_a, y_a) = c_a \iint\limits_{Q} \psi_Q(x_Q, y_Q) e^{-iks}\,dx_Q\,dy_Q \tag{59.39a}$$

with

$$s := |r_i^{(0)} - r_A| + W(r_A, r_Q) \tag{59.39b}$$

This wavefunction is now to be introduced into the integrand of (59.14), where the integration is taken with respect to x_a, y_a. The position r' is not restricted to the image plane, which means that a defocus $z' \neq z_I$ is still allowed. Once again, we collect all constant amplitude factors in a single constant c and assume small angle diffraction. The result is then

$$\psi(r') = \frac{c}{z'} \iint\limits_{A} \iint\limits_{Q} \psi_Q(x_Q, y_Q) e^{-ikD'}\,dx_Q\,dy_Q\,dx_a\,dy_a$$

with $D' := -|r' - r_A| + |r_i^{(0)} - r_A| + W$.

Apart from W, this expression for D' is the small difference of two large distances. This unfavourable form can be improved in the following way:

$$W - D' = \frac{|r' - r_A|^2 - |r_i^{(0)} - r_A|^2}{|r' - r_A| + |r_i^{(0)} - r_A|} = \frac{(r' - r_i^{(0)})(r' + r_i^{(0)} - 2r_A)}{|r' - r_A| + |r_i^{(0)} - r_A|}$$

which still holds exactly and lends itself to simplification.

A small defocus, $\Delta z := z' - z_I \neq 0$, could be considered by evaluation of (59.39a,b) in the corresponding out-of-focus plane. It is, however, usual and certainly more convenient to choose the image plane $z = z_I =: b$ and to compensate the ensuing error by adding a term

$$\Delta W = -\Delta z + (\Delta x^2 + \Delta y^2)/2\Delta z$$

to the wave aberration W, Δx and Δy being the lateral shifts due to the defocus Δz. The length b, which measures the distance from the aperture to the screen, is always much larger than all other geometrical parameters. We can thus replace the denominator by $2b$. Moreover, we can assume that the lateral extent $|r' - r_i^{(0)}|$ of the diffraction spot is very small, so that terms in $r'^2 - r_i^{(0)2}$ can be ignored. We thus arrive at

$$D' = W + \frac{1}{b}r_A \cdot (r' - r_i^{(0)})$$

$$= W(r_Q, r_a) + \frac{1}{b}\{x_a(x' - x_i^{(0)}) + y_a(y' - y_i^{(0)})\}$$

Introducing this into the integral expression for ψ and using the notation r_i instead of r' for position in the image plane, we obtain

$$\psi(r_i) = \frac{c}{b} \iint\limits_{A} \iint\limits_{Q} \psi_Q(x_Q, y_Q)e^{-ikD} \, dx_Q \, dy_Q \, dx_a \, dy_a \qquad (59.40a)$$

$$D := W(r_Q, r_a) + \frac{1}{b}\{x_a(x_i - x_i^{(0)}) + y_a(y_i - y_i^{(0)})\} \qquad (59.40b)$$

In practice this integral is never evaluated in this explicit form, since that would be far too laborious. Some further simplifications are possible for a point source Q located on the optic axis. These already allow a fairly accurate estimate of the resolution achieved in electron optical imaging to be made. This topic is dealt with in Sections 60.5 and 60.6.

A much more important aspect of equations (59.40) is that they play a fundamental role in the *transfer calculus* dealt with in Part XIII, especially in Section 65.3. We shall see that this calculus can be justified if W does not depend on r_Q. This results in such a far-reaching simplification that the assumption $W = W(r_a)$ is practically always made whenever this is not too misleading. We shall see that this is tantamount to limiting the discussion to isoplanatic systems.

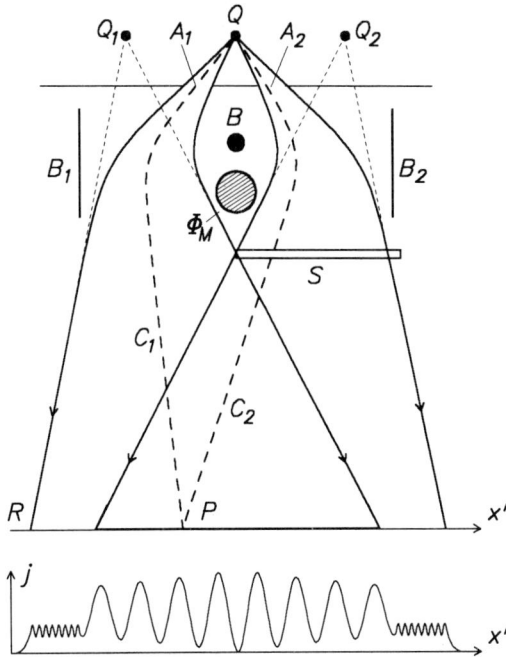

Fig. 59.6: Principal features of an electron biprism interferometer with point source Q, virtual sources Q_1, Q_2, effective entrance slits A_1, A_2, biprism wire B, electrostatic plates B_1, B_2, magnetic flux Φ_M, specimen S and recording screen R. C_1 and C_2 denote two classical electron trajectories starting at Q and ending at P on the screen R. The curve $j(x')$ below the ray diagram shows the recorded intensity. For clarity, the scales in the x-direction are drastically exaggerated. Further details are given in Chapter 62.

59.6 The observability of diffraction and interference fringes

Electron interference is mathematically included in the solutions of (59.37) or (59.40), the domain A of integration then consisting of several separate openings $A_1 \ldots A_N$. Thus diffraction and interference appear together as one wave-propagation process. The basic arrangement of an interferometer of biprism type is shown in Fig. 59.6. An electron source Q illuminates two slits A_1, A_2 *coherently*. For the moment, this just means that the incident wave emerges from a point source at Q and has a sharp energy E, so that a simple spherical wave can be used. The concept of coherence will be examined more carefully in later chapters, especially Chapter 78.

The two slits A_1 and A_2 act as apertures for the diffraction process. In the absence of any field, the diffraction fringes associated with the edges of the slits will not overlap. If a suitable electrostatic field is created by the electron optical element B (a biprism or a sequence of biprisms), the two initially separated electron waves will be deflected in such a manner that they overlap further downstream. An interference pattern can then be observed on the recording screen R.

This interference pattern can be altered by introducing a specimen S into one of the beams. Macroscopically, its action is due to the electromagnetic fields obtained by averaging over all atomic fields. A magnetic flux Φ_M, concentrated entirely in the non-overlap domain between the two beams, likewise has an influence on the interference pattern, as will be explained in detail below.

The intensity distribution, recorded on the screen R, is essentially the normal component j_z of the current density $\boldsymbol{j}$. According to (56.14) and with harmonic time-dependence, this is given by

$$j_z = \frac{\hbar}{2m_0 i}\left(\psi^* \frac{\partial \psi}{\partial z'} - \psi \frac{\partial \psi^*}{\partial z'}\right) + \frac{e}{m_0}A_z(\boldsymbol{r}')\psi^*\psi \qquad (59.41)$$

Differentiation of (59.37a) with respect to z' and multiplication of the result by $\hbar\psi^*/i$ gives

$$\frac{\hbar}{i}\psi^* \frac{\partial \psi}{\partial z'} = \frac{\hbar}{ib}\frac{db}{dz'}|\psi|^2 - ib\psi^* \iint \psi_0(x,y)\frac{\partial S}{\partial z'}\exp(\frac{i}{\hbar}S)\,dx\,dy$$

We now expand $\partial S/\partial z'$ in terms of x and y. The slopes of the classical trajectories are so small that this series expansion can be truncated after the zero-order term

$$S_{z0} := \left(\frac{\partial S}{\partial z'}\right)_{(0,0,z;x',y',z')}$$

This factor can be taken outside the integral, which then becomes proportional to ψ, hence

$$\frac{\hbar}{i}\psi^* \frac{\partial \psi}{\partial z'} = \frac{\hbar}{ib}\frac{db}{dz'}|\psi|^2 + S_{z0}|\psi|^2$$

Introducing this and its complex conjugate into (59.41), we obtain

$$j_z = \frac{1}{m_0}\{S_{z0} + eA_z(\boldsymbol{r}')\}|\psi|^2 = \frac{1}{m_0}g_z(\boldsymbol{r}')|\psi|^2$$

Again thanks to the very small slopes, there is little error in replacing g_z by $|\boldsymbol{g}|$, giving

$$j_z \approx j = \frac{1}{m_0}(2m_0 e\hat{\phi})^{\frac{1}{2}}|\psi(\boldsymbol{r}')|^2 \qquad (59.42)$$

To establish the *gauge invariance* of this result, it is sufficient to show that (56.15b) is valid for the time-dependence of (55.4). We assume (56.15a) and set

$$S(\mathbf{r}, \mathbf{r}') = S'(\mathbf{r}, \mathbf{r}') - eF(\mathbf{r}') + \hbar\alpha \tag{59.43}$$

as the generalization of (57.12). On introducing this into (59.37a), we see that we can take the exponential factor $\exp\{-ieF(\mathbf{r}')/\hbar + i\alpha\}$ outside the integral. Moreover, we see that the determinant (59.36) is unaltered, since F does not depend on x and y. Bringing all this together does indeed confirm the transformation (56.15b), from which the invariance of $|\Psi|^2 = |\psi|^2$ is immediately obvious. This shows that all the approximations were made consistently.

We now investigate the influence of a locally confined magnetic flux Φ_M on the interference pattern. Figure 59.6 shows a cross-section through this flux. The $\mathbf{B}$-lines are perpendicular to the plane of the drawing. Such a flux can be produced to a good approximation by a current-carrying solenoid with sufficiently small diameter and pitch, closed by magnetic yokes outside the electron beam. This arrangement does not produce a $\mathbf{B}$-field in the region of wave propagation; the corresponding vector potential $\mathbf{A}$ is such that curl $\mathbf{A} = 0$ (locally) and

$$\oint \mathbf{A} \cdot d\mathbf{r} = \Phi_M \tag{59.44}$$

The current density at any point P on the recording screen is determined by the phase *difference* between the two classical trajectories C_1 and C_2 starting from Q and intersecting again at P. We consider first the case $\mathbf{A}(\mathbf{r}) \equiv 0$, for which Φ_M is absent. The corresponding phase difference is then given by

$$\varphi_0 = \frac{1}{\hbar}(S_1^{(0)} - S_2^{(0)}) = \frac{1}{\hbar}\left(\overset{(C_1)}{\underset{Q}{\int}}{}^{P} g\, ds - \overset{(C_2)}{\underset{Q}{\int}}{}^{P} g\, ds \right) \tag{59.45}$$

A gauge transformation (59.43) with position vectors $\mathbf{r}$ for Q and $\mathbf{r}'$ for P does not alter this result.

We now suppose that the paths C_1 and C_2 enclose magnetic flux, as described above. Since the $\mathbf{B}$ field vanishes in the region of the electron beam, the classical trajectories remain unaffected and neither of the integrals in (59.45) is altered. The complete eikonals are, however, given by

$$S_j = \overset{(C_j)}{\underset{Q}{\int}}{}^{P} g\, ds - \overset{(C_j)}{\underset{Q}{\int}}{}^{P} e\mathbf{A} \cdot d\mathbf{r} \quad (j = 1, 2)$$

in which the second term *is* gauge-dependent. The phase difference is now given by

$$
\varphi = \frac{1}{\hbar}(S_1 - S_2) = \varphi_0 - \frac{e}{\hbar}\left(\int\limits_{Q}^{(C_1)\ P} \boldsymbol{A}\cdot d\boldsymbol{r} - \int\limits_{Q}^{(C_2)\ P} \boldsymbol{A}\cdot d\boldsymbol{r} \right)
$$

$$
= \varphi_0 - \frac{e}{\hbar}\left(\int\limits_{Q}^{(C_1)\ P} \boldsymbol{A}\cdot d\boldsymbol{r} + \int\limits_{P}^{(C_2)\ Q} \boldsymbol{A}\cdot d\boldsymbol{r} \right)
$$

$$
= \varphi_0 - \frac{e}{\hbar}\oint \boldsymbol{A}\cdot d\boldsymbol{r}
$$

in which the path of integration for the circuit integral is formed by C_1 from Q to P and by C_2 back from P to Q. This may be written as

$$
\varphi = \varphi_0 - \frac{e}{\hbar}\int \operatorname{curl}\boldsymbol{A}\cdot d\boldsymbol{S} = \varphi_0 - \frac{e}{\hbar}\int \boldsymbol{B}\cdot d\boldsymbol{S}
$$

in which the integral is taken over the surface between C_1 and C_2. We thus obtain the simple result (cf. p. 56)

$$
\varphi = \varphi_0 - \frac{e}{\hbar}\Phi_M \tag{59.46}
$$

Since the shift $\Delta\varphi = e\Phi_M/\hbar$ is independent of the position P, this causes all the maxima and minima in the interference pattern to be shifted by $\Delta\varphi/2$, and this is indeed observed.

The formula (59.46) was first derived by Ehrenberg and Siday (1949) and later found again by Aharonov and Bohm (1959, 1961), after whom the effect then became named. It is interesting to note that the reasoning of Ehrenberg and Siday was based almost wholly on classical mechanics, the only wave-optical notion being the elementary interference relation that phase difference $= k\times$ (path difference). It was, however, the paper by Aharonov and Bohm that attracted attention and the first experimental confirmation was published by Chambers (1960), followed shortly after by Fowler *et al.* (1961) and Boersch *et al.* (1961, 1962, see also 1981). In 1962, Möllenstedt and Bayh demonstrated the fringe-shift associated with enclosed magnetic flux in a way that was free of the objections to which the earlier experiments had given rise, the interference effect and other classical electromagnetic effects being well separated. A voluminous literature rapidly grew up, enlivened by vigorous controversy. Discussion of this is

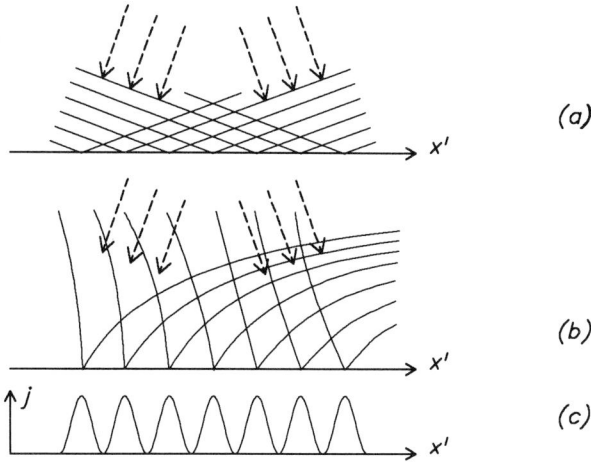

Fig. 59.7: Effect of a gauge transformation: (a) plane wavefronts and orthogonal trajectories; (b) transformed wavefronts, the trajectories (broken lines) remain unaltered; (c) corresponding current density distribution.

deferred to Chapter 62, in which electron interferometry is dealt with in more detail, and to the bibliography of Part XII.

We close this chapter with some remarks about the experimental significance of such wave-optical concepts as wavefronts, wavelength, phase, frequency and eikonal. We have seen that all these are gauge-dependent. The only invariant quantities are the particle and current density and related densities, and the expectation values of the classical kinematic functions. These wave-optical concepts serve only to illuminate vividly the otherwise very abstract quantum theory but do not correspond to any quantity that could be measured.

The influence of a gauge transformation is illustrated in Fig. 59.7. Case (a) shows two interfering plane waves with corresponding straight orthogonal trajectories. This picture is adequate in field-free space and represents the conventional ideas. Case (b) shows the *same* situation after the introduction of a vector potential $\boldsymbol{A}(\boldsymbol{r}) = \nabla F(\boldsymbol{r})$. The wavefronts can be *arbitrarily distorted*, as long as the intersection points on the x'-axis (indicating the screen) are unaltered, but the trajectories are no longer orthogonal to the wavefronts (cf. Section 5.4). The trajectories and the intensity distribution, likewise, remain unaltered: the wavefronts cannot have any real significance. This fact was very clearly pointed out by Lenz (1962), Rang (1964, 1977) and Boersch (1968). The ambiguity, introduced by gauge transformations, can of course be removed by adopting some *fixed* gauge, for instance by setting div $\boldsymbol{A} = 0$ and $\boldsymbol{A} \to 0$ as $|\boldsymbol{r}| \to \infty$, which seems to be the natural choice. There is, however, no real need to do this, and it is more a philosophical question whether or not one should sacrifice the freedom of gauge transformation.

60
Elementary Diffraction Patterns

In the previous chapter, we derived general diffraction formulae without offering any practical examples. We now examine some real situations that can be calculated in closed form. This is successful in practice only for diffraction in field-free space and we shall consider mainly the Fresnel and Fraunhofer diffraction treated in Section 58.3, therefore.

60.1 The object function

We consider the configuration shown in Fig. 60.1 and adopt the corresponding notation. A point source Q, located at the position $\boldsymbol{r}_s = (x_s, y_s, z_s)$ with $z_s =: -a < 0$, emits a spherical wave

$$\psi_s(\boldsymbol{r}) = \frac{C}{|\boldsymbol{r} - \boldsymbol{r}_s|} e^{ik|\boldsymbol{r} - \boldsymbol{r}_s|} \tag{60.1}$$

In the Fresnel approximation, this simplifies to

$$\psi_s(\boldsymbol{r}) = \frac{C}{|z - z_s|} \exp(ik|z - z_s|) \exp\left[\frac{ik}{2|z - z_s|}\{(x - x_s)^2 + (y - y_s)^2\}\right] \tag{60.2}$$

In the plane of the diffracting object, $z = 0$, and with object coordinates x_o, y_o, this formula reduces to

$$\hat{\psi}_o(x_o, y_o) = \frac{C}{a} \exp(ika) \exp\left[\frac{ik}{2a}\{(x_o - x_s)^2 + (y_o - y_s)^2\}\right] \tag{60.3}$$

This is the wavefunction incident on the object plane, on the far side with respect to the observer.

We now assume that the object is effectively plane and so thin that its extent in the z-direction can be ignored. It causes a modulation of any wave passing through it, which we can describe by a complex transmission function*

$$O(x_o, y_o) = e^{-\sigma(x_o, y_o) + i\eta(x_o, y_o)} \tag{60.4}$$

---

* Note that two conventions are in use for the sign of the phase term, corresponding to the choice of sign in (57.1).

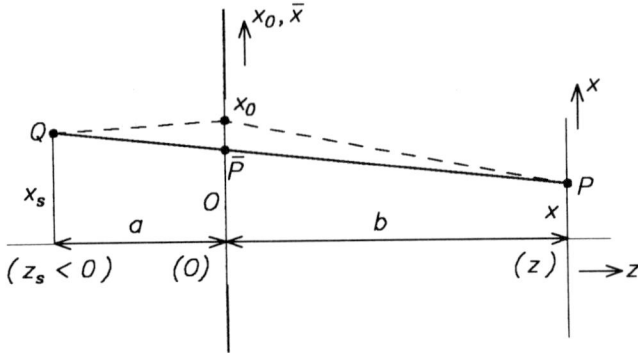

Fig. 60.1: Notation and choice of the coordinate system in a Fresnel-diffraction configuration; Q: point source, O: diffracting object, P: observation point.

This will be referred to as the *object function*. The factor $\exp(-\sigma)$ describes an attenuation and $\exp(i\eta)$ a phase shift, both locally; these are to
be understood as macroscopic effects. The following special cases are of
particular interest:

$$\left.\begin{array}{lll}
\sigma = 0 \quad , \quad \eta = 0 \quad : & \text{free opening} \\
\sigma = \infty \qquad\qquad : & \text{opaque screen} \\
\sigma = 0 \quad , \quad \eta \neq 0 \quad : & \text{pure phase object}
\end{array}\right\} \qquad (60.5)$$

The wavefunction on the exit side of the object is now given by

$$\psi_o(x_o, y_o) = \hat{\psi}_o(x_o, y_o) O(x_o, y_o) \qquad (60.6)$$

This is to be introduced into Fresnel's or Fraunhofer's formula as appropriate. At the observation point with coordinates $\mathbf{r} = (x, y, z)$, $z = b > 0$,
eqs. (59.17a) and (59.17b) with (60.3) and (60.6) now take the explicit form

$$\psi(\mathbf{r}) = \frac{C}{i\lambda ab} e^{ik(a+b)} \int\!\!\!\int_{-\infty}^{\infty} O(x_o, y_o) \exp\{iks/2\}\, dx_o\, dy_o$$

with the length

$$s = \frac{1}{a}\{(x_o - x_s)^2 + (y_o - y_s)^2\} + \frac{1}{b}\{(x_o - x)^2 + (y_o - y)^2\}$$

This can be rewritten as

$$s = \frac{1}{f}\{(x_o - \bar{x})^2 + (y_o - \bar{y})^2\} + \frac{1}{a+b}\{(x - x_s)^2 + (y - y_s)^2\}$$

Here we have introduced the abbreviations

$$\frac{1}{f} = \frac{1}{a} + \frac{1}{b} \qquad (60.7)$$

$$\overline{x} = \frac{ax + bx_s}{a + b} \quad , \quad \overline{y} = \frac{ay + by_s}{a + b} \tag{60.8}$$

The quantity f will sometimes be identified with a focal length. The coordinates $\overline{x}, \overline{y}$ are those of the point $\overline{P}$ at which the line $\overline{QP}$ intersects the screen (see Fig. 60.1). In the Fresnel approximation, the identification

$$D := a + b + \frac{(x - x_s)^2 + (y - y_s)^2}{2(a + b)} = \overline{QP} = |\mathbf{r} - \mathbf{r}_s| \tag{60.9}$$

can be made without introducing any additional error; D is thus the distance between the source and the observation point. The diffraction integral now becomes

$$\psi(\mathbf{r}) = \frac{C}{i\lambda ab} e^{ikD} \int\!\!\!\int\limits_{-\infty}^{\infty} O(x_o, y_o) \exp\left[\frac{ik}{2f}\{(x_o - \overline{x})^2 + (y_o - \overline{y})^2\}\right] dx_o \, dy_o$$

To simplify this integral further, we make the substitution

$$x_o := \overline{x} + Lu \quad , \quad y_o := \overline{y} + Lv \tag{60.10a}$$

with the diffraction length

$$L := \sqrt{\frac{\pi f}{k}} = \sqrt{\frac{\lambda f}{2}} \tag{60.10b}$$

whereupon the diffraction integral takes its final form

$$\psi(\mathbf{r}) = \psi_s(\mathbf{r})M(\mathbf{r}) \tag{60.11}$$

with

$$M(\mathbf{r}) = \frac{1}{2i} \int\!\!\!\int\limits_{-\infty}^{\infty} O(\overline{x} + Lu, \overline{y} + Lv) \exp\left\{\frac{i\pi}{2}(u^2 + v^2)\right\} du \, dv \tag{60.12}$$

The modulation factor $M(\mathbf{r})$ describes the influence of the object on the incident wave $\psi_s(\mathbf{r}) = C \exp(ikD)/(a + b)$ and is therefore of particular interest.

The formula includes the degenerate case of an incident *plane* wave with small transverse components of the wave vector. In this case we have

$$a \to \infty \quad , \quad b = f \quad , \quad \overline{x} = x - b\alpha_x \quad , \quad \overline{y} = y - b\alpha_y \tag{60.13}$$

α_x, α_y being the angles of incidence. The factor $\psi(\boldsymbol{r})$ is then, of course, a *plane* wave.

60.2 Rectangular structures

These are characterized by the property that the object function separates into two independent factors:

$$O(x_o, y_o) = O_1(x_o)O_2(y_o) \qquad (60.14)$$

whereupon the integral (60.12) separates also into two independent factors M_1, M_2 with

$$M = M_1 M_2 \quad , \quad M_k = \frac{1-\mathrm{i}}{2} \int\limits_{-\infty}^{\infty} O_k(\overline{x}_k + Lu) \exp\left(\frac{\mathrm{i}\pi}{2}u^2\right) du \qquad (60.15)$$

and $\overline{x}_1 = \overline{x}$, $\overline{x}_2 = \overline{y}$. This is a major simplification, since it suffices now to calculate two one-dimensional integrals.

The formulae simplify still further if the object function depends only on one coordinate, x_o say, while $O_2(y_o) \equiv 1$. Using the well-known formula

$$\int\limits_{-\infty}^{\infty} \exp\left(\frac{\mathrm{i}\pi}{2}v^2\right) dv = 1 + \mathrm{i} \qquad (60.16)$$

we see from (60.15) that $M_2 = 1$ and $M(\boldsymbol{r}) = M_1(x, z)$, the dependence on x and z being given by (60.7), (60.8) and (60.10).

If we have no diffracting object at all, $O_1(x_o)$ is also equal to unity everywhere; we can apply (60.16) a second time and find that $M(\boldsymbol{r}) = 1$, as it must be. This shows that the Fresnel approximation is self-consistent.

We consider now diffraction at the edge of an opaque half-plane. The object function is then

$$O_1(x_o) = \begin{cases} 0 & \text{for } x_o < 0 \\ 1 & \text{for } x_o \geq 0 \end{cases} \qquad (60.17)$$

We assume that the incident wave is plane and parallel to the object with unit amplitude,

$$\psi_s(\boldsymbol{r}) = \mathrm{e}^{\mathrm{i}kz} =: \psi_s(z)$$

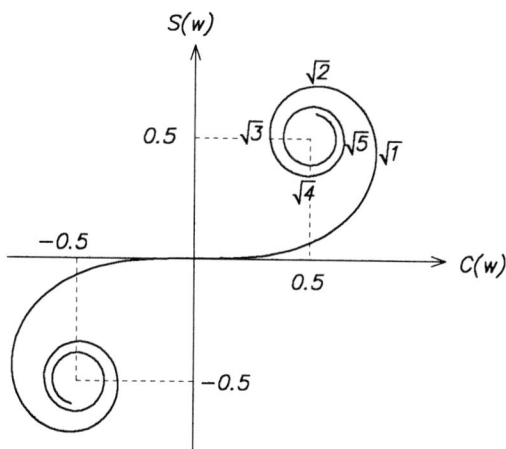

Fig. 60.2: Parametric representation of the Fresnel integrals $C(w)$ and $S(w)$.

the angles α_x, α_y in (60.13) are zero and so $\bar{x} = x$. With these assumptions (60.15) collapses to

$$M = M_1(x, z) = \frac{1 - i}{2} \int_{-x/L}^{\infty} \exp\left(\frac{i\pi}{2} u^2\right) du \qquad (60.18)$$

This integral can be expressed in terms of well-known analytic functions known as the *Fresnel integrals*. These are defined by

$$F(w) \equiv C(w) + iS(w) := \int_0^w \exp\left(\frac{i\pi}{2} v^2\right) dv \qquad (60.19)$$

The behaviour of this function is shown in Fig. 60.2, where the values of $C(w)$ and $S(w)$ are regarded as cartesian coordinates of points corresponding to all possible values of w. The resulting curve is the Cornu spiral. It is easy to show that

$$F(-w) = -F(w) \quad , \quad F(0) = 0 \qquad (60.20a)$$

$$F(\pm\infty) = \pm\frac{1 + i}{2} \qquad (60.20b)$$

The series expansion for $|w| \gg 1$ is

$$F(w) - F(\pm\infty) = \frac{1}{i\pi w} \exp\left(\frac{i\pi}{2} w^2\right) \left\{ 1 + \frac{1}{i\pi w^2} + \frac{1 \cdot 3}{(i\pi w^2)^2} + \frac{1 \cdot 3 \cdot 5}{(i\pi w^2)^3} + \cdots \right\} \qquad (60.21)$$

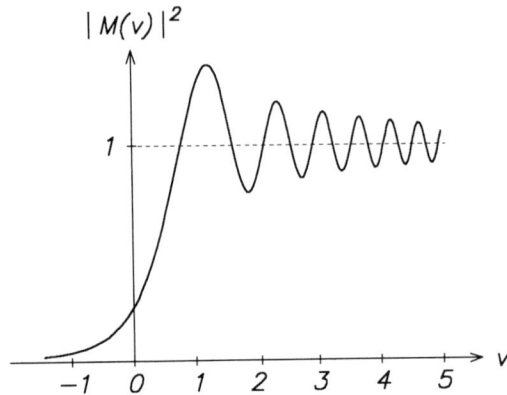

Fig. 60.3: Intensity distribution behind a straight edge of an opaque screen, $v = x/\sqrt{\lambda z/2}$.

but this series is asymptotic, which means that the most accurate result is obtained by halting the series after a certain number of terms, which depends on $|w|$. Beyond this, the error increases and the series finally diverges. For $|w| = 2$, the relative error is about 10^{-3} with the terms in (60.21), which is sufficient in practice.

For $|w| \leq 1$ the Taylor series expansion

$$F(w) = \sum_{m=0}^{\infty} \left(\frac{i\pi}{2}\right)^m \frac{w^{2m+1}}{m!(2m+1)} \tag{60.22}$$

can be evaluated. The accuracy is about 10^{-5} or better if the expansion is truncated after $m = 7$. Other approximations are given by Abramowitz and Stegun (1965).

In terms of Fresnel integrals, the solution (60.18) can be rewritten as

$$M(\boldsymbol{r}) = \frac{1}{2} + \frac{1-i}{2} F\left(\frac{x}{\sqrt{\lambda z/2}}\right) \tag{60.23}$$

The normalized intensity distribution is then given by

$$|M(\boldsymbol{r})|^2 = \frac{1}{4}\left|1 + (1-i)F\left(\frac{x}{\sqrt{\lambda z/2}}\right)\right|^2 \tag{60.24}$$

This intensity distribution is depicted in Fig. 60.3. For $x < 0$ (in the domain of the geometric shadow), the intensity decreases as $(|x| + L)^{-2}$ towards zero where $L := \sqrt{\lambda z/2}$. For $x > 0$, an interference pattern is observable in which the amplitude and fringe spacing gradually decrease. If we truncate the asymptotic expansion (60.21) after the first term, we obtain the simple approximation

$$|M(\boldsymbol{r})|^2 =: J(w) = 1 - \frac{\sqrt{8}}{\pi w} \cos \frac{\pi}{2}\left(w^2 + \frac{1}{2}\right) + O(w^{-2}) \tag{60.25}$$

From this it can be concluded that the maxima and minima are located at

$$w_n = \sqrt{2n - \frac{1}{2}} \quad , \quad x_n = \sqrt{\lambda z \left(n - \frac{1}{4}\right)} \quad , \quad n = 1, 2, 3, \ldots \tag{60.26}$$

with odd values of the integers n for maxima and even values for minima. These formulae are also valid for spherical incident waves if z is replaced by f of (57.9) with $b = z$ so that $1/f = 1/a + 1/z$.

On the basis of the solution (60.23), more sophisticated situations can be analysed. Often the screen is not completely opaque but is described by an object function of the general form (60.4). Lenz (1961) has investigated the case of a semitransparent half-plane, described by

$$O(x_o, y_o) = \begin{cases} \exp(-\sigma + i\eta) & x_o < 0 \\ 1 & x_o \geq 0 \end{cases} \tag{60.27}$$

The solution is then given by

$$M(\boldsymbol{r}) = \frac{1}{2}(1 + e^{i\eta - \sigma}) + \frac{1 - i}{2}(1 - e^{i\eta - \sigma}) F\left(\frac{x}{\sqrt{\lambda z/2}}\right) \tag{60.28}$$

where (60.20a) has been used. It is easy to verify that $M(\infty) = 1$ and $M(-\infty) = \exp(i\eta - \sigma)$ with the aid of (60.20b), as expected. Lenz has cast the relation (60.28) into a form that permits the complex vector representing the local wave excitation to be directly measured from a Cornu spiral. This construction is shown in Fig 60.4.

Another elementary case is a bar of breadth $2w$ illuminated by a plane wave propagating in the z-direction. The corresponding solution is given by

$$M_b(\boldsymbol{r}) = 1 + \frac{1 - i}{2}\left\{F\left(\frac{x - w}{\sqrt{\lambda z/2}}\right) - F\left(\frac{x + w}{\sqrt{\lambda z/2}}\right)\right\} \tag{60.29}$$

In the complementary case of a slit of breadth $2w$ we have:

$$M_s(\boldsymbol{r}) = \frac{1 - i}{2}\left\{F\left(\frac{x + w}{\sqrt{\lambda z/2}}\right) - F\left(\frac{x - w}{\sqrt{\lambda z/2}}\right)\right\} \tag{60.30}$$

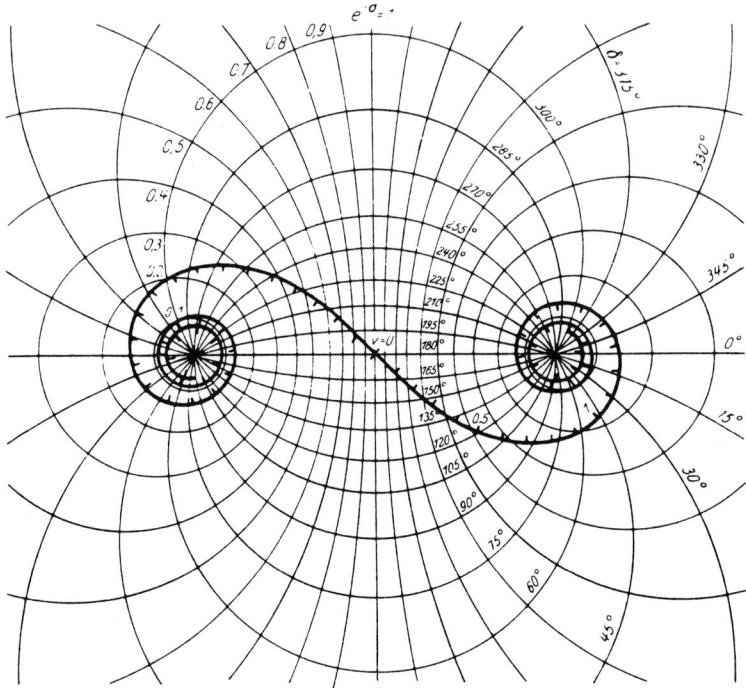

Fig. 60.4: Cornu spiral for reading the amplitude and phase of the wave diffracted by a semitransparent phase-shifting half-plane together with the circles $\sigma =$ const and $\eta =$ const. The length and direction of the vector from the point with coordinates σ and η to the current point with the parameter value w on the Cornu spiral represent the complex amplitude of the wave function in the plane of observation.

These two solutions satisfy $M_b + M_s = 1$ as they must do in agreement with Babinet's principle*.

Equation (60.30) simplifies considerably when the slit is very narrow. After some calculation, the approximation

$$M_s(\boldsymbol{r}) = \frac{(1+\mathrm{i})w}{\sqrt{\lambda z/2}} \exp\left(\frac{\mathrm{i}\pi x^2}{\lambda z}\right) \frac{\sin(2\pi x w/\lambda z)}{2\pi x w/\lambda z} \qquad (60.31)$$

is obtained, which represents essentially the Fraunhofer diffraction at the slit. The wave amplitude decreases as $z^{-1/2}$, like that of a cylindrical wave.

---

* In practice, the conditions in which Babinet's principle is valid are rarely satisfied. See Boersch (1951), Hosemann and Joerchel (1954) and Ditchburn (1963) and, for telling examples, Lipson and Walkley (1968).

This is a consequence of the assumption that the slit is infinitely long in the y-direction.

As an obvious generalization, systems of parallel slits or bars can be treated by appropriate superposition of Fresnel integrals. The final solution does of course become more complicated.

60.3 Circular structures

60.3.1 General expression for $M(\mathbf{r})$

Circular structures are described by a rotationally symmetric object function

$$O(x_o, y_o) = O(r_o) = e^{-\sigma(r_o) + i\delta(r_o)} \tag{60.32}$$

$$r_o = \sqrt{x_o^2 + y_o^2} \tag{60.33}$$

being the radial coordinate in the object plane. We examine a reasonably simple configuration, which leads to *rotationally symmetric* diffraction patterns, namely a point source on the axis of symmetry, the optic axis; thus $x_s = y_s = 0$. From (60.8), we obtain the simple proportionality relations

$$\bar{x} = \gamma x \quad , \quad \bar{y} = \gamma y \quad , \quad \gamma := \frac{a}{a + b} \tag{60.34}$$

x and y being the transverse coordinates in the plane of observation $z = b$. It is now advantageous to introduce polar coordinates (r_o, φ_o) in the object plane and (r, φ) in the plane of observation. From (60.10a) and (60.34), the relations

$$u = \frac{1}{L}(r_o \cos\varphi_o - \gamma r \cos\varphi)$$

$$v = \frac{1}{L}(r_o \sin\varphi_o - \gamma r \sin\varphi)$$

$$u^2 + v^2 = \frac{1}{L^2}\{r_o^2 + \gamma^2 r^2 - 2\gamma r_o r \cos(\varphi_o - \varphi)\} =: A$$

$$du\,dv = \frac{1}{L^2} r_o\,dr_o\,d\varphi_o$$

can then be derived. Introducing these into (60.12), we find

$$M = \frac{1}{2iL^2} \int\limits_0^\infty \int\limits_0^{2\pi} O(r_o) \exp\left(\frac{i\pi A}{2}\right) d\varphi_o\, r_o\, dr_o$$

or more explicitly

$$M = \frac{1}{2\mathrm{i}L^2} \exp\left(\frac{\mathrm{i}\pi\gamma^2 r^2}{2L^2}\right) \int\limits_0^\infty O(r_o) \exp\left(\frac{\mathrm{i}\pi r_o^2}{2L^2}\right) I(r_o) r_o \, dr_o$$

with

$$I(r_o) := \int\limits_0^{2\pi} \exp\left\{-\frac{\mathrm{i}\pi\gamma r_o r}{L^2} \cos(\varphi_o - \varphi)\right\} d\varphi_o$$

This latter integral may be written as a Bessel function of zero order, since

$$J_0(x) = \frac{1}{2\pi} \int\limits_0^{2\pi} \exp(\mathrm{i}x \cos\alpha) \, d\alpha \qquad (60.35)$$

Thus

$$M(r) = \frac{\pi}{\mathrm{i}L^2} \exp\left(\frac{\mathrm{i}\pi\gamma^2 r^2}{2L^2}\right) \int\limits_0^\infty O(r_o) \exp\left(\frac{\mathrm{i}\pi r_o^2}{2L^2}\right) J_0\left(\frac{\pi\gamma r_o r}{L^2}\right) r_o \, dr_o$$

$$(60.36a)$$

or explicitly (60.10, 60.34)

$$M(r) = \frac{2\pi}{\mathrm{i}\lambda}\left(\frac{1}{a} + \frac{1}{b}\right) \exp\left(\frac{\mathrm{i}\pi a r^2}{\lambda b(a + b)}\right)$$

$$\times \int\limits_0^\infty O(r_o) \exp\left\{\frac{\mathrm{i}\pi r_o^2}{\lambda}\left(\frac{1}{a} + \frac{1}{b}\right)\right\} J_0\left(\frac{2\pi r_o r}{\lambda b}\right) r_o \, dr_o$$

$$(60.36b)$$

The remaining integration can in general not be performed in closed form, but there are two important special cases that do allow further simplification. These are the *axial value* $M(0)$ and the *Fraunhofer approximation*.

60.3.2 Zone lenses
For $r = 0$ we have $J_0(0) = 1$ and consequently

$$M(0) = \frac{2\pi}{\mathrm{i}\lambda}\left(\frac{1}{a} + \frac{1}{b}\right) \int\limits_0^\infty O(r_o) \exp\left\{\frac{\mathrm{i}\pi r_o^2}{\lambda}\left(\frac{1}{a} + \frac{1}{b}\right)\right\} r_o \, dr_o$$

Using again $1/f = 1/a + 1/b$ and the substitution $u = r_o^2$, we obtain

$$M(0) = \frac{\pi}{\mathrm{i}\lambda f} \int\limits_0^\infty O(\sqrt{u}) \exp\left(\frac{\mathrm{i}\pi u}{\lambda f}\right) du \qquad (60.37)$$

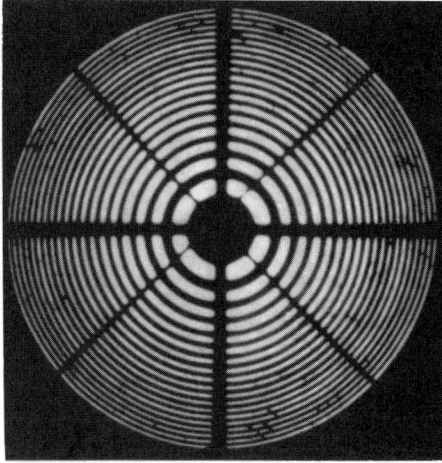

Fig. 60.5: A Fresnel zone plate.

This is essentially a *Fourier integral*. As an example, we consider now a system of concentric rings, specified by the object function

$$O(r_o) = \begin{cases} 1 & \text{for } r_{j1} \leq r_o \leq r_{j2} \quad, \quad j = 1 \ldots N \\ 0 & \text{for } r_{j2} < r_o < r_{j+1,1} \quad, \quad j = 1 \ldots N - 1 \end{cases} \tag{60.38}$$

Evaluation of (60.37) gives

$$M(0) = \sum_{j=1}^{N} \left\{ \exp\left(\frac{i\pi r_{j1}^2}{\lambda f}\right) - \exp\left(\frac{i\pi r_{j2}^2}{\lambda f}\right) \right\} \tag{60.39}$$

This formula helps to explain the focusing action of a Fresnel zone lens. Clearly, $M(0)$ assumes its maximum value if the phases of all the exponential terms are such that they sum up constructively. The simplest possible choice is

$$r_{j1} = \{(2j-2)\lambda f\}^{1/2} \quad, \quad r_{j2} = \{(2j-1)\lambda f\}^{1/2} \quad, \quad j = 1 \ldots N \tag{60.40}$$

so that all the open and opaque rings have the same area $A = \pi\lambda f$; the maximum is then $M(0) = 2N$. Such a ring system is shown in Fig. 60.5. It is also possible to interchange the open and the opaque rings; the result is then $M(0) = -2N$. Moreover, diffraction foci of higher orders can be obtained for

$$\frac{1}{a} + \frac{1}{b} = \frac{1}{f_m} = \frac{(2m+1)\lambda}{r_{1,2}^2} \tag{60.41}$$

m being a positive or negative integer.

Fresnel zone lenses have become of importance in X-ray optics, where they are useful in X-ray telescope and microscope design. In electron optics, there is no practical need for Fresnel lenses but a proposal has been made for reducing the effect of spherical aberration, from which all conventional lenses suffer, by means of a suitable ring-shaped diffraction grid.

This idea was first published by Hoppe (1961, 1963) and pursued by Lenz (1963, 1964, 1965). It is sketched in Fig. 60.6. The spherical aberration causes a deviation of the emergent wavefronts from an exactly spherical shape (cf. Fig. 24.5). For a perfect sphere, all the contributions at the image point (the centre of the sphere) would be equidistant from the surface of equal phase and hence all would have the same phase. For a surface that departs progressively farther from a sphere as we move away from the axis, the contributions arrive with different phases. The idea behind the zone-plate corrector is to suppress zones that (on average) would contribute with a phase difference of $\pi, 3\pi, \ldots (2n + 1)\pi, \ldots$ while allowing in-phase contributions $(2n\pi)$ to pass through and reach the image plane.

Lenz (1963, 1964) has suggested that the complementary ring system, for which the outer open zones interfere destructively with the central zone, might be useful to enhance the image contrast but subsequently withdrew this proposal.

Although all these zone plates should work in principle, they have not been successful in practice. The main reasons are that a grid within the electron beam causes a strong scattering background and is, moreover, soon deformed by strong heating. For attempts to use zone plates in the microscope, see Section 66.5.3.

60.3.3 Fraunhofer diffraction
The outer radius of the diffracting object is so small that the exponential term in the integrand of (60.36b) can be replaced by unity to a good approximation. This requires

$$r_o^2 \ll \frac{\lambda ab}{a + b}$$

which shows that the diffracting area must be much smaller than a Fresnel zone. We thus rewrite (60.36b) in the form

$$M(r) = \frac{2\pi}{i\lambda}\left(\frac{1}{a} + \frac{1}{b}\right)\exp\{i\delta(r)\}\int_0^\infty O(r_o)J_0\left(\frac{2\pi r_o r}{\lambda b}\right)r_o\,dr_o \qquad (60.42)$$

in which the unimportant exponential factor in front of the integral is denoted by $\exp(i\delta)$. Equation (60.42) is essentially a *Fourier–Bessel transform* of the object function $O(r_o)$.

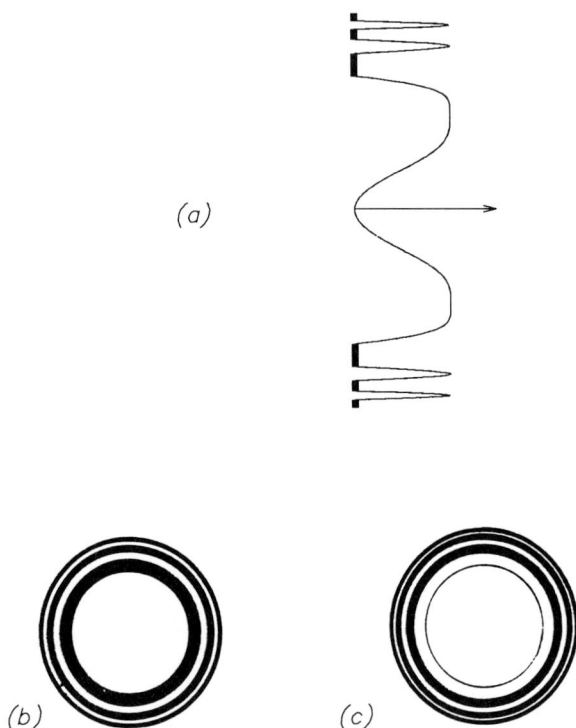

Fig. 60.6: Correction plates
(a) Construction, (b) Hoppe-plate, (c) Lenz-plate.

An example for which the result can be expressed in closed form is again the object function specified by (60.38). Using the familiar integral formula

$$\int_{x_1}^{x_2} J_0(x)x\,dx = x_2 J_1(x_2) - x_1 J_2(x_1) \qquad (60.43)$$

J_1 being the first-order Bessel function and identifying x with $2\pi r_o r/\lambda b$, we obtain

$$M(r) = \frac{a+b}{iar}\exp(i\delta)\sum_{j=1}^{N}\left\{r_{j2}J_1\left(\frac{2\pi r r_{j2}}{\lambda b}\right) - r_{j1}J_1\left(\frac{2\pi r r_{j1}}{\lambda b}\right)\right\} \qquad (60.44)$$

This formula has a finite value at $r = 0$; using $J_1(x) = x/2 + O(x^3)$, we

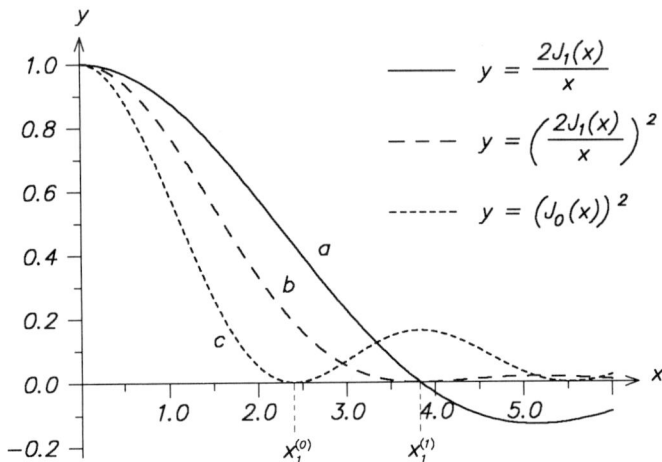

Fig. 60.7: Normalized distribution functions for the diffraction at a circular hole (Airy disc, a, b) and for a ring (c) of equal radius.

obtain the simple relation

$$M(0) = \frac{A}{i\lambda}\left(\frac{1}{a} + \frac{1}{b}\right) \quad , \qquad A := \pi \sum_{j=1}^{N}(r_{j2}^2 - r_{j1}^2) \qquad (60.45)$$

The factor A is the total open area of the ring system; $|M(0)|^2$ usually represents a maximum of the intensity distribution, which obviously increases with the *square* of the open area.

There are two special cases of importance. The first is a simple *circular hole* of radius R and area $A = \pi R^2$. We set $N = 1$, $r_{1,1} = 0$ $r_{1,2} = R$ and obtain the formula

$$M(r) = M(0)\exp\{i\delta(r)\}\frac{2J_1(x)}{x} \quad , \qquad x := \frac{2\pi R r}{\lambda b} \qquad (60.46)$$

The function $2J_1(x)/x$ and its square, which determines the intensity pattern, the Airy disc and rings, are depicted in Fig. 60.7, curves (a) and (b). The central peak extends to the zero at $x_1^{(1)} = 3.83$ and covers 83 per cent of the total intensity. This is of some importance in connection with elementary estimates of the resolution limit of imaging systems.

The other case of interest is a narrow ring with mean radius R and half-width $h \ll R$. We now specialize (60.44) to the case $N = 1$, $r_{1,1} = R - h$, $r_{1,2} = R + h$. Using the approximation

$$(x + \Delta x)J_1(x + \Delta x) - (x - \Delta x)J_1(x - \Delta x) = 2x\Delta x J_0(x) + O((\Delta x)^3)$$

and noting that the area is now $A = 2\pi Rh$, we can cast the result into the form

$$M(r) = M(0)e^{i\delta(r)} J_0(x) \quad , \quad x = \frac{2\pi Rr}{\lambda b} \tag{60.47}$$

This allows a comparison to be made with (60.46). The central peak is now bounded by the zero at $x_1^{(0)} = 2.40$, and is hence narrower than that of the Airy disc, see Fig. 60.7 curve (c), but the outer fringes contain much more intensity. This situation was investigated in detail by Lenz and Wilska (1966/67), who came to the conclusion that for the latter reason an electron microscope with an annular objective aperture will not give better resolution than a conventional instrument.

60.4 Caustic interferences

We consider now the situation sketched in Fig. 60.8. An electron beam passes through a slit aperture of width $2a$ in the object plane $z_o = -D$ and forms an imperfect line focus in the reference plane $z = 0$. In a purely geometrical description (Chapter 42) the rays form a caustic $x = x_c(z)$. This is a singular situation in which the approximations given so far must break down. The geometrical theory fails, since it predicts infinite electron intensity on the caustic itself. This is a consequence of the fact that the eikonal function then becomes singular. For the same reason the approximation of Chapter 57 also breaks down, even for very wide slits. It becomes obvious that only a wave mechanical treatment as a diffraction problem can be adequate. Since each point of the reference plane in the classically allowed region can be reached by two rays, we can expect that in the wave-optical formulation, two corresponding partial waves will interfere; we may hence anticipate that a characteristic interference phenomenon will be seen.

For conciseness, we assume that the waves propagate in field-free space. In the entrance plane $z_o = -D$, a unique definition of an eikonal is still possible:

$$L(x_o) := \frac{S(x)}{g} = -\frac{x_o^2}{2D} + \frac{1}{6C}\left(\frac{x_o}{D}\right)^3 \tag{60.48}$$

This is independent of y_o and thus represents a cylindrical wavefront; $g = \hbar k$ denotes the constant magnitude of the kinetic momentum.

The slope of the trajectory passing through the point $r_o = (x_o, 0, -D)$ is given by

$$x' = \frac{\partial L}{\partial x_o} = -\frac{x_o}{D} + \frac{x_o^2}{2CD^3}$$

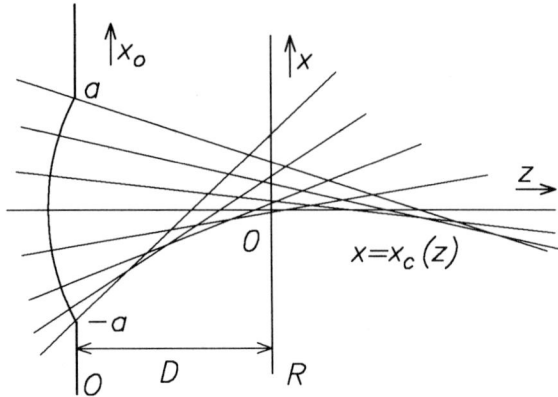

Fig. 60.8: Formation of a caustic $x = x_c(z)$ by rays passing through a slit in the object plane O; R denotes the plane chosen for recording the caustic interferences. The inclination of the rays with respect to the z-axis is exaggerated for clarity.

and the cartesian representation of this particular ray is hence

$$x(z) = x_o + x'(z + D) = -\frac{x_o z}{D} + \frac{x_o^2(z + D)}{2CD^3} \qquad (60.49)$$

The whole family of all such rays, sketched in Fig. 60.8, is obtained by regarding x_o as a *parameter*, defined in the interval $[a, -a]$. The caustic is obtained as the singularity of this family from the condition $\partial x/\partial x_o = 0$, giving

$$-\frac{z}{D} + \frac{x_o(z + D)}{CD^3} = 0 \qquad (60.50)$$

The cartesian representation of the caustic is found by elimination of x_o between (60.49) and (60.50), which results in

$$x_c(z) = -\frac{z^2 CD}{2(z + D)} \qquad (60.51)$$

This is valid for sufficiently small $|z|$, so that $|x_o| \leq a$; we see from this representation that the coefficient C is the *curvature* of the caustic in the observation plane $z = 0$.

The incident wave is now given by

$$\psi_o(x_o) = e^{ikL(x_o)} \qquad (60.52)$$

where we have dropped an unimportant amplitude term. After making the appropriate change of notation for the coordinates, the Fresnel integral

(58.17a,b) takes the form

$$\psi(x, y, 0) = \frac{\exp(ikD)}{i\lambda D} \int\limits_{x_o=-a}^{a} \int\limits_{y_o=-\infty}^{\infty} e^{i\Phi} \, dx_o \, dy_o$$

with

$$\Phi = \frac{k}{2D}\{(x_o - x)^2 + (y_o - y)^2 - x_o^2\} + \frac{k}{6C}\left(\frac{x_o}{D}\right)^3$$

The integration over y_o is again possible in closed form and results in a constant factor $(1 + i)(\pi D/k)^{1/2}$: ψ becomes independent of y, as it must. Combining all factors that do not depend on x_o in a common factor A, we obtain

$$\psi(x) = A \int_{-a}^{a} \exp\left\{\frac{ik}{6C}\left(\frac{x_o}{D}\right)^3 - \frac{ik}{D} x_o x\right\} dx_o$$

This integral is not soluble in closed form. It was first evaluated numerically by Lenz and Krimmel (1961, 1963). Following their paper, we cast this integral into a standard form by the introduction of new dimensionless variables:

$$t := \frac{x_o}{D}\left(\frac{\pi}{3C\lambda}\right)^{\frac{1}{3}}, \qquad b := \frac{a}{D}\left(\frac{\pi}{3C\lambda}\right)^{\frac{1}{3}}$$

$$X := 2x\left(\frac{3\pi^2 C}{\lambda^2}\right)^{\frac{1}{3}}$$

whereupon the diffraction integral becomes

$$\Psi(X) = \int_{0}^{b} \cos(t^3 - Xt) \, dt \tag{60.53a}$$

This integral can be evaluated by numerical quadrature, which is, however, very slow. Lenz and Krimmel therefore recommend solving the ordinary differential equation satisfied by $\Psi(X)$. Differentiation under the integral gives

$$\Psi'(X) = \int_{0}^{b} t \sin(t^3 - Xt) \, dt$$

$$\Psi''(X) = -\int_{0}^{b} t^2 \cos(t^3 - Xt) \, dt$$

We can easily verify that

$$\Psi''(X) + \frac{1}{3}X\Psi(X) = \int_0^b (\frac{1}{3}X - t^2)\cos(t^3 - Xt)\,dt$$

$$\text{(60.53b)}$$

$$= \frac{1}{3}\int_0^b \frac{d}{dt}\sin(Xt - t^3)\,dt = \frac{1}{3}\sin(Xb - b^3)$$

is the required differential equation. This is an inhomogeneous, linear second-order equation and can hence be solved once solutions of the corresponding homogeneous equation have been found. The latter is related to Bessel's equation (Eq. 2.162.11 in Part C of Kamke, 1977) and its solutions can be expressed in terms of the tabulated functions $Z_{1/3}(X)$, $Z_\nu := C_1 J_\nu + C_2 Y_\nu$ in which C_1, C_2 are arbitrary constants and J_ν, Y_ν are Bessel functions of the first and second kinds. Solutions in this form are tabulated under the name of Airy functions. Lenz and Krimmel solved (60.53b), after determining $\Psi(0)$ and $\Psi'(0)$ by numerical quadrature. The results for the corresponding intensity distribution $I(X, b) = |\Psi(X, b)|^2$ are presented in Fig. 60.9. They show a very clear interference pattern which is most pronounced in the classically accessible interval $0 \le X \le 3b^2$ (dashed lines in Fig. 60.9) The appearance of weak fringes in the classically forbidden domain is common to all diffraction phenomena. For $b \ll 1$, the solution $\Psi(X, b)$ converges to the Fraunhofer solution for a narrow slit, $\Psi(X, b) = X^{-1}\sin bX$, as it must.

Lenz and Krimmel (1961, 1963) observed such an interference pattern in the line focus of a spectrometer having an aperture defect of second order. This of course required a high degree of monochromatism of the incident beam for otherwise the interference frings would have been smeared out by chromatic effects.

This kind of diffraction pattern is characteristic of any cross-section through a *regular* part of the caustic, as is sketched in Fig. 60.8. It is unnecessary to assume wave propagation in field-free space — this was only done for reasons of conciseness; the same formulae are obtained for curved trajectories if X is identified with the *difference* between the curvature of the caustic and the rays tangent to it. The diffraction phenomenon naturally becomes more complicated if the caustic consists of several branches, which may even intersect. Such examples were, for instance, studied by Krimmel (1960, 1961). A highly complicated situation arises in the vicinity of caustic cusps formed by spherical aberration and astigmatism. Even the geometrical theory is then extremely complicated. The corresponding wave-optical problem is soluble only by numerical techniques or via crude estimates.

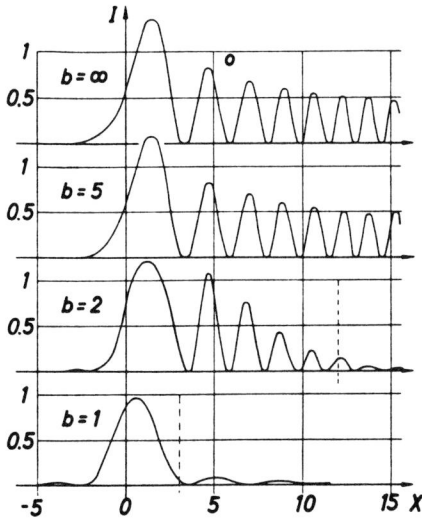

Fig. 60.9: Intensity distributions in caustic interferences for various values of the parameter b.

60.5 Diffraction disc with lens aberrations

We now study the situation sketched in Fig. 60.10. A point source is located near the object focal point of a round electron lens, which produces a crossover in a recording plane, usually thought of as the image plane. This crossover is affected by axial lens aberrations and by diffraction, the latter arising at the edge of the aperture confining the electron beam. In what conditions will the diameter of the crossover be smallest for given focal length, magnification and coefficient C_s of spherical aberration?

This is a standard problem in microscopy of all kinds. Its solution is of great importance for the resolution of the electron microscope. The results obtained, however, can furnish only a preliminary understanding of the resolution limit. The rigorous theory of image formation in the electron microscope is dealt with in later chapters.

In order to confine the topic in a reasonable manner, we consider only the spherical aberration of third order and a small defocus in addition to the diffraction. The aperture is located in the field-free space behind the lens, so that we need consider only the simpler Kirchhoff formula. The aperture angles ϑ_0 and ϑ_i (Fig. 60.10) are related by the Smith–Helmholtz

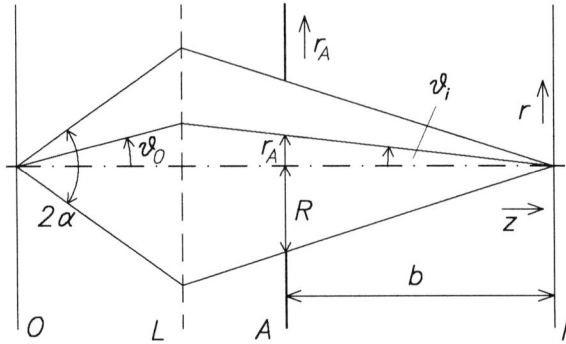

Fig. 60.10: Imaging system with notation: O: object plane, L: lens, A: aperture and I: image plane. The aperture angles are exaggerated.

formula (15.55):

$$\vartheta_i = \frac{r_A}{b} = -\frac{g_0 \vartheta_0}{g_i M} \tag{60.54}$$

g_0 and g_i being the corresponding kinetic momenta and M the lateral magnification. If the aperture is located in the lens field, then the corresponding asymptotic values for r_A and b are to be taken.

In the absence of aberrations, the wave incident on the aperture is spherical with its centre at the axial image point; the contribution of interest to the eikonal of the real wave therefore consists only of *aberration terms*. It is convenient to express these in terms of the aperture angle ϑ_o on the object side. The eikonal $S(\vartheta_o)$, to be introduced into the exponent of the incident wave, is then

$$S(\vartheta_o) = g_o \left(\frac{1}{2} \Delta_o \vartheta_o^2 - \frac{1}{4} C_s \vartheta_o^4 \right) \tag{60.55}$$

Δ_o being the axial defocus; all coefficients and factors refer to the object side. Apart from an unimportant amplitude factor, omitted here, the incident wave is given by

$$\psi_0(\vartheta_o) = \exp\left\{ \frac{i\pi}{\lambda_o} (\Delta_o \vartheta_o^2 - \frac{1}{2} C_s \vartheta_o^4) \right\} \tag{60.56}$$

with the wavelength $\lambda_o = h/g_o$.

This function is now to be identified with the 'object function' $O(r_o)$ in (60.42). Our diffracting object is now the aperture of the system (Fig. 60.10), and consequently the variable (r_o) is to be identified with r_A in (60.54), which can be expressed in terms of ϑ_o. Apart from factors of no interest, the wavefunction in the image plane is then

$$\psi(r) \propto \int_0^\alpha \exp\left\{\frac{i\pi}{\lambda_o}\left(\Delta_o\vartheta_o^2 - \frac{1}{2}C_s\vartheta_o^4\right)\right\} J_0\left(\frac{2\pi r\vartheta_o}{\lambda_o|M|}\right)\vartheta_o\, d\vartheta_o \qquad (60.57)$$

where we have made use of $\lambda \equiv \lambda_i = \lambda_o g_o/g_i$.

In order to cast this integral into a more concise form, we introduce dimensionless variables and coefficients as follows (cf. 65.31 and 66.85):

$$\gamma := \vartheta_o(C_s/\lambda_o)^{\frac{1}{4}} \qquad\qquad (a)$$
$$\Gamma := \alpha(C_s/\lambda_o)^{\frac{1}{4}} \qquad\qquad (b)$$
$$D := \Delta_o(C_s\lambda_o)^{-\frac{1}{2}} \qquad\qquad (c) \qquad (60.58)$$
$$\varrho := 2\pi r(C_s\lambda_o^3)^{-\frac{1}{4}}/|M| \qquad (d)$$

whereupon (60.57) becomes

$$\psi(r) \propto \int_0^\Gamma J_0(\gamma\varrho)\exp\left\{i\pi\left(D\gamma^2 - \frac{\gamma^4}{2}\right)\right\}\gamma\, d\gamma$$

With the substitution $u := \gamma^2 - D$ and omission of an unimportant constant factor, $\exp(-i\pi D^2/2)$, the complex conjugate wavefunction $\hat{\psi}(\varrho) \propto \psi^*(r)$ then takes the final form

$$\hat{\psi}(\varrho) = \int_{-D}^{\Gamma^2-D} J_0(\varrho\sqrt{u+D})\exp\left(\frac{i\pi}{2}u^2\right) du$$

For $D > 0$ and $\Gamma^2 > D$ this can be written as

$$\hat{\psi}(\varrho) = \int_0^{\Gamma^2-D} J_0(\varrho\sqrt{u+D})\exp\left(\frac{i\pi}{2}u^2\right) du + \int_0^D J_0(\varrho\sqrt{D-u})\exp\left(\frac{i\pi}{2}u^2\right) du$$

$$(60.59)$$

We shall deal only with this case, since the case $D < 0$ is of little practical interest; the reasons will be obvious later.

Apart from the axial value $\hat{\psi}(0)$, these integrals cannot be evaluated in closed form. The axial value is given in terms of Fresnel integrals:

$$\hat{\psi}(0) = F(\Gamma^2 - D) + F(D) \qquad (60.60)$$

Another useful result can be derived from the conservation of integrated intensity:

$$\int_0^\infty \varrho |\hat\psi(\varrho)|^2 \, d\varrho = 2\Gamma^2 \qquad (60.61)$$

This simply means that the integrated intensity increases as the area of the aperture.

These two results enable us to deduce optimum working conditions for the lens. It is tempting to make $|\hat\psi(0)|^2$ as large as possible. This gives a unique answer:

$$D_m = 1.2 \quad , \quad \Gamma_m = D_m\sqrt{2} = 1.7 \quad , \quad |\hat\psi_m(0)|^2 = 3.6 \qquad (60.62)$$

and corresponds to the point on the Cornu spiral (Fig. 60.2) that is farthest from the origin. This is, however, *not* necessarily the best choice; perhaps a slightly smaller value of $|\hat\psi(0)|^2$ would give a higher concentration of intensity in the central disc relative to the outer rings. We therefore seek the maximum of $|\hat\psi(0)|^2/\Gamma^2$. This quest again has a unique outcome:

$$D_{opt} = 1 \quad , \quad \Gamma_{opt} = \sqrt{2} \quad , \quad |\hat\psi(0)|^2_{opt} = 3.2 \qquad (60.63)$$

and $|\hat\psi(0)|^2_{opt}/\Gamma^2_{opt} = 1.6$ is indeed larger than $|\hat\psi_m(0)|^2/\Gamma^2_m = 1.25$ from (60.62): the choice (60.63) does lead to a better intensity concentration in the central disc. This is hence the required optimum. The quantity $|\hat\psi(0)|^2/\Gamma^2$ is very similar to the Strehl intensity ratio (*Definitionshelligkeit*) used in optical instrument design.

Introducing (60.63) into (60.58b,c), we find the aperture angle

$$\alpha_{opt} = \left(\frac{4\lambda_o}{C_s}\right)^{\frac{1}{4}} \qquad (60.64)$$

and the object defocus

$$\Delta_{opt} = \sqrt{\lambda_o C_s} \qquad (60.65)$$

Moreover, it is possible to estimate a mean square radius from (60.61). To find this, we replace $\hat\psi(\varrho)$ by a piecewise constant function:

$$\hat\psi(\varrho) \approx \frac{1}{2}\hat\psi(0) \text{ for } \varrho \le \bar\varrho \quad , \quad 0 \text{ otherwise}$$

Equation (60.61) then gives $|\hat\psi(0)|^2\bar\varrho^2/8 = 2\Gamma^2$, and so $\bar\varrho = 4\Gamma/|\hat\psi(0)| = 3.16 \approx \pi$. Substitution of this into (60.58d) gives the average disc radius

$$\bar r = 0.5|M|(C_s\lambda_o^3)^{\frac{1}{4}} \qquad (60.66)$$

$|\hat{\psi}(\rho)|^2/4$

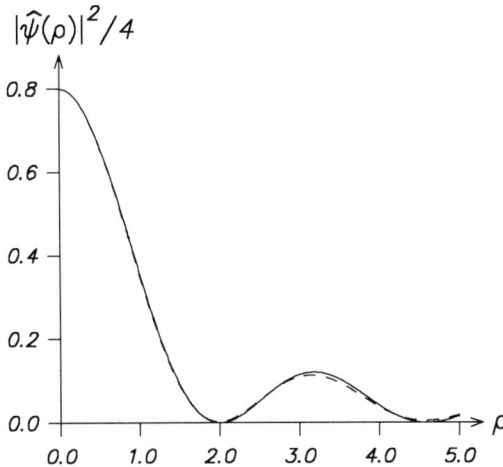

Fig. 60.11: Intensity distribution in a focus with optimum defocus. Broken lines: exact numerical solution; full lines: approximation by formula (60.67).

which agrees fairly well with experimental findings.

The intensity distribution $|\hat{\psi}(\varrho)|^2$ for the optimum case is plotted in Fig. 60.11. This function has a first minimum at $\varrho_m = 2$, which is distinctly smaller then $\bar{\varrho}$. For sufficiently small values of ϱ this function is represented approximately by

$$|\psi(\varrho)|^2 = \frac{3.2 J_0^2(1.2\varrho)}{1 + \varrho^2/144} \qquad (60.67)$$

which satisfies (60.61) acceptably for $\Gamma = 2$.

60.6 The Rayleigh rule and criterion

Rayleigh's quarter-wavelength rule is well-established in optics and frequently used to estimate the parameters of complicated aberration patterns. It states that the quality of an image is not seriously affected if the distance between the spherical wave surface and the true wave surface, distorted by aberrations, nowhere exceeds $\lambda/4$ and, hence, if the maximum absolute phase difference relative to the same point nowhere exceeds $\pi/2$.

We shall now reconsider the aberration disc produced by diffraction, spherical aberration and defocus, which we have investigated in the previous section, in the light of this rule. The reference pattern for the comparison is the Airy disc for aberration-free diffraction, given by (60.46).

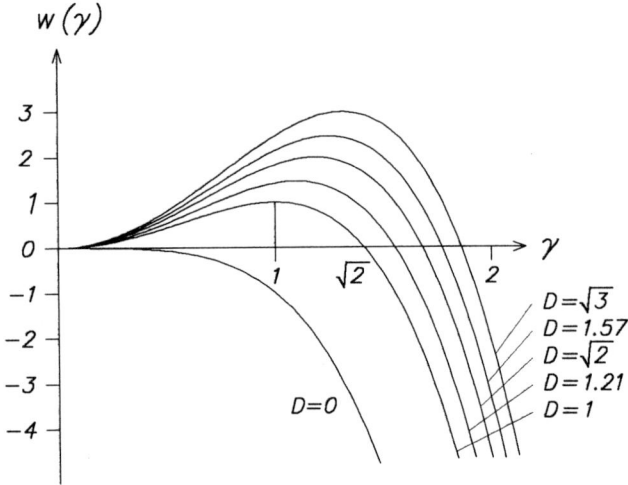

Fig. 60.12: Reduced wave aberration $w(\gamma)$ for various values of the defocus parameter D.

The Airy disc is produced by a constant object function $O(r_o) = \psi_o(\vartheta_o) = 1$. Hence (60.56) would lead to essentially the same pattern if, according to the Rayleigh rule,

$$|\Delta_o \vartheta_o^2 - C_s \vartheta_o^4/2| \leq \lambda_o/2 \tag{60.68}$$

is satisfied for all $\vartheta_o \leq \alpha^2$. In the reduced variables (60.58) this is equivalent to

$$|2D\gamma^2 - \gamma^4| \leq 1 \qquad \forall \gamma \leq \Gamma \tag{60.69}$$

The task of optimization is now to find those pairs of values Γ and D for which the greatest intensity is concentrated in the central disc subject to the constraint (60.69). The answer can be found from Fig. 60.12, which shows the function $w(\gamma) = 2D\gamma^2 - \gamma^4$ for various values of the parameter D.

The optimal situation corresponds to the curve that just reaches unity at its maximum. For this, $D = \gamma_M^2$, $w_M = \gamma_M^4 = 1$ and hence $\gamma_M = 1$, $D_{opt} = 1$. The maximum allowed aperture is then determined by the next zero of $w(\gamma)$: $w(\gamma) = 0$ gives $\Gamma_{opt} = \sqrt{2}$. These are the same as the parameters given in (60.63), which are thus confirmed by a different argument.

The Rayleigh rule also allows us to estimate the radius of the central diffraction spot. The radius of the Airy disc is traditionally chosen as the first zero of the Bessel function in (60.46); setting $x = 3.83$ gives

$$\bar{r} = 3.83\lambda b/2\pi R = 0.61\lambda b/R$$

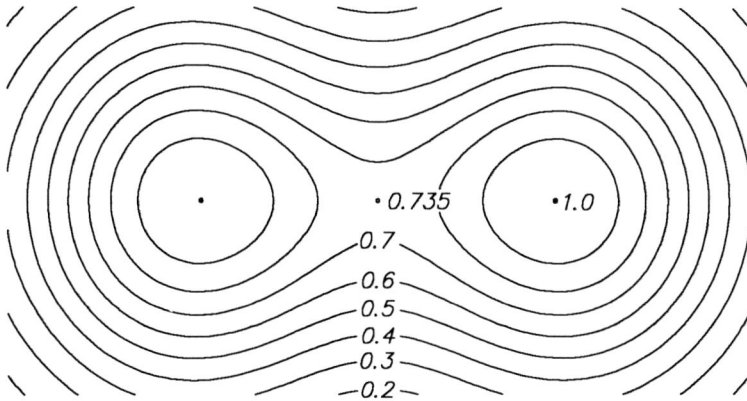

Fig. 60.13: Superposition of the intensities of two Airy discs at a mutual distance of $\delta = 0.61\lambda_o/\sin\alpha$.

If we use $\vartheta_i = R/b$, $\vartheta_o = \alpha$ and $r_A = R$ in (60.54) and recall that $g_o\lambda_o = g_i\lambda$, we obtain

$$\bar{r} = |M|\, 0.61\lambda_o/\alpha \tag{60.70}$$

which will be recognized as Abbe's resolution limit for an ideal lens. Inserting now the optimum aperture angle from (60.64), we arrive at

$$\bar{r} = 0.43|M|(C_s\lambda_o^3)^{\frac{1}{4}} \tag{60.71}$$

which agrees rather well with (60.66). An error of 15 per cent in such an estimate is quite acceptable since the accuracy of measurements is generally not better.

Traditionally, these formulae are used to determine the resolution limit of the electron microscope or any other imaging instrument. This is an everyday procedure in conventional light optics and has been transferred into electron optics by analogy. According to Rayleigh's criterion, two overlapping diffraction patterns are considered to be separated if the intensity at the saddle point is at most 75 per cent of that at the two maxima (see Fig. 60.13). For Airy discs, the distance $\bar{r}$ between the two maxima is then given by (60.70) and (60.71). This is then regarded as the classical resolution limit of the instrument.

All these arguments are based on the tacit assumption that the object can be treated as a collection of individual point sources that radiate incoherently. This means that it is the intensities and not the wavefunctions, produced by the individual sources, that superimpose linearly, as is shown in Fig. 60.13. This assumption is frequently unjustified in electron optics and the reasoning leading to the estimate (60.71) for the resolution limit is extremely over-simplified. Moreover, the incoherent asssumption is particularly unlikely to be true in high-resolution conditions. Much of the present volume is devoted to the task of developing a more satisfactory approach to image formation. It is, perhaps, reassuring that (60.64) and (60.65) also prove to be optimum conditions when we discuss resolution in terms of the contrast transfer function.

We close this section with an estimate of the maximum allowable astigmatism for which there is no significant image blurring. With axial astigmatism, the wavefunction in the aperture plane is

$$\psi_o(\vartheta_o, \varrho_o) = \exp\left[\frac{i\pi}{\lambda_o}\left\{\Delta_o\vartheta_o^2 - \frac{1}{2}C_s\vartheta_o^4 + A_o\vartheta_o^2 \cos 2(\varphi_o - \beta)\right\}\right] \quad (60.72)$$

instead of (60.56), $A_o > 0$ being the astimatism coefficient referred to the object side. In principle the astigmatism can be compensated but this does not always succeed perfectly and an estimate of its influence on the resolution is hence needed.

The corresponding diffraction integral cannot be evaluated in closed form; even the integration over the azimuth φ_o cannot be carried out analytically. We can, however, again apply Rayleigh's quarter-wave rule. Taking the diffraction pattern without astigmatism as reference for comparison, we see that the astigmatism term in (60.72) must satisfy

$$A_o\vartheta_o^2 \leq \frac{\lambda_o}{4}$$

since the phase difference between the value at $\varphi_o = \beta$ and the value at $\varphi_o = \beta + \pi$ must not exceed $\pi/2$. Introducing α_{opt} from (60.64) for ϑ_o, we obtain the estimate

$$A_o \leq \frac{1}{8}\sqrt{\lambda_o C_s} = \frac{\Delta_{opt}}{8} \quad (60.73)$$

This little example demonstrates clearly that Rayleigh's rule may be quite helpful when seeking preliminary estimates in complicated situations.

PART XII

Electron Interference and
Electron Holography

61

General Introduction

When an electron wavefront is divided into separate parts, which traverse regions of different refractive index and are subsequently recombined, interference fringes will be created. This situation may arise in many ways. In conventional electron microscopy, the electron beam may traverse a wedge-shaped crystal (Fig. 61.1a), in which case fringes will be formed by interference between the part of the beam that has passed through the specimen and the part that has not. Alternatively, the beam may be deliberately divided into two parts, typically by means of an electron biprism (Fig. 61.1b). The bright-field image in a conventional electron microscope may even be regarded as the interference pattern created by the undiffracted beam and the electrons scattered by the specimen (Fig. 61.1a).

Interference patterns are of course intensity distributions but their detailed structure is governed by the phase differences between the waves that interfere. A particularly important family of interference phenomena is now known as *holography*. This two-stage process was proposed by Gabor (1948, 1949a,b, 1950, 1951a,b) as a way of eliminating the adverse effects of lens aberrations on electron optical image formation. The basic idea was as follows.

A thin specimen, transparent to electrons, is irradiated and traversed by an electron wave, the *object wave*. The latter is modulated by the electric field inside the specimen and thus contains information about the object structure. This object wave is then brought into coincidence with a second wave, which has not been affected by the specimen and is known as the *reference wave* (Fig. 61.2a). The resulting interference pattern is then recorded on film. It is this pattern that is known as a *hologram* and it is of course affected by the properties of the device used to create the interference.

In a second quite separate step, the hologram is trans-illuminated by a coherent wave, which need not be an electron wave and will in practice be coherent light from a laser. As we shall see in more detail later, diffraction in the hologram generates *three* waves. One has the structure of the reference wave. A second wave is a reconstruction of the object wave and seems to emanate from a virtual object O' (Fig. 61.2b). The third wave, the *conjugate wave*, converges to a 'real' object O'', which has a different orientation from the specimen O. These three waves must now be sepa-

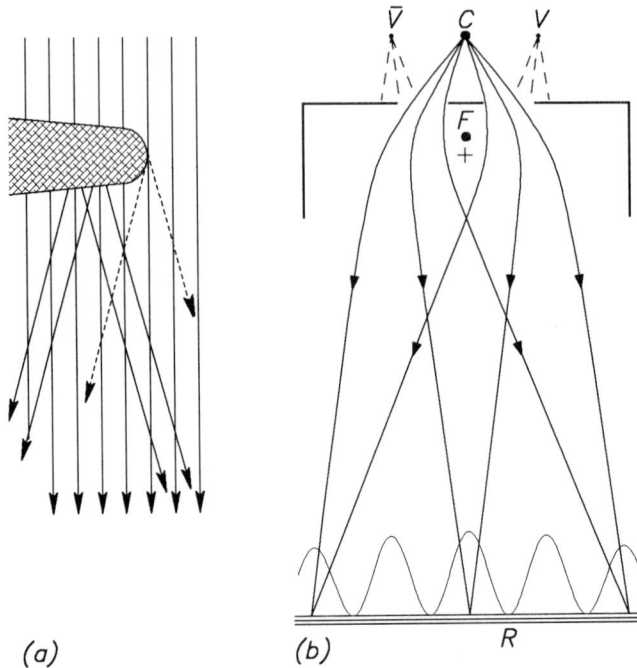

(a)   (b)

Fig. 61.1: Possible ways of producing electron interference: (a) *Diffraction* in a crystalline specimen and at its edges; the diffracted waves interfere with the undiffracted ones. The resulting fringes are visible in a defocused image. (b) *Electron biprism* with real source C, virtual sources V, $\overline{V}$, entrance slits, filament F and photographic plate R.

rated from one another as completely as possible so that one of the object waves can be isolated.

This stage is known as *reconstruction*. An advantage of this procedure in light optics is that we have complete control over the basic geometrical aberrations, and in particular over the spherical aberration; in principle, therefore, the electron optical aberrations can be compensated by suitable choice of the glass lenses employed in the reconstruction.

The principle of holography in this simple form is illustrated in Fig. 61.2, where light is used for both the recording of the hologram and for the reconstruction. Technical details (the laser, lenses, diaphragms) are omitted. In the reconstruction step, irradiation of the hologram H generates the virtual (O') and real (O'') images shown.

Holography is indeed a two-stage process, therefore. In the first stage, a reference wave and an object wave interfere, producing a hologram and in the second stage, information about the object wave is extracted from this hologram. There are numerous ways of performing each step and in practice

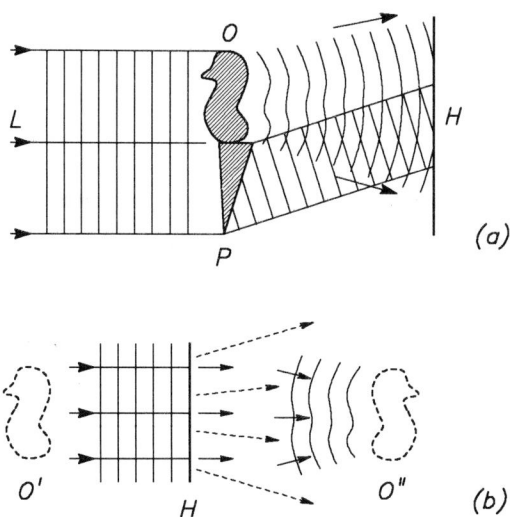

Fig. 61.2: Simplified presentation of light-optical holography: (a) formation of the hologram; (b) object reconstruction. L: laser-beam; O: physical object (specimen); P: prism; H: hologram; O', O'': virtual and real reconstructed object, respectively.

it is convenient to distinguish between *in-line* holography (Fig. 61.3a,b) and *off-axis* holography (Fig. 61.3c,d). Each technique has specific advantages and also disadvantages.

In the *in-line* arrangement, the optic axis is a symmetry axis and passes through the points O, O' and O'' of Figs 61.3a,b. The aberrations of the lenses (not shown here) have less effect than in the off-axis arrangement but it is not easy to separate the wavefronts R, W' and W'' sufficiently in the reconstruction stage.

There is no such difficulty in the off-axis arrangement (Fig. 61.3d) but the effect of the lens aberrations in the recording stage is greater, owing to the highly asymmetric disposition of the wavefronts (Fig. 61.3c). The choice between the two arrangements may well be governed by the availability and properties of the appropriate optical elements: sources, beam splitters, apertures, recording media.

The two wavefronts W' and W'' of Figs 61.3b and 61.3d are completely equivalent so far as the reconstruction is concerned. It is the subsequent arrangement of lenses and apertures that determines which of them is finally recorded. In Chapter 63, we shall see that the three waves R, W' and W'' can be considered as three orders, zero and ± 1, in the Fresnel diffraction pattern of the illuminating wave at the hologram film H. (The

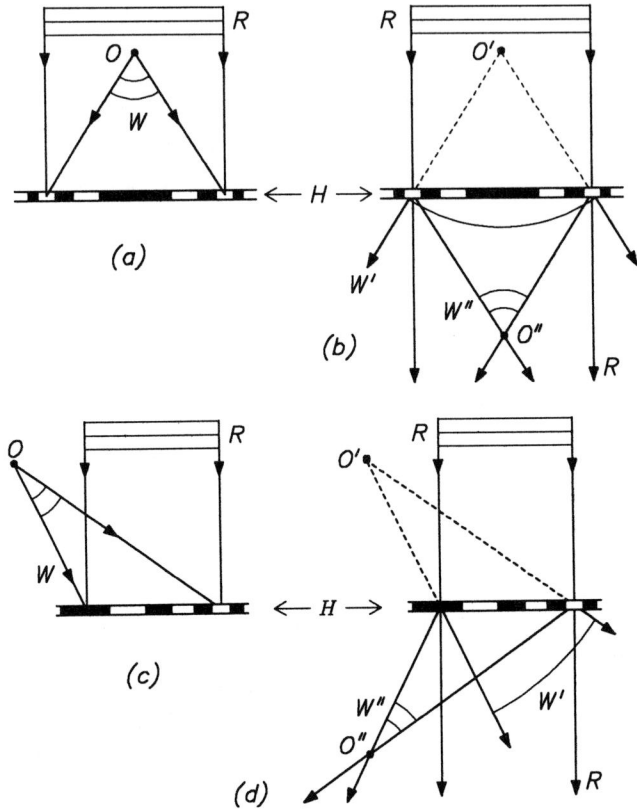

Fig. 61.3: Principal types of holography: (a,b): *in-line* set-up; (c,d): *off-axis* set-up; (a,c): recording; (b,d): reconstruction; R: reference wave; H: hologram; W: recorded object wave; O: physical specimen; W', W'': diverging and converging reconstructed waves; O', O'': reconstructed virtual and real objects.

diffraction angles in Fig. 61.3 are highly exaggerated —in reality, they will be extremely small.) In a first very simple approximation, the waves may be regarded as practically plane. The necessary focusing by a lens and beam selection by an aperture are then as shown in Fig. 61.4. The lens is adjusted in such a way that all the beams belonging to the same diffraction order are focused onto the aperture plane. Simultaneously, it must image the hologram with all the beamlets that pass through the aperture onto the image plane. This particular arrangement is known as *image plane holography*. Other arrangements for producing and reconstructing holograms are described in later sections.

We have followed the traditional development of holography in the

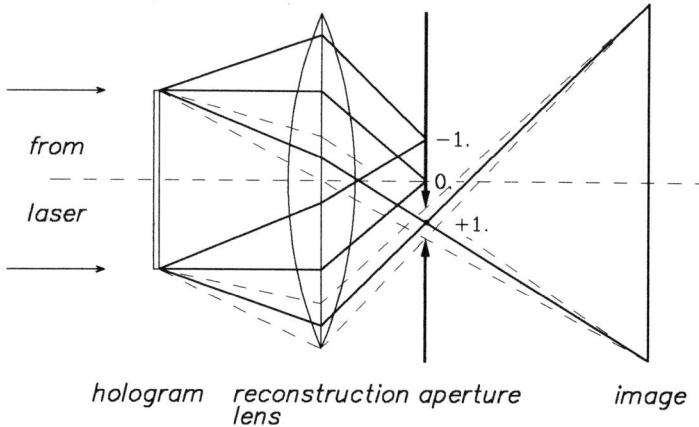

from

laser

hologram   reconstruction aperture   image
             lens

Fig. 61.4: Light-optical reconstruction: the foci of the reference wave (0) and of the diffraction order (-1) are excluded by an aperture-containing diaphragm. The rays that pass through the aperture produce a sharp image of the object (image-plane holography).

foregoing introduction: electron-optical hologram formation, light-optical reconstruction. It has long been apparent, however, that many of the practical difficulties of the reconstruction step could be avoided by digitizing the hologram and performing the reconstruction step in the computer. The flexibility would thereby be increased and the only noise present would be that due to the digitization and that of the hologram. Correction of the high spherical aberration of an electron lens by means of a glass lens with equal and opposite aberration is far from easy (see Rogers, 1978, 1980, for example, who had a special lens with abnormally high primary aberrations made for this purpose) but is trivial digitally. We shall frequently describe reconstruction in light-optical terms but all of these procedures can be and in practice are being replaced by digital routines once the hologram has been digitized.

It should be clear from this introduction that electron interferometry, electron holography and the bright-field image-forming process are all intimately related. We now attempt to disentangle the various strands by considering the main stages in the development of these themes. We stress that only the most important publications are mentioned here. Further references are included in the list for this Part.

The impossibility of designing a rotationally symmetric electron lens free of third-order spherical aberration (see Section 24.3, Scherzer's theorem) and the experimental difficulty of introducing correctors such as those proposed by Scherzer (Chapter 41) and by himself led Gabor to enquire whether electron images degraded by aberrations could not be reconstituted optically. He argued that the object information in an image obtained with a lens with aberrations was not lost but shuffled—wrongly presented to the eye—and suggested that a correct image could be reconstructed by the two-stage holography process we have described (Gabor, 1948, 1949a,b, 1950, 1951a,b). His publications provoked other suggestions and comments (Dyson, 1950; Rogers, 1950a,b, 1952; Haine and Dyson, 1950) and attempts to test the idea experimentally were made by his colleagues at the Associated Electrical Industries' Research Laboratory in Aldermaston (Haine and Mulvey, 1950, 1951, 1952). These could not succeed owing to the lack of coherence* of both the electron source used to form the hologram and the light source employed in the reconstruction (the laser had not yet been invented) and the idea lay fallow for many years.

At the same period, Marton and colleagues at the U.S. National Bureau of Standards built the first electron interferometer, an amplitude-division design in which the wavefront was divided and recombined by Bragg reflection within a thin crystalline film (Marton, 1952, 1954; Marton et al., 1953, 1954; Simpson, 1954, 1956). The problems of alignment and lack of intensity of their device were considerable; it has been re-examined by Matteucci et al. (1981). Real progress in interferometry was not made until Möllenstedt and Düker in the University of Tübingen introduced the electron biprism, in which a fine, charged wire mounted between earthed plates splits the incident beam into two parts; these are subsequently reunited by a lens (Möllenstedt and Düker, 1955, 1956; Düker, 1955; Möllenstedt, 1991). These early biprism experiments were performed with electrostatic lenses; similar work was undertaken with a magnetic column very soon after by Faget and Fert in the Toulouse Laboratory of Electron Optics (Faget and Fert, 1956, 1957; Faget et al., 1958; Faget, 1961; Fert, 1961, 1962). Figure 61.5 shows how the fringes build up as the biprism voltage is increased.

With the introduction of the biprism by Möllenstedt and Düker, interference microscopes were constructed in many laboratories and ingenious combinations of biprisms and other optical elements were tested. Arrangements using more than one biprism were employed by Möllenstedt and Bayh (1961a,b) and Schaal et al. (1966/67) and by Lichte who built an electron mirror interference microscope as shown in Fig. 18.2 (Lichte et al., 1972; Lichte, 1979, 1980a; Lichte and Möllenstedt, 1977, 1979). In Japan, the gun with a pointed filament developed by Hibi (1956) enabled Hibi and Takahashi (1963) to obtain many sharp fringes thanks to its brightness and several versions were subsequently built, with high mag-

---

* The term 'coherence' is used loosely throughout this Part in connection with non-vanishing source-size and energy spread. The related effects are discussed in Section 66.2 (contrast-transfer function) and more formally in Part XVI.

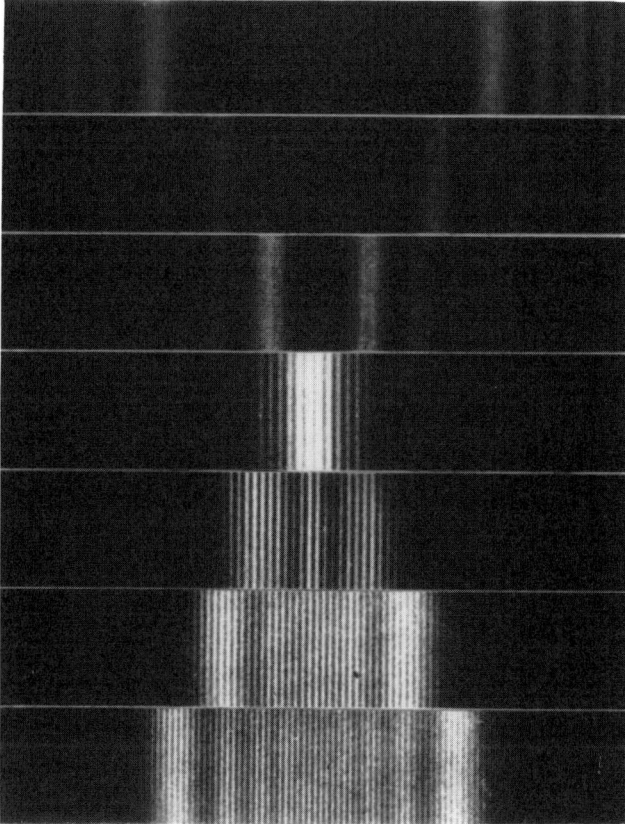

Fig. 61.5: The influence of the filament potential of an electron biprism. As the potential is increased from zero (top) to 7 V (bottom), the two sets of Fresnel fringes of the filament approach each another. When they overlap, the Fresnel fringes are superimposed and equidistant interference fringes become visible. In this early experiment of Möllenstedt and Düker (1956), the filament was 2 μm in diameter.

nification in mind (Tomita *et al.*, 1970a,b, 1972; Yada *et al.*, 1973). The advent of the field-emission gun brought further improvements (Brünger, 1968, 1972; Crewe and Saxon, 1971). Other interference microscopes were developed by Boersch *et al.* (1960, 1962a), who used anti-parallel magnetic domains in a ferromagnetic foil as a beam splitter, Buhl (1958, 1959), Keller (1958, 1961), Krimmel (1960), Feltynowski (1963), Drahoš and De-long (1963, 1964), Anaskin *et al.* (1966), Kerschbaumer (1967), Komrska *et al.* (1967) and Sonier (1968, 1971). The natural extension to multiple-beam interference was made by Möllenstedt and Jönsson (1959), using several slits and by Anaskin and Stoyanova (1967, 1968a-c) and Anaskin *et al.* (1968), using several wires in the 'biprism'. (See also Jönsson, 1961; Buhl, 1961a,b; and Jönsson *et al.*, 1974.) A magnetic biprism in which no part of the incident beam is obscured was examined by Krimmel (1960, 1961). A mixed interferometer incorporating both a crystalline film and a biprism is used by Matteucci and Pozzi (1980). These instruments have been used to measure the degree of coherence of electron sources (see Hibi and Takahashi, 1969; Hibi and Yada, 1976, Section IV.B of Hawkes, 1978, for many early references and also Braun, 1972; Speidel and Kurz, 1977, Möllenstedt and Wohland, 1980; Lenz and Wohland, 1984; Schmid, 1984; Lenz, 1987; Medina and Pozzi, 1990) and the mean inner potential (Chapter 68) of thin specimens (Möllenstedt and Buhl, 1957; Möllenstedt and Keller, 1957; Durand *et al.*, 1958; Langbein, 1958; Buhl, 1959; Keller, 1961; Hoffmann and Jönsson, 1965; Jönsson *et al.*, 1965; Kerschbaumer, 1967; Tomita and Savelli, 1968; Anaskin and Stoyanova, 1968b; Sonier, 1970, 1971; Yada *et al.*, 1973; Herring *et al.*, 1992). The use of an interferometer for mapping electric field distributions has been considered by Borzjak *et al.*(1977) and Kulyupin *et al.* (1978/79) and, in a different context, by Frabboni *et al.* (1987), Matteucci *et al.* (1987, 1988a, 1989, 1992b), Missiroli *et al.* (1991) and by Zhang *et al.* (1992, 1993, cf. Spence *et al.*, 1993). Contact potential differences have been studied by Brünger (1972), Brünger and Klein (1977) and Krimmel *et al.* (1964). The fields at p–n junctions are examined by Frabboni *et al.* (1985, 1986), Matteucci *et al.* (1988b), Merli and Pozzi (1978), Merli *et al.* (1976b) and Pozzi and Vanzi (1982). The electron analogue of the Michelson interferometer, in which high temporal coherence is vital, has been studied by Lichte *et al.* (1972), Lenz (1972) and Lichte (1980a,b) while Ohtsuki and Zeitler (1977) have repeated Young's experiment, using electrons. Stoyanova and Anaskin (1968) have attempted to estimate the ratio of coherently scattered electrons to those scattered incoherently in a specimen by interferometry and Menu and Evrard (1971) have studied the decline in fringe visibility as the specimen thickness is increased. The book by Stoyanova and Anaskin (1972) has chapters on interference and holography; a full review of the earlier work on interferometry and interference microscopy is given in Missiroli *et al.* (1981), which includes some applications not considered here. A later review by Matteucci *et al.* (1984) deals with the extensive work on interferometry with magnetic specimens, which became a major preoccupation in holography as we shall see. More isolated studies have been made of glass knives for ultramicrotomy (Lauer and Lickfeld, 1988), effects in crystals (Ade, 1986; Kawasaki *et al.*, 1991a,b, 1992b), edges (Subbarao, 1991), interfaces (Weiss *et al.*, 1991) and of possible uses in biology (Lichte and Weierstall, 1988; Matsumoto *et al.*, 1991).

Two other interference effects must also be mentioned: the Aharonov–Bohm effect (Ehrenberg and Siday, 1949; Aharonov and Bohm, 1959, 1961), around which a lively polemic developed, discussed in Section 62.4, and the attempts to detect the Sagnac effect, using electrons. The latter effect is a phase shift caused by rotation of interfering beams and has been demonstrated for electrons by Hasselbach and Nicklaus (1990); for a very detailed account, see Nicklaus (1989). We return to this briefly in Section 62.5.1.

Apart from an attempt by Hibi (1956) to repeat the early experiments of Haine and Mulvey with a pointed filament, electron holography attracted no interest until the late 1960s. In 1968, Möllenstedt and Wahl in Tübingen and Tonomura *et al.* in Tokyo described off-axis and on-axis electron hologram formation and the subject has been extensively developed in the following decades, largely in Germany and Japan with some work in the United States. At about the same time, Hanszen and colleagues in Braunschweig explored in great detail a holographic interpretation of bright-field electron microscope image formation. Such images may indeed be regarded as fringe patterns formed by interference between the unscattered wave and the wave describing electrons scattered within the specimen; these two waves may be regarded as a reference wave and an object wave so that the bright-field image may legitimately be described as an in-line hologram. Irradiation of this image with laser light and a suitable filter may thus be identified with the reconstruction step, which has led Hanszen (1971, 1982a, 1986a) and Stroke *et al.* (1971a,b, 1973, 1974, 1977; Stroke and Halioua, 1972, 1973) to identify many of the linear reconstruction algorithms described in Part XV with holography. We have to say that this is really a matter of personal preference and is helpful only in so far as one is more familiar with holography than with electron image formation.

The reconstruction stage of many holographic processes makes severe demands on the light optical arrangement employed and it was realized soon after Wahl described his experimental difficulties in detail (1974, 1975) that digital reconstruction would solve many of these (e.g. Hawkes, 1980: Preface); the only drawback would be the size of the matrices corresponding to holograms digitized on a fine enough sampling grid. Digital reconstruction has now been successfully implemented by Lichte (1988, 1991b) and Fu *et al.* (1991) and for interference holography by Ohshita *et al.* (1990). See also Hanszen (1982b), Franke *et al.* (1986, 1987), Ade (1988), Daberkow *et al.* (1988) and Ade and Lauer (1992a,b).

Owing to the complexity of the subject, we have divided the material into two chapters. Chapter 62 is devoted to the classical wave mechanical aspects of interferometry, in which the roles played by diffraction in the specimen and by lens aberrations are unimportant. A certain amount of space is devoted to the optics of biprisms, since these are the most important electron optical elements involved and were not treated in Volumes 1 and 2.

In Chapter 63, on holography in the proper sense of the term, diffraction in the object plays a central role. We begin by showing that in-line holography can be understood in terms of Fresnel zone lenses, as shown in Fig. 61.3a,b. We then study in some detail hologram formation in microscopes equipped with a biprism and digital reconstruction in the computer.

62
Principles of Interferometry

62.1 The electrostatic biprism

Together with a sufficiently coherent source and lenses to provide the necessary magnification, the electrostatic biprism is an important component of electron interference instruments of any kind. Its task is to split the coherent wave emitted by the electron source into two partial waves that are still coherent, as shown schematically in Fig. 61.1b. This separation can be made so large that a specimen can be brought into the path of one of the partial beams without loss of the necessary coherence.

The deflection of the electrons in the electrostatic biprism can be treated entirely within the frame of classical mechanics. The reason for this is that the important diffraction and interference events occur up- and downstream from the field of the biprism.

Electron biprism design has scarcely changed since the device was invented by Möllenstedt and Düker in 1955. The basic arrangement is shown schematically in Fig. 62.1a. A straight and very thin metallic filament is situated midway between two plates serving as screening electrodes. These are held at the potential of the instrument column, so that the asymptotic field vanishes. We set this potential equal to zero, and in practice the plates are earthed. This is not the same choice of potential origin as in Volumes 1 and 2, where zero voltage corresponded to zero electron velocity and hence to the potential of the filament of the gun. The filament must be *positively* charged in order to make the partial beams overlap further downstream.

Field model

Accurate calculation of the electrostatic field in this configuration is very complicated though possible in principle. In order to display the principal deflecting effect of the field on the electrons reasonably straightforwardly, we employ a simple model, which nevertheless describes the essentials correctly.

The central filament has a finite length $2a'$ and is *uniformly* charged with a total charge $Q > 0$. The charge per unit length, or line-charge density, is hence $q = Q/2a'$. The two electrodes are simulated by two parallel line charge distributions of equal length $2a'$ and line-charge density

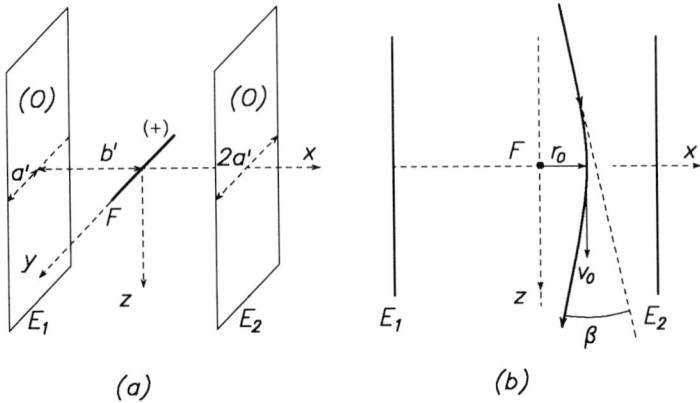

Fig. 62.1: Basic shape and physical action of an electrostatic biprism. (a) Simplified perspective view. F: filament (length $2a'$, charge $2a'q > 0$, potential $V_0 > 0$. E_1, E_2: screening electrodes (charge $-a'q$, each at zero potential; the connection to the voltage supply is omitted. (b) Deflection of a trajectory with apex distance r_0, apex velocity v_0 and total deflection angle β.

$-q/2$. This is necessary to achieve *neutrality* of the total system. Without loss of generality we can choose the coordinate system such that the central filament lies along the y-axis and the two screening lines lie in the x-y plane at positions $x = \pm b'$.

The potential produced by this charge distribution is given by

$$V(x,y,z) = \sum_{n=-1}^{1} \frac{q(1 - 3n^2/2)}{4\pi\epsilon_0} \int_{-a'}^{a'} \frac{dy'}{\{(y' - y)^2 + (x + nb')^2 + z^2\}^{\frac{1}{2}}}$$

The integral can be evaluated in closed form, giving

$$V(x,y,z) = \sum_{n=-1}^{1} \frac{q(1 - 3n^2/2)}{4\pi\epsilon_0} \ln\left|\frac{y + a' + \{(y + a')^2 + (x + nb')^2 + z^2\}^{\frac{1}{2}}}{y - a' + \{(y - a')^2 + (x + nb')^2 + z^2\}^{\frac{1}{2}}}\right|$$

$$(62.1)$$

From this general but still fairly complicated solution, expressions for two important special cases can be derived: the *asymptotic* behaviour and the potential in the vicinity of the origin.

An expansion for large distances

$$R = \sqrt{x^2 + y^2 + z^2} \gg \max(a', b') \tag{62.2a}$$

shows that a *quadrupole field* is the lowest non-vanishing multipole term:

$$V(x,y,z) = -\frac{q}{2\pi\epsilon_0} \frac{a'b'^2}{R^3} \left(\frac{3}{2}\frac{x^2}{R^2} - \frac{1}{2}\right) \tag{62.2b}$$

The quadrupole, characterizing the electrostatic fringe field, is orientated in the x-direction.

In the immediate vicinity of the central filament, which is the most important region for the deflection effect, we see from (62.1) that the potential, and consequently the field strength, depend on the coordinate y. This is a perturbation which causes electron optical aberrations, since the interference fringes will be straight only if the field does *not* depend on y. In a more realistically calculated field, this perturbation is less pronounced, since the filament surface is then an equipotential. Even so, the effect cannot be completely eliminated, being an inevitable consequence of the finite length of the filament and the surrounding screening plates.

It is hence necessary to confine the domain traversed by the electron beam in the y direction to a slit, the length of which is considerably smaller than a'. We can then set $y = 0$ in (62.1) and in the corresponding formulae for the field strength to a good approximation.

In the practical realization of biprisms, the radius ϱ of the central filament is much smaller than the lengths a' and b'; the electric field strength is therefore high only in the immediate vicinity of this filament. This implies that we can set

$$r := \sqrt{x^2 + y^2} \ll \min(a', b') \tag{62.3}$$

throughout the domain in which the electrons are effectively deflected. In the terms for which $n \neq 0$ in (62.1), we may therefore set $x = z = 0$ and the general formula simplifies to

$$V(x, 0, z) = \frac{q}{2\pi\epsilon_0} \left\{ \ln\left(\frac{2a'}{r}\right) - \ln\left(\frac{a' + \sqrt{a'^2 + b'^2}}{b'}\right) \right\} \tag{62.4a}$$

The *filament potential* V_0 is obtained by setting $r = \varrho$, giving

$$V_0 = \frac{q}{2\pi\epsilon_0} \ln\left[\frac{2b'}{\varrho\{1 + \sqrt{1 + (b'/a')^2}\}} \right] \tag{62.4b}$$

Usually this potential is fixed by the experimental conditions; (62.4b) can then be used to calculate the line-charge density if the above field model is accepted. We can also use the constant V_0 to simplify (62.4a), giving finally

$$V(r) = V_0 - \frac{q}{2\pi\epsilon_0} \ln\left(\frac{r}{\varrho}\right) \tag{62.4c}$$

We shall now use this ideal cylindrical potential, obtained after extreme simplification, to calculate the beam deflection in interferometers. We note that (62.4c) has the same form as that of a cylindrical condenser, in agreement with the observations of Möllenstedt and Düker (1956) who studied the biprism field with the aid of an analogue device commonly used at that time, the electrolytic tank. For good agreement, the outer radius of the condenser needed to be slightly smaller than the distance between the plates. Later, Septier (1959) obtained an expression for the field of a wire of negligible radius between earthed plates.

Asymptotic deflection of electron trajectories

In view of the many simplifications already made, there seems little hope of obtaining an accurate solution but, surprisingly, this is not so (Kasper, 1992). The only indispensable approximation is the reduction of the problem to two dimensions, which means that we continue to ignore the confinement of the field in the y-direction.

We start from the ray equations derived in Volume 1 (3.22). Since there is no magnetic field and no motion in the y-direction, these equations reduce to

$$\frac{x''(z)}{1+x'^2} = \frac{1}{2\hat{\phi}}\left(\frac{\partial\hat{\phi}}{\partial x} - x'\frac{\partial\hat{\phi}}{\partial z}\right)$$

On linearizing the accelerating potential, $\hat{\phi} = (U+V)\{1+\epsilon(U+V)\}$ with respect to V, which is permissible since $|V| \ll U$ and writing $\gamma = 1 + 2\epsilon U$ as usual, we obtain

$$\frac{x''}{1+x'^2} = \frac{\gamma}{2\hat{U}}\left(\frac{\partial V}{\partial x} - x'\frac{\partial V}{\partial z}\right) \tag{62.5}$$

Since $V(z,x)$ satisfies the two-dimensional Laplace equation, there must be an associated potential $W(z,x)$, related to $V(z,x)$ via the Cauchy–Riemann equations,

$$\frac{\partial V}{\partial x} = \frac{\partial W}{\partial z} \quad, \quad \frac{\partial V}{\partial z} = -\frac{\partial W}{\partial x}$$

If we use these to replace V by W in (62.5), we find that both sides can be rewritten as total derivatives with respect to z:

$$\frac{x''}{1+x'^2} = \frac{d}{dz}\{\arctan x'(z)\} = \frac{\gamma}{2\hat{U}}\left(\frac{\partial W}{\partial z} + x'\frac{\partial W}{\partial x}\right) = \frac{\gamma}{2\hat{U}}\frac{dW(z,x(z))}{dz}$$

We can integrate this immediately and obtain

$$\alpha(z) := \arctan x'(z) = \frac{\gamma}{2\hat{U}}W(z,x(z)) + \text{const} \tag{62.6}$$

This equation is not limited to the biprism but is true for any two-dimensional electric field; nor is it necessary to assume that the slopes are small.

The potential $W(z, x)$ must satisfy the Neumann condition $\partial W/\partial n = 0$ on all electrode surfaces and has to be made unique by suitable choice of cuts in the z-x plane. Let us now consider the specific case of the biprism. For a trajectory passing on one side of the central filament, we consider the closed loop consisting of the trajectory from the field-free domain on the entrance side to that on the exit side and a path outside the biprism in the positive sense. In this way, we have encircled the line-singularity $-q/4\pi\epsilon_0$ of the shielding plate on the same side of the filament as the trajectory. Stokes' integral theorem then tells us that the line integral of grad W around this loop is equal to $-q/2\epsilon_0$. This value is thus the potential difference between the two sides of the cut running from the singularity to infinity. We can hence conclude that

$$W(L, x(L)) - W(-L, x(-L)) = -q/2\epsilon_0 \quad , \quad L \to \infty$$

and from (62.6), that

$$\beta := \alpha(L) - \alpha(-L) = -\frac{\gamma q}{4\epsilon_0 \hat{U}} \quad \text{for} \quad L \to \infty \qquad (62.7)$$

The deflection angle β is hence *exactly* proportional to the line-charge density q of the central filament, if we adopt the earlier assumptions that we can neglect the fringe fields in the transverse (y) direction and the nonlinear terms in V. These latter simplifications are, indeed, not severe.

It is important to notice that the deflection angle β depends neither on the distance from the apex, the impact parameter r_0, nor on the asymptotic initial slope. It has the same value for all trajectories passing on the same side of the biprism filament and differs only in sign on the other side. An immediate consequence is that an axial point source above the biprism will be seen as a pair of *virtual* point sources, symmetrically placed about the z-axis, from below the biprism. These are labelled V and $\overline{V}$ in Fig. 62.2. This is the fundamental reason for the success of biprism interferometers.

Applications with real interferometers

The simple configuration shown in Fig. 62.2 is not suitable for electron beams. A more realistic, though still schematic, arrangement is shown in Fig. 62.3. A highly coherent electron wave is emitted by an electron source G, which should be as small as possible; in practice, a very bright field-emission gun with a fine cathode tip is used. Nowadays, tip radii smaller than 10 nm are attainable (see Part IX).

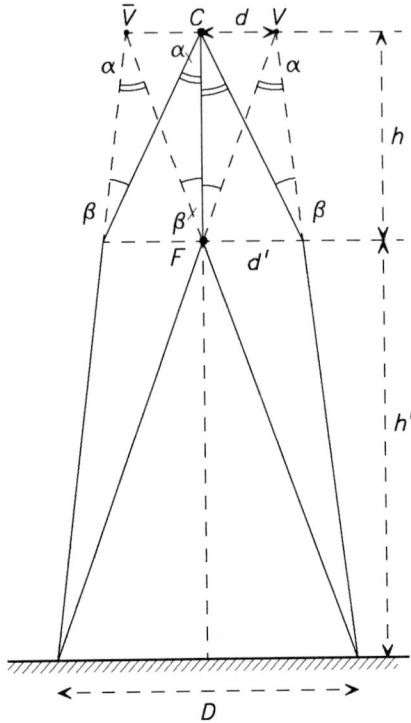

Fig. 62.2: Geometrical construction of the two virtual sources V and $\overline{V}$, located symmetrically at a distance d from the real source C. The rays (full lines) are schematically refracted in the plane of the filament F. The deflection angle β and the aperture angle α are much exaggerated for clarity. With $|\alpha| \ll 1$, $|\beta| \ll 1$ we have $d = h\beta$, $d' = h\alpha$, $D = (h + h')\alpha$.

The real object S, that is, the specimen to be investigated, is situated on one side of the axis in front of the objective lens L. The latter produces a highly demagnified crossover C, to which all the rays from the source are focused. This point is now the real source denoted by C in Fig. 62.2. The biprism is placed further downstream (as in Fig. 62.2) and forms two virtual sources V and $\overline{V}$; their positions are found by backward continuation of the deflected rays. To a good approximation, the asymptotes to *all* the deflected rays intersect in these points.

In the absence of deflection, the objective lens would image the specimen as an 'object' O; it will be convenient to regard this below as a 'real object', though it is of course only an intermediate image.

An interference pattern is now formed in the plane H at the bottom of Fig. 62.3 and often called the primary hologram plane. In principle this

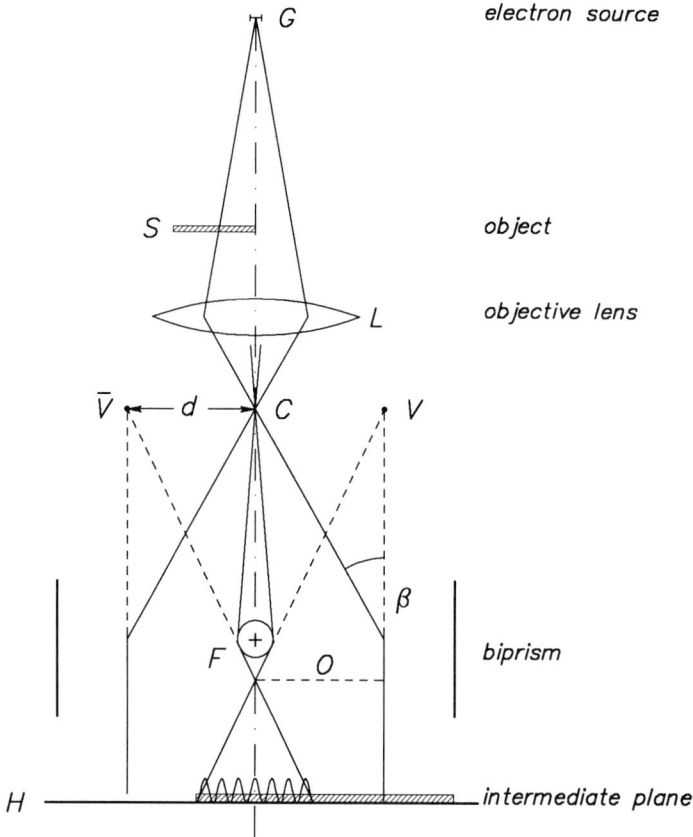

Fig. 62.3: Typical arrangement of an electron interferometer, simplified presentation. G: electron source, S: real object (specimen), L: objective lens, C: real crossover of G. V, $\overline{V}$: virtual sources (see also Fig. 62.2), F: filament of the biprism, β: deflection angle (62.7), O: object in the form of the image of S, H: primary (intermediate) hologram plane. The entrance slits shown in Figs. 61.1b and 62.4 are omitted here for clarity.

intensity distribution could be recorded, but the interference fringes are in fact far too fine to be resolved. A lens system is therefore placed beyond this plane to provide sufficient magnification to resolve the fringes. This lens system and the true recording plate are not shown here. So far as the final intensity distribution is concerned, it is of no importance whether the waves emanated from real or virtual sources nor whether they were diffracted in real or virtual objects. The only important requirement is that the wave amplitudes and phases be correctly transferred in the imaging process.

The electron waves that interfere below the biprism, are *coherent* if C is a coherent point source. This follows from the symmetry of the device:

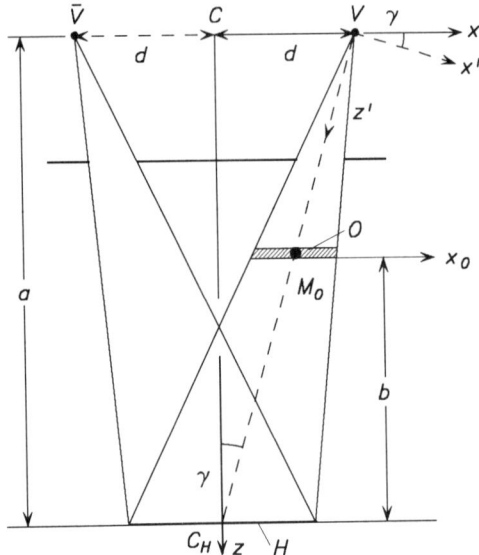

Fig. 62.4: Schematic presentation of a two-beam interferometer. V and $\overline{V}$: virtual sources; C: origin of the (x, y, z) coordinate system; z', x': rotated coordinate system; O: object (specimen), H: interferogram; M_0: midpoint of the object.

for instance, along all pairs of mirror-symmetric rays, the electron optical path lengths between point C and the plane H are exactly equal in zero diffraction order.

62.2 Quasi-homogeneous interference fringes

The recorded intensity pattern can become extremely complicated if an arbitrary specimen is brought into the interferometer, so complicated indeed that a lengthy computation is needed to calculate it. We therefore set out from a quite simple configuration and then gradually generalize.

We first consider the two-beam configuration shown in Fig. 62.4. Let us assume that two monochromatic spherical waves $\psi_V(\boldsymbol{r})$ and $\psi_R(\boldsymbol{r})$ emerge from the sources V and $\overline{V}$, respectively and that they are perfectly *coherent*. This simply means that the two sharply defined wavelengths and amplitudes are exactly equal and that there is a *constant* phase difference between these waves. Without loss of generality, we may set this phase difference equal to zero, and we shall also omit a common time-dependent factor $\exp(-i\omega t)$. Adopting the cartesian (x, y, z) system with origin at C (Fig. 62.4), we then have

$$\psi_V(\boldsymbol{r}) = Ar^{-1}\exp(ikr) \quad, \qquad r = \{(x-d)^2 + y^2 + z^2\}^{\frac{1}{2}}$$
$$\psi_R(\boldsymbol{r}) = A\bar{r}^{-1}\exp(ik\bar{r}) \quad, \qquad \bar{r} = \{(x+d)^2 + y^2 + z^2\}^{\frac{1}{2}}$$

This is still exactly true in the domains located in front of the two entrance slits of the interferometer. In practice the angle between the wave normal and the z-axis is always so small that the Fresnel approximation is justified, which is a simplification analogous to the step leading from (60.1) to (60.2). In the present case we obtain

$$\psi_V(\boldsymbol{r}) = Az^{-1}\exp\left(ikz + \frac{ik}{2z}y^2\right)\exp\left\{\frac{ik}{2z}(x-d)^2\right\} \qquad (62.8a)$$

$$\psi_R(\boldsymbol{r}) = Az^{-1}\exp\left(ikz + \frac{ik}{2z}y^2\right)\exp\left\{\frac{ik}{2z}(x+d)^2\right\} \qquad (62.8b)$$

No object present

Let us now suppose that the object O in Fig. 62.4 is absent. If we also disregard diffraction at the edges of the entrance slits, we achieve the utmost simplification: the intensity distribution in the recording plane $z = a$ is then given by

$$J_H(x,y) = |\psi_V(x,y,a) + \psi_R(x,y,a)|^2 = 4J_R\cos^2(kdx/a) \qquad (62.9)$$

with the constant reference value

$$J_R = |\psi_R(x,y,a)|^2 = A^2/a^2$$

This describes a perfect pattern of straight and equidistant fringes, in which the intensity minima fall to zero. The distance between adjacent maxima or minima is simply

$$\Delta x = \lambda a/2d \qquad (62.10)$$

In practice, this formula holds only in the *central* part of the fringe system. Near the boundaries, which are approximately determined by the geometrical shadows of the slit edges, the diffraction of waves at these edges must be taken into account. The mathematical tools for this are given in Section 60.2.

Here however the situation has become much more complicated owing to the fact that *two* slit functions for ψ_V and ψ_R are to be superimposed

Fig. 62.5: Above, an early example of biprism fringes, obtained in a special laboratory (below), in "a meadow far from all municipal mains". Building 1 housed the interferometer; the other huts contained the high-voltage source and the vacuum equipment.

prior to forming $|\psi|^2$. Moreover, in a real interferometer, the two shadow boundaries will not coincide perfectly, as was tacitly assumed in the construction of Fig. 62.4. Thus the intensity calculation becomes exceedingly complicated, even in this simple configuration.

An example of such an intensity distribution is given in Fig. 62.5. In practice, the shadow zones in which Fresnel fringes are formed are disregarded if the recorded intensity patterns are to be evaluated numerically.

Quasi-homogeneous object

We now consider the introduction of a specimen S into the interferometer, as in Fig. 62.3. This corresponds to the object O in Fig. 62.4. Its optical properties are described by a transparency or object function.

$$O(x_o, y_o) = \exp\{-\sigma(x_o, y_o) + i\eta(x_o, y_o)\} \qquad (62.11)$$

which is the same as that defined by (60.4). The origin of the (x_o, y_o) system is taken to be the midpoint M_0 of the object, as indicated in Fig. 62.4. This point is projected into the centre C_H of the plane H.

We now make the simplifying assumption that the function $O(x_o, y_o)$ varies so slowly with x_o and y_o that diffraction in the object does *not* need to be taken into account; this is what we mean by the term 'quasi-homogeneous object'. The object function is then transferred to the recording plane by a simple central projection with magnification M:

$$O_H(x, y) = O(x_o, y_o) = O(x/M, y/M) \qquad (62.12a)$$
$$M = a/(a - b) \qquad (62.12b)$$

The *object wave* at the screen H now becomes

$$\psi_o(x, y, a) = O(x/M, y/M)\psi_V(x, y, a) \qquad (62.13)$$

with ψ_V given by (62.8a), while the *reference wave* ψ_R remains unaffected. Instead of (62.9), we find

$$J_H(x, y) = |\psi_R + O\psi_V|^2$$

Evaluation of this expression using (62.8) and (62.11) results in

$$J_H(x, y) = J_R\{1 + e^{-2\sigma} + 2e^{-\sigma}\cos(2kxd/a - \eta)\} \qquad (62.14)$$

with $\sigma = \sigma(x/M, y/M)$ and $\eta = \eta(x/M, y/M)$. It is obvious from this formula that the object phase $\eta(x_o, y_o)$ causes a corresponding shift of the interference fringes in the recording plane:

$$\delta x_\eta = \frac{\lambda a}{4\pi d}\eta\left(\frac{x}{M}, \frac{y}{M}\right) \qquad (62.15)$$

For $\eta = 2\pi$ we find $\delta x_\eta = \Delta x$, as expected.

The attenuation factor $\exp(-\sigma)$ in (62.11) describes two effects at once, a decrease of the mean intensity and a loss of contrast.

The local *mean intensity* $\overline{J}(x, y)$ is obtained by averaging $J_H(x, y)$ over x in intervals that are large enough for the cosine term to cancel out, but also so small that σ and η do not alter significantly. We then find:

$$\overline{J}(x, y) = J_R[1 + \exp\{-2\sigma(x/M, y/M)\}] \qquad (62.16)$$

The local *contrast* is defined by

$$C(x, y) = \frac{J_{\max} - J_{\min}}{J_{\max} + J_{\min}} \qquad (62.17)$$

where we again consider σ and η to be slowly varying functions. From (62.14) we find

$$J_{max} = J_R(1 + e^{-2\sigma} + 2e^{-\sigma}) = J_R(1 + e^{-\sigma})^2$$
$$J_{min} = J_R(1 + e^{-2\sigma} - 2e^{-\sigma}) = J_R(1 - e^{-\sigma})^2$$

and hence

$$C(x, y) = \frac{2e^{-\sigma}}{1 + e^{-2\sigma}} < 1 \tag{62.18}$$

This contrast formula is still incomplete. There are stronger effects that reduce the contrast, as we shall see in the next sections.

The fringe shift $\delta x_\eta(x, y)$ can be immediately measured from recorded intensity patterns and thus enables us to deduce the phase shift $\eta(x_o, y_o)$. The latter results from changes in the wavenumber $k(\boldsymbol{r})$ associated with local variations of the electrostatic potential in the solid specimen:

$$\eta(x_o, y_o) = \int_{z_1}^{z_2} \{k(x_o, y_o, z) - k_v\} \, dz \tag{62.19}$$

Here k_v denotes the vacuum value of k and z_1, z_2 the local coordinates of the upper and lower specimen surfaces. If the potential is $U = $ const outside and $U + \Phi(\boldsymbol{r})$ with $|\Phi| \ll U$ inside the specimen, we obtain (with $k_v = 2\pi/\lambda$)

$$\eta(x_o, y_o) \approx \frac{\gamma\pi}{\lambda\hat{U}} \int_{z_1}^{z_2} \Phi \, dz \tag{62.20}$$

On introducing (62.20) into (62.15), we see that the wavelength λ cancels out and thus

$$\delta x_\eta = \frac{\gamma a}{4d\hat{U}} \int_{z_1}^{z_2} \Phi \, dz$$

In practice it is commonly assumed that the physical properties of the specimen do not alter significantly with the local depth z; we may then replace the integral by $\Phi\tau$, τ being the local *thickness* of the specimen, and finally arrive at

$$\delta x_\eta = \frac{a\gamma\tau}{4d\hat{U}} \Phi \tag{62.21}$$

Provided that the thickness τ is known from other information, the measurement of δx_η from interferograms gives us a method of determining the quantity Φ. This is the *'mean inner potential'* of the solid material (see

Fig. 62.6: Interference fringes; the shift is caused by the 'inner potential' of a solid specimen, here a MgO crystal (Völkl, 1991).

Section 69.3). It is positive and $e\Phi$ $(e > 0)$ is a measure of the binding energy of the electrons in the solid.

A typical example of the fringe shift by a mean inner potential is shown in Fig. 62.6. Such measurements and evaluation of Φ were first published by Möllenstedt and Keller (1957); many other references are listed in Chapter 61. Since the wavelength λ cancels out from (62.21), the temporal coherence requirements are less stringent than in modern holography. This is one of the reasons for the success of these early measurements.

The inner potential is not the only source of phase shifts between the two interfering beams. Variations in specimen thickness can also be detected as fringe shifts, and we shall return to this aspect when we consider interference holography (see Section 63.3.2), since this permits very small thickness differences to be detected. A rich field of application of interference microscopy is concerned with magnetic specimens; we refer to Missiroli *et al.* (1981), Matteucci *et al.* (1984) and Tonomura (1986a, 1992a) for detailed accounts.

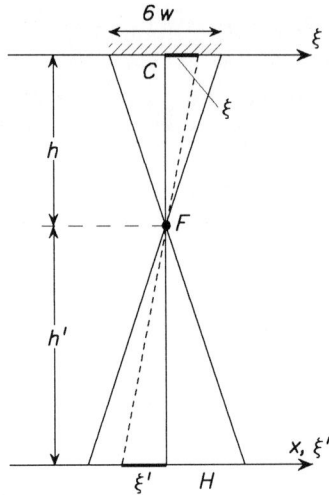

Fig. 62.7: Tilt of the local beam axis when the crossover is not vanishingly small. C: Centre of the crossover, F: filament axis, H: interferogram plane, ξ, ξ': displacement coordinates; dashed line: axis corresponding to a coordinate ξ. The breadth $6w$ corresponds to $\exp(-9) \approx 10^{-4}$.

62.3 Coherence problems

In the theory presented so far, we have explicitly assumed that the two sources V and $\overline{V}$ are point emitters of coherent waves. This coherence is a consequence of the symmetrical splitting of *one* spherical wave by the biprism and could not be achieved by placing two real sources at V and $\overline{V}$. But even the assumption of one ideally monochromatic point source, located at the point C (see Fig. 62.3), is unrealistic and will now be abandoned. There are two new features, which lead to a loss of contrast: the *lateral extent* of the real electron source and the width of the *energy spectrum* do not vanish. In reality both effects occur simultaneously, but for ease of presentation we study them separately. We shall see, in Section 66.2, in a different context, that this is usually legitimate.

62.3.1 Partial lateral coherence (spatial coherence)
In order to obtain a very small effective electron source, the cathode tip G (Fig. 62.3) is demagnified into the crossover C, but even with the best field electron emission sources, the lateral extent of C, small though it is, cannot be ignored. The consequence of this is shown in Fig. 62.7: lateral displacement of a point source in combination with the biprism causes a

tilt of the corresponding beam axis.

How will this affect the interference pattern? Different areas of the cathode surface emit electron waves that are completely *uncorrelated* in phase, since the emission of single particles is a stochastic process. We can hence consider the small cross-section in C, from which the electron beam emanates, as a large number of *incoherent* point sources. The tilted axis (dashed line in Fig. 62.7) then refers to one of these point sources.

We now introduce new transverse cartesian coordinates (ξ, ϱ) and an intensity distribution $\hat{F}(\xi, \varrho)$, such that $\hat{F}(\xi, \varrho) \, d\xi \, d\varrho$ is the *relative* intensity coming from the element of area $d\xi \, d\varrho$ at position (ξ, ϱ). This is expressed by the normalization condition

$$\iint\limits_{-\infty}^{\infty} \hat{F}(\xi, \varrho) \, d\xi \, d\varrho = 1 \tag{62.22}$$

Since the deflecting field in the essential part of the biprism is cylindrical, this deflection is independent of the direction of incidence: the whole electron bundle starting from a common off-axis point (ξ, ϱ) has an axis which is simply tilted round the filament F, as is shown in Fig. 62.7. This causes a corresponding shift

$$\xi' = -\mu\xi \quad , \quad \varrho' = 0 \quad , \quad \mu := h'/h \tag{62.23}$$

in the recording plane. (The distances h and h' are the same as in Fig. 62.4.)

The total intensity distribution is obtained by superposing all the partial intensities, which gives:

$$\hat{J}(x,y) = \iint J_H(x+\xi', y+\varrho') \hat{F}(\xi, \varrho) \, d\xi \, d\varrho = \iint J_H(x-\mu\xi, y) \hat{F}(\xi, \varrho) \, d\xi \, d\varrho$$

Any shift in the ϱ-direction is clearly immaterial and we therefore define a one-dimensional distribution function

$$F(\xi) := \int\limits_{-\infty}^{\infty} \hat{F}(\xi, \varrho) \, d\varrho \quad , \quad \int\limits_{-\infty}^{\infty} F(\xi) \, d\xi = 1 \tag{62.24}$$

whereupon

$$\hat{J}(x,y) = \int\limits_{-\infty}^{\infty} J_H(x - \mu\xi, y) F(\xi) \, d\xi \tag{62.25}$$

Once the distribution function $F(\xi)$ is known, the remaining task is reduced to an integration; this has the form of a convolution of the distribution $F(\xi)$

with the 'coherent' intensity J_H. The situation becomes even simpler if the electron source is *symmetric*, which implies that

$$F(-\xi) = F(\xi) \tag{62.26}$$

In practice, perfect mirror symmetry cannot be reached by adjustment of the instrument, but the coordinate system at the final recording screen can always be laterally shifted so that the centre of the fringe system in the absence of any specimen coincides with the origin. This is always done in practice and is equivalent to the assumption (62.26).

The convolution becomes fairly simple in the case of a quasi-homogeneous object. The object functions σ and η then vary so slowly that they may be considered as constant with respect to the convolution. Using (62.14) for J_H and (62.24) and (62.26) for F, we find

$$\hat{J}(x,y) = J_R\{1 + e^{-2\sigma} + 2e^{-\sigma}\cos(2kxd/a - \eta)K_T\} \tag{62.27}$$

which now includes a *lateral contrast factor*

$$K_T := \int_{-\infty}^{\infty} F(\xi)\cos(2kd\mu\xi/a)\,d\xi \tag{62.28}$$

The effect of this on fringe contrast is explored further in Section 62.3.3. (Note that although (62.28) appears at first sight to be a cosine Fourier transform, it is in fact a constant since x is not present in the argument of the cosine.)

62.3.2 Partial longitudinal coherence (temporal coherence)
We now take into account the fact that the electrons in the interfering beams are not exactly monoenergetic, as we have hitherto always assumed, but have a finite —albeit narrow— *energy spectrum*. The wavenumber $k = 2\pi/\lambda$ also has a spectrum, therefore, which we describe by a distribution $G(k)$, normalized so that

$$\int_{0}^{\infty} G(k)\,dk = 1 \tag{62.29}$$

We make the simplifying assumption that this function $G(k)$ is the same for all surface elements of the electron source, and is thus independent of the coordinates ξ and ϱ introduced above. This assumption is not at all trivial or self-evident but is a reasonably good approximation if the emitting area remains small.

The intensity $\hat{J}(x,y)$ obtained after averaging over the lateral source coordinates will now be a function of the wavenumber k and should therefore be denoted by $\hat{J}(x,y,k)$. In order to obtain the total intensity distribution $J(x,y)$ on the screen, we have to integrate over k:

$$J(x,y) = \int_0^\infty \hat{J}(x,y,k)G(k)\,dk \qquad (62.30)$$

This formula and all the intensity formulae already discussed imply that the intensity is really *recorded* in the plane H. If this is not the case, some modifications are necessary, which will be examined below.

The integration in (62.30) can be simplified and finally carried out in closed form if we treat the spectral distribution as a very sharp peak in a small interval $|k - \bar{k}| < 3\kappa \ll \bar{k}$ outside which it effectively vanishes. The value $\bar{k}$ refers to the intensity maximum, and κ is a measure of the spectral width. We now substitute $\hat{J}(x,y)$ given by (62.27) into the integral in (62.30). We use the fact that the distribution $F(\xi)$ is very narrow and its Fourier transform $K_T(k)$ is hence a very slowly varying function of k. ($K_T(k)$ is defined by (62.28), in which k is now regarded as a variable instead of a constant.) It is then quite sufficient to replace k by $\bar{k}$ in (62.28), which again becomes a constant and can hence be taken outside the resulting integral. With the normalization condition (62.29) we obtain the intermediate result

$$J(x,y) = J_R\{1 + e^{-2\sigma} + 2K_T e^{-\sigma}\int_0^\infty \cos(2kxd/a - \eta)G(k)\,dk\}$$

With the substitution $k = \bar{k} + u$, the remaining integral can be rewritten as

$$\int_0^\infty \cos(2kxd/a - \eta)G(k)\,dk$$

$$= \cos(2\bar{k}xd/a - \eta)\int_{-\infty}^\infty \cos(2uxd/a)G(\bar{k} + u)\,du$$

$$- \sin(2\bar{k}xd/a - \eta)\int_{-\infty}^\infty \sin(2uxd/a)G(\bar{k} + u)\,du$$

In practice the assumption that the spectral function is symmetric is well justified:

$$G(\bar{k} - u) = G(\bar{k} + u) \qquad (62.31)$$

The sine terms then vanish in the integration and we are led to define a *longitudinal* contrast factor

$$K_L := \int\limits_{-\infty}^{\infty} G(\overline{k} + u)\cos(2uxd/a)\,du \qquad (62.32)$$

62.3.3 Superposition

Definition of a *total* contrast factor is straightforward:

$$K(x) = K_T K_L(x) \qquad (62.33)$$

The intensity distribution in the fringe system can then be cast into the very compact form

$$J(x,y) = J_R\{1 + e^{-2\sigma} + 2Ke^{-\sigma}\cos(2\overline{k}xd/a - \eta)\} \qquad (62.34)$$

Note that the familiar assumption of weakly scattering objects, $\sigma \ll 1$ and $|\eta| \ll 1$, has *not* been necessary in the derivation of this formula. A careful investigation shows that the fringe shift δx_η in (62.15) and the mean intensity $\overline{J}$ in (62.16) remain unaltered by the factor K, while the contrast becomes

$$C(x,y) = 2K(x)\frac{e^{-\sigma}}{1 + e^{-2\sigma}} \qquad (62.35)$$

The factors K_L and K_T, determining the decrease of contrast, are both Fourier-like transforms of emission intensity distributions. In order to estimate their effect, we consider the case in which the latter are Gaussian:

$$F(\xi) = \frac{1}{w\sqrt{\pi}}\exp(-\xi^2/w^2) \qquad (62.36a)$$

$$G(k) = \frac{1}{\kappa\sqrt{\pi}}\exp\{-(k - \overline{k})^2/\kappa^2\} \qquad (62.36b)$$

w and κ are the widths at which these functions have decreased to $e^{-1} = 0.37$ of their maximum values. Evaluation of the corresponding integrals gives

$$K = K_T K_L(x) = \exp\left\{-\left(\frac{\overline{k}d\mu w}{a}\right)^2\right\}\exp\left\{-\left(\frac{\kappa xd}{a}\right)^2\right\} \qquad (62.37)$$

The first exponential factor K_T describes a *uniform* loss of contrast, while the second factor $K_L(x)$ describes an attenuation, with full contrast $K_L =$

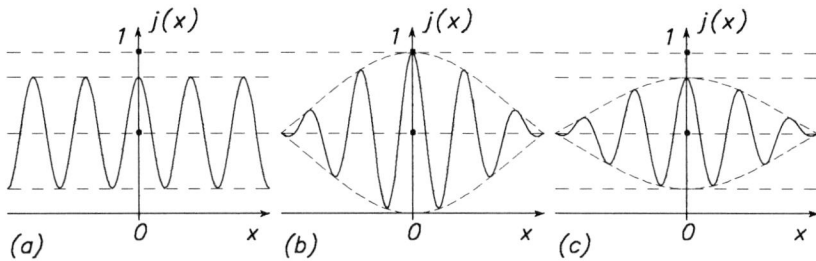

Fig. 62.8: Normalized intensity distributions $j(x) = J(x)/2\bar{J}$. (a) Extended source only; (b) polyenergetic source only; (c) superposition of both effects.

1 in the centre and a decay to invisibility as $|x| \to \infty$. This is shown in Fig. 62.8, which corresponds to the case in which no specimen is present, $\sigma \equiv 0$.

In practice, this theory contains unacceptable simplifications. The crossover C is produced by a system of electron lenses and is hence affected by their aberrations. In particular, the lateral distribution function $F(\xi)$ is broadened by transverse lens aberrations and by the Boersch effect. Furthermore, the plane of the source C, tacitly assumed to be at $z = 0$, is subject to a defocus and longitudinal chromatic aberrations.

Moreover, the intensity is not recorded in the plane H, since the fringes are far too narrow there. This plane is, instead, highly magnified onto the final recording screen. The necessary projector system is not shown in the figures. It is evident that these lenses must cause some further loss of fringe-contrast as a consequence of their inevitable aberrations. In particular, the lateral chromatic aberrations, both isotropic and anisotropic cause a loss of contrast for large values of $|x|$ and $|y|$, respectively. These effects have been studied in detail by Lenz and Wohland (1984).

Other sources of deterioration of the fringe contrast are the noise of the electron beam and mechanical vibrations. The superposition of all the lateral aberrations gives rise to a decrease of the maximum contrast at $x = 0$, while the combination of longitudinal or chromatic aberrations reduces the number of fringes visible. A rough estimate of this number is given by $N = U_r/\Delta U$, where ΔU is the effective energy spread. Möllenstedt and Wohland (1980) and Schmid (1984) used this to determine the spread ΔU. This quantity is not just determined by the energy spectrum of the emission process at the cathode, but is also affected by the numerous aberrations, as explained above. Nowadays values $N > 1000$ are feasible in well-designed interferometers.

62.4 The Aharonov–Bohm effect

We have already briefly mentioned this effect in Section 5.6, where we showed that if an electron beam is divided and passes on either side of a local magnetic field, the lengths of the optical paths will be different by an amount that depends on the magnitude of the enclosed field, even if the field is negligibly small in the regions traversed by the electrons. The electrons are affected by the vector potential, A, which is different in the two regions. The difference in path length can be rendered visible as an interference effect.

We shall not enter into the details of the extensive literature of this subject here, but simply recall that the effect was first noticed by Ehrenberg and Siday in 1949 but attracted no attention. It was rediscovered by Aharonov and Bohm, who used very different arguments, in 1959 and first demonstrated by Chambers (1960), followed soon after by Fowler et al. (1961), Boersch et al. (1961, 1962b), Möllenstedt and Bayh (1962a,b) and Bayh (1962). These experiments are discussed in detail in the monograph of Peshkin and Tonomura (1989). In Fig. 62.9, the Möllenstedt and Bayh experiment and the resulting fringe shift are reproduced. A voluminous literature grew up during the 1960s and 1970s, consisting essentially of three groups of papers: by authors who believed that the effect did not exist, theoretical defences and experimental justifications of the effect and finally, abstract discussion of its theoretical implications. The first group is now of largely historic interest, since Tonomura has shown that the effect persists even when the beams are completely protected from any residual magnetic field (Tonomura et al., 1982a, 1984). The experiment uses the shielding properties of the superconducting state (Kuper, 1980) and is conclusive. First, however, we consider a ferromagnet in the normal state. An electron beam is divided into two partial beams, one of which passes through the central hole in a toroidal ferromagnet while the other passes outside (Fig. 62.10a). The resulting interferogram (Fig. 62.10b) shows that there is a fringe shift, and hence a phase difference, between these two regions. In the improved form of the experiment, described in full detail in Tonomura et al. (1986, 1987, 1990) and Osakabe et al. (1986), the toroid consisted of a permalloy core enclosed in niobium, which becomes superconducting at 9.2 K. The interference fringes obtained with the niobium in the normal and superconducting states (Fig. 62.11) are consistent with the phase differences predicted by the Aharonov–Bohm effect. The magnetic field is far too small in this experiment to cause the fringe shift.

For extremely full discussion of the literature on this effect, we refer to the monograph of Peshkin and Tonomura (1989). Subsequent develop-

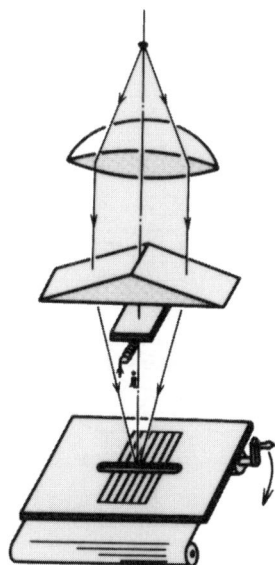

(a)

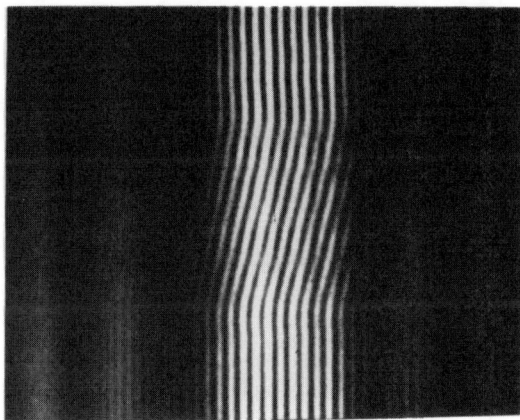

(b)

Fig. 62.9: The Möllenstedt–Bayh experiment. As the current in the solenoid is increased, the film is advanced and the fringes that pass through the slit move laterally.

ments are recorded in Selleri (1992) and the proceedings of the International Symposia on the Foundations of Quantum Mechanics (see Kamefuchi *et al.*, 1984; Namiki *et al.*,1987; and Kobayashi *et al.*, 1990). See also O'Raifeartaigh *et al.*(1991), Herman (1992), Home and Selleri (1992), Kobe

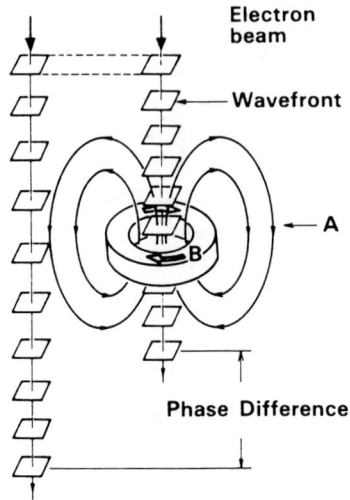

Fig. 62.10a: Vector potentials near ferromagnetic particle with hole (*A*: vector potential; *B*: magnetic flux).

et al. (1992) and Tonomura (1992c). The Aharonov–Bohm effect has an electrostatic analogue; we refer to Matteucci *et al.* (1992a) for details and references.

62.5 Other electron interference studies

62.5.1 The Sagnac effect

If a coherent beam of light or electrons, or indeed other particles is divided into two coherent partial beams, which later recombine to produce interference fringes, a fringe shift can be caused by rotating the whole structure about an axis perpendicular to the area enclosed by the separate beams. If the area enclosed is A, the angular velocity Ω and the particle energy E, the phase shift $\Delta\phi$ will be given by

$$\Delta\phi = \frac{2}{\hbar c^2} E A \Omega$$

and for electrons of relativistic mass m,

$$\Delta\phi = \frac{2m}{\hbar} A \Omega$$

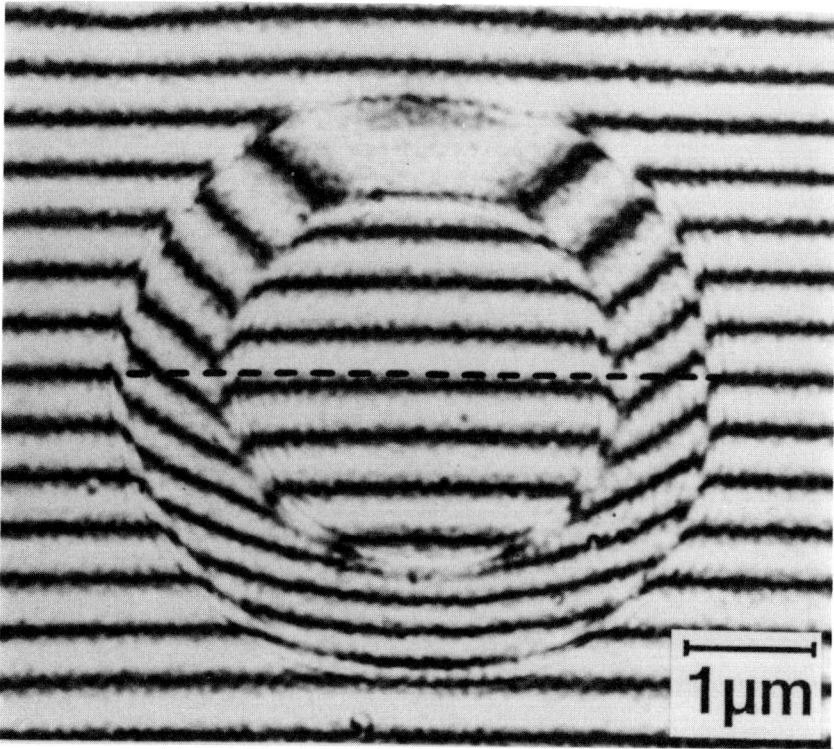

Fig. 62.10b: Interferogram showing phase distribution of electron beam passing through toroidal ferromagnet.

(See Nicklaus, 1989 for a full derivation and an account of earlier work.) The technical problems to be overcome are severe (Hasselbach and Nicklaus, 1966), for an entire electron optical bench with field-emission gun, biprisms, Wien filter and detection unit has to be rotated at a frequency of the order of a revolution per second. The Wien filter is included owing to its ability to shift the partial wave-packets longitudinally, in the direction of the optic axis, and hence to ensure that they superimpose satisfactorily (Hasselbach and Nicklaus, 1988, Nicklaus and Hasselbach, 1993, cf. Möllenstedt and Wohland, 1980). An interferometer of rugged design was built for the purpose by Hasselbach (1988). The effect was successfully detected and the magnitude of the phase shift agreed with the theoretical prediction to within the experimental error (Nicklaus, 1989; Hasselbach and Nicklaus, 1990, 1993; Hasselbach, 1992).

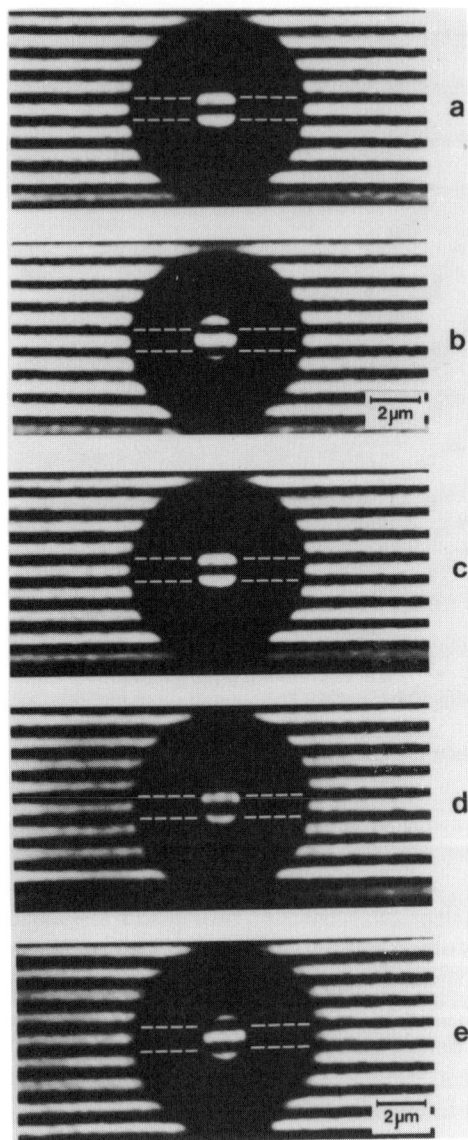

Fig. 62.11: Interferograms of toroidal samples. (a) and (b), different samples at 4.5 K. In (a), the phase difference between outside and inside is $\pi(\mod 2\pi)$ corresponding to the quantized flux $nh/2e$ in the superconducting niobium (n odd). In (b) n is even and so the phase difference is $2\pi(\mod 2\pi)$. (c) $T = 4.5$ K. Fringe displacement between inside and outside is half a fringe-spacing. (d) $T = 15$ K (above the critical temperature). Fringe displacement can now vary continuously and is here equal to 0.4 of the fringe-spacing. (e) Room temperature. Further change in fringe displacement, due to a small decrease in the magnetization of the permalloy core.

62.5.2 Modification of the phase of diffraction spots

Considerable effort has been devoted to the problem of establishing values of phase from intensity records. A large part of Chapter 74 is concerned with this 'phase problem' and the uncertainty about the phase of the electron wave at diffraction spots from crystalline specimens is the principal difficulty in electron crystallography. Among the many attempts to solve the phase problem, which are examined in detail in Chapter 74, there is one that is more appropriate here: by introducing a very small coil in the back focal plane of the objective lens of a transmission microscope, it should be possible to create a controlled, variable phase shift between diffracted beams from periodic specimens. Coils of suitable size, of the order of 18 μm or less in diameter, have been made and preliminary tests showed that the method is "feasible, although its application is difficult" (Valdrè, 1979); for a few more details, see Valdrè (1974).

62.5.3 Convergent-beam electron diffraction interferometry

A recent development in the use of the biprism consists in bringing together different regions of the convergent-beam electron diffraction (CBED) pattern with the aid of a biprism. Thus Herring and Tanji (1993) and Herring et al. (1993a,b) have succeeded in recording fringes resulting from interference between the main beam and a set of diffracted beams and between two different sets of diffracted beams. Effects due to the wave aberration have been detected in these patterns.

63
Principles of Holography

Electron interference and holography are intimately related, as we saw in the general introduction (Chapter 61). One major difference is that the formation of interference fringes in a two-beam interferometer can be understood without considering diffraction whereas the reconstruction of object waves from a hologram is an intrinsic diffraction process. It is this distinction that has guided our organization of the subject material.

Before discussing in-line and off-axis holography in detail, we recall some general features of the holographic process and the associated vocabulary. The essence of holography is the formation of an interference pattern by adding a reference beam and a beam modulated by the specimen. In the case of in-line holography (where reference beam and image-forming beam coincide), the hologram may be recorded in or close to a plane conjugate to the specimen, in which case it is called a *Fresnel hologram*; alternatively, it may be recorded in some other plane, distant from an image plane, in which case it is called a *Fraunhofer hologram*. The proximity to an image plane is measured by the Fresnel number, ν, defined by

$$\nu = \frac{\pi s^2}{\lambda \Delta}$$

in which λ is the electron wavelength, Δ the distance from the hologram plane to the image plane referred back to object space and s is the size of the specimen detail of interest. For high resolution work, $s = 0.2$ nm for example, and operation at 300 kV ($\lambda \approx 2$ pm), we have $\nu \approx 60/\Delta$ (Δ in nm). Values of ν of the order of unity or less correspond to Fraunhofer conditions, while values appreciably greater than unity are associated with Fresnel hologram formation. For values of Δ less than about 10 nm, therefore, Fresnel conditions prevail, but otherwise we shall be involved in Fraunhofer holography. The same distinction is made in off-axis holography, introduced by Leith and Upatnieks in 1962 (see also Leith and Upatnieks, 1963, 1967). Here, image-plane holograms (Fresnel holograms with $\Delta \approx 0$) are attractive because finite source-size proves to have least effect on resolution in this case.

A further ramification is *interference holography* (Tonomura *et al.*, 1979d, see Tonomura, 1984 and 1987a), which is particularly attractive for displaying phase variations and hence thickness distributions or magnetization patterns. Here, the laser beam used in the reconstruction is split into two beams by a half-silvered mirror, one of which falls on the hologram and generates the usual reconstructed images. One of these is selected by an aperture and allowed to interfere with the other partial beam (Fig. 63.1a). A further refinement allows us to amplify the phase differences involved by arranging that a reconstructed image and its conjugate interfere (Fig. 63.1b). We illustrate these possibilites below. This by no means exhausts the many forms of electron holography. For a very full list, see Cowley (1992) and for a complete account, see the monograph of Tonomura (1993).

63.1 In-line holography

We have already mentioned that, owing to the difficulty of separating the two images in the reconstruction step, the in-line technique has lost much of its original attraction; nevertheless, it will be considered briefly here since it provides a simple access to the mathematical formalism.

Hologram recording

We start from the configuration shown in Fig. 63.2a. The axial point source S at $z = -a$ emits a monoenergetic spherical reference wave denoted $\psi_R(x, y, z)$. We adopt the Fresnel approximation and the complex wave amplitude in the hologram plane $H(z = 0)$ is then

$$\psi_R(x, y, 0) = \frac{A}{a} \exp\left\{ \frac{ik}{2a}(x^2 + y^2) \right\} \tag{63.1}$$

Quite analogously, a point object O at $z = -b$ produces an object wave

$$\psi_W(x, y, 0) = \frac{B}{b} \exp\left\{ \frac{ik}{2b}(x^2 + y^2) \right\} \tag{63.2}$$

The intensity of the superimposed waves is then

$$\begin{aligned} J(x, y) &= |\psi_R(x, y, 0) + \psi_W(x, y, 0)|^2 \\ &= \frac{|A^2|}{a^2} + \frac{|B^2|}{b^2} + \frac{A^* B}{ab} \exp\left\{ \frac{ik}{2}\left(\frac{1}{b} - \frac{1}{a} \right)(x^2 + y^2) \right\} + \text{c.c.} \end{aligned} \tag{63.3}$$

where c.c. denotes the complex conjugate of the last term. Apart from an unimportant scaling factor, this intensity pattern is recorded on the hologram film. It is obvious that $J(x, y)$ is *nonlinear* in ψ_W, since a term

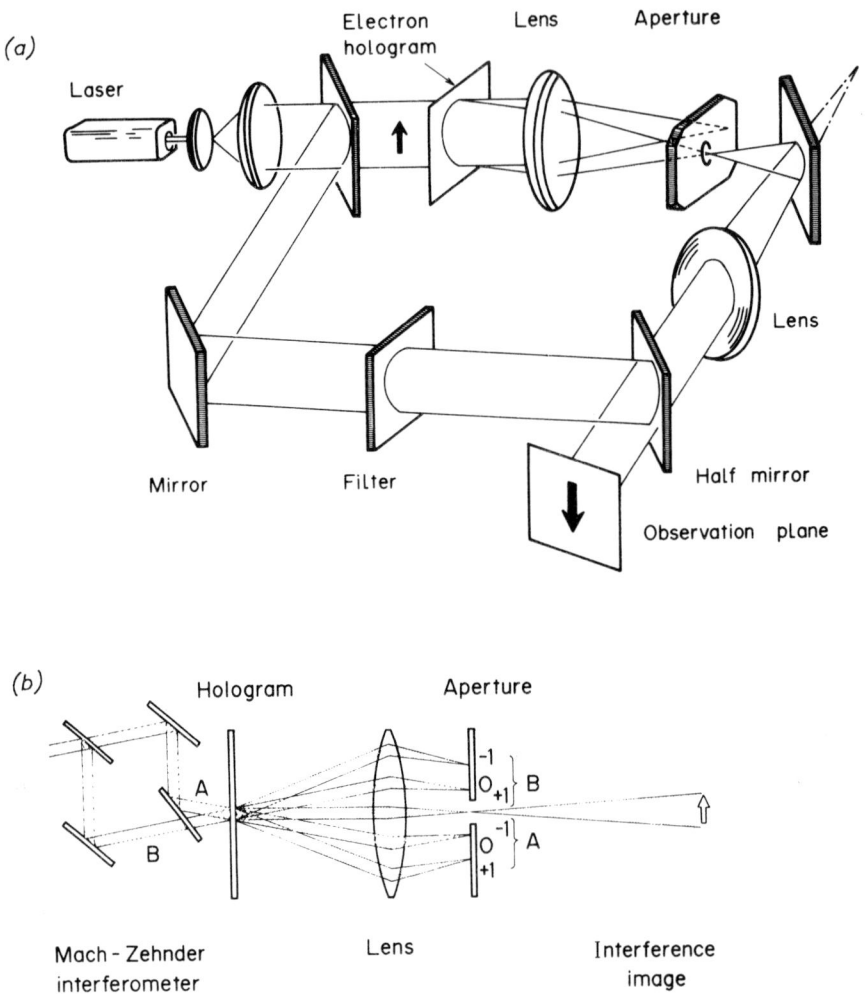

Fig. 63.1: (a) Optical reconstruction system for interference microscopy. (b) Optical reconstruction system for interference microscopy with phase amplification ($\times 2$).

$|\psi_W|^2 = |B|^2/b^2$ appears. When object waves emanating from many object points are superposed, such a term leads to nonlinear interactions, which make direct reconstruction impossible. It is hence necessary to assume that the object waves are *weak*, which means that $|B/b|^2 \ll |A/a|^2$.

The hologram must be processed in such a way that its *transparency*

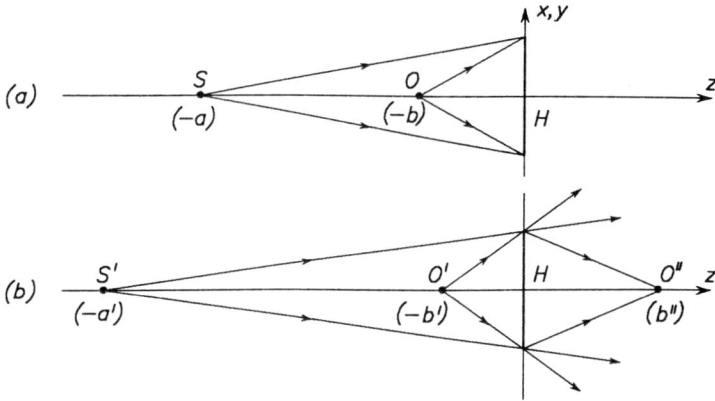

Fig. 63.2: Simplified ray constructions with geometric parameters. (a) Recording step; (b) reconstruction. (In these configurations, a, b and a', b', b'' are all positive.)

$T(x, y)$ becomes proportional to $J(x, y)$:

$$T(x, y) = T_0 J(x, y)$$
$$= T_0 \frac{|A|^2}{a^2} + \frac{T_0 A^* B}{ab} \exp\left\{\frac{ik}{2f}(x^2 + y^2)\right\} \tag{63.4a}$$
$$+ \frac{T_0 AB^*}{ab} \exp\left\{-\frac{ik}{2f}(x^2 + y^2)\right\}$$

with

$$\frac{1}{f} := \frac{1}{b} - \frac{1}{a} > 0 \tag{63.4b}$$

Object reconstruction

This step is sketched in Fig. 63.2b. The processed hologram is illuminated with a coherent wave ψ'_R emanating from an axial point source S' at $z = -a'$. The wavenumber k' may be different from that of the recording step. On the entrance side of the hologram we have the wave excitation

$$\psi'_R(x, y, 0) = \frac{A'}{a'} \exp\left\{\frac{ik'}{2a'}(x^2 + y^2)\right\} \tag{63.5}$$

and on the exit side

$$\chi(x, y) = T(x, y)\psi'_R(x, y, 0) = \sum_{j=-1}^{1} \chi_j(x, y) \tag{63.6}$$

The three partial waves χ_{-1}, χ_0 and χ_1 result from the three contributions to $T(x, y)$ in (63.4a). Introducing these into (63.6) and using (63.5), we find

$$\chi_0(x, y) = \frac{T_0 |A^2| A'}{a^2 a'} \exp\left\{\frac{ik'}{2a'}(x^2 + y^2)\right\} \tag{63.7a}$$

$$\chi_1(x, y) = \frac{T_0 A^* B A'}{aba'} \exp\left\{\frac{i}{2}\left(\frac{k'}{a'} + \frac{k}{f}\right)(x^2 + y^2)\right\} \tag{63.7b}$$

and

$$\chi_{-1}(x, y) = \frac{T_0 A B^* A'}{aba'} \exp\left\{\frac{i}{2}\left(\frac{k'}{a'} - \frac{k}{f}\right)(x^2 + y^2)\right\} \tag{63.7c}$$

The wave structure behind the hologram is a consequence of the diffraction in the hologram. This diffraction is described by employing χ_{-1}, χ_0 and χ_{+1} as *start-functions* or one-sided boundary values in the diffraction integral. In the Fresnel approximation, the latter is given by

$$\psi'_j(x, y, z) = \frac{k' e^{ik'z}}{2\pi i z} \int\!\!\int\limits_{-\infty}^{\infty} \chi_j(u, v) \exp\left[\frac{ik'}{2z}\{(x - u)^2 + (y - v)^2\}\right] du\, dv$$

$$(j = -1, 0, 1; z \gg D)$$

$$\tag{63.8}$$

D being the diameter of the hologram. The resulting integrals, containing quadratic forms of u and v in the exponents, can all be evaluated in closed form by means of the general formula

$$\int\!\!\int\limits_{-\infty}^{\infty} \exp\left\{\frac{i}{2}p(u^2 + v^2) - iq(xu + yv) + \frac{i}{2}r(x^2 + y^2)\right\} du\, dv$$

$$= \frac{2\pi i}{p} \exp\left\{\frac{i}{2}(x^2 + y^2)(r - q^2/p)\right\}$$

After some elementary calculations, we find

$$\psi'_0(x, y, z) = \frac{T_0 |A^2|}{a^2} \frac{e^{ik'z}}{z + a'} \exp\left\{\frac{ik'(x^2 + y^2)}{2(z + a')}\right\} \tag{63.9}$$

$$\psi'_1(x, y, z) = \frac{T_0 A^* B A' b'}{aba'} \frac{e^{ik'z}}{z + b'} \exp\left\{\frac{ik'(x^2 + y^2)}{2(z + b')}\right\} \tag{63.10a}$$

with

$$\frac{1}{b'} := \frac{1}{a'} + \frac{k}{k'f} \tag{63.10b}$$

and finally

$$\psi'_{-1}(x,y,z) = -\frac{T_0 AB^* A' b''}{aba'} \frac{e^{ik'z}}{z - b''} \exp\left\{\frac{ik'(x^2 + y^2)}{2(z - b'')}\right\} \quad (63.11a)$$

with

$$\frac{1}{b''} := -\frac{1}{a'} + \frac{k}{k'f} \quad (63.11b)$$

Interpretation of the results

Comparison of (63.9) with (63.5) shows that

$$\psi'_0(x,y,z) = \frac{T_0 |A|^2}{a^2} \psi'_R(x,y,z) \quad (63.12)$$

if we generalize (63.5) for $z \neq 0$ by replacing a' by $a' + z$; ψ'_0 is hence the *reconstructed reference wave* of Fig. 63.2 and is of little importance, since this wave is not the required one and must be excluded.

The next partial wave ψ'_1 can be regarded as a *reconstructed object wave*. The corresponding object point is, however, shifted from the position O to O', as sketched in Fig. 63.2. This is a consequence of the altered illumination conditions. In fact if we specialize to $k = k'$, $a = a'$ in (63.10a,b) we find $b = b'$ and then

$$\psi'_1(x,y,z) = \frac{T_0 A^* A'}{a^2} \psi_W(x,y,z) \quad (63.13)$$

(again by the generalization $b \to z + b$ in (63.2)). This is usually the required wave.

The third partial wave $\psi'_{-1}(x,y,z)$ converges to an object point at a position $z = b''$ *behind* the hologram, as sketched in Fig. 63.2b. It can likewise be used as a reconstructed object wave, if some phases change their sign. This becomes easier to understand, if we again specialize to $a = a'$, $k = k'$, giving $1/b'' = 1/b - 2/a$. A further specialization is the case of parallel illumination: $a \to \infty$ and $A = A' \to \infty$ such that $C := A/a$ remains finite. We then find $b'' = b$ and hence from (63.11a):

$$\psi'_{-1}(x,y,z) = -\frac{T_0 |C|^2 B^*}{z - b} e^{ikz} \exp\left\{\frac{ik(x^2 + y^2)}{2(z - b)}\right\}$$

$$= \psi'^*_1(x,y,-z) \quad (63.14)$$

The focus $z = +b$ of this wave is mirror-symmetric with respect to the original object position at $z = -b$. The presence of the factor B^* instead of B shows that the phase has reversed its sign. This wave is therefore often called the *conjugate wave*.

We have now shown that the method, sketched in Fig. 61.3 and briefly described in the introduction, must indeed work and we could even deal with the more general configuration shown in Fig. 63.2.

Focal equations

From (63.10b) and (63.11b), we can immediately derive the focal equations

$$\frac{1}{b'} - \frac{1}{b''} = \frac{2}{a'} \tag{63.15}$$

and

$$\frac{1}{b'} + \frac{1}{b''} = \frac{2k}{k'f} =: \frac{1}{F} \tag{63.16}$$

The latter relation resembles the focal equation of a thin lens, and we may hence associate a focal length

$$F = \frac{\lambda ab}{2\lambda'(a - b)} = \frac{\lambda f}{2\lambda'} \tag{63.17}$$

with the hologram.

From this formula it is immediately obvious that light-optical reconstruction from a hologram produced with electron waves is *not* immediately possible for technical reasons. The difference $a - b$ in the denominator cannot be made arbitrarily small, since there must be space enough to bring the object into the electron beam; we have roughly $F \approx \lambda b/2\lambda'$. Now the wavelength λ' of the laser light, used for reconstruction, is very much greater than the wavelength λ of electrons at some kilovolts. This implies that $F \ll b$, which is technically not feasible. The hologram must therefore be *highly magnified* in the electron optical system before being recorded. This situation is quite analogous to the technical conditions for the operation of an interferometer. Magnification of the hologram is equivalent to increasing the ratio λ/λ'.

The focal equations are exactly the same as for the lowest diffraction orders obtained with Fresnel *zone plates*; there must hence be a close relation between holography and zone plates (Rogers, 1950a). This relation is depicted in Fig. 63.3; the variable r is here the radial coordinate $r = (x^2 + y^2)^{\frac{1}{2}}$. The average transparency is $\overline{T} = T_0|A|^2 a^{-2}$ in agreement with the first term of (63.4). In this example we have chosen the phase relations such that $\Re(A^*B) = 0$, hence

$$T(r) = \overline{T}\{1 - 2\sin(kr^2/2f)\} \tag{63.18}$$

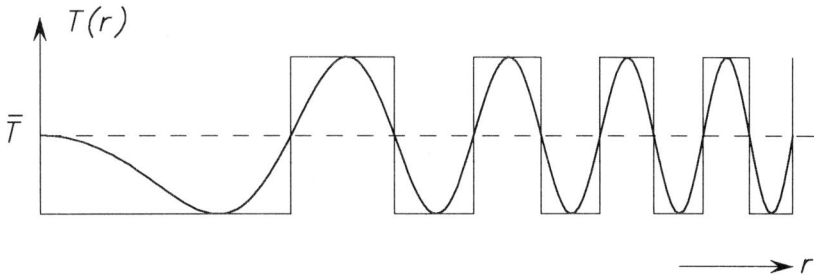

Fig. 63.3: Holographic transparency distribution (smooth curve) and equivalent zone lens (steps); $\overline{T}$ is the average transparency.

The hologram and the zone plate have essentially the same fringe structure. The only important difference lies in the fact that the holographic transparency is *smooth*, while the zone-plate distribution has sharp edges. An important consequence is that in holography there appear just *three* different diffraction orders, while a zone plate produces an infinite (or at least very large) number of them.

We can now understand qualitatively how the reconstruction of an extended object from its hologram is possible. For this we consider the object to be built up from a large number of very small surface elements. Each of these produces its concentric ring system on the hologram, the centre being appropriately shifted. In the reconstruction process, each ring system acts as a tiny Fresnel lens and produces three different waves. Their superposition gives the *correct* total wave, since the amplitude and phase relations are maintained by virtue of the linearization.

It will be clear from the foregoing account and the detailed analysis of Section 66.1 that in-line electron holography and high-resolution image formation of weakly scattering specimens have much in common. We shall not pursue this interpretation of bright-field electron image formation here; the interested reader is referred to the accounts of Hanszen (1971, 1982a). An early in-line holographic reconstruction using a large enough defocus value ($\nu \approx 0.04$) to reduce the conjugate image to a blurred background behind the principal image was made by Tonomura and Watanabe in 1968 (Tonomura and Watanabe, 1968; Tonomura *et al.*, 1968a,b; Watanabe and Tonomura, 1969). Other early attempts were made by Voronin *et al.* (1972) and Munch (1975; Munch and Zeitler, 1974).

63.2 Off-axis holography: hologram formation in a two-beam interferometer

63.2.1 Expression for the two waves
Electron holography is nowadays performed using the off-axis configuration with the aid of a two-beam interferometer. We considered fringe formation in this device in Sections 62.2–3 and for simplicity we return to the arrangement shown in Fig. 62.4. Now, however, the object O may have a very fine structure and we can therefore no longer disregard diffraction effects within it. The reasoning becomes considerably more complicated, as we shall see.

It is convenient to express the object wave $\psi_{\overline{V}}$, emanating from the virtual source $\overline{V}$ (see Fig. 62.4), in tilted coordinate system (x', y', z'), defined by

$$x' = (x - d)\cos\gamma + z\sin\gamma \qquad y' = y$$
$$z' = -(x - d)\sin\gamma + z\cos\gamma \qquad \tan\gamma = d/a \tag{63.19}$$

The coordinates of the midpoint (M_0) of the object are then

$$z_M = a - b \qquad\qquad z'_M = (a - b)\sec\gamma =: D_0$$
$$x_M = d - (a - b)\tan\gamma \qquad x'_M = 0$$

The object may be slightly tilted with respect to the coordinate plane $z' = D_0$ or even warped. We then describe the position of an arbitrary object point by

$$x_o = x' \quad , \quad y_o = y' \quad , \quad z_o = D_0 - h(x_o, y_o) \tag{63.20}$$

where $h(x_o, y_o)$ describes the very small vertical displacement $|h| \ll D_0$.

Determination of the wavefunction

The wavefunction ψ_V on the entrance side of the object can now be written as

$$\psi_V(\mathbf{r}_o) = A{z'}^{-1}\exp\left\{ikz' + \frac{ik}{2z'}(x'^2 + y'^2)\right\}$$
$$= AD_0^{-1}\exp\left\{ikD_0 - ikh + \frac{ik}{2D_0}(x_o^2 + y_o^2)\right\} \tag{63.21a}$$

The wavefunction on the exit side of the object is obtained by multiplying ψ_V by the specimen transparency $O(x_o, y_o)$:

$$\psi_W(\mathbf{r}_o) = O(x_o, y_o)\psi_V(\mathbf{r}_o) \tag{63.21b}$$

This function represents the boundary value for the diffraction integral. In the far zone, the distance $z' - z_o$ from the object may be replaced by $z' - D_0$ in the denominator, thus neglecting the small variation $h(x_o, y_o)$. In the Fresnel approximation, the diffraction integral becomes

$$\psi_W(\mathbf{r}) = -\frac{ik}{2\pi(z' - D_0)} \iint\limits_{-\infty}^{\infty} \psi_W(\mathbf{r}_o) \exp[ik\{z' - D_0 + h(x_o, y_o)\}]$$

$$\times \exp\left[\frac{ik}{2(z' - D_0)}\{(x' - x_o)^2 + (y' - y_o)^2\}\right] dx_o \, dy_o$$

On substituting expressions (63.21a,b) for ψ_W and ψ_V into this formula, we find that the terms involving $h(x_o, y_o)$ cancel out. This is of great practical importance, as it shows that small vertical displacements do not degrade the quality of holograms.

The diffracted wave can now be written in the form

$$\psi_W(\mathbf{r}) = -\frac{ikA}{2\pi D_0(z' - D_0)} \exp(ikz') \iint\limits_{-\infty}^{\infty} O(x_o, y_o) \exp \frac{ik}{2D_0}\left[(x_o^2 + y_o^2)\right.$$

$$\left. + \frac{D_0}{z' - D_0}\{(x' - x_o)^2 + (y' - y_o)^2\}\right] dx_o \, dy_o$$

Before adding this wave to the reference wave ψ_R, the tilted coordinates must be eliminated. In the plane of superposition, $z = a$, (63.19) give

$$x' = x \cos\gamma \quad , \quad y' = y \quad , \quad z' = a \sec\gamma - x \sin\gamma$$

In the denominator, we make the approximation $z' \approx a$ and $D_0 \approx a - b$. As $\mathbf{r}_H$ then has the components (x, y, a) we obtain

$$\psi_W(\mathbf{r}_H) = -\frac{ikA}{2\pi(a - b)b} \exp\{ik(a \sec\gamma - x \sin\gamma)\}$$

$$\times \iint\limits_{-\infty}^{\infty} O(x_o, y_o) \exp\left[\frac{ik}{2}\left\{\frac{x_o^2 + y_o^2}{a - b} + \frac{(x - x_o)^2 + (y - y_o)^2}{b}\right\}\right] dx_o \, dy_o$$

$$(63.22)$$

The small-angle approximation, $\gamma \ll 1$, $\sin\gamma \approx \gamma$ and $\cos\gamma \approx 1$ can be made except in the term in $ka/\cos\gamma$, since $ka \gg 1$. In the same approximation, the reference wave is given by

$$\psi_R(\mathbf{r}_H) = \frac{A}{a} \exp\left\{ik\left(a \sec\gamma + x \sin\gamma + \frac{x^2 + y^2}{2a}\right)\right\} \qquad (63.23)$$

63.2.2 Addition of object wave and reference wave

The intensity in the hologram plane is now given by $|\psi_W + \psi_R|^2$, but it is no longer necessary to assume that the object scatters weakly. We shall see that the diffraction pattern of the hologram separates into three distinct regions, which can in principle be made far enough apart to avoid any overlapping: (i) a central region and (ii) two sidebands in which information about the amplitude and phase of the specimen is coded *linearly*. Nevertheless, high-resolution information is frequently sought from weakly scattering specimens and we shall therefore consider both situations.

We first consider the general case, in which the specimen is not necessarily weak, and for simplicity we disregard the curvature of the two waves ψ_R and ψ_W. The latter are thus assumed to be plane waves, with wave vectors $\boldsymbol{k} = (\pm q_s/\pi, 0, k)$, where $q_s = 2\gamma/\lambda$. The reference wave ψ_R thus has the form

$$\psi_R(\boldsymbol{r}) = \exp(\mathrm{i}\pi q_s x)e^{\mathrm{i}kz} \tag{63.24}$$

while the object wave is given by

$$\psi_W(\boldsymbol{r}) = A(\boldsymbol{r})\exp\{-\mathrm{i}\pi q_s x + \varphi(\boldsymbol{r})\}e^{\mathrm{i}kz} \tag{63.25}$$

The functions A and φ will in general not be the same as the amplitude and phase of the specimen transparency, since (63.25) is the result of a diffraction process in the Fraunhofer approximation and phase-shifts due to aberrations will hence have an effect.

In the hologram plane, where $z = a$, the intensity will be given by

$$J_H(x,y) = 1 + A^2(x,y) + 2A\cos\{2\pi q_s x - \varphi(x,y)\} \tag{63.26}$$

If we consider the Fourier transform of this intensity distribution, we obtain three groups of terms. The unit term and $A^2(x,y)$ yield a central peak and the autocorrelation function of the object wave, which is also centred on the origin. The cosine term yields two sidebands, shifted owing to the presence of the term $2\pi q_s x$ in the argument of the cosine:

$$\mathcal{F}\{2A\cos(2\pi q_s x - \varphi)\} = \int A\exp(2\mathrm{i}\pi q_s x)\exp(-\mathrm{i}\varphi)\exp(2\pi\mathrm{i}\boldsymbol{q}\cdot\boldsymbol{r})\,d\boldsymbol{r}$$

$$+ \int A\exp(-2\mathrm{i}\pi q_s x)\exp(\mathrm{i}\varphi)\exp(2\pi\mathrm{i}\boldsymbol{q}\cdot\boldsymbol{r})\,d\boldsymbol{r}$$

$$= \int A e^{-\mathrm{i}\varphi}\exp\{2\pi\mathrm{i}(\boldsymbol{q}+\boldsymbol{q}_s)\cdot\boldsymbol{r}\}\,d\boldsymbol{r}$$

$$+ \int A e^{\mathrm{i}\varphi}\exp\{2\pi\mathrm{i}(\boldsymbol{q}-\boldsymbol{q}_s)\cdot\boldsymbol{r}\}\,d\boldsymbol{r} \tag{63.27}$$

Fig. 63.4: Fourier transform of a hologram showing the central autocorrelation distribution and the two sidebands. The figure demonstrates the symmetry $|R(-g)| = |R(g)|$.

in which $r = (x, y)$, $q = (q_x, q_y)$ and $q_s = (q_s, 0)$.

In reality, the specimen will have a spatial frequency spectrum of finite extent: beyond some cutoff frequency $|q| = q_c$, the spectrum contains no useful information but only noise. Since the bandwidth of the autocorrelation function will be twice this cutoff frequency, the sidebands will not overlap the central band (Fig. 63.4) provided that the frequency q_s is high enough,

$$q_s \geq 3q_c \tag{63.28}$$

This in turn implies that the angle of inclination of the wavefronts must be steep enough:

$$\gamma = \frac{1}{2}q_s\lambda \geq 1.5q_c\lambda \tag{63.29}$$

By isolating one sideband, therefore, we have direct access to the Fourier transform of $Ae^{\pm i\varphi}$ and should hence be able to recover the specimen phase and amplitude, perturbed by aberrations, directly. We return to this in

Section 63.3, after considering the more realistic case of spherical reference and object waves.

We now form $|\psi_W + \psi_R|^2$ in which the expressions (63.22) and (63.23) are used for ψ_W and ψ_R respectively. We shall later consider the weakly scattering object and we therefore write

$$O =: 1 + S(x_o, y_o) = 1 - \sigma(x_o, y_o) + i\eta(x_o, y_o) \qquad (63.30)$$

For weak objects, S, σ and η will all be small.

Substituting $O = 1 + S$ in (63.22), the expression for ψ_W divides into two parts, one corresponding to the unity and the other to S. The Fresnel integral for the former can be evaluated in closed form, since

$$\int\!\!\!\int_{-\infty}^{\infty} \exp\left[\frac{ik}{2}\left\{\frac{x_o^2 + y_o^2}{a - b} + \frac{(x - x_o)^2 + (y - y_o)^2}{b}\right\}\right] dx_o dy_o$$
$$= \frac{2\pi i b(a - b)}{ak} \exp\left\{\frac{ik}{2a}(x^2 + y^2)\right\} \qquad (63.31)$$

The resulting partial wave is then

$$\psi_V(\boldsymbol{r}_H) = \frac{A}{a} \exp\left\{ik\left(a \sec\gamma - x \sin\gamma + \frac{x^2 + y^2}{2a}\right)\right\}$$

which is identical with expression (63.23) for ψ_R except that the sign of the linear term in x is reversed. This was to be expected because ψ_V represents a spherical wave that emanated from the virtual source V and passed through the object *without interaction*. The other partial wave, the scattered wave $\psi_s(\boldsymbol{r}_H)$, is obtained by replacing $O(x_o, y_o)$ in (63.22) by $S(x_o, y_o)$.

It is advantageous to combine the two strong contributions, ψ_R and ψ_V, into one background wave $\psi_B(\boldsymbol{r}_M)$; this will represent the wave pattern created by the interferometer in the absence of a specimen, its intensity being the familiar fringe pattern. In the hologram plane H, where $\boldsymbol{r}_H = (x, y, a)$, we obtain

$$\psi_B(x, y) =: \psi_R(x, y) + \psi_V(x, y)$$
$$= \frac{2A}{a} \exp\left\{ik\left(a \sec\gamma + \frac{x^2 + y^2}{2a}\right)\right\} \cos x k_x \qquad (63.32)$$

in which a is omitted from the arguments of the functions and we have written

$$k_x := k \sin\gamma \qquad (63.33)$$

this quantity is exactly the x-component of the $\boldsymbol{k}$ vector at the midpoint of the hologram. The mean intensity is thus $\bar{J} = 2|A|^2/a^2$ and the background intensity may be written

$$J_B(x, y) = 2\bar{J} \cos^2(k_x x) \qquad \forall y \tag{63.34}$$

in agreement with (62.15) for $|\gamma| \ll 1$. The large terms in $ika \sec\gamma$ cancel out exactly from (63.34).

The hologram intensity is given exactly by

$$J_H(x, y) = |\psi_R + \psi_V + \psi_s|^2 = |\psi_B + \psi_s|^2 \tag{63.35}$$

and we could proceed as we did in the case of plane waves and separate the various regions of the spatial spectrum of the hologram. Here, however, we consider a different situation, the weakly scattering object, for which $|S| \ll 1$. This implies that $|\psi_s| \ll |\psi_R|$ and $|\psi_s| \ll |\psi_V|$ everywhere but, near the zeros of J_B ($k_x x$ an odd multiple of $\pi/2$ in (63.22)), it will not be true that $|\psi_s|^2 \ll J_B$. Nevertheless, if we wish to obtain a hologram intensity that is always linear in ψ_s, we are obliged to omit $|\psi_s|^2$. As we saw earlier in this section, the great advantage of off-axis holography is that part of the spectrum of the hologram contains information about ψ_s and we shall show that this part can be isolated by suitable techniques. Only if we require linearity everywhere must we neglect $|\psi_s|^2$ in J_H. When we do impose this requirement, J_H takes the form

$$J_H(x, y) = J_B(x) + 2\Re\{\psi_B^*(x, y)\psi_s(x, y)\}$$

As usual, we obtain ψ_s by replacing $O(x_o, y_o)$ with $S(x_o, y_o)$ in (63.22), whereupon J_H becomes

$$J_H = 2\bar{J} \cos^2(k_x x) + \bar{J} \frac{ka}{\pi b(a - b)} \cos(k_x x) F_s \tag{63.36a}$$

in which F_s is the scattering amplitude,

$$F_s := \Im\left\{ e^{-ik_x x} \iint\limits_{-\infty}^{\infty} S(x_o, y_o) \exp(ikL - ikW) \, dx_o \, dy_o \right\} \tag{63.36b}$$

The variable L in the exponent denotes the quadratic function

$$L := -\frac{x^2 + y^2}{2a} + \frac{x_o^2 + y_o^2}{2(a - b)} + \frac{(x - x_o)^2 + (y - y_o)^2}{2b} \tag{63.36c}$$

while W is the perturbation eikonal for the geometric aberrations.

A number of special cases of (63.36a) are of interest. We examine these briefly before turning to the various reconstruction procedures.

(a) In-line holography as a degenerate case

When the distance $2d$ between the virtual sources V and $\overline{V}$ vanishes, $\gamma = 0$ and hence $k_x = 0$. The waves ψ_R and ψ_V coincide and form *one* coherent wave of amplitude $\hat{A} = 2A$; this is to be identified with the amplitude in Section 63.1. The midpoint of the object moves onto the optic axis. If we now set $\overline{J} = |\hat{A}|^2/a^2$, we must likewise replace $S(x_0, y_0)$ by $2S$, since the wave amplitude is twice that used earlier. Hence

$$J_H = \overline{J}\left[1 + \frac{ka}{\pi b(a-b)}\Im\left\{\iint S(x_o, y_o)e^{ik(L-W)}\,dx_o\,dy_o\right\}\right] \qquad (63.37)$$

where L and W are as defined above. Clearly, the separation of the hologram spectrum into regions does not occur and the advantage of off-axis operation for the reconstruction is lost.

(b) Quasi-homogeneous objects

As in Section 62.3, we assume that the functions $\sigma(x_o, y_o)$ and $\eta(x_o, y_o)$ alter very slowly with x_o and y_o. The aberration term will be omitted here, since it has little significance.

The factor S can be taken outside the integral, but before we can do this, we must transform the quadratic function L in order to find the appropriate arguments of S. We rearrange the expression (63.36c) in such a manner that x_o and y_o appear only in quadratic terms:

$$L = \frac{a}{2b(a-b)}\left\{\left(x_o - \frac{a-b}{a}x\right)^2 + \left(y_o - \frac{a-b}{a}y\right)^2\right\}$$

$$= \frac{a}{2b(a-b)}\left\{\left(x_o - \frac{x}{M}\right)^2 + \left(y_o - \frac{x}{M}\right)^2\right\}$$

with $M = a/(a-b)$, as in (62.12b). It is now obvious that the oscillations of the integrand are slow only in the vicinity of the point $(x/M, y/M)$; we may thus simplify (63.36b) to

$$F_s = \Im\left\{e^{-ik_x x}S(x/M, y/M)\int\limits_{-\infty}^{\infty}\!\!\int e^{ikL}\,dx_o dy_o\right\}$$

The remaining integral can be evaluated analytically and gives $2\pi i b(a - b)/ak$ and hence

$$F_s = \frac{2\pi b(a - b)}{ak} \Re\left\{ e^{-ik_x x} S(x/M, y/M) \right\}$$

After introducing this into (63.36a), we find

$$J_H(x, y) = 2\overline{J}\cos^2 k_x x + 2\overline{J}\cos(k_x x)\Re\left\{ e^{-ik_x x} S(x/M, y/M) \right\}$$

or with $S = -\sigma + i\eta$:

$$J_H(x, y) = 2\overline{J}\left\{ 1 - \sigma(x/M, y/M) \right\} \cos^2 k_x x + \overline{J}\eta(x/M, y/M)\sin 2k_x x$$

$$(63.38)$$

This is in agreement with (63.20) if we linearize with respect to σ and η and identify $\overline{J}$ with $2J_R$ and k_x with $kd/a = k\gamma$. This confirms that our calculations are self-consistent.

(c) *Very small objects*

The diameters of the specimens investigated in holography experiments at high resolution are very much smaller than those of the holograms finally recorded, since a high intermediate magnification is necessary to achieve the required resolution. This implies that we can ignore all terms in $x_o^2 + y_o^2$ in (63.36c). The integral in (63.36b) then reduces essentially to a Fourier integral since (63.36c) simplifies to

$$L = \frac{x^2 + y^2}{2f} - \frac{xx_o + yy_o}{b} \qquad (63.39)$$

with $1/f := 1/b - 1/a$, as in (63.4b). The last term in (63.39) shows that it is convenient to define a *spatial frequency vector*

$$\boldsymbol{q} = (q_x, q_y) = \left(\frac{x}{\lambda b}, \frac{y}{\lambda b}\right) = \frac{1}{\lambda}(\theta_x, \theta_y)$$

$\theta_x = x/b$ and $\theta_y = y/b$ being essentially the *scattering angles* in the object. Equation (63.39) can then be rewritten as

$$kL = \pi\lambda\boldsymbol{q}^2 b/M - 2\pi\boldsymbol{q}\cdot\boldsymbol{r}_o \qquad (63.41)$$

Owing to the very small lateral extent of the specimen, it is sufficient to consider only those terms in the wave aberration W that depend on the angles θ_x and θ_y (isoplanatic approximation, see Chapter 65). It is convenient to express θ_x and θ_y in terms of $\boldsymbol{q}$ and to incorporate the first term in (63.41), a *defocus*, into the wave aberration. We thus write

$$w(\boldsymbol{q}) := 2\pi W(\lambda\boldsymbol{q})/\lambda - \pi\lambda\boldsymbol{q}^2 b/M \qquad (63.42)$$

With $k_x = k\sin\gamma \approx 2\pi\gamma/\lambda$, $x = \lambda b q_x$ from (63.40) and $s := \gamma b$ we obtain

$$k_x x = 2\pi q_x s \qquad (63.43)$$

s being the *off-axis* distance of the very small object detail being investigated. Putting all this together and returning to (63.36b), we see that the scattering amplitude simplifies to

$$F_s = \Im\left[\exp\{-2\pi i s q_x - iw(\boldsymbol{q})\}\tilde{S}(\boldsymbol{q})\right] \qquad (63.44)$$

with the two-dimensional *Fourier transform*

$$\tilde{S}(\boldsymbol{q}) = \mathcal{F}^-(S) = \iint S(x_o, y_o)e^{-2\pi i(x_o q_x + y_o q_y)}\, dx_o dy_o \qquad (63.45)$$

It is now easy to express the hologram intensity J_H, given by (63.36a), in terms of q_x and q_y. We shall not pursue this, since the formulation of holography for weak objects has nowadays lost much of its earlier interest. One result of the present considerations is, however, of general interest: the conditions for the application of Fourier transforms do not need to be enforced by a special optical arrangement, but are simply a consequence of the very high magnification needed for small object detail.

63.3 Reconstruction procedures

The numerous reconstruction techniques differ essentially in the choice of the waves that are caused to interfere. They differ too in the choice between optical and digital reconstruction but, although this choice has far-reaching practical implications, it has little bearing on the principles. The arguments on both sides are well known and will not be rehearsed here; in short, optical reconstruction is fast (once the optical setup has been chosen and aligned) and large areas can be processed with ease while numerical reconstruction requires digitization and powerful computing facilities but enables us to manipulate waves easily that would be difficult to generate optically.

63.3.1 Aberration correction
We recall that this was the purpose of Gabor's early investigations. Equations (63.36) show that the aberrations of the electron microscope lenses

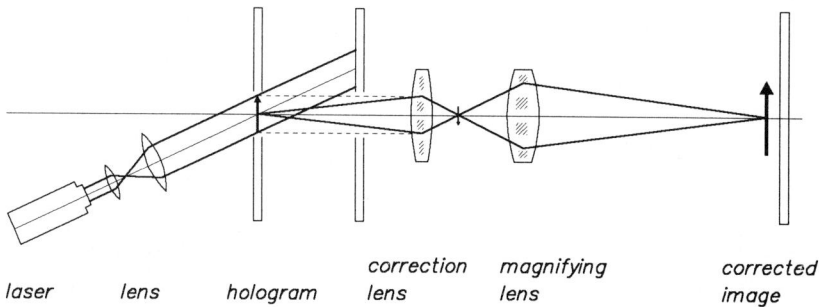

Fig. 63.5: Optical reconstruction system for spherical aberration correction.

can be removed by selecting one sideband of the hologram spectrum and cancelling the wave aberration, either with a suitably chosen glass lens (Fig. 63.5) or, more straightforwardly, by digitizing the hologram and adding the appropriate phase shifts. The optical approach has been used by Tonomura *et al.* (1979b) and Rogers (1978, 1980). Numerical correction has been successfully demonstrated by Lichte and colleagues (Franke *et al.*, 1987; Fu *et al.*, 1991; Lichte, 1991a–e, 1993) who even included a quadratic term in q^2 of the form $\exp(cq^2)$ to reduce the attenuation due to chromatic aberration, and by Harada *et al.* (1990a) and Ishizuka *et al.* (1991). A major problem with aberration correction is the need for very accurate knowledge of the aberration coefficients of the image-forming lenses. We show in Chapter 77 that it is possible, though not easy, to establish these quantities with the necessary precision. The theoretical studies of Hanszen *et al.* (1972a,b), Hanszen (1972a, 1973b) and Ade (1973) are relevant here. An ingenious way of relaxing the requirements for extremely fine sampling has been proposed by Ishizuka (1993), who has gone on to discuss correction (Ishizuka, 1994). See also Kawasaki and Rodenburg (1993).

The final task is to derive the distribution of atoms in the specimen from the functions $\sigma(r_o)$ and $\eta(r_o)$. This question of image interpretation cannot be answered within the context of holography.

A simulation of the correction sequence calculated by Joy *et al.* (1993) is shown in Fig. 63.6. The images of Fig. 63.6a show the amplitude and phase of the wavefunction emerging from a silicon carbide crystal, illuminated by a coherent plane wave, as calculated by the multislice simulation program MacTempas* (see Section 69.8). Figure 63.6b shows the same quantities modified by the objective lens. In Figure 63.6c, the hologram is shown above and its Fourier transform below; in the latter, the central disc

---

* Available from Total Resolution, 20 Florida Ave., Berkley CA 94707, USA.

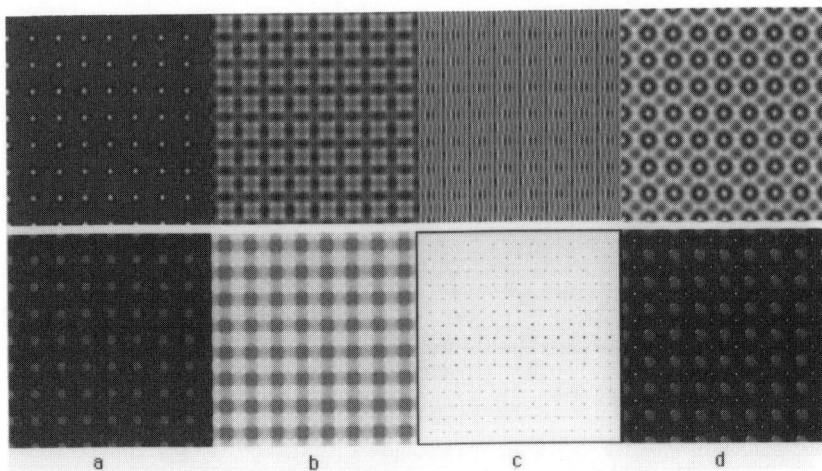

Fig. 63.6: Amplitude (above) and phase (below) of the wavefunction emerging from a silicon carbide crystal. (a) Simulation. (b) Simulation with objective lens aberrations. (c) Hologram (above) and its Fourier transform (below). (d) Reconstruction.

and the two sidebands just touch: the fringe spacing is one quarter of the atomic spacing. The final pair of figures, 63.3d, show the reconstruction, which is good but not perfect.

63.3.2 Interference holography

If we add a plane wave to one of the reconstructed images, as illustrated in Fig. 63.1a, we can produce a contour map of specimen thickness. An example is shown in Fig. 63.7 which also illustrates the principle of phase amplification; a shearing interferogram, obtained as shown in Fig. 63.8 is also included. Here the reconstructed image and its conjugate are brought into coincidence, to give double the phase difference. Higher multiples can be obtained by using higher-order diffracted beams. For extensive discussion and many examples, see Tonomura (1987a). Recently, numerical methods have been introduced into interference holography by Ohshita *et al.* (1990), Ru *et al.* (1991a,b) and Matsuda *et al.* (1991).

A slightly different approach to the reconstruction stage has been introduced by Ade and Lauer (1988, 1990, 1991; Lauer and Ade, 1990). In their (numerical) procedure, the two sidebands of (63.26) are retained and said to form a 'filtered' hologram, with the general form

$$J_H^{(f)} = b + 2A\cos\{2\pi \boldsymbol{q} \cdot \boldsymbol{r} - \varphi(x,y)\} \qquad (63.46)$$

in which b is a constant bias term; the filtering consists in removing the term $A^2(x,y)$ and modifying the bias. This filtered hologram is now multiplied

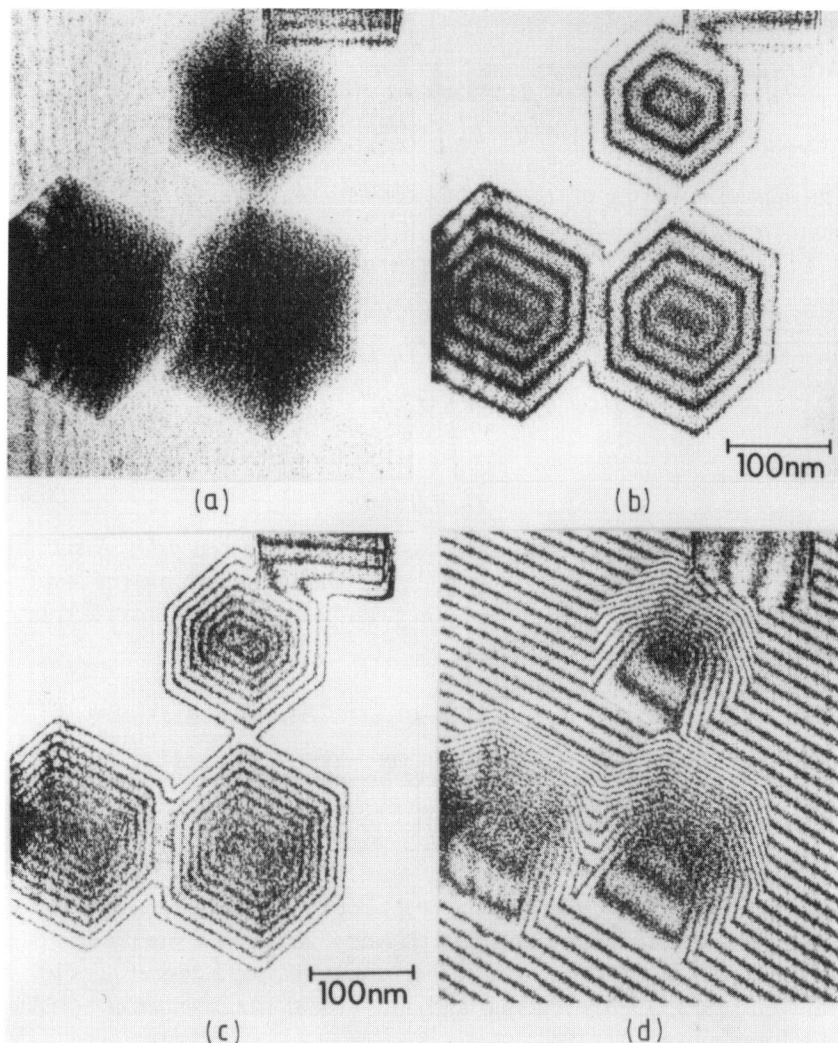

Fig. 63.7: Interference micrographs of MgO particles. (a) Reconstructed image. (b) Contour map. (c) Contour map (with phase amplification). (d) Interferogram (with plane wave at oblique incidence as in Fig. 63.8b).

(Ade and Lauer, 1991) by a raised sinusoid of the form

$$s(\boldsymbol{r}) = 1 + 2B\cos(2\pi \boldsymbol{q}_b \cdot \boldsymbol{r}) \tag{63.47}$$

giving

$$
\begin{aligned}
J_H^{(f)}s = {} & b + 2bB\cos(2\pi \boldsymbol{q}_b \cdot r) + 2A\cos(2\pi \boldsymbol{q}\cdot \boldsymbol{r} - \varphi)\\
& + 2AB[\cos\{2\pi(\boldsymbol{q}+\boldsymbol{q}_b)\cdot \boldsymbol{r} - \varphi\} + \cos\{2\pi(\boldsymbol{q}-\boldsymbol{q}_b)\cdot \boldsymbol{r} - \varphi\}]
\end{aligned}
\tag{63.48}
$$

The spatial spectrum of this product consists of several terms:

$$
\begin{aligned}
\mathcal{F}(J_H^{(f)}s) = {} & b\delta(\boldsymbol{q}') + bB\{\delta(\boldsymbol{q}-\boldsymbol{q}_s+\boldsymbol{q}') + \delta(\boldsymbol{q}-\boldsymbol{q}_s-\boldsymbol{q}')\}\\
& + \tilde{O}(\boldsymbol{q}+\boldsymbol{q}') + \tilde{O}^*(\boldsymbol{q}-\boldsymbol{q}')\\
& + B\{\tilde{O}(2\boldsymbol{q}-\boldsymbol{q}_s+\boldsymbol{q}') + \tilde{O}^*(2\boldsymbol{q}-\boldsymbol{q}_s-\boldsymbol{q}')\}\\
& + B\{\tilde{O}(\boldsymbol{q}_s+\boldsymbol{q}') + \tilde{O}^*(\boldsymbol{q}_s-\boldsymbol{q}')\}
\end{aligned}
\tag{63.49}
$$

in which

$$\boldsymbol{q}_s := \boldsymbol{q} - \boldsymbol{q}_b \tag{63.50}$$

and $\tilde{O}$ is the Fourier transform of the object transparency O. A suitable choice of the difference frequency $\boldsymbol{q}_s$ brings the terms forming the last line of (63.49) close to the origin, where they can be selected and inverse Fourier tranformed to give the reconstruction $r(\boldsymbol{r})$,

$$r(\boldsymbol{r}) := \mathcal{F}^-[B\{\tilde{O}(\boldsymbol{q}_s+\boldsymbol{q}') + \tilde{O}^*(\boldsymbol{q}_s-\boldsymbol{q}')\}] = 2AB\cos(2\pi \boldsymbol{q}_s \cdot \boldsymbol{r} - \varphi) \tag{63.51}$$

For visual presentation, this can either be set against a bias, b', giving

$$r(\boldsymbol{r}) + b' = b' + 2AB\cos(2\pi \boldsymbol{q}_s \cdot \boldsymbol{r} - \varphi) \tag{63.52}$$

or its modulus can be displayed, $|r(\boldsymbol{r})|$. For examples of such reconstructions, see Ade and Lauer (1990, 1991), Ade (1992) and a similar approach by Harada *et al.* (1990b). For earlier proposals, see Endo *et al.* (1979), Hanszen (1983, 1985), Hanszen and Ade (1983), Hanszen *et al.* (1983a) and Tonomura *et al.* (1985).

63.3.3 Other aspects of electron holography

(a) *Statistical considerations*

Although the reasoning in this Chapter has been expressed in terms of wavefunctions, we have said nothing about the uncertainties introduced by sampling, finiteness and the low electron dose often essential to obviate radiation damage. These questions have begun to be studied and we refer to the work of Ade (1980), Hanszen (1982d, 1987), Ade *et al.* (1984), Hanszen *et al.* (1984), Lauer and Hanszen (1986), Herrmann and Lichte (1986), Lichte *et al.* (1987, 1988), Lenz (1988), Lenz and Völkl (1990) and de Ruijter and Weiss (1993) for further details.

(b) *Holography in the scanning transmission electron microscope (STEM)*

With the successful incorporation of the electron biprism in the transmission electron microscope*, it was natural to enquire whether the presence of such a biprism in a scanning transmission instrument would offer any new possibilities. A preliminary study by Leuthner *et al.* (1988, 1989, 1990) shows that the flexibility at the detector level that is characteristic of the STEM (and extensively studied by Chapter 67) can be exploited to separate the phase and amplitude of the specimen. The experimental arrangement is shown in Fig. 63.9, which shows that each pixel of the object will create an interference pattern in the detector plane as the scanning probe passes over it. Clearly, this pattern can be averaged in many ways, depending on the detector geometry, as explained in connection with routine STEM operation in Chapter 67. Leuthner *et al.* suggest that two signals should be extracted from each pixel, one obtained by summing the incident intensity, the other obtained by modulating the incident signal by means of a grating before detection and summing. These two signals should be sufficient to yield the specimen amplitude and phase separately. Other types of modulation would certainly merit investigation and numerical simulation of such modulated detectors would undoubtedly be advantageous. Leuthner *et al.* also suggest an 'interferometric mode' in which the electron interference pattern is scanned over the detector in synchronism with the probe. Preliminary results are presented in Leuthner *et al.* (1989, 1990).

Shortly afterwards, STEM holography with a biprism was examined by Cowley (1990) and pursued in considerable detail by Gribelyuk and Cowley (1991, 1992a,b, 1993), Wang and Cowley (1991), Mankos *et al.* (1992) and by Konnert and d'Antonio (1992). We refer to those studies for a thorough analysis.

Before leaving this topic, we should just mention that, just as brightfield TEM image formation has been interpreted in the language of in-line holography, the STEM signal in the image plane has also been discussed in the language of holography. The pioneering work of Veneklasen (1975) and the later discussion of Cowley and Walker (1981) and Lin and Cowley

---

* A commercial version became available in 1990 from Hitachi.

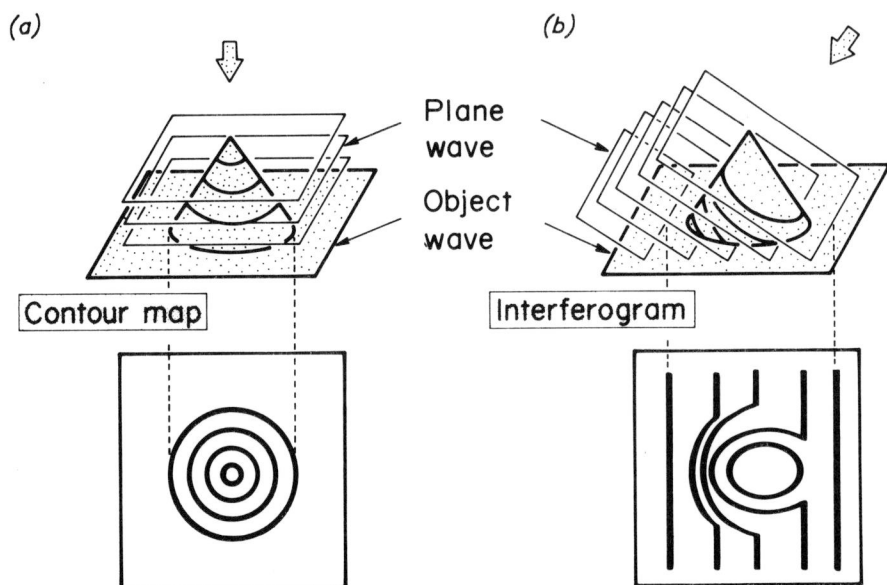

Fig. 63.8: Two kinds of interference micrographs. (a) Contour map obtained with the plane wave and the reconstructed object wave in the same direction. (b) Interferogram obtained with the plane wave inclined to the reconstructed object wave. The latter shows whether the wavefronts have been advanced or retarded.

(1986) are examples of this. Veneklasen in particular discussed the use of structured detectors in this context.

(c) *Choice of defocus*

It has been pointed out by Lichte (1990, 1991b, 1992a,b) that the best choice of defocus value for bright-field high-resolution imaging in the transmission electron microscope is not necessarily the most suitable for electron holography. The influence of the defocus value in the former case is examined in great detail in Chapter 66, where it emerges that the value known as 'Scherzer focus', $\Delta = (C_s \lambda)^{\frac{1}{2}}$, is particularly suitable for converting the phase variations in the electron wave into intensity variations at the image plane. Lichte argues that, in holography, it is more important to prevent unwanted exchange between amplitude and phase information and hence seeks the value of defocus for which the *amplitude* contrast transfer function (see 66.19a below) satisfies a particular condition. This function

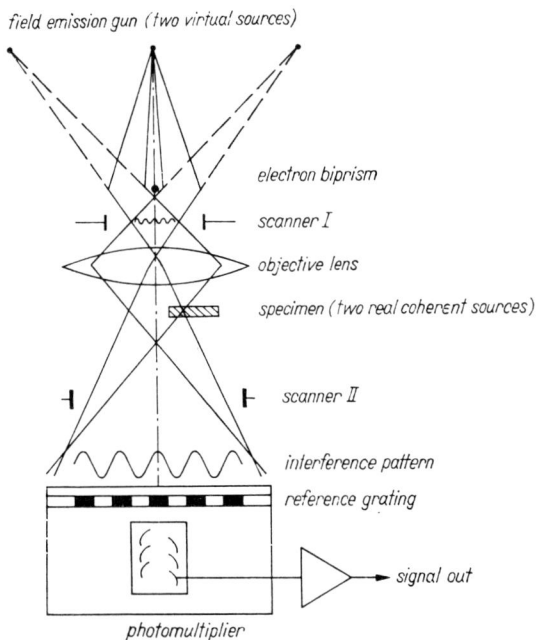

Fig. 63.9: Ray diagram illustrating hologram formation in the STEM. One virtual source forms a probe that explores the specimen while the other forms a probe in free space. The resulting waves interfere in the detector plane.

is oscillatory, and is given by

$$\hat{K}_a(Q) = \cos \pi \left(DQ^2 - \frac{1}{2}Q^4 \right) \tag{63.53}$$

in terms of reduced quantities (65.31-32). It first reaches zero for $Q = Q_0$, where

$$Q_0^4 - 2DQ_0^2 = 1$$

or

$$Q_0^2 = D \pm (D^2 + 1)^{\frac{1}{2}} \tag{63.54}$$

It passes through a minimum at Q_m, where $d\hat{K}_a(Q)/dQ = 0$, giving $Q_m^2 = D$ and hence $\hat{K}_a(Q_m) = \cos(\pi D^2/2)$. The phase contrast transfer function (66.19b), given by

$$\hat{K}_p(Q) = \sin \pi \left(DQ^2 - \frac{1}{2}Q^4 \right) \tag{63.55}$$

also passes through a minimum for this value of Q and transfer of amplitude specimen information to the amplitude variations of the wavefunction at the image and of phase information to phase variations will be greatest when the depths of the phase and amplitude minima are the same. Correspondingly, 'cross-talk', which maps specimen amplitude into image phase and vice versa, will be least. In order to find the defocus value for which this situation obtains, we equate $\hat{K}_p(Q_m)$ and $\hat{K}_a(Q_m)$:

$$\sin \frac{\pi}{2} D^2 = \cos \frac{\pi}{2} D^2 \qquad (63.56)$$

which has the obvious solution $D^2 = \frac{1}{2}$. It is therefore proposed that this value of defocus should be called the *Gabor focus**, D_G:

$$D_G := \frac{1}{\sqrt{2}} \qquad (63.57)$$

Clearly,
$$D_G = D_S/\sqrt{2} = 0.707 D_S \qquad (63.58)$$

where D_S is the Scherzer focus, $D_S = 1$. This value of defocus is optimal when aberration correction is not envisaged. Lichte shows that when the spherical aberration is to be cancelled, a different value is preferable, for which

$$D = 0.75 Q^2_{\max} \qquad (63.59)$$

where $Q_{\max}$ is the greatest reduced spatial frequency to be included in the correction procedure. This is essentially the plane of the disc of least confusion, which we found in Section 24.3 by geometrical arguments (24.50).

(d) Reflection holography

For surface studies, a reflection mode is of considerable interest. For details of this technique, see Banzhof et al. (1988, 1992), Osakabe (1992), Osakabe et al. (1988, 1989, 1993), Takeguchi et al. (1990, 1992) and Banzhof and Herrmann (1993).

(e) Applications

The most numerous applications of holography are concerned with magnetic specimens. The following list contains a representative selection of this work: Arii et al. (1981), Fabbri et al. (1987), Frost and Lichte (1988),

---

* This is not the same as the definition of Gabor focus adopted by Lichte (1991), who gives $D_G = \{(2/\pi) \arccos 0.75\}^{\frac{1}{2}} \approx 0.68$. The above definition is suggested by Hawkes (1992).

Fukuhara *et al.* (1983), Hasegawa *et al.* (1989), Lau and Pozzi (1978), Matsuda *et al.* (1982), Matteucci *et al.* (1982a, 1988a,c), Matteucci and Muccini (1992), Migliori and Pozzi (1992), Olivei (1969), Osakabe *et al.* (1983), Pozzi (1980a), Pozzi and Missiroli (1973), Ru *et al.* (1991b), Tonomura (1972, 1983, 1987b, 1992b), Tonomura *et al.* (1980a,b, 1982b,c,d), Wahl and Lau (1979), Yoshida *et al.* (1983). Trapped flux in superconductors and fluxons are particularly examined by Boersch and Lischke (1970a,b), Boersch *et al.* (1974), Harada *et al.* (1992), Lischke (1969, 1970), Matsuda *et al.* (1989, 1990), Migliori and Pozzi (1991), Migliori *et al.* (1993), Tonomura *et al.* (1984, 1990) and Wahl (1968/9, 1970a,b). Ferroelectrics have also been studied, by Zhang *et al.* (1992, 1993a,b).

63.4 Further reading

The references cited in this Part represent only a small fraction of the voluminous literature of electron interferometry and holography. We have therefore grouped here further papers, of a general nature or dealing with aspects of the subject not considered here.

63.4.1 Electron interferometry

Instrumental aspects are considered by Buhl (1961b), Costa *et al.* (1989), Harada *et al.* (1988), Hibi and Takahashi (1963), Kawasaki *et al.* (1990a,b, 1992a), Matteucci (1978), Matteucci *et al.* (1982b), Merli *et al.* (1974), Möllenstedt and Krimmel (1964), Ogai *et al.* (1991), Pozzi (1977, 1983a,b, 1992), Rau *et al.* (1991a,b), Ru *et al.* (1991a, 1992a,b), Schaal (1971), Stumpp (1984), Stumpp *et al.* (1984) and Tanji *et al.* (1991).

For further details of electron interferometry, see Donati *et al.* (1973), Faget *et al.* (1960), Fagot *et al.* (1961), Fert and Faget (1958), Fert *et al.* (1962a,b), Komrska (1971, 1975), Komrska and Lenc (1970), Komrska and Vlachová (1973), Komrska *et al.* (1964a,b, 1967), Fischer and Lischke (1967), Gabor (1956), Hasselbach (1992), Lenz (1965), Lichte (1984b, 1986d), Matsuda *et al.* (1978), Matteucci *et al.* (1979), Merli *et al.* (1976a), Möllenstedt (1960, 1962, 1987), Möllenstedt and Bayh (1961a,b), Möllenstedt and Lenz (1962), Möllenstedt and Lichte (1978a,b, 1979), Pozzi (1975, 1980b), Rang (1953), Rogers (1970), Stroke (1967), Takeda and Ru (1985), Tonomura *et al.* (1978c, 1989), Fu *et al.* (1987) and Yatagai *et al.* (1987). The light shed on the concept of electron phase by interference experiments is discussed notably by Möllenstedt (1988) and Möllenstedt and Lichte (1989).

63.4.2 Electron holography

The following papers are all concerned with some aspect of the technique: Ade (1982, 1994), Allard *et al.* (1992), Anaskin and Stoyanova (1972), Boersch (1967), Bonhomme and Beorchia (1980), Bonnet *et al.* (1978), Boseck *et al.* (1986), Buhl (1961a), Chen *et al.* (1987, 1993), Cowley (1991), Endo and Tonomura (1990), Endo *et al.* (1986, 1989), Estrada *et al.* (1991), Gabor (1968/9), Gabor *et al.* (1965), Gajdardziska-Josifovska *et al.* (1993), Greenaway and Huiser (1976), Hanszen (1969, 1970a, 1980, 1982c, 1984, 1986b) Hanszen and Ade (1976a,b, 1977, 1984), Hanszen and Lauer (1980), Hanszen *et al.* (1980, 1981, 1982, 1985, 1986), Harada and Shimizu (1991), Herrmann *et al.* (1978), Laberrigue *et al.* (1980), Lannes (1978, 1980, 1982), Lauer (1982, 1984a,b), Lauer and Ade (1992), Lichte (1982, 1984a, 1985, 1986a,b,c, 1989, 1991c,d,e, 1992a,b,c), Lichte and Völkl (1988, 1991), Lichte *et al.* (1992), Matteucci *et al.* (1982c), Menzel *et al.* (1973), Plass and Marks (1992), Pozzi (1992, 1993), Pozzi and Prola (1987), Saxon (1972a,b), Tonomura (1969, 1986b, 1987a–c, 1989, 1990, 1991a,b, 1992d), Tonomura *et al.* (1978a,b, 1979a, 1979c), Völkl and Lichte (1990a,b), Wade (1974, 1975, 1980), Weingärtner *et al.* (1969a,b, 1970, 1971), Weiss *et al.* (1993), Zeitler (1979) and Zhang and Joy (1991).

Some unconventional forms of holography involving electrons are discussed by Bartell (1972, 1975), Bartell and Johnson (1977), Bartell and Ritz (1974) and Gabor (1980); by Bates and Lewitt (1975); by Garcia (1989); by Fink *et al.* (1991), Fink and Kreuzer (1992) and Spence *et al.* (1992); by Lichte and Hornstein (1982); by Tong *et al.* (1991); by Tonomura and Matsuda (1980); by Saldin (1991); and by Qian *et al.* (1993), Scheinfein *et al.* (1993) and Spence and Qian (1992).

For further discussion of the relation between contrast-transfer theory and holography, see the following papers, from Hanszen's group: Hanszen (1970b, 1971, 1972b, 1973a, 1974, 1976), Hanszen and Ade (1974), Hanszen *et al.* (1983b) and the historical accounts by Hanszen (1990a,b).

Holographic display of scanning electron microscope images is proposed by Kulick *et al.* (1987).

Part XIII

Theory of Image Formation

64
General Introduction

This Part is concerned with the theory of image formation in the types of transmission electron microscope capable of very high resolution imaging, the (conventional) transmission electron microscope (TEM, or CTEM if it is necessary to stress the fact that the conventional instrument is meant) and the scanning transmission electron microscope (STEM). Scanning electron microscopes (SEMs) of the latest generation are also capable of providing high-resolution information but the image-forming processes are very different. In this introductory chapter, we describe the principal features of these various types of microscope, limiting the account to those that affect image formation directly. In practice, a single instrument may be capable of operating in several modes: transmission electron microscopes may be designed to operate as either conventional or scanning instruments; a 'dedicated' STEM (which does not provide TEM images) may furnish some of the signals routinely collected by a SEM. We disregard these aspects of instrument design.

A (conventional) transmission electron microscope consists of a source, the electron gun; condenser lenses to direct the beam onto the specimen and to control the area illuminated and the angles at which the electrons are incident; an objective lens to provide the first stage of magnification; projector lenses to provide further magnification; a fluorescent screen for direct observation of the image and a photographic recording unit (Fig 64.1). Apertures are included at several levels: in the condenser system, in the back-focal plane of the objective lens and in the plane of the first intermediate image. There are many other elements that do not concern us here (notably alignment coils and stigmators). In modern microscopes, direct image read-out to computer memory and an electron energy-loss spectrometry unit (EELS) may be fitted below the image plane.

Two modes of operation are of everyday interest in TEM and a number of others are important in special circumstances. The two common modes are illustrated in Fig. 64.1. In bright-field imagery, a small source provides a beam of electrons, which are directed onto the specimen by condenser lenses; frequently the object is immersed in the magnetic field of the objective, the prefield of which thus acts as a last condenser lens. These condensers and the associated apertures ensure that the beam covers the desired area of the specimen and that the electrons arrive with the appropriate angular spread. In much of what follows, we shall assume that the

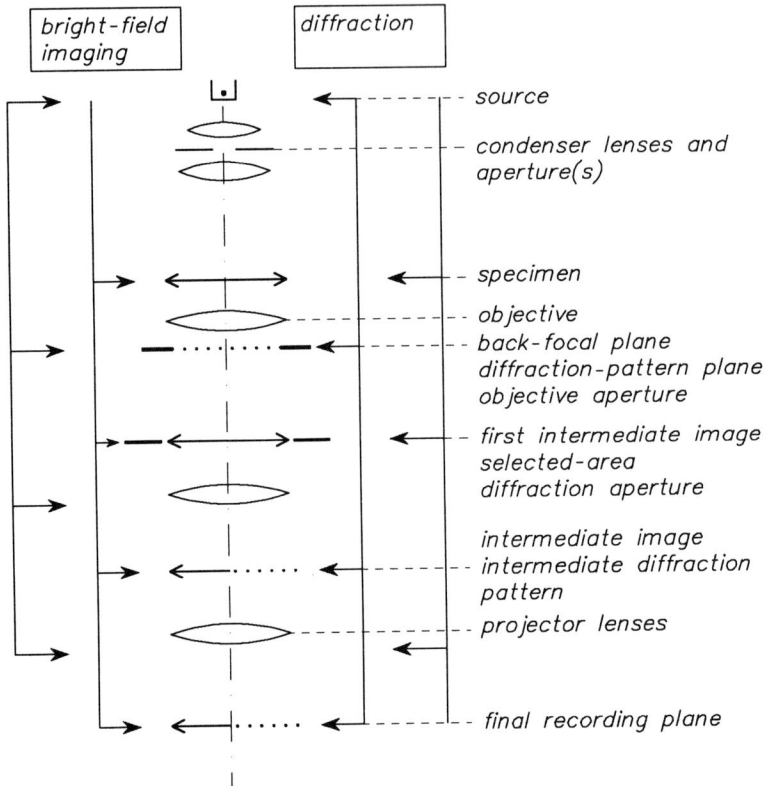

Fig. 64.1: Conjugate planes in a TEM in the bright-field imaging mode and in the diffraction mode. Sets of conjugate planes are indicated by arrows. The condenser aperture may be regarded as conjugate to the specimen though in practice this is not always the case.

condensers have been adjusted so that the electrons are travelling parallel to the optic axis when they strike the specimen and hence correspond to a (truncated) plane incident wave. In fact, this would be possible (neglecting condenser lens aberrations) only if the source size were vanishingly small. From (16.12) with $z_1 = z_{Fo}$, we see that $x_2' = -x_1/f$, where x_1 is the source width and x_2' is the corresponding beam tilt at the specimen plane; f is the focal length of the complete condenser system.

If the beam is indeed parallel to the axis when it strikes the specimen[*], the source will be conjugate to the back-focal plane of the objective lens and

---

[*] This is known as Köhler illumination (Köhler, 1893, 1899) and its merits for electron microscopy have been underlined by Probst *et al.* (1991a,b) and Benner *et al.* (1990, 1991). It cannot, however, be achieved exactly when the specimen is immersed in the field of the objective (Christenson and Eades, 1986).

to planes close to the back-focal planes of the intermediate and projector lenses if these are operating at relatively high magnification. The specimen will be conjugate to a sequence of planes beyond the objective, culminating in the final image plane, the recording plane. The first intermediate image lies in the plane of the selected-area diffraction aperture and if all lenses are operating at relatively high magnification, these intermediate image planes will not be far from the object focal planes of successive lenses. The object plane is also frequently but not always conjugate to a condenser aperture.

In the standard diffraction mode, the strengths of the condenser and objective lenses are unaltered so that nothing in Fig. 64.1 changes down to the first intermediate image. The intermediate and projector lenses are now adjusted so that not this intermediate image but the objective back-focal plane is conjugate to the final recording plane. The roles of the various planes downstream are thus exchanged. (Depending on the way in which the projector lens currents are changed, these planes may be shifted as well.)

On the basis of these simple diagrams, the many other modes may be easily understood. Convergent-beam diffraction, for example, is achieved by focusing the source on the specimen. The tilted-illumination modes are obtained with the aid of coils in the space upstream from the specimen. These can be used to illuminate the object with a plane tilted beam or with a hollow-cone beam, in which the tilt angle is kept constant but the azimuth is changed so that the beam sweeps out a hollow conical volume. Tilted illumination is employed in crystal imaging and in principle improves bright-field resolution anisotropically; hollow-cone illumination should provide the same improvement isotropically but, in both cases, the gain is offset by a loss of absolute contrast. We return to this in Section 66.5. The various dark-field modes are obtained by preventing the electrons unscattered in the specimen from reaching the image. This is usually achieved by tilting the beam incident on the specimen so that the unscattered beam is intercepted by the objective aperture or by displacing the latter so that only selected scattered electrons, typically those concentrated in diffraction spots when the specimen is crystalline, reach the image.

The STEM is a microscope in which a small probe explores the specimen and information collected by the beam electrons as they traverse each small picture-element (pixel) of the specimen is extracted by various detectors downstream from the latter. In the basic instrument, two modes of operation are routinely used but the situation is more complex than for the TEM. The microscope consists of a source (in practice only a field-emission gun is suitable); condenser lenses; sets of scanning coils; the probe-forming lens, which is commonly referred to as the objective; intermediate lenses; an annular dark-field detector; an electron energy analysis unit with fur-

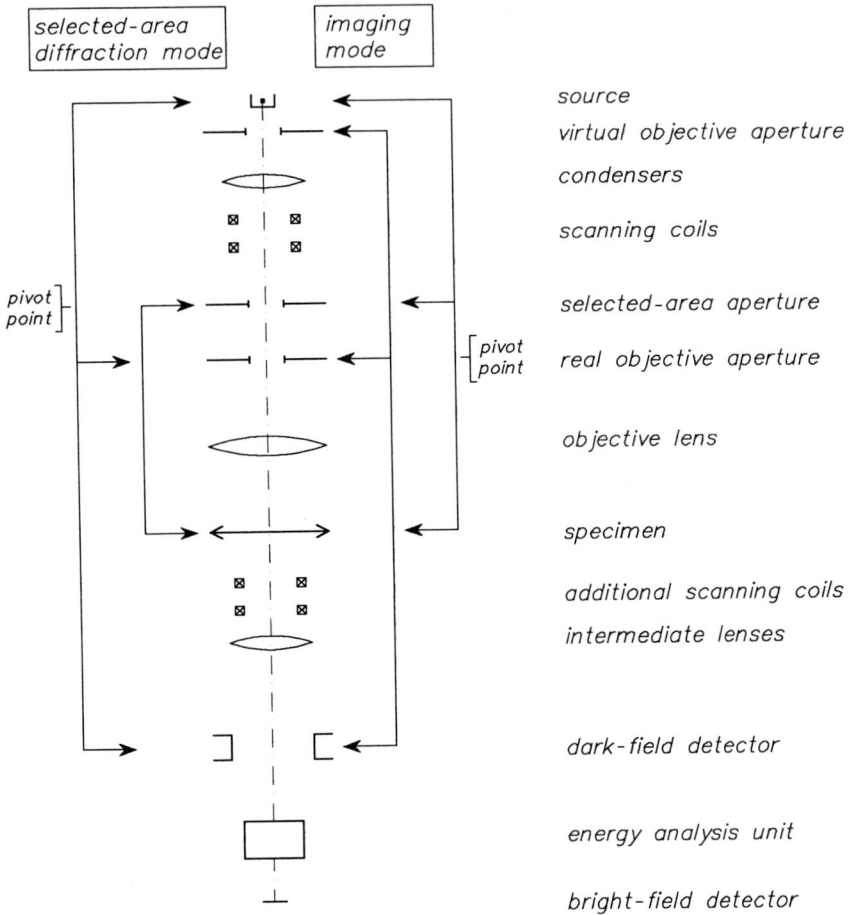

Fig. 64.2: Conjugate planes in a STEM in the bright-field imaging mode and in the diffraction mode. Sets of conjugate planes are indicated by arrows.

ther detectors. There are in addition alignment coils, stigmators and apertures.

In the standard imaging mode (Fig. 64.2), the source is demagnified by the condenser and objective lenses, so that crossover, a selected-area aperture and specimen are conjugates. The pivot point of the scanning coils coincides with the front-focal plane of the objective so that, as the beam is rocked, the probe explores the specimen. This plane may also contain a real objective aperture or the latter may be replaced by a 'virtual' objective aperture in a conjugate plane closer to the source. Both of these planes will be conjugate to the plane of the dark-field detector.

In the selected-area diffraction mode, the source is no longer conjugate to the specimen but is instead conjugate to the real objective aperture (front-focal plane of the objective). By raising the pivot point of the scanning coils to the selected-area aperture plane, which is conjugate to the specimen, a parallel beam will be rocked about the axis while illuminating an (approximately) fixed area of the specimen.

The various images are thus formed sequentially and are of different kinds. Owing to the general rule that inelastically scattered electrons are deflected, on average, through smaller angles than elastically scattered electrons (see Part XIV), a crude but often very effective separation is performed by the annular dark-field detector in the imaging mode; the central opening is chosen to allow the electrons undeflected by the specimen to pass through, together with a substantial fraction of those scattered inelastically. The dark-field detector then collects mostly electrons that have been scattered elastically by each pixel in turn; the resulting dark-field signal is used to modulate the intensity of a monitor scanned in synchronism with the STEM. The electrons that pass through the opening in the dark-field detector are dispersed by a prism and can then be used to form energy-filtered images or for EELS. If a very small bright-field detector is employed, the image formation will prove to be closely analogous to that in the TEM. We shall also see that the fact that the detector geometry can be chosen freely offers new and interesting possibilities in the bright-field mode.

We note, without further comment, that the design of the lens system between the specimen and the detectors is very important, both to match the information emerging from the object to the detector geometry and, more recently, to enable the electrons scattered through large angles to be collected and exploited efficiently.

We conclude this short chapter with a reminder that the scanning electron microscope, like the STEM, produces the information that is used to generate its numerous images sequentially. We shall not need to refer to its optics, which is in any case basically very simple; the problems for the designer arise from the difficulty of situating so many detectors in so confined a space. We shall, however, often have occasion to mention image processing techniques that are of particular interest for the SEM in Part XV.

65

Fundamentals of Transfer Theory

In earlier chapters, we have developed the tools that are needed to construct a theory of image formation: a wave equation that describes electron propagation and the corresponding diffraction theory. We now reconsider electron wave propagation from a rather different standpoint. Throughout this chapter, we assume that the waves are monoenergetic; the effect of abandoning this assumption will be examined in Chapter 66.

65.1 The integral transformation

The most important feature of the wave equation in all its various forms is its *linearity*. This implies that, for any pair of particular solutions $\psi_1(\mathbf{r})$ and $\psi_2(\mathbf{r})$, the linear combination

$$\psi(\mathbf{r}) = c_1\psi_1(\mathbf{r}) + c_2\psi_2(\mathbf{r}) \tag{65.1a}$$

with arbitrary complex coefficients c_1 and c_2 is also a solution. This *superposition principle* can be cast into the following more general forms:

$$\psi(\mathbf{r}) = \sum_{n=1}^{N} c_n\psi_n(\mathbf{r}) \tag{65.1b}$$

$$\psi(\mathbf{r}) = \sum_{m=1}^{M}\sum_{n=1}^{N} c_{mn}\psi_{mn}(\mathbf{r}) \tag{65.1c}$$

$$\psi(\mathbf{r}) = \int_{\alpha_1}^{\alpha_2} c(\alpha)\hat{\psi}(\mathbf{r},\alpha)\,d\alpha \tag{65.2a}$$

$$\psi(\mathbf{r}) = \int_{\alpha_1}^{\alpha_2}\int_{\beta_1}^{\beta_2} c(\alpha,\beta)\hat{\psi}(\mathbf{r},\alpha,\beta)\,d\alpha\,d\beta \tag{65.2b}$$

The variables of integration α and β are parameters with respect to all spatial operations.

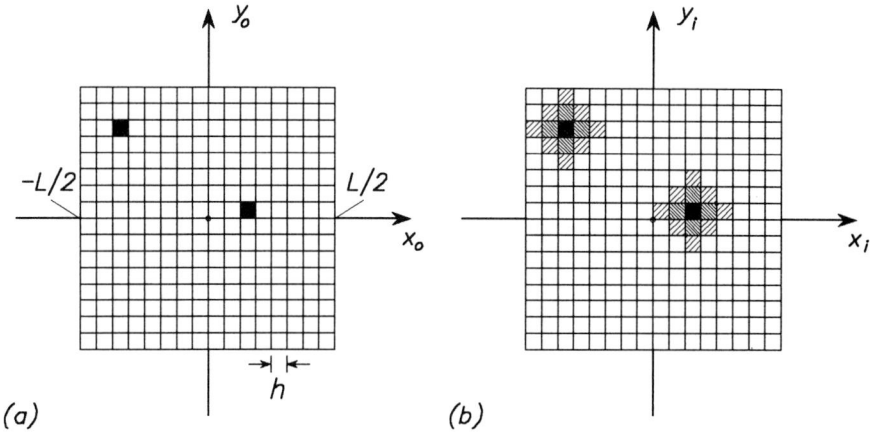

Fig. 65.1: Concepts of image processing. The object (a) and the image (b) are divided into a large number (here $N^2 = 256$) of square elements of area (pixels). The image is transformed back to object size and orientation. The information relating to each object pixel is not only stored in the corresponding image pixel but is also spread over its neighbourhood.

These superposition rules all refer to the same position $\boldsymbol{r}$. Our aim now is to find superposition rules that tell us how contributions from different points $\boldsymbol{r}_o$ add up to produce the wavefunction at a point $\boldsymbol{r}_i$.

It proves convenient to approach this problem by dividing the object and image areas into discrete zones, as shown in Fig. 65.1, thereby introducing the first stage of digital image processing, sampling. Consider a square of side $L \times L$ in the object plane, $z = z_o$, and divide it into N^2 identical subsquares of side $h = L/N$. These are labelled with two indices m and n running from 1 to N; the coordinates of the centres of the subsquares are thus

$$x_{mn}^{(0)} = -\frac{L}{2} + \left(m - \frac{1}{2}\right)h, \quad y_{mn}^{(0)} = -\frac{L}{2} + \left(n - \frac{1}{2}\right)h \qquad (65.3)$$

Next, we consider wavefunctions $\psi_{mn}(\boldsymbol{r})$ with the following values:

$$\psi_{mn}(x_o, y_o, z_o) = \begin{cases} 1 & \text{for } |x_o - x_{mn}^{(0)}| < h/2 \ |y_o - y_{mn}^{(0)}| < h/2 \\ 0 & \text{otherwise} \end{cases} \qquad (65.4)$$

Each wavefunction is thus normalized to unity within the element of area (m,n) and vanishes in the remainder of the object plane. This could be regarded as the starting condition of a wave propagation problem. Together with a reasonable assumption about the normal derivative,

$$\left(\frac{\partial \psi_{mn}}{\partial z}\right)_{r_o} = ik_{mn}\psi_{mn}(r_o) \tag{65.5}$$

for example, we have a fully defined problem that could be solved by means of diffraction integrals. This is not, however, the most favourable way, as we shall see later.

We can superimpose all these particular solutions, as in (65.1c), with weights

$$c_{mn} = \psi(x_{mn}^{(0)}, y_{mn}^{(0)}, z_o) \tag{65.6}$$

and $M = N$, thereby obtaining an approximate solution, which is piecewise constant in the object plane. This is indeed how image processing programs operate; the value of N is usually a power of 2, typically $N = 256 = 2^8$ or $N = 512 = 2^9$, since this suits both computer architectures and various widely used algorithms.

The representation (65.1c) together with (65.6) is an approximate form of a *continuous* superposition of the general form (65.2b), to which it tends if we keep L fixed and let N increase indefinitely. We may then identify α and β with x_o and y_o, $\alpha_1 = \beta_1 = -L$ and $\alpha_2 = \beta_2 = L$ and $c_{mn} \to \psi(x_o, y_o, z_o)$. The particular solution $\hat{\psi}$ is then the *integral kernel* $\hat{G}$ in

$$\psi(x, y, z) = \iint \hat{G}(x, y, z; x_o, y_o, z_o)\psi(x_o, y_o, z_o)\, dx_o\, dy_o \tag{65.7a}$$

This integral kernel can be identified with the one that appears in the diffraction theory but, in practice, (65.7a) and its special case (65.7b) below are rarely used for calculations and serve only as basis for the theoretical framework.

The plane $z = z_o$ is usually identified with the object plane, as we have already done, and $z = z_i$ is the image plane. Nevertheless, (65.7a) can of course be used to relate any two planes and we shall see that a form of it is needed for other pairs of planes when we consider coherence theory in Part XVI. It is often unnecessary to include z explicitly among the arguments of the wavefunction; specific planes are then indicated by a suffix, ψ_o and ψ_i for example. Equation (65.7a) then takes the form

$$\hat{\psi}(\hat{x}_i, \hat{y}_i) = \iint_{-\infty}^{\infty} \hat{G}(\hat{x}_i, \hat{y}_i; x_o, y_o)\psi_o(x_o, y_o)\, dx_o\, dy_o \tag{65.7b}$$

We can extend the limits of integration formally to infinity by assuming that $\psi_o \equiv 0$ downstream from the specimen outside the region occupied by the latter.

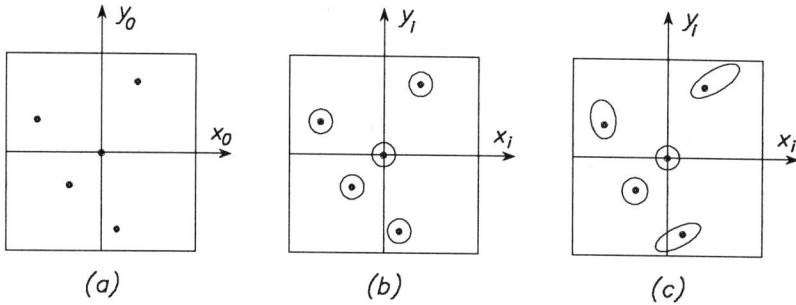

Fig. 65.2: Illustration of the concept of isoplanatism. (a) Object consisting of five points. (b) Image in the isoplanatic approximation: the aberration disks are all equal in size and shape, their centres are unshifted. (c) Image with violation of the isoplanatism: the aberration disks may alter in size, shape and orientation, their centres are shifted (distortion). These effects increase with increasing distance from the optic axis (where $x_i = y_i = 0$).

65.2 Isoplanatism and Fourier transforms

In certain conditions, the double integral in (65.7b) takes the form of a convolution. This suggests that we should examine its Fourier transform since the latter converts convolution products into direct products. For our present purposes, we shall not consider any other approach but we note in passing that when the functions ψ_o and $\hat{G}$ have appreciably different supports (that is, $\hat{G}$ is effectively non-zero over a much smaller area than ψ_o), direct convolution may be preferable to two Fourier transforms and a direct product.

The condition in which (65.7b) reduces to a convolution is that the system be *isoplanatic*. This implies in our case that the object is so small that all off-axis lens aberrations can be ignored in the image-forming process and that only the aberrations that do not vanish for $x_o = y_o = 0$ need be retained (Fig. 65.2). This is not the original definition of isoplanatism, but is a consequence of the more fundamental requirement that $\hat{G}$ may be written in the form

$$\hat{G}(\hat{x}_i, \hat{y}_i; x_o, y_o) = G(x_i - x_o, y_i - y_o) \tag{65.8}$$

which we now impose. For a careful discussion, see Dumontet (1955a). The coordinates $\hat{x}_i$, $\hat{y}_i$ are the natural image coordinates, measured in the same laboratory frame as x_o, y_o, while x_i, y_i are *scaled* image coordinates, related to $\hat{x}_i$, $\hat{y}_i$, by

$$x_i + iy_i = (\hat{x}_i + i\hat{y}_i)/M(z_i, z_o) \tag{65.9}$$

in which $M(z_i, z_o)$ is the complex magnification (15.45–46). We recall that

$$M(z_i, z_o) = M \exp\{i(\theta_i - \theta_o)\}$$

and that M is an algebraic quantity (negative for single-stage imaging). Equation (65.9) means that the image is referred back to the object scale and orientation.

When $\hat{G}$ may be replaced by G (65.8–9), Equation (65.7b) acquires the form of a *convolution* integral or *convolution product*. Such integrals are mapped into *direct products* by Fourier transformation and it is for this reason that Fourier transform techniques are ubiquitous in image analysis. The Fourier transform is not, however, the only operator that maps convolutions into a simpler type of product and we shall mention the alternatives briefly in Section 66.5.5.

The Fourier transforms are of course all two-dimensional. We first recall the one-dimensional form with 2π in the exponent, since this extends immediately to two dimensions for any pair of functions $u(x)$ and $U(q)$ satisfying the necessary mathematical requirements (essentially that the integral of their squared modulus remains finite, $u, U \in L^2$).

We have

$$u(x) =: \mathcal{F}(U) = \int_{-\infty}^{\infty} U(q) \exp(2\pi i q x)\, dq \qquad (65.10a)$$

with inverse*

$$U(q) =: \mathcal{F}^-(u) = \int_{-\infty}^{\infty} u(x) \exp(-2\pi i q x)\, dx \qquad (65.10b)$$

Then for any two functions $u(x)$, $v(x)$ with transforms $U(q)$ and $V(q)$, the convolution theorem tells us that for

$$w(x) := \int_{-\infty}^{\infty} u(x - x')v(x')\, dx' \equiv \int_{-\infty}^{\infty} u(x')v(x - x')\, dx'$$
$$\qquad (65.11a)$$
$$=: u(x) * v(x) \equiv v(x) * u(x)$$
or
$$=: (u * v)(x) \equiv (v * u)(x)$$

we have

$$W(q) = U(q)V(q) \qquad (65.11b)$$

---

* The inverse Fourier transform operator is the adjoint of the direct operator but as a mnemonic aid, we denote the inverse operator by $FT^-(\cdot)$ or $\mathcal{F}^-(\cdot)$ as a reminder that a minus sign appears in the exponent.

In imaging applications, in electron optics as elsewhere, all the Fourier and convolution integrals become two-dimensional. The role of the coordinate x is taken over by the position vector in the object plane, $\boldsymbol{u}_o = (x_o, y_o)$ and that of the conjugate variable q by the spatial frequency vector $\boldsymbol{q}$ with components (q_x, q_y) or occasionally (p, q) when we need to add further suffixes. Like q, q_x and q_y have the dimensions of reciprocal length. The simple product qx in the Fourier transform exponents is replaced by the scalar product $\boldsymbol{q} \cdot \boldsymbol{u}_o = q_x x_o + q_y y_o$.

It is convenient to introduce the *scaled* image coordinates (x_i, y_i), defined by (65.9) not only in the integrand of (65.7b) but also as arguments on the left-hand side. Summarizing, we write

$$
\begin{aligned}
\boldsymbol{u}_o &= (x_o, y_o) \\
\boldsymbol{q} &= (q_x, q_y) = (p, q) \\
\boldsymbol{u}_i &= (x_i, y_i)
\end{aligned}
\tag{65.12}
$$

The convolution integral then takes the compact form

$$
\begin{aligned}
\psi_i(\boldsymbol{u}_i) &= \int_{-\infty}^{\infty} G(\boldsymbol{u}_i - \boldsymbol{u}_o)\psi_o(\boldsymbol{u}_o)\,d\boldsymbol{u}_o \\
&= (G * \psi_o)(\boldsymbol{u}_i)
\end{aligned}
\tag{65.13}
$$

in which

$$
\psi(\boldsymbol{u}_i) := \hat{\psi}(\hat{\boldsymbol{u}}_i)
\tag{65.14}
$$

It is important to note the scale change implicit in this definition, which must not be forgotten when deriving explicit object–image relations.

Several Fourier transforms will occur so frequently that we introduce a special notation for them. In particular, we define the *object spectrum*

$$
\begin{aligned}
S_o(\boldsymbol{q}) :&= \mathcal{F}^-(\psi_o) \\
&= \int_{-\infty}^{\infty} \psi_o(\boldsymbol{u}_o)\exp(-2\pi i \boldsymbol{q} \cdot \boldsymbol{u}_o)\,d\boldsymbol{u}_o
\end{aligned}
\tag{65.15a}
$$

the *image spectrum*

$$
S_i(\boldsymbol{q}) := \mathcal{F}^-(\psi_i)
\tag{65.15b}
$$

and the *instrumental wave transfer function*

$$
T(\boldsymbol{q}) := \mathcal{F}^-(G)
\tag{65.15c}
$$

From (65.13), therefore, we have

$$S_i(\boldsymbol{q}) = T(\boldsymbol{q})S_o(\boldsymbol{q}) \tag{65.16}$$

in which we have used the two-dimensional form of the convolution theorem (65.11a–b).

The intensity distribution in the image plane may now be calculated efficiently by the following sequence:

(i) We define an object wavefunction $\psi_o(\boldsymbol{u}_o)$, usually in terms of a theoretical model (Part XIV), and perform the first Fourier transform (65.15a) with the aid of one of the various fast-Fourier-transform (FFT) algorithms.

(ii) The functions T and S_o are then multiplied (65.16). The transfer function $T(\boldsymbol{q})$ is not obtained by transforming G numerically since it can be established directly in closed form by theoretical arguments, as we shall see in the next section.

(iii) We form the inverse of (65.15b),

$$\psi_i(\boldsymbol{u}_i) = \int\limits_{-\infty}^{\infty} S_i(\boldsymbol{q})\exp(2\pi i\boldsymbol{q}\cdot\boldsymbol{u}_i)\,d\boldsymbol{q} \tag{65.17}$$

and hence obtain $|\psi_i(\boldsymbol{u}_i)|^2$, which is proportional to the intensity distribution in the image.

This sequence is shown in Fig. 65.3, which brings out the analogy with electronic signal processing. The transfer function $T(\boldsymbol{q})$ acts as a *complex filter*, which modifies the object spectrum prior to the recording step. Provided that $T(\boldsymbol{q})$ nowhere vanishes, a condition that will be investigated later (and shown not to be satisfied in important cases), the whole procedure from ψ_o to ψ_i is *invertible*. It is in the last step, in which $|\psi_i|^2$ is recorded, that the wave phase, $\arg(\psi_i)$, is irretrievably lost. This makes direct interpretation of recorded images unreliable whenever the phase is important. We shall see in the following chapters how this difficulty can be overcome or at least palliated. The holographic solution has already been presented in Chapter 63.

65.3 The wave transfer function

The transfer calculus presented schematically in Fig. 65.3 stands or falls with the assumption (65.8), namely, that the arguments of $\hat{G}$ only appear as the differences $\boldsymbol{u}_i - \boldsymbol{u}_o$ (with the scaling of 65.9). This assumption must now be justified. An extremely useful by-product of this calculation will

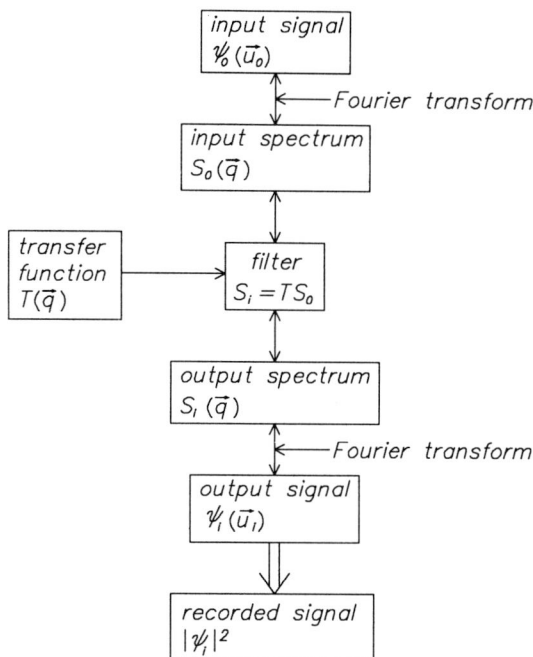

Fig. 65.3: Image formation as a linear filter process with a nonlinear recording step.

be a simple analytical formula for the instrumental wave transfer function, $T(\boldsymbol{q})$.

We set out from the situation shown in Fig. 59.5, where a specimen is located close to the object focal plane of an electron lens and hence highly magnified. We study the diffraction at the asymptotic aperture (O, A), described by (59.40). The notation of the present chapter is related to that of (59.40) as follows:

$$\psi_Q(x_Q, y_Q) \equiv \psi_o(x_o, y_o) \equiv \psi_o(\boldsymbol{u}_o)$$
$$\psi(\boldsymbol{r}_i) \equiv \hat{\psi}(\hat{x}_i, \hat{y}_i) \tag{65.18}$$
$$(x_a, y_a) \equiv (\hat{x}_a, \hat{y}_a) \equiv \hat{\boldsymbol{u}}_a$$

The use of the original (unscaled) image coordinates $\hat{\boldsymbol{u}}_i$ is unfavourable and we therefore introduce the scaled, rotated coordinates $\boldsymbol{u}_i$ with the aid of (65.9). Since $(\hat{x}_i^{(0)}, \hat{y}_i^{(0)})$ are the coordinates of the Gaussian image point, the introduction of (65.9) gives $x_i^{(0)} = x_o$, $y_i^{(0)} = y_o$. It is likewise advantageous to define *rotated* aperture coordinates,

$$x_a + iy_a := (\hat{x}_a + i\hat{y}_a)\exp(-i\theta) \tag{65.19a}$$

and a corresponding two-dimensional vector

$$\boldsymbol{u}_a = (x_a, y_a) \tag{65.19b}$$

The scalar product in (59.40b) is invariant with respect to this rotation and the diffraction integral takes the more compact form

$$\psi_i(\boldsymbol{u}_i) = \frac{c}{b} \int_O \int_A \psi_o(\boldsymbol{u}_o) \exp(-ikD) \, d\boldsymbol{u}_o \, d\boldsymbol{u}_a \tag{65.20a}$$

with

$$D = \overline{W}(\boldsymbol{u}_o, \boldsymbol{u}_a) - \frac{|M|}{b} \boldsymbol{u}_a \cdot (\boldsymbol{u}_i - \boldsymbol{u}_o) \tag{65.20b}$$

(In order not to overburden the notation, we have simply written $\overline{W}(\boldsymbol{u}_o, \boldsymbol{u}_a)$ here. Formally, we should have stated that $\overline{W}(\boldsymbol{r}_Q, \boldsymbol{r}_a)$ occurring in (59.40b) becomes $\overline{W}(\boldsymbol{u}_o, \hat{\boldsymbol{u}}_a)$ in our present notation and that this in turn becomes $\hat{W}(\boldsymbol{u}_o, \boldsymbol{u}_a)$ after making the scale change (65.19a).)

This expression must now be reconciled with (65.17). Comparison of the terms containing $\boldsymbol{u}_i$ in the exponent gives

$$2\pi i \boldsymbol{q} \cdot \boldsymbol{u}_i = \frac{ik|M|}{b} \boldsymbol{u}_a \cdot \boldsymbol{u}_i$$

Since this must hold for any pair of vectors $\boldsymbol{u}_a$ and $\boldsymbol{u}_i$, the spatial frequency vector $\boldsymbol{q}$ must be related to $\boldsymbol{u}_a$ by

$$\boldsymbol{q} = \frac{|M|}{\lambda b} \boldsymbol{u}_a \tag{65.21a}$$

in which we have used $k = 2\pi/\lambda$. At high magnification, $b/|M| \approx f$ and so

$$\boldsymbol{q} = \boldsymbol{u}_a / \lambda f \tag{65.21b}$$

We have $|\boldsymbol{u}_a| = r_a$ and usually $M < 0$ so that with the aid of (60.54) and the relation $g_o \lambda_o = g_i \lambda$ we can express $\boldsymbol{q}$ in terms of angular coordinates at the specimen. We find

$$\boldsymbol{q} = \frac{\theta_o}{\lambda_o} (\cos \varphi_o, \sin \varphi_o) \tag{65.21c}$$

and $q = |\boldsymbol{q}| = \theta_o/\lambda_o$. Formulae (65.21b,c) have the merit of simplicity and are almost invariably used in the literature of the subject.

It must also be possible to reorganize all terms in (65.20a,b) that depend on $\boldsymbol{u}_o$ in the form of (65.15a). Clearly, this cannot be done unless $\overline{W}$ is independent of $\boldsymbol{u}_o$ and depends only on $\boldsymbol{u}_a$:

$$\overline{W} =: W(\boldsymbol{u}_a) =: W'(\boldsymbol{q}) \tag{65.22}$$

which is the appropriate form of the *isoplanatic approximation*. Equation (65.20a) thus becomes

$$\psi_i(\boldsymbol{u}_i) = \frac{cb\lambda^2}{M^2} \int S_o(\boldsymbol{q}) \exp\{-ikW'(\boldsymbol{q})\} \exp(2\pi i \boldsymbol{q} \cdot \boldsymbol{u}_i) \, d\boldsymbol{q} \tag{65.23}$$

which is in agreement with (65.16) and (65.17) if we identify the transfer function $T(\boldsymbol{q})$ with the term in $W'(\boldsymbol{q})$:

$$T(\boldsymbol{q}) = T_o \exp\{-ikW'(\boldsymbol{q})\} \tag{65.24}$$

The value of the constant T_o will be determined later but we note that for a completely open diaphragm, it will simply be unity.

The exponential term in (65.24) can be cast into a more convenient form: the constant $k = 2\pi/\lambda$ refers to the image whereas the wavelength λ_o in (65.21) refers to the object. We therefore introduce a final change of scale and define the wave aberration by

$$W'(\boldsymbol{q}) =: \frac{\lambda}{\lambda_o} W(\lambda_o \boldsymbol{q}) \equiv \frac{\lambda}{\lambda_o} W(\theta_o \cos\varphi_o, \theta_o \sin\varphi_o) \tag{65.25}$$

so that, for the completely transparent diaphragm, we obtain the standard form

$$T_L(\boldsymbol{q}) := \exp\left\{-\frac{2\pi i}{\lambda_o} W(\lambda_o \boldsymbol{q})\right\} \tag{65.26}$$

We now determine the value of T_o for this case. A pure phase factor of the form $\exp(i\alpha)$ being irrelevant, we can assume from the outset that T_o is a positive constant. From the continuity equation, it can be seen that the total (integrated) intensity,

$$I = \int |\psi_o|^2 \, d\boldsymbol{u}_o = \frac{\lambda_o}{\lambda} \iint |\hat{\psi}_i|^2 d\hat{x}_i \, d\hat{y}_i$$

is conserved for a wholly transparent diaphragm. The factor λ_o/λ is inconvenient and since we have already made one scale change (65.9), we introduce an intensity scale that eliminates this factor and hence normalizes the intensity integral I; we then have the very convenient relation

$$\int |\psi_o|^2 \, d\boldsymbol{u}_o = \int |\psi_i|^2 \, d\boldsymbol{u}_i = 1 \tag{65.27}$$

Parseval's theorem then tells us that

$$\int |S_o(\boldsymbol{q})|^2 \, d\boldsymbol{q} = \int |S_i(\boldsymbol{q})|^2 \, d\boldsymbol{q} = 1 \qquad (65.28)$$

Recalling that $S_i = T_L S_o$ and hence $|S_i| = |T_L| \cdot |S_o|$, this can only be generally true if $|T_L(\boldsymbol{q})| = 1$. We thus recover (65.26) and confirm that $T_o = 1$ for a completely open diaphragm.

We have retained the distinction between λ and λ_o for generality but, in practice, the lenses used for high-resolution imaging are virtually always magnetic. We then have $\lambda = \lambda_o$ and the conservation of the norm of ψ follows directly, without an additional change of scale.

65.4 Explicit formulae

The wave aberration terms that are to be considered in the isoplanatic approximation for a lens system with at worst small departures from rotational symmetry are as follows:

$$W = \tfrac{1}{4} C_s \theta_o^4 \qquad\qquad \text{spherical aberration}$$

$$-\tfrac{1}{2} \Delta_o \theta_o^2 \qquad\qquad \text{defocus}$$

$$+\theta_o (d_1 \cos \varphi_o + d_2 \sin \varphi_o) \qquad\qquad \text{deflection axial coma}$$

$$+\tfrac{1}{3} \theta_o^3 (c_1 \cos \varphi_o + c_2 \sin \varphi_o) \qquad\qquad \text{cubic axial coma}$$

$$+\tfrac{1}{2} \theta_o^2 (a_1^{(2)} \cos 2\varphi_o + a_2^{(2)} \sin 2\varphi_o) \qquad \text{twofold axial astigmatism}$$

$$+\tfrac{1}{3} \theta_o^3 (a_1^{(3)} \cos 3\varphi_o + a_2^{(3)} \sin 3\varphi_o) \qquad \text{threefold axial astigmatism}$$

$$(65.29)$$

(cf. 60.55 and 60.72). With the aid of (65.21), we obtain $W(\lambda_o \boldsymbol{q})$ and hence $T_L(\boldsymbol{q})$ from (65.26).

The parasitic terms, those depending on φ_o, can be made arbitrarily small by careful alignment of the microscope, at least in principle. In practice, they are reduced as far as technically possible but the practical limits are not known beforehand and depend heavily on the skill and experience of the microscopist. Computer-aided methods of alignment are gradually coming into use (Section 77.3) and, even if these are not perfect, they can be expected to give reproducible results at least. Be this as it may, we can hardly hope to draw any general conclusions from these terms and we hence disregard them forthwith, though we shall occasionally return to the axial astigmatism term.

The surviving contributions are therefore the spherical aberration and the defocus, corresponding to the distance Δ_o between the specimen and the plane conjugate to the (fixed) image plane. For these, $T_L(\boldsymbol{q})$ becomes

$$T_L(\boldsymbol{q}) = \exp\left(-\frac{\mathrm{i}\pi\lambda_o^3}{2}C_s q^4 + \mathrm{i}\pi\lambda_o\Delta_o q^4\right)$$

$$= \exp 22\left\{-\frac{\mathrm{i}\pi}{2}C_s\lambda_o^3(q_x^2 + q_y^2)^2 + \mathrm{i}\pi\lambda_o\Delta_o(q_x^2 + q_y^2)\right\}$$

$$= \exp\left\{-\frac{2\pi\mathrm{i}}{\lambda_o}\left(\frac{1}{4}C_s\lambda_o^4 q^4 - \frac{1}{2}\Delta_o\lambda_o^2 q^2\right)\right\}$$

$$=: \exp\{-\mathrm{i}\chi(\boldsymbol{q})\} \tag{65.30}$$

$$\chi(\boldsymbol{q}):= \pi\left(\frac{1}{2}C_s\lambda_o^3 q^4 - \Delta_o\lambda_o q^2\right)$$

This function depends on three parameters, C_s, λ_o and Δ_o. It is often convenient* to scale distances perpendicular to the axis relative to $(C_s\lambda_o^3)^{\frac{1}{4}}$ and those along the axis relative to $(C_s\lambda_o)^{\frac{1}{2}}$, thereby reducing the number of degrees of freedom and permitting a simpler graphical representation of the various functions that we shall be encountering. We introduce the reduced spatial frequency

$$\boldsymbol{Q}:= (C_s\lambda_o^3)^{\frac{1}{4}}\boldsymbol{q} \quad , \quad Q = |\boldsymbol{Q}| \tag{65.31}$$

and the reduced defocus

$$D:= \Delta_o/(C_s\lambda_o)^{\frac{1}{2}} \tag{65.32}$$

The function $T_L(\boldsymbol{q})$ simplifies to

$$T_L(\boldsymbol{q}) \rightarrow \hat{T}_L(Q) = \exp\left\{\mathrm{i}\frac{\pi}{2}(2DQ^2 - Q^4)\right\} \tag{65.33}$$

The simplest special case, $D = 0$, is shown in Fig. 65.4; we shall see later that this is not an experimentally attractive value of D. For $Q \leq 1$, the real and imaginary parts of $\hat{T}_L$ vary fairly slowly but, for $Q > 1$, they start to oscillate with increasing speed. The practical consequences of this

---

* The use of this scaling is invaluable in practice since results obtained for a particular microscope (given C_s and accelerating voltage) can be easily transferred to any other. It has been recommended (Hawkes, 1980b) that the scaling factors be given names, and this usage is gradually being adopted. If we write $(C_s\lambda)^{1/2} =: 1\,\text{scherzer} =: 1\,\text{Sch}$, $(C_s\lambda^3)^{1/4} =: 1\,\text{glaser} =: 1\,\text{Gl}$ then we can speak of a defocus of 1.5 Sch, for example or a spatial frequency cutoff at 0.5 Gl.

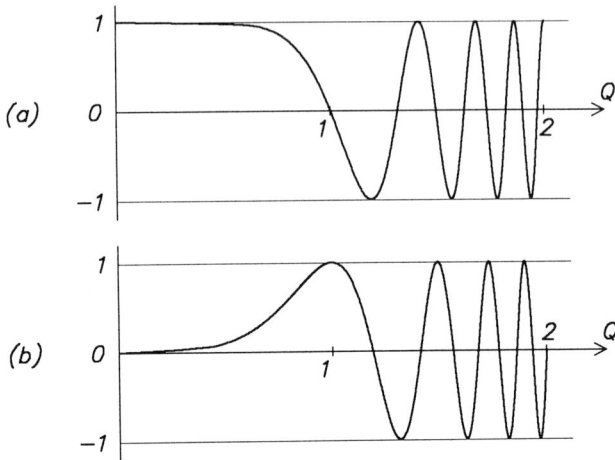

Fig. 65.4: (a) Real and (b) imaginary parts of the wave transfer function for vanishing defocus, $\hat{T}_L(Q) = \exp(-\mathrm{i}\pi Q^4/2)$. Curve (b) shows $-\Im(\hat{T}_L)$.

and the behaviour of $\hat{T}_L$ for other choices of D will be examined in the next chapter.

So far, we have dealt exclusively with a completely transparent aperture plane, which has no effect on the electrons. This is unrealistic for, in all practical devices, apertures with a finite bore radius are introduced to confine the beam. What factor must we include in $T(\boldsymbol{q})$ to characterize this situation?

This factor will have the same structure as the object function given by (60.4): the laws of electron propagation that we have derived relate any two planes and we are at liberty to place the 'object' plane at the aperture and regard the aperture transparency as the 'object function'. The object coordinates (x_o, y_o) are thus to be replaced by the aperture coordinates transformed with the aid of (65.21a). We thus obtain an *aperture factor*

$$T_A(\boldsymbol{q}) = \exp\left\{-\sigma_A(\boldsymbol{q}) + \mathrm{i}\eta_A(\boldsymbol{q})\right\}$$
$$= \exp\left\{-\sigma_A\left(\frac{|M|}{\lambda b}\boldsymbol{u}_A\right) + \mathrm{i}\eta_A\left(\frac{|M|}{\lambda b}\boldsymbol{u}_A\right)\right\} \qquad (65.34)$$

several examples of which are shown in Figs. 65.5–7. The complete transfer function is then

$$T(\boldsymbol{q}) = T_L(\boldsymbol{q})T_A(\boldsymbol{q}) \qquad (65.35)$$

We have already considered the case in which no aperture is present, for which σ_A and η_A are zero everywhere and hence $T_A(\boldsymbol{q}) \equiv 1$. The most

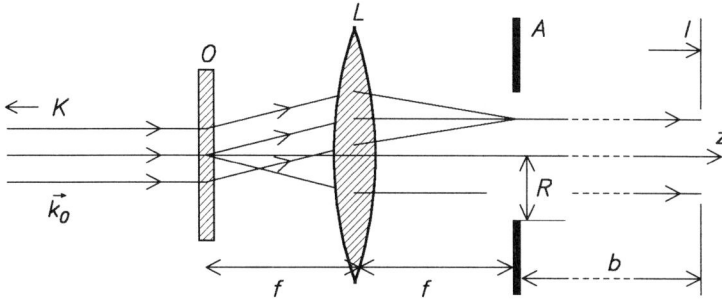

Fig. 65.5: Effect of a simple hole as an aperture
K: condenser (far left), $\mathbf{k}_0$: wave vector of the incident beam, O: object plane, L: lens,
A: aperture, I: image screen, far to the right, f: focal length of the lens, b: distance
$\overline{AI}, b \gg f$, R: radius of the circular opening.

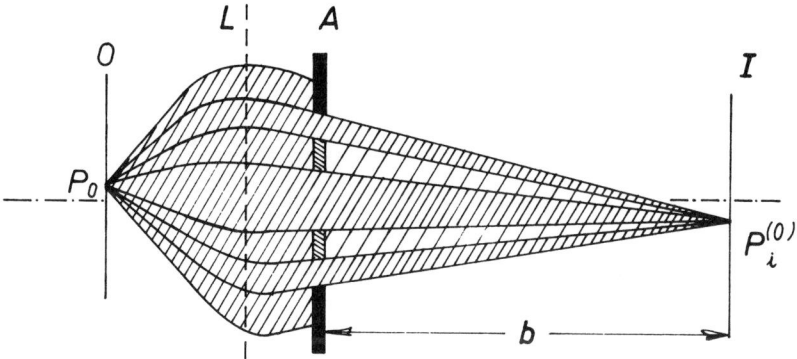

Fig. 65.6: The general possibilities of beam modification by an aperture diaphragm
having open zones and semitransparent zones enclosed in an opaque screen. For simplicity,
only one object point P_o and the corresponding image point $P_i^{(o)}$ are considered.

important special case is a round hole in a completely opaque screen, to
which the apertures in an electron microscope correspond closely. We then
write

$$\sigma_A = 0, \quad \eta_A = 0 \qquad \to T_A = 1 \quad : \text{open parts}$$
$$\sigma_A = \infty, \quad \eta_A \text{ arbitrary} \quad \to T_A = 0 \quad : \text{opaque parts}$$

For a circular aperture of radius R centred on the axis, we obtain (Fig. 65.5)

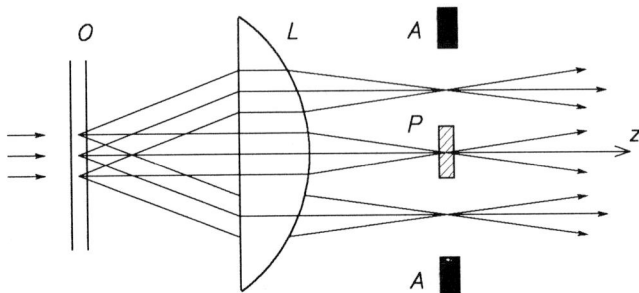

Fig. 65.7: Zernike's phase-contrast procedure in light microscopy as an example of a nontrivial aperture function $T_A(q)$. The object O is a pure 'phase object', common in biology, which means that the specimen shifts only the phase of the incident wave and does not attenuate the amplitude of the light. The contrast plate P in the aperture plane diminishes the intensity of the central undiffracted beam and shifts its phase by $\pi/2$, thereby turning the phase differences into amplitude differences. The aperture function is: $T_A(q) = \begin{cases} -\mathrm{i}\exp(-\sigma_A) & \text{in } P \\ 1 & \text{in the open ring} \\ 0 & \text{on the opaque screen} \end{cases}$ The eyepiece is located far to the right and is not shown in the figure.

$$T_A(\boldsymbol{q}) = \begin{cases} 1 & q \le |M|R/\lambda b = R/\lambda f \\ 0 & q > |M|R/\lambda b = R/\lambda f \end{cases} \qquad (65.36)$$

The function $T_L(\boldsymbol{q})$ of (65.30) is simply cutoff at the value $q_c = |M|R/\lambda b$. In Chapter 66, we shall discuss the optimum value of the cutoff value, which can be chosen freely by selecting the aperture radius appropriately.

66
The Theory of Bright-field Imaging

The transfer theory established in Chapter 65 is incomplete in several respects. We have studied the linear relation between the wavefunctions in the object plane and in the image plane, whereas only the intensity, the squared modulus of the wavefunction, can be detected. Furthermore, we have considered only perfectly monochromatic waves, but in practice the illumination will always have a finite energy spectrum. Finally, we have invariably regarded the incident wave as plane or spherical. Although this is not an obvious limitation, for an appropriate set of plane waves can always be superimposed to reproduce more complicated incident wave patterns, we need to know how such a superposition interacts with the effects due to a finite energy spectrum. We now consider these various points in turn.

In the general case, the loss of the phase of the wavefunction in the recording step, which is only sensitive to $|\psi_i|^2$, makes it impossible to reconstruct the object wavefunction from its measured image intensity distribution, as we saw in Section 65.2. There are, however, situations in which reconstruction is, at least approximately, possible. Some of these involve unconventional imaging modes, of which the most successful is off-axis holography using an electron biprism; this is described in detail in Section 63.2. A very important case is bright-field imaging in a conventional transmission electron microscope of a certain class of specimen, and we now examine this in detail. We first consider the simpler case in which the waves are assumed to be monoenergetic and plane, travelling in the axial direction. The effects of finite energy spectrum and a spread of incident directions are examined in Sections 66.2–3. At the end of Section 66.2, we show that these calculations would have been much simpler if we could have assumed at the outset that the effects were independent. In the closing section of this chapter, we consider the representation of tilted illumination in transfer theory, hollow-cone illumination and some other less important aspects of transfer function theory.

66.1 Image contrast for weak specimens

Figure 66.1 shows a transparent foil in the object plane of a transmission electron microscope being imaged at high magnification. In this

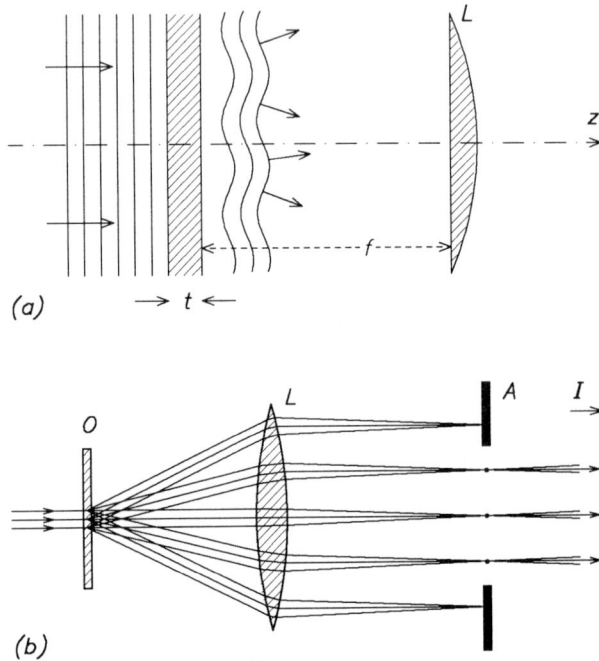

Fig. 66.1: Principle of bright-field imaging. (a) Distorted waves behind a thin transparent foil located close to the front focal plane of the lens L. The effects are highly exaggerated; the thickness $t \ll f$. (b) Wave propagation as Fraunhofer diffraction, only the trajectories corresponding to a few directions are drawn. The aperture A acts selectively on the diffracted wavelets. Here only the 'zero-order' and first two 'side reflections' can pass through it.

section, we assume that a *plane wave* is incident in the axial direction from the condenser system. The Fourier transform of the wave emerging from the specimen may be pictured vividly by recalling that its squared modulus is the same as the Fraunhofer diffraction pattern of the specimen. The diffracted, locally plane, partial waves are focused in the aperture plane, where all waves travelling in the same direction are brought to a point. The aperture then selects some wavelets and excludes those beyond the spatial-frequency cutoff value (65.36).

With no loss of generality, we may normalize the amplitude of the incident plane wave to unity and set the z-coordinate origin at the entrance plane of the specimen; in this plane, the incident wave is then simply $\psi = 1$. Its passage through the thin specimen causes a decrease $\exp\{-\sigma_o(x_o, y_o)\}$ of the amplitude and a phase shift $\exp\{i\eta_o(x_o, y_o\}$, as in (60.4). These are *total* interaction effects, obtained by integration along the classical electron paths through the specimen. More details will be found in Part XIV. A *weakly scattering specimen* is characterized by $\sigma_o \ll 1$ and $|\eta_o| \ll 1$, so that

$$\psi_o(x_o, y_o) = \exp(i\eta_o - \sigma_o) \approx 1 + i\eta_o(x_o, y_o) - \sigma_o(x_o, y_o) \qquad (66.1)$$

is the wave emerging from the specimen. The 'distorted waves' of Fig. 66.1a then correspond to surfaces of equal phase and hence represent the variations of η_o. The object spectrum S_o corresponding to (66.1) is

$$S_o(\boldsymbol{q}) = \delta(\boldsymbol{q}) + i\tilde{\eta}(\boldsymbol{q}) - \tilde{\sigma}(\boldsymbol{q}) \qquad (66.2)$$

in which $\delta(\boldsymbol{q})$ denotes the Dirac δ-function in two dimensions, $\delta(\boldsymbol{q}) = \delta(q_x)\delta(q_y)$; $\tilde{\eta}$ denotes the Fourier transform of η_o and is hence the phase shift spectrum:

$$\tilde{\eta}(\boldsymbol{q}) = \mathcal{F}^-(\eta_o) = \int \eta_o(\boldsymbol{u}_o) \exp(-2\pi i \boldsymbol{q} \cdot \boldsymbol{u}_o) \, d\boldsymbol{u}_o \qquad (66.3)$$

while $\tilde{\sigma}$, the transform of σ_o, is the 'absorption' or amplitude spectrum*,

$$\tilde{\sigma}(\boldsymbol{q}) = \mathcal{F}^-(\sigma_o) = \int_{-\infty}^{\infty} \sigma_o(\boldsymbol{u}_o) \exp(-2\pi i \boldsymbol{q} \cdot \boldsymbol{u}_o) \, d\boldsymbol{u}_o \qquad (66.4)$$

For the moment, we retain the most general form of the aperture factor T_A (65.34). The spectrum of the image wavefunction is thus

$$S_i(\boldsymbol{q}) = \exp\left\{ i\eta_A(\boldsymbol{q}) - \sigma_A(\boldsymbol{q}) - \frac{2\pi i}{\lambda_o} W(\lambda_o \boldsymbol{q}) \right\} \{\delta(\boldsymbol{q}) + i\tilde{\eta}(\boldsymbol{q}) - \tilde{\sigma}(\boldsymbol{q})\} \quad (66.5)$$

On evaluating the image wavefunction $\psi_i(\boldsymbol{u}_i) = \mathcal{F}(S_i)$, we notice that the term in $\delta(\boldsymbol{q})$ simply gives the integrand at $\boldsymbol{q} = 0$. Since $W(0) = 0$, we obtain

$$\begin{aligned} \psi_i(\boldsymbol{u}_i) = {} & \exp\left\{ i\eta_A(0) - \sigma_A(0) \right\} \\ & + \mathcal{F}\{(i\tilde{\eta} - \tilde{\sigma}) \exp(i\eta_A - \sigma_A - 2\pi i W/\lambda_o)\} \end{aligned} \qquad (66.6)$$

---

* The physical interpretation of this term is not straightforward. In weakly scattering specimens, there is virtually no absorption in the sense that no incident electron is brought to a halt within the specimen. Electrons may, however, be 'lost' to the image-forming process in various ways and the absorption term is intended to account for these. For very weak specimens, the amplitude term can usually be neglected but we retain it in the theory because of its importance for thicker but still fairly weak specimens.

This formula is still general enough to include dark-field imaging ($\sigma_A(0) \to$ ∞ for central-stop dark-field) and the presence of phase-shifting devices in the aperture plane such as the Zernike phase plate ($\eta_A(0) = \pi/2$, $0 \leq \sigma_A(0) < 1$). These and other examples of intervention in the back-focal plane are examined briefly in Section 66.5. An essential feature of conventional bright-field imaging is that the neighbourhood of the optic axis is an *open* part of the aperture, so that the undiffracted wave, the partial wave that has passed through the specimen without any interaction, can reach the image plane unmodified. In these conditions, we have

$$\eta_A(0) = 0, \quad \sigma_A(0) = 0, \quad T_A(0) = 1 \tag{66.7}$$

Moreover, we now assume that the aperture is a simple circular hole centred on the optic axis (65.36). We denote this aperture by $\mathcal{A}$,

$$\mathcal{A} : q^2 = q_x^2 + q_y^2 \leq q_c^2 \tag{66.8}$$

Expression (66.6) for ψ_i now simplifies to

$$\psi_i(\boldsymbol{u}_i) =: 1 + \mu(\boldsymbol{u}_i) \tag{66.9a}$$

$$= 1 + \int_{\mathcal{A}} \{i\tilde{\eta}(\boldsymbol{q}) - \tilde{\sigma}(\boldsymbol{q})\} \exp\{-2\pi i W(\lambda_o \boldsymbol{q})/\lambda_o\}$$

$$\times \exp(2\pi i \boldsymbol{q} \cdot \boldsymbol{u}_i) d\boldsymbol{q} \tag{66.9b}$$

A consequence of our earlier assumption that σ_o and $|\eta_o|$ are both much smaller than unity is that $|\mu| \ll 1$. It is this that enables us to simplify the expression for the image intensity $|\psi_i|^2$ in such a way that the *image contrast*[*] $C(\boldsymbol{u}_i)$ becomes *linearly related* to the object functions σ_o and η_o. The contrast, here defined by

$$C(\boldsymbol{u}_i) := \frac{-|\psi_i(\boldsymbol{u}_i)|^2 + 1}{|\psi_i(\boldsymbol{u}_i)|^2 + 1}, \tag{66.10a}$$

is the most direct of the various image quality measures that we shall meet. Neglecting quadratic terms in μ, we see immediately that

$$C(\boldsymbol{u}_i) = -\Re\mu(\boldsymbol{u}_i) \tag{66.10b}$$

The image contrast for weakly scattering specimens is clearly *very low*, which is an intrinsic drawback of the bright-field technique.

---

[*] In the literature, the contrast is frequently defined as the numerator only of (66.10a). A factor 2 then appears, notably in the contrast transfer functions.

Explicit evaluation of (66.10b) with the aid of (66.9b) yields two contributions, the *amplitude contrast*

$$C_a(\boldsymbol{u}_i) = \Re \int_{\mathcal{A}} \tilde{\sigma}(\boldsymbol{q}) \exp\{2\pi\mathrm{i}\boldsymbol{q}\cdot\boldsymbol{u}_a - \frac{2\pi\mathrm{i}}{\lambda_o}W(\lambda_o\boldsymbol{q})\}\,d\boldsymbol{q} \qquad (66.11a)$$

and the *phase contrast*

$$C_p(\boldsymbol{u}_i) = \Im \int_{\mathcal{A}} \tilde{\eta}(\boldsymbol{q}) \exp\{2\pi\mathrm{i}\boldsymbol{q}\cdot\boldsymbol{u}_a - \frac{2\pi\mathrm{i}}{\lambda_o}W(\lambda_o\boldsymbol{q})\}\,d\boldsymbol{q} \qquad (66.11b)$$

The total contrast is $C = C_a + C_p$.

In the most general case, no further progress can be made. These integral expressions are valid even for a slightly excentric or oval aperture, provided of course that the undiffracted beam can pass through it. We next consider the case of a round aperture centred on the axis, for which (66.11a,b) can be further simplified.

First, however, we reconsider the roles of σ and η and hence the relative importance of amplitude and phase contrast. We have already observed that, in the very thin specimens employed in high resolution electron microscopy, there is no absorption and hence amplitude contrast may be expected to be much weaker than phase contrast. For specimens that are thin but not excessively so, however, other phenomena appear that contribute indirectly to the 'absorption' term. Some electrons may be scattered inelastically, which means that they can no longer be included in a theory in which the wavelength is fixed; they are therefore 'lost' as far as the present form of the transfer theory is concerned and contribute to σ. As the specimen thickness is increased, the next term in the expansion (66.1) will become more important; this then becomes (with $|\eta_o| \gg \sigma_o$)

$$\psi_o(x_o, y_o) \approx 1 + \mathrm{i}\eta_o - \sigma_o - \frac{1}{2}\eta_o^2 + \cdots \qquad (66.1')$$

again giving a contribution to the real part. The relative importance of these various effects is not well understood but there is experimental evidence that the 'amplitude image' can sometimes give a faithful representation of the object (Saxton, 1986, 1987). We reconsider this question in Part XIV.

Returning to (66.11a,b), we replace the factor $\exp(-2\pi\mathrm{i}W/\lambda_o)$ by

$T_L(\boldsymbol{q})$ as in (65.26), giving

$$C_a(\boldsymbol{u}_i) = \frac{1}{2} \int_{\mathcal{A}} \tilde{\sigma}(\boldsymbol{q}) T_L(\boldsymbol{q}) \exp(2\pi i \boldsymbol{q} \cdot \boldsymbol{u}_a) \, d\boldsymbol{q}$$

$$+ \frac{1}{2} \int_{\mathcal{A}} \tilde{\sigma}^*(\boldsymbol{q}) T_L^*(\boldsymbol{q}) \exp(-2\pi i \boldsymbol{q} \cdot \boldsymbol{u}_a) \, d\boldsymbol{q}$$

$$C_p(\boldsymbol{u}_i) = \frac{1}{2i} \int_{\mathcal{A}} \tilde{\eta}(\boldsymbol{q}) T_L(\boldsymbol{q}) \exp(2\pi i \boldsymbol{q} \cdot \boldsymbol{u}_a) \, d\boldsymbol{q}$$

$$- \frac{1}{2i} \int_{\mathcal{A}} \tilde{\eta}^*(\boldsymbol{q}) T_L^*(\boldsymbol{q}) \exp(-2\pi i \boldsymbol{q} \cdot \boldsymbol{u}_a) \, d\boldsymbol{q}$$

On changing the sign of the variable $\boldsymbol{q}$ in the second integral of each expression and using the fact that $\mathcal{A}$ is symmetric about the axis, we obtain

$$C_a(\boldsymbol{u}_i) = \frac{1}{2} \int_{\mathcal{A}} \{\tilde{\sigma}(\boldsymbol{q}) T_L(\boldsymbol{q}) + \tilde{\sigma}^*(-\boldsymbol{q}) T_L^*(-\boldsymbol{q})\}$$

$$\times \exp(2\pi i \boldsymbol{q} \cdot \boldsymbol{u}_a) \, d\boldsymbol{q} \tag{66.12a}$$

$$C_p(\boldsymbol{u}_i) = \frac{1}{2i} \int_{\mathcal{A}} \{\tilde{\eta}(\boldsymbol{q}) T_L(\boldsymbol{q}) - \tilde{\eta}^*(-\boldsymbol{gq})\{T_L^*(-q)\}$$

$$\times \exp(2\pi i \boldsymbol{q} \cdot \boldsymbol{u}_a) \, d\boldsymbol{q} \tag{66.12b}$$

These can be further simplified by means of various symmetry relations. First, we note that $\tilde{\sigma}^*(-\boldsymbol{q}) = \tilde{\sigma}(\boldsymbol{q})$ and $\tilde{\eta}^*(-\boldsymbol{q}) = \tilde{\eta}(\boldsymbol{q})$ since both $\sigma_o(\boldsymbol{u}_o)$ and $\eta_o(\boldsymbol{u}_o)$ are real functions. In addition, (65.30) shows that

$$T_L(-\boldsymbol{q}) = T_L(\boldsymbol{q}) \tag{66.13}$$

and consequently $T_L^*(-\boldsymbol{q}) = T_L^*(\boldsymbol{q})$. This remains true if *twofold axial astigmatism* is included but not in the general case when other terms of (65.29) are retained. With these symmetry relations, (66.12a,b) become

$$C_a(\boldsymbol{u}_i) = \int_{\mathcal{A}} \tilde{\sigma}(\boldsymbol{q}) \Re T_L(\boldsymbol{q}) \exp(2\pi i \boldsymbol{q} \cdot \boldsymbol{u}_a) \, d\boldsymbol{q} \tag{66.14a}$$

$$C_p(\boldsymbol{u}_i) = \int_{\mathcal{A}} \tilde{\eta}(\boldsymbol{q}) \Im T_L(\boldsymbol{q}) \exp(2\pi i \boldsymbol{q} \cdot \boldsymbol{u}_a) \, d\boldsymbol{q} \tag{66.14b}$$

Each of these has the form of a Fourier transform and we therefore introduce their *contrast spectra*,

$$C_a(\boldsymbol{u}_i) =: \mathcal{F}(S_a), \qquad S_a := \mathcal{F}^-(C_a) \tag{66.15a}$$

$$C_p(\boldsymbol{u}_i) =: \mathcal{F}(S_p), \qquad S_p := \mathcal{F}^-(C_p) \tag{66.15b}$$

and the associated *contrast transfer functions,*

$$K_a(\boldsymbol{q}) := \begin{cases} \Re T_L(\boldsymbol{q}) & \text{if } \boldsymbol{q} \in \mathcal{A} \\ 0 & \text{otherwise} \end{cases} \qquad (66.16a)$$

$$K_p(\boldsymbol{q}) := \begin{cases} \Im T_L(\boldsymbol{q}) & \text{if } \boldsymbol{q} \in \mathcal{A} \\ 0 & \text{otherwise} \end{cases} \qquad (66.16b)$$

We may then write (66.14 a, b) in the form

$$S_a(\boldsymbol{q}) = K_a(\boldsymbol{q})\tilde{\sigma}(\boldsymbol{q}) \qquad (66.17a)$$
$$S_p(\boldsymbol{q}) = K_p(\boldsymbol{q})\tilde{\eta}(\boldsymbol{q}) \qquad (66.17b)$$

Recalling that (66.10b, 66.11a,b)

$$C(\boldsymbol{u}_i) = C_a(\boldsymbol{u}_i) + C_p(\boldsymbol{u}_i)$$

and hence that

$$S_c(\boldsymbol{q}) = S_a(\boldsymbol{q}) + S_p(\boldsymbol{q})$$

we see that

$$S_c(\boldsymbol{q}) = K_a(\boldsymbol{q})\tilde{\sigma}(\boldsymbol{q}) + K_p(\boldsymbol{q})\tilde{\eta}(\boldsymbol{q}) \qquad (66.17c)$$

Equation (66.17c) is analogous to (65.16) but with important differences. The spectrum on the left-hand side is now given by the observed contrast and is hence a measurable quantity. There are, however, two terms on the right-hand side so that, even if the transfer functions could be inverted, a single image would not be sufficient to yield $\tilde{\sigma}$ and $\tilde{\eta}$. If σ_o is negligible, of course, this objection vanishes.

Figure 66.2 shows the various stages of the reasoning for a weak *phase* object, and this is to be compared with Fig. 65.3. The analogy is not exact: a minor difference is that it is no longer necessary to form the squared modulus in Fig. 66.2, since the contrast C is already the required quantity; more important is the fact that $S_p = K_p\tilde{\eta}$ is *not invertible.* We may not simply write $\tilde{\eta} = S_p/K_p$ because $K_p(\boldsymbol{q})$ has zeros and inevitably vanishes outside $\mathcal{A}$. The possibilities of object reconstruction from a recorded image are therefore limited. The same is true if we retain both the phase and the amplitude terms and record two images in different conditions, so that K_a and K_p are altered (typically by changing the defocus Δ_o). Since it is just such object reconstruction that is of paramount interest in the practical applications of electron microscopy, the theory has led us to a very unfortunate conclusion! Many of the later chapters are devoted to attempts to overcome this obstacle.   Electron holography (Chapter 63) seems to offer a particularly attractive solution.

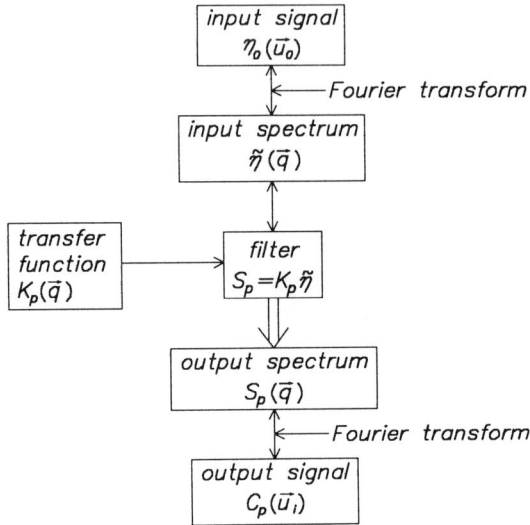

Fig. 66.2: Bright-field image formation in terms of transfer functions. Only the phase term is shown.

We conclude this section with graphs of the phase shift χ and the contrast-transfer functions K_p and K_a in reduced coordinates for various values of the reduced defocus (D). Introducing (65.31–33) into (66.16) and denoting the cutoff value of $q = |\mathbf{q}|$ imposed by the aperture radius by q_c, we write

$$Q_c = (C_s \lambda_o^3)^{1/4} q_c \qquad (66.18)$$

and the transfer functions in terms of the reduced spatial frequency Q become

$$\hat{K}_a(Q) = \begin{cases} \cos \pi (DQ^2 - \frac{1}{2}Q^4) & \text{for } Q \leq Q_c \\ 0 & \text{for } Q > Q_c \end{cases} \qquad (66.19a)$$

$$\hat{K}_p(Q) = \begin{cases} \sin \pi (DQ^2 - \frac{1}{2}Q^4) & \text{for } Q \leq Q_c \\ 0 & \text{for } Q > Q_c \end{cases} \qquad (66.19b)$$

Figure 66.3 shows χ/π as a function of Q for a range of values of D. Figures 66.4 a–d show the behaviour of $K_p(Q)$ for $D = 0$, 1, $\sqrt{2}$ and $\sqrt{3}$ respectively; similar curves for $K_a(Q)$ are shown in Fig. 66.5. No cutoff value Q_c is shown. A common feature of all these curves is the onset of rapid oscillations shortly beyond the first or second zero. How are we to interpret these curves and what values of the parameters should we choose to obtain good image contrast? We defer consideration of these questions to Section 66.4.

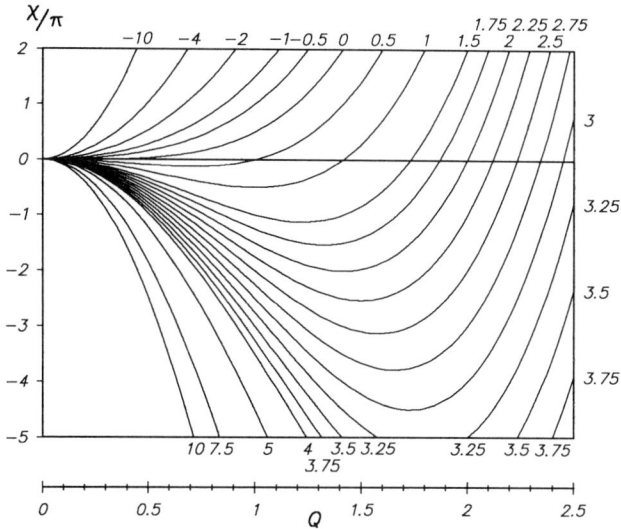

Fig. 66.3: The phase shift $\chi(Q)/\pi$ as a function of Q. From (65.30) we have $\chi(Q)/\pi = Q^4/2 - DQ^2$. Each curve corresponds to the reduced defocus D indicated.

Discussion

The phase and amplitude transfer functions for axial illumination were gradually developed during the 1960s, in particular by Hanszen and colleagues at the Physikalisch-Technische Bundesanstalt in Braunschweig but before tracing their work more closely, we point out that Fourier electron optics was nearly discovered by Glaser: equation (47.10) of Glaser (1956), which gives the image wavefunction, is already in the form of a convolution and the Fourier transform of this is at the heart of transfer theory!

The earliest discussion of contrast-transfer functions in electron optics is to be found in Hanszen *et al.* (1964) but the notion of such instrumental functions, although present (e.g. Eq. 24b and Eq. 26), does not yet emerge clearly. Another paper by Hanszen *et al.* (1965) dealt with contrast transfer for incoherent electron imaging but the expressions for the phase and amplitude contrast-transfer functions with coherent illumination (monochromatic point source) first appear clearly in Hanszen and Morgenstern (1965) as Eq. (32) and Eq. (32b). The same expressions are presented more transparently in the paper by Hanszen (1966a), after which the notion of linear transfer between image contrast and object transparency (σ_o and η_o) could be considered well-established. Further publications (Hanszen, 1967, 1969, 1971) and a conference paper (Hanszen, 1966b) helped to make the theory better known but it was the dramatic optical diffraction pat-

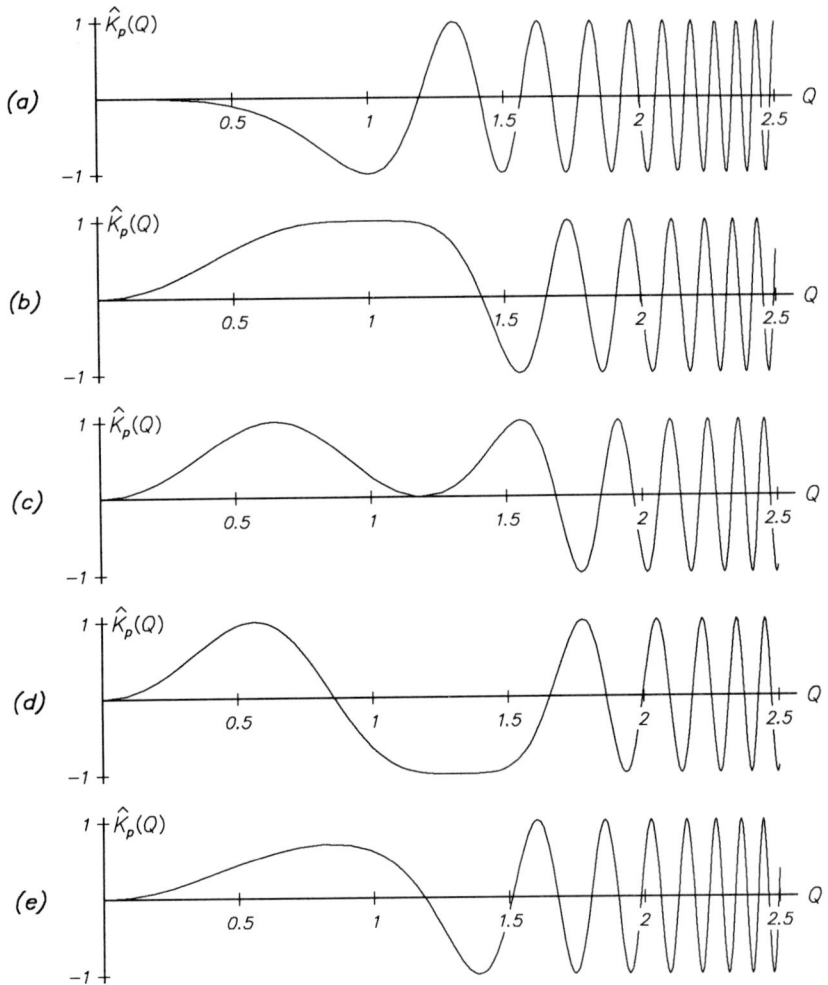

Fig. 66.4: Phase contrast transfer function $\hat{K}_p(Q)$. (a) $D = 0$. (b) $D = 1$. (c) $D = \sqrt{2}$. (d) $D = \sqrt{3}$. (e) $D = 1/\sqrt{2}$; this is Gabor focus, defined in Chapter 63, see (63.57).

terns of amorphous specimens obtained by Thon (1966a, 1967, 1968a) in the University of Tübingen and presented at International and European electron microscopy conferences in Kyoto (Thon, 1966b) and Rome (Thon, 1968b) that excited widespread interest. If the specimen is a weak phase object with a fairly 'white' spatial frequency spectrum, so that $\tilde{\eta} \approx$ const. for a good range of frequencies, the Fourier transform of the image contrast will show $\sin \chi$ directly; the light-optical diffraction pattern of the micro-

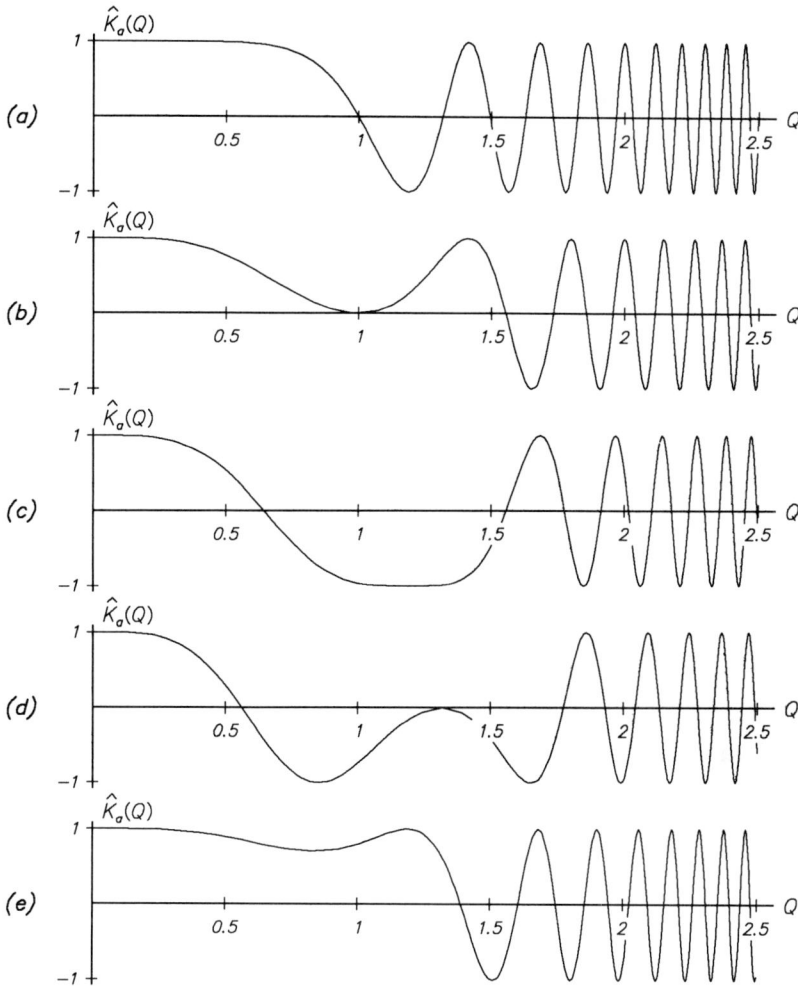

Fig. 66.5: Amplitude contrast transfer function $\hat{K}_a(Q)$. (a) $D = 0$, $\hat{K}_a = \cos(\pi Q^4/2)$. (b) $D = 1$, $\hat{K}_a = \cos\{\pi(2Q^2 - Q^4)/2\}$. (c) $D = \sqrt{2}$, $\hat{K}_a = \cos\{\pi(2\sqrt{2}Q^2 - Q^4)/2\}$. (d) $D = \sqrt{3}$, $\hat{K}_a = \cos\{\pi(2\sqrt{3}Q^2 - Q^4)/2\}$. (e) $D = 1/\sqrt{2}$, $\hat{K}_a = \cos\{\pi(\sqrt{2}Q^2 - Q^4)/2\}$.

graph will thus represent $\sin^2 \chi$. Thon's ring patterns (Fig. 66.6), which reveal this sinusoidal variation arrestingly, were an easily grasped proof of the correctness of the approach. For an extended account, see Thon (1971) or Lenz (1971a,b), Hawkes (1973b) or Hanszen and Ade (1977).

Once established, the theory was soon applied to many other imaging modes and the effects of partial source coherence were included. References are given in later sections. Hanszen and colleagues explored in very

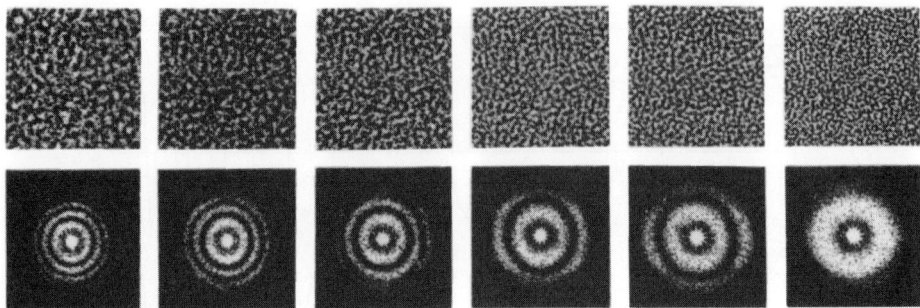

(a)

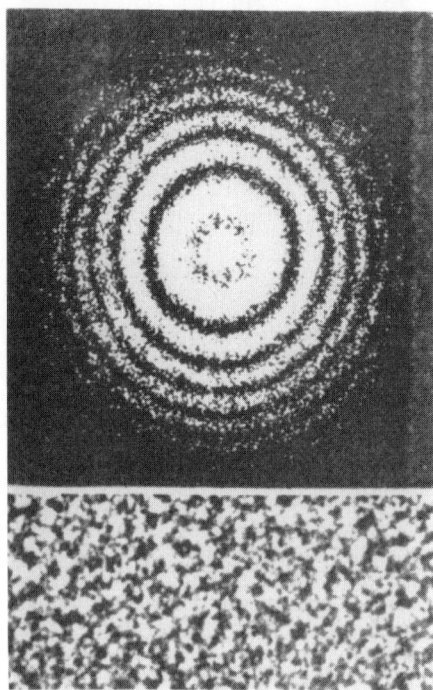

(b)

Fig. 66.6: Diffractograms of a thin carbon film. (a) Focal series for $D = 134$ (right), 191, 229, 324, 419 and 519 (left) nm. (b) $D = 659 \pm 19$ nm. The effect of the astigmatism is seen on the right. $C_s = 4$ mm, accelerating voltage = 100 kV. (Owing to residual astigmatism D is 30 nm greater in the vertical direction.)

great detail the relation between in-line holography and transfer theory and their work may be traced with the aid of the fully referenced survey by Hanszen (1982). A very convenient classified bibliography of the contributions from the Physikalisch-Technische Bundesanstalt in Braunschweig has been prepared by Hanszen (1990).

The transfer functions did not, of course, emerge fully-fledged in the mid-sixties; they had a long prehistory, which goes back to an early paper by Scherzer (1949) giving the wave aberration (χ of (65.30)) in terms of spherical aberration and defocus. This was used on several occasions to interpret electron image formation close to the classical limit of resolution and the suggestions of Hoppe (1961, 1963) and Lenz (1963) for improving the electron image by means of zone plates, the rings of which correspond to the regions between successive zeros of $\sin \chi$, were inspired by it. Von Borries and Lenz (1956) and, more particularly, Lenz and Scheffels (1958) attempted to interpret the fine structure seen in high-resolution micrographs of supposedly amorphous specimens in the language of phase contrast but retained only the defocus term in the wave aberration, neglecting the spherical aberration (cf. Thon, 1965). Until the nature of phase contrast transfer was fully elucidated (and indeed for some time afterwards), these fine structures were used by electron microscope manufacturers as a measure of the resolution of their instruments.

During the 1970s, the use of transfer functions became common and they began to make their appearance in commercial brochures. An imaging mode employed for studying magnetic specimens that was analysed by Guigay *et al.* (1971) reminds us of the early explorations of Lenz and Scheffels; here the defocus has to be large and the spherical aberration can indeed be neglected. Moreover, the weak phase constraint is considerably less stringent.

66.2 Spectral distributions of the illumination

The situation considered above is highly idealized in various ways: the condenser system has been assumed to furnish an ideal plane and monoenergetic wave, propagating exactly in the axial (z) direction. This is clearly unrealistic for, even ignoring the condenser lens properties, it would require a vanishingly small source emitting electrons at exactly the same energy. Electron sources do approximate to this ideal but the small emitting area and the energy spread have some effect, which may be regarded as a perturbation of the ideal case so far considered. In this section, we establish formulae describing this perturbation.* Non-vanishing source-size and non-vanishing energy spread may be regarded as *spectral effects*: the former is characterized by a spectrum of wavevectors $\boldsymbol{k} = (k_x, k_y, k_z)$, with $k_x, k_y \ll k_z$, instead of a single component $\boldsymbol{k} = (0, 0, k_z)$ and the latter by a range of values of $|\boldsymbol{k}|$.

(i) *Non-vanishing source-size.* The wavevector $\boldsymbol{k}$ now has small transverse components. It will be convenient to introduce the *two-dimensional* vector $\boldsymbol{\kappa}$,

$$\boldsymbol{\kappa} = (\kappa_x, \kappa_y) := \frac{1}{2\pi}(k_x, k_y) \tag{66.20}$$

and we shall always have $\kappa^2 = \kappa_x^2 + \kappa_y^2 \ll \lambda^{-2}$, since the slopes of the electron trajectories are very small, typically $\kappa\lambda \lesssim 10^{-3}$ rad. We saw in Chapter 64 how the angles at the specimen (θ) are related to the transverse dimensions of the source: for parallel illumination, $|\theta| = r_s/f_i$ in which r_s is the crossover radius and f_i is the image focal length of the entire condenser system.

(ii) *Non-vanishing energy spread.* The wavenumber $k = |\boldsymbol{k}|$ now has a spectrum of non-vanishing width. The value $k_o = 2\pi/\lambda_o$ that we have used so far refers to the maximum of this spectrum. Extensive discussion of the various forms of this spectrum is to be found in Chapter 44.

We recall that k is given (56.12) by $k(\boldsymbol{r}) = \{2m_o e\hat{\Phi}(\boldsymbol{r}) + 2Em(\boldsymbol{r})\}^{1/2}/\hbar$. The specimen may well be situated within the field of a magnetic lens but there will certainly be no external electrostatic field and we may therefore replace $\hat{\Phi}$ by the constant value $\hat{U}$ and regard $m = \gamma m_0$ as a constant. With no loss of generality, we may choose the energy scale in such a way that $E = 0$ corresponds to the maximum of the energy spectrum; $|E|$ will hence be very small and the above expression for k can be expanded as a Taylor series:

$$k_o = \frac{2\pi}{\lambda_o} = \frac{1}{\hbar}\sqrt{2m_o e\hat{U}} \tag{66.21}$$

$$k = k_o + \frac{mE}{\hbar^2 k_o} + O(E^2)$$

The energy E is not very convenient as a parameter in the subsequent calculations and we therefore introduce its *volt equivalent* v, $E =: ev$, whereupon

---

* It is usual to discuss these perturbing influences in the language of partial coherence: non-vanishing source-size is assimilated to partial spatial coherence and energy spread to partial temporal coherence. (The terms lateral and longitudinal coherence are also used.) Coherence is discussed more formally in Part XVI; in this Chapter, we shall usually speak of source-size and energy spread rather than of partial coherence.

the expression for k in (66.21) becomes

$$k = k_o + \frac{em}{\hbar^2 k_o} v \equiv k_o \left(1 + \frac{\gamma v}{2\hat{U}}\right) \qquad (66.22)$$

where as usual (Section 2.3) $\gamma = 1 + 2\epsilon U_o$, $\hat{U} = U_o(1 + \epsilon U_o)$ and $\epsilon = e/2m_oc^2$. The quantity v that characterizes the energy spread is usually such that $|v| \lesssim 1$ V; see Part IX for extensive discussion.

The incident electrons are now *wave packets*. It is not advantageous to use the general form for these, which would be unnecessarily complicated. Instead, we consider the *partial waves* building them up, which offer an easily understood and physically correct model. Each partial wave is given by

$$\psi_p(\boldsymbol{r}) = A(\boldsymbol{\kappa}, v) \exp\{2\pi i(x\kappa_x + y\kappa_y)\} \exp\left\{i\left(k_o + \frac{emv}{\hbar^2 k_o} - \frac{2\pi^2\kappa^2}{k_o}\right)z\right\} \qquad (66.23)$$

from which the unimportant time factor $\exp(-iEt/\hbar)$ has been omitted. This expression is valid in the half-space $z \leq 0$ upstream from the specimen. The amplitude A will be specified below. For the moment, we just impose the normalization

$$\iint |A(\boldsymbol{k}, v)|^2 \, d\boldsymbol{\kappa} \, dv = 1 \qquad (66.24)$$

In the entrance plane of the specimen where $z = 0$, (66.23) simplifies to

$$\psi_p(\boldsymbol{u}_o, 0) = A(\boldsymbol{\kappa}, v) \exp(2\pi i \boldsymbol{\kappa} \cdot \boldsymbol{u}_o) \qquad (66.25)$$

Note that the energy does not appear in the exponential here. The specimen is assumed to be so thin that, in a first-order approximation, the phase shift η_o is unaffected by v; at the exit surface of the specimen, therefore, the wavefunction is given by

$$\psi_o(\boldsymbol{u}_o, \boldsymbol{\kappa}, v) = \{1 + i\eta_o(\boldsymbol{u}_o)\} \exp(2\pi i \boldsymbol{\kappa} \cdot \boldsymbol{u}_o) A(\boldsymbol{\kappa}, v) \qquad (66.26)$$

We shall not consider the absorption term $\sigma_o(\boldsymbol{u}_o)$ here; the calculation can be straightforwardly extended to include it if required.

The object spectrum S_o now becomes

$$S_o(\boldsymbol{q}, \boldsymbol{\kappa}, v) = \int \psi_o(\boldsymbol{u}_o, \boldsymbol{\kappa}, v) \exp(-2\pi i \boldsymbol{q} \cdot \boldsymbol{u}_o) \, d\boldsymbol{u}_o$$

$$= A(\boldsymbol{\kappa}, v) \int \{1 + i\eta(\boldsymbol{u}_o)\} \exp\{2\pi i(\boldsymbol{\kappa} - \boldsymbol{q}) \cdot \boldsymbol{u}_o\} \, d\boldsymbol{u}_o$$

which, recalling (66.3) and taking into account a new effect, namely, the shift of spatial frequency from $\boldsymbol{q}$ to $\boldsymbol{q} - \boldsymbol{\kappa}$, may be written as follows:

$$S_o(\boldsymbol{q}, \boldsymbol{\kappa}, v) = A(\boldsymbol{\kappa}, v)\{\delta(\boldsymbol{q} - \boldsymbol{\kappa}) + i\tilde{\eta}(\boldsymbol{q} - \boldsymbol{\kappa})\} \qquad (66.27)$$

This spectrum is again transferred to the image as shown in Fig. 65.3 but the wave transfer function T now depends explicitly on v, so that $T = T(\boldsymbol{q}, v)$. The main effect is due to the *axial chromatic aberration* C_c of the objective lens. We give the explicit formula later; in the general calculation, we simply need to recall that v is now an argument of T.

For the image spectrum, we have

$$S_i(\boldsymbol{q}, \boldsymbol{\kappa}, v) = A(\boldsymbol{\kappa}, v)T(\boldsymbol{q}, v)\{\delta(\boldsymbol{q} - \boldsymbol{\kappa}) + i\tilde{\eta}(\boldsymbol{q} - \boldsymbol{\kappa})\}$$

which is inverted to give

$$\psi_i(\boldsymbol{u}_i, \boldsymbol{\kappa}, v) = \int S_i(\boldsymbol{q}, \boldsymbol{\kappa}, v)\exp(2\pi i \boldsymbol{q} \cdot \boldsymbol{u}_i)\, d\boldsymbol{q}.$$

The term involving the δ-function is easily integrated and we find

$$\psi_i(\boldsymbol{u}_i, \boldsymbol{\kappa}, v) = A(\boldsymbol{\kappa}, v)T(\boldsymbol{\kappa}, v)\exp(2\pi i \boldsymbol{\kappa} \cdot \boldsymbol{u}_i)$$
$$+ iA(\boldsymbol{\kappa}, v)\int T(\boldsymbol{q}, v)\tilde{\eta}(\boldsymbol{q} - \boldsymbol{\kappa})\exp(2\pi i \boldsymbol{q} \cdot \boldsymbol{u}_i)\, d\boldsymbol{q}$$

As usual, we neglect quadratic terms in $\tilde{\eta}$ when calculating $|\psi_i|^2$ and we also assume that $|T(\boldsymbol{\kappa}, v)|^2 = 1$. Then

$$|\psi_i(\boldsymbol{u}_i, \boldsymbol{\kappa}, v)|^2 = |A(\boldsymbol{\kappa}, v)|^2(1 - 2\Im\mu_i) \qquad (66.28a)$$

where

$$\mu_i := \int T(\boldsymbol{q}, v)T^*(\boldsymbol{\kappa}, v)\tilde{\eta}(\boldsymbol{q} - \boldsymbol{\kappa})\exp\{2\pi i(\boldsymbol{q} - \boldsymbol{\kappa}) \cdot \boldsymbol{u}_i\}\, d\boldsymbol{q} \qquad (66.28b)$$

This is the image intensity for a single partial wave. In order to obtain the contrast, we must integrate over the angular and energy spreads, $\boldsymbol{\kappa}$ and v, for a definition such as (66.10a) involves the total intensity. By integrating the partial intensity (66.28a) and not the partial wave $\psi_i(\boldsymbol{u}_i, \boldsymbol{\kappa}, v)$ over $\boldsymbol{\kappa}$ and v, we are assuming that the partial waves are completely independent, that there is no relation (on average) between electron emission at different energies and from different points on the source. We shall return to this question in the last Part of the book.

The integrated intensity, in which the contributions of all the partial intensities are added, is given by

$$
\begin{aligned}
I &= \iint |\psi_i(\boldsymbol{u}_i, \boldsymbol{\kappa}, v)|^2 \, d\boldsymbol{\kappa} \, dv \\
&= 1 - 2 \int_q \int_v \int_\kappa |A(\boldsymbol{\kappa}, v)|^2 \, \Im[T(\boldsymbol{q}, v) T^*(\boldsymbol{\kappa}, v) \tilde{\eta}(\boldsymbol{q} - \boldsymbol{\kappa}) \\
&\qquad \times \exp\{2\pi \mathrm{i}(\boldsymbol{q} - \boldsymbol{\kappa}) \cdot \boldsymbol{u}_i\}] \, d\boldsymbol{q} \, dv \, d\boldsymbol{\kappa}
\end{aligned}
\tag{66.29}
$$

The image (phase) contrast is now

$$
C_p(\boldsymbol{u}_i) = \frac{1 - I(\boldsymbol{u}_i)}{1 + I(\boldsymbol{u}_i)}
\tag{66.30}
$$

(cf. 66.10a) and, in the linear approximation, this can be written in the form

$$
\begin{aligned}
C_p(\boldsymbol{u}_i) = \Im\Bigg[\int_q \bigg\{ \int_\kappa \int_v |A(\boldsymbol{\kappa}, v)|^2 T(\boldsymbol{q} + \boldsymbol{\kappa}, v) T^*(\boldsymbol{\kappa}, v) \, dv \, d\boldsymbol{\kappa} \bigg\} \\
\times \tilde{\eta}(\boldsymbol{q}) \exp(2\pi \mathrm{i} \boldsymbol{q} \cdot \boldsymbol{u}_i) \, d\boldsymbol{q} \Bigg]
\end{aligned}
\tag{66.31}
$$

after a change in the order of integration and a change of variable from $\boldsymbol{q}$ to $\boldsymbol{q} - \boldsymbol{\kappa}$.

We now seek to rearrange this expression in such a way that it has the form of a Fourier integral with respect to $\boldsymbol{q}$. This will enable us to write down the contrast *spectrum*, $S_c(\boldsymbol{q})$, immediately and hence to extract a contrast-transfer function $K_p(\boldsymbol{q})$. We first write out the imaginary part explictly and make the double substitution, $\boldsymbol{q} \to -\boldsymbol{q}$ and $\boldsymbol{\kappa} \to -\boldsymbol{\kappa}$ in the final term, as in (66.12). We obtain

$$
\begin{aligned}
C_p(\boldsymbol{u}_i) = \frac{1}{2} \int_q \bigg\{ \int_\kappa \int_v |A(\boldsymbol{\kappa}, v)|^2 T(\boldsymbol{q} + \boldsymbol{\kappa}, v) T^*(\boldsymbol{\kappa}, v) \, dv \, d\boldsymbol{\kappa} \bigg\} \\
\times \tilde{\eta}(\boldsymbol{q}) \exp(2\pi \mathrm{i} \boldsymbol{q} \cdot \boldsymbol{u}_i) \, d\boldsymbol{q} \\
- \frac{1}{2\mathrm{i}} \int_q \bigg\{ \int_\kappa \int_v |A(-\boldsymbol{\kappa}, v)|^2 T^*(-\boldsymbol{q} - \boldsymbol{\kappa}, v) T(-\boldsymbol{\kappa}, v) \, dv \, d\boldsymbol{\kappa} \bigg\} \\
\times \tilde{\eta}^*(-\boldsymbol{q}) \exp(2\pi \mathrm{i} \boldsymbol{q} \cdot \boldsymbol{u}_i) \, d\boldsymbol{q}
\end{aligned}
$$

We must again assume that the aperture is symmetric about the axis. Even in the presence of axial chromatic aberration, the wave transfer function T satisfies

$$
T(-\boldsymbol{q}') = T(\boldsymbol{q}')
\tag{66.32}
$$

for all $\boldsymbol{q}'$, and in particular for $\boldsymbol{q}' = \boldsymbol{q} + \boldsymbol{\kappa}$ and for $\boldsymbol{q}' = \boldsymbol{\kappa}$. We must also assume that the illumination is rotationally symmetric, so that

$$|A(-\boldsymbol{\kappa}, v)|^2 = |A(\boldsymbol{\kappa}, v)|^2 \tag{66.33}$$

Finally, we know that $\tilde{\eta}^*(-\boldsymbol{q}) = \tilde{\eta}(\boldsymbol{q})$ since η is real. Bringing all these symmetries together, we find

$$
\begin{aligned}
C_p(\boldsymbol{u}_i) &= \int_q \left[\int \! \cdot \! \int_{\kappa} \int_v |A(\boldsymbol{\kappa}, v)|^2 \frac{1}{2i} \left\{ T(\boldsymbol{q} + \boldsymbol{\kappa}, v) T^*(\boldsymbol{\kappa}, v) - \text{c.c.} \right\} dv \, d\boldsymbol{\kappa} \right] \\
&\quad \times \tilde{\eta}(\boldsymbol{q}) \exp(2\pi i \boldsymbol{q} \cdot \boldsymbol{u}_i) \, d\boldsymbol{q} \\
&=: \int_q K(\boldsymbol{q}) \tilde{\eta}(\boldsymbol{q}) \exp(2\pi i \boldsymbol{q} \cdot \boldsymbol{u}_i) \, d\boldsymbol{q}
\end{aligned}
\tag{66.34}
$$

(c.c. signifies 'complex conjugate.') This has the desired appearance, and we see that

$$K_p(\boldsymbol{q}) = \int_{\kappa} \! \cdot \! \int_v |A(\boldsymbol{\kappa}, v)|^2 \Im \left\{ T(\boldsymbol{q} + \boldsymbol{\kappa}, v) T^*(\boldsymbol{\kappa}, v) \right\} dv \, d\boldsymbol{\kappa} \tag{66.35}$$

is the generalized form of the transfer function. It should contain our earlier result (66.16) as a special case. The latter corresponds to infinitely narrow spectra:

$$|A(\boldsymbol{\kappa}, v)|^2 \to \delta(\kappa_x) \delta(\kappa_y) \delta(v) \tag{66.36}$$

which satisfies (66.24); the integration in (66.35) may then be performed, giving

$$
\begin{aligned}
K_p(\boldsymbol{q}) &= \Im \left\{ T(\boldsymbol{q}, 0) T^*(0, 0) \right\} \\
&= \Im T(\boldsymbol{q}, 0)
\end{aligned}
$$

as we expect, since $T(0, 0) = 1$.

Examination of the expression (66.35) for the transfer function $K_p(\boldsymbol{q})$ shows that both instrumental parameters associated with the image-forming system (T) and parameters describing the illumination (A) are involved. We now enquire to what extent these can be separated before examining specific forms of $A(\boldsymbol{\kappa}, v)$ in Section 66.3.

For this, we must study the function $T(\boldsymbol{q}, v)$ in more detail. The associated wave aberration W including a chromatic aberration term is now

$$W = \frac{1}{4} C_s \theta_o^4 - \frac{1}{2} \Delta_o \theta_o^2 - \frac{1}{2} \frac{\gamma v}{\hat{U}} \theta_o^2 \tag{66.37}$$

in which C_c is the familiar (axial) chromatic aberration coefficient (Chapter 26) and $\hat{U}$ is again the relativistic accelerating voltage. In the relation between θ_o and q, $\theta_o = \lambda_o q$ (65.18), we use the nominal value λ_o given by (66.21) even though there is now a spread of wavelengths. This is permissible since W consists of small aberration terms and if we retained terms in $\lambda - \lambda_o$ in W we should obtain aberrations of higher order. The function $T(q, v)$, which is the generalization of (65.30), is therefore given by

$$T(\boldsymbol{q}, v) = \exp\left[i\pi\left\{-\frac{1}{2}C_s\lambda_o^3(q_x^2 + q_y^2)^2 + \lambda_o\Delta_o(q_x^2 + q_y^2)\right.\right.$$
$$\left.\left. + \frac{\lambda_o C_c\gamma}{\hat{U}}v(q_x^2 + q_y^2)\right\}\right] \qquad (66.38)$$

and it will be convenient to write the combination $T(\boldsymbol{q} + \boldsymbol{\kappa}, v)T^*(\boldsymbol{\kappa}, v)$ in the following form:

$$T(\boldsymbol{q} + \boldsymbol{\kappa}, v)T^*(\boldsymbol{\kappa}, v) =: \exp\left\{i(-T_s + T_\Delta + T_v v\right\} \qquad (66.39)$$

with

$$T_s := -\frac{\pi}{2}C_s\lambda_o^3\left[\left\{(\boldsymbol{q} + \boldsymbol{\kappa}) \cdot (\boldsymbol{q} + \boldsymbol{\kappa})\right\}^2 - (\boldsymbol{\kappa} \cdot \boldsymbol{\kappa})^2\right]$$
$$= -\frac{\pi}{2}C_s\lambda_o^3\left\{(\boldsymbol{q} \cdot \boldsymbol{q})^2 + 4(\boldsymbol{q} \cdot \boldsymbol{\kappa})^2 + 4(\boldsymbol{q} \cdot \boldsymbol{q})(\boldsymbol{q} \cdot \boldsymbol{\kappa})\right.$$
$$\left. + 2(\boldsymbol{q} \cdot \boldsymbol{q})(\boldsymbol{\kappa} \cdot \boldsymbol{\kappa}) + 4(\boldsymbol{q} \cdot \boldsymbol{\kappa})(\boldsymbol{\kappa} \cdot \boldsymbol{\kappa})\right\} \qquad (66.40)$$
$$T_\Delta := \pi\Delta_o\lambda_o\left\{(\boldsymbol{q} + \boldsymbol{\kappa}) \cdot (\boldsymbol{q} + \boldsymbol{\kappa}) - \boldsymbol{\kappa} \cdot \boldsymbol{\kappa}\right\}$$
$$= \Delta_o\lambda_o(\boldsymbol{q} \cdot \boldsymbol{q} + 2\boldsymbol{q} \cdot \boldsymbol{\kappa})$$
$$T_v := \pi\frac{\lambda_o C_c\gamma}{\hat{U}}(\boldsymbol{q} \cdot \boldsymbol{q} + \boldsymbol{q} \cdot \boldsymbol{u}) =: \pi\lambda_o C'_c(\boldsymbol{q} \cdot \boldsymbol{q} + 2\boldsymbol{q} \cdot \boldsymbol{\kappa})$$

and

$$C'_c := \gamma C_c/\hat{U}.$$

Returning to (66.35), we see that $K_p(\boldsymbol{q})$ can be written in the form

$$K_p(\boldsymbol{q}) = \frac{1}{2i}\int\!\!\!\int_{\boldsymbol{\kappa}}\int_v |A(\boldsymbol{\kappa}, v)|^2\left[\exp\left\{i(-T_s + T_\Delta + T_v v)\right\}\right.$$
$$\left. - \exp\left\{-i(-T_s + T_\Delta + T_v v)\right\}\right] d\boldsymbol{\kappa}\, dv$$
$$= \frac{1}{2i}\int\!\!\!\int_{\boldsymbol{\kappa}}\int_v\left[|A(\boldsymbol{\kappa}, v)|^2 \exp\left\{-i(T_s - T_\Delta)\right\}\right. \qquad (66.41)$$
$$\left. - |A(\boldsymbol{\kappa}, -v)|^2 \exp\left\{i(T_s - T_\Delta)\right\}\right] \exp(iT_v v)\, d\boldsymbol{\kappa}\, dv$$

The distribution $A(\boldsymbol{\kappa}, v)$ is unlikely to be exactly even in v (though it is often assumed to be and we shall model it by an even function in the next Section) and we therefore write

$$|A(\boldsymbol{\kappa}, v)|^2 =: a_e + i a_o \qquad (66.42)$$

in which the even and odd contributions are given by

$$
\begin{aligned}
a_e & := \frac{1}{2} \left\{ |A(\boldsymbol{\kappa}, v)|^2 + |A(\boldsymbol{\kappa}, -v)|^2 \right\} \\
a_o & := \frac{1}{2i} \left\{ |A(\boldsymbol{\kappa}, v)|^2 - |A(\boldsymbol{\kappa}, -v)|^2 \right\}
\end{aligned}
\qquad (66.43)
$$

so that $K_p(\boldsymbol{q})$ becomes

$$
\begin{aligned}
K_p(\boldsymbol{q}) &= \frac{1}{2i} \int_{\kappa}\!\!\int_v a_e \left[e^{-i(T_s - T_\Delta)} - e^{i(T_s - T_\Delta)} \right] e^{iT_v v} \, d\boldsymbol{\kappa} \, dv \\
&\quad + \frac{1}{2} \int_{\kappa}\!\!\int_v a_o \left[e^{-i(T_s - T_\Delta)} + e^{i(T_s - T_\Delta)} \right] e^{iT_v v} \, d\boldsymbol{\kappa} \, dv \qquad (66.44) \\
&= \int_{\kappa} \left\{ \tilde{a}_e \sin(T_\Delta - T_s) + \tilde{a}_o \cos(T_\Delta - T_s) \right\} d\boldsymbol{\kappa}
\end{aligned}
$$

The quantities $\tilde{a}_e$ and $\tilde{a}_o$ are, with the appropriate scaling, the Fourier transforms with respect to v of the even and odd parts of the source distribution:

$$\tilde{a}_e := \int a_e \exp(iT_v v) \, dv \quad , \quad \tilde{a}_o = \int a_o \exp(iT_v v) \, dv \qquad (66.45)$$

Equation (66.44) is quite close to the representation we are seeking but requires further manipulation, particularly as $\boldsymbol{\kappa}$ appears in the arguments of $\tilde{a}_e$ and $\tilde{a}_o$. Let us now examine the terms in $T_\Delta - T_s$ in more detail.

From the definitions (66.40), we see that

$$
\begin{aligned}
T_\Delta - T_s &= \pi \Delta_o \lambda_o (\boldsymbol{q} \cdot \boldsymbol{q} + 2\boldsymbol{q} \cdot \boldsymbol{\kappa}) \\
&\quad - \frac{\pi}{2} C_s \lambda_o^3 \{ (\boldsymbol{q} \cdot \boldsymbol{q})^2 + 4(\boldsymbol{q} \cdot \boldsymbol{\kappa})^2 + 4(\boldsymbol{q} \cdot \boldsymbol{q})(\boldsymbol{q} \cdot \boldsymbol{\kappa}) \\
&\quad + 2(\boldsymbol{q} \cdot \boldsymbol{q})(\boldsymbol{\kappa} \cdot \boldsymbol{\kappa}) + 4(\boldsymbol{\kappa} \cdot \boldsymbol{\kappa})(\boldsymbol{q} \cdot \boldsymbol{\kappa}) \} \\
&=: -\chi(\boldsymbol{q}) + T(\boldsymbol{q}, \boldsymbol{\kappa}) \qquad (66.46)
\end{aligned}
$$

with

$$
\begin{aligned}
\chi(\boldsymbol{q}) &= \frac{\pi}{2} C_s \lambda_o^3 q^4 - \pi \Delta_o \lambda_o q^2 \\
T(\boldsymbol{q}, \boldsymbol{\kappa}) &= 2\pi \Delta_o \lambda_o \boldsymbol{q} \cdot \boldsymbol{\kappa} - \frac{\pi}{2} C_s \lambda_o^3 \\
&\quad \times \{ 4q^2 \boldsymbol{q} \cdot \boldsymbol{\kappa} + 4(\boldsymbol{q} \cdot \boldsymbol{\kappa})^2 + 2q^2 \kappa^2 + 4\kappa^2 (\boldsymbol{q} \cdot \boldsymbol{\kappa}) \}
\end{aligned}
\qquad (66.47)
$$

Hence $K_p(\boldsymbol{q})$ may be written

$$
\begin{aligned}
K_p(\boldsymbol{q}) = & - \sin \chi(\boldsymbol{q}) \int_\kappa \{\tilde{a}_e \cos T(\boldsymbol{q}, \boldsymbol{\kappa}) - \tilde{a}_o \sin T(\boldsymbol{q}, \boldsymbol{\kappa})\} \, d\boldsymbol{\kappa} \\
& + \cos \chi(\boldsymbol{q}) \int_\kappa \{\tilde{a}_e \sin T(\boldsymbol{q}, \boldsymbol{\kappa}) + \tilde{a}_o \cos T(\boldsymbol{q}, \boldsymbol{\kappa})\} \, d\boldsymbol{\kappa}
\end{aligned}
\tag{66.48}
$$

and we have thus succeeded in showing that non-vanishing source-size and energy spread lead to a *modulation* of the ideal transfer functions, $\sin \chi$ and $\cos \chi$; if the energy spread is not even in v, there will be some intermixing of the ideal functions or, in other words, a shift of the zeros. If $|A(\boldsymbol{\kappa}, v)|^2$ is even in v, then of course a_o, $\tilde{a}_o$ vanish, leaving

$$
K_p(\boldsymbol{q}) = - \sin \chi(\boldsymbol{q}) \int_\kappa \tilde{a}_e \cos T(\boldsymbol{q}, \boldsymbol{\kappa}) \, d\boldsymbol{\kappa} + \cos \chi(\boldsymbol{q}) \int_\kappa \tilde{a}_e \sin T(\boldsymbol{q}, \boldsymbol{\kappa}) \, d\boldsymbol{\kappa}
\tag{66.49}
$$

The definition of $T(\boldsymbol{q}, \kappa)$, (66.47), shows that both odd and even terms in $\boldsymbol{\kappa}$ are present and we could write $T(\boldsymbol{q}, \boldsymbol{\kappa}) =: T_e + T_o$, expand the terms in $\cos T(\boldsymbol{q}, \boldsymbol{\kappa})$ and $\sin T(\boldsymbol{q}, \boldsymbol{\kappa})$, and analyse the integrals further. This is fairly laborious, since even if (as has been assumed) $|A(\boldsymbol{\kappa}, v)|^2$ is even in $\boldsymbol{\kappa}$ (symmetric illumination spread) the symmetry of a_e with respect to κ is more complicated $(a_e(\boldsymbol{\kappa}, \boldsymbol{q}) = a_e(-\boldsymbol{\kappa}, -\boldsymbol{q}))$. We shall not pursue this here. The final question concerns the interaction between energy spread, characterized by v, and source-size, measured by $\boldsymbol{\kappa}$. In what circumstances can we write the integral over $\boldsymbol{\kappa}$ (66.48 or 66.49) as the product of a term arising from $\boldsymbol{\kappa}$ and another term involving v? As before, we assume for simplicity that $|A(\boldsymbol{\kappa}, v)|^2$ is even in v; a more complicated reply can be found if this assumption is not made.

We first require that the roles of $\boldsymbol{\kappa}$ and v in $A(\boldsymbol{\kappa}, v)$ be separated:

$$
A(\boldsymbol{\kappa}, v) =: A_\kappa(\boldsymbol{\kappa}) A_v(v) = A_\kappa(\boldsymbol{\kappa}) A_v(-v)
\tag{66.50}
$$

Hence

$$
a_e = |A_\kappa(\boldsymbol{\kappa})|^2 |A_v(v)|^2, \quad a_o = 0
\tag{66.51}
$$

and

$$
\tilde{a}_e = |A_\kappa(\boldsymbol{\kappa})|^2 \int |A_v(v)|^2 \exp\{i\pi C_c'(q^2 + 2\boldsymbol{q} \cdot \boldsymbol{\kappa})v\} \, dv
\tag{66.52}
$$

If we can neglect $\boldsymbol{q} \cdot \boldsymbol{\kappa}$ in the exponent, then,

$$
\tilde{a}_e \approx |A_\kappa(\boldsymbol{\kappa})|^2 t_v(\boldsymbol{q})
\tag{66.53}
$$

with

$$t_v(\boldsymbol{q}) := \int |A_v(v)|^2 \exp(i\pi C_c' q^2 v)\, dv \qquad (66.54)$$

and as required $t_v(\boldsymbol{q})$ is independent of $\boldsymbol{\kappa}$. Thus $K_p(\boldsymbol{q})$ now reduces to

$$K_p = -\sin\{\chi(\boldsymbol{q})\}t_v(\boldsymbol{q})t_\kappa^{(c)}(\boldsymbol{q}) + \cos\{\chi(\boldsymbol{q})\}t_v(\boldsymbol{q})t_\kappa^{(s)}(\boldsymbol{q}) \qquad (66.55)$$

with

$$t_\kappa^{(c)}(q) := \int_\kappa |A_\kappa(\boldsymbol{\kappa})|^2 \cos\{T(\boldsymbol{q},\boldsymbol{\kappa})\}\, d\boldsymbol{\kappa} \qquad (66.56a)$$

$$t_\kappa^{(s)}(q) := \int_\kappa |A_\kappa(\boldsymbol{\kappa})|^2 \sin\{T(\boldsymbol{q},\boldsymbol{\kappa})\}\, d\boldsymbol{\kappa} \qquad (66.56b)$$

If $T(\boldsymbol{q},\boldsymbol{\kappa})$ is truncated beyond the linear terms in $\boldsymbol{\kappa}$, so that

$$T(\boldsymbol{q},\boldsymbol{\kappa}) \approx T'(\boldsymbol{q},\kappa) := 2\boldsymbol{q}\cdot\boldsymbol{\kappa}(\Delta_o\lambda_o - C_s\lambda_o^3 q^2) \qquad (66.57)$$

the function $t_\kappa^{(s)}(\boldsymbol{q})$ vanishes and

$$K_p(\boldsymbol{q}) \approx -t_v(\boldsymbol{q})t_\kappa(\boldsymbol{q})\sin\chi(\boldsymbol{q}) \qquad (66.58)$$

with

$$t_\kappa(\boldsymbol{q}) := \int_\kappa |A_\kappa(\kappa)|^2 \cos\{T'(\boldsymbol{q},\boldsymbol{\kappa})\}\, d\boldsymbol{\kappa} \qquad (66.56c)$$

Discussion

The envelope functions that modulate $\sin\chi(\boldsymbol{q})$ and $\cos\chi(\boldsymbol{q})$ in (66.55) are both real. If we return to the discussion of (65.35) in the ideal case, we see that they occupy the same role as the aperture function, $T_A(\boldsymbol{q})$: the effect of non-vanishing source-size and energy spread is analogous to that of an objective aperture with a transparency given for (66.58) by $t_v t_\kappa$. But this seems to mean that electrons are 'absorbed'! Where have they gone? Electrons have of course not been lost but their capacity to convey specimen information to the image has been impaired. A convenient set of measures of this reduction in information transfer are the image quality criteria introduced by Linfoot (1956, 1957, 1960, 1964) and discussed on several occasions (O'Neill, 1963; Franke, 1966; Frieden, 1966; Röhler, 1967; Frank et al., 1970), as pointed out by Frank (1975b; Beer et al., 1975). These are the *dissimilarity*, K, and the *structural resolving power*, Ξ, and two derived quantities: the *fidelity*, $\Phi := 1 - K$ and the *correlation quality*

$\Psi = -(\Phi + \Xi)/2$, which is the same as the Strehl intensity ratio *(Definitionshelligkeit)* of traditional microscopy (Strehl, 1902; see Born and Wolf, 1980 Section 9.1.1). The measures K and Ξ are defined, for modulation by finite source-size only, as follows:

$$K := \frac{\int |1 - t_{\boldsymbol{\kappa}}(\boldsymbol{q})|^2 d\boldsymbol{q}}{\int d\boldsymbol{q}} \tag{66.59a}$$

(Frank, 1975a), in which the integration is taken over the aperture, and

$$\Xi := \frac{\int |t_{\kappa}(\boldsymbol{q})|^2 d\boldsymbol{q}}{\int d\boldsymbol{q}} \tag{66.59b}$$

The latter is interesting in that the signal-to-noise ratio at the image can be shown to remain constant as the source-size is increased if the exposure, and hence the electron dose n, is increased so that $n^2\Xi = 1$ (signal-to-noise ratio limited by the statistics of the photographic recording medium) or so that $n^3\Xi = 1$ (ratio limited by the electron statistics, at high magnification and in minimum exposure conditions, for example). Figure 66.7 shows the behaviour of K and Ξ as a function of source-size and defocus in reduced coordinates for the Gaussian source model introduced in the next section.

The effects of non-vanishing source-size and energy spread on the transfer functions were established in the early 1970s. The product representation (66.48) did not emerge from the earliest attempts to include these effects (Erickson and Klug, 1970a,b, 1971, see Erickson, 1973; Hanszen and Trepte, 1971b) but Hanszen and Trepte (1970, 1971a) did show that the effect of an energy spread (or temporal partial coherence) can be represented by an envelope. Frank (1973) noticed that the zeros of the curves of Hanszen and Trepte (1971b) and those of Hirt and Hoppe (1972), representing the effect of source-size on the ideal transfer functions, remained approximately stationary for a wide range of source radii and concluded that it should be possible to derive an envelope representation for this case also. Many studies followed, notably Hahn (1973), Hahn and Seredynski (1974), Bonhomme *et al.* (1973), Beorchia and Bonhomme (1974), Misell (1973), Misell and Atkins (1973), Fejes (1977), Saxton (1977), Frank *et al.* (1978/79), Wade (1978), O'Keefe (1979), Bonnet and Bonhomme (1980), Zemlin and Schiske (1980), Lannes *et al.* (1981) and Krakow (1982).

In these early studies, source-size and energy spread were studied separately but the legitimacy of simply multiplying the two envelopes was soon questioned (Frank, 1976a): was it certain that there was no interaction between the two? A careful study by Wade and Frank (1977) showed that the interaction is usually negligible for axial illumination but could become important for tilted illumination.

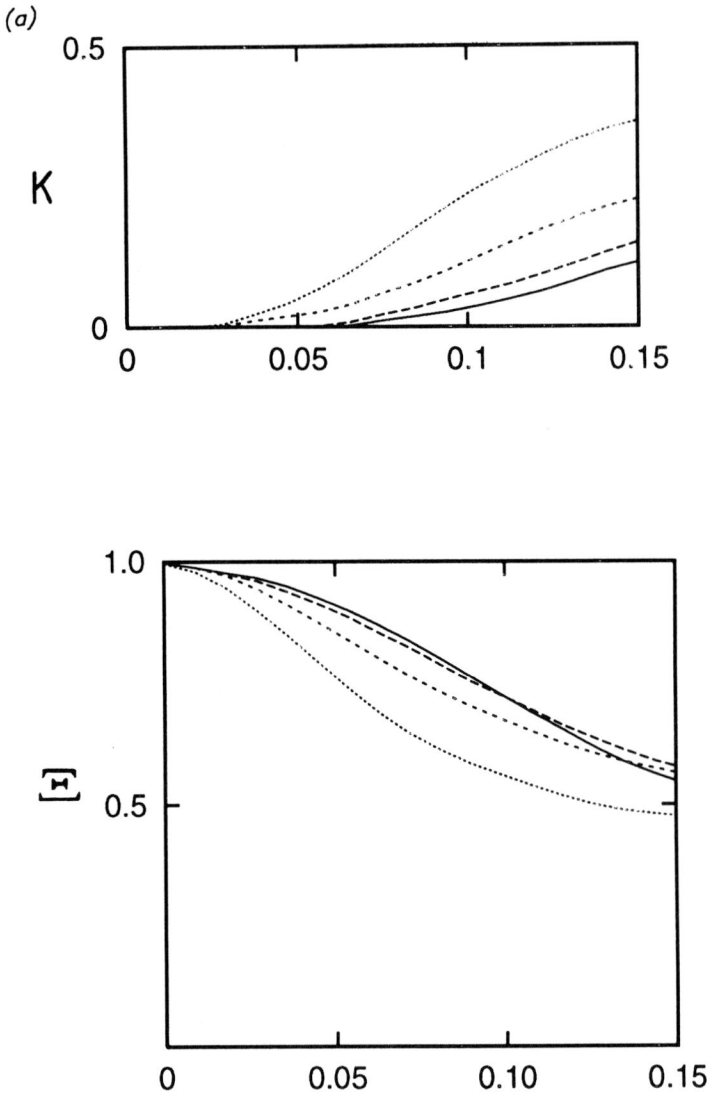

Fig. 66.7: (a) Dissimilarity K, and (b) structural resolving power, Ξ, functions of generalized source-size for four values of defocus: full curve, $\Delta = \sqrt{7}$; long dashes, $\Delta = \sqrt{5}$; short dashes, $\Delta = \sqrt{3}$; dots, $\Delta = 1$.

If the two effects can be safely assumed to be independent, the envelope representations can be obtained much more simply. For non-vanishing energy spread, for example, we associate a probability distribution $H(\Delta)$ with the variation in defocus caused by the distribution of energy. The current density at the image is thus of the form

$$\frac{dj}{j} = H(\Delta)d\Delta \quad \text{and} \quad \int H(\Delta)d\Delta = 1 \qquad (66.60)$$

In the ideal case,

$$j \propto |\psi_i \psi_i^*| = 1 + 2\int (\tilde{\eta}\sin\chi - \tilde{\sigma}\cos\chi)\exp(2\pi i \mathbf{q}\cdot\mathbf{u}_i)\,d\mathbf{q}$$

and so

$$\frac{dj}{d\Delta} = jH(\Delta) \propto H(\Delta) + 2\int (\tilde{\eta}\sin\chi - \tilde{\sigma}\cos\chi)H(\Delta)\exp(2\pi i \mathbf{q}\cdot\mathbf{u}_i)\,d\mathbf{q}.$$

Integration over Δ then gives

$$C(\mathbf{u}_i) = -\iint \tilde{\eta}\{\sin\chi(\Delta)H(\Delta)d\Delta\}\exp(2\pi i \mathbf{q}\cdot\mathbf{u}_i)\,d\mathbf{q}$$

$$+ \iint \tilde{\sigma}\{\cos\chi(\Delta)H(\Delta)d\Delta\}\exp(2\pi i \mathbf{q}\cdot\mathbf{u}_i)\,d\mathbf{q} \qquad (66.61)$$

The defocus Δ will differ by only a small amount from the nominal defocus Δ_o:

$$\Delta = \Delta_o + \delta, \quad , \quad \delta \ll \Delta_o \qquad (66.62)$$

and hence

$$\int \sin\chi(\Delta)H(\Delta)\,d\Delta = \sin\chi(\Delta_o)\int H(\delta)\cos(\pi\lambda q^2\delta)\,d\delta$$

$$- \cos\chi(\Delta_o)\int H(\delta)\sin(\pi\lambda q^2\delta)\,d\delta$$

$$\int \cos\chi(\Delta)H(\Delta)\,d\Delta = \sin\chi(\Delta_o)\int H(\delta)\sin(\pi\lambda q^2\delta)\,d\delta$$

$$+ \cos\chi(\Delta_o)\int H(\delta)\cos(\pi\lambda q^2\delta)\,d\delta \qquad (66.63)$$

If $H(\Delta)$ is even, then the integrals in $\sin(\pi\lambda q^2\delta)$ vanish, leaving

$$\int \sin\chi(\Delta)H(\Delta)\,d\Delta = \sin\chi(\Delta_o)H_c(\lambda q^2)$$

$$\int \cos\chi(\Delta)H(\Delta)\,d\Delta = \cos\chi(\Delta_o)H_c(\lambda q^2)$$

$$(66.64)$$

where

$$H_c(\lambda q^2) = \int H(\delta)\cos(\pi\lambda q^2\delta)\,d\delta \qquad (66.65)$$

We have thus recovered an alternative form of the envelope function (66.54); this approach is convenient if other perturbing effects (lens current fluctuations, in particular) are to be included. A similar study may be made of the envelope corresponding to the source-size; see Frank (1973) or the account in Hawkes (1978c) for details. The most useful result is that for a symmetric source, the phase and amplitude contrast-transfer functions $\sin\chi$ and $\cos\chi$ are modulated by an envelope function of the form

$$t(\boldsymbol{q}) \approx \int \tilde{\alpha}(\boldsymbol{\kappa})\cos\{\boldsymbol{\kappa}\cdot\operatorname{grad}\chi(\boldsymbol{q})\}\,d\boldsymbol{\kappa}$$
$$= \alpha(\operatorname{grad}\chi(\boldsymbol{q}))$$

with $\tilde{\alpha}(\boldsymbol{\kappa}) := |A(\boldsymbol{\kappa},0)|^2$.

66.3 Particular forms of the spectra

There are only a few forms of the function $A(\boldsymbol{\kappa},v)$ for which the integrations in the various approximations for $K_p(\boldsymbol{q})$ can be performed analytically. In practice, only two expressions have been examined, Gaussian distributions and top-hat functions. We shall not consider the latter here ($A(\boldsymbol{\kappa},v) = \text{const}$, $|\kappa| \leq \kappa_o$, $|v| < v_o$, $A(\boldsymbol{\kappa},v) = 0$ elsewhere) since the Gaussian model often matches the experimental conditions rather well. Almost all estimates of the source-size and energy spread based on the measurement of the form of $K_p(\boldsymbol{q})$ have made the assumption that the distributions are Gaussian. Moreover, the rare attempts to distinguish between these two models by studying the attenuation of the transfer function suggest that, given the experimental errors and uncertainties, both are compatible with the measurements.

In the Gaussian model, we write

$$|A(\boldsymbol{\kappa},v)|^2 = C_v \exp\{-C_\kappa(\kappa_x^2 + \kappa_y^2) - v^2/v_m^2\} \qquad (66.66)$$

in which

$$C_v := \frac{\lambda_o^2}{\pi^{3/2}v_m\alpha_c^2} \quad , \quad C_\kappa := \lambda_o^2/\alpha_c^2 \qquad (66.67)$$

where α_c is a measure of the aperture angle of the beam furnished by the condenser system and hence of the source-size while v_m is the mean energy spread. We see from the very outset that the contributions of $\boldsymbol{\kappa}$ and v

to $|A|^2$ are separable; we recall that this was one of the conditions for obtaining $K_p(\boldsymbol{q})$ in the product form (66.58). We note too that $|A|^2$ is even in $\boldsymbol{\kappa}$ and in v.

Let us now reconsider the integral $\tilde{a}_e$ (66.45):

$$
\begin{aligned}
\tilde{a}_e &= \int a_e \exp(\mathrm{i}T_v)\, dv \\
&= C_v \exp(-C_\kappa \kappa^2) \int \exp(-v^2/v_m^2) \exp(\mathrm{i}T_v v)\, dv
\end{aligned}
\tag{66.68}
$$

After introducing the expression for T_v (66.40) and rearranging the exponent, we can integrate over v, giving

$$
\begin{aligned}
\int_{-\infty}^{\infty} & \exp(-v^2/v_m^2) \exp(\mathrm{i}T_v v)\, dv \\
&= \pi^{1/2} v_m \exp\left(-\frac{1}{4}T_v^2 v_m^2\right) \\
&= \pi^{1/2} v_m \exp\left\{-\frac{1}{4}\pi^2 v_m^2 C_c'^2 \lambda_o^2 (q^2 + 2\boldsymbol{q}\cdot\boldsymbol{\kappa})^2\right\}
\end{aligned}
\tag{66.69}
$$

We now have to evaluate

$$
\int \tilde{a}_e \cos T(\boldsymbol{q},\boldsymbol{\kappa})\, d\boldsymbol{\kappa} = \Re \int \tilde{a}_e \exp\{\mathrm{i}T(\boldsymbol{q},\boldsymbol{\kappa})\}\, d\boldsymbol{\kappa}
\tag{66.70}
$$

The terms involving $\boldsymbol{\kappa}$ are of the form

$$
\exp\left[-\left\{L\kappa^2 + 2M\boldsymbol{q}\cdot\boldsymbol{\kappa} + N(\boldsymbol{q}\cdot\boldsymbol{\kappa})^2\right\}\right]
\tag{66.71}
$$

neglecting a small term in $\kappa^2(\boldsymbol{q}\cdot\boldsymbol{\kappa})$; we give the explicit forms of L, M and N below. Since the result of integrating over $\boldsymbol{\kappa}$ in (66.70) is the same for all values of $|\boldsymbol{q}|$, irrespective of azimuth, we may consider the special case $\boldsymbol{q} = (q,0)$ without introducing any error. The integral can then be performed analytically; it takes the form

$$
\int \exp\left\{-(L\kappa^2 + 2Mq\kappa_x + Nq^2\kappa_x^2)\right\}\, d\boldsymbol{\kappa}
\tag{66.72}
$$

The general formula

$$
\int_{-\infty}^{\infty} \exp(-ax^2 - 2bx)\, dx = (\pi/a)^{1/2} \exp(b^2/a^2) \qquad \Re a > 0
\tag{66.73}
$$

which, for $b = 0$, reduces to $(\pi/a)^{1/2}$, enables us to integrate over κ_y immediately and (66.72) becomes

$$(\pi/L)^{1/2} \int \exp -\{(L + Nq^2)\kappa_x^2 + 2Mq\kappa_x\} \, d\kappa_x$$

which, again using (66.73), is equal to

$$\left(\frac{\pi}{L}\right)^{1/2} \left(\frac{\pi}{L + Nq^2}\right)^{1/2} \exp\left(\frac{M^2 q^2}{L + Nq^2}\right) \tag{66.74}$$

The orders of magnitude of the many quantities occurring in this expression are very different and it can be replaced by an approximate formula that is still highly accurate and shows much more clearly the influence of the various parameters. In order to establish this, we examine L, M and N more closely.

$$L = C_\kappa + i\pi C_s \lambda_o^3 q^2 = \frac{\lambda_o^3}{\alpha_c^2}(1 + i\pi C_s \lambda_o \alpha_c^2 q^2)$$

$$M = \pi\lambda_o \left\{ \frac{1}{2}\pi v_m^2 C_c'^2 \lambda_o q^2 + i(C_s \lambda_o^2 q^2 - \Delta_o) \right\}$$

$$N = \pi\lambda_o^2(\pi v_m^2 C_c'^2 + 2iC_s\lambda_o)$$

$$L + Nq^2 = C_\kappa + 3i\pi C_s \lambda_o^3 + \pi^2 v_m^2 C_c'^2 \lambda_o^2 q^2$$

$$= \frac{\lambda_o^2}{\alpha_c^2}\left(1 + 3i\pi C_s \lambda_o \alpha_c^2 q^2 + \pi^2 v_m^2 C_c'^2 \alpha_c^2 q^2\right) \tag{66.75}$$

For typical values of the parameters ($C_s \approx 10^{-3}$ m, $C_c' \approx 10^{-8}$ m/V, $\lambda \approx 4 \times 10^{-12}$ m, $\alpha_c \lesssim 10^{-4}$ rad, $v_m \approx 1$ V, $q \approx 10^8$ m^{-1}), the second term of L and the second and third terms of $L + Nq^2$ are found to be much smaller than the first and we therefore write

$$L =: \frac{\lambda_o^2}{\alpha_c^2}(1 + ir_1) \qquad\qquad r_1 := \pi C_s \lambda_o \alpha_c^2 q^2 \ll 1$$

$$L + Nq^2 = \frac{\lambda_o^2}{\alpha_c^2}(1 + 3ir_1 + r_2) \qquad r_2 := \pi^2 v_m^2 C_c'^2 \lambda_o \alpha_c^2 q^2 \ll 1 \tag{66.76}$$

The first two factors in (66.74) may thus be written

$$\left(\frac{\pi}{L}\right)^{\frac{1}{2}} \left(\frac{\pi}{L + Nq^2}\right)^{\frac{1}{2}} \approx \frac{\pi\alpha_c^2}{\lambda_o^2}\left(1 - 2ir_1 - \frac{1}{2}r_2\right) \tag{66.77}$$

and it will be convenient, if artificial, to regard $(1 - 2ir_1 - \frac{1}{2}r_2)$ as the first term in the expansion of an exponential, so that

$$\left(\frac{\pi}{L}\right)^{\frac{1}{2}} \left(\frac{\pi}{L + Nq^2}\right)^{\frac{1}{2}} \approx \frac{\pi \alpha_c^2}{\lambda_o^2} \exp\left(-2ir_1 - \frac{1}{2}r_2\right) \tag{66.78}$$

The argument of the exponential in (66.74) may be written as $M^2 q^2 \alpha_c^2 / \lambda_o^2$,

$$\frac{M^2 q^2 \alpha_c^2}{\lambda_o^2} = \pi^2 q^2 \alpha_c^2 \left\{ \frac{1}{2} \pi v_m^2 C_c'^2 \lambda_o q^2 + i(C_s \lambda_o^2 q^2 - \Delta_o) \right\}^2$$

$$\approx -\pi^2 q^2 \alpha_c^2 (C_s \lambda_o^2 q^2 - \Delta_o)^2 \tag{66.79}$$

Collecting up all the factors, we finally obtain

$$C_v(\pi^{\frac{1}{2}} v_m) \left(\frac{\pi \alpha_c^2}{\lambda_o^2}\right) \exp\left(-2ir_1 - \frac{1}{2}r_2\right) \exp\{iT(q)\}$$

$$\times \exp\left\{-\pi^2 q^2 \alpha_c^2 (C_s \lambda_o^2 q^2 - \Delta_o)^2\right\}$$

But $C_v = \lambda_o^2 / \pi^{\frac{3}{2}} v_m \alpha_c^2$ and so the phase contrast-transfer function becomes

$$K_p(\mathbf{q}) = \sin\{-\chi(\mathbf{q}) - 2r_1\} t_v t_\kappa \exp\left(-\frac{1}{2}r_2\right)$$

$$= \sin\left\{\pi \lambda_o q^2 \left(\Delta_o - \frac{1}{2}C_s \lambda_o^2 q^2 - 2C_s \alpha_c^2\right)\right\} t_v t_\kappa \tag{66.80a}$$

$$\times \exp\left(-\frac{1}{2}\pi^2 v_m^2 C_c'^2 \alpha_c^2 q^2\right)$$

with

$$t_v = \exp\left(-\frac{1}{4}\pi^2 v_m^2 C_c'^2 \lambda_o^2 q^4\right) \tag{66.81a}$$

$$t_\kappa = \exp\left\{-\pi^2 q^2 \alpha_c^2 (C_s \lambda_o^2 q^2 - \Delta_o)^2\right\} \tag{66.81b}$$

This expression reveals very strikingly all the features of interest of the modulation of $\sin T(\mathbf{q})$ by the source-size and energy spread: modulation by a source-size term (t_κ) and an energy spread term t_v; further modulation (usually neglected and indeed usually negligible) by a mixed term $\exp(-\pi^2 v_m^2 C_c'^2 \alpha_c^2 q^2 / 2)$; and a small shift in the zeros of $\sin \chi(\mathbf{q})$, which is likewise usually neglected.

It can be shown that a small twofold axial astigmatism, which is to be expected in practice, can be included to a good approximation simply by replacing Δ_o by an azimuth-dependent quantity:

$$\Delta_o \rightarrow \Delta_o + C_a \cos 2\varphi_o,$$

C_a being the astigmatic length constant referred back to the object (like Δ_o) and φ the azimuth at the exit plane of the object.

Formula (66.80a) for $K_p(q)$ contains numerous system constants. The situation can be improved slightly by the introduction of reduced quantities (65.31–32) but it is reasonable to look for a set of system parameters that are optimal in some well-defined sense by considering the working conditions that are advantageous from the point of view of the ideal transfer function, $\sin \chi(q)$ and then examine the influence of the envelope functions on these. This will be the subject of the next section. First, however, we extract the formulae that are most useful in practice from among the foregoing calculations and give the equivalent expressions in reduced coordinates.

Phase contrast-transfer function

(i) for vanishing source-size and energy spread

$$K_p(q) = \sin\left\{\pi q^2 \lambda \left(\Delta_o - \frac{1}{2}C_s\lambda^2 q^2\right)\right\}$$

$$\hat{K}_p(Q) = \sin\left\{\pi\left(DQ^2 - \frac{1}{2}Q^4\right)\right\} \qquad (66.19b)$$

(ii) for Gaussian energy spread and angular distribution (general formula)

$$K_p(q) = t_v(q)t_\kappa(q)\exp\left(-\frac{1}{2}\pi^2 v_m^2 C_c'^2 \alpha_c^2 q^2\right)$$

$$\times \sin\left\{\pi q^2 \lambda \left(\Delta_o - \frac{1}{2}C_s\lambda^2 q^2 - 2C_s\alpha_c^2\right)\right\} \qquad (66.80b)$$

$$\hat{K}_p(Q) = \hat{t}_v(Q)\hat{t}_\kappa(Q)\exp\left(-\frac{\pi}{2}B^2\alpha^2 Q^2\right)$$

$$\times \sin\left\{\pi Q^2 \left(D - \frac{1}{2}Q^2 - 2\alpha^2\right)\right\} \qquad (66.80c)$$

(iii) for Gaussian energy spread and angular distribution (approximate formula, usually adopted)

$$K_p(q) = t_v(q)t_\kappa(q)\sin\left\{\pi q^2 \lambda \left(\Delta_o - \frac{1}{2}C_s\lambda^2 q^2\right)\right\}$$

$$\hat{K}_p(Q) = \hat{t}_v(Q)\hat{t}_\kappa(Q)\sin\left\{\pi Q^2 \left(D - \frac{1}{2}Q^2\right)\right\} \qquad (66.82)$$

in which

$$Q := (C_s \lambda_o^3)^{\frac{1}{4}} q \tag{65.31}$$

$$D := \Delta_o / (C_s \lambda_o)^{\frac{1}{2}} \tag{65.32}$$

$$\alpha := \alpha_c (C_s / \lambda_o)^{\frac{1}{4}} \tag{66.83a}$$

$$B := \frac{C_c' v_m}{(C_s \lambda_o)^{\frac{1}{2}}} \tag{66.83b}$$

and

$$t_v(q) = \exp\left(-\frac{1}{4}\pi^2 v_m^2 C_c'^2 \lambda_o^2 q^4\right) \tag{66.81a}$$

$$\hat{t}_v(Q) = \exp\left(-\frac{1}{4}\pi^2 \frac{v_m^2 C_c'^2}{C_s \lambda} Q^4\right) = \exp\left(-\frac{\pi^2}{4} B^2 Q^4\right) \tag{66.81a'}$$

$$t_\kappa(q) = \exp\left\{-\pi^2 q^2 \alpha_c^2 (C_s \lambda_o^2 q^2 - \Delta_o)^2\right\} \tag{66.81b}$$

$$\hat{t}_\kappa(Q) = \exp\left\{-\pi^2 Q^2 \alpha^2 (Q^2 - D)^2\right\} \tag{66.81b'}$$

We recall that

$$C_c' := \frac{\gamma C_c}{\hat{U}}$$

66.4 Optimum defocus and resolution limit

The complete expression (66.80) for the phase contrast-transfer function depends on bewilderingly many instrumental parameters. We therefore consider first the ideal transfer function $\sin\chi(q)$, for which the influence of the spectral distributions is neglected, and then see how the latter affect it.

A common feature of all the graphs in Figs. 66.4–5 is the increasingly rapid oscillation of $\hat{K}(Q)$ for values of Q beyond about 1.25. Changes in the sign of $\hat{K}(Q)$ make the interpretation of the image complicated or even impossible. We must therefore choose the free parameters D (defocus of the microscope) and q_c (radius of the objective aperture) in $\sin\chi(Q)$ in such a way that we obtain as wide an interval with no change of sign of K as possible.

The graphs of Fig. 66.4 suggest that curve (b), $D = 1$ with $Q_c = \sqrt{2}$, is a good choice. For this, most of the interval $0 \leq Q \leq \sqrt{2}$ is available and the function $\hat{K}(Q)$ has a broad maximum within it. Equation (66.19b) now becomes

$$\hat{K}(Q) = \begin{cases} \sin\left\{\pi\left(DQ^2 - \frac{1}{2}Q^4\right)\right\} & \text{for } Q \leq \sqrt{2} \\ 0 & \text{for } Q > \sqrt{2} \end{cases} \tag{66.84}$$

The optimum defocus $D = 1$ corresponds to $\Delta_{opt} = (\lambda_o C_s)^{1/2}$ (65.28) in agreement with the value obtained in Chapter 60 (60.65), the Scherzer defocus. The cutoff $Q_c = \sqrt{2}$ corresponds to the spatial frequency q_c (66.18)

$$q_c = \left(4/C_s \lambda_o^3\right)^{\frac{1}{4}} \qquad (66.85a)$$

and with $\theta = \lambda_o q$, the maximum aperture angle on the object side becomes

$$\theta_c = \lambda_o q_c = (4\lambda_o/C_s)^{\frac{1}{4}} \qquad (66.85b)$$

in accord with expression (60.64) for α_{opt}.

This agreement with our earlier estimates of optimal conditions shows that these are indeed realistic; reasoning based on the transfer function shows clearly what is meant by 'optimal'. We can also see that although the choice $D = 1$ is a good general value, there may well be conditions in which other choices are better. For very high resolution work, $D = \sqrt{3}$ or even $D = \sqrt{5}$ could be preferable; with $D = \sqrt{3}$, for example, spatial frequencies in the second band would be interpreted directly and the lower resolution information discarded (or processed). The cutoff in the second band is better than $Q = \sqrt{2}$. Images have even been obtained using the information in several peaks of the oscillations, after tuning the microscope very carefully to the spatial frequency to be detected in the specimen (Hashimoto et al., 1977; Endoh and Hashimoto, 1977; Hashimoto and Endoh, 1978); this 'aberration-free focusing' is, however, rather a tour de force than a practical technique, for the microscopist is choosing a priori what is to be seen.

Returning to the case $D = 1$ (66.84), we now estimate the resolution limit with the aid of the uncertainty principle, $\Delta x \cdot \Delta p_x \geq h = 2\pi\hbar$. The result will not be over-optimistic if we identify Δx with the resolution limit d_{min} and set $\Delta p_x = hq_c$, giving

$$d_{min} = 1/q_c = 0.7 \left(C_s \lambda_o^3\right)^{\frac{1}{4}} \qquad (66.86)$$

This is slightly larger than $\bar{r}/|M|$ as given by (60.66) or (60.71) and is probably more realistic.

The contrast transfer function changes quite rapidly in the neighbourhood of the optimum defocus $D = 1$, as can be seen in Fig. 66.8. If we regard a contrast of 80% as acceptable in practice, too small a value of defocus is clearly worse than too large. This is in accordance with the arguments of Section 60.5, leading to (60.62).

We now examine more closely the effects of non-vanishing source-size and energy spread. Equation (66.80) can be cast into a much simpler

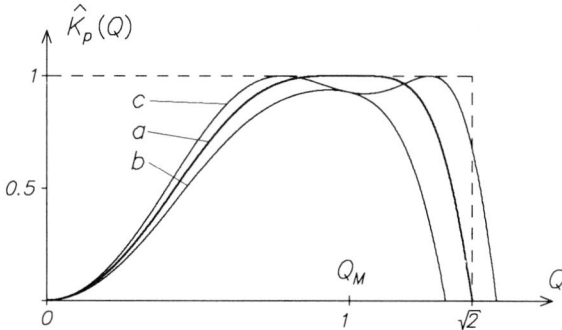

Fig. 66.8: Undamped contrast transfer function $\hat{K}_p(Q)$ for various values of the defocus D, close to $D = 1$. *Line a*: best choice $D = 1$; this has a broad maximum at $Q_M = 1$. *Line b*: too *small* a value of D, the maximum of $\hat{K}_p$ is less than unity and the zero is reached for $Q < \sqrt{2}$. *Line c*: too *large* a value of D; the curve has a local minimum at Q_M but the limiting zero is shifted beyond $Q = \sqrt{2}$.

form by introducing the reduced spatial frequency Q (65.31) and two more dimensionless parameters (cf. 66.83), A and B, which have a clear physical meaning:

$$A := \frac{2\alpha_c}{\theta_c} = \alpha_c(4C_s/\lambda_o)^{\frac{1}{4}} = \alpha\sqrt{2} \qquad (66.87a)$$

is the ratio of the apertures on either side of the specimen and

$$B := \frac{1}{d_{min}}\frac{\gamma C_c v_m \theta_c}{2\hat{U}} = \frac{\gamma C_c v_m}{\hat{U}(\lambda_o C_s)^{\frac{1}{2}}} = \frac{C_c' v_m}{(\lambda_o C_s)^{\frac{1}{2}}} \qquad (66.87b)$$

is the ratio of the radius of the chromatic aberration disc and the resolution limit. Both parameters should be made as small as is experimentally possible in order to come close to the theoretical limit of resolution: $A \ll 1$ and $B \ll 1$.

Introducing A and B into (66.80), the transfer function takes the form

$$\hat{K}(Q) = \sin\left\{\pi Q^2\left(D - \frac{1}{2}Q^2 - A^2\right)\right\}\exp\left\{-\frac{\pi^2}{2}A^2 Q^2(D - Q^2)^2\right\}$$

$$\times \exp\left\{-\frac{\pi^2}{4}B^2 Q^2(A^2 + Q^2)\right\} \qquad (66.88a)$$

which reduces to

$$\hat{K}(Q) = \sin\left\{\pi Q^2\left(1 - \frac{1}{2}Q^2 - A^2\right)\right\}\exp\left\{-\frac{\pi^2}{2}A^2 Q^2(1 - Q^2)^2\right\}$$

$$\times \exp\left\{-\frac{\pi^2}{4}B^2 Q^2(A^2 + Q^2)\right\} \qquad (66.88b)$$

when $D = 1$. Careful inspection of these formulae shows that the *axial chromatic aberration* will in general have a *more deleterious effect* than the source-size, since the exponential term involving B (energy spread) decreases monotonically with Q whereas the first exponential factor, which involves A (source size) returns to unity at $Q = D^{1/2}$; this coincides with the maximum of the sine function for $D = 1$, $\sqrt{3}$, $\sqrt{5}$, ... for negligible values of A and is always close to the maximum. Beyond $Q = D^{1/2}$, this factor falls off monotonically. The separate effects of the two envelope functions are shown in Figs. 66.9 and 66.10; if B is negligible (no energy spread) and $D = 1$, we have

$$\hat{K}(Q) = \sin\left\{\pi Q^2\left(1 - A^2 - \frac{1}{2}Q^2\right)\right\}\exp\left\{-\frac{\pi^2}{2}A^2 Q^2(1 - Q^2)^2\right\}$$

$$(66.89a)$$

whereas if A is set to zero (vanishing source-size), we find

$$\hat{K}(Q) = \sin\left\{\pi Q^2\left(1 - \frac{1}{2}Q^2\right)\right\}\exp\left(-\frac{\pi^2}{4}B^2 Q^4\right) \qquad (66.89b)$$

These functions are plotted in Figs. 66.9a–d ($B = 0$) and 66.10a–d ($A = 0$) for a range of values of A and B. The corresponding curves for $D^2 = 3$ are shown in Figs. 66.9e–h and 66.10e–h.

For values of Q beyond $\sqrt{2}$, the damping by the exponential factors may become so strong that the oscillations of $\hat{K}(Q)$ soon become insignificant. No useful information then survives in the image. This is shown by the graphs of Fig. 66.9, which confirm that a cutoff at $Q_c = \sqrt{2}$ is a prudent choice. These curves also justify some of the approximations made in deriving (66.80) and hence (66.88): the integrals were evaluated on the assumption that the upper limits were infinite and hence that $Q_c \to \infty$. Formally, therefore, we should introduce finite limits if we wish to include the effect of setting the cutoff at $Q_c = \sqrt{2}$, which would give a much more complicated result involving the Gaussian error function, Fresnel integrals and Bessel functions. The strong damping, together with the many approximations made during the derivation, show that it is unnecessary to pursue such complicated reasoning: (66.80) will in practice be a perfectly adequate approximation.

When the ultimate resolution of the microscope is required, exceptional efforts must be made to keep A and B small and information will then survive in the image from several oscillations of the sinusoidal function. It is then usual to distinguish between the instrumental resolution and the information limit. The *instrumental resolution* indicates the capacity of the microscope to transfer spatial frequencies faithfully to the image and is

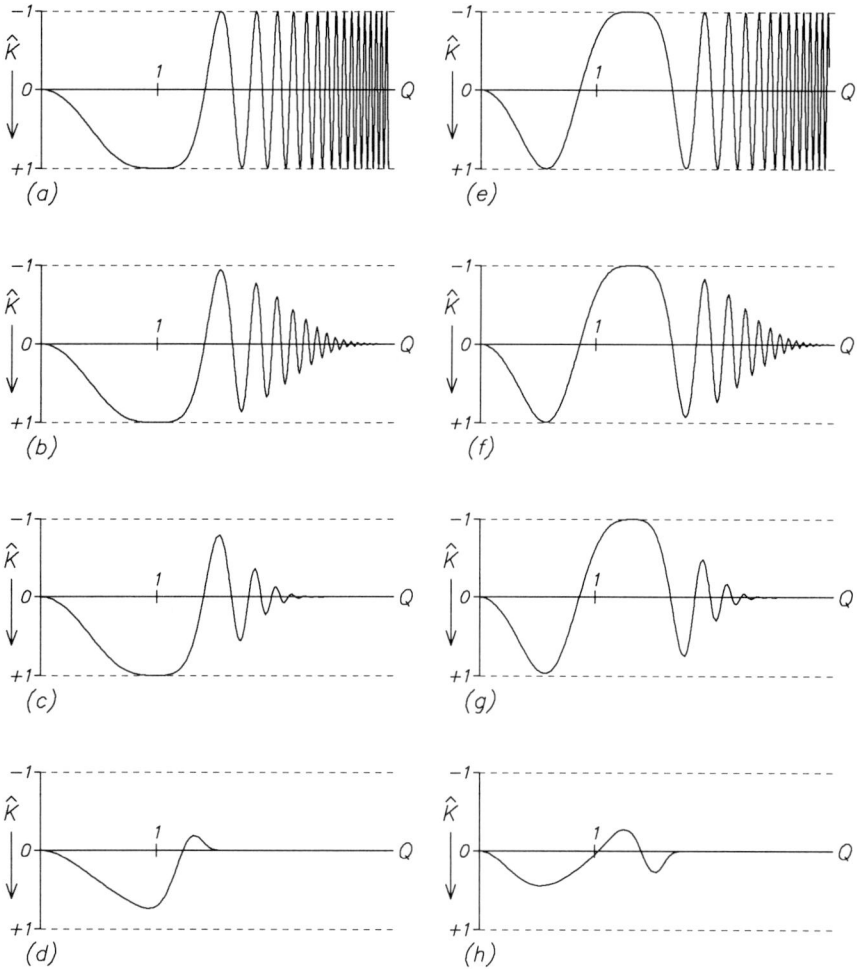

Fig. 66.9: Damping effect of a non-vanishing condenser aperture $A \neq 0, B = 0$ on the phase contrast transfer function for $D = 1$ (a–d) and $D = \sqrt{3}$ (e–h).
(a) and (e) Undamped curve, $A = B = 0$, (b) and (f) $A = 0.05$, (c) and (g) $A = 0.1$, (d) and (h) $A = 0.5$.

usually defined to be proportional to $(C_s \lambda^3)^{1/4}$ where the constant of proportionality takes one of the values we have found earlier. The *information limit* is defined in terms of the rate of fall-off of the exponential functions, typically, the value of Q for which the product of the two exponentials in (66.88a) reaches $1/e$. For careful discussion of these resolution limits, see van Dyck (1989, 1991, 1992), van Dyck and de Jong (1992), de Jong and

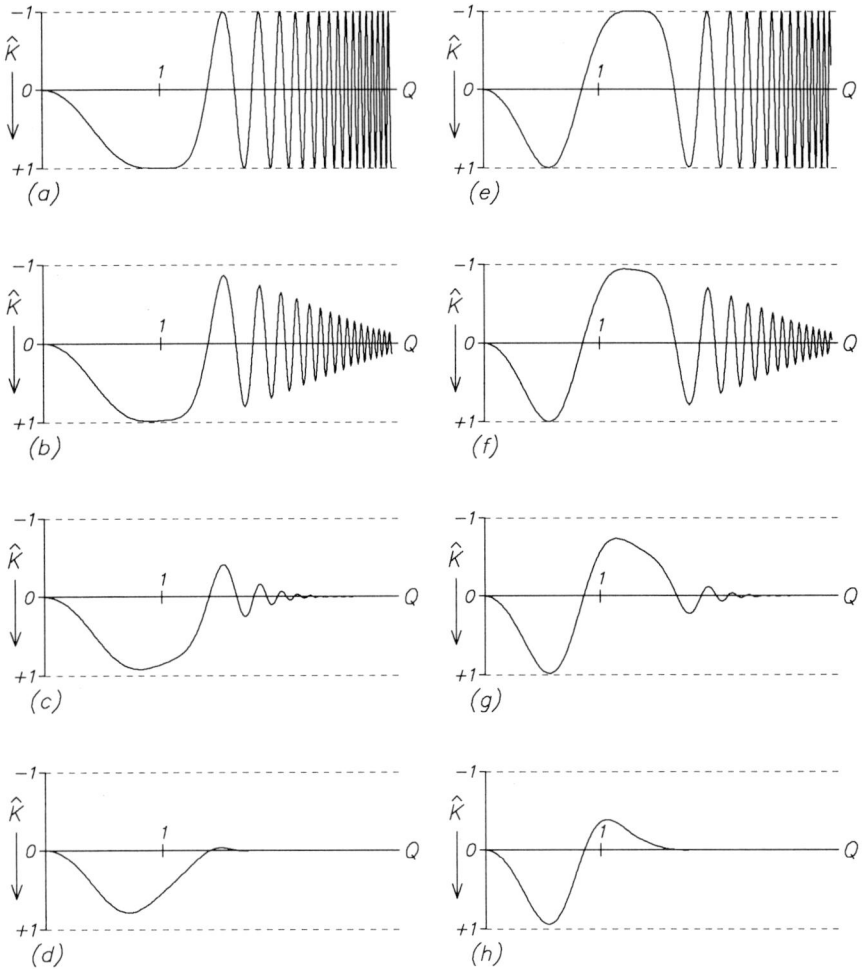

Fig. 66.10: Damping effect of a non-vanishing energy spread $B \neq 0, A = 0$ on the phase contrast transfer function for $D = 1$ (a–d) and $D = \sqrt{3}$ (e–h).
(a) and (e) Undamped curve, $A = B = 0$, (b) and (f) $B = 0.1$, (c) and (g) $B = 0.25$, (d) and (h) $B = 0.5$.

van Dyck (1993), Sarikaya and Howe (1992), O'Keefe and Spence (1991), O'Keefe (1992), van den Bos (1991, 1992) and Tsuno (1993).

A convenient way of assessing the limit of information transfer has been introduced by Frank (1975a), based on his earlier work on the effect of specimen drift (1969). Two images of a specimen with a reasonably uniform spatial frequency spectrum, the second shifted laterally with respect to the first, are superimposed. In the diffractogram, the familiar ring pattern will be seen, modulated by a sinusoidal fringe pattern, the extent of which gives a good estimate of the practical resolution. The effect of source-size and energy spread and the possibility of estimating these from the fringe pattern are examined in a later paper (Frank, 1976a). See Zemlin and Weiss (1993) for a careful examination of this idea.

The use of the spectral signal-to-noise ratio as a measure of the resolution, of correlation-averaged images in particular (though not exclusively), is very fully analysed by Unser *et al.* (1987a,b).

66.5 Extensions of the theory

The foregoing sections have been devoted to the imaging of weakly scattering specimens in axial bright-field imaging conditions. Although this is by far the most widely-used mode, there are others that are also encountered and the present section will be concerned with these. In particular, we consider bright-field imaging with a tilted incident beam and its natural extension, hollow-cone imaging (Section 66.5.1). We also devote a short section to the effect on transfer theory of aberrations, coma in particular, that render the conditions anisoplanatic (Section 66.5.2). Unusual forms of the aperture function T_A are considered briefly in Section 66.5.3. We then examine phase contrast for crystalline specimens that do not necessarily scatter weakly (Section 66.5.4). Finally, as promised earlier, we comment on transforms other than the Fourier transform that map convolution products into direct products (Section 66.5.5).

66.5.1 Tilted and hollow-cone illumination
We first consider the case of a plane monochromatic illuminating wave no longer travelling in the direction of the optic axis (Hawkes, 1980a). This incident wave is now written

$$\psi = \exp(2\pi i \boldsymbol{m} \cdot \boldsymbol{x}_o) \tag{66.90}$$

and with the specimen transparency represented as before (66.1) by $1 - \sigma_o + i\eta_o$, the emergent object wave will be

$$\psi_o(x_o, y_o) = \{1 - \sigma_o(x_o, y_o) + i\eta_o(x_o, y_o)\} \exp(2\pi i \boldsymbol{m} \cdot \boldsymbol{x}_o) \tag{66.91a}$$

with spectrum

$$S_o = \delta(\boldsymbol{q} - \boldsymbol{m}) - \tilde{\sigma}(\boldsymbol{q} - \boldsymbol{m}) + i\tilde{\eta}(\boldsymbol{q} - \boldsymbol{m}) \tag{66.91b}$$

In the image plane, therefore,

$$S_i(\boldsymbol{q}) = T_A(\boldsymbol{q})T_L(\boldsymbol{q})\{\delta(\boldsymbol{q}-\boldsymbol{m}) - \tilde{\sigma}(\boldsymbol{q}-\boldsymbol{m}) + i\tilde{\eta}(\boldsymbol{q}-\boldsymbol{m})\} \qquad (66.92)$$

and hence

$$\psi_i(\boldsymbol{u}_i) = \int T_A(\boldsymbol{q})T_L(\boldsymbol{q})\{\delta(\boldsymbol{q}-\boldsymbol{m}) - \tilde{\sigma}(\boldsymbol{q}-\boldsymbol{m}) + i\tilde{\eta}(\boldsymbol{q}-\boldsymbol{m})\}$$
$$\times \exp(2\pi i \boldsymbol{u}_i \cdot \boldsymbol{q})\,d\boldsymbol{q} \qquad (66.93a)$$

or with

$$\boldsymbol{p} := \boldsymbol{q} - \boldsymbol{m} \qquad (66.94)$$

$$\psi_i(\boldsymbol{u}_i) = \exp(2\pi i \boldsymbol{u}_i \cdot \boldsymbol{m})\int T_A(\boldsymbol{p}+\boldsymbol{m})T_L(\boldsymbol{p}+\boldsymbol{m})\{\delta(\boldsymbol{p}) - \tilde{\sigma}(\boldsymbol{p}) + i\tilde{\eta}(\boldsymbol{p})\}$$
$$\times \exp(2\pi i \boldsymbol{u}_i \cdot \boldsymbol{p})\,d\boldsymbol{p} \qquad (66.93b)$$

The image contrast may again be written as the sum of an amplitude term and a phase term,

$$C(\boldsymbol{u}_i) := \frac{1 - |\psi_i|^2}{1 + |\psi_i|^2} =: C_a(\boldsymbol{u}_i) + C_p(\boldsymbol{u}_i) \qquad (66.95)$$

and as before, we introduce the phase and amplitude spectra:

$$S_a = \mathcal{F}^-(C_a) = \frac{1}{2}\tilde{\sigma}(\boldsymbol{p})\{T_A^*(\boldsymbol{m})T_L^*(\boldsymbol{m})T_A(\boldsymbol{m}+\boldsymbol{p})T_L(\boldsymbol{m}+\boldsymbol{p})$$
$$+ T_A(\boldsymbol{m})T_L(\boldsymbol{m})T_A^*(\boldsymbol{m}-\boldsymbol{p})T_L^*(\boldsymbol{m}-\boldsymbol{p})\}$$

$$S_p = \mathcal{F}^-(C_p) = -\frac{1}{2}i\tilde{\eta}(\boldsymbol{p})\{T_A^*(\boldsymbol{m})T_L^*(\boldsymbol{m})T_A(\boldsymbol{m}+\boldsymbol{p})T_L(\boldsymbol{m}+\boldsymbol{p})$$
$$-T_A(\boldsymbol{m})T_L(\boldsymbol{m})T_A^*(\boldsymbol{m}-\boldsymbol{p})T_L^*(\boldsymbol{m}-\boldsymbol{p})\} (66.96)$$

We must pause here to consider the limits of integration. In the presence of a real aperture, small enough to have appreciable effects on the integration, the above expressions are not correct, since the limits for the first and second terms in C_a and C_p will be different and hence the above expressions for S_a and S_p cannot be obtained by Fourier transformation. We shall therefore assume that there is no physical aperture, or in practice an aperture so large that the (tilted) main beam does not come near to it and hence that the integrals can be taken to infinity. Conversely, we leave the term T_A in the formulae because, as we saw earlier, the effects of finite energy spread and source-size can be incorporated in this term.

The phase and amplitude transfer functions now have the form

$$K_a(p) = \frac{1}{2}T_A(m)\{T_A(m+p)T_L^*(m)T_L(m+p)$$
$$+ T_A(m-p)T_L(m)T_L^*(m-p)\} \quad (66.97a)$$

$$K_p(p) = -\frac{1}{2}iT_A(m)\{T_A(m+p)T_L^*(m)T_L(m+p)$$
$$- T_A(m-p)T_L(m)T_L^*(m-p)\} \quad (66.97b)$$

in which we have taken T_A to be real. With $T_L(q) = \exp\{-i\chi(q)\}$ (65.30), these become

$$K_a(p) = \frac{1}{2}T_A(m)\,[T_A(m+p)\exp i\{\chi(m)-\chi(m+p)\}$$
$$+ T_A(m-p)\exp -i\{\chi(m)-\chi(m-p)\}]$$

$$K_p(p) = -\frac{1}{2}iT_A(m)\,[T_A(m+p)\exp i\{\chi(m)-\chi(m+p)\}$$
$$- T_A(m-p)\exp -i\{\chi(m)-\chi(m-p)\}](66.98)$$

We now consider the special case of vanishing source-size and energy spread, $T_A = 1$. We introduce the reduced coordinates

$$\boldsymbol{P} = (C_s\lambda^3)^{\frac{1}{4}}\boldsymbol{p} \qquad \boldsymbol{M} = (C_s\lambda^3)^{\frac{1}{4}}\boldsymbol{m}$$
$$D = \Delta/(C_s\lambda)^{\frac{1}{2}} \qquad (66.99)$$

as in (60.31–32) and rewrite the transfer functions as follows:

$$K_a(p) \rightarrow \hat{K}_a(\boldsymbol{P}) = \exp(-2\pi i\alpha)\cos\pi\beta \qquad (66.100a)$$
$$K_p(p) \rightarrow \hat{K}_p(\boldsymbol{P}) = \exp(-2\pi i\alpha)\sin\pi\beta \qquad (66.100b)$$

in which

$$\alpha(\boldsymbol{P}) = \boldsymbol{P}\cdot\boldsymbol{M}\,(P^2 + M^2 - D)$$
$$\beta(\boldsymbol{P}) = DP^2 - \frac{1}{2}\{P^4 + 2P^2M^2 + 4(\boldsymbol{P}\cdot\boldsymbol{M})^2\} \qquad (66.101)$$

The function $\hat{K}_p(\boldsymbol{P})$ has zeros wherever $\beta(\boldsymbol{P}) = n$ (n an integer or zero) and hence when

$$P^4 + 2(M^2 - D)P^2 + 4(\boldsymbol{P}\cdot\boldsymbol{M})^2 = 2n \qquad (66.102a)$$

Setting $P^2 =: r$ and $\boldsymbol{P} \cdot \boldsymbol{M} = PM \cos\theta$, this becomes

$$r^2 + 2(M^2 + 2M^2 \cos^2\theta - D)r = 2n \qquad (66.102b)$$

which is the equation in polar form (r, θ) for the zeros of the phase contrast-transfer function. The form of the corresponding curves is governed by the relative magnitude of $\boldsymbol{M}$ and $\boldsymbol{D}$. For small tilt and non-vanishing defocus, the influence of $\cos\theta$ will be small and the curves resemble those for axial illumination. For large tilt and modest defocus, the linear term dominates and (66.100b) resembles the polar form of the equation for an ellipse.

Returning to (66.102b), it is interesting to calculate the number of times the function returns to zero along the axes ($\theta = 0$ or π, $\theta = \pi/2$ or $3\pi/2$). For these values of θ, (66.102b) becomes

$$r^2 + 2(3M^3 - D)r = 2n$$
$$r^2 + 2(M^2 - D)r = 2n$$

and so

$$
\begin{aligned}
r = D - 3M^2 \pm \left\{(D - 3M^2)^2 + 2n\right\}^{\frac{1}{2}} \quad & \theta = 0, \pi \\
r = D - M^2 \pm \left\{(D - M^2)^2 + 2n\right\}^{\frac{1}{2}} \quad & \theta = \pi/2, 3\pi/2
\end{aligned}
\qquad (66.103)
$$

For negative values of n, therefore, r and hence P^2 may have two positive values. The curves then consist of closed loops with separate branches. Some examples are shown in Fig. 66.11.

It is, however, essential to consider the effect of source-size and energy spread in the case of tilted illumination since their effect is more complicated than the simple attenuation that we have encountered for axial illumination. These effects can be included straightforwardly provided that the envelope representation (66.58) is acceptable. We consider only the phase contrast-transfer function, $K_p(\boldsymbol{p})$ of (66.97b); the amplitude term $K_a(\boldsymbol{p})$ can be analysed in a similar fashion if required.

Instead of (66.97b), we find (with 66.99)

$$
\begin{aligned}
\hat{K}_p(\boldsymbol{P}) = -\frac{1}{2}\mathrm{i}\,\big\{ & T_L(\boldsymbol{M} + \boldsymbol{P})T_L^*(\boldsymbol{M})t_\kappa^+ t_v^+ \\
& - T_L^*(\boldsymbol{M} - \boldsymbol{P})T_L(\boldsymbol{M})t_\kappa^- t_v^- \big\}
\end{aligned}
\qquad (66.104)
$$

in which the envelopes corresponding to source-size ($\boldsymbol{\kappa}$) and energy spread (v) are given by

$$
\begin{aligned}
t_\kappa^\pm &= \exp\left\{-\left(\frac{\alpha_c \tau^\pm}{\lambda_o}\right)^2\right\} \\
t_v^\pm &= \exp\left\{-\pi^2(P^2 \pm 2\boldsymbol{P} \cdot \boldsymbol{M})^2 v_m^2/4\right\}
\end{aligned}
\qquad (66.105)
$$

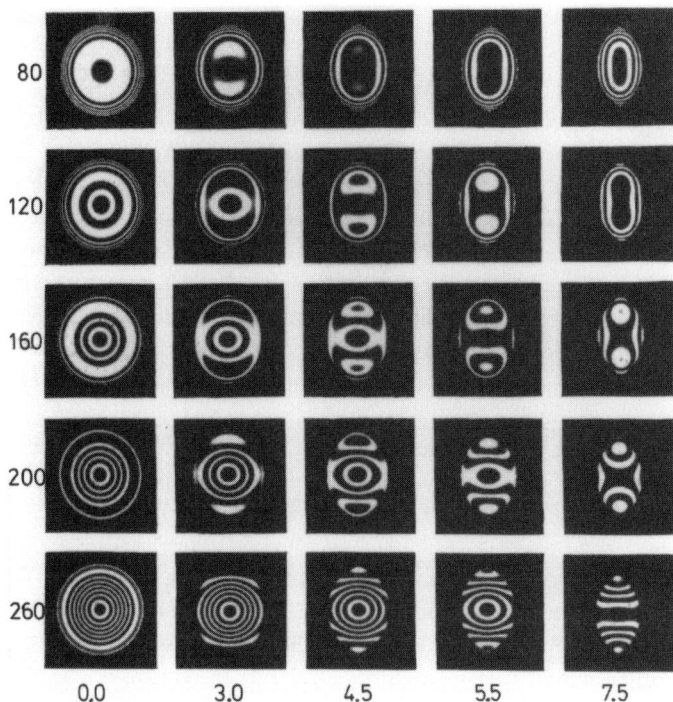

Fig. 66.11: Calculated diffractograms for vanishing energy spread. The ordinate is defocus in nanometres and the abscissa is beam tilt in mrad. The angular aperture of the illumination is 15 μrad, $C_s = 1.35$ mm and the accelerating voltage is 100 kV.

in which

$$\tau^{\pm} = 2\pi[\{(M \pm P)^2 - D\}(\pm P) + (P^2 \pm 2M \cdot P)M] \qquad (66.106)$$

The Gaussian expressions for the source-size and energy spread (66.66–67) have been used here.

The envelope $t_v^{\pm}$ is equal to unity when $P^2 \pm 2P \cdot M = 0$, that is, when

$$(P_x \pm M_x)^2 + (P_y \pm M_y)^2 = M_x^2 + M_y^2 \qquad (66.107)$$

These are circles of radius $|M|$ centred on the points $(\pm M_x, \pm M_y)$ around which the energy spread has no effect: they are the *achromatic circles* described by Hoppe (1974), Hoppe *et al.* (1974), Willasch (1976) and Frank (1976b).

How does the energy-spread envelope vary in the neighbourhood of an achromatic circle? The envelope falls to $1/e$ of its maximum value when

$$\exp\left\{-\pi^2(P^2 + 2P \cdot M)^2 D^2/2\right\} = 1/e \qquad (66.108a)$$

and hence when

$$P^2 \pm 2\boldsymbol{P} \cdot \boldsymbol{M} = \pm \frac{2^{\frac{1}{2}}}{D\pi} \qquad (66.108b)$$

which represents circles with radii $(M^2 \pm \sqrt{2}/D\pi)^{\frac{1}{2}}$.

The larger the defocus D, the narrower will be the envelope around the achromatic circle (for values of $M^2 \geq \sqrt{2}/D\pi$). We shall see that the source-size has an analogous effect but it is already clear that the expressions (66.100) are likely to be very misleading because the envelopes $t_v^{\pm}$ and $t_\kappa^{\pm}$ may well attenuate their terms in (66.104) before these two terms overlap: we must expect three contributions to $\hat{K}_p(\boldsymbol{P})$, one corresponding to a central region in which the two terms do overlap and hence interfere and individual contributions near to $\pm\boldsymbol{M}$ for which the interference is much smaller. The effects of course vary in magnitude with the choice of parameters.

We now analyse $t_\kappa^{\pm}$ in a similar way. The term $\tau^{\pm}$ may be written

$$\tau^{\pm} = 2\pi \left\{ (P^2 \pm 2\boldsymbol{P} \cdot \boldsymbol{M})(\boldsymbol{M} \pm \boldsymbol{P}) \pm (M^2 - D)\boldsymbol{P} \right\} \qquad (66.109)$$

which brings out the importance of the defocus value $D = M^2$. The envelopes are then symmetric about $\boldsymbol{P} = \pm\boldsymbol{M}$ in the sense that they take the same value at $\boldsymbol{P} = \boldsymbol{M} \pm \boldsymbol{\mu}$ (or $\boldsymbol{P} = -\boldsymbol{M} \pm \boldsymbol{\mu}$). The envelope function returns to unity when

$$(P^2 \pm 2\boldsymbol{P} \cdot \boldsymbol{M})(\boldsymbol{M} \pm \boldsymbol{P}) = 0 \qquad (66.110)$$

and hence at $\boldsymbol{P} = \pm\boldsymbol{M}$ and on the achromatic circle. The defocus value $D = M^2$ is therefore optimal in the sense that there is no attenuation around the achromatic circle and at the tilt points $(\pm\boldsymbol{M})$.

By far the most detailed study of the interaction between beam tilt, source-size and energy spread is to be found in the work of Jenkins (1979), Wade and Jenkins (1978) and Jenkins and Wade (1977). Earlier work by Hoppe et al. (1975), Downing (1975), Hanszen (1976), Hanszen and Ade (1976c), Wade (1976a,b), Krakow et al. (1976), Typke and Köstler (1976, 1977), Rose (1977), Krakow (1976a,b), Kiselev and Sherman (1976) and Hoppe and Köstler (1976) is of direct relevance; see also Kunath (1979). The experimental observation by Parsons and Hoelke (1974) of wings in the optical diffractogram of images obtained with tilted illumination led to much discussion, as did the genuineness of fringes seen in images of supposedly amorphous films of germanium and silicon (see Howie et al., 1972, 1973; McFarlane, 1975; Krivanek, 1975, 1976, 1978; Krivanek and Howie, 1975; Goldfarb et al., 1975; Krivanek et al., 1976; Krakow, 1976a, b; Wade, 1976a,b; Saxton et al., 1977; and Howie,1978, 1983).

The analysis given above is only approximate but the terms neglected can be shown to be very small; for discussion of these, see McFarlane and Cochrane (1975), Wade and Jenkins (1978) and especially Jenkins (1979).

An interesting concept that emerged from this work was the *effective coherent aperture* introduced by Downing (1975) and subsequently investigated and employed by Wade (1976a,b) and Wade and Jenkins (1978). The idea here is that the integration over κ and v should be performed on the filtered object spectrum $S_o T_L$ instead of on the image contrast. The neglect of the quadratic terms in $|\psi_i|^2$ ensures that the results are the same, even though we seem to be adding wavefunctions and not intensities. We shall not describe this further here but simply point out that the filtered object spectrum can in this way be written as the product of the usual term $T_L(\boldsymbol{q})$, the amplitude term $T_A(\boldsymbol{q})$ defining the radius of any aperture, the (shifted) object spectrum $\tilde{\eta}(\boldsymbol{q} - \boldsymbol{m})$ and envelope functions corresponding to energy spread and source-size; these have the advantage that the resulting effects at the image can be understood in terms of the behaviour of a single function for each spread function and not two for each as is the case for (66.104).

Hollow-cone illumination

The anisotropy of image formation with tilted illumination disappears if all tilt angles are present simultaneously, as they are when the electrons incident on the specimen form a hollow cone. The associated transfer functions can in principle be obtained by integrating the tilted-beam expressions over $\boldsymbol{m}$, keeping $|\boldsymbol{m}|$ constant (thin cone) or over both the angular part of $\boldsymbol{m}$ and over a small range of values of $|\boldsymbol{m}|$ as well (thick cone). The effects of source-size and energy spread complicate the integrals, but not unduly, and in any case these have to be evaluated numerically. The fullest study of hollow-cone transfer is to be found in the work of Jenkins (1979). We follow this closely.

We set out from the expression (66.104) for $\hat{K}_p(\boldsymbol{P})$, and note that the second term is the complex conjugate of the first if $\boldsymbol{P}$ is replaced by $-\boldsymbol{P}$. This means that when we integrate over $\boldsymbol{M}$, keeping the tilt-angle fixed, contributions that are themselves complex conjugates will be added and the transfer function will be real: there will be no shift in the zeros of the transfer function. Considerable insight into hollow-cone transfer can be gained by the effective aperture approach and although we confine this account to the results, we shall use this concept. It is not difficult to show that the real part of the transfer function for tilted illumination is given by

$$t_\kappa t_v \sin\left\{\frac{1}{2}\pi(P^2 - M^2)^2\right\} \tag{66.111}$$

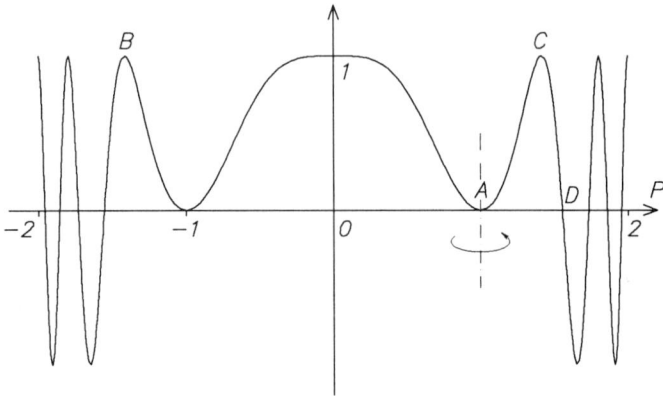

Fig. 66.12: The unattenuated real part of the transfer function, $\sin\{\pi(P^2 - M^2)^2/2\}$ as a function of P for $M^2 = 1$.

where

$$t_\kappa = \exp\left\{-\pi^2(P^2 - M^2)P^2/C_\kappa\right\}$$

$$t_v = \exp\left\{-\frac{1}{4}\pi^2(P^2 - M^2)D^2 v_m^2\right\} \tag{66.112}$$

The function $\sin\frac{\pi}{2}(P^2 - M^2)^2$ is shown in Fig. 66.12 for the optimum defocus $M^2 = D = 1$ and in Fig. 66.13 for a series of optimum defocus values, $D = M^2 = 0.5$, 0.75,... 2.75. After rotational averaging, the bright-field hollow-cone phase contrast-transfer functions of Fig. 66.14 are obtained: for each, $D = M^2$; the reduced source-size is 0.12 and the defocus corresponding to the energy spread is 0.15 Sch. As a general conclusion, we may say that hollow-cone illumination offers the possibility of extending the resolution of a given microscope beyond the cutoff for axial illumination but at the cost of lowering the contrast, which is no longer close to unity anywhere in the passband but at best 0.5.

For further discussion, see Hanszen and Trepte (1971b), Niehrs (1973), Krakow and Howland (1976), Kunath (1976), Krakow (1977, 1978), Rose and Fertig (1977), Rose (1977), Freeman *et al.* (1977, 1980), Bonnet *et al.* (1978), Saxton *et al.* (1978), Fertig and Rose (1978a,b, 1979), Saxton and Smith (1979), Kunath and Weiss (1980), Kunath *et al.* (1981, 1985, 1986), Zemlin *et al.* (1982), Balossier and Thomas (1984), Kunath and Gross (1985) and Herring (1991).

A procedure for simulating images and diffraction patterns in various less conventional modes is described and abundantly illustrated by Krakow (1984); this includes convergent-beam diffraction, hollow-cone bright-field and dark-field illumination and diffraction with 'virtual apertures'. In

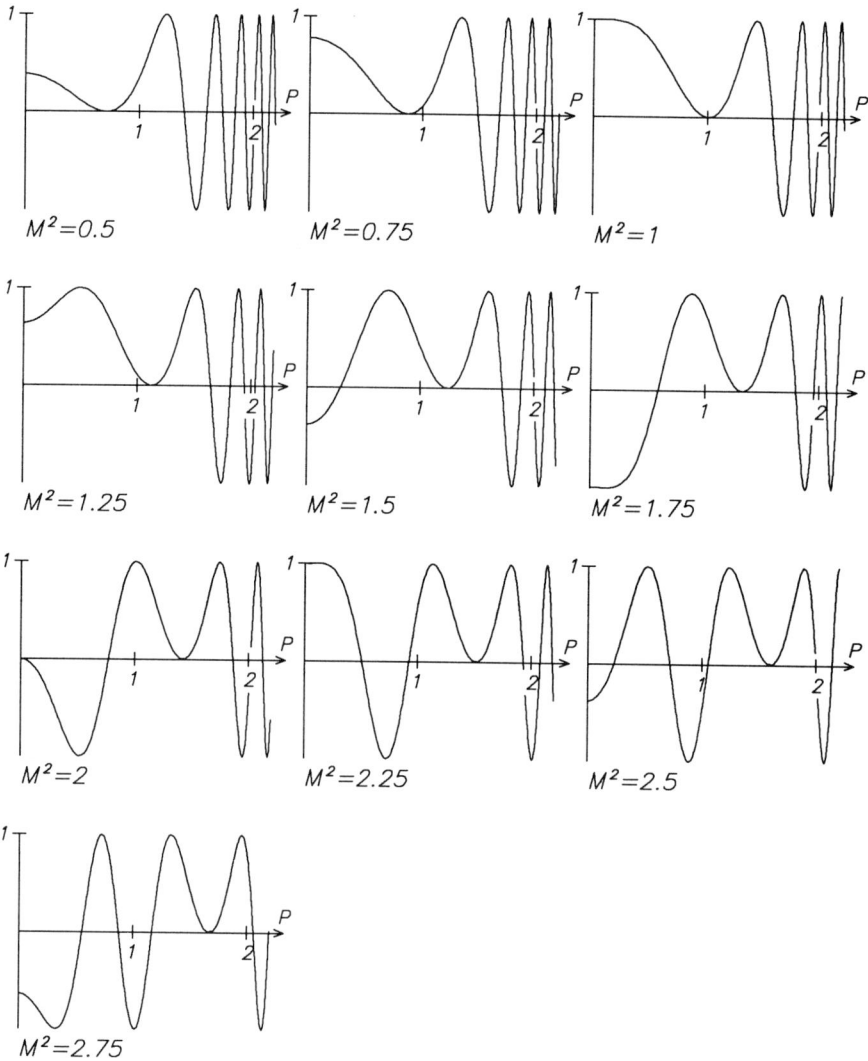

Fig. 66.13: The function $\sin\{\pi(P^2 - M^2)^2/2\}$ as a function of P for the values of M^2 shown.

the dark-field hollow-cone mode, the cone angle is so large that the direct (conical) beam is intercepted by the objective aperture. In the virtual-aperture mode, the direction and intensity of the incident beam are under computer control and describe a more complicated pattern.

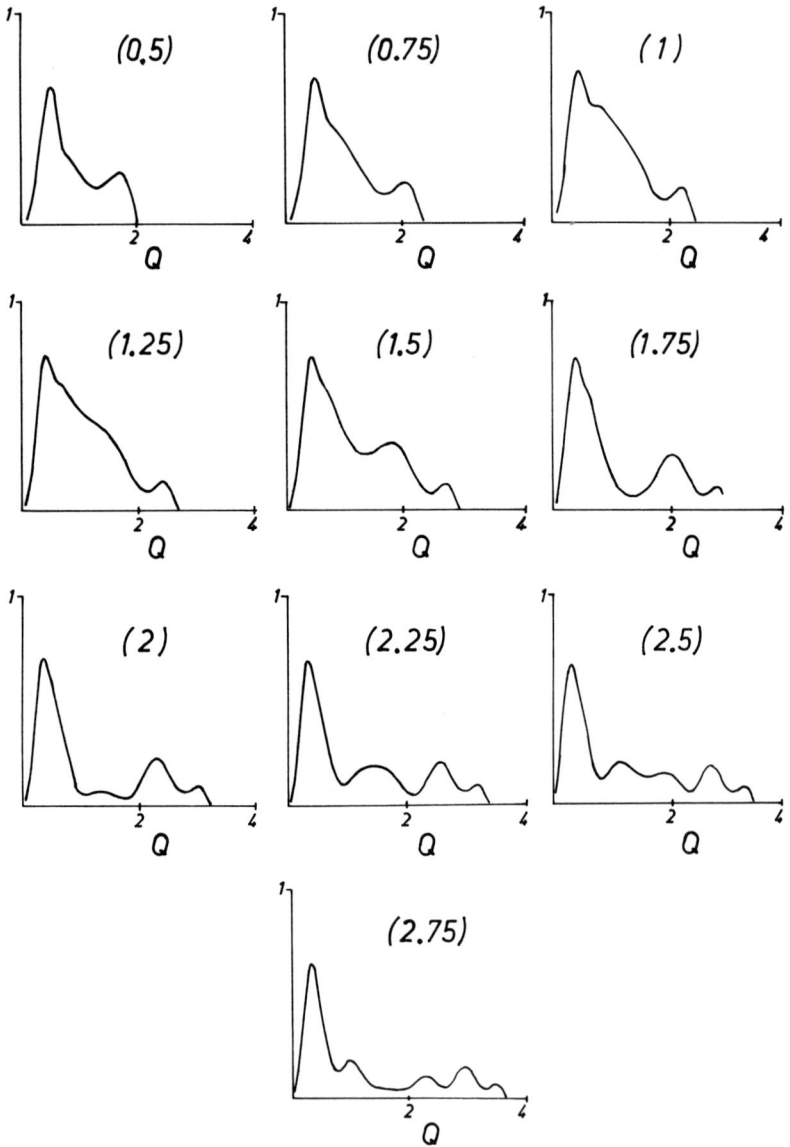

Fig. 66.14: Bright-field hollow-cone transfer functions at optimum focus for various values of Γ^2.

66.5.2 Anisoplanatism

The development of transfer theory, as we have presented it, was contingent on the convolutional structure of (65.13). This limited the geometrical aberrations that could be included to spherical aberration and axial astigmatism. At very high resolution, we may wish to consider other aberrations, however, if only to show that their effect is negligible. The coma is a case in point, being the next most important third-order aberration after spherical aberration since it is quadratic in angle and linear in position in the object plane. Can we say anything about the contrast transfer in the presence of coma and perhaps other aberrations that destroy isoplanasy? We now show that a filter function analogous to $T_L(\boldsymbol{q})$ can be defined but that this varies with position in the image plane; it thus enables us to make predictions about the quality of the image but cannot be inverted (Hawkes, 1971). We limit the present analysis to the isotropic coma K (Section 24.4).

Coma contributes a term of the form $K\lambda^3 q^2 \boldsymbol{q}\cdot\boldsymbol{u}_o$ to the wave aberration W (65.29). This can be forced into a form involving $\boldsymbol{u}_i - \boldsymbol{u}_o$ by writing

$$\boldsymbol{q}\cdot\boldsymbol{u}_o = -\boldsymbol{q}\cdot(\boldsymbol{u}_i - \boldsymbol{u}_o) + \boldsymbol{q}\cdot\boldsymbol{u}_i$$

so that the kernel $G(\boldsymbol{u}_i, \boldsymbol{u}_o)$ in

$$\psi_i(\boldsymbol{u}_i) = \int G(\boldsymbol{u}_i, \boldsymbol{u}_o)\psi_o(\boldsymbol{u}_o)\, d\boldsymbol{u}_o$$

(65.7) has the form

$$G(\boldsymbol{u}_i, \boldsymbol{u}_o) = G(\boldsymbol{u}_i - \boldsymbol{u}_o; \boldsymbol{u}_i) \tag{66.113}$$

and

$$\psi_i(\boldsymbol{u}_i) = \int G(\boldsymbol{u}_i - \boldsymbol{u}_o; \boldsymbol{u}_i)\psi_o(\boldsymbol{u}_o)\, d\boldsymbol{u}_o$$

Setting $\boldsymbol{u}_s := \boldsymbol{u}_i - \boldsymbol{u}_o$, we have

$$\psi_i = \int G(\boldsymbol{u}_s; \boldsymbol{u}_i)\psi_o(\boldsymbol{u}_i - \boldsymbol{u}_s)\, d\boldsymbol{u}_s \tag{66.114}$$

This is again a convolution and hence

$$\psi_i = \int \tilde{G}(\boldsymbol{q}; \boldsymbol{u}_i)S_o(\boldsymbol{q})\exp(2\pi i \boldsymbol{u}_s \cdot \boldsymbol{q})\, d\boldsymbol{q} \tag{66.115}$$

As usual, we write $\psi_o(\boldsymbol{u}_o) = 1 - \sigma_o + i\eta_o$, σ_o, $|\eta_o| \ll 1$ and obtain, in the conditions assumed above for the isoplanatic situation,

$$\begin{aligned} C(\boldsymbol{u}_i) = \frac{1}{2} \int \tilde{\sigma} \left\{ \tilde{G}(\boldsymbol{q}; \boldsymbol{u}_i) + \tilde{G}^*(-\boldsymbol{q}; \boldsymbol{u}_i) \right\} \exp(-2\pi i \boldsymbol{q}\cdot\boldsymbol{u}_i)\, d\boldsymbol{q} \\ - \frac{i}{2} \int \tilde{\eta} \left\{ \tilde{G}(\boldsymbol{q}; \boldsymbol{u}_i) - \tilde{G}^*(-\boldsymbol{q}; \boldsymbol{u}_i) \right\} \exp(-2\pi i \boldsymbol{q}\cdot\boldsymbol{u}_i)\, d\boldsymbol{q} \end{aligned} \tag{66.116}$$

where

$$S_o(\boldsymbol{q}) = \int \psi_o(\boldsymbol{u}_o) \exp(-2\pi i \boldsymbol{q} \cdot \boldsymbol{u}_o) \, d\boldsymbol{u}_o$$

$$\tilde{G}(\boldsymbol{q}; \boldsymbol{u}_i) = \int G(\boldsymbol{u}_s; \boldsymbol{u}_i) \exp(-2\pi i \boldsymbol{q} \cdot \boldsymbol{u}_s) \, d\boldsymbol{u}_s \qquad (66.117)$$

The spatial frequency spectra of the object, characterized by $\tilde{\sigma}$ and $\tilde{\eta}$, will therefore be filtered in a predictable fashion but the filter functions, $\tilde{G}(\boldsymbol{q}; \boldsymbol{u}_i) \pm \tilde{G}^*(-\boldsymbol{q}; \boldsymbol{u}_i)$, now depend not only on the defocus and lens aberrations but also on the image point considered: we can still calculate how contrast is transferred from object to image but we cannot invert the transformation.

For further discussion of anisoplanatic transfer, see the later work of Hawkes (1972, 1973a), Schiske (1973) and Ade (1978).

66.5.3 Forms of the aperture function T_A

Although the standard round aperture centred on the optic axis is by far the most important case, a number of other aperture shapes and opacities (η_A in 65.34) have attracted sporadic attention. Thus Hoppe (1961, 1963, 1970, 1971) and Lenz (1963, 1964, 1965) considered the use of zone plates, in which the open and opaque (or phase-shifting) parts of the plate matched the positive and negative regions of the function $\sin \chi$. Attempts to make and test such plates are described by Langer and Hoppe (1966/7, 1967), Möllenstedt et al. (1968) and Hoppe et al. (1970). A natural extension, even more difficult to put into practice, consists in using a *phase-shifting* plate, the film thickness adding phase shifts that will render the contrast-transfer function as uniform as possible. For discussion, see Boersch (1947), Agar et al. (1949), Locquin (1954, 1955, 1956), Kanaya et al. (1954, 1957, 1958a,b), Kanaya and Kawakatsu (1958), Faget et al. (1960a,b, 1962), Eisenhandler and Siegel (1966b), Siegel et al. (1966), Badde and Reimer (1970), Reimer and Badde (1970), Thon and Willasch (1970, 1971, 1972a,b), Tochigi et al. (1970), Müller (1971), Müller and Rindfleisch (1971), Parsons and Johnson (1972), Willasch (1973), Anaskin and Ageev (1974), Thon (1974), Willasch (1975a,b), Müller (1976), Balossier et al. (1980), Laberrigue et al. (1980).

A more complicated modification to the wavefunction that was once considered promising involves the introduction of an electrically charged element with a view to creating a local phase shift. For this, a fine thread is stretched across the objective aperture and charge is allowed to build up on it, thereby creating a potential variation. The difficulties and uncertainties of these devices were, however, such that they have been abandoned. For details, see the very full study by Unwin (1970a,b, 1971, 1972, 1974) and the work of Krakow and Siegel (1975), Balossier et al. (1980) and Balossier and Bonnet (1981).

A very ingenious attempt to modify the wave function in the back-focal plane for crystalline specimens has been made by Valdrè and colleagues. Their aim was to insert a very tiny coil in the aperture plane and position it around a selected diffraction spot. It was hoped that, by changing the current in the coil, information about the phase of the wave at the spot could be deduced. Brief accounts are to be found in Valdrè (1974, 1979).

66.5.4 Transfer theory and crystalline specimens

The linear transfer theory depends essentially on the assumption that the unscattered beam is considerably stronger than the scattered beam but this may well not be true for crystalline specimens, where the electrons are concentrated in spots in the back-focal plane of the objective. If several diffracted spots have intensities that are not markedly weaker than the central spot, interference between the diffracted waves may be significant, thereby rendering the direct interpretation of the image difficult or indeed misleading. This observation led Ishizuka (1980) to investigate the nonlinear term in the image intensity in detail, for crystalline specimens in particular, and we summarize his analysis here. The main purpose of this study is to indicate the limits that must be imposed on the credibility of the transfer theory for objects for which the weak scattering approximation may break down.

We set out from the general formula (65.16), from which we derive the spectrum of the image intensity as the weighted convolution of the spectra of ψ_i and ψ_i^*. The weighting describes the effect of non-vanishing source-size and energy spread on the assumption that these can be treated separately. Equation (65.16) tells us that

$$S_i(\boldsymbol{q}) = T(\boldsymbol{q})S_o(\boldsymbol{q}) \qquad (66.118)$$

and hence

$$\mathcal{F}(\psi_i\psi_i^*) =: S_I(\boldsymbol{q};\Delta)$$
$$= \int X(\boldsymbol{q}',\boldsymbol{q};\Delta)S_o(\boldsymbol{q}'+\boldsymbol{q})S_o^*(\boldsymbol{q}')\,d\boldsymbol{q}' \qquad (66.119)$$

in which $X(q',q;\Delta)$ is the transmission cross-coefficient, which we shall meet again in Part XVI. It involves the auto-correlation of $T(\boldsymbol{q})$, weighted by the source functions

$$H_1(\Delta) := \frac{1}{\pi^{\frac{1}{2}}\Delta_o}\exp\left(-\frac{\Delta^2}{\Delta_o^2}\right) \qquad \text{(energy spread)} \qquad (66.120a)$$

(66.60) and

$$H_2(\boldsymbol{q}) := \frac{1}{\pi q_o^2} \exp\left(-\frac{q^2}{q_o^2}\right) \qquad \text{(source-size)} \qquad (66.120b)$$

Equations (66.120) are in conformity with (66.66–67) if we write the latter in the form

$$|A(\boldsymbol{\kappa}, v)|^2 = \frac{1}{\pi^{\frac{1}{2}} v_m} \exp\left(-\frac{v^2}{v_m^2}\right) \times \frac{\lambda_o^2}{\pi \alpha_c^2} \exp\left(-\frac{\lambda_o^2 \kappa^2}{\alpha_c^2}\right)$$

and identify q_o with α_c/λ_o in the second factor. In the first we have chosen to characterize the effect of chromatic aberration in terms of the resulting focal shift.

Explicitly,

$$X(\boldsymbol{q}', \boldsymbol{q}; \Delta) = \int \cdot \int H_2(\boldsymbol{q}'') H_1(\Delta) T(\boldsymbol{q}' + \boldsymbol{q}''; \Delta + \Delta') \\ \times T^*(\boldsymbol{q} + \boldsymbol{q}''; \Delta + \Delta') \, d\boldsymbol{q}'' \, d\Delta' \qquad (66.121)$$

The object function corresponding to the spectrum $S_o(\boldsymbol{q})$ in (66.118) is now formally separated into a constant part and a variable part but the latter is now not necessarily small compared with the former (taken to be unity):

$$S_o(\boldsymbol{q}) =: \delta(\boldsymbol{q}) + s(\boldsymbol{q}) \qquad (66.122)$$

After some calculation, we find that the image intensity spectrum, S_I, is composed of three parts:

$$S_I(\boldsymbol{q}; \Delta) = S_I^{(0)}(\boldsymbol{q}; \Delta) S_I^{(1)}(\boldsymbol{q}; \Delta) S_I^{(2)}(\boldsymbol{q}; \Delta) \qquad (66.123)$$

The first two terms we have already met in the linear theory; it is the final term that is new and will enable us to assess the impact of the nonlinear interactions. We find

$$S_I^{(0)}(\boldsymbol{q}; \Delta) = X(q, 0; \Delta)\delta(\boldsymbol{q})$$
$$S_I^{(1)}(\boldsymbol{q}; \Delta) = X(\boldsymbol{q}, 0; \Delta)s(\boldsymbol{q}) + X(0, -\boldsymbol{q}; \Delta)s^*(-\boldsymbol{q})$$
$$\qquad \Longrightarrow X(\boldsymbol{q}, 0; \Delta)s(\boldsymbol{q}) + X^*(\boldsymbol{q}, 0; \Delta)s^*(-\boldsymbol{q}) \quad \text{(symmetric source)}$$
$$S_I^{(2)}(\boldsymbol{q}; \Delta) = \int X(\boldsymbol{q}' + \boldsymbol{q}, \boldsymbol{q}'; \Delta)s(\boldsymbol{q}' + \boldsymbol{q})s^*(\boldsymbol{q}') \, d\boldsymbol{q}' \qquad (66.124)$$

A lengthy calculation, given in detail by Ishizuka, shows that the function X can be written as the product of five factors:

$$X(\boldsymbol{q}', \boldsymbol{q}; \Delta) =: X_L X_p X_s X_e X_x \qquad (66.125)$$

where

$$X_L := T_A(\boldsymbol{q}')T_A(\boldsymbol{q})\exp\left[-\mathrm{i}\left\{\chi(\boldsymbol{q}',\Delta) - \chi(\boldsymbol{q},\Delta)\right\}\right]$$

(effect of aperture and lens properties, $T_A T_L$)

$$X_p(\boldsymbol{q}',\boldsymbol{q};\Delta) := \exp\{2\pi\mathrm{i}(\pi q_o \Delta_o)^2(\chi'_\Delta - \chi_\Delta)(\chi'_q - \chi_q, \chi'_{q\Delta} - \chi_{q\Delta})/u\}$$

$$X_s(\boldsymbol{q}',\boldsymbol{q};\Delta) := \exp\left\{-\pi q_o^2(\chi'_q - \chi_q)^2/u\right\}$$

$$X_e(\boldsymbol{q}',\boldsymbol{q}) = \exp\left\{-(\pi\Delta_o)^2(\chi'_\Delta - \chi_\Delta)^2/u\right\}$$

$$X_x(\boldsymbol{q}',\boldsymbol{q}) = \exp\left[-(\pi^2 q_o^2 \Delta_o)^2\left\{(\chi'_q - \chi_q)^2(\chi'_{q\Delta} - \chi_{q\Delta})^2\right.\right.$$
$$\left.\left. - (\chi'_q - \chi'_q, \chi'_{q\Delta} - \chi_{q\Delta})^2\right\}/u\right]u^{-1/2}$$
$$= \exp\left[(\pi^2 q_o^2 \Delta_o)^2\left\{(\chi'_q - \chi_q)(\chi'_{q\Delta} - \chi'_{q\Delta})^2\right\}/u\right]u^{-1/2}$$

$$(66.126)$$

in which

$$u := 1 + \xi(\boldsymbol{q}' - \boldsymbol{q})^2, \qquad \xi = (\pi\lambda q_o \Delta_o)^2 \qquad (66.127)$$

and

$$\chi_q = (C_s\lambda^2|q^2| - \Delta')\lambda\boldsymbol{q}$$
$$\chi_\Delta = -\frac{1}{2}\lambda|q^2|$$
$$\chi_{q\Delta} = -\lambda\boldsymbol{q} \qquad (66.128)$$

and analogously for χ'_q, χ'_Δ and $\chi'_{q\Delta}$.

This product representation enables us to assess in detail the various effects that may occur when the nonlinear terms cannot be neglected, a situation that is particularly likely to occur with crystals. Ishizuka gives curves and perspective plots of each term in turn and then of the cross-coefficient X. From these, it emerges that the envelopes X_e and X_s are the most important. Figure 66.15 shows $X_e X_s$ for various defocus values for a microscope operating at 100 kV, $C_s = 0.7$ mm and $q_o = 0.01$ Å^{-1}, $\Delta_o = 100$ Å. Clearly, the nonlinear term can have a very significant effect and Ishizuka suggests that such effects were important in the related experimental observations of Hashimoto et al. (1977), Izui et al. (1977, 1978), Sieber and Tonar (1975, 1976) and Desseaux et al. (1977).

The expressions (66.128) are in fact only first-order approximations to the correct quantities and Ishizuka has later examined the validity of this approximation (1989, 1990). We refer to these papers for further details.

Another important effect that may be encountered when imaging crystal defects at the highest resolution is *delocalization* (Coene and Janssen, 1991). If the image is formed with very coherent illumination, so that the transfer function begins to oscillate well before the information limit is

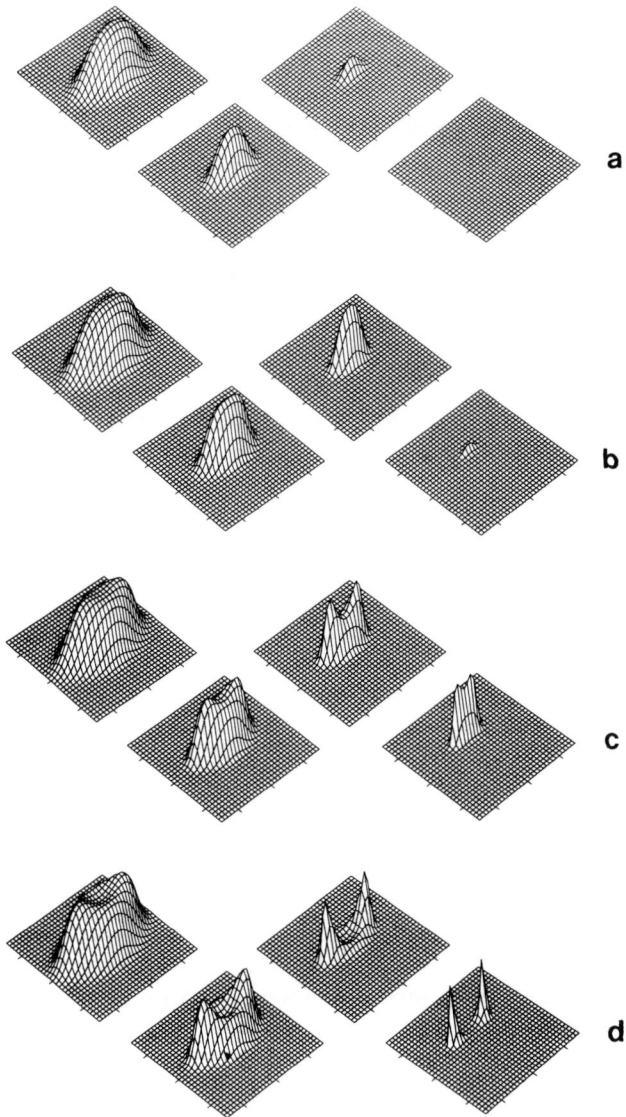

Fig. 66.15: Perspective views of the overall envelope function $X_e X_s$ for four values of the defocus. For each value, the four envelopes correspond from left to right to spatial frequencies $q_x = 1, 2, 4, 8 \text{ nm}^{-1}$, $q_y = 0$. The marks on the q'_x and q'_x axes correspond to $-1, 0$ and 1. (a) $\Delta = -200$ nm. (b) $\Delta = 0$ nm (c) $\Delta = 200$ nm. (d) $\Delta = 400$ nm.

reached, the position of a crystal defect may appear to be delocalized in the image. The associated analysis is given in full by Coene and Janssen.

66.5.5 Other transforms having a convolution theorem

Much of the theory presented in this chapter has yielded formulae of practical importance thanks to the fact that the Fourier transform maps convolution products into direct products. The Fourier transform is not alone in possessing this property and some of its rivals exhibit desirable features that it does not possess. We shall see that the principal attraction is speed and precision of computation. Against this, we must recall that the very simple form of the filter functions $(\sin \chi(\boldsymbol{q})$ and $\cos \chi(\boldsymbol{q}))$ is a consequence of the fact that it is the *Fourier* transform of the kernel function G (65.13) that has a simple explicit form. There is a risk that any advantage gained by applying other transforms to $C(\boldsymbol{u}_i)$ will be offset by the need to evaluate the transform of G; this remains to be investigated.

A rather general type of transform that maps convolution products into a simpler form is the polynomial transform; to understand this, some familiarity with the manipulation of rings and fields of polynomials is required. An account of this branch of algebra would lead us too far afield and we therefore give only a very superficial description before turning to the more easily understood number-theoretic transforms, which form a special case of the polynomial transforms.

We can see why polynomials are of interest in connection with convolutions by considering an elementary one-dimensional case,

$$y_\ell = \sum_{j=0}^{N-1} h_j x_{\ell-j} \tag{66.129}$$

in which ℓ runs from 0 to $2N - 2$. Consider now the polynomials in z whose coefficients are the same as h_j and x_j:

$$H(z) := \sum_{j=0}^{N-1} h_j z^j \qquad X(z) := \sum_{j=0}^{N-1} x_j z^j \tag{66.130}$$

Multiplication of H by X will give a polynomial $Y(z)$ of degree $2(N-1)$ and the coefficient of z^j in $Y(z)$ will be given by

$$y_j = \sum_{k=0}^{j} h_k x_{j-k} \tag{66.131}$$

There is thus a close relation between polynomial multiplication and convolution and this can be exploited to generate a very powerful family of transforms.

We now turn to the special case of the number-theoretic transforms and first consider the simple situation, in which arithmetic is performed modulo a prime q in the Galois field $\mathrm{GF}(q)$. We again set out from (66.129) and scale so that all the coefficients h_j and x_j are integers. We now define

$$H_k := \sum_{j=0}^{N-1} h_j g^{jk} \text{ modulo } q \qquad (66.132a)$$

where g is an integer and likewise for x. We can also define an inverse

$$a_j = N^{-1} \sum_{k=0}^{N-1} H_k g^{-jk} \text{ modulo } q \qquad (66.132b)$$

with $NN^{-1} = 1$ modulo q. These definitions bear a strong resemblance to those for a Fourier transform and its inverse except that an integer g replaces $\exp(2\pi i)$. Let us now examine the product of two transformed arrays and see whether their inverse has the form of a convolution. The product of the transforms of $\{h\}$ and $\{x\}$ is given by

$$H_k X_k = \sum_{j=0}^{N-1} \sum_{\ell=0}^{N-1} h_j x_\ell g^{(j+\ell)k} \text{ modulo } q \qquad (66.133a)$$

with inverse

$$a_m = N^{-1} \sum_{k=0}^{N-1} \sum_{j=0}^{N-1} \sum_{\ell=0}^{N-1} h_j x_\ell g^{(j+\ell-m)k} \text{ modulo } q \qquad (66.133b)$$

This will have the same form as (66.129), with $a_m \rightarrow y_m$, if

$$\sum_{k=0}^{N-1} g^{(j+\ell-m)k} \begin{cases} \equiv N & \text{for } j+\ell-m \equiv 0 \text{ modulo } N \\ \equiv 0 & \text{for } j+\ell-m \not\equiv 0 \text{ modulo } N \end{cases}$$

The first condition implies that we must have

$$g^N \equiv 1 \text{ modulo } q \qquad (66.134a)$$

and the second, that

$$(g^{j+\ell-m} - 1) \sum_{k=0}^{N-1} g^{(j+\ell-m)k} \equiv g^{N(j+\ell-m)} - 1 \equiv 0 \text{ modulo } q \qquad (66.134b)$$

The integer g, which has hitherto been left undefined, must therefore be a root of unity of order N modulo q, that is $g^N \equiv 1$ modulo q. For this choice of g, the transform (66.132) maps convolution products into direct products. For given values of N and the prime q, we can always find a transform provided that N divides $q - 1$.

An analogous conclusion is reached if q is not prime. It can be shown by similar reasoning that a number-theoretic transform with the convolution mapping can be found if and only if

$$g^N \equiv 1 \text{ modulo } q \qquad (66.135a)$$

$$NN^{-1} \equiv 1 \text{ modulo } q \qquad (66.135b)$$

$g^{j+\ell-m}$ and q are mutually prime for

$$j + \ell - m = 1, 2 \ldots N - 1 \qquad (66.135c)$$

The various forms of the number-theoretic transform effectively correspond to different choices of the quantities g and q. Thus for $q = 2^p - 1$, p prime, the integers q are the Mersenne numbers and the transforms are likewise known as *Mersenne transforms*. Another choice is $q = 2^p + 1, p = 2^t$, which means that the q coincide with the Fermat numbers and we obtain the *Fermat number transforms*. Mersenne transforms have the attraction that multiplication is unnecessary but there is no fast algorithm and word-length and transform-length are connected by a very restrictive relation. Fermat transforms can be calculated using addition and multiplication by a power of two for values of N up to $2^t + 2$; the connection between word-length and transform-length is less rigid than in the Mersenne case.

The real advantage of these transforms is only apparent when we go beyond these simple choices for q and consider the pseudo-Mersenne and pseudo-Fermat transforms. Here, the transform is defined modulo q_i, where q_i is a factor of a pseudo-Mersenne number, $q = 2^p - 1$ (p is now composite) or of a pseudo-Fermat number, $q = 2^p + 1$ ($p \neq 2^t$). Both of these families of transforms can be extended to complex-valued arrays $\{x\}$ and $\{h\}$ and highly efficient fast algorithms can now be found.

Number-theoretic transforms can therefore rival or surpass the Fourier transform in certain circumstances, depending on the transform length involved, and have the great attraction that since integers replace the exponentials of the Fourier transform, there is no rounding-error. The calculations are exact. For a recent improvement, see Boussakta and Holt (1992).

This account can give no more than the flavour of these alternatives to the Fourier transform and we refer to specialized texts for extensive discussion, and in particular to Nussbaumer (1982), which we have followed closely, and to Hawkes (1974a,b, 1975), where they were first examined in connection with electron microscopy.

Other types of convolution

We mention in passing that the numerous discrete transforms that are commonly discussed in connection with image-processing—Walsh–Hadamard, Haar, slant are the best known—are not relevant here because they do not map convolutions of the cyclic type (66.129) into direct products. The Walsh transform does map a different type of convolution (dyadic convolution) into a simpler form but this dyadic product does not arise naturally in image formation; these transforms may, however, prove valuable for image coding (Section 71.4).

67

Image Formation in the Scanning Transmission Electron Microscope

67.1 Introduction

It was realized soon after the first scanning transmission instrument (STEM) was built that its optical properties could be interpreted in terms of those of a conventional fixed-beam transmission electron microscope running backwards: a small axial STEM detector corresponds to a very small crossover in TEM; the probe-forming lens that precedes the specimen in STEM corresponds to the objective that follows it in TEM; it is not quite so easy to see what corresponds to the large image plane (fluorescent screen, film) in the TEM but some reflection shows that it is analogous to the large area swept out by the scanning probe *projected back to the source plane.* Many of the phenomena observed in STEM operation can be easily understood by considering the corresponding TEM situation. In practice, however, there are ways of using the STEM for which the TEM analogue would be difficult to implement and which are hence not discussed in the TEM literature. We therefore provide a short account of STEM image formation and in particular, of the new flexibility offered at the detector level, which is readily accessible, whereas the illumination in TEM can only be altered in a limited number of simple ways; the more complicated modifications to the incident beam that have been proposed usually involve introducing beam-shaping or phase-shifting devices into the path of the electrons in the TEM and the adverse effects of these outweigh any potential benefits.

67.2 Wave propagation in STEM

The wave incident on the specimen is not even approximately plane, as it is in the normal imaging modes in TEM, since the source and specimen planes are conjugate and the probe-forming lens operates at a high demagnification. To a good approximation, the wave surface at the specimen is spherical, perturbed by the aberrations of the lenses between the source and the specimen and essentially by those of the final, probe-forming lens. The perturbation is described by the point-spread function of this lens ($\hat{G}$ or G in Chapter 65), which is itself the Fourier transform (65.15c) of the product $T = T_A T_L$ or $T_A \exp(-\mathrm{i}\chi)$ that we have met in the discussion of image formation in TEM (65.35). The wave incident on the specimen will be denoted by $\psi_Q(x_o, y_o, z_o)$ when the probe is centred on the axis, in the absence of any deflecting field therefore.

The illumination is assumed to be coherent (vanishing source-size, monochromatic source); we mention the effect of partial coherence briefly in Part XVI. The intensity corresponding to this incident wave is concentrated within a very small zone, the size of which defines the pixel-size and resolution of the instrument. Since probe-sizes of the order of a few ångströms can be attained, we may expect to attain resolutions comparable with those of a TEM.

We assume that, as the probe is swept over the specimen surface, the wavefunction ψ_Q is translated bodily with no appreciable change of shape or direction. When the probe is no longer centred on the axis but on a point with coordinates $\boldsymbol{u}_o = \boldsymbol{\xi}$, where we have written $\boldsymbol{u}_o := (x_o, y_o)$, the incident wave will just be $\psi_Q(\boldsymbol{u}_o - \boldsymbol{\xi})$ therefore. For a specimen transparency $O(\boldsymbol{u}_o)$, the emergent wave will be

$$\psi_o(\boldsymbol{u}_o; \boldsymbol{\xi}) = O(\boldsymbol{u}_o)\psi_Q(\boldsymbol{u}_o - \boldsymbol{\xi}) \tag{67.1}$$

and if this is allowed to propagate freely to a detector plane distant R from the object, the wavefunction there will be

$$\hat{\psi}(\boldsymbol{u}_d; \boldsymbol{\xi}) = \int O(\boldsymbol{u}_o)\psi_Q(\boldsymbol{u}_o - \boldsymbol{\xi}) \exp\left(\frac{2\pi\mathrm{i}}{\lambda R}\boldsymbol{u}_d \cdot \boldsymbol{u}_o\right) d\boldsymbol{u}_o \tag{67.2}$$

The extensive use of the spatial frequency in Chapters 65 and 66 has accustomed us to the idea that a quantity with the dimensions of length is not always the most suitable measure of off-axial distance in wavefunctions and intensity functions. Thus in a set of conjugate planes that include the object and image planes of the TEM, the corresponding coordinates, scaled perhaps by the magnification, are usually appropriate $(x_o, y_o; x_i, y_i; \ldots)$; in the set of conjugate planes in which certain beam properties are related to those in the first set via a Fourier transform, however—the 'diffraction planes'—it is convenient to introduce 'reciprocal' quantities with the dimensions $[\text{length}]^{-1}$. The spatial frequency (65.21), which measures off-axis distance in (or close to) the plane of the objective aperture in the TEM but divided by λf, is the obvious example. The same is true in the STEM, where the detector plane is a 'reciprocal' plane relative to the specimen: at any instant, the far-field diffraction pattern of the area illuminated by the probe is formed in the detector plane. We therefore introduce a position coordinate in that plane that has dimensions $[L]^{-1}$:

$$\boldsymbol{q}_d := \boldsymbol{u}_d / \lambda R \qquad (67.3a)$$

and write

$$\psi_d(\boldsymbol{q}_d; \boldsymbol{\xi}) := \hat{\psi}_d(\boldsymbol{u}_d; \boldsymbol{\xi}) \qquad (67.3b)$$

giving

$$\psi_d(\boldsymbol{q}_d; \boldsymbol{\xi}) = \int O(\boldsymbol{u}_o)\phi_Q(\boldsymbol{u}_o - \boldsymbol{\xi})\exp(2\pi i \boldsymbol{q}_d \cdot \boldsymbol{u}_o)\, d\boldsymbol{u}_o \qquad (67.4)$$

The current density distribution in the detector plane, when the probe is centred on the point $\boldsymbol{u}_o = \boldsymbol{\xi}$ is denoted by $J_d(\boldsymbol{q}_d; \boldsymbol{\xi})$:

$$
\begin{aligned}
J_d(\boldsymbol{q}_d; \boldsymbol{\xi}) = \iint & O(\boldsymbol{u}_o)O^*(\boldsymbol{u}_o')\psi_Q(\boldsymbol{u}_o - \boldsymbol{\xi})\psi_Q^*(\boldsymbol{u}_o' - \boldsymbol{\xi}) \\
& \times \exp\left\{2\pi i \boldsymbol{q}_d \cdot (\boldsymbol{u}_o - \boldsymbol{u}_o')\right\}\, d\boldsymbol{u}_o \, d\boldsymbol{u}_o'
\end{aligned}
\qquad (67.5)
$$

As in Chapter 60 (60.4), we write $O(\boldsymbol{u}_o)$ in the form

$$
\begin{aligned}
O(\boldsymbol{u}_o) =: & \exp\{-\sigma(\boldsymbol{u}_o) + i\eta(\boldsymbol{u}_o)\} \\
\equiv & 1 + \left(e^{-\sigma}\cos\eta - 1\right) + i e^{-\sigma}\sin\eta \\
=: & 1 - s + i\varphi
\end{aligned}
\qquad (67.6)
$$

in which s and φ are not necessarily small. We shall make the weak-scattering approximation later on. The intensity at the detector, $J_d(\boldsymbol{q}_d; \boldsymbol{\xi})$, is now composed of three groups of terms: those independent of the specimen functions s and φ, ($J_d^{(0)}$), those linear in s or φ, ($J_d^{(1)}$) and quadratic terms, ($J_d^{(2)}$). Thus

$$J_d(\boldsymbol{q}_d; \boldsymbol{\xi}) = \sum_{j=0}^{2} J_d^{(j)}(\boldsymbol{q}_d; \boldsymbol{\xi}) \qquad (67.7)$$

The first term is easily seen to be

$$
\begin{aligned}
J_d^{(0)}(\boldsymbol{q}_d; \boldsymbol{\xi}) &= \iint \psi_Q(\boldsymbol{u}_o - \boldsymbol{\xi})\psi_Q^*(\boldsymbol{u}_o' - \boldsymbol{\xi})\exp\{2\pi i \boldsymbol{q}_d \cdot (\boldsymbol{u}_o - \boldsymbol{u}_o')\}\, d\boldsymbol{u}_o \, d\boldsymbol{u}_o' \\
&= |T_A|^2
\end{aligned}
\qquad (67.8)
$$

and is independent of $\boldsymbol{\xi}$.

The second group of terms, in which s and φ appear linearly, is of great interest. It consists of four terms, each of the form

$$\iint o(\boldsymbol{u}_o \text{ or } \boldsymbol{u}_o')\psi_Q(\boldsymbol{u}_o - \boldsymbol{\xi})\psi_Q^*(\boldsymbol{u}_o' - \boldsymbol{\xi})\exp\{2\pi i \boldsymbol{q}_d \cdot (\boldsymbol{u}_o - \boldsymbol{u}_o')\}\,d\boldsymbol{u}_o\,d\boldsymbol{u}_o'$$

in which o may be s or φ. We examine one such term in detail, with $o(\boldsymbol{u}_o)$ as the specimen function. On introducing the spectra of o and ψ_Q (66.2, 66.3), namely

$$\tilde{o}(\boldsymbol{q}) := \mathcal{F}^-(o) = \int o(\boldsymbol{u}_o)\exp(-2\pi i \boldsymbol{q} \cdot \boldsymbol{u}_o)\,d\boldsymbol{u}_o$$

and likewise for ψ_Q, the above integral becomes

$$\iiiint \tilde{o}(\boldsymbol{q})\tilde{\psi}_Q(\boldsymbol{m})\tilde{\psi}_Q^*(\boldsymbol{n})\exp\{\boldsymbol{q}_d \cdot \boldsymbol{u}_o - \boldsymbol{q}_d \cdot \boldsymbol{u}_o'$$
$$+ \boldsymbol{q}\boldsymbol{u}_o + \boldsymbol{m} \cdot (\boldsymbol{u}_o - \boldsymbol{\xi}) - \boldsymbol{n} \cdot (\boldsymbol{u}_o' - \boldsymbol{\xi})\}\,d\boldsymbol{u}_o\,d\boldsymbol{u}_o'\,d\boldsymbol{m}\,d\boldsymbol{n}\,d\boldsymbol{q}$$
$$= \iiint \tilde{o}(\boldsymbol{q})\tilde{\psi}_Q(\boldsymbol{m})\tilde{\psi}_Q^*\,\delta(\boldsymbol{q}_d + \boldsymbol{q} + \boldsymbol{m})\,\delta(-\boldsymbol{q}_d - \boldsymbol{n})$$
$$\times \exp\{2\pi i \boldsymbol{\xi} \cdot (\boldsymbol{n} - \boldsymbol{m})\}\,d\boldsymbol{m}\,d\boldsymbol{n}\,d\boldsymbol{q}$$
$$= \int \tilde{o}(\boldsymbol{q})\tilde{\psi}_Q(\boldsymbol{q} + \boldsymbol{q}_d)\tilde{\psi}_Q^*(\boldsymbol{q}_d)\exp(2\pi i \boldsymbol{\xi} \cdot \boldsymbol{q})\,d\boldsymbol{q}$$

in which we have assumed that $\tilde{\psi}_Q(-\boldsymbol{q}) = \tilde{\psi}_Q(\boldsymbol{q})$. An analogous calculation for the case $o(\boldsymbol{u}_o')$ gives

$$\int \tilde{o}(\boldsymbol{q})\tilde{\psi}_Q(\boldsymbol{q}_d)\tilde{\psi}_Q^*(\boldsymbol{q}_d - \boldsymbol{q})\exp(2\pi i \boldsymbol{\xi} \cdot \boldsymbol{q})\,d\boldsymbol{q}$$

Collecting up the four terms, we obtain in all

$$J_d^{(1)}(\boldsymbol{q}_d; \boldsymbol{\xi}) = -\int \tilde{s}(\boldsymbol{q})\left\{\tilde{\psi}_Q(\boldsymbol{q}_d + \boldsymbol{q})\tilde{\psi}_Q^*(\boldsymbol{q}_d) + \tilde{\psi}_Q(\boldsymbol{q}_d)\tilde{\psi}_Q^*(\boldsymbol{q}_d - \boldsymbol{q})\right\}$$
$$\times \exp(2\pi i \boldsymbol{\xi} \cdot \boldsymbol{q})\,d\boldsymbol{q}$$
$$+ i\int \tilde{\varphi}(\boldsymbol{q})\left\{\tilde{\psi}_Q(\boldsymbol{q}_d + \boldsymbol{q})\tilde{\psi}_Q^*(\boldsymbol{q}_d) - \tilde{\psi}_Q(\boldsymbol{q}_d)\tilde{\psi}_Q^*(\boldsymbol{q}_d - \boldsymbol{q})\right\}$$
$$\times \exp(2\pi i \boldsymbol{\xi} \cdot \boldsymbol{q})\,d\boldsymbol{q}$$

$$(67.9)$$

The quadratic terms cannot be simplified in any useful way. We have

$$J_d^{(2)}(\boldsymbol{q}_d; \boldsymbol{\xi}) = \left|\int \{s(\boldsymbol{u}_o) - i\varphi(\boldsymbol{u}_o)\}\psi_Q(\boldsymbol{u}_o - \boldsymbol{\xi})\exp(2\pi i \boldsymbol{q}_d \cdot \boldsymbol{u}_o)\,d\boldsymbol{u}_o\right|^2$$

into which the spectra of s, φ and ψ_Q can be introduced straightforwardly.

The current density recorded per picture element will depend on the shape and response of the detector. These are characterized by means of a *detector response function*, $D(\boldsymbol{q}_d)$. If the recorded current density is denoted by $I_d(\boldsymbol{\xi})$, then

$$I_d(\boldsymbol{\xi}) = \int J_d(\boldsymbol{q};\boldsymbol{\xi})D(\boldsymbol{q}_d)\,d\boldsymbol{q} \qquad (67.10)$$

The form of (67.5) suggests that the spatial frequency response of the detector will also be useful. We define this as in (65.15):

$$S_D(\boldsymbol{u}_d) := \mathcal{F}^-(D) = \int D(\boldsymbol{q}_d)\exp(-2\pi i\boldsymbol{q}_d \cdot \boldsymbol{u}_d)\,d\boldsymbol{q}_d \qquad (67.11)$$

(Since $\boldsymbol{q}_d$ has dimension $[L]^{-1}$, $\boldsymbol{u}_d$ is a length.) The recorded current density $I_d(\boldsymbol{\xi})$ may thus be written

$$I_d(\boldsymbol{\xi}) = \iint O(\boldsymbol{u}_o)O^*(\boldsymbol{u}'_o)\psi_Q(\boldsymbol{u}_o - \boldsymbol{\xi})\psi_Q^*(\boldsymbol{u}'_o - \boldsymbol{\xi})S_D(\boldsymbol{u}'_o - \boldsymbol{u}_o)\,d\boldsymbol{u}_o\,d\boldsymbol{u}'_o$$

$$(67.12)$$

Like $J_d(\boldsymbol{q}_d;\boldsymbol{\xi})$, the recorded current density $I_d(\boldsymbol{\xi})$ may be written as three contributions,

$$I_d^{(j)}(\boldsymbol{\xi}) =: \int J_d^{(j)}(\boldsymbol{q}_d;\boldsymbol{\xi})D(\boldsymbol{q}_d)\,d\boldsymbol{q}_d \qquad j = 0,1,2 \qquad (67.13)$$

of which the linear term $I_d^{(1)}(\boldsymbol{\xi})$ is of most interest here. We have

$$I_d^{(1)}(\boldsymbol{\xi}) = -\iint \{s(\boldsymbol{u}_o) + s(\boldsymbol{u}'_o)\}\psi_Q(\boldsymbol{u}_o - \boldsymbol{\xi})\psi_Q^*(\boldsymbol{u}'_o - \boldsymbol{\xi})S_D(\boldsymbol{u}'_o - \boldsymbol{u}_o)\,d\boldsymbol{u}_o\,d\boldsymbol{u}'_o$$

$$+ i\iint \{\varphi(\boldsymbol{u}_o) - \varphi(\boldsymbol{u}'_o)\}\psi_Q(\boldsymbol{u}_o - \boldsymbol{\xi})\psi_Q^*(\boldsymbol{u}'_o - \boldsymbol{\xi})S_D(\boldsymbol{u}'_o - \boldsymbol{u}_o)\,d\boldsymbol{u}_o\,d\boldsymbol{u}'_o$$

$$(67.14)$$

or

$$I_d^{(1)}(\boldsymbol{\xi}) = -\iint \tilde{s}(\boldsymbol{q})\{\tilde{\psi}_Q(\boldsymbol{q}_d + \boldsymbol{q})\tilde{\psi}_Q^*(\boldsymbol{q}_d) + \tilde{\psi}_Q(\boldsymbol{q}_d)\tilde{\psi}_Q^*(\boldsymbol{q}_d - \boldsymbol{q})\}$$

$$\times D(\boldsymbol{q})\exp(2\pi i\boldsymbol{\xi}\cdot\boldsymbol{q})\,d\boldsymbol{q}\,d\boldsymbol{q}_d$$

$$+ i\iint \tilde{\varphi}(\boldsymbol{q})\{\tilde{\psi}_Q(\boldsymbol{q}_d + \boldsymbol{q})\tilde{\psi}_Q^*(\boldsymbol{q}_d) - \tilde{\psi}_Q(\boldsymbol{q}_d)\tilde{\psi}_Q^*(\boldsymbol{q}_d - \boldsymbol{q})\}$$

$$\times D(\boldsymbol{q})\exp(2\pi i\boldsymbol{\xi}\cdot\boldsymbol{q})\,d\boldsymbol{q}\,d\boldsymbol{q}_d$$

$$(67.15)$$

The form of this expression suggests that we should examine the spectrum of $I_d^{(1)}$, which we denote $S_I(\boldsymbol{q})$:

$$
\begin{aligned}
S_I(\boldsymbol{q}) &:= \mathcal{F}^-(I_d^{(1)}) \\
&= -\tilde{s}(\boldsymbol{q})K_a(\boldsymbol{q}) - \mathrm{i}\tilde{\varphi}(\boldsymbol{q})K_p(\boldsymbol{q})
\end{aligned}
\tag{67.16}
$$

in which

$$
\begin{aligned}
K_a(\boldsymbol{q}) &:= \int \{\tilde{\psi}_Q(\boldsymbol{q}_d + \boldsymbol{q})\tilde{\psi}_Q^*(\boldsymbol{q}_d) + \tilde{\psi}_Q(\boldsymbol{q}_d)\tilde{\psi}_Q^*(\boldsymbol{q}_d - \boldsymbol{q})\} \\
&\quad \times D(\boldsymbol{q}_d)\, d\boldsymbol{q}_d
\end{aligned}
\tag{67.17a}
$$

$$
\begin{aligned}
K_p(\boldsymbol{q}) &:= \int \{\tilde{\psi}_Q(\boldsymbol{q}_d)\tilde{\psi}_Q^*(\boldsymbol{q}_d - \boldsymbol{q}) - \tilde{\psi}_Q(\boldsymbol{q}_d + \boldsymbol{q})\tilde{\psi}_Q^*(\boldsymbol{q}_d)\} \\
&\quad \times D(\boldsymbol{q}_d)\, d\boldsymbol{q}_d
\end{aligned}
\tag{67.17b}
$$

This term will be small compared with $\tilde{I}_d^{(2)}$ only if the specimen scatters weakly and hence s and φ are small. For this situation the image contrast[*] is described by *transfer functions* $K_a(\boldsymbol{q})$ and $K_p(\boldsymbol{q})$, which closely resemble those encountered in the discussion of image formation in TEM.

67.3 Detector geometry

Within the bright-field cone, therefore, the transfer of contrast to the detector plane is similar to the mechanism of linear image formation in the TEM. This observation is at the heart of suggestions by Rose (1974a,b, 1977) to divide the bright-field detection area into zones and form a weighted sum of the signals, for specimens for which the phase component is dominant.

This was not, however, the first subdivision of the STEM detector to be proposed. In 1974, Dekkers and de Lang showed that by dividing the circular bright-field detector into two semicircular parts, the phase gradient in one direction could be detected directly; see too Dekkers *et al.* (1976), Dekkers and de Lang (1977, 1978a), de Lang and Dekkers (1979), Dekkers (1979) and Bouwhuis and Dekkers (1980). The natural extension to division into four quadrants to yield both components of the phase gradient was mentioned shortly after (Rose, 1975) and explored in detail by Hawkes (1978b).

---

[*] Note that these transfer functions differ by a factor of two from those for the TEM because we have not defined STEM contrast in quite the same way, see (66.10a).

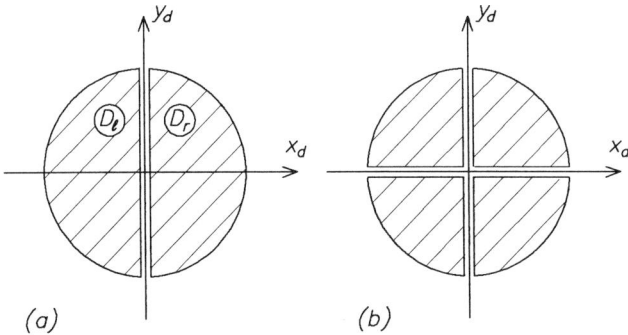

Fig. 67.1: (a) The semicircular detectors of Dekkers and de Lang. (b) Quadrant detectors.

The principle is easily explained with the aid of (67.14). For a detector consisting of two semicircular regions (Fig 67.1a), we have

$$D_r(\boldsymbol{q}_d) = \begin{cases} 1 & q_{xd} > 0 \\ 0 & q_{xd} < 0 \end{cases} \quad \forall q_{yd} \qquad \text{right detector}$$

$$D_l(\boldsymbol{q}_d) = \begin{cases} 1 & q_{xd} < 0 \\ 0 & q_{xd} > 0 \end{cases} \quad \forall q_{yd} \qquad \text{left detector}$$

(67.18)

with $\boldsymbol{q}_d =: (q_{xd}, q_{yd})$. Hence

$$S_{Dr}(\boldsymbol{u}_d) = \frac{1}{2} \left\{ \delta(x_d) - \frac{\mathrm{i}}{\pi x_d} \right\} \delta(y_d)$$

$$S_{Dl}(\boldsymbol{u}_d) = \frac{1}{2} \left\{ \delta(x_d) + \frac{\mathrm{i}}{\pi x_d} \right\} \delta(y_l)$$

(67.19)

A straightforward calculation set out in detail in Hawkes (1978b) shows that the sum of the two signals yields information about $s(\boldsymbol{u}_o)$ to a good approximation while the difference provides the derivative of the phase function, $\partial\varphi/\partial x_o$. Extension of the argument to a *quadrant* detector (Fig. 67.1b) shows that, as we should expect, both $\partial\varphi/\partial x$ and $\partial\varphi/\partial y$ can be extracted. Moreover the detailed analysis of Waddell and Chapman (1979) shows that the restriction to weak phase variations of the specimen is not so severe as in the conventional imaging modes. Split detectors are hence very suitable for studying magnetic specimens and have been used extensively for this purpose by Chapman and colleagues (Chapman *et al.*, 1978a,b, 1983, 1990; Morrison *et al.*, 1979; Chapman and Morrison, 1983; Morrison and Chapman, 1981, 1982, 1983; Chapman, 1989).

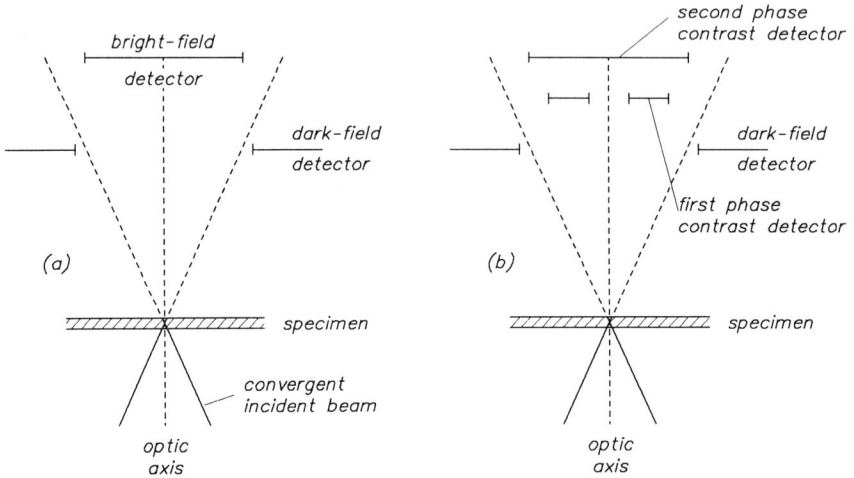

Fig. 67.2: (a) Standard arrangement of detectors in STEM. (b) Subdivision of the bright-field cone to improve phase contrast.

The split detector of Dekkers and de Lang was the first of several proposals designed to exploit the freedom of choice of detector geometry. Waddell *et al.* (1977, 1978) described a 'first-moment' detector, also analysed by Waddell and Chapman (1979), which involves weighting the current $J_d(\boldsymbol{q}_d; \boldsymbol{\xi})$ before adding the contributions from the points in the detector plane: $D(\boldsymbol{q})$ now has a more complicated variation than the simple binary response we have so far considered, $D(\boldsymbol{q}_d) = |\boldsymbol{q}_d|$.

The form of the function $K_p(\boldsymbol{q})$ led Rose (1974a) to investigate the signals that could be collected by subdividing the bright-field cone into complementary rings (Fig. 67.2b). We say no more about this here as the idea has been superseded by the more recent possibility of dividing the cone into a large number of fine rings, with which a very desirable form of the contrast transfer function can be attained (Hammel *et al.*, 1990).

It was soon realized, thanks largely to the work of Rose (1974a,b, 1975, 1977) and Rose and Fertig (1976), that flexible detectors, permitting a variety of forms of $D(\boldsymbol{q}_d)$ to be generated, were desirable and several were discussed (Cowley, 1976b, cf. Dekkers and de Lang, 1978b; Cowley and Au, 1978a,b; Lackovic *et al.*, 1979; Burge and van Toorn, 1979, 1980; Burge *et al.*, 1979a,b, 1982; Robinson, 1979; van Toorn and Robinson 1980; Ward *et al.*, 1979; Browne and Ward, 1982; Hawkes, 1980c; Cowley, 1993). Figure 67.3 shows the subdivisions recommended by Burge and van Toorn (1980).

More recently, a version with four quadrants and 32 rings has been designed (Haider *et al.*, 1988). Meanwhile, Smith and Erasmus (1982)

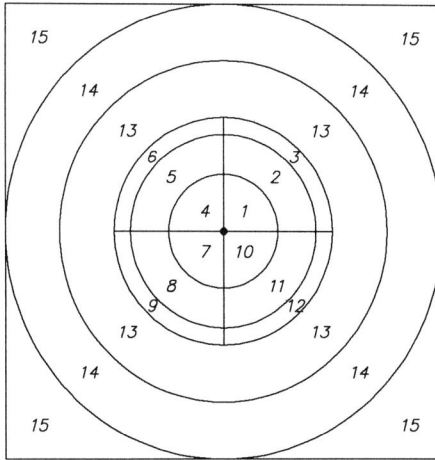

Fig. 67.3: Division of the detector plane with which two-ring detectors, split and quadrant detectors and other geometries can be synthesized.

pointed out that any desired configuration could be generated by reading the image from each pixel into computer memory and multiplying it by the appropriate pattern of zeros and ones (or weights), adding the result and storing these sums. This inevitably slowed down the acquisition, however. The idea has been revived by Daberkow and Herrmann (1984, 1988) and Daberkow *et al.* (1993) in a more modest form, in which only a few patterns are allowed. Hammel *et al.* (1990) discuss the use of detectors with many rings inside the bright-field cone to combat the adverse effect of the contrast-transfer function in the high-resolution study of weakly-scattering specimens. See too Wang *et al.* (1992).

If we are prepared to collect *all* the information for every pixel, however, a different form of the foregoing analysis is more appropriate (Bates and Rodenburg, 1989; Rodenburg, 1989a,b; Rodenburg and McCallum, 1991; Friedman *et al.*, 1991; Rodenburg and Bates, 1992; Plamann and Rodenburg, 1992; McCallum and Rodenburg, 1992, 1993a,b; Rodenburg *et al.*, 1993). This involves recording not $I_d(\boldsymbol{\xi})$ but $J_d(\boldsymbol{q}_d; \boldsymbol{\xi})$ for each point $\boldsymbol{\xi}$, giving not a (two-dimensional) image but a four-dimensional data set. We again set out from (67.5) and consider the transform of $J_d(\boldsymbol{q}_d; \boldsymbol{\xi})$ with respect to $\boldsymbol{q}_d$ as well as to $\boldsymbol{\xi}$. We write

$$\tilde{J}_d(\boldsymbol{u}_d; \boldsymbol{q}) := \iint J_d(\boldsymbol{q}_d; \boldsymbol{\xi}) \exp\{-2\pi \mathrm{i}(\boldsymbol{q}_d \cdot \boldsymbol{u}_d + \boldsymbol{\xi} \cdot \boldsymbol{q})\} \, d\boldsymbol{q}_d \, d\boldsymbol{\xi} \qquad (67.20)$$

giving

$$
\tilde{J}_d(\boldsymbol{u}_d; \boldsymbol{q}) = \iiiint O(\boldsymbol{u}_o)O^*(\boldsymbol{u}'_o)\psi_Q(\boldsymbol{u}_o - \boldsymbol{\xi})\psi_Q^*(\boldsymbol{u}'_o - \boldsymbol{\xi})
$$
$$
\times \exp\left[-2\pi\mathrm{i}\{\boldsymbol{q}_d \cdot (\boldsymbol{u}'_o - \boldsymbol{u}_o + \boldsymbol{u}_d) + \boldsymbol{\xi} \cdot \boldsymbol{q}\}\right] d\boldsymbol{u}_o \, d\boldsymbol{u}'_o \, d\boldsymbol{q}_d \, d\boldsymbol{\xi}
$$
$$
= \iint O(\boldsymbol{u}'_o + \boldsymbol{u}_d)O^*(\boldsymbol{u}'_o)\psi_Q(\boldsymbol{u}'_o + \boldsymbol{u}_d - \boldsymbol{\xi})\psi_Q^*(\boldsymbol{u}'_o - \boldsymbol{\xi})
$$
$$
\times \exp(-2\pi\mathrm{i}\boldsymbol{\xi} \cdot \boldsymbol{q}) \, d\boldsymbol{u}'_o \, d\boldsymbol{\xi}
$$

$$(67.21)$$

This separates very conveniently into terms in O and terms in ψ_Q if we introduce a new variable $\boldsymbol{v}$,

$$
\boldsymbol{v} = \boldsymbol{u}'_o - \boldsymbol{\xi} \tag{67.22}
$$

whereupon $\tilde{J}_d$ becomes

$$
\tilde{J}_d(\boldsymbol{u}_d; \boldsymbol{q}) = \iint O(\boldsymbol{u}'_o + \boldsymbol{u}_d)O^*(\boldsymbol{u}'_o)\psi_Q(\boldsymbol{v} + \boldsymbol{u}_d)\psi_Q^*(\boldsymbol{v})
$$
$$
\times \exp\{-2\pi\mathrm{i}(\boldsymbol{u}'_o - \boldsymbol{v}) \cdot \boldsymbol{q}\} \, d\boldsymbol{u}'_o \, d\boldsymbol{v} \tag{67.23}
$$
$$
=: A_O(\boldsymbol{u}_d, \boldsymbol{q})A_\psi(\boldsymbol{u}_d, -\boldsymbol{q})
$$

where

$$
A_O(\boldsymbol{u}_d, \boldsymbol{q}) = \int O(\boldsymbol{u}'_o + \boldsymbol{u}_d)O^*(\boldsymbol{u}'_o)\exp(-2\pi\mathrm{i}\boldsymbol{u}'_o \cdot \boldsymbol{q}) \, d\boldsymbol{u}'_o \tag{67.24}
$$

$$
A_\psi(\boldsymbol{u}_d, -\boldsymbol{q}) = \int \psi_Q(\boldsymbol{v} + \boldsymbol{u}_d)\psi^*(\boldsymbol{v})\exp(2\pi\mathrm{i}\boldsymbol{v} \cdot \boldsymbol{q}) \, d\boldsymbol{v} \tag{67.25}
$$

A_O and A_ψ are identical in form, apart from the change from O to ψ_Q, since $\boldsymbol{v}$ and $\boldsymbol{u}'_o$ are essentially dummy variables:

$$
A_f(\boldsymbol{u}_d, \boldsymbol{q}) = \int f(\boldsymbol{u}_d + \boldsymbol{v})f^*(\boldsymbol{v})\exp(-2\pi\mathrm{i}\boldsymbol{q} \cdot \boldsymbol{v}) \, d\boldsymbol{v} \tag{67.26}
$$

In principle, ψ_Q and hence A_ψ are known and the numerous methods described in Chapter 73 will yield estimates of $A_O(\boldsymbol{u}_d, \boldsymbol{q})$ from the transformed measured distribution $\tilde{J}_d(\boldsymbol{u}_d, \boldsymbol{q})$. Fourier transformation of A_O with respect to $\boldsymbol{q}$ yields S_O,

$$
S_O(\boldsymbol{u}'_o, \boldsymbol{u}_d) := O(\boldsymbol{u}'_o + \boldsymbol{u}_d)O^*(\boldsymbol{u}'_o) \tag{67.27}
$$

We set the origin of $\boldsymbol{u}'_o$ at the point where $S_O(\boldsymbol{u}'_o, 0)$ is greatest and hence obtain

$$
S_O(0, \boldsymbol{u}_d) = O(\boldsymbol{u}_d)O^*(0) \tag{67.28}
$$

We are at liberty to choose the phase origin at $S_O(0,0)$ so that finally

$$O(\boldsymbol{u}_d) = S_O(0, \boldsymbol{u}_d)/O(0) \qquad (67.29)$$

The method is thus capable of yielding the complex object transparency directly, without iteration, but at the cost of recording a huge volume of data. Experiments to establish the practical usefulness of the technique are continuing, but the electron dose that the specimen must support is clearly high if the individual diffraction patterns $J_d(\boldsymbol{q}_d; \boldsymbol{\xi})$ are not to be excessively noisy. We note that a form of this approach was adumbrated by Hoppe long before (1982), in a discussion of STEM ptychography.

The foregoing analysis does not, however, correspond to the most common way of forming an image in the STEM. As we have already mentioned, the electrons scattered within the specimen may be separated into three groups, those scattered inelastically and elastically and the unscattered beam. On average, the elastically scattered electrons will be deflected through larger angles than those scattered inelastically and a crude but effective selection can be made by collecting the former on an annular detector, the central opening of which permits all the unscattered and most of the inelastically scattered electrons to pass unimpeded. This central beam will then be separated into beams of different energy with the aid of a spectrometer.

A somewhat primitive procedure for estimating the signal collected by such an annular dark-field detector is often adequate. The specimen transparency is described by two terms, one of which characterizes scattering inside the bright-field cone, the other outside it:

$$O(\boldsymbol{u}_o) =: s_1 + s_2(\boldsymbol{u}_o) \qquad (67.30)$$

and from (67.6), we have

$$I_d(\boldsymbol{\xi}) = \iiint I^{(0)}(\boldsymbol{\xi})(1 + I^{(1)} + I^{(2)})\, d\boldsymbol{u}_d\, d\boldsymbol{u}_o\, d\boldsymbol{u}'_o \qquad (67.31)$$

in which

$$I^{(0)}(\boldsymbol{\xi}) := s_1 s_1^* \psi_Q(\boldsymbol{u}_o - \boldsymbol{\xi})\psi_Q^*(\boldsymbol{u}'_o - \boldsymbol{\xi})D(\boldsymbol{u}_d)\exp\{2\pi i \boldsymbol{u}_d \cdot (\boldsymbol{u}_o - \boldsymbol{u}'_o)\} \qquad (67.32)$$

The integral of this term vanishes since $D(\boldsymbol{u}_d) = 0$ in the central area occupied by s_1. The term $I^{(1)}$ has the form

$$I^{(1)} = s_2(\boldsymbol{u}_o) + s_2^*(\boldsymbol{u}'_o) \qquad (67.33)$$

and the contribution to $I_d(\boldsymbol{\xi})$ is thus

$$I_d^{(1)}(\boldsymbol{\xi}) = \int |s_1|^2 |\psi_Q(\boldsymbol{u}_o - \boldsymbol{\xi})|^2 \{s_2(\boldsymbol{u}_o) + s_2^*(\boldsymbol{u}_o)\} \, d\boldsymbol{u}_o \qquad (67.34)$$

The final term is quadratic in s_2:

$$I^{(2)} = s_2(\boldsymbol{u}_o) + s_2^*(\boldsymbol{u}_o)$$

and

$$I_d^{(2)}(\boldsymbol{\xi}) = \int |s_1|^2 |\psi_Q(\boldsymbol{u}_o - \boldsymbol{\xi})|^2 |s_2|^2 \, d\boldsymbol{u}_o \qquad (67.35)$$

The last term is frequently dominant and dark-field STEM image formation in these circumstances is hence said to be *incoherent*.

Examination of the dependence of the elastic and inelastic cross-sections on atomic number Z shows that, for non-crystalline material, it is advantageous to form the ratio of the dark-field and bright-field signals in the STEM. The resulting image exhibits 'Z-contrast' and this simple combination of two signals, introduced by Crewe et al. (1970a,b), has been found useful for biological specimens especially.

For crystalline material, this procedure is not useful since collective diffraction effects replace the single-atom scattering that makes the ratio technique effective. In order to create Z-dependent contrast with crystalline specimens, the opening in the annular dark-field detector must be made so large that the Bragg-diffracted electrons pass through as well as the central beam, only the electrons scattered through very high angles (Rutherford scattering) being collected. In practice, the unwanted electrons are focused through the central hole by means of a lens, which also helps to bring the electrons scattered through large angles back onto the detector.

For further details of Z-contrast for non-crystalline specimens, see Isaacson et al. (1980) and for the crystalline case, see Treacy et al. (1978), Howie (1979), Pennycook (1981), Treacy (1981) and especially Pennycook (1989, 1992), Jesson and Pennycook (1990), Pennycook and Jesson (1990, 1992), Pennycook et al. (1991) and Pennycook and Boatner (1988).

For extensive discussion of the practical aspects of the various imaging modes in the STEM, we refer to recent texts on high-resolution microscopy, notably Spence (1988) and Spence and Zuo (1992). One such mode that we have not considered here permits lattice images to be formed in the STEM (Spence and Cowley, 1978). A survey by Spence (1992), in which the close relation between this technique and other TEM and STEM operating modes is emphasized, is particularly illuminating.

67.4 Concluding remarks

The foregoing account of image formation in the STEM is intended to be no more than an introduction to the subject. We have neglected the effects of finite source-size and energy spread in the probe (Burge and Dainty, 1976; Zeitler, 1975). A much more important omission is the lack of a thorough study of the interaction between the various scattering mechanisms in the specimen and the signals detected. This is essential for correct interpretation of the images provided by the STEM; many of the papers by Rose, Spence and Cowley listed below are devoted to this topic.

The first full analysis of image formation in the STEM is to be found in the remarkable study by Zeitler and Thomson (1970), which remains invaluable. The nature of STEM imagery was investigated in great detail during the 1970s, both in research laboratories and in the development groups of the companies that produced commercial instruments (VG, Siemens and AEI) or hybrid TEM-STEM systems. The publications from Chicago, where the first STEM was built (Crewe, 1966, 1970, 1973, 1974, 1979, 1980a-c; Crewe and Wall, 1970a,b; Crewe et al., 1969, 1979; Crewe and Groves, 1974; Groves, 1975; Beck and Crewe, 1975; Beck, 1977; Crewe and Ohtsuki, 1980a,b; Crewe and Kopf, 1980; Ohtsuki and Crewe, 1980; Thomson, 1973), Darmstadt (Rose, 1974a,b, 1975, 1977, 1978; Rose and Fertig, 1976, 1977; Fertig and Rose, 1977), Arizona (Cowley, 1969, 1970, 1975, 1976a,b,c, 1978, 1978/9, 1980; Cowley and Jap, 1976a,b; Spence and Cowley, 1978; Spence, 1978; Cowley and Au, 1978a,b, 1980; Cowley and Spence, 1979), London (Welford, 1972; Barnett, 1973, 1974; Misell et al., 1974; Browne et al., 1975; Burge and Derome, 1976; Burge, 1977; Misell, 1977; Burge et al., 1976, 1979a,b; Burge and van Toorn, 1979, 1980) and Braunschweig (Hanszen and Ade, 1974, 1976a, 1978; Hanszen, 1974; Ade, 1977a,b) give a good idea of the way the subject developed. See too Colliex et al. (1977), Engel (1974), Engel et al. (1974), Howie (1972, 1974), Kermisch (1977), Misell et al. (1974), Reimer et al. (1975), Reimer and Hagemann (1977) and Veneklasen (1975). (Many other papers are listed in the relevant sections of Hawkes, 1978a, 1982b, 1992.)

The published coverage of the commercial instruments—the short-lived Siemens ST100F and AEI STEM and the thriving VG HB series—is less full. On the Siemens microscope, see Krisch et al. (1976, 1977), Hubert et al. (1978) and an anonymous account in *Microscopica Acta* **77** (1976) 455–458. For the AEI instrument, see Banbury and Bance (1973a,b), Banbury (1974), Banbury et al. (1975) and Ray et al. (1975). The VG HB5 is discussed by Wardell et al. (1973); however, many improvements to the early version were made by users and these are reported in the conference proceedings of the period, many of which had a section on STEM, see for example Browne *et al.* (1975), Waddell *et al.* (1978), Pennycook *et al.* (1977), Treacy *et al.* (1979). The early work on hybrid instruments is described by Thompson (1973a,b, 1975) and Kuypers *et al.* (1973) in the case of Philips, by Koike *et al.* (1974), Someya *et al.* (1974) and Harada *et al.* (1975) for JEOL and by Tochigi *et al.* (1974) for Akashi.

By the following decade, the STEM was in routine use in the rather few laboratories that possessed one and only VG continued to supply such instruments commercially. Most of the manufacturers of transmission microscopes offered scanning transmission as one of the modes of operation of the TEM though it was some years before reliable field-emission sources became available. The following papers give an idea of both theoretical and practical developments: Brown (1981), Colliex (1985), Colliex *et al.* (1984, 1989), Mory and Colliex (1985), Mory *et al.* (1987a,b); Cowley (1981, 1983, 1984, 1985, 1986, 1987), Cowley and Disko (1980), Cowley and Walker (1981), Cowley *et al.* (1980), Cowley and Spence (1981); Crewe (1980a, 1983, 1985), Crewe and Ohtsuki (1981a,b); Crewe and Salzman (1982); Fertig and Rose (1981); Rose (1984), Kohl and Rose (1985), Kohl (1986); Hammel *et al.* (1990); Haider *et al.* (1988), Jones (1988), Jones and Haider (1989); Hawkes (1982a,b, 1985); Humphreys (1981); Kanaya *et al.* (1985, 1986); Reichelt and Engel (1985, 1986); Li and Dorignac (1986); Bovey and Nicholls (1987); Hammel and Rose (1993), Cowley (1993).

Part XIV

Electron Interactions in Thin Specimens

68
Electron Interactions in Amorphous Specimens

68.1 Introduction

The many facets of the electron microscope imaging process have been studied in great detail in the preceding chapters. The effect of the specimen on the electron waves traversing it has been described by an 'object function' or its spectrum and the relation between this and the specimen structure was regarded as known. In this Part, we show how the interaction of the electron beam with a specimen can be understood in terms of basic quantum mechanical principles.

We have divided this topic into two chapters, one on *amorphous* and the other on *crystalline* specimens. The essential difference between these lies in the fact that the irregular atomic structure in the former causes practically random scattering whereas crystalline structures generate strong, regular interferences, which modify the electron waves in a characteristic manner.

This account is intended merely to introduce the reader to this vast subject and we therefore consider only interactions in *thin* transparent foils, typical of the specimens studied in transmission microscopy. The numerous physical processes that occur when electrons are scattered in bulk material, which are of central importance for analytical electron microscopy and spectroscopy, have had to be excluded; they form a whole subject of their own.

We shall be mainly concerned with *elastic* electron scattering at higher energies, above 30 keV say. This is the most important process so far as the formation of the so-called 'object wave' is concerned. We shall treat the three most important methods of calculation: Born's approximation in first order, Molière's high-energy approximation and the partial wave method.

The standard approach at higher particle energies is based on *Born's approximation.* This is attractive in that fairly simple analytic formulae are obtained and some useful general insight can be gained from these. Its main drawback is that it is not always very accurate.

The high-energy approximation of Molière gives distinctly more reliable results and we shall describe it in some detail, since its formalism comes closest to the concepts used in the definition of 'weakly scattering' and 'phase' objects. The method furnishes an approximate expression for the wavefunction immediately behind a scattering atom.

The third method, partial wave analysis, is dealt with only very briefly. It is the only exact method and must be employed at lower energies, below about 10 keV. This energy range is not considered here. At higher energies, the partial wave analysis provides a way of checking the accuracy of the other methods, from which it emerges that Molière's procedure gives much more accurate results than Born's approximation.

In specimens consisting of elements of lower atomic number, typically organic material, *inelastic* electron scattering becomes important. The corresponding theory is briefly sketched, on the basis of Born's approximation. A generalization of Molière's method would be better but becomes unduly complicated. We emphasize that our presentation is far from being complete, and is highly simplified. Our main goal here is to point out the differences between the angular distributions of elastically and inelastically scattered electrons and the dependence on atomic number, important for understanding the imaging of biological specimens.

We conclude this chapter with a brief account of plural and multiple electron scattering, in which we present a theory developed by Schwertfeger (1974) in an attempt to generalize the standard treatment of Lenz (1954). This theory yields the incoherent intensity distribution on the exit side of a specimen as a function of position and direction. From this, the image contrast can be calculated in the low-resolution limit.

The physical description of scattering processes is to be found in any textbook on quantum mechanics; our presentation will hence be very concise but, in order to keep these chapters reasonably self-contained, we do include definitions of the various cross-sections and recapitulate the basic relations. For more detail, the reader is referred to such specialized works on scattering as Wu and Ohmura (1962) and Mott and Massey (1965); the books of Reimer (1985, 1993) and Egerton (1989) deal extensively with the application of the scattering theories in electron optics. The International Tables for Crystallography (Wilson, 1992) are an invaluable source of information.

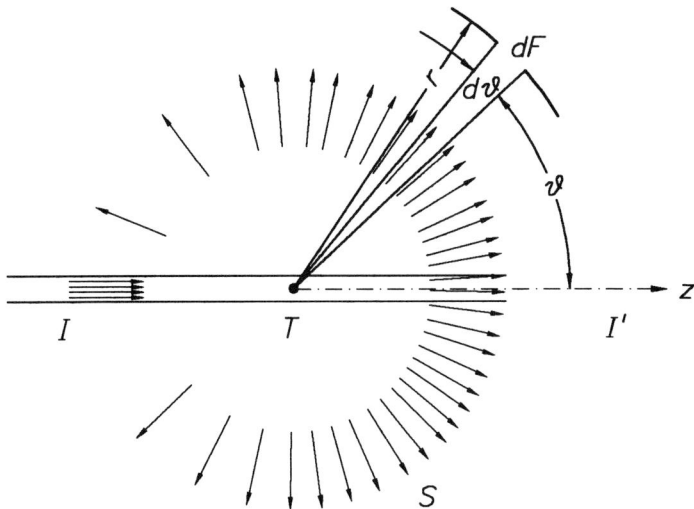

Fig. 68.1: Essential feature of a scattering experiment: I: incident or primary particle beam; I': unscattered beam; S: scattered particles; dF: detector aperture at a distance r from the target T at an angle ϑ.

68.2 Definition of the elastic cross-sections

Let us now examine the situation shown in Fig. 68.1. A very small target T is brought into the path of a parallel and monoenergetic beam of electrons. Without loss of generality, a spherical coordinate system (z, θ, φ) can be chosen such that the origin coincides with the centre of the target and that the beam travels in the positive z-direction. The lateral extent of the beam is assumed to be very much greater than the diameter of the target, which permits us to describe the incident electrons by a plane wave $\psi_i = \exp(\mathrm{i}kz)$. The electrostatic potential of the target produces an additional outgoing wave ψ_s. In the close vicinity of the target itself, this wave will have a complicated structure, but this is of little interest. It is the *asymptotic* behaviour of ψ_s, for large values of r, that will prove to be of greatest importance. This is characterized by *outgoing spherical wavefronts*, described by a wave term of the form $r^{-1}\exp(\mathrm{i}kr)$. This will be weighted by an additional factor f, which will usually depend on the angles θ and φ. Altogether, we thus arrive at the *asymptotic* formula

$$\psi(r, \theta, \varphi) = \psi_i + \psi_s = \exp(\mathrm{i}kz) + f(\theta, \varphi)\frac{\mathrm{e}^{\mathrm{i}kr}}{r} \qquad (68.1)$$

The amplitude of the incident plane wave has been normalized to unity, which can always be done without loss of generality.

The current density in the incident beam is given by

$$j_i = -i\frac{\hbar}{m}\Im\left(\psi_i^*\frac{\partial\psi_i}{\partial z}\right) = \frac{\hbar k}{m} = \text{const} \qquad (68.2a)$$

which is of the familiar form $j_i = \rho v$ with $\rho \equiv 1$ and $v = g/m = \hbar k/m$. The current density in the scattered beam, flowing in the radial direction, is given by

$$j_s = -i\frac{\hbar}{m}\Im\left(\psi_s^*\frac{\partial\psi_s}{\partial r}\right) = \frac{\hbar k}{mr^2}|f(\theta,\varphi)|^2 \qquad (68.2b)$$

We now define the *differential cross-section* as the ratio

$$\sigma(\theta,\varphi) := r^2 j_s/j_i = |f(\theta,\varphi)|^2 \qquad (68.3)$$

This can be measured, as sketched in Fig. 68.1: a small detector with aperture area $\Delta F = r^2\Delta\Omega = r^2\sin\theta\Delta\theta\Delta\varphi$ is brought into the path of the scattered beam and the number ΔN of particles that pass through it during a sufficiently long time span Δt is counted. We can then determine the quantity $j_s r^2 = \Delta N/\Delta\Omega\Delta t$. The current density in the incident beam is determined in a similar manner, after which the ratio is formed, as required by (68.3). All absolute intensities cancel out from this ratio, so that the normalization, made in (68.1) has indeed led to no loss of generality.

The *total cross-section* is defined by integration over the full unit sphere:

$$\sigma_{tot} := \oint \sigma\, d\Omega \equiv \int\limits_{\theta=0}^{\pi}\int\limits_{\varphi=0}^{2\pi}\sigma(\theta,\varphi)\sin\theta\, d\theta\, d\varphi \qquad (68.4)$$

This quantity can be pictured as the area of a disc, which is brought into the incident beam with its surface normal to the latter. All particles that hit the disc are then removed from the beam by scattering. Quite generally the so-called *optical theorem*

$$\sigma_{tot} = \frac{4\pi}{k}\Im f(0,0) \qquad (68.5)$$

can be derived. The proof can be found in textbooks on quantum mechanics. From (68.5) it is clear that the scattering amplitude $f(\theta,\varphi)$ is necessarily *complex* (if it does not vanish identically).

68.3 The first-order Born approximation for elastic scattering

In order to calculate the differential cross-section σ for a given target, we have to solve the Schrödinger equation in its time-independent form. The target is here characterized by its electrostatic potential $V(\boldsymbol{r})$, which acts on the incident particles. We confine our consideration to *screened* potentials, which implies *neutral* targets. Scattering at charged particles such as free electrons, ions or even bare nuclei may be important but not in electron microscopy, where the scattering atoms are all embedded in a neutral specimen.

A consequence of this assumption is that the potential decreases rapidly, typically exponentially, as $|\boldsymbol{r}| \to \infty$, so that is makes sense to speak of an asymptotically *field-free* space. In such field-free domains, the wavenumber k and the electron mass m satisfy (56.17) and (56.18), respectively. Here we shall drop the suffix '∞' and identify the resulting quantities k and m with those in (68.1–5).

Moreover we make the stronger assumption that the term in V^2 can be neglected in the relativistically corrected Schrödinger equation. This assumption is usually satisfied in practice; conditions in which this is too inaccurate are briefly mentioned in Section 68.5. After this linearization with respect to $V(\boldsymbol{r})$, the Schrödinger equation takes the form (56.19), which is more concisely rewritten as

$$\nabla^2 \psi(\boldsymbol{r}) + k^2 \psi(\boldsymbol{r}) + \frac{2me}{\hbar^2} V(\boldsymbol{r})\psi(\boldsymbol{r}) = 0$$

It is inconvenient to retain the constant factors in the potential term and we hence define a new function with dimension $[\mathrm{L}]^{-2}$

$$v(\boldsymbol{r}) := \frac{2me}{\hbar^2} V(\boldsymbol{r}) \tag{68.6}$$

whereupon the Schrödinger equation takes the compact form

$$\nabla^2 \psi(\boldsymbol{r}) + k^2 \psi(\boldsymbol{r}) = -v(\boldsymbol{r})\psi(\boldsymbol{r}) \tag{68.7}$$

A formal solution is readily obtained by treating the right-hand side as a source term. By recalling that the Green's function of free-wave propagation is given by

$$G(\boldsymbol{r}, \boldsymbol{r}') = \frac{1}{4\pi|\boldsymbol{r} - \boldsymbol{r}'|} \exp(ik|\boldsymbol{r} - \boldsymbol{r}'|) \tag{68.8}$$

which satisfies the inhomogeneous equation

$$\nabla^2 G + k^2 G = -\delta(r - r') \tag{68.9}$$

we can easily verify that the expression

$$\psi(r) = \exp(ikz) + \int G(r, r')v(r')\psi(r')\, dr' \tag{68.10}$$

satisfies (68.7). The term $\exp(ikz)$, which we are at liberty to include, is a solution of the homogeneous equation, here the field-free wave equation. In this way, we satisfy the asymptotic boundary conditions in (68.1).

Equation (68.10) is well-known as the Lippmann–Schwinger equation. It is an inhomogeneous *integral* equation for ψ, since ψ appears in the integral as well as on the left-hand side. In practice, therefore, it is no easier to solve than (68.7) but suitable approximations can be found more straightforwardly.

A first simplification arises from the fact that only the asymptotic form of the solutions of (68.10) is needed for the calculation of the cross-section σ. It is therefore sufficient to make use of the asymptotic approximation to the Green's function for $|r| \gg |r'|$. With $r = |r|$ and the unit vector $e := r/r$ in the direction of observation, this takes the form

$$G(r, r') \rightarrow \frac{1}{4\pi r} \exp(ikr)\exp(-ike \cdot r') \tag{68.11}$$

Substituting this in (68.10), we find

$$\psi(r) = \exp(ikz) + \frac{\exp(ikr)}{r} \frac{1}{4\pi} \int v(r')\psi(r')\exp(-ike \cdot r')\, dr'$$

This does indeed have the basic form of (68.1) and, by comparison, we obtain the exact formula

$$f(\theta, \varphi) = \frac{1}{4\pi} \int v(r')\psi(r')\exp(-ike \cdot r')\, dr' \tag{68.12}$$

This expression still does not have the required form, since the unknown ψ appears under the integral.

In the *Born approximation* in its first-order form, the scattering process is assumed to be very *weak*, which means $|\psi_s| \ll |\psi_i| = 1$ in (68.1); it is hence sufficient to approximate ψ by $\psi_i = \exp(ikz')$ in (68.12). We have later to discuss whether this simplification is acceptable but, if we adopt

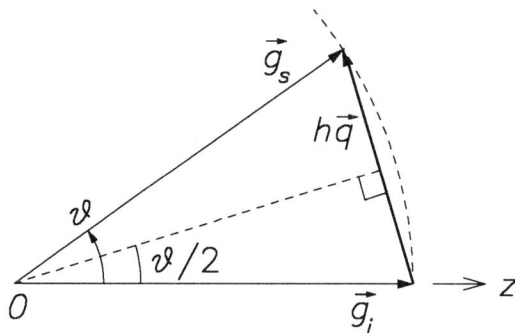

Fig. 68.2: Momentum transfer diagram for elastic scattering, showing the momentum g_i of the incident particle, g_s of the scattered particle and transfer vector $hq = g_s - g_i$. The latter is a chord of a circle of radius $g = \hbar k$. Note that the direction of this vector is often reversed in other presentations.

it for the moment, we obtain a result that can be cast into the form of an inverse three-dimensional Fourier integral:

$$
\begin{aligned}
f(\theta, \varphi) =: \hat{f}(\boldsymbol{q}) &= \frac{1}{4\pi} \mathcal{F}^-(v) \\
&= \frac{1}{4\pi} \int v(\boldsymbol{r}') \exp(-2\pi i \boldsymbol{q} \cdot \boldsymbol{r}') \, d\boldsymbol{r}'
\end{aligned}
\tag{68.13}
$$

with $k = 2\pi/\lambda$; the vector $\boldsymbol{q}$ is defined by

$$
\boldsymbol{q} = \frac{1}{\lambda}(\sin\theta\cos\varphi, \sin\theta\sin\varphi, \cos\theta - 1)
\tag{68.14}
$$

Apart from a constant factor, this describes the *transfer of momentum* to the scattered electron; in particular we have

$$
\Delta\boldsymbol{g} := \boldsymbol{g}_s - \boldsymbol{g}_i = \hbar k(\boldsymbol{e} - \boldsymbol{e}_z) = h\boldsymbol{q}
\tag{68.15}
$$

This is illustrated in Fig. 68.2.

The first-order Born approximation violates the optical theorem (68.5). This is immediately obvious from (68.13), which predicts that $f(0,0)$ is real:

$$
f(0,0) = \frac{1}{4\pi} \int v(\boldsymbol{r}') \, d\boldsymbol{r}'
$$

and hence that $\Im f(0,0) = 0$. Conversely, $\sigma_{tot} > 0$, since $f(\theta, \varphi)$ does not vanish identically. This deficiency could be remedied by evaluating the Born approximations of higher order; but in practice, even the second order is so complicated that it cannot compete with other methods and is therefore almost nowhere employed. In spite of its evidently approximate character, formula (68.13) is frequently used, doubtless because the techniques for performing Fourier transforms are so highly advanced.

The theory given above holds for fairly general potentials. We now specialize to spherically symmetric potentials $V(r)$ and write these as

$$V(r) =: \frac{Ze}{4\pi\varepsilon_0 r} u(r) \tag{68.16}$$

where, as usual, Z denotes the atomic number and $u(r)$ the *screening function*. The latter is bounded by the conditions

$$u(r) \geq 0, \quad u(0) = 1, \quad \lim_{r\to\infty} u(r) = 0 \tag{68.17}$$

which imply *neutrality* of the scattering atom. Moreover it is advantageous to introduce the Bohr radius

$$a_0 := \frac{4\pi\varepsilon_0 \hbar^2}{m_0 e^2} = 0.0529 \text{ nm} \tag{68.18}$$

as the natural length unit. With $\gamma := m/m_0$ we obtain

$$v(r) = \frac{2\gamma Z}{a_0 r} u(r) \tag{68.19}$$

It is common practice to introduce the *particle density* $\rho(\boldsymbol{r})$ of the shell electrons and normalize it by writing

$$\int \rho(\boldsymbol{r}) \, d\boldsymbol{r} = Z \tag{68.20}$$

For spherically symmetric shells, Poisson's equation takes the form

$$u''(r) = 4\pi Z^{-1} r \rho(r) \tag{68.21}$$

This differential equation can be used to relate the scattering amplitude $f(\theta)$ to the atomic form factor, which is the Fourier transform of $\rho(r)$:

$$F(q) := \mathcal{F}(\rho) = 4\pi \int_0^\infty \rho(r) \frac{\sin(2\pi q r)}{2\pi q r} r^2 \, dr \tag{68.22}$$

The variable

$$q = 2\lambda^{-1} \sin\theta/2 \tag{68.23}$$

is here the absolute value of the momentum transfer vector $\boldsymbol{q}$ of (68.14). A longer calculation, given in most textbooks on quantum mechanics (see e.g. Mott and Massey 1965, Chapter V), results in

$$\hat{f}(q) := f(\theta) = \frac{\gamma\{Z - F(q)\}}{2\pi^2 a_0 q^2} \qquad (68.24)$$

For very large values of q, the oscillations in the integrand of $F(q)$ become so rapid that $F(q)$ becomes negligibly small. We then obtain the familiar Rutherford formula

$$\sigma(\theta) = |f(\theta)|^2 = \frac{\gamma^2 Z^2}{4k^4 a_0^2 \sin^4 \theta/2} \qquad (68.25)$$

as a degenerate special case. This means that the incident electron has such a high energy that it can approach the nucleus quite closely; the screening by the shell is then ineffective.

So far, we have described Born's first-order approximation in its general form. We now come to the special features and requirements of electron optics. Once again, we exclude all low-energy applications which are characteristic of analytic electron microscopy and SEM operation. We limit the discussion to electron energies eU above about 30 keV, for which it can be shown that Born's approximation is valid with an accuracy that, even if not very high, is still reasonable. A consequence of such high energies is that the scattering angles θ are small, of the order of milliradians. The approximation $2\sin\theta/2 = \theta$ is then perfectly adequate.

Elastic electron scattering was first investigated in this way by Lenz (1954). By expanding the sine factor in (68.22) up to third order, he found the approximate expression

$$F(q) = Z - \Theta\frac{2\pi^2}{3}q^2 + O(q^4) \qquad (68.25a)$$

with

$$\Theta := Z\bar{r}^2 = 4\pi \int_0^\infty \rho(r)r^4\, dr \qquad (68.25b)$$

Equation (68.24) then gives

$$\sigma_{max} := \sigma(0) = \left(\frac{\gamma\Theta}{3a_0}\right)^2 \qquad (68.26)$$

for the maximum value of the cross-section. It now becomes obvious that scattering into small angles mostly occurs in the *outer* zones of the atomic shell, since these give the largest contributions to the integral Θ. This result is of interest in connection with scattering contrast, to which we return in Section 68.7.

In order to pursue the calculations further, a knowledge of the potential $V(r)$ or, equivalently, of the particle density $\rho(r)$ is required. But here we face a real difficulty, since these functions are not known exactly, even for free atoms. Their incorporation into solid amorphous specimens causes such a violent perturbation that we cannot expect to obtain more than crude estimates.

For *free atoms*, the self-consistent atomic shell models of Hartree, Fock and Slater and the statistical models of Thomas, Fermi and Dirac in their various degrees of sophistication are in frequent use. These can be fairly accurately approximated by screening functions of the general form

$$u(r) = \sum_{v=1}^{N} \alpha_v(Z)\exp\{-\beta_v(Z)r/a_0\} \tag{68.27}$$

Different coefficient sets for $N = 3$ have been published, by Byatt (1956), Bonham and Strand (1963) and Haase (1966, 1968, 1970). Such simple model functions have the great advantage that simple closed formulae are obtained for $\sigma(\theta)$ and σ_{tot}: we find

$$F(q) = Z\sum_{v=1}^{N} \frac{\alpha_v\beta_v^2}{\beta_v^2 + (2\pi a_0 q)^2}$$

$$\hat{f}(q) = 2a_0\gamma Z\sum_{v=1}^{N} \frac{\alpha_v}{\beta_v^2 + (2\pi a_0 q)^2},$$

$$\Theta = 6a_0^2 Z\sum_{v=1}^{N} \alpha_v/\beta_v^2, \tag{68.28}$$

$$\sigma_{tot} = \frac{1}{\pi}\sum_{v=1}^{N}\left(\frac{\gamma Z\lambda\alpha_v}{\beta_v}\right)^2 + \frac{2}{\pi}(\gamma Z\lambda)^2\sum_{\mu=1}^{N}\sum_{v=1}^{\mu-1} \frac{\alpha_\mu\alpha_v\ln(\beta_\mu/\beta_v)}{\beta_\mu - \beta_v}$$

For atoms embedded in an amorphous solid target, a so-called 'muffin tin' model is in use. For each particular atom a spherical domain is constructed, within which the potential V is spherically symmetric, while it vanishes outside. The radius a is taken to be the mean half-distance to the neighbouring atoms. Raith (1968) suggests the formula

$$V_c(r) = \begin{cases} V(r) + V(2a-r) - 2V(a) & r \leq a \\ 0 & r \geq a \end{cases}$$

This potential satisfies $V'_c(a) = 0$, so that the continuation into the outside domain is smooth. Schwertfeger (1974) constructs an average Wigner–Seitz cell of radius a on the inscribed sphere. The charge of the atom is now redistributed in such a manner that its total value is still $-Ze$ and that $\rho(r)$ is smooth inside the whole Wigner–Seitz cell. A fraction of charge remains homogeneously distributed in the space inside the cell but outside the sphere. This implies that the screening effect is less pronounced than in Raith's model, but $\rho \neq 0$ conflicts with $V = 0$ for $r \geq a$. Nevertheless, the results obtained with this model come slightly closer to reality for, in Raith's model, the compression of the charge distribution into a sphere of radius a is too strong.

Since none of these more complicated models is really convincing, we adopt the Wentzel or Yukawa potential, this being the simplest possible model, for our further considerations. This potential is defined by

$$u(r) = \exp(-r/R) \tag{68.29}$$

and we use the screening radius R as a free-fit parameter. Formally this is a special case of (68.27) and (68.28) with $N = 1$, $\alpha_1 = 1$, $\beta_1 = a_0/R$ and so from (68.23) we obtain

$$|\hat{f}(q)|^2 = \sigma(\theta) = \frac{4\gamma^2 Z^2}{a_0^2(4k^2 \sin^2 \theta/2 + R^{-2})^2} \tag{68.30a}$$

$$\Theta = 6ZR^2 \tag{68.30b}$$

and

$$\sigma_{tot} = \frac{1}{\pi}\left(\frac{\gamma Z \lambda R}{a_0}\right)^2 \tag{68.30c}$$

A fairly good approximation for the radius R, valid for all atomic numbers Z, is simply $R \approx a_0 Z^{-1/3}$. A factor close to unity may appear in this relation, but we shall neglect it here. We then obtain

$$\Theta = 6a_0^2 Z^{\frac{1}{3}}, \quad \sigma_{max} = 4\gamma^2 a_0^2 Z^{\frac{2}{3}},$$
$$\sigma_{tot} = \frac{1}{\pi}\gamma^2 \lambda^2 Z^{\frac{4}{3}} = 4\pi a_0^2 \gamma^2 \frac{E_0}{e\hat{U}} Z^{\frac{4}{3}} \tag{68.31}$$

$E_0 = 13.6$ eV denoting the Rydberg energy constant. These results show that the total cross-section increases more strongly with atomic number than does the maximum. The decrease $\sigma_{tot} \propto U^{-1}$ for not too high energies means that the scattering contrast decreases correspondingly.

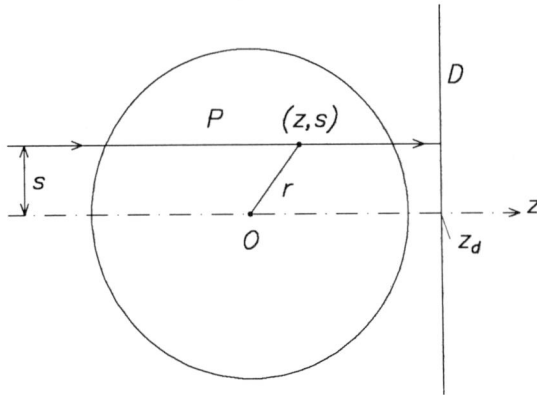

Fig. 68.3: Application of the high-energy approximation. D: diffraction plane; P: a particular straight path of integration and an arbitrary position (z, s) on it. The circle represents a rough estimate of the range of the atomic potential $V(r)$.

It is favourable to define a mean solid angle $\overline{\Omega}$ of scattering by

$$\overline{\Omega} = \pi\overline{\theta}^2 := \sigma_{tot}/\sigma_{max} \tag{68.32}$$

From (68.31), we obtain

$$\overline{\theta} = Z^{\frac{1}{3}}\sqrt{E_0/e\hat{U}} \tag{68.33}$$

This shows that the mean scattering angle $\overline{\theta}$ increases only very slowly with the atomic number. From carbon $(Z = 6)$ to lead $(Z = 82)$, there is only a factor of 2.4. The dependence on $\hat{U}$ is more pronounced. For $U = 100$ kV the square-root factor gives 0.011. The effects of these results on image contrast in the electron microscope will be dealt with in Section 68.6.

68.4 The high-energy approximation

We shall present this theory in some detail since it offers a way of determining the phase shift of the electron wave after it has passed through a target. Moreover it gives more accurate results than the first-order Born approximation. In its original form, the high-energy approximation was developed by Molière (1947) but it is also often associated with the name of Glauber (1959). Here we shall incorporate some useful modifications which remove a singularity.

The basic idea is to apply the *eikonal approximation* to the propagation of the electron wave. Since the scattering in the forward direction is very pronounced for higher kinetic energy (≥ 10 keV), we can assume that the paths of integration are straight in the direction of incidence, as shown in Fig. 68.3. The potential $v(\boldsymbol{r})$ is spatially confined, so that practically no error is caused by assuming infinite limits of integration for z. As it is not necessary to assume that the potentials are spherically symmetrical from the beginning, we start with the more general case. The results may then also be useful in cases where a target consisting of several kinds of atoms is to be studied.

We denote the lateral coordinates by $\boldsymbol{s} = (x, y)$. The *phase shift* $\eta(\boldsymbol{s})$ in some plane $z = $ const behind the target is then

$$\eta(\boldsymbol{s}) = \int_{-\infty}^{\infty} \left(\sqrt{k^2 + v(z, \boldsymbol{s})} - k \right) dz \qquad (68.34a)$$

It is common practice to assume that the potential is *weak*, $|v| \ll k^2$. The square-root term can then be expanded and only the linear term retained, giving

$$\eta(\boldsymbol{s}) = \frac{1}{2k} \int_{-\infty}^{\infty} v(z, \boldsymbol{s}) \, dz \qquad (68.34b)$$

Though quite widely used, this formula is unsatisfactory. For instance, it gives a logarithmic singularity at $\boldsymbol{s} = 0$, when applied to the potential v of (68.19). As far as we know, this difficulty is never considered in the literature. Since the more correct formula (68.34a) would give an even stronger square root singularity, we are forced to make an essential modification.

In domains where $|v| \ll k^2$ is valid, let us evaluate the next few terms in the series expansion:

$$\sqrt{k^2 + v} - k = \frac{v}{2k} - \frac{v^2}{8k^3} + \frac{v^3}{16k^5} \mp \cdots$$

An easy way of removing the singularity, no matter how strong it is, is to replace the integrand by a *rational function*:

$$\sqrt{k^2 + v} - k \Rightarrow \frac{v}{2k + \mu v/k} = \frac{v}{2k} - \frac{\mu v^2}{4k^2} + \frac{\mu^2 v^3}{8k^3} \pm \cdots$$

For $\mu = 1/2$ the quadratic term is reproduced correctly in the far zone. We therefore choose this value and obtain the simple formula

$$\eta(\boldsymbol{s}) = \int_{-\infty}^{\infty} \frac{2kv(z, \boldsymbol{s}) \, dz}{v(z, \boldsymbol{s}) + 4k^2} \qquad (68.35)$$

This expression always remains well-behaved for $s = 0$, as it should.

We now choose a suitable plane $z = z_d = \text{const}$ as a diffraction plane (see Fig. 68.3). The incident wave is here

$$\psi(s, z_d) = \exp\{ikz_d + i\eta(s)\}$$

We decompose the wavefunction into the unmodified incident wave ψ_i and the scattered wave ψ_s by the following simple trick:

$$\psi = \exp(ikz_d) + \exp(ikz_d)(e^{i\eta} - 1) \tag{68.36}$$

The first term is evidently the value of $\exp(ikz)$ at $z = z_d$ in (68.1); the second term must hence be ψ_s.

We now apply the diffraction theory in the Fraunhofer approximation to the wave propagation in the half-space $z > z_d$. The plane wave ψ_i already has the correct form $\exp(ikz)$. The application of Kirchhoff's diffraction formula to the scattered wave ψ_s in the far zone gives

$$\psi_s(\boldsymbol{r}) = -\frac{ik}{2\pi r} \exp\{ik(r - z_d)\} \int \psi_s(s, z_d) \exp(-2\pi i \boldsymbol{q} \cdot s) \, ds$$

with $\boldsymbol{q}$ given by (68.14); we have ignored the very small quantity $z_d(1-\cos\theta)$ in deriving this formula. On introducing the boundary values ψ_s from (68.36), we see that the terms in z_d cancel out completely; we obtain a first factor $\exp(ikr)/r$ for the outgoing spherical wave, and by comparison with (68.1) we conclude that

$$f(\theta, \varphi) = \hat{f}(\boldsymbol{q}) = \frac{k}{2\pi i} \int \{\exp i\eta(s) - 1\} \exp(-2\pi i \boldsymbol{q} \cdot s) \, ds \tag{68.37a}$$

This is evidently a Fourier integral; we can rewrite it more concisely as

$$\hat{f}(\boldsymbol{q}) = \frac{1}{i\lambda} \mathcal{F}^-(e^{i\eta} - 1) \tag{68.37b}$$

We have derived this formula explicitly here since it demonstrates the close relation between the scattering amplitude $\hat{f}$ and the *object spectrum* (65.15a and 66.2) in the theory of image formation.

Formula (68.37b) is certainly more correct than the familiar first-order Born approximation (68.13). The latter is obtained by making the *weak phase* approximation $\exp(i\eta) = 1 + i\eta$ and the still more restrictive *linearization* leading to (68.34b). The reason why the high-energy approximation is less frequently used in practice is to be found in the fact that simple closed formulae like those given in the previous section cannot be derived. With the widespread availability of powerful computers, there is no obstacle to carrying out all the necessary integrations numerically.

The high-energy approximation satisfies the optical theorem to a high degree of accuracy. It is not necessary to assume small scattering angles θ, as do Wu and Ohmura (1962, p. 50), although the scattering angles will be small in practice. We see from (68.37a) that

$$\Im \hat{f}(0) = \frac{2}{\lambda} \int\limits_0^\infty \sin^2\{\eta(\boldsymbol{s})/2\}\, d\boldsymbol{s}$$

The evaluation of σ_{tot} is straightforward: by making use of Parseval's theorem, which is suggested by the form of (68.37b), we obtain

$$\int\limits^\infty |\hat{f}(\boldsymbol{q})|^2\, d\boldsymbol{q} = \frac{1}{\lambda^2} \int\limits^\infty |e^{i\eta} - 1|^2\, d\boldsymbol{s}$$

This is exactly valid, but a small error is introduced on the left-hand side by setting the upper limit of integration equal to infinity instead of the correct value $q \leq 2k$. However, $|\hat{f}|^2$ decreases so rapidly for large values of q that for $q \approx 2k$ we can already ignore the contributions to the integral. Recalling that

$$d\Omega = \sin\theta\, d\theta\, d\varphi = \lambda^2\, d\boldsymbol{q}$$

we see that

$$\sigma_{tot} = \oint |f|^2\, d\Omega = \lambda^2 \int |\hat{f}|^2\, d\boldsymbol{q}$$

$$= \int |e^{i\eta} - 1|^2\, d\boldsymbol{s} = 4 \int \sin^2(\eta/2)\, d\boldsymbol{s} = \frac{2}{\lambda}\Im f(0,0)$$

The optical theorem is hence satisfied.

In the case of a spherical target potential $V(r)$, some simplifications are possible. On introducing (68.19) into (68.35), we find

$$\eta(s) = \int\limits_0^\infty \frac{4k u(r)\, dr}{u(r) + 2k^2 a_0 r/\gamma Z} \tag{68.38}$$

with $s = (x^2 + y^2)^{\frac{1}{2}}$, $r = (s^2 + z^2)^{\frac{1}{2}}$. Since this phase shift is rotationally symmetric about the z-axis, the integration over the azimuth in (68.37a) can be carried out in closed form, giving

$$f(\theta) = -\mathrm{i}k \int\limits_0^\infty \left\{ e^{\mathrm{i}\eta(s)} - 1 \right\} J_0\left(2ks \sin\frac{\theta}{2} \right) s\, ds \qquad (68.39)$$

In the familiar small-angle approximation, we replace $2\sin\theta/2$ by θ but the integration still has to be performed numerically. The latter calculation can be made reasonably fast by executing the inner integration (68.38) only once for a suitable set of discrete s-values and then storing the results in a table for subsequent interpolation.

Values of cross-sections and phase shifts obtained with the high-energy approximation have been published by Ferwerda and Visser (1973), Reimer and Sommer (1968), Reimer and Gilde (1973), Reimer and Lödding (1984) and by Zeitler and Olsen (1964, 1966, 1967). For more details of these calculations, especially concerning the atom models and approximations, we must refer to the original publications. Typke and Radermacher (1982) have evaluated the phase of the complex scattering factor from the associated shifts of the zeros of the transfer function.

We have tested the novel modification (68.35) (E. Kasper, unpublished) and compared it with the results obtained from (68.34b). For $U = 10$ kV and a Thomas–Fermi–Dirac model for the potential, we obtained only quite insignificant alterations in the results. This confirms that very large phases $\eta(s)$ for very small values of s have practically no influence on the value of the integral (68.39). The factor $\exp\{\mathrm{i}\eta(s)\}$ oscillates so rapidly that the integration over $s\exp\{\mathrm{i}\eta(s)\}$ makes no appreciable contribution for small values of s. The formula (68.38) has the advantage of removing this unphysical behaviour and thus facilitates the integration. On the other hand, this also shows that it is very difficult to derive phase shifts from recorded scattering information. The value $\eta(0)$ *cannot* be calculated by Fourier inversion of (68.37) with any real reliability. We always find that $|\eta| > 0.3$ in an appreciable zone.

68.5 Partial wave analysis

The methods outlined so far fail for low electron energy. Their range of application depends also on the atomic number Z, but a safe estimate of the lower limit is 10 keV. Scattering of fairly slow electrons is of importance in the operating conditions of scanning electron microscopes (SEM), in which bulk specimens are investigated. The calculation of elastic cross-sections is then a fundamental step in the calculation of multiple scattering processes.

The standard method of investigating low-energy scattering is partial wave analysis. This is extensively studied in the literature of quantum mechanics and the following discussion is therefore very brief. We begin with the theory of Faxén and Holtsmark.

Schrödinger's equation (68.7) may be reduced to a sequence of uncoupled ordinary differential equations by separation of variables in spherical coordinates. From

$$\psi(\boldsymbol{r}) = \frac{1}{r} \sum_{l=0}^{\infty} \sum_{m=-l}^{l} \chi_l(r) Y_{\ell m}(\theta, \varphi) \tag{68.40}$$

we thus obtain the *radial* equations

$$\chi_l''(r) + \{k^2 + v(r) - l(l+1)/r^2\}\chi_l(r) = 0 \tag{68.41}$$

In the theory of Faxén and Holtsmark, the *phases* η_l of the *asymptotic* solutions are to be determined, which means that we have to find them from solutions of the form

$$\chi_l(r) \rightarrow \sin(kr - l\pi/2 + \eta_l) \quad \text{as} \quad r \rightarrow \infty \tag{68.42}$$

The amplitudes are not needed here. It can be shown that

$$f(\theta) = \frac{1}{2ik} \sum_{l=0}^{\infty} (2l+1)(e^{2i\eta_l} - 1) P_l(\cos\theta) \tag{68.43a}$$

$$\sigma = |f(\theta)|^2 = \frac{1}{4k^2} \left| \sum_{l=0}^{\infty} (2l+1)(e^{2i\eta_l} - 1) P_l(\cos\theta) \right|^2 \tag{68.43b}$$

and

$$\sigma_{tot} = \frac{4\pi}{k^2} \sum_{l=0}^{\infty} (2l+1) \sin^2 \eta_l \tag{68.44}$$

These series expansions are always convergent, although the convergence may become very slow for high electron energies.

For very low electron energies, direct numerical solution of (68.41) with the appropriate starting conditions where $r \rightarrow 0$ is the best approach. We have to trace this solution until it becomes practically harmonic, by which we mean that the deviations from (68.42) can be neglected. There are formulae in the literature (Yates, 1971; Oswald, 1992) for the practical performance of this task, but for reasons of space we shall omit them here.

With increasing electron energy, this direct method becomes more and more laborious, since we have to integrate numerically over an increasingly large number of oscillations. The Wentzel–Kramers–Brillouin (WKB) method then becomes advantageous. This involves finding the eikonal approximation to the solutions of (68.41). The result is that the whole argument of the sine function in (68.42) is given by the integral expression

$$kr - l\pi/2 + \eta_l = \int_{r_l}^{r} \left\{ k^2 + v(r') - (l + 1/2)^2 \big/ r'^2 \right\}^{\frac{1}{2}} dr' + \frac{\pi}{4}, \quad (68.45)$$

r_l denoting the (largest) zero of the radicand. Replacement of $l(l + 1)$ by $(l + 1/2)^2$ has been justified by Langer (1937). That it is correct can be verified for the special case $v(r) \equiv 0$. We then find, with $L := l + 1/2$,

$$\frac{\pi}{4} + \int_{L/k}^{r} \sqrt{k^2 - L^2/r'^2} dr' = \sqrt{k^2 r^2 - L^2} - L \arccos\left(\frac{L}{kr}\right) + \frac{\pi}{4}$$

$$= kr - l\pi/2 + O(r^{-1}) \quad (68.46)$$

which is the correct asymptotic phase of the spherical Bessel functions.

The domain of validity of the WKB method is difficult to assess. Apart from the eikonal approximation, which requires a slowly varying potential $v(r)$, all the information about $v(r)$ for $r < r_l$ is ignored and should therefore be inessential. It is obvious that this cannot always be true! For sufficiently low energies, the WKB method must inevitably break down, since r_l then becomes very large. But for very large energies too, it cannot remain valid since r_l then becomes so small that the potential gradient becomes too large.

For large values of $L = l + 1/2$, the determination of the phase shift η_l from (68.45) and (68.46) becomes numerically unstable, being given by a small difference between large quantities. Since $v(r)$ decreases exponentially, it is then permissible to assume $|v(r)| \ll L^2/r^2$ and we can eliminate the difference of square root terms; the result is now

$$\eta_l = \int_{(l+1/2)/k}^{\infty} \frac{v(r)\, dr}{\sqrt{k^2 - (l + 1/2)^2/r^2}} \quad (68.47)$$

This formula is sometimes called Born's approximation for the scattering phase.

For still larger values of k , even the series expansion (68.43) becomes numerically unstable (although we might know the phases η_l exactly in some special cases). It can then be shown (see e.g. Wu and Ohmura, 1962, Section C) that this series expansion is equivalent to Born's approximation, or better to Molière's approximation, which should be used instead. The appropriate domains of application of the different methods are difficult to assess and require some experience, since they depend on the electron energy and the atomic number and—within the partial wave analysis itself—on the angular momentum l as well.

Remaining problems

In the theory of scattering, outlined above, it is tacitly assumed that the electrostatic target potential is unaffected by the incident electrons. This assumption is fairly well justified for fast electrons, and hence for the usual operating conditions of a CTEM or a STEM, but becomes questionable at low electron energies (below about 10 keV). We then have to take into account effects caused by *exchange* of an incident electron with one of the shell electrons, which are indistinguishable particles. Moreover, the atomic shell does not remain static but is *polarized* by the field of the incident electron. Other effects are *relativistic* in nature; they can be understood by solution of the more correct Dirac equation and are known collectively as *Mott scattering*. The reason why this is more noticeable at low energies is that the incident particle then has a much higher probability of coming very close to the nucleus, where the potential *gradient* becomes extremely large.

We cannot devote space to these problems and refer to the corresponding literature. Detailed calculations can be found in comprehensive textbooks on the quantum mechanics of scattering, for instance in Mott and Massey (1965) and Wu and Ohmura (1962).

68.6 Inelastic electron scattering

The process of elastic electron scattering is characterized by the fact that the quantum state of the target remains unaltered. This implies that the kinetic electron energy is conserved; the energy transferred to the target as a whole, associated with the vector q, can be completely ignored, owing to the large mass of the target atoms. For recent discussion, see Marks and Zhang (1992).

In contrast to this, inelastic electron scattering is characterized by transition to another quantum state in the target during the scattering process. For simplicity, we assume that the target—a single atom, a group

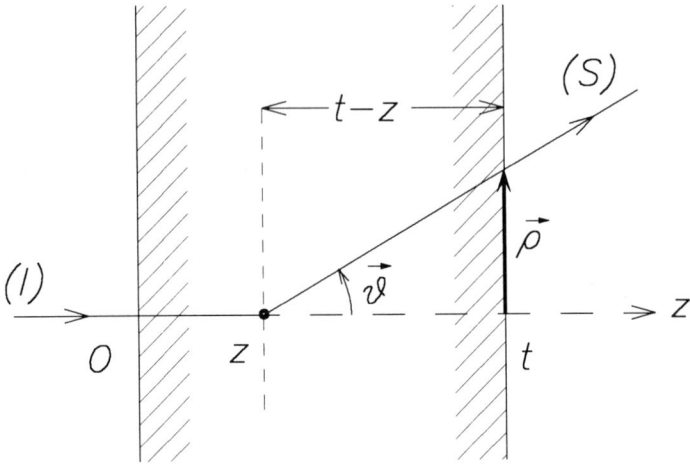

Fig. 68.4: Momentum transfer diagrams for inelastic scattering; (a) a general vector diagram for one particular excitation; (b) degenerate case corresponding to a very small scattering angle ϑ: the point O of diagram (a) has shifted to minus infinity; all transfer vectors $2\pi q_n$ effectively terminate on a straight line at a distance $k\vartheta$ from the z-axis.

of atoms or a layer of solid matter—is in its ground state (with label 0) prior to the scattering process. The incident electron then causes a *transition* into one of the discrete or continuous excited states (label n). Since the energy of the complete system is conserved, the scattered electron must lose an amount $E_n - E_0$ of energy, which corresponds to an alteration of the wavenumber from $k =: k_0$ to $k_n < k_0$.

These kinematic reactions are shown in Fig. 68.4. The laws of conservation of energy and momentum tell us that

$$\Delta k_n := k_0 - k_n = \overline{m}(E_n - E_0)/\hbar^2 \overline{k}_n \qquad (68.48)$$

with $\overline{k}_n = (k_n + k)/2$ and $\overline{m} = (m + m_0)/2 = m_0(1 + \gamma)/2$; this extremely small mass change is ignored. The momentum transfer is given by

$$2\pi q_n := |\boldsymbol{k}_n - \boldsymbol{k}| = \left\{ 4k k_n \sin^2 \theta/2 + (\Delta k_n)^2 \right\}^{\frac{1}{2}} \qquad (68.49)$$

Note that $2\pi q_n > \Delta k_n$ for $\theta \neq 0$.

The determination of cross-sections requires a rigorous quantum mechanical calculation, in which the wavefunction of the complete system is expanded in terms of the target eigenfunctions, the coefficients being asymptotically the scattering amplitudes f_n of the electron. We refer to the literature (e.g. Reimer, 1985, Chapter 3 or Mott and Massey, 1965,

Chapter XIII) for an account of this (see also Fanidis $et\ al.$, 1992, 1993). As a first result, (68.1) takes the more general form

$$\psi_n(\boldsymbol{r}) = \exp(ikz)\delta_{n0} + f_n(\theta, \varphi)\frac{e^{ik_n r}}{r} \tag{68.50a}$$

from which the differential cross-section σ_n is obtained by the slight modification

$$\sigma_n(\theta, \varphi) = |f_n|^2 k_n/k_0 \tag{68.50b}$$

The correct form of the optical theorem is now

$$\sum_{n=0}^{\infty} \int |f_n(\theta, \varphi)|^2 \, d\Omega = \frac{4\pi}{k_0}\Im f_0(0,0) \tag{68.50c}$$

In the high-energy approximation ($U \geq 30$ kV), we may set $k_n/k_0 \to 1$ and use the total cross-sections on the left-hand side.

So far, this theory is almost exactly valid and includes elastic scattering as the special case with label 0. The difficult task is now the calculation of the scattering amplitudes f_n. We shall outline this very briefly for $inner$-$shell\ excitation$ and $ionization$. At higher electron energies, Born's approximation is quite acceptable and gives exactly

$$f_n(\theta, \varphi) = \frac{\gamma\{Z\delta_{n0} - F_{n0}(\boldsymbol{q}_n)\}}{2\pi^2 a_0 q_n^2} \tag{68.51}$$

which is the generalization of (68.24). Here we have introduced $\boldsymbol{q}_n :=$ $(\boldsymbol{k}_n - \boldsymbol{k})/2\pi$ (see Fig. 68.4), while $q_n = |\boldsymbol{q}_n|$ is given by (68.49). The transition matrix elements in (68.51) are given by

$$F_{n0}(\boldsymbol{q}_n) = \left\langle n \left| \sum_{j=0}^{Z} \exp(-2\pi i \boldsymbol{q}_n \cdot \boldsymbol{r}_j) \right| 0 \right\rangle \tag{68.52}$$

$\boldsymbol{r}_j$ denoting the positions of the shell electrons. Since these matrix elements are often unknown in practice, some essential simplifications are now necessary.

(i) We consider only the sum of all the inelastic cross-sections, not individual members; only then can the sum rules of quantum mechanics be employed.

(ii) Inelastic scattering has appreciable intensity only for very small angles θ; for $\theta < 10^{-2}$ rad, we have $q_n a_0 \ll 1$. Equation (68.49) then simplifies to

$$4\pi^2 q_n^2 = k^2\theta^2 + \Delta k_n^2, \quad n = 0, 1, 2\ldots \tag{68.53a}$$

We then have the practically orthogonal contributions

$$2\pi q_n^{\perp} = 2\pi q_0 = k\theta, \quad 2\pi q_z = -\Delta k_n \tag{68.53b}$$

We see that, in practice, an energy transfer alters only the *longitudinal* component.

(iii) The summation over an infinite number of excited eigenstates is feasible only if the different energy transfers $E_n - E_0$, $n = 1, 2, 3 \ldots$ are replaced by one average value $\Delta \overline{E}$. Following a proposal of Koppe (1947), the half-ionization energy of the atom, $J/2$, is commonly chosen. In (68.48) it is permissible to write $k_n \approx \overline{k}_n \approx k$ and we then have

$$\kappa := \Delta \overline{k} = \overline{m} J / 2\hbar^2 k \tag{68.54}$$

(iv) The ground state of the atom is spherically symmetric. Making use of quantum sum rules, approximate formulae for the differential inelastic cross-section $\sigma_{in}(\theta)$ can be derived. The results depend on the assumptions made for the atomic eigenfunctions and on the way in which the approximation (68.54) is used during the calculation. Various proposals are discussed by Lenz (1954). Here we give only the simplest formula, which is quite often used:

$$\sigma_{in}(\theta) := \sum_{n=1}^{\infty} \sigma_n(\theta) = \frac{4\gamma^2 \{ Z - F^2(q)/Z \}}{a_0^2 (2\pi q)^4} \tag{68.55a}$$

with

$$2\pi q = (k^2 \theta^2 + \kappa^2)^{\frac{1}{2}} \tag{68.55b}$$

The form-factor $F(q) \equiv F_{00}(q)$ is defined by (68.22) but with q as given in (68.55b) instead of the simple value $k\theta/2\pi$.

Here we see the great advantage of employing sum rules: it is possible to derive a reasonable estimate for σ_{in} without knowing the excited states explicitly!

General properties of inelastic scattering

We consider here only the Wentzel model, the simplest possible expression, since even this gives qualitatively correct results. The appropriate atomic form-factor is

$$F(q) = Z/(1 + 4\pi^2 q^2 R^2)$$

R being the screening radius of (68.29). From (68.55a) we then obtain, after a short calculation:

$$\sigma_{in}(\theta) = \frac{\gamma^2 Z}{a_0^2} \frac{2 + 4\pi^2 q^2 R^2}{(1 + 4\pi^2 q^2 R^2)^2} \frac{R^2}{\pi^2 q^2} \tag{68.56}$$

with q given by (68.55b). The maximum value is reached for $\theta = 0$, $2\pi q = \kappa$:

$$\sigma_{in}^{max} = \frac{4\gamma^2 Z}{a_0^2} \frac{2 + \kappa^2 R^2}{(1 + \kappa^2 R^2)^2} \frac{R^2}{\kappa^2} \tag{68.57a}$$

It is interesting to compare this with the corresponding maximum of the elastic cross-section

$$\sigma_{el}^{max} = 4\gamma^2 Z^2 R^4 / a_0^2 \tag{68.57b}$$

If we introduce (68.54), $R = a_0 Z^{-1/3}$ and the Rydberg energy $E_0 = \hbar^2/2m_0 a_0^2$, we find

$$\frac{\sigma_{in}^{max}}{\sigma_{el}^{max}} \approx \frac{2}{Z\kappa^2 R^2} \approx \frac{32 E_0 eU}{J^2 Z^{\frac{1}{3}}} \tag{68.57c}$$

with $\kappa R \ll 1$. Since $J \approx E_0 \ll eU$, this clearly demonstrates that inelastic scattering is in practice always dominant in the forward direction.

We may likewise compare the total cross-sections. Integration over $\sigma_{in}(\theta)$ given by (68.56) with small-angle approximations leads to

$$\sigma_{in}^{tot} := \oint \sigma_{in} \, d\Omega \equiv \int_0^\pi 2\pi\sigma_{in}(\theta) \sin\theta \, d\theta$$

$$\approx \frac{8\pi\gamma^2 R^2 Z}{k^2 a_0^2} \int_{\kappa/2\pi}^\infty \frac{(2 + 4\pi^2 q^2 R^2)}{(1 + 4\pi^2 q^2 R^2)^2} \frac{dq}{q}$$

Evaluation of the integral results in

$$\sigma_{in}^{tot} = \frac{4\pi\gamma^2 R^2 Z}{k^2 a_0^2} \left\{ 2\ln\left(1 + \frac{1}{\kappa^2 R^2}\right) - \frac{1}{1 + \kappa^2 R^2} \right\}$$

With $\kappa R \ll 1$, we may use the second part of (63.57c) to obtain

$$\sigma_{in}^{tot} \approx 16\pi Z \left(\frac{\gamma R}{k a_0}\right) \ln\frac{1}{\kappa R}$$

$$= 16\pi Z^{\frac{1}{3}} \gamma^2 a_0^2 \frac{E_0}{e\hat{U}} \ln(4Z^{\frac{1}{3}} \sqrt{E_0 eU}/J) \tag{68.58}$$

With the aid of (68.31), we may form the ratio $\sigma_{in}^{tot}/\sigma_{el}^{tot}$:

$$\sigma_{in}^{tot}/\sigma_{el}^{tot} = \frac{4}{Z} \ln\left(\frac{1}{\kappa R}\right) \tag{68.59}$$

in which we have used $\kappa R = J/4Z^{\frac{1}{3}}(E_0 eU)^{\frac{1}{2}}$. From this result it is obvious that inelastic scattering is dominant for light elements ($Z \lesssim 6$), while elastic scattering prevails for heavy elements.

Another characteristic property of inelastic scattering is the behaviour of the mean scattering angle, $\bar{\theta}_{in}$, which is defined in the same way as $\bar{\theta}_{el}$ (68.32); the evaluation gives

$$\bar{\theta}_{in} = \left(\sigma_{in}^{tot}/\pi\sigma_{in}^{max}\right)^{\frac{1}{2}} = \frac{J}{4e\hat{U}}\left(\ln\frac{1}{\kappa^2 R^2}\right)^{\frac{1}{2}} \tag{68.60}$$

The accelerating voltage $\hat{U}$ appears here raised to a higher negative power that in the expression for $\bar{\theta}_{el}$ (68.33); $\bar{\theta}_{in}$ is hence very small, usually less than a milliradian. This confirms our earlier assumption (ii).

Another quite familiar observation has found a quantitative explanation: since inelastic scattering generally involves the *ionization* of atoms, and since this often destroys the local atomic configuration, (68.59) tells us that specimens with light elements are those most likely to suffer specimen destruction. This *radiation damage* makes electron microscopy in biology so difficult.

Further remarks

The theory presented here is highly simplified and only atomic ionization has been considered explicitly. In reality, of course, the target does not consist of individual free atoms but of atoms incorporated in a solid. In a full study, the many cooperative effects, notably the excitation of phonons, of spatial and surface plasmons and of intraband transitions and many others, have to be taken into account. This would oblige us to include a full account of solid state physics! We merely state, without derivation, that the (unshifted) frequency of spatial plasmons (plasma oscillations) is given by

$$\omega_p = \sqrt{\frac{N_v e^2}{m^*\varepsilon_0}}$$

N_v being the number of plasma electrons per unit volume and m^* the effective mass. The frequency of surface plasmons is then given by $\omega_s = \omega_p(1 + \varepsilon)^{-1/2}$, ε being the relative dielectric constant of the conductor. Plasmons can be excited with energies $E_n = n\hbar\omega_p$ or $n\hbar\omega_s$, where n is an integer. The corresponding energy losses in the scattered electron beam can be recognized as sharp equidistant peaks in the energy-loss spectrum. Reviews of the corresponding techniques and detailed presentations of the underlying theories can be found in more specialized textbooks of Reimer

(1985, 1993) and Egerton (1989) and in Disko *et al.* (1992). For information about the theory of electron interactions in solids, see the current series on Solid State Physics edited by H. Ehrenreich and D. Turnbull (Academic Press, San Diego; vol. 46 in 1992).

68.7 Plural and multiple electron scattering

Although the specimens studied in very high resolution electron microscopy are often so thin that none of the incident electrons is scattered more than once, this is by no means the most usual situation. In specimens that are not extremely thin, the electrons may well undergo several interactions or 'scattering events' before emerging from the far side. We speak of *plural* scattering if the number of interactions is less than about 10, otherwise of *multiple* scattering. This number is only a rough guide!

The statistics of plural and multiple scattering can be investigated in various ways. We confine this account to a very simple version. Thus we assume that the target is homogeneous and amorphous and consists of only one kind of atom, having locally spherical symmetry. Here we do not distinguish between elastic and inelastic processes, but only between the alternatives, collision or no collision. The corresponding cross-section is then $\sigma_T = \sigma_{el}^{tot} + \sigma_{in}^{tot}$ and the *mean free path* Λ with ν atoms per unit volume is given by $\Lambda = 1/\nu\sigma_T$. This quantity can be interpreted as the distance over which an electron can travel, on average, without a collision.

We assume that the target film has a constant thickness t and make the small-angle approximation for the scattering process, so that the total path-length of all electrons in the target is practically equal to t; scattering may thus occur in the interval $0 \leq z \leq t$.

Scattering can be considered as a rather *improbable* process. In an amorphous target successive scattering events can be assumed to be *independent* of each other, since there is no constructive interference which could couple them. In these conditions, we have to employ Poisson statistics. This tells us that the probability that an electron will reach the penetration depth z with just n collisions is given by

$$p_n(z) = \frac{1}{n!} \left(\frac{z}{\Lambda}\right) \exp(-z/\Lambda) \qquad (n = 0, 1, 2, 3, \ldots) \qquad (68.61)$$

We see that $\sum_0^\infty p_n(z) = 1$.

A second class of probabilities concerns the distribution of directions after a collision. In the small-angle approximation, introduced above, it is convenient to describe the direction of a path by a two-dimensional vector

$$\boldsymbol{\theta} = (\theta_x, \theta_y) = (\theta \cos \varphi, \theta \sin \varphi) \qquad (68.62)$$

θ and φ being the conventional angles relative to the z-axis and $|\theta| \ll 1$. The element of solid angle is then approximated by

$$d\Omega = \theta \, d\theta \, d\varphi = d\boldsymbol{\theta} = d\theta_x \, d\theta_y$$

The required probability functions are now *continuous*; the probability *densities* for n-fold scattering are in turn:

$$
\begin{aligned}
w^{(0)}(\boldsymbol{\theta}) &= \delta(\boldsymbol{\theta}) \quad \text{(primary beam)} \\
w^{(1)}(\boldsymbol{\theta}) &= \sigma(\theta)/\sigma_T \\
w^{(2)}(\boldsymbol{\theta}) &= (w^{(1)} * w^{(1)})(\boldsymbol{\theta}) \\
w^{(n)}(\boldsymbol{\theta}) &= (w^{(n-1)} * w^{(1)})(\boldsymbol{\theta}) \quad n \geq 1
\end{aligned}
\tag{68.63}
$$

Here $\sigma(\theta) = \sigma_{in}(\theta) + \sigma_{el}(\theta)$ denotes the differential cross-section and the asterisk a two-dimensional convolution. We verify that $\int W^{(n)}(\boldsymbol{\theta}) \, d\Omega = 1$, as it must be. The repeated convolutions are unfavourable for practical calculations, and we therefore introduce the *spectrum* $\phi(\boldsymbol{u})$ of $w^{(1)}(\boldsymbol{\theta})$:

$$\phi(\boldsymbol{u}) := \mathcal{F}^-(w^{(1)}) = \frac{1}{\sigma_T} \int \sigma(\theta) \exp\left(-2\pi i \boldsymbol{u} \cdot \boldsymbol{\theta}\right) d\Omega$$

or

$$\phi(u) = \frac{2\pi}{\sigma_T} \int_0^\infty \sigma(\theta) J_0(2\pi u \theta)\theta \, d\theta \qquad u = |\boldsymbol{u}| \tag{68.64}$$

The higher convolution powers are then given by transforming ordinary powers of ϕ:

$$w^{(n)}(\boldsymbol{\theta}) = \mathcal{F}(\phi^n(\boldsymbol{u})) \tag{68.65}$$

for $n = 0, 1, 2, 3, \ldots$ The combination of these formulae with the probabilites $p_n(z)$ of (68.61) is known as Bothe's theory (Bothe, 1921, 1933). The order of summation and Fourier transform can be reversed and we then obtain the following expressions for the angular distribution density $W_A(z, \theta)$ at the penetration depth z:

$$
\begin{aligned}
W_A(z, \theta) &= \sum_{n=0}^\infty p_n(z) w^{(n)}(\boldsymbol{\theta}) \\
&= \sum_{n=0}^\infty \frac{1}{n!} \left(\frac{z}{\Lambda}\right)^n \exp\left(-\frac{z}{\Lambda}\right) \mathcal{F}(\phi^n(\boldsymbol{u})) \\
&= \exp\left(-\frac{z}{\Lambda}\right) \mathcal{F}\left[\sum_{n=0}^\infty \left\{\frac{1}{n!}\left(\frac{z}{\Lambda}\phi(\boldsymbol{u})\right)^n\right\}\right] \\
&= \mathcal{F}\left[\exp\left\{\frac{z}{\Lambda}(\phi(\boldsymbol{u}) - 1)\right\}\right]
\end{aligned}
$$

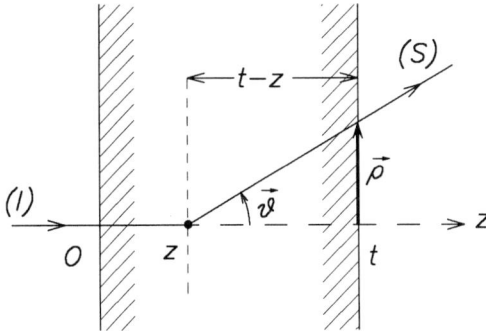

Fig. 68.5: Relation between transverse vectors in the case of single scattering; z: position of scattering event, $\vartheta = |\vec{\vartheta}|$: scattering angle, $\rho = |\vec{\rho}|$: off-axis distance in the exit plane $z = t$. The figure shows a part of the specimen with incident ray I and scattered ray S. For simplicity, only a planar construction is drawn.

Owing to the assumed rotational symmetry of the cross-section $\sigma(\theta)$, this simplifies to

$$W_A(z, \theta) = 2\pi \int_0^\infty \exp\left\{\frac{z}{\Lambda}(\phi(u) - 1)\right\} J_0(2\pi u \theta) u \, du \qquad (68.66)$$

The practical evaluation of this formula with the familiar models for the cross-section was carried out by Lenz (1954). For detailed results, we refer to his paper.

The restriction to very small angles θ is adequate for the needs of transmission electron microscopy but is not really necessary. In the more correct theory of Goudsmit and Saunderson (1940) the convolutions are carried out over the full unit sphere. The analogue of the two-dimensional Fourier transform is then series expansion with respect to Legendre polynomials. At higher electron energies, the scattering through very small angles is highly pronounced, which means that very many terms are necessary and the summation becomes tedious, slow and numerically unstable. This theory is hence advantageous only in the case of slow electrons. Important though this may be, it is beyond the scope of this book.

The foregoing method can be further extended to yield the combined angular and surface distribution of the scattered electrons at the exit side of the target. This calculation first appeared in the thesis of Schwertfeger (1974). Since this is in German and the contents were not published in one of the familiar journals, it is little known. We feel that this theory is useful and therefore present it briefly.

In addition to the two-dimensional vector $\boldsymbol{\theta}$, which characterizes direction, we now introduce a two-dimensional vector $\boldsymbol{\rho} = (x, y)$, which describes position in the exit plane $z = t$, as indicated in Fig. 68.5. As before, we consider an electron that enters the target along the z-axis from the left-hand side. After a single scattering event in which the electron is deflected through an angular vector $\boldsymbol{\theta}$ at the position z, it reaches the exit position $\boldsymbol{\rho} = (t - z)\boldsymbol{\theta}$. The *differential* probability that this process occurs in the depth increment dz, the element of solid angle $d\Omega = d\boldsymbol{\theta}$ and the element of surface area $d\boldsymbol{\rho}$, is given by

$$dp^{(1)} = f(\boldsymbol{\rho}, \boldsymbol{\theta}, z) \, dz \, d\boldsymbol{\rho} \, d\boldsymbol{\theta}$$

with

$$f(\boldsymbol{\rho}, \boldsymbol{\theta}, z) = w^{(1)}(\theta)\delta(\boldsymbol{\rho} - (t - z)\boldsymbol{\theta}) \qquad (68.67a)$$

This function is normalized to unity in the sense that

$$\iint f(\boldsymbol{\rho}, \boldsymbol{\theta}, z) \, d\boldsymbol{\rho} \, d\boldsymbol{\theta} = 1 \qquad (68.67b)$$

is true for all values of z, as can be verified easily. Formally, we obtain the probability density when the depth of scattering z can be anywhere between $z = 0$ and $z = t$ by *averaging* and hence integrating over z, according to

$$F^{(1)}(\boldsymbol{\rho}, \boldsymbol{\theta}, t) = \frac{1}{t} \int\limits_0^t f(\boldsymbol{\rho}, \boldsymbol{\theta}, z) \, dz \qquad (68.68a)$$

It is easy to verify that

$$\iint F^{(1)}(\boldsymbol{\rho}, \boldsymbol{\theta}, t) \, d\boldsymbol{\rho} \, d\boldsymbol{\theta} = 1 \qquad (68.68b)$$

is true, as it must be. It is not advantageous to attempt to eliminate the delta functions by explicit integration. Instead, we introduce the double Fourier transform

$$\begin{aligned}
\tilde{F}(\boldsymbol{q}, \boldsymbol{u}, t) :&= \mathcal{F}^-(F^{(1)}) \\
&= \iint F^{(1)}(\boldsymbol{\rho}, \boldsymbol{\theta}, t) \exp\{-2\pi\mathrm{i}(\boldsymbol{q} \cdot \boldsymbol{\rho} + \boldsymbol{u} \cdot \boldsymbol{\theta})\} \, d\boldsymbol{\rho} \, d\boldsymbol{\theta} \\
&= \frac{1}{t} \int\limits_0^t \int\limits_{\boldsymbol{\theta}} w^{(1)}(\theta) \exp\{-2\pi\mathrm{i}(\boldsymbol{u} + z\boldsymbol{q}) \cdot \boldsymbol{\theta}\} \, d\boldsymbol{\theta} \, dz
\end{aligned} \qquad (68.69)$$

This is a function from which all the delta functions have disappeared.

The distributions for plural scattering are again obtained by successive convolutions, but now over both vectors ρ and q. This implies that we have to raise the function $\tilde{F}$ of (68.69) to powers. By steps similar to those leading to (68.66), we now obtain

$$
F(\rho, \theta, t) = \mathcal{F}\left[\exp\left\{\frac{t}{\Lambda}\tilde{F}(q, u, t) - \frac{t}{\Lambda}\right\}\right]
$$

$$
= \iint \exp\left\{2\pi i(q \cdot \rho + u \cdot \theta) + \frac{t}{\Lambda}\tilde{F}(q, u, t) - \frac{t}{\Lambda}\right\} dq\, du
$$

$$
\tag{68.70}
$$

for the combined distribution $F(\rho, \theta, t)$ in the exit plane. For rotationally symmetric configurations, all azimuthal integrations can be carried out analytically, leading to expressions containing Bessel functions; these are not given here.

Some interesting special cases can be derived by integration:

(i) If we integrate over ρ before attempting to perform any other integrations, we obtain a factor $\delta(q)$, whereupon the integration over q becomes trivial, and we are led to evaluate the function $\tilde{F}(0, u, t)$. By comparison of (68.69) with (68.64), we see that this degenerate function is just $\phi(u)$ (the z-integration has here become trivial). Hence Bothe's theory is a special case of (68.70).

(ii) We may likewise seek the surface distribution of the scattered electrons at the exit plane. We then have to integrate first over all directions θ, obtaining a factor $\delta(u)$. Integration over u leads us to set $u = 0$ in $\tilde{F}$ and we obtain from (68.69) a new function

$$
\chi(q) := \tilde{F}(q, 0, t) = \frac{1}{t}\int\limits_0^t \int\limits_\theta w^{(1)}(\theta)\exp(2\pi i z q \cdot \theta)\, d\theta\, dz
$$

$$
= \frac{2\pi}{t}\int\limits_0^t \int\limits_0^\infty w^{(1)}(\theta)J_0(2\pi z q\theta)\,\theta\, d\theta\, dz
\tag{68.71}
$$

Here the integration over z is non-trivial. Evaluation of (68.70) now leads to the lateral or surface distribution formula of Jost and Kessler (1963):

$$
W_L(t, \rho) = 2\pi\int\limits_0^\infty \exp\left[\frac{t}{\Lambda}\{\chi(q) - 1\}\right] J_0(2\pi q\rho)q\, dq
\tag{68.72}
$$

which is quite analogous to (68.66).

So far, we have presented the theory in its most simplified form. We mention only one of the various extensions that have been studied. The conditions of incidence shown in Fig. 68.5 correspond to delta functions with respect to the starting position $\rho_0 = (x_0, y_0)$ in the plane $z = 0$ and to the angle θ_0 of incidence: $\rho_0 = 0$, $\theta_0 = 0$. If, however, the electron beam already has an initial distribution, described by a normalized function $f_0(\rho_0, \theta_0)$, then the exit distribution is obtained by convolution with F given by (68.70):

$$f(\rho, \theta, t) = (f_0 * F)(\rho, \theta) \qquad (68.73)$$

Since F itself has the form of a Fourier transform, it is most favourable to make use also of the Fourier transform $\tilde{f}_0 := \mathcal{F}^-(f_0)$; we then have the simple expression

$$f(\rho, \theta, t) = \mathcal{F}\left[\tilde{f}_0 \exp\left\{\frac{t}{\Lambda}(\tilde{F} - 1)\right\}\right] \qquad (68.74)$$

It is essentially this intensity distribution that gives us information about the specimen in an electron microscope, if the resolution is low enough for interference effects in the electron wave to be completely negligible. Here we have not distinguished between elastic and inelastic scattering, which is certainly unrealistic. The case in which *all* inelastically scattered electrons are filtered out is readily obtained by a minor modification: by replacing Λ by $\Lambda_{el} = 1/v\sigma_{el}^{tot}$ and $\sigma(\theta)$ in (68.63) by $\sigma_{el}(\theta)$, all further expressions will refer to elastic scattering only. The inelastic processes are now represented by an *attenuation factor* $\exp(-vt\sigma_{in}^{tot})$ at the end:

$$f(\rho, \theta, t) = \mathcal{F}\left[\tilde{f}_0 \exp\left\{\frac{t}{\Lambda_{el}}(\tilde{F}_{el} - 1)\right\}\right] \exp(-vt\sigma_{in}^{tot}) \qquad (68.75)$$

This distribution is no longer normalized to unity. The case of partial filtering, which would be more realistic, is not considered here.

The scattering contrast

We are now in a position to understand how image contrast is formed in the transmission electron microscope at lower resolution. This contrast can be understood directly in terms of scattering processes within the specimen; the interference and diffraction effects that are so very important in transfer theory, which enables us to understand contrast formation close to the limit of resolution of the microscope, can be ignored.

In the simplest approach, for a uniform specimen, we note that electrons scattered by atoms in the specimen will be intercepted by the objective aperture located in the back-focal plane of the objective lens if they are deflected through a large enough angle. If the scattering cross-sections corresponding to an objective aperture θ_0 are denoted by $\sigma_{el}(\theta_0)$ and $\sigma_{in}(\theta_0)$, then the decrease in the number of electrons beyond the aperture will be given by

$$\frac{dn}{n} = -N_0\{\sigma_{el}(\theta_0) + \sigma_{inel}(\theta_0)\}\nu\,dt =: -dt/t_0$$

or

$$\frac{n}{n_0} = \exp(-t/t_0)$$

where ν is the density of the specimen material and N_0 is the number of atoms per unit weight.

A fuller understanding of the scattering contrast can be obtained with the aid of the function F (68.70). We assume that all the specimen properties such as the cross-sections, the nature and density ν of the atoms and hence the mean free paths Λ_{el} and Λ_{in} and also the thickness t are *slowly varying* functions of position, $\rho_0 = (x_0, y_0)$. The vector ρ in all equations up to (68.72) should correctly be replaced by $\rho - \rho_0$, since only *relative* coordinates determine the local scattering process. The vector ρ_0 now appears as an additional material parameter. We thus obtain a modified distribution $F(\rho_0; \rho - \rho_0; \theta - \theta_0; t(\rho_0))$.

This is to be convolved with the illumination distribution $f_0(\rho_0, \theta_0)$, as stated in (68.73), but the Fourier method used in (68.74), can no longer be applied. If we integrate over values of θ corresponding to the solid angle of acceptance of the objective aperture, we shall obtain the intensity distribution $I(x, y)$ that is subsequently magnified. The image contrast $C(x, y)$ due to scattering is then simply obtained by normalization.

Although scattering contrast can be understood without recourse to the wave theory of image formation, we should expect to be able to explain it in terms of this theory as well. This can be shown to be possible, at least for a single scattering event; we refer to Niehrs (1969), Reimer (1969) and Reimer and Gilde (1973) for details, recapitulated in Section 6.3.3 of Reimer (1993).

69
Electron Interactions in Crystalline Specimens

69.1 Introduction

Crystalline specimens are characterized by the fact that the atoms are located on a lattice, which is periodic in all three directions. In real crystals, this symmetric ordering is inevitably finite in extent and is frequently destroyed, locally at least, by imperfections of various kinds. Nevertheless, there are usually sufficiently large domains in which the lattice is so nearly perfect that the potential distribution that acts on the electron waves propagating through it can be taken to be periodic. A consequence of this periodicity is that the phases of scattered electrons are not distributed freely, as they are in amorphous specimens, but may take preferred values; strong interference patterns can hence build up and it is no longer safe to assume that the unscattered beam will be substantially stronger than any of the individual scattered beams. This implies that the scattering theory of Chapter 68 is no longer directly useful and we must develop a new approach.

We first recapitulate briefly the fundamentals of crystallography. This is merely intended to provide the tools that will be needed here. Next, the properties of the periodic potential are investigated. This potential is built up by simply summing the contributions of free atoms, situated at their appropriate, crystallographically determined, positions. Incorrect though this procedure is, since it ignores all modifications due to chemical bonding, it is widely employed and gives fairly useful results.

The *kinematic theory*, which is the subject of the next section, is essentially an application of the Born approximation to scattering from a (large) array of atoms. This simple theory is already capable of explaining the most striking difference between scattering in crystals and in amorphous materials: the formation of sharp and regularly ordered diffraction spots. However, the kinematic theory is not always appropriate, since it requires that the primary (unscattered) wave be much stronger than the total of the scattered waves. This theory will therefore necessarily fail for thicker specimens, in which many and sometimes almost all of the electrons may be scattered somewhere for large penetration depths.

This weakness is removed in the *dynamical theory* by the introduction of reversible coupling of all elastically scattered waves. These waves can be described in two different but equivalent ways: as plane waves with slowly varying amplitudes or as Bloch waves. The general form of the dynamical theory is very complex, involving large systems of equations. In this chapter, we can deal only with the simplest non-trivial case, the *two-beam approximation* but this is sufficient to explain many phenomena that cannot be analysed using the kinematic theory, the anomalous transmission effect for example.

The literature on the dynamical theory is very extensive and the subject has many ramifications. Here, we present the general theory for *perfect* crystals in a fairly compact and self-contained form. The two-beam case is then examined more closely; the possible extensions, many of which are necessary in practice, are sketched only briefly. The numerous investigations on imperfect crystals are therefore excluded, important though they certainly are, for it is one of the great strengths of the electron microscope that one can study directly departures from crystallinity, dislocations and interfaces in particular. For these and many other practical situations, we must refer to the specialist literature (the classic Hirsch *et al.*, 1965, for example, and the more recent treatise of Spence, 1988; see too Amelinckx *et al.*, 1970; Howie, 1971, 1978, 1984; Thomas and Goringe, 1979; Forwood and Clarebrough, 1991; Amelinckx, 1992; Humphreys and Bithell, 1992; and Amelinckx and van Dyck, 1993). Another major omission is convergent-beam diffraction, a technique that is capable of furnishing crystallographic information directly. We refer to the surveys by Steeds (1984) and Steeds and Carlino (1992), the book by Spence and Zuo (1992) and the chapters by Eades, Spence and Cowley in Cowley (1992); there is a significant difference between the analysis when the area under investigation is illuminated essentially coherently (in practice, in STEM) and incoherently (TEM) and these situations are discussed separately by Cowley (1992) and Eades (1992) respectively.

The notation employed below is that in widespread use, wherever that is convenient and does not conflict with the remainder of the book, but in some cases we shall adopt a distinctly simpler notation. This concerns essentially the wave amplitudes and the labels of the Bloch waves.

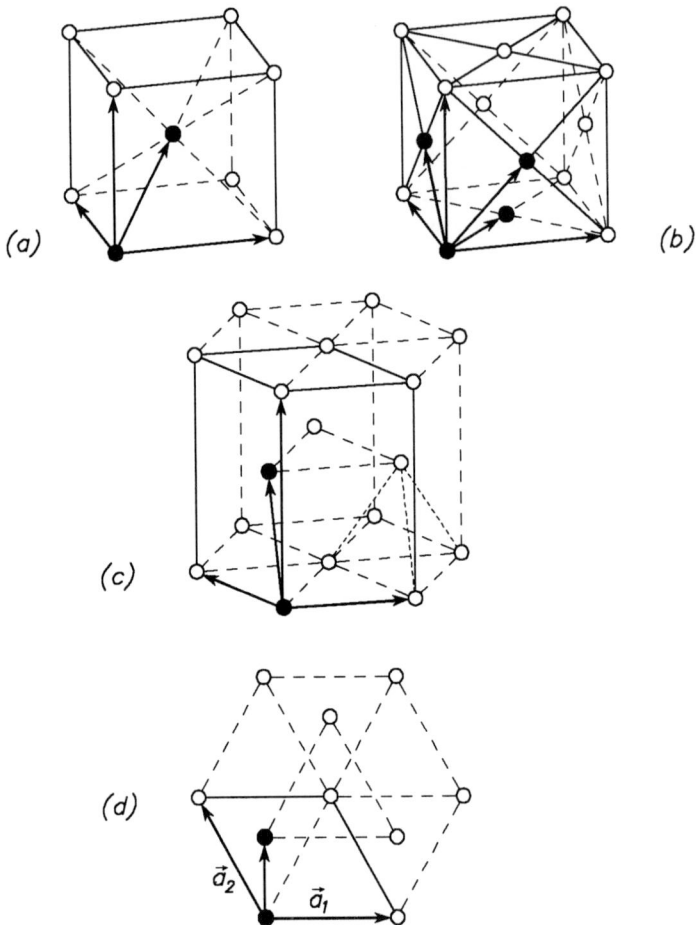

Fig. 69.1: Some common lattice types: (a) body-centred cubic lattice; (b) face-centred cubic lattice; (c), (d) hexagonal close-packed lattice, (c) perspective view, (d) top view.

69.2 Fundamentals of crystallography

An ideal crystal is a three-dimensional periodic array of atoms. Strictly speaking, this structure extends indefinitely, which is of course unrealistic, but if we assume that there are very many layers of atoms in each direction, 100 at least, then the perturbations caused by surface effects can be ignored in the interior. This simplification is adopted here.

The lattice of atoms can be built up by periodic repetition of an elementary cell in all three spatial directions; some common forms of cell are shown in Fig. 69.1. There are different ways of introducing such cells, each of which has specific advantages and disadvantages. We shall not deal with these different possibilities, which are not needed here, and we say nothing about Bravais and Wigner–Seitz cells. The *extended* elementary cells illustrated here are adapted to particular geometric symmetries, orthogonal or hexagonal for example, as shown in Fig. 69.1. Each cell will hence contain *more than one atom* if we count consistently and some positions must be specified by fractions of cell dimensions, often $1/2$ or $1/3$, $2/3$. Apart from the obvious convenience of choosing orthogonal basis vectors in orthogonal crystal systems, the extended elementary cells have the advantage that the classification of structures composed of different chemical elements is straightforward.

We specify the shape of a particular (generally parallelepipedal) cell by three linearly independent *basis vectors* (a_1, a_2, a_3). Periodic repetition of this cell generates the set of *lattice vectors*

$$a(n_1, n_2, n_3) = n_1 a_1 + n_2 a_2 + n_3 a_3 \qquad (69.1)$$

from the origin of the first cell to the origins of all the others. Here, the quantities n_i in the triplet n_1, n_2, n_3 must be *integers*. The positions of atoms, if needed explicitly, are then given by the vectors

$$r(m, n_1, n_2, n_3) = \sum_{j=1}^{3} (n_j + r_{mj}) a_j \equiv a + r_m \qquad (69.2)$$

Here, the integer $m \leq M$ serves to distinguish between the M different kinds of atoms in a cell and the components r_{mj} will generally be fractional, $0 \leq r_{mj} \leq 1$.

The basis (a_1, a_2, a_3) will not be normalized and is often not even orthogonal (in hexagonal structures, for example). The corresponding reciprocal basis $(a_1^\dagger, a_2^\dagger, a_3^\dagger)$ is therefore defined by

$$a_1^\dagger := a_2 \times a_3 / \mathcal{V}_E \quad , \quad a_2^\dagger := a_3 \times a_1 / \mathcal{V}_E \quad , \quad a_3^\dagger := a_1 \times a_2 / \mathcal{V}_E \quad (69.3)$$

with

$$\mathcal{V}_E := \det(a_1, a_2, a_3) = (a_1 \times a_2) \cdot a_3 \qquad (69.4)$$

$\mathcal{V}_E$ denotes the cell volume in real space. It is easy to verify the orthonormality relations

$$a_j \cdot a_k^\dagger = \delta_{jk} \quad (j, k = 1, 2, 3) \qquad (69.5)$$

The cell volume formed with the reciprocal basis vectors is then just $\mathcal{V}_E^{-1}$.

It is now appropriate to introduce the ideas of *reciprocal space* and the *reciprocal crystal lattice*. Just as in (69.1), the set of reciprocal lattice vectors is given by

$$\mathbf{g}(g_1, g_2, g_3) = g_1 \mathbf{a}_1^\dagger + g_2 \mathbf{a}_2^\dagger + g_3 \mathbf{a}_3^\dagger \tag{69.6}$$

where the g_i, $i = 1, 2, 3$ are integers, which may be negative. The Miller indices are a subset of these triplets, as they must be non-negative and have no common divisor. They serve to specify lattice planes in real space by means of the parametric equation

$$\mathbf{g}(h, k, l) \cdot \mathbf{r} = n = \text{const} \tag{69.7}$$

from which it is obvious that $\mathbf{g}$ is the normal to such a plane. The lattice points $\mathbf{a}(n_1, n_2, n_3)$ that lie on the plane satisfy $n = hn_1 + kn_2 + ln_3$, as can easily be verified. The distance of the plane from the origin is evidently given by

$$d = nd_{hkl} \tag{69.8}$$

where $d_{hkl} = 1/|\mathbf{g}|$ is the corresponding spacing between the planes. This relation is of importance in the description of Bragg reflections.

We conclude this Section with a list of the seven familiar crystal systems. These are classified in terms of the three lattice parameters $a = |\mathbf{a}_1|$, $b = |\mathbf{a}_2|$ and $c = |\mathbf{a}_3|$ and of the three angles $\alpha = (\mathbf{a}_2, \mathbf{a}_3) = \arccos(\mathbf{a}_2 \cdot \mathbf{a}_3/bc)$, $\beta = (\mathbf{a}_3, \mathbf{a}_1) = \arccos(\mathbf{a}_3 \cdot \mathbf{a}_1/ca)$ and $\gamma = (\mathbf{a}_1, \mathbf{a}_2) = \arccos(\mathbf{a}_1 \cdot \mathbf{a}_2/ab)$. For the first four classes, the vector basis is given in cartesian components. The quantity d_{hkl} is the distance to the lattice plane with Miller indices (hkl), see (69.8). This list is not complete, since composite lattice systems are not considered.

1. **Cubic lattices**: $a = b = c$, $\alpha = \beta = \gamma = 90°$,

$\mathbf{a}_1 = (a, 0, 0)$, $\mathbf{a}_2 = (0, a, 0)$, $\mathbf{a}_3 = (0, 0, a)$

$\mathbf{a}_1^\dagger = \left(\dfrac{1}{a}, 0, 0\right)$, $\mathbf{a}_2^\dagger = \left(0, \dfrac{1}{a}, 0\right)$, $\mathbf{a}_3^\dagger = \left(0, 0, \dfrac{1}{a}\right)$

$V_E = a^3$, $d_{hkl} = a/\sqrt{h^2 + k^2 + l^2}$

(a) *Body-centred lattice*: $M = 2$ (Fig. 69.1a)

$\mathbf{r}_1 = (0, 0, 0)$, $\mathbf{r}_2 = \left(\dfrac{a}{2}, \dfrac{a}{2}, \dfrac{a}{2}\right)$

(b) *Face-centred lattice*: $M = 4$ (Fig. 69.1b)

$(0, 0, 0)$, $\left(\dfrac{a}{2}, \dfrac{a}{2}, 0\right)$ $\left(\dfrac{a}{2}, 0, \dfrac{a}{2}\right)$ $\left(0, \dfrac{a}{2}, \dfrac{a}{2}\right)$

2. Hexagonal lattices $a = b \neq c$, $\alpha = \beta = 90°$, $\gamma = 120°$ (one third of the complete hexagonal cell),

$$\boldsymbol{a}_1 = (a,0,0),\ \boldsymbol{a}_2 = \left(-\frac{a}{2}, \frac{a}{2}\sqrt{3}, 0\right),\ \boldsymbol{a}_3 = (0,0,c)$$

$$\boldsymbol{a}_1^\dagger = \left(\frac{1}{a}, \frac{1}{a\sqrt{3}}, 0\right),\ \boldsymbol{a}_2^\dagger = \left(0, \frac{2}{a\sqrt{3}}, 0\right),\ \boldsymbol{a}_3^\dagger = \left(0,0,\frac{1}{c}\right)$$

$$V_E = \frac{1}{2}\sqrt{3}a^2 c,\ d_{hkl} = a\left\{\frac{4}{3}(h^2 + hk + k^2) + \frac{a^2}{c^2}l^2\right\}^{-\frac{1}{2}}$$

Close-packed lattice: $M = 2$, $c = a\sqrt{8/3}$ (Fig. 69.1c,d)

$$\boldsymbol{r}_1 = (0,0,0),\ \boldsymbol{r}_2 = \frac{1}{3}\boldsymbol{a}_1 + \frac{2}{3}\boldsymbol{a}_2 + \frac{1}{2}\boldsymbol{a}_3 = \left(0, \frac{a}{\sqrt{3}}, \frac{c}{2}\right)$$

3. Tetragonal lattices $a = b \neq c$, $\alpha = \beta = \gamma = 90°$,

$$\boldsymbol{a}_1 = (a,0,0),\ \boldsymbol{a}_2 = (0,a,0),\ \boldsymbol{a}_3 = (0,0,c)$$

$$\boldsymbol{a}_1^\dagger = \left(\frac{1}{a}, 0, 0\right),\ \boldsymbol{a}_2^\dagger = \left(0, \frac{1}{a}, 0\right),\ \boldsymbol{a}_3^\dagger = \left(0,0,\frac{1}{c}\right)$$

$$V_E = a^2 c,\ d_{hkl} = a/\sqrt{h^2 + k^2 + a^2 l^2/c^2}$$

4. Orthorhombic lattices $a \neq b \neq c$, $\alpha = \beta = \gamma = 90°$,

$$\boldsymbol{a}_1 = (a,0,0),\ \boldsymbol{a}_2 = (0,b,0),\ \boldsymbol{a}_3 = (0,0,c)$$

$$\boldsymbol{a}_1^\dagger = \left(\frac{1}{a}, 0, 0\right),\ \boldsymbol{a}_2^\dagger = \left(0, \frac{1}{b}, 0\right),\ \boldsymbol{a}_3^\dagger = \left(0,0,\frac{1}{c}\right)$$

$$V_E = abc,\ d_{hkl} = \left\{\left(\frac{h}{a}\right)^2 + \left(\frac{k}{b}\right)^2 + \left(\frac{l}{c}\right)^2\right\}^{-\frac{1}{2}}$$

5. Trigonal (rhombohedral) lattices $a \neq b \neq c$, $\alpha = \beta = \gamma \neq 120°$,

$$V_E = a^3\sqrt{1 - 3\cos^3\alpha + 2\cos^3\alpha}$$

$$d_{hkl} = \frac{a\sqrt{1 + \cos\alpha - 2\cos^2\alpha}}{\sqrt{(1 + \cos\alpha)(h^2 + k^2 + l^2) - 2\cos\alpha(hk + kl + lk)}}$$

6. Monoclinic lattices $a \neq b \neq c$, $\alpha = \gamma = 90°$, $\gamma \neq 90°$,

$$V_E = abc\sin\beta,\ d_{hkl} = \sin\beta\left\{\frac{h^2}{a^2} + \frac{l^2}{c^2} + \frac{k^2}{b^2}\sin^2\beta - \frac{2hl}{ac}\cos\beta\right\}^{-\frac{1}{2}}$$

7. Triclinic lattices $a \neq b \neq c$, $\alpha \neq \beta \neq \gamma \neq 90°$

General parallelepipedal cells, evaluation of vector formulae more favourable.

69.3 The periodic potential

Electrons that traverse a neutral crystalline layer are influenced by an effective electrostatic potential $V(\mathbf{r})$. This is built up by superposition of contributions from all the nuclei and shell electrons and must therefore exhibit the same periodicity as the crystal. This implies that the potential must satisfy a shift relation

$$V(\mathbf{r}) = V(\mathbf{r} + \mathbf{a}) \tag{69.9}$$

for *all* lattice vectors $\mathbf{a}$. The same must be true of all functions that can in principle be observed, the charge density for example, so that $\rho(\mathbf{r}) = \rho(\mathbf{r} + \mathbf{a})$.

A consequence of (69.9) is that the potential can be expanded as a Fourier series, most concisely expressed in terms of *reciprocal* lattice vectors. This expansion is given explicitly by

$$V(\mathbf{r}) = \sum_{g_1,g_2,g_3=-\infty}^{\infty} V_{g_1 g_2 g_3} \exp\{2\pi i(g_1 \mathbf{a}_1^\dagger + g_2 \mathbf{a}_2^\dagger + g_3 \mathbf{a}_3^\dagger) \cdot \mathbf{r}\}$$

or more concisely

$$V(\mathbf{r}) = \sum_g V_g \exp(2\pi i \mathbf{g} \cdot \mathbf{r}) \tag{69.10}$$

The Fourier coefficients V_g are known as *structure potentials*. The fact that $V(\mathbf{r})$ is real, $V(\mathbf{r}) = V^*(\mathbf{r})$, implies that

$$V_{-g} = V_g^* \tag{69.11}$$

for all triplets $\mathbf{g}$, where $-\mathbf{g} := (-g_1, -g_2, -g_3)$. If the crystal structure is also *centrosymmetric*, which means that $V(\mathbf{r}) = V(-\mathbf{r})$ for a suitable choice of origin, we can conclude that $V_{-g} = V_g$. Together with (69.11), this implies that all the coefficients are real, so that the series expansion (69.10) consists only of cosine terms for this situation. The same must hold for the expansion of the charge density $\rho(\mathbf{r})$.

The structure potentials are obtained formally by integration of $V(\mathbf{r})$ over one elementary cell (E):

$$V_g = V_E^{-1} \int_E V(\mathbf{r}) \exp(-2\pi i \mathbf{g} \cdot \mathbf{r}) \, d\mathbf{r} \tag{69.12}$$

In particular, the coefficient V_0 of lowest order is simply the spatial average of $V(\mathbf{r})$ and is therefore called the *mean inner potential*. It is this quantity

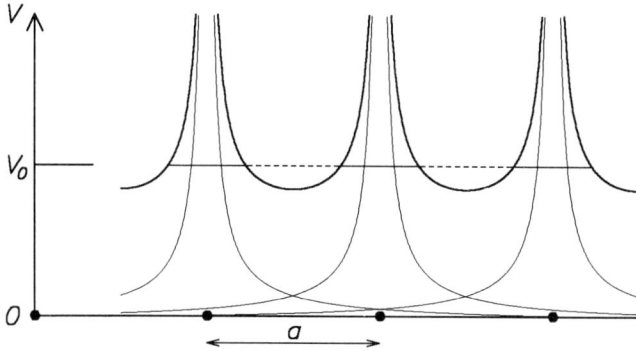

Fig. 69.2: One-dimensional superposition of single-atom potentials to create a periodic potential; the positions of the atoms are indicated by dots. V_0 denotes the mean inner potential. Here, all the Fourier coefficients are made positive by locating one atom at the origin.

that was determined in the early interference experiments, see Section 62.2.

The full quantum mechanical calculation of the structure potentials is a highly complicated task, which goes far beyond the theme of this book. A first—and very rough—approximation for the crystal potential is obtained by superposing the spherical potentials U_m of free atoms located at the correct, experimentally observable, lattice sites (Fig. 69.2):

$$V(\boldsymbol{r}) = \sum_{\boldsymbol{a}} \sum_{m=1}^{M} U_m(|\boldsymbol{r} - \boldsymbol{a} - \boldsymbol{r}_m|) \qquad (69.13)$$

Here, $\boldsymbol{r}$ is given by (69.2) and $\boldsymbol{a} = (n_1, n_2, n_3)$. On introducing this into (69.12) and writing $\boldsymbol{r} =: \boldsymbol{r}' + \boldsymbol{r}_m$, we obtain

$$V_g = V_E^{-1} \sum_m \exp(-2\pi i \boldsymbol{g} \cdot \boldsymbol{r}_m) \sum_{\boldsymbol{a}} \int_E U_m(|\boldsymbol{r}' - \boldsymbol{a}|) \exp(-2\pi i \boldsymbol{g} \cdot \boldsymbol{r}') \, d\boldsymbol{r}'.$$

The integration over one elementary cell together with summation over all shifted positions is equivalent to integration over the whole crystal and, owing to the exponential decrease of the atomic potential at large distances, this integration may even be extended to infinity without significant error. Equations (69.1), (69.6) and (69.5) allow us to conclude that $\exp(-2\pi i \boldsymbol{g} \cdot \boldsymbol{a}) = 1$, whereupon we obtain the spherically symmetric *spec-*

tral function of a single atom:

$$\tilde{U}_m = \mathcal{F}^{-1}(U_m) = \sum_a \int_E U_m(|\boldsymbol{r}' - \boldsymbol{a}|) \exp\{-2\pi i \boldsymbol{g} \cdot (\boldsymbol{r}' - \boldsymbol{a})\} \, d^3 \boldsymbol{r}'$$

giving

$$\tilde{U}_m(|\boldsymbol{g}|) = \int_0^\infty U_m(r) \frac{\sin(2\pi|\boldsymbol{g}|r)}{2\pi|\boldsymbol{g}|r} 4\pi r^2 \, dr \qquad (69.14)$$

The structure potential V_g can now be written as

$$V_g = V_E^{-1} \sum_{m=1}^M \tilde{U}_m(|\boldsymbol{g}|) \exp(-2\pi i \boldsymbol{g} \cdot \boldsymbol{r}_m) \qquad (69.15)$$

These coefficients are, however, hardly ever used in practice since the Schrödinger equation to be solved takes the form of (68.7) with (68.6). It is then more convenient to employ the structure potentials v_g of $v(\boldsymbol{r})$. For a single atom located at the origin, it can be shown that the required spectrum is given essentially by (68.24); we have simply to replace q by $|\boldsymbol{g}|$ and include a factor 4π, as is obvious from (68.13). If we make the corresponding substitution into (69.15) and include all the appropriate factors, we finally obtain

$$v_g = \frac{4\pi}{V_E} \sum_{m=1}^M f_m(|\boldsymbol{g}|) \exp(-2\pi i \boldsymbol{g} \cdot \boldsymbol{r}_m) \qquad (69.16)$$

This expression is now evaluated for two important crystal classes.

(i) In *body-centred cubic* lattices (caesium chloride, for example), one kind of atom is situated at all the corners of the cubes, the other at all their centres. We choose the basis $\boldsymbol{a}_j = a\boldsymbol{e}_j$ ($j = 1, 2, 3$), $\boldsymbol{e}_j$ denoting the familiar unit vectors and then find that $\boldsymbol{a}_j^\dagger = \boldsymbol{e}_j/a$. With no loss of generality, the reference cell can be chosen so that $\boldsymbol{r}_1 = 0$ for atoms of the first kind. The second relative position vector is then

$$\boldsymbol{r}_2 = a(\boldsymbol{e}_1 + \boldsymbol{e}_2 + \boldsymbol{e}_3)/2$$

(see Section 69.2 and Fig. 69.1a). If we substitute this in (69.16), we obtain

$$v_g = \frac{4\pi}{a^3} \{f_1(|\boldsymbol{g}|) + f_2(|\boldsymbol{g}|) \cos(g_1 + g_2 + g_3)\pi\} \qquad (69.17)$$

For even $(+)$ and odd $(-)$ values of $g_1 + g_2 + g_3$, this yields

$$v_g = 4\pi(f_1 \pm f_2)/a^3$$

For some metals, notably for Cr, Fe, Mo and W, $f_1 = f_2$.

(ii) In *face-centred cubic* lattices (with the chemical composition AB_3, for example), atoms of the first kind (A) again occupy all the corners of the cubes. We select one of these and again choose $r_1 = 0$. Atoms of the second kind (B) occupy the midpoints of the faces of the cubes. We select three of these, $r_2 = a(e_1 + e_2)/2$, $r_3 = a(e_1 + e_3)/2$ and $r_4 = a(e_2 + e_3)/2$ (Section 69.2 and Fig. 69.1b). With $M = 4$ in (69.16) and two different values of the amplitudes f_i, $f_1 =: f_A$, $f_2 = f_3 = f_4 =: f_B$, we now obtain

$$v_g = \frac{4\pi}{a^3}\left[f_A + f_B\{\cos(g_1 + g_2)\pi + \cos(g_1 + g_3)\pi + \cos(g_2 + g_3)\pi\}\right]$$
(69.18)

In the crystals of some pure metals (Al, Ni, Cu, Ag, Au and Pt), we have $A = B$ and $f_A = f_B$. The approximation (69.13), on which all these formulae are based, is fairly accurate in the vicinity of the atomic nuclei, which correspond to large values of $|g| \gg a^{-1}$; inevitably, it becomes poor for small values of $|g|$, which correspond to the interstices between the atoms, where the chemical bonds are ignored.

69.4 Kinematic theory of electron scattering

If the crystal layer is so thin that the total intensity of the scattered electrons is always much smaller than the primary intensity, we may apply the Born approximation. When scattering in crystals is in question, the result of using this approximation is traditionally known as the *kinematic* theory (e.g. Gevers, 1970). We now have to solve the Schrödinger equation with the potential $v(r)$ examined in the preceding section.

The solution could be cast into the form of (68.13-15), where the spherical symmetry has not yet been exploited, but this is unfavourable since the polar angles θ and ϕ are not adapted to the crystal symmetry. Conversely, the representation involving the scattering factor $\hat{f}(q)$ might well be advantageous if we could find a better representation of the momentum transfer, q.

It is now necessary to assume that the crystal has a finite volume. For simplicity, we consider a parallelepipedal shape adapted to the crystal symmetry, so that all six surfaces are lattice planes. The greatest side-length must be much smaller than the distance to the detector, so that the general geometrical considerations for scattering experiments are satisfied.

At the same time, the length of the shortest side must be substantially greater than the lattice spacing, so that the crystal contains many atomic layers and it is reasonable to assume that the structure is periodic. These two requirements can almost always be reconciled.

It is easier to perform the necessary integrations if we express the position vector r in terms of the basis vectors a_1, a_2 and a_3:

$$r = \sum_{i=1}^{3} u_i a_i \quad , \quad -N_i/2 \leq u_i \leq N_i/2 \gg 1 \qquad (69.19)$$

The coordinates u_1, u_2 and u_3 are dimensionless real variables and the (even) integers N_1, N_2 and N_3 are the numbers of atomic layers in the corresponding directions. The momentum transfer q is best represented in the reciprocal basis

$$q = q_1 a_1^\dagger + q_2 a_2^\dagger + q_3 a_3^\dagger \qquad (69.20)$$

in which the components q_1, q_2 and q_3 are again dimensionless; they take integral values at the reciprocal lattice points. From (69.10), we obtain

$$v(r) = \sum_g v_g \exp(2\pi i g \cdot r)$$

$$= \sum_g v_g \prod_{k=1}^{3} \exp(2\pi i g_k u_k) \qquad (69.21)$$

The evaluation of $\hat{f}(q)$ is now straightforward; we find

$$\hat{f}(q) = \frac{\mathcal{V}_c}{4\pi} \sum_g v_g \prod_{k=1}^{3} \frac{\sin\{\pi N_k(q_k - g_k)\}}{\pi N_k(q_k - g_k)} \qquad (69.22)$$

where $\mathcal{V}_c$ denotes the volume of the crystal.

69.4.1 Geometrical rules and scattering

We now establish simple rules for the positions of the intensity maxima; these are readily obtained from (69.22). Since the N_k are always much greater than unity, large values of $|\hat{f}|^2$ can be expected only if $|q_k - g_k| \approx 1/2N_k$ is true for $k = 1, 2$ and 3 simultaneously. When this is the case, the intensity exhibits a sharp peak. The highest peak occurs when $q = g$. If we combine this with the relation $q = (k' - k)/2\pi$, we obtain a form of the Laue conditions. The definition of the wavevector that we have been using, for which $|k| = 2\pi/\lambda$, is favourable in many branches of optics and quantum mechanics but not here. In many presentations of the subject,

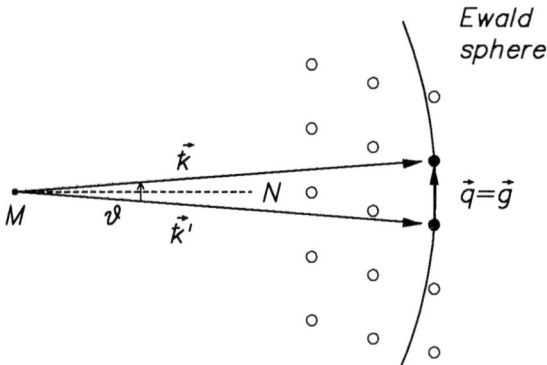

Fig. 69.3: Momentum transfer diagram for (69.24) in the reciprocal lattice: two-dimensional construction in which M is the centre of Ewald sphere and the reflecting planes lie along M–N.

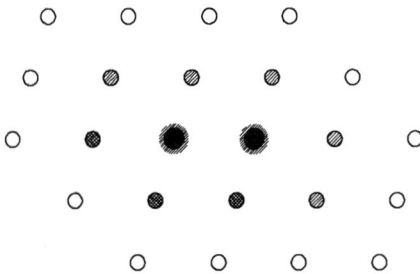

Fig. 69.4: A diffraction pattern with two strongly excited reflexes.

notably the authoritative books of Hirsch *et al.* (1965) and Spence (1988), the alternative normalization $|k| = 1/\lambda$ is adopted. To prevent confusion with earlier chapters, we introduce the symbol

$$\boldsymbol{\mathchar'26\mkern-9mu k} := k/2\pi \qquad (69.23a)$$

so that

$$|\boldsymbol{\mathchar'26\mkern-9mu k}| = 1/\lambda \qquad (69.23b)$$

This quantity is proportional to the momentum and will be loosely referred to by this name, though of course it has the dimensions of reciprocal length, like the spatial frequency. Laue's rule now takes the compact form

$$q = \boldsymbol{\mathchar'26\mkern-9mu k}' - \boldsymbol{\mathchar'26\mkern-9mu k} = g \qquad (69.24)$$

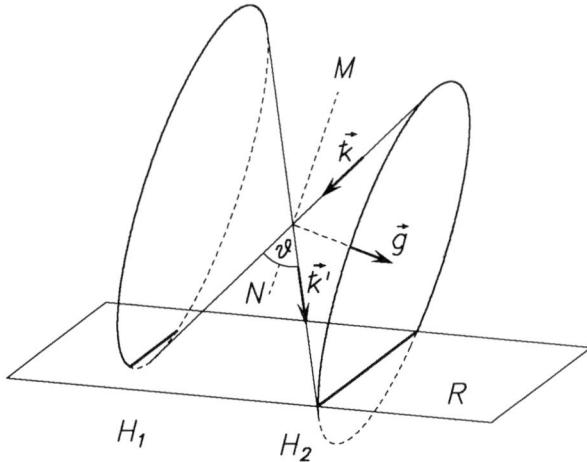

Fig. 69.5: The two Kossel cones and the two branches H_1 and H_2 of the hyperbola in which the cones intersect the recording plane R.

The momentum transfer must thus be a reciprocal lattice vector. If the starting points of the vectors $\boldsymbol{k}$ and $\boldsymbol{k}'$ are kept fixed, then the endpoints must be located on a spherical surface known as the *Ewald sphere*, in parametric form $|\boldsymbol{k}'| = 1/\lambda = \text{const}$. Equation (69.24) then states that reflections are excited only for points of the reciprocal lattice that lie on this Ewald sphere, as shown in Fig. 69.3. A corresponding two-dimensional intensity pattern is shown in Fig. 69.4.

Suppose now that we consider a particular lattice vector $\boldsymbol{g}$ and allow the direction of the incident electrons, and hence the momentum vector $\boldsymbol{k}$, to vary—this would be the case for convergent illumination, for instance. All pairs of vectors $\boldsymbol{k}'$ and $\boldsymbol{k}$ for which (69.24) remains satisfied are located on a double cone, known as the *Kossel cones* (Fig. 69.5). These intersect a plane parallel to the image plane along the two branches of a hyperbola but since the curvature of the latter is very small, the intensity pattern is virtually indistinguishable from a pair of straight lines. This too is shown in Fig. 69.5.

A third form of the geometrical condition can be obtained from (69.24) by considering the absolute lengths on both sides. Using (69.8), we find

$$|\boldsymbol{q}| = \frac{2}{\lambda}\sin(\theta/2) = |\boldsymbol{g}| = n/d$$

or after rearrangement

$$2d\sin(\theta/2) = n\lambda \tag{69.25}$$

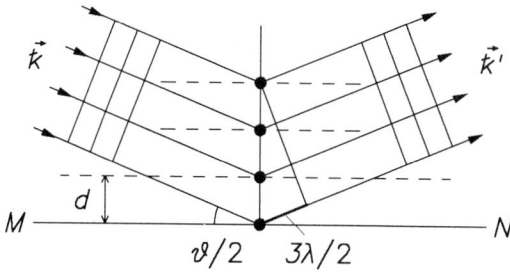

Fig. 69.6: Graphical expression of Bragg's condition in the real lattice. This shows the incident wave and a wave scattered at real atoms. M–N and the broken parallel lines denote real lattice planes.

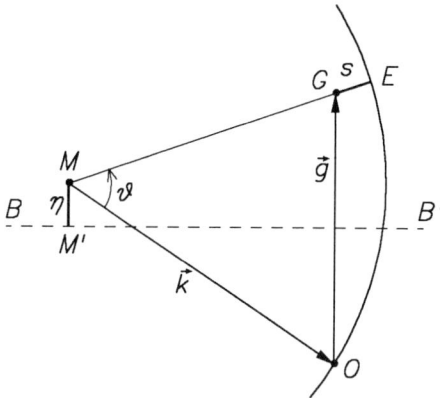

Fig. 69.7: Definition of the excitation error $s = \overline{GE}$ and the resonance error $\eta = \overline{M'M}$ for two points O and G in the reciprocal lattice. The broken line $\overline{BB'}$ indicates the position of the reflecting plane. (Note that $\vec{k}' = \mathbf{ME} \neq \vec{k} + g$ here!)

This is the *Bragg condition*. Its form is interesting for it shows that the diffraction pattern can be interpreted as constructive interference of reflections at lattice planes, which are perpendicular to the vector $\boldsymbol{g}$. This interpretation is illustrated in Fig. 69.6.

69.4.2 Intensity formulae for diffraction peaks

The intensity peaks are of course never infinitely sharp and Laue's condition refers only to maxima in ideal conditions of excitation. We now enquire what happens when (69.24) is nearly, but not exactly, satisfied. This situation is shown in Fig. 69.7, which represents the two most common kinds of deviation from the ideal condition, characterized by s and η.

(i) The *excitation error* s is the algebraic distance of the Ewald sphere from the Laue position:

$$s \equiv s_g := |\mathbf{k}'| - |\mathbf{k} + \mathbf{g}|$$
$$\approx (\mathbf{k} \cdot \mathbf{g} + g^2/2)\lambda \tag{69.26a}$$

The last expression is valid only for very small values of s, which will be assumed henceforth to be the case.

(ii) The *resonance error* η is the departure of the midpoint M from the ideal symmetrical Bragg position. Again for small values, this may be written

$$\eta = -(\mathbf{k} \cdot \mathbf{g} + g^2/2)/|\mathbf{g}|$$
$$= s/|\lambda \mathbf{g}| \tag{69.26b}$$

Since η is proportional to s, it does not contain any new information; we shall therefore use only the measure s. In the conditions shown in Fig. 69.7, s is positive.

We show how the intensity is calculated for a simple example only, a crystalline sheet of cubic structure, fairly thin in the z-direction ($t = N_3 a$) but of much larger extent in the x- and y-directions. The incident electron beam is tilted at a very small angle to the z-axis (the optic axis) and the scattering angles are assumed to remain small, as is typical for electron optical imaging.

In such conditions, only Bragg reflections with vanishing longitudinal component can be excited, $g_3 = q_z = 0$. The individual reflections are so sharp that they do not overlap and hence we do not need to consider interference terms when calculating $|\hat{f}|^2$ from (69.22). Thus for the reflection with reciprocal lattice vector $\mathbf{g}$, we find

$$|\hat{f}_g|^2 = \left|\frac{V_c}{4\pi} v_g\right|^2 \prod_{j=1}^{2} \frac{\sin^2\{\pi N_j(q_j - g_j)\}}{\pi^2 N_j^2 (q_j - g_j)^2} \cdot \frac{\sin^2(\pi N_3 q_3)}{\pi^2 N_3^2 q_3^2}$$

The total intensity scattered into this spot is given by integrating over all the corresponding scattering angles. This is easily done by making the substitution

$$d\Omega = \lambda^2 \, dk'_x \, dk'_y = (\lambda/a)^2 \, dq_1 \, dq_2$$

in which we have used (69.20). Recalling that

$$\int_{-\infty}^{\infty} \frac{\sin^2 \alpha}{\alpha^2} \, d\alpha = \pi$$

we find

$$J_g := \int |\hat{f}_g|^2 \, d\Omega$$

$$= \left| \frac{\mathcal{V}_c}{4\pi} \lambda v_g \right|^2 \frac{1}{N_1 N_2 a^2} \frac{\sin^2(\pi N_3 q_3)}{(\pi N_3 q_3)^2}$$

This expression can be cast into a more convenient form. First, we note that $\mathcal{A} := N_1 N_2 a^2$ is the irradiated area in the $x-y$ plane and consequently $\mathcal{V}_c = \mathcal{A}t$. Furthermore, we see that $s = q_3/a$ for beams inclined at very small angles to the z-axis. This relation is easily obtained from (69.26) together with $\mathbf{k'} = \mathbf{k} + \mathbf{g} + s\mathbf{e}_z = \mathbf{k} + \mathbf{q}$, $g_z = 0$ and $q_z = q_3/a$ (69.20).

Altogether, we may set

$$\pi N_3 q_3 = \pi s t$$

and hence obtain

$$J_g = \mathcal{A} \left| \frac{\lambda}{4\pi} v_g \right|^2 \frac{\sin^2(\pi s t)}{\pi^2 s^2}$$

The intensity J_g is thus proportional to the area irradiated, as we should expect, but varies with the thickness t; the variation is described by the oscillatory factor $\sin^2(\pi s t)$. It is usual to collect up to the factors not containing s and hence to introduce the *extinction distance*, ξ_g. We write

$$\xi_g := \frac{4\pi^2}{\lambda |v_g|} \equiv \frac{\pi \hbar u_E}{e |V_g|} \tag{69.27}$$

where

$$u_E = \gamma^{-1} (2e\hat{U}/m_0)^{\frac{1}{2}} \tag{69.28}$$

here denotes the velocity of the electron (2.19, with $u_E = g/m_0\gamma$ as in 2.22). The intensity per unit area irradiated can then be written

$$j_g = J_g/\mathcal{A} = \frac{\sin^2(\pi s t)}{s^2 \xi_g^2} \tag{69.29}$$

Since the primary intensity j_0 is equal to unity, we see that a sufficient condition for the kinematic approximation to be valid is $|s| \gg \xi_g$, which means far from the corresponding Laue positions. This is, however, a rather uninteresting situation. If s becomes very small, the sine factor can be replaced by its argument, whereupon the small quantity s^2 cancels out. A sufficient criterion for $j_g \ll 1$ is then that $\pi t \ll \xi_g$. A reasonable value is $t \lesssim \zeta_g/20$, which tells us that the crystalline layer must be sufficiently thin;

otherwise, we have to employ the dynamical theory, which is presented in Section 69.5.

69.4.3 Energy spread in the illumination

The oscillations of the intensity with st are hardly ever observed in practice, as it is difficult to ensure that the incident beam is sufficiently monochromatic. Let us now suppose that the energy spectrum of the primary beam is quite narrow but still wide enough to cover the relevant s interval. Since the latter is itself small, we can extend the limit of integration to infinity without introducing any appreciable error. The relation between E and s is found from $E = eU$, $k = 1/\lambda$ and $|ds| = |dk|$; an elementary calculation then gives

$$\left| \frac{dE}{ds} \right| = 2\pi\hbar u_E \equiv hu_E \tag{69.30}$$

where u_E is given by (69.28).

We describe the energy spread in terms of a normalized spectrum with density $S(E) \geq 0$ and note that

$$\int_{-\infty}^{\infty} S(E)\, dE = 1$$

The intensity of the primary beam is thus again equal to unity. For high velocities u_E and fairly thin layers, the relevant interval of integration,

$$\Delta E = \left| \frac{dE}{ds} \right| \Delta s = (2\pi\hbar u_E)(2/t)$$

becomes so small that the factors $|dE/ds|$ and $S(E)$ are practically constant and may be replaced by their values at E_0, corresponding to $s = 0$. From (69.29), we then obtain

$$I_g := \int_{-\infty}^{\infty} S(E)J_g(s(E))\, dE$$

$$= S(E_0) \left| \frac{dE}{ds} \right|_0 \int_{-\infty}^{\infty} J_g(s)\, ds$$

With

$$\int_{-\infty}^{\infty} J_g(s)\, ds = \mathcal{A}t\pi^2/\xi_g^2 = V_c\pi^2/\xi_g^2$$

we finally find

$$I_g = \mathcal{V}_c \frac{\pi^2}{\xi_g^2} h u_{E_0} S(E_0) \qquad (69.31)$$

This tells us that the scattered intensity increases as the thickness t and not as its square. This form of the result, in which only the crystal volume $\mathcal{V}_c$ appears, is applicable even for more general shapes of crystal, always provided that the maximum thickness in the z-direction remains sufficiently small.

69.5 General formulation of the dynamical theory

The kinematic theory can be valid only for very thin crystals, since (69.31) suggests that the scattered intensity can become larger than that of the primary beam if we merely make the crystalline layer thick enough. This evidently wrong finding is a consequence of the fact that we have considered only energy transfer from the primary to the scattered wave, ignoring the reverse process. Moreover, we have to allow for the fact that such coupling may exist not just between two waves but, more generally, between several. Finally, we must recognize that *inelastic* processes cannot be ignored in thicker crystalline specimens.

In this section, we shall briefly sketch two different but demonstrably equivalent approaches to the dynamical theory (e.g. Metherell, 1976; Whelan, 1970). In the next section, we examine the practical application of the theory to the important two-beam case.

69.5.1 The method of oscillating amplitudes
This form of the dynamic theory was first introduced by Darwin (1914), who applied it to the diffraction of X-rays. Later Howie and Whelan (1961) and Hashimoto *et al.* (1962) modified it to make it applicable to electron scattering. A detailed presentation can be found in the book of Hirsch *et al.* (1965) and in many later texts, notably Cowley (1981), Reimer (1993), Buseck *et al.* (1988) and Spence (1988).

A characteristic feature of this method is the assumption that all partial electron waves $\psi_g(\boldsymbol{r})$ can be regarded as plane waves with amplitudes $A_g(z)$ that vary slowly with the penetration depth z:

$$\psi_g(\boldsymbol{r}) = A_g(z) \exp\{2\pi i(\boldsymbol{k} + \boldsymbol{g}) \cdot \boldsymbol{r}\} \exp(2\pi i s_g z) \qquad (69.32)$$

The parameter $s_g := s$ is the excitation error defined in (69.26) and sketched in Fig. 69.7. Such a representation also includes the primary wave, which is specified by $\boldsymbol{g} = 0$ and $s_0 = 0$.

The amplitudes $A_g(z)$ are coupled by a system of ordinary differential equations, *linear* and *of first order*. If we disregard absorption for the moment, we can show that the conservation of total intensity requires that the system of differential equations be Hermitian. Since $-\mathrm{i}d/dz$ is a Hermitian differential operator, we find that

$$\frac{1}{2\pi\mathrm{i}}\frac{d}{dz}A_g(z) = \sum_{g'} K_{gg'} A_{g'}(z)\exp\{2\pi\mathrm{i}(s_{g'} - s_g)\} \qquad (69.33)$$

is an appropriate form provided that the coefficient matrix $K_{gg'}$ is Hermitian, which means that

$$K_{gg'} = K_{g'g}^* \qquad (69.34)$$

for all reflections. All reciprocal lattice vectors $\boldsymbol{g}$, $\boldsymbol{g}'$ with triplet indices g_i, g_i' ($i = 1, 2, 3$) must be located in a plane perpendicular to the z-axis, a characteristic of the high-energy approximation with small scattering angles. It is easy to verify that the conservation law

$$\sum_g |A_g(z)|^2 = \text{constant} = 1 \qquad (69.35)$$

is satisfied.

This normalization to unity, which can be made with no loss of generality, can be interpreted as the study of the propagation of a single particle, in which case the individual terms $|A_g(z)|^2$ represent the probabilities of finding this particle in the corresponding Bragg reflection state. The initial conditions are then

$$|A_g(0)|^2 = \delta_{0g} \qquad (69.36)$$

The conservation law can also be interpreted in a different way. The total wavefunction $\psi(\boldsymbol{r})$ is obtained by summation of all the partial waves:

$$\psi(\boldsymbol{r}) = \sum_g \psi_g(\boldsymbol{r})$$
$$= \sum_g A_g \exp\left[2\pi\mathrm{i}\{(\boldsymbol{k} + \boldsymbol{g}) \cdot \boldsymbol{r} + s_g z\}\right] \qquad (69.37)$$

The corresponding particle density $\rho(\boldsymbol{r})$ is then

$$\rho(\boldsymbol{r}) := |\psi(\boldsymbol{r})|^2$$
$$= \sum_g \sum_{g'} \psi_g^*(\boldsymbol{r})\psi_{g'}(\boldsymbol{r}) \qquad (69.38)$$

which leads to

$$\rho(\boldsymbol{r}) = \sum_g \sum_{g'} A_g^*(z) A_{g'}(z) \exp\{2\pi\mathrm{i}(\boldsymbol{g'} - \boldsymbol{g}) \cdot \boldsymbol{r}\} \exp\{2\pi\mathrm{i}(s_{g'} - s_g)z\}$$

The first exponential factor involves only the coordinates x and y and has the same periodicity as the real lattice. If we average over an area $\mathcal{A}_c$ that covers an integral number of periods, therefore, all terms for which $\boldsymbol{g} \neq \boldsymbol{g'}$ cancel out and we find

$$\overline{\rho} := \frac{1}{\mathcal{A}_c} \int\int_{\mathcal{A}_c} \rho(x, y, z)\, dx\, dy$$

$$= \sum_g \sum_{g'} A_g^*(z) A_{g'}(z) \exp\{2\pi\mathrm{i}(s_{g'} - s_g)\} \delta_{gg'}$$

$$= \sum_g |A_g(z)|^2 = 1$$

Equation (69.35) is hence the familiar normalization to unity for one free (unbounded) particle.

This presentation of the method is, however, still incomplete: the definition of the coefficients $K_{gg'}$ is missing and will be given in Section 69.5.3.

69.5.2 Formulation as an eigenvalue problem

This formulation was first introduced by Bethe (1928, 1933) and is extensively discussed by Hirsch et al. (1965) and by Cowley (1981). The characteristic feature of this approach is the attempt to find a direct, approximate solution of the Schrödinger equation (68.7) in terms of *Bloch waves*. The latter, a well-known tool in solid-state physics, here take the general form

$$b_\mu(\boldsymbol{k}, \boldsymbol{r}) = \sum_g C_{\mu g}(\boldsymbol{k}) \exp[2\pi\mathrm{i}\{(\boldsymbol{k} + \boldsymbol{g}) \cdot \boldsymbol{r} + q_\mu z\}] \qquad (69.39)$$

As before, Roman subscripts $\boldsymbol{g}$, $\boldsymbol{g'}$ denote the index triplets of reciprocal lattice vectors. The additional Greek subscripts (μ here) will distinguish between different quantum states in the solutions of the Schrödinger equation. (They are often written as superscripts in parentheses but our notation is more convenient here and causes no confusion.)

On introducing (69.39) into the Schrödinger equation and cancelling out all exponential factors after some manipulation of the indices, we obtain a homogenous, linear system of equations for the series expansion coefficients, denoted by C in (69.39):

$$\sum_{g'} \left[\{-(\boldsymbol{k} + \boldsymbol{g} + q_\mu \boldsymbol{e}_z)^2 + \boldsymbol{k}^2\} \delta_{gg'} + \frac{1}{4\pi^2} v_{g-g'} \right] C_{\mu g'}(\boldsymbol{k}) = 0$$

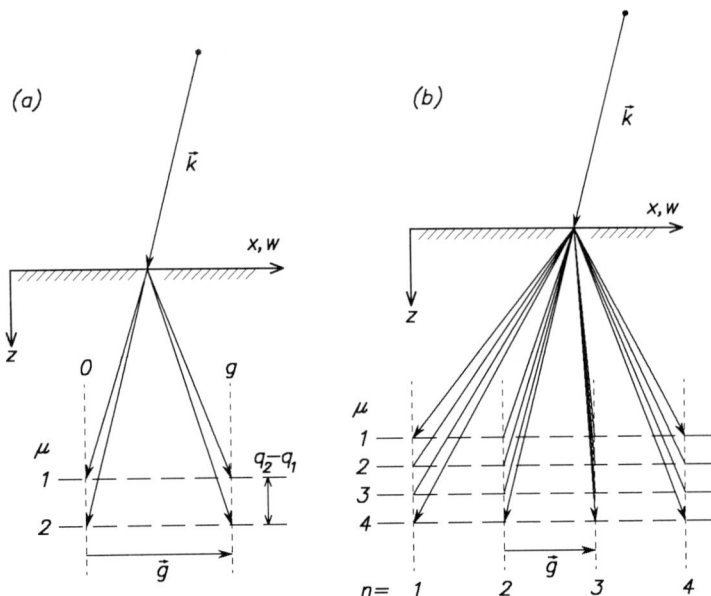

Fig. 69.8: Vector diagrams of N-beam cases for (a) $N = 2$ and (b) $N = 4$. The momentum $\bar{k}$ of the incident wave splits up into N^2 vectors $k_{\mu n}$. The label μ refers to the z-component, while n refers to the transverse components. In the two-beam case the labels 0 and g are employed instead of $n = 1, 2$.

This is still exactly valid but can be simplifed as before by making the approximation of high energies and small scattering angles:

$$g_z = 0, \quad |\boldsymbol{g}| \ll k, \quad |q_\mu| \ll k \equiv 1/\lambda$$

If we consider the excitation errors s_g given by (69.26) and define new matrix elements $K_{gg'}$,

$$K_{gg'} := \frac{\lambda}{8\pi^2} v_{g-g'} \tag{69.40}$$

for all index triplets $\boldsymbol{g}$ and $\boldsymbol{g}'$, we obtain an essential simplification:

$$\sum_{g'} \{K_{gg'} + (s_g - q_\mu)\delta_{gg'}\} C_{\mu g'}(\boldsymbol{k}) = 0 \tag{69.41}$$

This is a Hermitian eigenvalue problem. In order to solve it, we have to choose a sufficiently large set $\boldsymbol{g}_1, \boldsymbol{g}_2, \ldots, \boldsymbol{g}_N$ of reciprocal lattice vectors with $g_z = 0$, one of which must vanish (primary beam). This configuration is then called an N-beam case (Fig. 69.8) and the system (69.41) has rank N. The characteristic equation

$$\det[K_{gg'} + (s_g - q_\mu)\delta_{gg'}] = 0 \qquad (69.42)$$

gives exactly N real eigenvalues, q_1, q_2, ... , q_N; we use the additional subscript μ, with $1 \leq \mu \leq N$, to label them. Strictly speaking, the triplet indices g and g' should now have labels too: $g_1,\ldots,g_N, g'_1,\ldots,g'_N$; we shall however omit these labels as before, whenever this causes no confusion. Since $s_g = s_g(\boldsymbol{k})$ in (69.26), the eigenvalues q_μ will be functions of k, the momentum of the incident particle; the latter is thus a free parameter. It is advantageous to normalize the coefficients $C_{\mu g}$ to unity, and since they are complex, we require that

$$\sum_g C^*_{\mu g}(\boldsymbol{k})C_{\nu g}(\boldsymbol{k}) = \delta_{\mu\nu} \qquad (69.43)$$

The coefficients therefore form a *unitary* matrix C

$$C := \{C_{\mu g_n}(\boldsymbol{k}) \mid \mu = 1,\ldots,N; \; n = 1,\ldots,N\}$$

From this property, a second set of orthogonality relations can be derived:

$$\sum_\mu C_{\mu g}(\boldsymbol{k})C^*_{\mu g'}(\boldsymbol{k}) = \delta_{gg'} \qquad (69.44)$$

Once all the Bloch waves $b_\mu(\boldsymbol{k},\boldsymbol{r})$ have been calculated by solving this eigenvalue problem, the entire wavefunction $\psi(\boldsymbol{r})$ is obtained by linear superposition with the appropriate weights $\alpha_\mu(\boldsymbol{k})$:

$$\psi(\boldsymbol{r}) = \sum_{\mu=1}^{N} \alpha_\mu(\boldsymbol{k})b_\mu(\boldsymbol{k},\boldsymbol{r}) \qquad (69.45)$$

These weights α_μ are determined uniquely by the initial boundary condition,

$$\psi(x,y,0) = \exp\{2\pi i(x k_x + y k_y)\} \qquad (69.46)$$

at the entrance plane, the surface of the crystal on which the electrons are incident. This condition generates the set of linear equations

$$\sum_\mu \alpha_\mu(\boldsymbol{k})C_{\mu g}(\boldsymbol{k}) = \delta_{g0}$$

with solutions

$$\alpha_\mu(\boldsymbol{k}) = C^*_{\mu 0}(\boldsymbol{k}) \quad , \quad \mu = 1,\ldots,N \qquad (69.47)$$

This can be easily verified by setting $g' = 0$ in (69.44). The full solution can now be written in the compact form

$$\psi(r) = \sum_\mu \sum_g C^*_{\mu 0}(k) C_{\mu g}(k) \exp[2\pi i\{r \cdot (k + g) + q_\mu(k)z\}] \qquad (69.48)$$

The problem of wave propagation in perfect crystals can in principle be completely solved in this way. The practical calculation remains to be performed and the results to be interpreted.

69.5.3 The equivalence of the two methods

Since we set out from a single well-defined expression of the problem, it must be possible to show that the two, superficially different, solutions (69.37) and (69.48) are in fact identical. The quantities $\exp\{2\pi i(k + g) \cdot r\}$ are linearly independent; each coefficient must therefore agree, which implies that

$$A_g(z) = \sum_\mu C^*_{\mu 0}(k) C_{\mu g}(k) \exp[2\pi i\{q_\mu(k) - s_g\}z] \qquad (69.49)$$

Differentiation of this relation gives

$$\frac{1}{2\pi i} A'_g(z) = \sum_\mu C^*_{\mu 0} C_{\mu g}(q_\mu - s_g) \exp\{2\pi i(q_\mu - s_g)z\} \qquad (69.50)$$

We can now proceed in two different ways.

(i) We regard the differential equation (69.33) as given. We then replace g by g' in (69.49) and substitute these expressions for $A_{g'}$ in the right-hand side R of (69.33). The terms in $s_{g'}$ cancel out and we find

$$R = \sum_\mu C^*_{\mu 0} \sum_{g'} K_{gg'} C_{\mu g'} \exp\{2\pi i(q_\mu - s_g)z\} \qquad (69.51)$$

This will be in agreement with (69.50) if the term $C_{\mu g}(q_\mu - s_g)$ in the latter is the same as the term $\sum_{g'} K_{gg'} C_{\mu g'}$ in (69.51). On setting them equal, we recover (69.41), which is the expression of the eigenvalue problem.

(ii) We now regard (69.41) as given and attempt to construct an equivalent system of linear, first-order, differential equations. We first replace the term $C_{\mu g}(q_\mu - s_g)$ in the derivative of $A_g(z)$ given by (69.50) with the aid of (69.41), thereby obtaining the expression R of (69.51). This must now be rewritten in such a way that the entire expressions $A_{g'}$ appear; this can be achieved by writing

$$\exp(-2\pi i s_g z) \equiv \exp(-2\pi i s_{g'} z) \exp\{2\pi i(s_{g'} - s_g)z\}$$

and taking the final factor into the coefficients of the differential equations. We thus arrive at (69.33).

The method involving oscillating amplitudes is not self-contained: to make it complete, the information in (69.40) has to be added. For perfect crystals, it can be abandoned but its real advantages emerge when we consider electron propagation in imperfect crystals. It is then much easier to solve linear differential equations *of first order* than it is to solve the eigenvalue problem. We refer to Hirsch *et al.* (1965) or Buseck *et al.* (1988) for further discussion.

69.5.4 Absorption

So far, we have completely ignored inelastic processes but, in thicker crystal layers, these become very appreciable and must be taken into consideration. Inelastic scattering of an electron destroys its phase relation with the Bloch waves initially excited. The latter can gain in intensity only by constructive interference, however, so that each inelastic event weakens them; this has the phenomenological appearance of an absorption, although the total number of electrons is of course conserved. We have encountered a similar situation in Chapter 66, in the discussion of weakly scattering specimens. Even for phase objects, we retained an amplitude term in the specimen transparency and one of the reasons for this was that inelastically scattered electrons are excluded from an essentially monochromatic theory.

The easiest way of describing this apparent absorption is to incorporate an *imaginary* potential into the Schrödinger equation. Many quantities that were real in the preceding reasoning then become complex. We shall retain the notation already employed for the real part and denote the imaginary part by a circumflex; the corresponding complex quantity is given a tilde. The new form of Schrödinger's equation is thus

$$\nabla^2 \tilde{\psi}(\boldsymbol{r}) + 4\pi k^2 \tilde{\psi}(\boldsymbol{r}) + \{v(\boldsymbol{r}) + i\hat{v}(\boldsymbol{r})\}\tilde{\psi}(\boldsymbol{r}) = 0$$

in which we have distinguished the new wavefunction by a tilde, even though this function is complex with and without absorption.

This is a general approach to the description of inelastic processes. In crystals, the imaginary potential $\hat{v}(\boldsymbol{r})$, like the real potential $v(\boldsymbol{r})$, must be a periodic function:

$$\hat{v}(\boldsymbol{r}) = \sum_g \hat{v}_g \exp(2\pi i \boldsymbol{g} \cdot \boldsymbol{r}) \geq 0 \tag{69.53}$$

We can again seek a solution of the wave equation (69.57) in the form of Bloch waves but there must now be exponential damping in the z-direction. We have

$$\tilde{b}_\mu(\boldsymbol{k}, \boldsymbol{r}) = \sum_g \tilde{C}_{\mu g}(\boldsymbol{k}) \exp\{2\pi i (\boldsymbol{k} + \boldsymbol{g}) \cdot \boldsymbol{r}\} \exp\{2\pi (iq_\mu - \hat{q}_\mu)z\} \tag{69.54}$$

The algebraic techniques described earlier are then successful but the expressions finally obtained are unnecessarily complicated. A considerable simplification is obtained if we assume that the absorption processes are very weak, which is justified by experiment. All terms of second or higher order in $\hat{v}_g$ and $\hat{q}_\mu$ can then be ignored. It is permissible to set $\hat{C}_{\mu g} = 0$, so that $\tilde{C}_{\mu g}$ is the same as $C_{\mu g}$.

If we now introduce (69.54) into (69.52) and cancel out all the exponential factors, we find that (69.41) with (69.40) and all the equations up to (69.44) remain valid without modification. In addition, we find that the imaginary parts of the coefficients $\hat{K}_{gg'}$ are given by

$$\hat{K}_{gg'} = \frac{\lambda}{8\pi^2}\hat{v}_{g-g'} = \hat{K}^*_{g'g} \tag{69.55}$$

as we might expect, and these appear in a second set of linear equations for the unknowns $\hat{q}_\mu$:

$$\hat{q}_\mu C_{\mu g} = \sum_{g'} \hat{K}_{gg'} C_{\mu g'} \tag{69.56}$$

These cannot be quite correct, since they imply that the set of C-coefficients are eigenvectors of the matrices $\{K_{gg'}\}$ and $\{\hat{K}_{gg'}\}$ simultaneously, which will not in general be true. We can minimize the associated error by multiplying (69.56) throughout by $C^*_{\mu g}$ and summing over g. With the normalization condition (69.44), we find

$$\hat{q}_\mu = \sum_g \sum_{g'} C^*_{\mu g} \hat{K}_{gg'} C_{\mu g'} \tag{69.57}$$

(even when $\hat{C} \neq 0$). This is the *expectation value* of the $\hat{K}$ matrix in the state with label μ. We can now write the corresponding Bloch wave in the compact form

$$\tilde{b}_\mu(\boldsymbol{k},\boldsymbol{r}) = b_\mu(\boldsymbol{k},\boldsymbol{r})\exp\{-2\pi\hat{q}_\mu(\boldsymbol{k})z\} \tag{69.58}$$

This expression shows that each Bloch wave is slowly attenuated without distortion. However, the attenuation is different for each Bloch wave, whereas a single parameter is sufficient to characterize ordinary absorption. This is therefore known as *anomalous absorption*.

The representation of $\hat{q}_\mu$ as an expectation value is at first sight surprising; it can, however, be written in a more convincing and more concise form as follows:

$$8\pi^2 k \hat{q}_\mu = \langle \hat{v}(\boldsymbol{r}) \rangle_\mu$$

$$= \frac{\displaystyle\int_E |\tilde{b}_\mu(\boldsymbol{k},\boldsymbol{r})|^2 \hat{v}(\boldsymbol{r})\,d\boldsymbol{r}}{\displaystyle\int_E |\tilde{b}_\mu(k,\boldsymbol{r})|^2\,d\boldsymbol{r}} \tag{69.59}$$

In these integrations over one elementary cell E, the exponential attenuation factors vary so very slowly that they can be taken outside the integrals and cancel out from the numerator and denominator. The tildes may then be dropped and $\langle \hat{v}(\boldsymbol{r}) \rangle_\mu$ becomes the familiar quantum mechanical expectation value of $\hat{v}(\boldsymbol{r})$, which can be calculated by standard perturbation techniques. Such calculations are presented by Yoshioka (1957), Howie (1963), Humphreys and Hirsch (1968) and Radi (1968).

The absorption of electron waves in crystals can also be described in the language of the Darwin theory. There are no essentially new aspects. The basic form of the wave representation, (69.32) and (69.37) with the initial condition (69.36), remains valid without modification. The general form of the differential equations (69.33) is unaltered provided that the matrix elements $K_{gg'}$ are replaced by

$$\tilde{K}_{gg'} = K_{gg} + i\hat{K}_{gg'} \equiv \frac{\lambda}{8\pi^2}(v_{g-g'} + i\hat{v}_{g-g'})$$

The differential equations then cease to be Hermitian.

It can easily be shown that this form of the Darwin theory is equivalent to Bethe's theory with absorption if q_μ is replaced by

$$\tilde{q}_\mu = q_\mu + i\hat{q}_\mu \tag{69.60}$$

in (69.49–51) and the same approximations as those for (69.56) are made. With the aid of the modified form of (69.49) for A_g, it is a straightforward task to derive the intensity loss formula:

$$
\begin{aligned}
I'(z) : & = \frac{d}{dz} \sum_g |A_g(z)|^2 \\
& = -4\pi \sum_{g,g'} A_g^*(z) \hat{K}_{gg'} A_g'(z) \exp\{2\pi i(s_{g'} - s_g)z\} \\
& = -4\pi \sum_\mu |C_{\mu 0}|^2 \hat{q}_\mu \exp(-4\pi\hat{q}_\mu z) \tag{69.61}
\end{aligned}
$$

which evidently satisfies all the basic requirements.

69.6 The two-beam case

The foregoing form of the dynamical theory is self-contained in the sense that it is ready to be applied to practical situations. The resulting calculations are, however, usually so extensive that simple results cannot be obtained and we must have recourse to numerical techniques.

An exception to this disappointing situation is the 'two-beam case', for which results can be obtained in closed form. These tell us a great deal about the differences between kinematic and dynamical wave propagation.

In a *two-beam case*, we consider only two strong reflections, the primary beam with reciprocal lattice vector $g_1 = 0$ and a second beam with some other vector $g_2 =: g \neq 0$. For $N = 2$, we construct two Bloch waves with labels $\mu = 1, 2$. For simplicity, we assume here that the crystals are centrosymmetric, whereupon the coefficients v_g become real. By a suitable choice of the coordinate system, they can even be made positive and we assume that this is the case.

69.6.1 General calculations

We return to the definitions of s_g and ξ_g (69.26 and 69.29, respectively), in which the absolute-value signs can now be omitted. Since we have only one non-zero vector g, we can also drop the label g in s_g and ξ_g. We now define

$$w := s\xi = \cot 2\beta \quad , \quad 0 \leq \beta \leq \pi/2 \qquad (69.62)$$

which differs by a factor two from the definition adopted in Hirsch *et al.* (1965) and many subsequent accounts of the theory. We also write

$$q_0 := K_{00} = \frac{\lambda v_0}{8\pi^2} \equiv \frac{eV_0}{hu_E} \qquad (69.63)$$

u_E being given by (69.28). If we introduce the extinction distance $\xi (= \xi_g)$ from (69.27) into (69.40) and recall that $V_g > 0$, we can cast the characteristic equation (69.42) into the form

$$\begin{vmatrix} K_{00} - q_\mu & K_{0g} \\ K_{0g} & K_{gg} + s_g - q_\mu \end{vmatrix} = \begin{vmatrix} q_0 - q_\mu & 1/2\xi \\ 1/2\xi & q_0 - q_\mu + s \end{vmatrix} = 0$$

The two solutions can be written

$$q_{1,2} = q_0 + \frac{1}{2\xi}(w \mp \sqrt{1 + w^2}) \qquad (69.64)$$

in which w is defined by (69.62). The constant term q_0 is a scaled measure of the mean inner potential V_0 of the crystal; it thus causes a classical *refraction* of the electron waves at the entrance and exit planes. This is a minor effect of little significance and will therefore not be discussed further. The coefficient matrix C can be cast into the form

$$C = \begin{pmatrix} C_{10} & C_{1g} \\ C_{20} & C_{2g} \end{pmatrix} = \begin{pmatrix} \cos\beta & -\sin\beta \\ \sin\beta & \cos\beta \end{pmatrix} \qquad (69.65)$$

which is clearly orthogonal.

In this first step, we have solved the eigenvalue problem and obtained a quite simple result. We are now ready to evaluate the absorption coefficients, $\hat{q}_\mu$, using (69.57). There are only two relevant matrix elements, which can be taken to be positive, and hence

$$\hat{K} = \frac{\lambda}{8\pi^2} \begin{pmatrix} \hat{v}_0 & \hat{v}_g \\ \hat{v}_g & \hat{v}_0 \end{pmatrix} \tag{69.66}$$

This expression, together with (69.65), yields

$$\hat{q}_{1,2} = \frac{\lambda}{8\pi^2}(\hat{v}_0 \mp \hat{v}_g \sin 2\beta) \equiv \frac{\lambda}{8\pi^2}\left(\hat{v}_0 \mp \frac{\hat{v}_g}{\sqrt{1+w^2}}\right) \tag{69.67}$$

Since both values of $\hat{q}$ must be positive, $\hat{v}_g$ must satisfy the condition $\hat{v}_g < \hat{v}_0$.

The formulae for the wavefunctions and intensities take a convenient form if we introduce four new parameters, $\bar{\gamma}$, γ_a, κ and $\hat{\xi}$ as follows:

$$\bar{\gamma} := 2\pi(\hat{q}_2 + \hat{q}_1) = \hat{v}_0/k$$

$$\gamma_a := 2\pi(\hat{q}_2 - \hat{q}_1) = \frac{\hat{v}_g}{k\sqrt{1+w^2}} =: \frac{2\pi}{\hat{\xi}\sqrt{1+w^2}}$$

$$\kappa := 2\pi(q_2 - q_1) = \frac{v_g\sqrt{1+w^2}}{k} = \frac{2\pi\sqrt{1+w^2}}{\xi} \tag{69.68}$$

Here, $k = 2\pi/\lambda$ is the electron wavenumber *in vacuo*. A slightly better approximation could be achieved by considering the effect of the mean inner potential on the wavenumber but the gain is negligible. The parameter $\bar{\gamma}$ is the *mean absorption coefficient*, which is obviously independent of w. The quantity γ_a is the *anomalous absorption coefficient*, the measurement of which will be considered below. It is not really necessary to introduce $\hat{\xi}$ but we give the definition here because it is often employed in the literature of the subject. Finally, the parameter κ will prove to be the *beat frequency* of the intensity oscillations. Note that these quantities depend in different ways on the factor $\sqrt{1+w^2}$: γ_a vanishes as $w \to \infty$ whereas κ increases monotonically.

We are now ready to express the wavefunction in closed form. Owing to the fact that $g_z = 0$ and hence that $\boldsymbol{g} \cdot \boldsymbol{r} = g_x x + g_y y$, we obtain product separations, which simplify the expressions for the amplitudes and intensities considerably. The two damped Bloch waves are given by

$$\tilde{b}_1(\boldsymbol{k}, \boldsymbol{r}) = \exp\{2\pi \mathrm{i}(\boldsymbol{k} \cdot \boldsymbol{g}r + \tilde{q}_1 z)\}\{\cos\beta - \exp(2\pi \mathrm{i}\boldsymbol{g} \cdot \boldsymbol{r})\sin\beta\}$$

$$\tilde{b}_2(\boldsymbol{k}, \boldsymbol{r}) = \exp\{2\pi \mathrm{i}(\boldsymbol{k} \cdot \boldsymbol{g}r + \tilde{q}_2 z)\}\{\sin\beta + \exp(2\pi \mathrm{i}\boldsymbol{g} \cdot \boldsymbol{r})\cos\beta\} \tag{69.69}$$

The corresponding intensities are then

$$|\tilde{b}_{1,2}|^2 = \exp\{-(\overline{\gamma} \mp \gamma_a)z\}\left\{1 \mp \frac{\cos 2\pi(xg_x + yg_y)}{\sqrt{1+w^2}}\right\} \qquad (69.70)$$

Superposition of the two Bloch waves as in (69.45) here gives

$$\psi(\boldsymbol{r}) = \tilde{b}_1(\boldsymbol{k}, \boldsymbol{r})\cos\beta + \tilde{b}_2(\boldsymbol{k}, \boldsymbol{r})\sin\beta$$
$$= \exp(2\pi i\boldsymbol{k} \cdot \boldsymbol{r} + \tilde{q}_1 z)\{\cos^2\beta - \exp(2\pi i\boldsymbol{g} \cdot \boldsymbol{r})\sin\beta\cos\beta\} \qquad (69.71)$$
$$+ \exp(2\pi i\boldsymbol{k} \cdot \boldsymbol{r} + \tilde{q}_2 z)\{\sin^2\beta + \exp(2\pi i\boldsymbol{g} \cdot \boldsymbol{r})\sin\beta\cos\beta\}$$

The amplitudes are commonly denoted

$$T(z) := A_0(z) \quad , \quad S(z) := A_g(z) \qquad (69.72)$$

and refer to the 'transmitted' primary wave and the 'scattered' wave, respectively. They are easily found by writing out (69.37) explicitly, which here gives

$$\psi(\boldsymbol{r}) = \exp(2\pi i\boldsymbol{k} \cdot \boldsymbol{r})[T(z) + \exp\{2\pi i(sz + \boldsymbol{g} \cdot \boldsymbol{r})\}S(z)] \qquad (69.73)$$

and casting (69.71) into this form. A short calculation yields

$$T(z) = \exp(2\pi i\tilde{q}_1 z)\cos^2\beta + \exp(2\pi i\tilde{q}_2 z)\sin^2\beta$$
$$S(z) = \exp(-2\pi isz)\cos\beta\sin\beta\{\exp(2\pi i\tilde{q}_2 z) - \exp(2\pi i\tilde{q}_1 z)\} \qquad (69.74)$$

After calculation of the parameter β, the corresponding intensities can be written

$$|S(z)|^2 = \frac{\exp(-\overline{\gamma}z)}{2(1+w^2)}(\cosh\gamma_a z - \cos\kappa z) \qquad (69.75)$$

and

$$|T(z)|^2 = \frac{\exp(-\overline{\gamma}z)}{2(1+w^2)}\left\{(1 + 2w^2)\cosh\gamma_a z + \cos\kappa z + 2w\sqrt{1+w^2}\sinh\gamma_a z\right\} \qquad (69.76)$$

The terms in $\cos\kappa z$ describe an *oscillation* of the intensities between the two wave fields, an effect to which we return later; it disappears from the total intensity, $I(z)$:

$$I(z) := |T(z)|^2 + |S(z)|^2 = e^{-\overline{\gamma}z}\left(\cosh\gamma_a z + \frac{w}{\sqrt{1+w^2}}\sinh\gamma_a z\right) \qquad (69.77)$$

This expression is a generalization of (69.35), to include absorption.

69.6.2 Interpretation of the results

The expression (69.71) describes the linear superposition of four exponentially damped plane waves with real reduced wavevectors

$$\mathbf{k}_{\mu 0} = \mathbf{k} + q_\mu \mathbf{e}_z \quad , \quad \mathbf{k}_{\mu g} = \mathbf{k} + \mathbf{g} + q_\mu \mathbf{e}_z \quad (\mu = 1, 2)$$

These vectors must be located on a surface known as the *dispersion surface* in reciprocal $\mathbf{k}$-space, as illustrated in Fig. 69.9b. This surface must exhibit the three-dimensional periodicity of the reciprocal lattice and consists of many separate branches. An approximate construction with only one-dimensional periodicity in the direction of the vector $\mathbf{g}$ is shown in Fig. 69.9a; it is obtained in the following way. We first draw all the circles

$$k_z^2 + (k_x - n|\mathbf{g}|)^2 = (k + q_0)^2 = \text{const}$$

the centres of which are shifted by the distances $n|\mathbf{g}|$. The integer n may be positive or negative. The term q_0 on the right-hand side corresponds to refraction by the mean inner potential. These circles are the curves along which planes cut the Ewald sphere and hence correspond to ideal Laue positions.

From the quantum mechanical standpoint, the Laue positions are *degenerate states*, as each point can be reached by two wavevectors. The propagation of an electron wave through the periodic potential can now be understood in terms of a quantum mechanical interaction that removes the degeneracy: the dispersion surface, being a surface of constant total energy, splits into two smooth, separate branches, which correspond to the labels $\mu = 1$ and $\mu = 2$. The shortest distance between the two branches is just ξ^{-1}. This interpretation extends to the N-beam case, the number of $\mathbf{g}$-vectors and branches increasing accordingly.

We now study two simple special cases of some importance (Figs. 69.10a,b).

(i) *Kinematic case* $w \gg 1$. We are now far from the resonance position. Equations (69.64) give $q_1 \approx q_0$ and $q_2 \approx q_0 + w/\xi = q_0 + s$ and from (69.68), we find $\kappa \approx 2\pi/\xi$ and $\gamma_a \approx 0$. There is virtually no anomalous absorption. The Bloch wave representation becomes unfavourable but (69.75–77) remain quite useful and simplify to

$$I(z) = \exp(-\bar{\gamma}z), \quad |T(z)|^2 = I(z) - |S(z)|^2 \tag{69.78}$$

with

$$|S(z)|^2 = \frac{\exp(-\bar{\gamma}z)}{\xi^2 s^2} \sin^2(\pi s z) \tag{69.79}$$

This is similar to the kinematic formula (69.29), now improved by the presence of the attenuation factor, which is easily interpreted. A less drastic

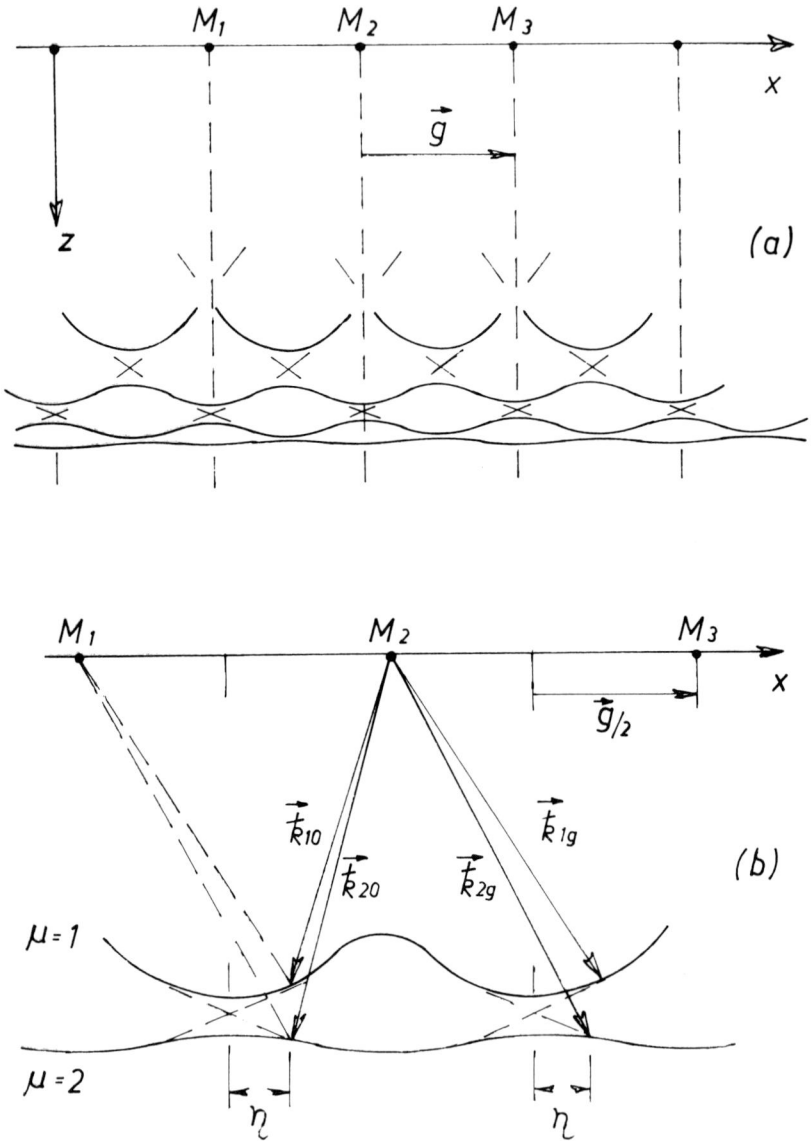

Fig. 69.9: Dispersion surfaces. (a) Construction based on Ewald spheres with equidistant centres M_n and continuous splitting in the neighbourhood of each point of intersection. (b) Location of the four wavevectors $\vec{k}_{\mu 0}$ and $\vec{k}_{\mu g}$, $\mu = 1, 2$ in the two-beam case relative to the dispersion surfaces; the related constructions differ from this one by vector shift (broken lines); η is the resonance error, given by (69.26b).

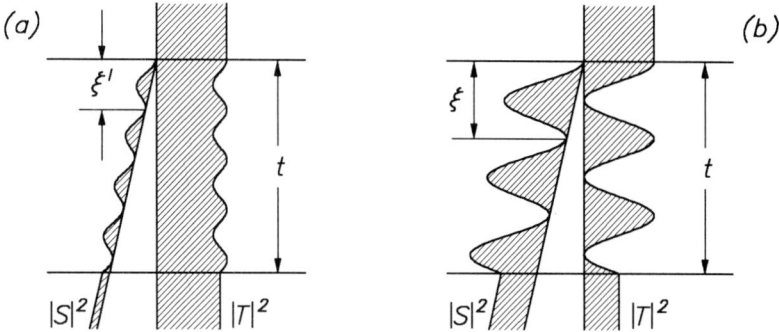

Fig. 69.10: Difference between the intensity distributions in a kinematic case (a) and a resonant dynamic case, $w = 0$ (b); $\xi' = \xi\sqrt{1 + w^2}$ denotes the reduced extinction distance.

simplification is obtained if $|w|$ is large enough for γ_a to be negligible but the other terms in (69.75–77) are retained. We again obtain (69.78) but $|S(z)|^2$ is now given by

$$|S(z)|^2 = \frac{\exp(-\bar{\gamma}z)}{1 + w^2} \sin^2(\pi z\sqrt{1 + w^2}/\xi) \qquad (69.80)$$

This function is plotted in Fig. 69.11. It can be shown, and is indeed obvious from the last results, that the case $w \ll -1$ again leads to (69.78–80).

(ii) *Resonant case* $w = 0$. There are now two positions for which the separation between the two branches of the dispersion surface is smallest. We find $q_{1,2} = q_0 \mp 1/2\xi$, $\beta = \pi/4$ and $\kappa = 2\pi/\xi$. In direct contrast to the case discussed above, $\gamma_a = 2\pi/\hat{\xi}$ now assumes its greatest value, so that anomalous absorption becomes significant.

Equations (69.75) and (69.76) now specialize to

$$\begin{aligned}
|S(z)|^2 &= \frac{1}{2}e^{-\bar{\gamma}z}\{\cosh(2\pi z/\hat{\xi}) - \cos(2\pi z/\xi)\} \\
|T(z)|^2 &= \frac{1}{2}e^{-\bar{\gamma}z}\{\cosh(2\pi z/\hat{\xi}) + \cos(2\pi z/\xi)\}
\end{aligned} \qquad (69.81)$$

The corresponding graphs are plotted in Fig. 69.12. They reveal that the oscillations of both intensities are more strongly damped than their mean values. For $\hat{\xi} \to \infty$, this solution reduces to (69.80) with $w = 0$!

A more interesting conclusion can be drawn from (69.70). Without loss of generality, we can always orientate the (x, y) axes in such a way

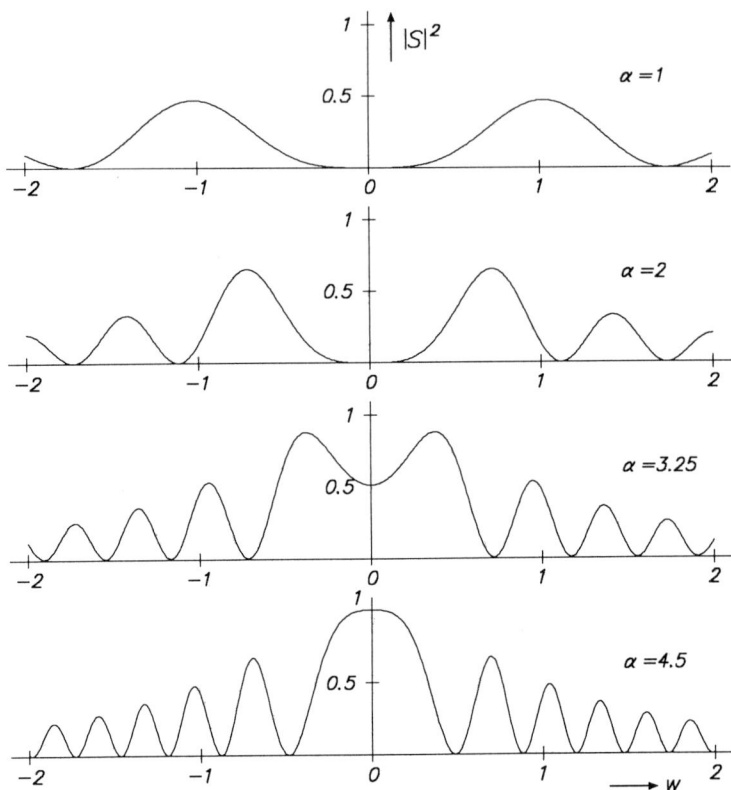

Fig. 69.11: Graph of the function $|S|^2 = (1+w^2)^{-1}\sin^2(\pi\alpha\sqrt{1+w^2})$ for various values of the parameter $\alpha = z/\xi$. (The function is normalized to unity; inclusion of absorption reduces to a mere change to scale.)

that $g_y = 0$. If we then set $g_x = 1/d$, d denoting the corresponding lattice-plane spacing, we have $\boldsymbol{g}\cdot\boldsymbol{r} = g_x x = x/d$. For $w = 0$, (69.70) take the form

$$
\begin{aligned}
|\tilde{b}_1|^2 &= 2\exp\{-(\overline{\gamma}-\gamma_a)z\}\sin^2(\pi x/d) \\
|\tilde{b}_2|^2 &= 2\exp\{-(\overline{\gamma}+\gamma_a)z\}\cos^2(\pi x/d)
\end{aligned}
\tag{69.82}
$$

Since $0 < \gamma_a \le \tilde{\gamma}$, the intensity $|\tilde{b}_2|^2$ is much more strongly damped than $|\tilde{b}_1|^2$. The maxima with respect to x of the former occur at the lattice planes, whereas those of the latter fall in the space between them. This effect is sketched in Fig. 69.13. The quantum mechanical explanation of this is based on the idea that the Bloch wave $\tilde{b}_2$ undergoes many more inner-shell interactions with the atomic electrons than $\tilde{b}_1$ since it is strongest in the regions occupied by them. These interactions have a strong inelastic

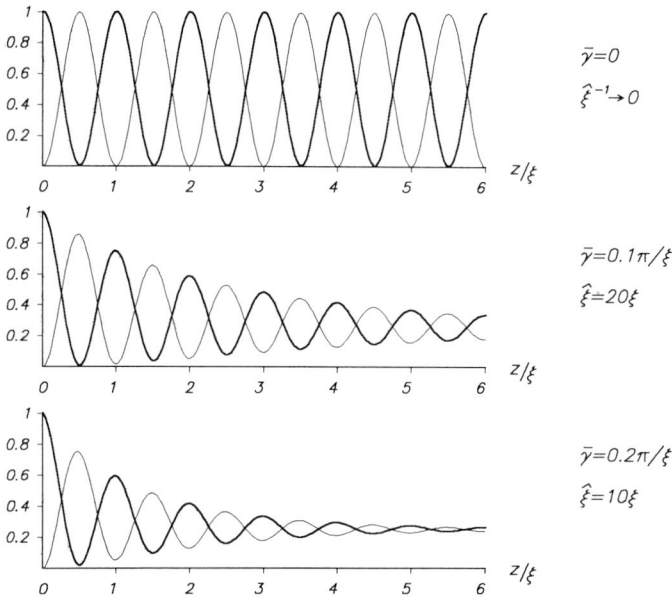

Fig. 69.12: Graphs of the functions of (69.81), $f(z/\xi) = \frac{1}{2}\exp(-\bar{\gamma}z)\{\cosh(2\pi z/\hat{\xi}) \pm \cos(2\pi z/\xi)\}$ for various values of the parameters $\bar{\gamma}$ and $\hat{\xi}$.

component, which leads to the anomalous part of the absorption. Non-localized interactions, plasmon excitations for example, affect both waves equally and cause an increase of $\bar{\gamma}$.

This effect has become known as *anomalous transmission*, since only the Bloch wave $\tilde{b}_1(\boldsymbol{k}, \boldsymbol{r})$ survives; if the crystalline layer is thick enough, its intensity is unexpectedly high. The electrons that do finally arrive have passed through channel-like domains in the space between the atoms. This is perhaps the most striking difference between the kinematic and dynamical descriptions of wave propagation.

The corresponding effect in X-ray physics was discovered long before by Borrmann (1941) and is named after him. It was exploited in the X-ray interferometer of Bonse and Hart (1965). An attempt to develop the counterpart of this in electron microscopy failed because the inelastic electron interactions in matter are very strong: such a device cannot compete with the biprism interferometer, in which the beam-splitting field is created in free space and not within a crystalline layer.

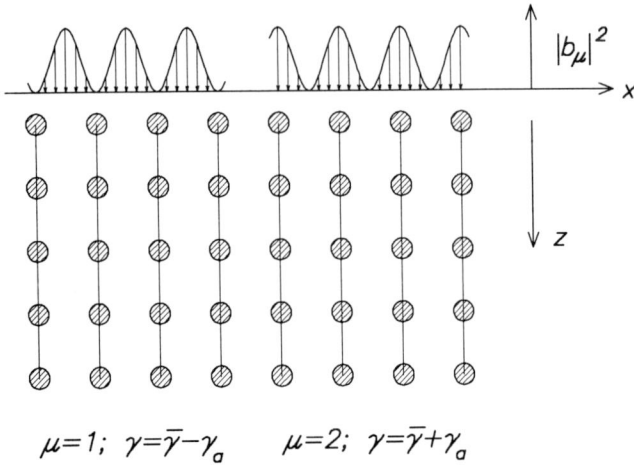

$\mu=1; \; \gamma=\bar{\gamma}-\gamma_a \qquad \mu=2; \; \gamma=\bar{\gamma}+\gamma_a$

Fig. 69.13: Sketch of the two Bloch wave intensities and their relation to the position of atoms.

69.7 Applications and extensions of the dynamical theory

Over the years, a vast amount of work on the use of dynamical theory in the study of materials has accumulated, so vast indeed that the whole subject can be covered only in specialized texts. The various experimental techniques consist essentially in combining appropriate illumination with a suitable imaging mode.

(i) The specimen can be illuminated with either a *parallel* electron beam (within experimental limits) or a *convergent* beam. In scanning instruments, the direction of incidence can be varied by *rocking* the beam (Fig. 69.14).

(ii) The electron waves emerging from the exit surface of the specimen can be transmitted to the recording plane in either the *imaging mode* or the *diffraction mode*. In the first case, an image of the information at the exit surface is seen, in the second, a diffraction pattern.

The situation is complicated by the fact that the ideal theory is practically never sufficient but only a first, over-simplified approximation. We conclude this chapter with a brief indication of the main extensions of the theory.

69.7.1 Nearly perfect crystals
(i) At higher electron energies, the curvature of the Ewald sphere may become so small that it is far from sufficient to consider only two lattice

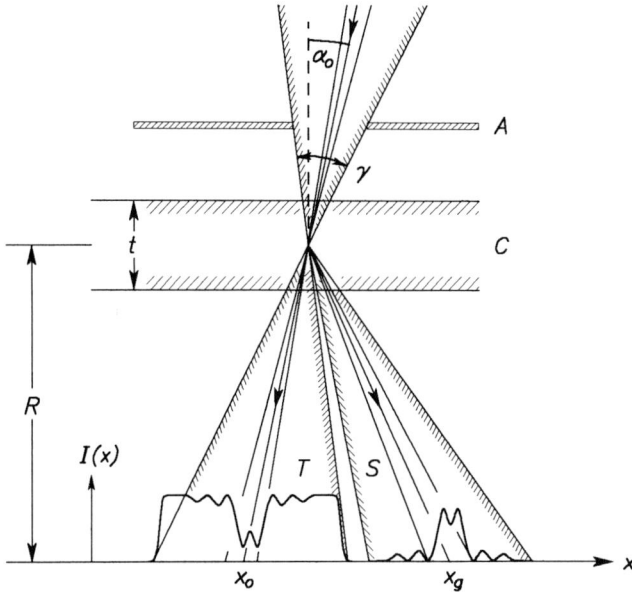

Fig. 69.14: Schematic presentation of a convergent-beam or rocking-beam configuration with aperture A, crystal layer C (thickness exaggerated) and recording plane. The position x_0 depends on the inclination angle α_0 of the beam axis; the width of the intensity pattern $I = |T(x)|$ is γR, γ being the rocking angle; and the separation $x_g - x_0 = R\lambda|\boldsymbol{g}|$.

vectors, $\boldsymbol{g}_1 = 0$ and $\boldsymbol{g}_2 = \boldsymbol{g}$. The next simplest generalization is to consider a *systematic row*, in which all the vectors $\boldsymbol{g}_n = n\boldsymbol{g}$ (where n is a positive or negative integer) are retained. Anomalous transmission again occurs but the simple sinusoidal factors must be replaced by more complicated Fourier series expansions.

(ii) At temperatures T far from zero, the *thermal motion* of the lattice atoms has to be taken into account. The effect of this motion is encapsulated in the *Debye–Waller factor*:

$$v_g(T) = v_g(0)\exp\{-M_g(T)\} \tag{69.83a}$$

in which

$$M_g(T) = 2\pi^2|\boldsymbol{g}|^2\langle\boldsymbol{u}^2(T)\rangle \tag{69.83b}$$

and $\langle\boldsymbol{u}^2\rangle$ denotes the mean squared amplitude of the vibrations. The Debye–Waller factor strongly reduces the high-order structure potentials.

(iii) *Dynamic potentials.* The structure potentials v_h that would otherwise be neglected in the N-beam case can be taken into account in an

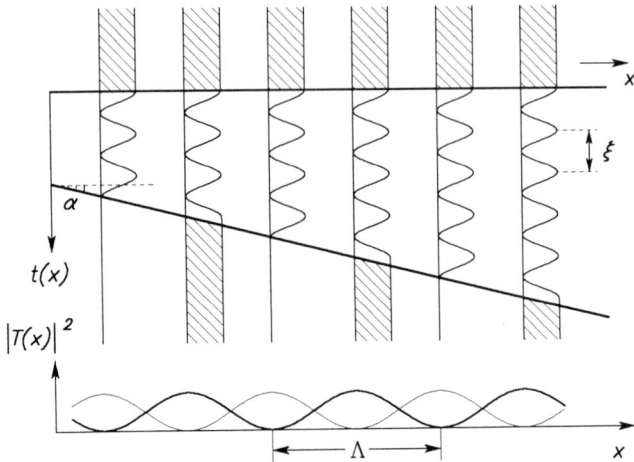

Fig. 69.15: Interference in a wedge-shaped crystal: the stripe distance is $\Lambda = \xi \cot \alpha$.

approximate fashion by a modification suggested by Bethe (1928): v_g is replaced by

$$v_g^B = v_g - \sum_{h'} \frac{1}{4\pi^2\gamma} \sum_h{}' \frac{v_h v_{g-h}}{k^2 - (\mathbf{k} + \mathbf{h})^2} \qquad (69.84)$$

The prime on the summation indicates that all triplets $\mathbf{h}$ already considered explicitly in the N-beam calculation are to be excluded. The formula is valid for centrosymmetric crystals, for which all the v-coefficients are real. The factor $\gamma = m/m_0$ must be present in the denominator since it appears in the numerator of (68.6). For a certain value of γ, which is reached for electron energies in the region of 1 MeV or more, it can happen that v_g^B vanishes and, at higher energies still, reverses its sign. This is known as the *critical voltage effect*. The critical voltage depends sensitively on the numerical values of the v-coefficients, including the correction described by the Debye–Waller factor.

(iv) *Kikuchi patterns.* Convenient though it is to describe inelastic scattering processes phenomenologically in terms of an imaginary potential, this can only give an incomplete description. The inelastically scattered electrons also travel through the crystal and are diffracted by the same mechanisms as the other electrons. The novel aspect is that the initial directions are not fixed, typically those of a parallel beam, but are diffusely distributed, rather as in convergent-beam illumination. After diffraction at the lattice planes, the inelastically scattered electrons cluster preferentially (though not exclusively) on *Kossel cones* (cf. Fig. 69.5). Kikuchi lines therefore bear a strong resemblance to Kossel lines.

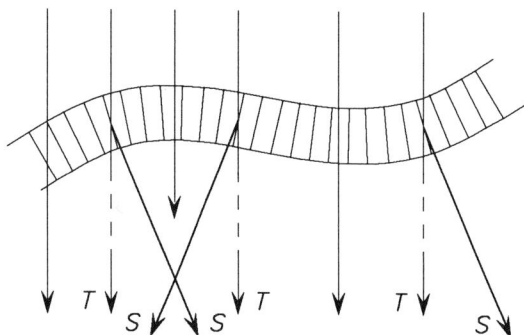

Fig. 69.16: Warped crystalline layer: at certain positions, the inclination of the lattice planes relative to the incident beam is such that S-waves can be excited with appreciable intensity.

(v) *Slowly varying crystal shape.* Perfect crystals are always an idealization. A first, simple approach to reality consists in recognizing that thin layers with nearly perfect lattices may be warped or vary in the thickness t. If these variations are slowly changing functions of the lateral coordinates x and y, the assumption that the lattice is *locally* perfect will be justified. We can still apply the dynamical theory, therefore, but must regard the lattice vectors $\boldsymbol{a}_j(x,y)$, $\boldsymbol{a}_j^\dagger(x,y)$ ($j = 1,2,3$) and the thickness $t(x,y)$ as only locally constant. The effect of the slow variation with x and y on the result of the calculation must then be investigated. Some possibilities are sketched in Figs. 69.15 and 69.16.

69.7.2 Imperfect crystals

Lattice imperfections can take very many different forms and a large literature has grown up around them. We give here only the briefest indication of the most important way of studying such specimens (Fig. 69.17), referring to the specialized texts listed in the bibliography for further information, notably Goringe (1976) and Gjønnes (1993).

Distorted structures are described by a *vector field* $\boldsymbol{R}(\boldsymbol{r})$ *of local displacements.* Equation (69.13) must then be replaced by

$$V(\boldsymbol{r}) = \sum_a \sum_m U_m(|\boldsymbol{r} - \boldsymbol{a} - \boldsymbol{r}_m - \boldsymbol{R}(\boldsymbol{r})|) \qquad (69.85)$$

In some cases, the displacement $\boldsymbol{R}(\boldsymbol{r})$ may be regarded as a slowly varying function and it then has an effect similar to that of the vectors $\boldsymbol{r}_m$: an exponential factor $\exp\{-2\pi i \boldsymbol{g} \cdot \boldsymbol{R}(r)\}$ appears in the structure potentials

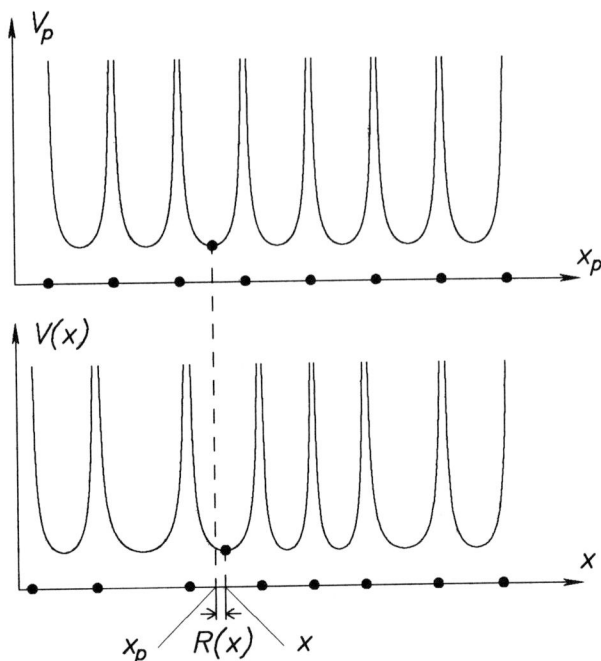

Fig. 69.17: One-dimensional sketch of the model of deformable atoms. Upper curve: undeformed periodic potential $V_p(x_p)$, lower curve: deformed potential $V(x)$. The value $V_p(x_p)$ is shifted to the position $x = x_p + R(x)$, $V(x) = V_p(x_p) = V_p(x - R(x))$.

and we now write

$$V(\boldsymbol{r}) = \sum_g V_g \exp[2\pi\mathrm{i}\{\boldsymbol{r} - \boldsymbol{R}(\boldsymbol{r})\}] \qquad (69.86)$$

instead of (69.10). As solution of Schrödinger's equation, we replace $A_g(z)$ of (69.32) by $A_g(\boldsymbol{r})$:

$$\psi(\boldsymbol{r}) = \sum_g A_g(\boldsymbol{r}) \exp\{2\pi\mathrm{i}(\boldsymbol{k} + \boldsymbol{g}) \cdot \boldsymbol{r}\} \exp(2\pi\mathrm{i}s_g z) \qquad (69.87)$$

Since $R(\boldsymbol{r})$ is slowly varying, the differential equations for A_g take the form

$$\frac{1}{2\pi\mathrm{i}} \frac{\partial}{\partial z} A_g(\boldsymbol{r}) = \sum_{g'} \tilde{K}_{gg'} A_g(\boldsymbol{g}\boldsymbol{r}) \exp\{2\pi\mathrm{i}(s_{g'} - s_g)z\}$$
$$\times \exp\{2\pi\mathrm{i}(\boldsymbol{g}' - \boldsymbol{g}) \cdot \boldsymbol{R}(\boldsymbol{r})\} \qquad (69.88)$$

The second exponential factor describes the influence of the distortions on the amplitudes. The transverse coordinates x and y are treated here as mere parameters but they cannot be ignored completely. Sometimes, $\boldsymbol{R}$ is taken to be a function of z only, $\boldsymbol{R} = \boldsymbol{R}(z)$, which would be true for stratified material. Solution of (69.88) then involves an integration over z, giving the 'column approximation'.

Equations of the general form (69.88) or special cases of them were derived as long ago as the early 1960s (Hirsch et al., 1960; Howie and Whelan, 1961, 1962; Takagi, 1962). More detailed studies, notably of dislocations, for which $\boldsymbol{R}(\boldsymbol{r})$ is not everywhere slowly varying, are to be found in the textbooks of Hirsch et al. (1965), Reimer (1993) and Buseck et al. (1988). Equation (69.88) is clearly not appropriate for such situations and the study of diffraction processes in crystals with severe imperfections is extremely complicated.

69.8 Image simulation

In this chapter, we have attempted to provide a concise introduction to the study of crystalline specimens by electron microscopy. Except for very thin and weakly scattering specimens, however, there is no direct relation that enables us to interpret the features of a structure image at high resolution in terms of the atoms in the lattice, the imperfections of which are usually of particular interest. It has therefore become usual to compare the observed image with a simulated image. The latter is calculated on the basis of the assumed structure, or of a range of possible structures, using all available information about the microscope parameters, illumination characteristics and scattering cross-sections.

Although several ways of performing such simulations have been thoroughly studied, using Bloch waves (Metherell, 1976), a real space formulation (van Dyck, 1980, 1983; van Dyck and Coene, 1984; Coene and van Dyck, 1984; see survey by van Dyck, 1985) and direct wave propagation through thin layers of material (Cowley and Moodie, 1957; van Dyck, 1985), most simulations in fact use this last technique, which is known as the *multislice method*. The principle is simple: the specimen is cut into a large number of thin slices, each of which is regarded as a two-dimensional phase object and the Fresnel approximation is used to calculate the wavefunction ψ^{n+1} at the entrance plane of the $(n + 1)$-th slice in terms of that at the exit plane of the preceding n-th slice. Constant factors apart, the wavefunction $\psi^{(n+1)}$ has the form of a convolution product, the first function being the specimen transmission $\exp(i\pi V/\lambda\hat{U})$ and the second the 'propagator' $\exp\{i\pi(x^2 + y^2)/\lambda\tau\}$ where τ is the slice thickness. The introduction of the FFT algorithm by Ishizuka and Uyeda (1977) increased the speed of the calculation, which is now used in a very wide range of applications including convergent-beam and STEM images and reflection electron microscopy. For many applications, see Krakow and O'Keefe (1989) and for an extremely clear presentation and full historical details, see van Dyck (1985) or more recently, Barry (1992). The 'combined basis algorithm' of Tochilin and Whelan (1993) offers a considerable improvement in the rate of convergence.

The multislice approach has recently been applied to wave propagation through electron lens fields by Pozzi. The paraxial theory is complete and work is in progress on extending this to include the effect of aberrations (Pozzi, 1989a,b, 1990, 1992; di Sebastiano and Pozzi, 1992).

Part XV

Digital Image Processing

70

Introduction

70.1 Organization of the subject

Not all the information present in an electron image can be directly appreciated by simple visual scrutiny, for a host of very different reasons. At the simplest extreme, the image may be too dark or too light for the eye to be able to discern fine detail. Or again, the information sought may be scattered over several different kinds of image, typically corresponding to the various signals generated by a scanning instrument. We may be interested not just in the appearance of the image but in measurements made on it: what is the area, or the total perimeter, of structures of a particular type, themselves characterized by size, or texture, or perhaps by the presence of a particular chemical element? At high resolution, we may need to *restore* the image, some detail of which has been lost or falsified in the image-forming process, characterized by an instrumental transfer function $T(q)$. In other circumstances, the quality of the raw image may be poor because the electron dose used was too small to generate a satisfactory image but could not be raised for fear of damaging the specimen. As a final example, we may be interested in combining a set of images for three-dimensional reconstruction.

All these tasks fall within the province of digital image processing, in which the image is first digitized and quantized and then manipulated in a computer. The subject may be conveniently divided into four broad sections: (i) image acquisition, sampling and coding; (ii) enhancement; (iii) restoration; (iv) analysis and pattern recognition. We shall discuss each of these in detail in this Part but we first say a few words about this classification and introduce the unifying notion of image algebra, with the aid of which the many different strands of image processing can be plaited together. We also introduce notation that will be used as consistently as possible throughout this Part of the book.

The first chapter, *acquisition, sampling and coding* (Chapter 71), is concerned with the steps that convert the raw image, or set of images, into an array of numbers suitable for the computer. The quasi-continuous image must be broken up into small picture-elements, or pixels(some authors speak of 'pels'), and the intensity of each must be measured and a numerical value attributed to each measurement. Several decisions must be taken here: how should we choose the size and shape of the pixel? When we measure the intensity, how many *grey levels* should be allowed between black and white and how should they be distributed? Finally, do we need to retain all the measured values or can we compress the volume of data in some way? This last question is prompted by the fact, that, without compression, a digitized image is typically very space-consuming: thus a square image consisting of 512×512 pixels, each characterized by an 8-bit number measuring the grey levels, occupies 2 Mbit. Moreover, 512×512 only represents about $50 \times 50 \ nm^2$ for pixels 1Å $\times$ 1Å.

In the second chapter, *enhancement* (Chapter 72), we present a range of techniques that improve the image but make little or no use of any knowledge we may have concerning the origin of such imperfections as may be apparent: in short, they are techniques that adapt the image to the visual response. Thus images that are very dark or very light or have an uneven background can be rendered easier to see without introducing any artefacts. Features of especial interest, such as small changes in contrast, can be highlighted. These simple aids to viewing are often extremely useful: some have been incorporated in SEMs for a number of years and are gradually finding their place in transmission microscopy as well.

The third field, *restoration* (Chapters 73–75), is of most direct interest in transmission electron microscopy, owing to the nature of the image-forming process at high resolution, to the desire for three-dimensional structural information and to the damage caused by bombarding fragile specimens with the number of electrons needed to form a clear image. In these chapters, we find methods of creating a directly interpretable image in which all available knowledge about the image-forming process is exploited: notably, but not exclusively, the instrumental transfer function and certain statistical properties of signal and noise.

The penultimate chapter, on *image analysis and pattern recognition* (Chapter 76), has a different purpose. Here, the aim is not to produce a clearer or more faithful image but to analyse features of it of particular interest: to identify structures, count them, measure some property of them, draw intensity contours, single out boundaries—we could give numerous other examples. Of the various methods that have been developed for performing individual tasks of this kind, many have now been incorporated into the subject known as *mathematical morphology*, which thus provides some degree of unification. Image *description* is also included in this chapter and a few words will be devoted to the syntactic approach to description.

We conclude Part XV with a chapter on instrumental applications of image processing. For some procedures, in restoration in particular, values of such instrumental parameters as the spherical aberration coefficient, the defocus, the astigmatism and the parameters characterizing the degree of coherence of the illumination must be estimated from the image itself. Ways of obtaining these estimates are described in Section 77.2. It is gradually becoming more common to use such measurements for automatic instrument control by means of direct feedback to the focus, alignment and stigmator controls of the microscope; some information and references on control are given in Section 77.3. In the STEM, the geometry of the detector is a free parameter that can be chosen to bring out specific features of the image—we discuss this and other aspects of image processing of particular interest for STEM in Section 77.4. Some techniques that fit into none of these categories are mentioned in Section 77.5.

At first sight, the material that comprises image processing appears to be highly diverse, with little in common apart from the fact that it all has something to do with images. One reason for this heterogeneity is that the methods for the most part involve operations on individual pixels or on small groups of pixels, whereas instinctively we should like to operate on whole images or even on image sets. In an attempt to remedy this, several *image algebras* have been developed, which differ in the nature of the images accepted (binary, with multiple grey levels, real, complex, matrix-valued, ...) and in the choice of basic elements (single-sorted, many-sorted*) but which all have in common the property of addressing entire images. We discuss such image algebras in some detail for, although the subject is young and in rapid evolution, it seems clear that such algebraic presentations will soon supplant the piecemeal description of image processing operations.

---

* These technical terms will be explained later but are not at all mysterious: in a single-sorted algebra, every array has the same status, so that masks for example or arrays of weights are all regarded as 'images'; in a many-sorted algebra, a distinction is made between images and other arrays, templates in particular, which include masks and other constructs.

70.2 Image algebra[*]

70.2.1 Introduction

Despite the stalwart efforts of the authors of the many textbooks on image processing to impose a pattern on their subject, the reader is all too likely to perceive it as a magpie collection of methods, gadgets, knacks and contrivances. One reason for this is that the various branches of the subject have been developed in very different fields, often with little awareness of one another, but a more fundamental cause is that most image processing operations have been described in terms not of whole images but of individual pixels or clusters of pixels. Attempts are being made to impose order on this chaotic, intellectually displeasing situation by the introduction of image algebras, in which the various image manipulations are expressed in terms of variables that are entire images and a small number of operators.

In the remainder of this chapter, we present the main ideas of image algebra and mention the differences between the algebras that are at present attracting most attention.

70.2.2 Images and templates

An algebra consists of *operands* (quantities on which we perform such operations as addition, exponentiation and more complicated functional operations) and *operators*, which allow us to manipulate the operands. The operands are typically *value sets* (the integers, $\mathbf{Z}$; real numbers excluding $\pm\infty$, $\mathbb{R}$; complex numbers, $\mathbf{C}$; binary numbers of length k, $\mathbf{Z}(2^k)$; and real numbers including $+\infty$ or $-\infty$, $\mathbb{R}_{+\infty}$ or $\mathbb{R}_{-\infty}$), *coordinate sets* and their elements, *images* and (perhaps) *templates*. Coordinate sets have their ordinary meaning, as subsets of n-dimensional Euclidean space $\mathbb{R}^n$. This brings us to the principal operand, the *image*.

Although the images that are manipulated by the user of image algebra are no different from the arrays (matrices) of pixel intensities that characterize any digitized image, it is also useful to have a definition in terms of value sets ($\mathbf{Z}$, $\mathbb{R}$, $\mathbf{C}$, etc.) and coordinate sets. For some value set $\boldsymbol{F}$ and coordinate set $\boldsymbol{X}$, we define an $\boldsymbol{F}$-valued image $\boldsymbol{a}$ on $\boldsymbol{X}$ to be the graph of a function $\boldsymbol{a} : \boldsymbol{X} \to \boldsymbol{F}$ so that

$$\boldsymbol{a} = \{(\boldsymbol{x}, \boldsymbol{a}(\boldsymbol{x})) | \boldsymbol{x} \in \boldsymbol{X}\} \tag{70.1}$$

where $\boldsymbol{a}(\boldsymbol{x}) \in \boldsymbol{F}$ and $\boldsymbol{x}(= x_1, x_2, \ldots)$ is an element of the coordinate set $\boldsymbol{X}$. In more everyday language, a typical pair $(\boldsymbol{x}, \boldsymbol{a}(\boldsymbol{x}))$ characterizes a pixel of

---

[*] This rather mathematical material is included in the belief that papers on image processing will soon be written on the assumption that their readers have some familiarity with image algebra. This section may be omitted, however, with almost no incidence on the remainder of Part XV.

the image a and, in particular, tells us that the grey level intensity at the pixel identified by its coordinates x is $a(x)$. All possible F-valued images on X form a set, which is denoted by F^X. In practice, the value set F will often be $Z(2^k)$, to which the 256 real grey levels that we regard as typical belong for $k = 8$; or C, when we are interested in both amplitude and phase, in coherent imagery, for example, or in the Fourier transform of an image. It is also useful to be able to work with sets of pixels simultaneously and, for this purpose, forms of algebra in which the elements of an image are themselves matrices are being developed (Wilson, 1989, 1990, 1991, 1992a, 1993). There are contexts in which it is convenient to regard the upper and lower grey levels of non-binary images as zero and one and in this case F is $\mathbb{R}$ (or perhaps Q, the set of rational numbers).

When a quasi-continuous image is digitized into an array of 512×512 pixels and the grey levels run from 0 to 255, the general F-valued image on X will be a real-valued image on (x, y), where x and y run in unit steps from 1 to 512; F is $Z(2^8)$. The large dynamic range of a diffraction pattern might oblige us to quantize the grey levels on 12 bits, 0—5095 (this range is adopted for the Fuji image plate, mentioned in Section 71.1). After Fourier transformation, the values in the array may well be complex and the 'F-valued image' will now be complex-valued. It is one of the strengths of a well-designed image algebra that the definitions of the operations involve the very general definition of an image but, in order to become familiar with this way of thinking, it is prudent to translate these general statements into the simpler forms corresponding to specific examples.

Of the operands, only *templates* remain to be discussed and we must begin with a warning. There are (at least) two large schools of thought in image algebra, one associated with Ritter and colleagues and the other with Dougherty and Giardina. The former are enthusiastic defenders of the notion of templates while the latter (see in particular Dougherty, 1989) regard them as superfluous and treat them on the same footing as images. We shall not attempt to compare the merits of these approaches but we essentially follow Ritter in the belief that, necessary or not, the use of templates often makes it easier to understand what is meant by the abstract reasoning. The masks and windows and other shapes that we use in image processing all belong to this category.

Formally, a template is defined in terms of a value set F (often Z) and *two* coordinate sets, which we denote Y and X, which may of course be the same or may be similar but different in scale; Y, for example, might be $y = (y_1, y_2)$ where y_1 and y_2 explore the square domain from $y_1, y_2 = 0$ to $y_1, y_2 = 63$ in unit steps whereas X is $x = (x_1, x_2)$, x_1 and x_2 running from zero to 1023. Again Y and X may be different, polar coordinates and cartesians for example. A generalized F-valued template t from the *target domain* Y of t to the *range space* X of t is a function $t : Y \rightarrow F^X$. In other words, applying t to y creates an F-valued image on X. It is convenient to use the notation t_y:

$$t_y := t(y) \tag{70.2}$$

and so

$$t(y) \equiv t_y = \{(x, t_y(x)) | x \in X\} \tag{70.3}$$

Thus a template is an image, the pixel values of which are themselves images. The pixel values $t_y(x)$ of the image t_y are known as the weights of the template t. By analogy with the notation F^X introduced above, we denote the set of all F-valued templates from Y to X by $(F^X)^Y$.

A certain amount of vocabulary is helpful in connection with these ideas. A template t that belongs to $(F^X)^Y$ may be invariant under translation; this will be the case if and only if, for every triplet $x^{(1)}, x^{(2)}, x^{(3)} \in X$, where $x^{(1)} + x^{(3)}$ and $x^{(2)} + x^{(3)} \in X$, it is true that $t_{x^{(2)}}(x^{(1)}) = t_{x^{(2)}+x^{(3)}}(x^{(1)} + x^{(3)})$.

A template may be *transposed*. If $t \in (F^X)^Y$, the transpose of t, denoted by t' and belonging to $(F^Y)^X$, is defined by

$$t'_x(y) := t_y(x) \tag{70.4}$$

Extended-real-valued templates have two kinds of *dual*. If $t \in (\mathbb{R}^X_{-\infty})^Y$ or $(\mathbb{R}^X_{+\infty})^Y$, then the *additive dual* of t is again a template, denoted by t^* and defined by

$$t^*_x(y) = -t_y(x) \tag{70.5}$$

where $t^* \in (\mathbb{R}^Y_\infty)^X$ or $(\mathbb{R}^Y_{-\infty})^X$. The *multiplicative dual* is likewise a template, denoted by $\bar{t}$ and defined by

$$\bar{t}_x(y) = \begin{cases} 1/t_y(x) & \text{if } t_y(x) \neq 0 \text{ and } \neq \infty \\ +\infty & \text{if } t_y(x) = 0 \\ 0 & \text{if } t_y(x) = +\infty \end{cases} \tag{70.6}$$

The permitted values of t and $\bar{t}$ are slightly more complicated here: $t \in (r^X)^Y$ and $\bar{t} \in (r^Y)^X$, where r is the set of non-negative real numbers including $+\infty$.

We have asserted that the notion of template will help us to manipulate the masks, windows and the like that occur frequently in image processing. This can be foreshadowed by considering a translation-invariant template with finite support, where the *support* S of a template is defined to be the set of values of x for which $t_y(x)$ does not vanish:

$$S(t_y) = \{x \in X | t_y(x) \neq 0\} \tag{70.7}$$

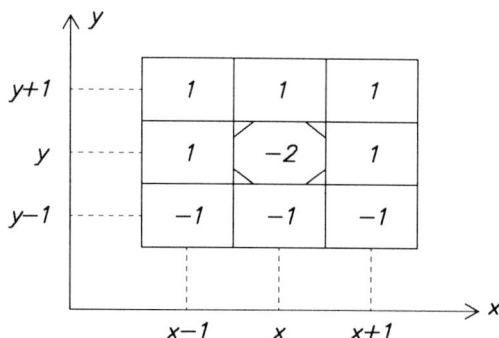

Fig. 70.1: Example of a shift-invariant template. In the shift-variant case, the pattern would not be the same for all target points y.

Suppose that the ordered pairs x belong to $\mathbf{Z} \times \mathbf{Z}$, that is, they are pairs of integers and hence form the nodes of a simple lattice. We define

$$\begin{aligned}
\boldsymbol{x}_1 &= (x-1, y+1) & \boldsymbol{x}_5 &= (x+1, y) \\
\boldsymbol{x}_2 &= (x, y+1) & \boldsymbol{x}_6 &= (x-1, y-1) \\
\boldsymbol{x}_3 &= (x+1, y+1) & \boldsymbol{x}_7 &= (x, y-1) \\
\boldsymbol{x}_4 &= (x-1, y) & \boldsymbol{x}_8 &= (x+1, y-1)
\end{aligned}$$

where $\boldsymbol{y} = (x, y)$ is an arbitrary point of $\boldsymbol{X}$. We must now complete the specification of the template by defining its weight for each point $\boldsymbol{y}$ belonging to $\boldsymbol{X}$. Looking ahead to (72.18), we choose

$$\begin{aligned}
& t_y(\boldsymbol{y}) = -2 \\
& t_y(\boldsymbol{x}_1) = t_y(\boldsymbol{x}_2) = t_y(\boldsymbol{x}_3) = t_y(\boldsymbol{x}_4) = t_y(\boldsymbol{x}_5) = 1 \\
& t_y(\boldsymbol{x}_6) = t_y(\boldsymbol{x}_7) = t_y(\boldsymbol{x}_8) = -1
\end{aligned}$$

(70.8a)

and

$$t_y(\boldsymbol{x}) = 0 \text{ if } \boldsymbol{x} \text{ is not } \boldsymbol{y} \text{ or any of the } \boldsymbol{x}_j \quad (j = 1\text{–}8)$$

(70.8b)

The support of t_y is thus

$$S(t_y) = \{\boldsymbol{y}, \boldsymbol{x}_j | j = 1 \ldots 8\}$$

(70.8c)

The template t can be pictured as in Fig. 70.1, in which the broken corners indicate the target point $\boldsymbol{y}$.

70.2.3 Operations

Section 70.2.2 contains definitions of the operands and, in particular, provides a framework for image description. We now list the *operations* that we include in the algebra. For real-valued images, addition, multiplication, choice of maximum are defined as usual. For images $a, b \in R^X$

$$a + b := \{(x, c(x))|c(x) = a(x) + b(x) \, , \, x \in X\} \qquad (70.9)$$
$$a \times b := \{(x, c(x))|c(x) = a(x)b(x) \, , \, x \in X\} \qquad (70.10)$$
$$a \vee b := \{(x, c(x))|c(x) = a(x) \vee b(x) \, , \, x \in X\} \qquad (70.11)$$

where the symbol $\vee$ between $a(x)$ and $b(x)$ means that the greater of the two is the result. Functions are defined in the same way,

$$f(a) = \{(x, c(x))|c(x) = f(a(x))\} \qquad (70.12)$$

with the exception of exponentiation and the logarithm, both of which require two quantities to be specified (e.g. the base of the logarithm and the operand); both of these may be images:

$$a^b := \left\{ (x, c(x)) \, \middle| \, \begin{array}{ll} c(x) = a(x)^{b(x)} & \text{if } a(x) \neq 0 \\ c(x) = 0 & \text{if } a(x) = 0 \end{array} \quad x \in X \right\} \qquad (70.13)$$

$$\log_a b := \left\{ (x, c(x))|c(x) = \log_{a(x)} b(x) \, , \quad x \in X \right\} \qquad (70.14)$$

In the work that we are following closely here (Ritter, 1991; Ritter *et al.*, 1990), these operations are confined to real quantities: a and b must be such that $a(x)^{b(x)} \in \mathbb{R}$ and in the logarithm, $a(x) > 0$ and $b(x) > 0$. Likewise, a and b in $a \vee b$ are required to be real. However, this reflects the types of image that the developers of these algebras had in mind and the extension into the complex domain has now begun. Subtraction, division and taking the minimum can be expressed in terms of the operators already defined but it is of course convenient to have symbols for them:

$$a - b := a + (-b) \quad , \quad -b = \{(x, -b(x))\} \qquad (70.15)$$
$$b/a := b \times a^{-1} \qquad (70.16)$$
$$a \wedge b := -(-a \vee -b) \qquad (70.17)$$

The reciprocal of a, denoted as usual by a^{-1}, is obtained from (70.13), in which the exponent b is a constant image, that is, an image all of whose pixels have the same grey level intensity g; to obtain the reciprocal, we set $g = -1$.

Before leaving these questions of elementary notation and vocabulary, we introduce three other terms, which are important by virtue of the fact that the number of pixels in an image and the number of grey levels are limited. In set theory, the *characteristic function* of a (measurable) set, S, denoted by χ_s, is in effect a label reminding us whether or not a quantity is a member of that set:

$$\chi_s(\boldsymbol{a}) = \left\{ (\boldsymbol{x}, \boldsymbol{c}(\boldsymbol{x})) \left| \begin{array}{l} \boldsymbol{c}(\boldsymbol{x}) = 1 \text{ if } \boldsymbol{a}(\boldsymbol{x}) \in S \\ \boldsymbol{c}(\boldsymbol{x}) = 0 \text{ otherwise} \end{array} \right. \right\} \tag{70.18}$$

The *domain* of an image $\boldsymbol{a}$ is the set over which $\boldsymbol{a}$ is defined and the *range* of $\boldsymbol{a}$ is the set of values that $\boldsymbol{a}$ may take.

70.2.4 Operations involving images and templates

We complete this summary of the ideas of image algebra with an account of the principal operations in which an image and a template are allowed to interact; these are to some extent analogous to the (linear) operations of convolution and cross-correlation and to such (nonlinear) operations as creation of a new function from two given functions by selecting the greater of their values at each point.

General definition

Consider an image $\boldsymbol{a}$ and a template $\boldsymbol{t}$, and value sets $\boldsymbol{F}_1$ and $\boldsymbol{F}_2$, with $\boldsymbol{a} \in \boldsymbol{F}_1^{\boldsymbol{X}}$ and $\boldsymbol{t} \in (\boldsymbol{F}_2^{\boldsymbol{X}})^{\boldsymbol{Y}}$. Let o be a binary operation, $o|\boldsymbol{F}_1 \times \boldsymbol{F}_2 \to \boldsymbol{F}$, such as addition or choice of maximum. Let γ be an associative binary operation on $\boldsymbol{F}$; we define the *generalized backward template operation* or *right product of $\boldsymbol{a}$ with $\boldsymbol{t}$*, denoted by $\oslash$, as follows:

$$\boldsymbol{a} \oslash \boldsymbol{t} := \{(\boldsymbol{y}, \boldsymbol{b}(\boldsymbol{y})) | \boldsymbol{b}(\boldsymbol{y}) = \Gamma_{\boldsymbol{x} \in \boldsymbol{X}} \boldsymbol{a}(\boldsymbol{x}) \, o \, \boldsymbol{t}_{\boldsymbol{y}}(\boldsymbol{x}), \boldsymbol{y} \in \boldsymbol{Y}\} \tag{70.19}$$

in which the 'global reduction' Γ on $\boldsymbol{F}^{\boldsymbol{X}}$ induced by γ (i.e. associated with γ) is obtained by forming the combination $\boldsymbol{a}(\boldsymbol{x}_1)\gamma\boldsymbol{a}(\boldsymbol{x}_2)\ldots\gamma\boldsymbol{a}(\boldsymbol{x}_m)$:

$$\Gamma\boldsymbol{a} = \Gamma_{\boldsymbol{x} \in \boldsymbol{X}} \boldsymbol{a}(\boldsymbol{x}) := \boldsymbol{a}(\boldsymbol{x}_1)\gamma\boldsymbol{a}(\boldsymbol{x}_2)\ldots\boldsymbol{a}(\boldsymbol{x}_m), \boldsymbol{a} \in \boldsymbol{F}^{\boldsymbol{X}} \tag{70.20}$$

$(\Gamma : \boldsymbol{F}^{\boldsymbol{X}} \to \boldsymbol{F})$.

In the same way, a *generalized forward template operation* or *left product of $\boldsymbol{a}$ with $\boldsymbol{t}$* may be defined, assuming now that $\boldsymbol{t} \in (\boldsymbol{F}_2^{\boldsymbol{Y}})^{\boldsymbol{X}}$:

$$\boldsymbol{t} \oslash \boldsymbol{a} := \{(\boldsymbol{y}, \boldsymbol{b}(\boldsymbol{y})) | \boldsymbol{b}(\boldsymbol{y}) = \Gamma_{\boldsymbol{x} \in \boldsymbol{X}} \boldsymbol{a}(\boldsymbol{x}) \, o \, \boldsymbol{t}_{\boldsymbol{x}}(\boldsymbol{y}), \boldsymbol{y} \in \boldsymbol{Y}\} \tag{70.21}$$

Special cases

In practice, Ritter *et al.* (1990) have found that extremely few binary operations γ are needed to express the image processing operations currently in use; three basic operations are very common and it is convenient to have symbols for them and for their duals, even though the latter are defined in terms of the former.

    Generalized convolution is generated by choosing γ to be 'plus' $(+)$ and o to be a product. Thus the Γ of (70.19–21) becomes the familiar summation $\sum$. Generalized convolution is hence denoted by $\oplus$ and we have

$$a \oplus t := \{(y, b(y)) | b(y) = \sum_{x \in X} a(x) t_y(x), y \in Y\} \tag{70.22}$$

$(a \in F^X$ and $t \in (F^X)^Y)$ and similarly

$$t \oplus a := \{(y, b(y)) | b(y) = \sum_{x \in X} a(x) t_x(y), y \in Y\} \tag{70.23}$$

$(t \in (F^Y)^X)$. We note that the set F may be $\mathbb{C}$ or $\mathbb{R}$ so that the image (or template) may be complex, an important feature of the definition for electron image processing.

    The two other operations commonly needed are *additive maximum* (and its dual, *additive minimum*) and *multiplicative maximum* (and its dual, *multiplicative minimum*). The *additive maximum* is denoted by $\boxdot$ and defined by

$$a \boxdot t := \{(y, b(y)) | b(y) = \bigvee_{x \in X} a(x) + t_y(x), y \in Y\} \tag{70.24}$$

where $a \in \mathbb{R}_{-\infty}^X$, $t \in (\mathbb{R}_{-\infty}^X)^Y$ and the symbol $\vee$ signifies 'take the maximum value of' (just as $\sum$ represents 'take the sum of'):

$$\bigvee_{x \in X} a(x) + t_y(x) = \max\{a(x) + t_y(x) | x \in X\} \tag{70.25}$$

Likewise

$$t \boxdot a := \{(y, b(y)) | b(y) = \bigvee_{x \in X} a(x) + t_x(y), y \in Y\} \tag{70.26}$$

Additive minimum, denoted by $\boxslash$ is defined in terms of the additive dual t^* (70.5):

$$a \boxslash t := (t^* \boxdot a^*)^*$$

The *multiplicative maximum* is denoted by $\otimes$ and defined by

$$a \otimes t := \{(y, b(y)) | b(y) = \bigvee_{x \in X} a(x) t_y(x), y \in Y\} \tag{70.27}$$

where $a \in (\mathbb{R}_{-\infty}^{\geq 0})^X$ and $t \in ((\mathbb{R}_{+\infty}^{\geq 0})^X)^Y$ or by

$$t \otimes a := \{(y, b(y)) | b(y) = \bigvee_{x \in X} a(x) t_x(y), y \in Y\} \tag{70.28}$$

where now $t \in ((\mathbb{R}_{+\infty}^{\geq 0})^Y)^X$; the dual multiplicative minimum $\oslash$ is defined by

$$a \oslash t := \overline{(\overline{t} \otimes \overline{a})} \tag{70.29}$$

where the multiplicative dual $\overline{t}$ is defined by (70.6).

We conclude this rather abstract introduction to the ideas of image algebra with a comment on the support of the generalized convolution, that is, the zone over which it is non-zero. By definition, $t_y(x)$ is zero whenever $x \notin S(t_y)$, $S(t_y)$ being the support of t_y (70.7); summation or maximization of $a(x) t_y(x)$ for all $x \in X$ is thus identical with the same operations for all $x \in S(t_y)$. The support $S(t_y)$ is thus closely analogous to a window or a mask. A similar analogy may be drawn between $S_{-\infty}(t_y)$ and a morphological structuring element, where

$$S_{-\infty}(t_y) = \{x \in X | t_y(x) \neq -\infty\} \tag{70.30}$$

70.2.5 Concluding remarks

This section is strongly influenced by the work of the school of G.X. Ritter. The distinction between Ritter's many-sorted algebra ('many-sorted' in the sense that images and templates are regarded as different in nature) and Dougherty's homogeneous algebra (homogeneous is synonymous with single-sorted: images and templates are regarded as members of the same family) is not, however, very serious when we confine our interest to the more straightforward operations, and in particular to convolutional filters. Newcomers to the subject may well find Dougherty's presentation of his homogeneous image algebra (Dougherty, 1989) easier to take in than the papers of Ritter and colleagues (Ritter *et al.*, 1990; Ritter, 1991; Davidson, 1992a), though it is essential to study the latter once the basic ideas have been grasped. In the following chapters, we have adopted the traditional divisions of image processing but we give the algebraic representation of many of the procedures described.

70.3 Notation

Many of the image processing routines to be presented in the following chapters have as starting point one or other of the relations between object and image wavefunction or between object wavefunction and image intensity derived in Chapter 65. There, the conventional notation for wavefunctions was employed but this is no longer so convenient when we need to be able to add multiple suffixes and other labels; we therefore introduce a parallel notation, closer to that often encountered in the literature of image processing. For convenience, we repeat the key equations of Chapters 65 and 66 here, with the alternative notation and equivalences.

(i) Object and image wavefunctions, general case

$$\psi_i(x_i, y_i) = \iint G(x_i, y_i; x_o, y_o)\psi_o(x_o, y_o)\, dx_o\, dy_o \qquad (70.31)$$

$$g(x_i, y_i) = \iint h(x_i, y_i; x_o, y_o)f(x_o, y_o)\, dx_o\, dy_o \qquad (70.32)$$

We assume for convenience that the discrete situation is generated by sampling the functions at regular intervals on a square lattice. The use of a rectangular lattice introduces very little complication. There are good arguments for considering seriously the use of a hexagonal pattern in some circumstances but these are not examined here; see Petersen and Middleton (1962), Mersereau (1979), Golay (1969), Koenderink and van Doorn (1979), Serra and Laÿ (1985), Bell et al.. (1989) and cf. Hawkes (1982a). Equation (70.32) then becomes

$$g_{ij} = \sum_k \sum_l h_{ijkl} f_{kl} \qquad (70.33)$$

in which the suffixes of the arrays $\{g_{ij}\}$ and $\{f_{ij}\}$ correspond to the lattice points

$$x^{(i)} = i\Delta x \quad \text{and} \quad y^{(j)} = j\Delta y \qquad (70.34)$$

where $\Delta x = \Delta y$ is the lattice spacing and the integers i, j run from 0 to $N - 1$. It may be convenient to replace the tensor-matrix products of (70.33) by a matrix–vector product, which can be done by introducing the stacking operator S (Bellman, 1970, at p. 245). Consider an $N \times N$ matrix G with columns $g^{(i)}(i = 0, 1, 2 \ldots N - 1)$. Then $\mathsf{S}(G)$ generates the vector $\boldsymbol{g}$ obtained by stacking the columns of G above one another:

$$\boldsymbol{g} = \mathsf{S}(G) := \begin{pmatrix} g^{(0)} \\ g^{(1)} \\ \vdots \\ g^{(N-1)} \end{pmatrix} \qquad (70.35)$$

or in other terms

$$g_p = G_{ij} \quad \text{with} \quad p = i + jN \tag{70.36}$$

If we apply the stacking operator to a product of the form (70.33), we find that

$$S(HF) = S(H)\,S(F)$$

in which H denotes $\{h_{ijkl}\}$ and F denotes $\{f_{ij}\}$. $S(F)$ is the stacked form of F as defined above and $S(H)$ is an $N^2 \times N^2$ matrix h with elements h_{pq} obtained from h_{ijkl} as follows:

$$\left.\begin{array}{l} \text{for} \quad p = i + jN \\ \qquad q = k + lN \\ \qquad p, q = 0, 1, 2 \ldots N^2 - 1 \end{array}\right\} h_{pq} := h_{ijkl} \tag{70.37}$$

(ii) Object and image wavefunctions, isoplanatic case

$$\psi(x_i, y_i) = \iint G(x_i - x_o, y_i - y_o)\psi_o(x_o, y_o)\, dx_o\, dy_o \tag{70.38}$$

$$g(x_i, y_i) = \iint h(x_i - x_o, y_i - y_o)\psi_o(x_o, y_o)\, dx_o\, dy_o \tag{70.39}$$

$$g_{ij} = \sum_k \sum_l h_{ijkl} f_{kl} \tag{70.40}$$

but now only N^2 of the elements h_{ijkl} are independent and we have

$$h_{ijkl} = \kappa_{pq} \begin{cases} p = i - k \\ q = j - l \end{cases} \tag{70.41}$$

If $0 \le p,\, q \le N - 1$, then $\kappa_{pq} = h_{p+k,q+l,k,l}$; otherwise $\kappa_{pq} = 0$. Alternatively, we may introduce the arrays c_{ijk},

$$c_{ijk} := \delta_{i-j,k} \tag{70.42}$$

whereupon

$$h_{ijkl} = \sum_r \sum_s \kappa_{rs} c_{irk} c_{jsl} \tag{70.43}$$

(iii) Object transparency and image intensity, weakly scattering specimen

$$S_c(\boldsymbol{q}) = K_p(\boldsymbol{q})\tilde{\eta}(\boldsymbol{q}) + K_a(\boldsymbol{q})\tilde{\sigma}(\boldsymbol{q}) \tag{70.44}$$

$$G_{ij} = K_{ij}^{(p)}\tilde{\eta}_{ij} + K_{ij}^{(a)}\tilde{\sigma}_{ij} \tag{70.45}$$

(iv) Diffraction plane wavefunction and image plane wavefunction

$$\psi_i(x_i, y_i) = \mathcal{F}(\psi_d(q_x, q_y))$$

$$= \iint \psi_d(q_x, q_y) \exp(2\pi i \mathbf{q} \cdot \mathbf{x}_i) \, dq_x \, dq_y \qquad (70.46)$$

$$g_{ij} = \sum_k \sum_l F_{ik} d_{kl} F_{lj} \qquad (70.47)$$

where

$$F_{jk} := N^{-\frac{1}{2}} W_n^{jk} \quad , \quad W_n := \exp(2\pi i/N) \qquad (70.48)$$

with inverse

$$d_{ij} = \sum_k \sum_l F_{ik}^{-1} g_{kl} F_{lj}^{-1} \qquad (70.49)$$

and of course

$$F_{jk}^{-1} = N^{-\frac{1}{2}} W_n^{-jk} \qquad (70.50)$$

71
Acquisition, Sampling and Coding

71.1 Acquisition

The electron current density distribution at the detector plane of an electron microscope may be 'acquired' in various ways, depending both on the type of microscope and on the accessories with which it is equipped. In the case of microscopes in which the image is formed sequentially, the SEM and the STEM, one or more signals are collected from each area illuminated by the electron probe that scans the specimen and these analogue signals may be sampled and quantized immediately and subsequently stored in computer memory. The digitized image thus created is immediately suitable for computer processing, though in practice the image generated by a single fast scan may well be of poor quality owing to the paucity of electrons per pixel. We return to this below but mention in passing that the standard and obvious remedy is to superimpose the signals from several successive frames, thereby accentuating the useful signal at the expense of the noise.

In the case of the TEM, the whole image is formed simultaneously and some further step is needed to dissect the continuously varying intensity into picture elements and to measure the mean intensity at each of these. In the past, it was necessary to record the image on a photographic emulsion and measure the opacity of the developed film with a microdensitometer, an expensive device with which few laboratories were equipped: digital image processing was a rare luxury. Microdensitometry was, moreover, a slow and inconvenient step and is gradually being supplanted by the direct methods of image readout now available. A television camera is situated behind the detector unit of the microscope and, together with a sampling, quantizing unit, sends a digitized image to computer memory. We shall not describe this microscope–computer link in any more detail here for not only is such a description beyond the scope of the book but the subject is also in a phase of such rapid evolution that the technical details would soon be obsolete. We simply mention that the requirements of the various parts of such systems have been examined in great detail by Herrmann and Krahl (1984); for more recent information, see Saxton and Chang (1988),

Herrmann (1990), Baumeister and Herrmann (1990), Krivanek *et al.* (1991) and Tietz (1992).

Commercial solutions aimed specifically at the electron microscopy community are offered by Tietz Video and Image Processing Systems (Tietz, 1990), Synoptics and Gatan (Fan and Krivanek, 1990; Krivanek and Fan, 1991).

A very recent alternative to the image-readout systems based on the traditional television camera is the Fuji image plate (TEM–IP). This is a flexible recording medium consisting of a phosphor layer and a protective film on a support. Electron impact deposits energy in the phosphor, which can be released as photostimulated luminescence by irradiating the image plate with light. In practice, the image plate is scanned with a laser beam and the resulting luminescence is guided to a photomultiplier by a light-pipe. The pixel size in the reader described by Mori *et al.* (1989) can be as small as 25×25 μm^2 and the grey levels are quantized on 12 bits (4096 levels). The sensitivity is high and the dynamic range over which the response is linear is very wide. For further details, see Hayakawa *et al.* (1987); Oikawa *et al.* (1989,a,b, 1990, 1991, 1992), Mori *et al.* (1988, 1989), Isoda *et al.* (1990, 1992), Shindo *et al.* (1990, 1991, 1992) and Burmester and Schröder (1992).

71.2 Sampling

71.2.1 The sampling theorem
The size of the picture element is intimately related to the resolution attainable. This must be interpreted in two ways: in cases in which we have no control over the pixel size once the image has been recorded—in the SEM, for example, where the probe diameter effectively defines the pixel—it must be realized that the useful resolution has already been determined; when, however, the pixel size can be chosen freely (within certain limits), in the microdensitometer, for example, where the window may be as small as $2.5\mu m$ square, the choice must be small enough to permit all genuine image detail to survive though not so small that random detail is retained (the grain-size of photographic emulsions is an obvious limit).

These qualitative remarks can be made quantitative with the aid of the *sampling theorem*. This theorem emerges in reply to the following question. Consider a two-dimensional continuous function $f(x, y)$, which we sample at a grid of points distance a apart (Fig. 71.1). In what circumstances are these sample values sufficient to reconstruct the original continuous function exactly? We shall also consider the practical extension of this procedure, in which the sampling consists not of hypothetical measure-

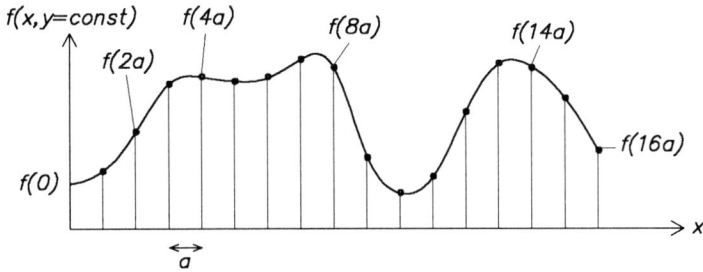

Fig. 71.1: Sampling: replacement of a continuous function $f(x,y)$ by an array of sample values. The one-dimensional example shown here may be regarded as the result of sampling for a single value of y.

ments at points (δ-function or comb sampling) but of averaging over small squares (pixels), which corresponds more closely to real analogue-to-digital conversion.

We require the reconstruction from the sample values to be linear and enquire whether we can find an interpolation function $g(x,y)$ such that the following equation is exactly true:

$$f(x,y) = \sum_{j=-\infty}^{\infty} \sum_{k=-\infty}^{\infty} f(ja,ka)g(x-ja,y-ka) \qquad (71.1)$$

On using the identity

$$f(\xi',\eta')g(x-\xi',y-\eta') \equiv \iint f(\xi,\eta)g(x-\xi,y-\eta)\delta(\xi-\xi',\eta-\eta')\,d\xi\,d\eta \qquad (71.2)$$

(71.1) becomes

$$f(x,y) = \iint f(\xi,\eta)g(x-\xi,y-\eta)\sum_{j}\sum_{k}\delta(\xi-ja,\eta-ka)\,d\xi\,d\eta \quad (71.3)$$

The delta-function after the double summation can be written as a Fourier series; it is easy to show that all the coefficients of the latter are equal to $1/a^2$ and hence that

$$\sum_{j}\sum_{k}\delta(\xi-ja,\eta-ka) = \frac{1}{a^2}\sum_{m}\sum_{n}\exp\left\{\frac{2\pi i}{a}(m\xi+n\eta)\right\} \qquad (71.4)$$

so that (71.1) is equivalent to

$$f(x,y) = \sum_{m}\sum_{n}\iint f(\xi,\eta)\exp\left\{\frac{2\pi i}{a}(m\xi+n\eta)\right\}\frac{g(x-\xi,y-\eta)}{a^2}\,d\xi\,d\eta \qquad (71.5)$$

The right-hand side has the form of a convolution and the Fourier transform of (71.5) is therefore

$$F(p,q) = \frac{G(p,q)}{a^2} \sum_m \sum_n F(p - \frac{m}{a}, q - \frac{n}{a}) \qquad (71.6)$$

in which

$$F(p,q) := \mathcal{F}^{-1}(f) = \iint f(x,y) \exp\left\{-2\pi i(px + qy)\right\} dx\, dy \qquad (71.7)$$

and likewise for $G(p,q)$. If, therefore, $f(x,y)$ can be written as the sum (71.1), then its Fourier transform $F(p,q)$ must be expressible in the form (71.6). The summation in (71.6) involves only *shifted* versions of $F(p,q)$ so that, if the latter is of finite support, in the sense that

$$F(p,q) = 0 \text{ for } |p| \geq q_0 , |q| \geq q_0 \qquad (71.8)$$

we can clearly satisfy (71.6) by choosing the distance a so small that the shifted versions do not overlap (Fig. 71.2):

$$a \leq \frac{1}{2q_0} \qquad (71.9)$$

By setting

$$G(p,q) = \begin{cases} a^2 & |p| < q_0, |q| < q_0 \\ 0 & \text{elsewhere} \end{cases} \qquad (71.10)$$

(71.6) becomes identically true. The sampling function $g(x,y)$, the transform of a rectangular function, is thus a sinc function:

$$g(x,y) = \frac{\sin(2q_0 \pi x) \sin(2q_0 \pi y)}{4q_0^2 \pi^2 xy}$$
$$= \text{sinc}(2q_0 x)\, \text{sinc}(2q_0 y) \qquad (71.11)$$

where[*]

$$\text{sinc}\, x := \frac{\sin(\pi x)}{\pi x} \qquad (71.12)$$

(Woodward, 1953).

The foregoing analysis tells us that if an image is bandlimited in the sense that its Fourier transform contains no useful information beyond the

---

[*] The more common definition of sinc x is that of (71.12) but sinc $x = \sin x / x$ is also in use. Some authors even use both, presumably inadvertently.

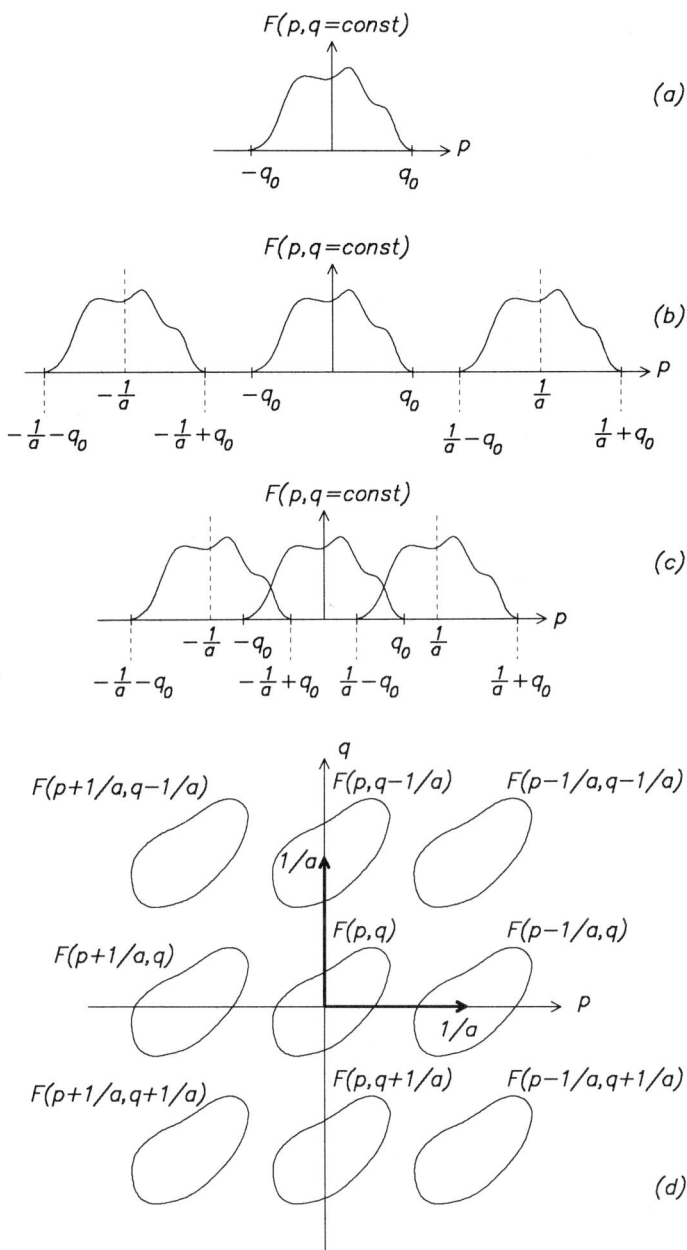

Fig. 71.2: Choice of sampling interval a. (a) $F(p, q = \text{const})$, bandlimited case; (b) shifted copies of $F(p,q)$, $a < 1/2q_0$; (c) shifted copies of $F(p,q)$, a chosen too large; (d) two-dimensional version of (b).

spatial frequencies $p, q = q_0$, it can be reconstructed *exactly* from a lattice of point samples, distance a apart, by means of the interpolation formula (71.1) with $a \leq 1/2q_0$:

$$f(x,y) = \sum_j \sum_k f(ja, ka) \operatorname{sinc}\left(\frac{x}{a} - j\right) \operatorname{sinc}\left(\frac{y}{a} - k\right)$$

$$= \sum_j \sum_k f\left(\frac{j}{2q_0}, \frac{k}{2q_0}\right) \operatorname{sinc}(2q_0 x - j) \operatorname{sinc}(2q_0 y - k) \tag{71.1'}$$

This is the two-dimensional form of the (Whittaker–Shannon–Kotel'nikov) sampling theorem. The distances $1/2q_0$ are sometimes known as the Nyquist intervals. If the condition (71.8) is not satisfied, spurious features will appear in the reconstruction (aliasing).

In practice, the image is more likely to be sampled by averaging over a small area, square or circular, than by a δ-function or even a close approximation to the latter. Thus a microdensitometer typically measures the mean transmittance of the photographic film through a small square window. Likewise, reconstruction is in practice more likely to be achieved by multiplying the sample values by a square uniform window function: filmwriters, for example, illuminate small areas (typically square) of film with a light intensity proportional to each sample value. How can we model this situation? Let us define a set of orthogonal normal functions $\{\varphi_{mn}(x, y)\}$ such that

$$\varphi_{mn}(x,y) = \begin{cases} \dfrac{N^2}{A^2} & \text{for} & \begin{aligned} \frac{mA}{N} \leq x \leq \frac{(m+1)A}{N} \\[2mm] \frac{nA}{N} \leq y \leq \frac{(n+1)A}{N} \end{aligned} \\[4mm] 0 & \text{elsewhere} \end{cases} \tag{71.13}$$

(for $m = 0, 1 \ldots N - 1$, $n = 0, 1 \ldots N - 1$)

If we picture the image area as a chessboard with $N \times N$ squares, total area A^2, each function takes the constant value N^2/A^2 in one of the squares.

We now enquire how we can approximate a continuous function $f(x, y)$ defined on the chessboard $\mathcal{A}$ by an expression of the form

$$\hat{f}(x,y) = \sum_{m=0}^{N-1} \sum_{n=0}^{N-1} a_{mn} \varphi_{mn}(x,y) \tag{71.14}$$

as accurately as possible. A reasonable measure of the accuracy is the mean-squares error $\iint_{\mathcal{A}} |f - \hat{f}|^2 \, dx \, dy$ and it is not difficult to show that this error is smallest when the coefficients a_{mn} are chosen such that

$$a_{mn} = \iint f(x,y)\varphi_{mn}(x,y)\,dxdy \qquad (71.15)$$

We thus have the reassuring result that if a continuous image is sampled by dividing it up into small squares and measuring the mean transmittance through each, the best way of reconstructing it is by illuminating small squares with an intensity proportional to the corresponding transmittance. We might say that replacement of the δ-function spikes by small squares has simultaneously transformed the inconvenient sinc functions (which have positive and negative parts) into small squares. This is of course a crude picture, for in the case of the small squares the error involved has only been minimized, not eliminated.

The sampling theorem continues to attract attention. For sampling schemes based on polar and other coordinates, see the work of Jerri (1977) and for a study of errors arising from imperfect bandlimitation, see Stens (1983). A major study of sampling for morphological processing has been made by Haralick *et al.* (1989). For a thorough study of sampling see the recent books of Marks (1991, 1993).

71.2.2 Degrees of freedom

The conditions of the sampling theorem do not conform to reality: in practice, all the transverse dimensions of an imaging system are not merely finite but are not even very different in size, neglecting scaling factors. The objective aperture in an electron microscope is typically a few tens of micrometres in diameter and the specimen area illuminated at a magnification of 100 000, a few micrometres across. The functions that we are manipulating are to all intents and purposes bandlimited in both real and Fourier space, which is not allowed! This inconsistency is even more disconcerting when we come to consider the nonlinear algorithms that iterate between image and diffraction plane on the assumption that a Fourier transform relates the wavefunctions in these planes.

The incongruity of studying finite signals in two domains related by a Fourier transform is not of course limited to imaging systems and is resolved by recognizing that a finite number of effective *degrees of freedom* is associated with such systems. This notion, introduced by Toraldo di Francia (1951, 1955, 1956a,b, 1969), is most clearly expressed with the aid of the prolate spheroidal functions, studied at length by Slepian and colleagues (Slepian, 1964, 1965, 1976, 1978; Slepian and Pollack, 1961; Landau and Pollack, 1961, 1962; Slepian and Sonnenblick, 1965). Its ramifications have been explored by Gori and colleagues (Gori and Guattari, 1973, 1974, 1975; Gori, 1974; Gori *et al.*, 1975; Bendinelli *et al.*, 1974) and by Saleh (1977).

An extremely clear account of the role of the prolate spheroidal functions in optics, in which related applications are also described, has been prepared by Frieden (1971). We present the idea in its simplest form, referring to the above-mentioned papers, especially those of Gori and Guattari, Bendinelli *et al.* and the Frieden review, for more thorough treatment of the subject.

We consider an image-forming system in which the image and object are related as usual by

$$g(x_i) = \int_{\mathcal{D}} h(x_i, x_o) f(x_o) \, dx_o \tag{71.16}$$

and the imaging is bandlimited: the spectrum of $g(x_i)$ is of finite support. We consider the one-dimensional case, which is sufficient to illustrate the principles. The domain of integration is likewise finite in reality, $\mathcal{D} = \{x_o \| |x_o| \le X_o\}$. The sampling theorem enables us to expand $g(x_i)$ in terms of its samples (71.1'):

$$g(x_i) = \sum_{j=-\infty}^{\infty} g(ja) \operatorname{sinc}(x/a - j) \tag{71.17}$$

in which $q_0 = \pm 1/2a$ are the extreme points of the spectrum. If the aperture in the diffraction plane is such that $|x_d| \le X_d$, then $q_0 = X_d/\lambda f$, $a = \lambda f/2X_d$, where f is the focal length. The range of values of the image coordinate x_i at which the intensity is actually recorded will be limited, $|x_i| \le X_i$; we might hence define the *number of degrees of freedom*, S, as the ratio of the image width $2X_i$ to the number of sampling points in this zone:

$$S := \frac{2X_i}{a} = 4q_0 X_i =: \frac{2b}{\pi}$$
$$b := 2\pi q_0 X_i = 2\pi \frac{X_d X_i}{\lambda f} \tag{71.18}$$

The quantity b is thus a (dimensionless) space–bandwidth product. Toraldo di Francia has called the number of degrees of freedom S the *Shannon number*.

In the simplest case, with all scaling factors omitted, the function $h(x_i, x_o)$ in (71.16) has the form

$$h(x_i, x_o) = \frac{\sin 2\pi q_0 (x_i - x_o)}{\pi (x_i - x_o)} \tag{71.19}$$

What are the eigenfunctions and eigenvalues of $h(x_i, x_o)$, regarded as an integral kernel? The solutions to

$$\int_{-X_o}^{X_o} \frac{\sin\{2\pi q_0 (x_i - x_o)\}}{\pi (x_i - x_o)} \psi_n(x_o) \, dx_o = \lambda_n \psi_n(x_i) \tag{71.20}$$

are the (linear) prolate spheroidal functions, defined by

$$\int_{-X_o}^{X_o} \exp(2\pi i q x)\psi_n(x)\, dx = i^n \left(\frac{\lambda_n X_o}{q_0}\right)\psi_n\left(\frac{q X_o}{q_0}\right) \tag{71.21}$$

and satisfying the remarkable double orthogonality:

$$\int_{-X_o}^{X_o} \psi_m \psi_n\, dx = \lambda_n \delta_{mn} \quad , \quad \int_{-\infty}^{\infty} \psi_m \psi_n\, dx = \delta_{mn} \tag{71.22}$$

so that

$$\int_{-\infty}^{\infty} \psi_n(x)\exp(-2\pi i q x)\, dx = \begin{cases} 0 & |q| > q_0 \\ \dfrac{1}{i^n}\left(\dfrac{X_o}{q_0 \lambda_n}\right)^{\frac{1}{2}}\psi_n\left(\dfrac{X_o q}{q_0}\right) & |q| \le q_0 \end{cases} \tag{71.23}$$

The eigenvalues have an important property, which is the reason why they are of interest in the present context: for a given value of the space–bandwidth product b, the λ_n remain close to unity for $n < 2b/\pi$ and fall abruptly to values close to zero for $n > 2b/\pi$ (Fig. 71.3).

Since the ψ_n are complete and orthonormal, we can expand $g(x_i)$ and $f(x_o)$ as series:

$$f(x_o) = \sum_l a_l \psi_l(x_o) \text{ with } a_l = \int f(x_o)\psi_l^*(x_o)\, dx_o$$
$$g(x_i) = \sum_l b_l \psi_l(x_i) \text{ with } b_l = \int g(x_i)\psi_l^*(x_i)\, dx_i \tag{71.24}$$

Substituting these expansions into (71.16) and recalling (71.22), we see immediately that

$$\sum_l b_l \psi_l(x_i) = \int h(x_i, x_o)\sum_l a_l \psi_l(x_o)\, dx_o = \sum_l a_l \lambda_l \psi_l \tag{71.25}$$

and hence

$$b_l = \lambda_l a_l \quad , \quad l = 0, 1, 2\ldots \tag{71.26}$$

But the λ_l are very small for all $l > 2b/\pi$: information contained in the object coefficients a_l beyond this cutoff value does not reach the image. We thus have an explanation of the paradox with which we began, namely, that

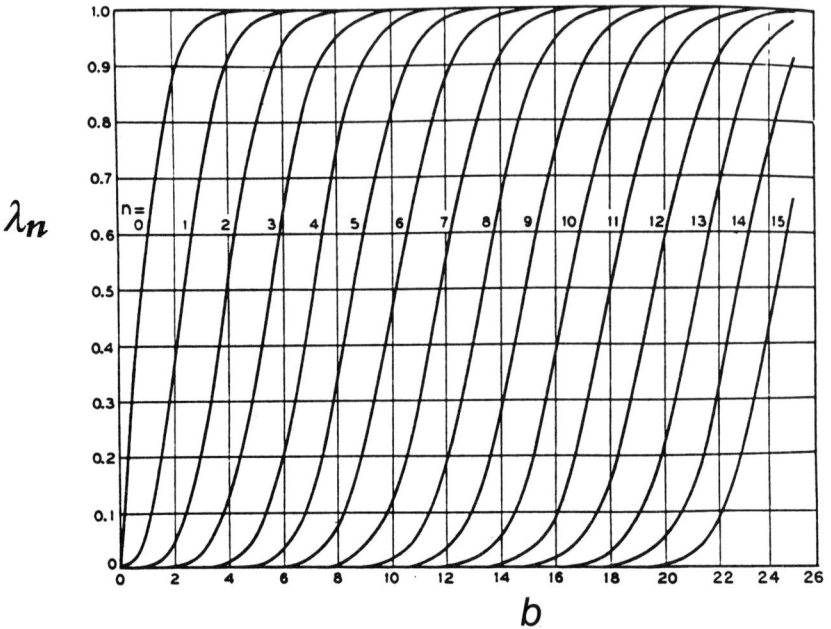

Fig. 71.3: Dependence of the eigenvalues λ_n of the prolate spheroidal functions on the space–bandwidth product b.

finite spectrum and finite image (or object) can in practice coexist. It can be shown that the number of effective or 'useful' degrees of freedom for coherent imagery is unaffected by noise (Toraldo di Francia, 1969; Bendinelli *et al.*, 1974) and virtually unaltered by lens aberrations (Gori *et al.*, 1975).

For an interesting relation between space–bandwidth products and wavelets, see Daubechies (1990).

71.3 Quantization

The intensity of each sample, which is a measure of the number of electrons incident on the corresponding pixel in the case of a direct electron image or of some specimen-related property in the case of many SEM signals, is quasi-continuous. In some situations, where electrons or photons are counted directly, the intensity may be discrete but in most cases the continuous variation of intensity must be *quantized* into a number of levels

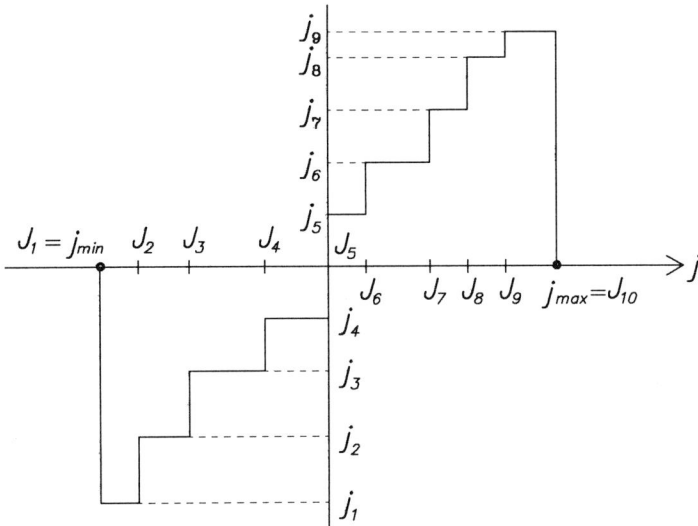

Fig. 71.4: Quantization. The quasi-continuous values j are mapped into N discrete values, j_i. The input values lie on the abscissa and the output values are given by the ordinate, so that a range of input values correspond to single output value. The steps need not be equal in height or width.

that is large enough to reflect any significant variations in intensity but not so large as to be unmanageable. In practice, for images at least, this number is often 256, for the technical reason that $256 = 2^8$ and computers are designed to work with words that are eight (or a multiple of eight) bits long.

Formally, we may say that a quantizer maps the quasi-continuous intensity values j into a set $\hat{j}$ of N discrete values $\{j_i | i = 1 \ldots N\}$. If j_{max} and j_{min} are the largest and smallest values of the intensity, then we define a set of steps or transition values $\{J_k | k = 1 \ldots N+1\}$ where $J_1 = j_{min}$ and $J_{N+1} = j_{max}$ (Fig. 71.4). All values of the (continuous) variable j lying in the range between J_k and J_{k+1} will be ascribed to the quantized value j_k.

In the simplest situation, the quantization is uniform; for 256 grey levels, we should have

$$J_k = \frac{j_{max} - j_{min}}{256}(k-1) \qquad k = 1, \ldots, 257$$
$$j_k = J_k \frac{0.5(j_{max} - j_{min})}{256} \qquad k = 1, \ldots, 256$$

$$(71.27)$$

If, however, we had some idea of the probability distribution $p(j)$ that described the likelihood of occurrence of the intensity values, we could improve the quantization by minimizing the mean-squares error between the continuous and discrete values,

$$
\begin{aligned}
\mathcal{E}(j - \hat{j})^2 &= \int_{J_1}^{J_{N+1}} (j - \hat{j})^2 p(j)\, dj \\
&= \sum_{k=1}^{N} \int_{J_k}^{J_{k+1}} (j - j_k)^2 p(j)\, dj
\end{aligned}
\tag{71.28}
$$

The minimum is found by setting the derivatives of this quantity with respect to J_k and j_k equal to zero, which gives

$$
\begin{aligned}
J_k &= \frac{1}{2}(j_k + j_{k+1}) \\
j_k &= \frac{\displaystyle\int_{J_k}^{J_{k+1}} j\, p(j)\, dj}{\displaystyle\int_{J_k}^{J_{k+1}} p(j)\, dj}
\end{aligned}
\tag{71.29}
$$

This is the *optimum mean-squares quantizer* also known as the Lloyd–Max quantizer. For further details and convenient approximations, see Jain (1989). In practice, the user rarely has much control over the choice of quantization levels.

71.4 Coding

The sampling and quantization of a continuously varying image represent the latter as a two-dimensional array of grey-level values. The size of the array is determined by the area measured and the sampling interval and may range from 16×16 for simple processing up to 1024×1024 or even bigger for very demanding tasks. Images invariably contain much redundant information, however, and it is therefore of interest to ask whether they can be coded more efficiently, for archival purposes and transfer between users in particular. It is found that very considerable savings can be made in this way and we mention briefly the more efficient methods, once again referring to specialized texts for details, and in particular for proofs of many of the following statements (Clarke, 1985; Netravali and Haskell, 1988; Gersho and Gray, 1992).

71.4.1 Use of image transforms

When we look at the Fourier spectrum of a one-dimensional signal, we associate the lower frequency components with coarse structures in the original signal, high frequency components with fine detail and very high frequency components with noise. This pattern is repeated for images and it therefore seems reasonable to take the Fourier (or some other) transform of an image and discard the coefficients beyond a certain value, which is believed to represent the frontier between genuine image detail and noise. Coding of the image is therefore achieved by retaining these transform coefficients instead of all the grey-level values of the original image, which can be reconstructed when needed by an inverse transform. The Fourier transform is only one of many possible choices of transform, however, and it can be shown that there exists an optimum transform, for which there is, on average, no correlation between the coefficients of the individual members. This optimum transform is established by regarding any image as a member of a set of images forming a random field $\{f(x, y)\}$. Can we find orthonormal matrices $G^{pq}(m, n)$ such that

$$f = \sum_p \sum_q G^{pq}(m, n) F(p, q) \tag{71.30}$$

and that[*]

$$\mathcal{E}\{F(p, q)F^*(p', q')\} = \mathcal{E}\{F(p, q)\}\mathcal{E}(F^*(p,' q'))\} \tag{71.31}$$

unless $p = p$ and $q = q'$? A calculation, given in detail in Rosenfeld and Kak (1982) for example, shows that such matrices do exist and are the solutions of the eigenvalue equation

$$\sum_k \sum_l R(m, n, k, l) G^{pq}(k, l) = \gamma_{pq} G^{pq}(m, n) \tag{71.32}$$

in which R is the auto-correlation function of the set of images $\{f\}$,

$$R(m, n, k, l) := \mathcal{E}\{f(m, n)f(k, l)\} \tag{71.33}$$

---

[*] The symbol $\mathcal{E}\{\cdot\}$ denotes the *expectation* of the quantity within braces, taken over the set of images forming the random field $\{f\}$. Under the pressure of circumstances, this expectation is occasionally replaced by the mean value of the quantity within braces for a single image; this is tantamount to assuming that the ergodic hypothesis is true for the images in question, which is almost never likely to be legitimate. It is more reasonable to use the hypothesis for small areas of an image; this is the basis of certain adaptive filters, see Section 73.5 and in particular the discussion in Kuan *et al.* (1985).

and γ_{pq} is given by

$$\mathcal{E}\{F(p,q)F^*(p,q)\} \tag{71.34}$$

This transform, known as the Karhunen–Loève transform, is not convenient in practice because the matrices G^{pq} (which play the same role as the exponential functions of the Fourier transform) are determined by $R(m,n,k,l)$ and hence by the family of images being studied. Fortunately, its performance for image coding is very closely rivalled by a transform that does not suffer from this disadvantage, namely the discrete cosine transform (DCT), defined by

$$F(p,q) = \frac{4c}{N^2} \sum_{m=0}^{N-1} \sum_{n=0}^{N-1} f(m,n) \cos \frac{(2m+1)\pi p}{2N} \cos \frac{(2n+1)\pi q}{2N}$$

$$p,q = 0 \ldots N-1$$

$$\tag{71.35}$$

$$f(m,n) = \sum_{p=0}^{N-1} \sum_{q=0}^{N-1} cF(p,q) \cos \frac{(2m+1)\pi p}{2N} \cos \frac{(2n+1)\pi q}{2N}$$

$$m,n = 0 \ldots N-1$$

in which $c = 1/2$ for $p = q = 0$ and $c = 1$ otherwise (Ahmed *et al.*, 1974; Ahmed and Rao, 1975; Rao and Yip, 1990).

This belongs to the family of transforms that possess fast algorithms and is hence easy and rapid to implement (Ahmed and Rao, 1975; Nussbaumer, 1982; Clarke, 1985). Comparisons of the performances of the Karhunen–Loève, cosine and Fourier transforms are to be found in image processing treatises. These are invariably presented with block-size as a parameter, the image to be coded being divided into sub-images (16×16 or 64×64 pixels, for example) before coding. An advantage of the cosine transform over the Fourier transform is the absence of artefacts (Gibbs phenomenon) between blocks but, as increased computing power becomes widely available, the separation into blocks will doubtless fall into disuse, except for extremely large arrays. For all further information about this transform, see the treatise of Rao and Yip (1990).

71.4.2 Predictive coding

We merely mention the principle of predictive coding, which we shall meet again in a different form in the section on image restoration. Generally speaking, an image will contain 'quiet' areas, in which the intensity values vary little or not at all, and other 'busy' areas, where the values vary much more rapidly. Intuitively, it seems that less information should be needed to characterize the quiet zones than the busy ones. Predictive methods exploit this idea in the following way. Suppose that we scan the pixels of

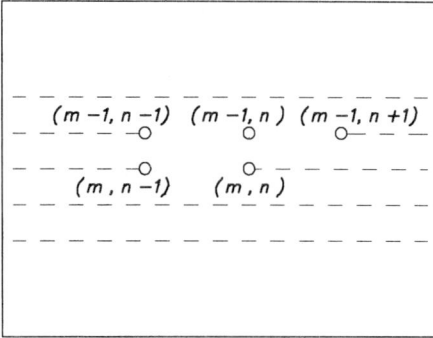

Fig. 71.5: A common choice of nearest neighbours of the point (m, n) for predictive coding.

an image in a raster pattern and devise a rule that predicts the intensity value of each pixel as we reach it from the intensity values of its nearest neighbours. We then need to code only the error between this predicted value and the true value; for a well-chosen prediction rule, this error will be very much smaller than the value itself and can hence be represented by substantially fewer bits. The predictor is usually a linear combination of the intensity values of nearest neighbours of the form

$$\hat{\jmath}(m,n) = \sum_k \sum_l a(k,l)j(m-k, n-l) \qquad (71.36)$$

The choice of values of k, l defines the nearest neighbours included; these are commonly those shown in Fig. 71.5, for which $k = 0$, $l = 1$ and $k = 1$, $l = -1, 0, 1$. The weights $a(k,l)$ are obtained by minimizing $\mathcal{E}\{|j(m,n) - \hat{\jmath}(m,n)|^2\}$ with respect to these unknowns. The conditions for a minimum generate four simultaneous equations for $a(k,l)$:

$$\begin{pmatrix} r(1,0) \\ r(0,1) \\ r(1,1) \\ r(1,-1) \end{pmatrix} = \begin{pmatrix} r(0,0) & r(1,-1) & r(0,1) & r(0,1) \\ r(1,-1) & r(0,0) & r(1,0) & r(1,-2) \\ r(0,1) & r(1,0) & r(0,0) & r(0,2) \\ r(0,1) & r(1,-2) & r(0,2) & r(0,0) \end{pmatrix} \begin{pmatrix} a(1,0) \\ a(0,1) \\ a(1,1) \\ a(1,-1) \end{pmatrix}$$
$$(71.37)$$

in which $r(m, n)$ is the covariance of the $N \times N$ image:

$$r(m,n) = \mathcal{E}[\{j(m'+m, n'+n) - \bar{\jmath}\}\{j(m',n') - \bar{\jmath}\}]$$
$$\approx \frac{1}{N^2} \sum_{m'=1}^{N-m} \sum_{n'=1}^{N-n} \{j(m+m', n+n') - \bar{\jmath}\}\{(j(m',n') - \bar{\jmath}\} \qquad (71.38)$$

and $\bar{\jmath}$ is the mean,

$$\bar{\jmath} = \frac{1}{N^2} \sum_{m=1}^{N} \sum_{n=1}^{N} j(m,n) \qquad (71.39)$$

Inversion of (71.37) yields the weights $a(m,n)$.

71.4.3 Huffman and vector codes

A much simpler code is obtained by attributing short codewords to frequently occurring grey level values and *vice versa*. A systematic procedure for this is known as Huffman coding (Huffman, 1962).

At the other extreme are the vector codes, capable in principle of the very highest compression rates but suffering from the drawback of generating very large codebooks. The idea here is based on Shannon's coding theorem, which implies that it is more economical to ascribe a single codeword to a cluster of grey-level values than to use (say) eight bits for each possible grey level. In the extreme case, every image would have its own codeword but even for 16×16 images with 8 bits/pixel this would require $2^{2048} \approx 10^{614}$ codewords! Work on these codes is in active progress (Gibson and Sayood, 1988) and it is already clear that high data compression rates can indeed be obtained. The size of the codebook is usually not as large as might be feared owing to the fact that, in practice, only a small fraction of all the possible codewords occurs other than with negligible frequency. The codebook can thus be limited to words that are needed in practice with only a very small loss of efficiency. For a comprehensive study, see Gersho and Gray (1992).

71.5 Electron optical considerations

Acquisition

The microscope–computer link has a large literature, for the scanning microscope especially. Readers interested in the earlier work may consult MacDonald (1968, 1969), White *et al.* (1968, 1972b), Kluge (1970), Herzog and Everhart (1973), Joy and Verney (1973), Herzog *et al.* (1974), Schmiesser (1974, 1975), Borus (1975), Rohr (1975), Ekelund (1975), Ekelund and Werlefors (1976a,b), Unitt and Smith (1976), Oron and Gilbert (1976), Ekelund *et al.* (1977), Hinterberger *et al.* (1980), Kerzendorf and Hoppe (1980), Jones and Unitt (1980, 1982), Smith (1981, 1982), Erasmus and Smith (1981), Miyokawa *et al.* (1988), Wisse and Zanger (1988), Yoshihara *et al.* (1988).

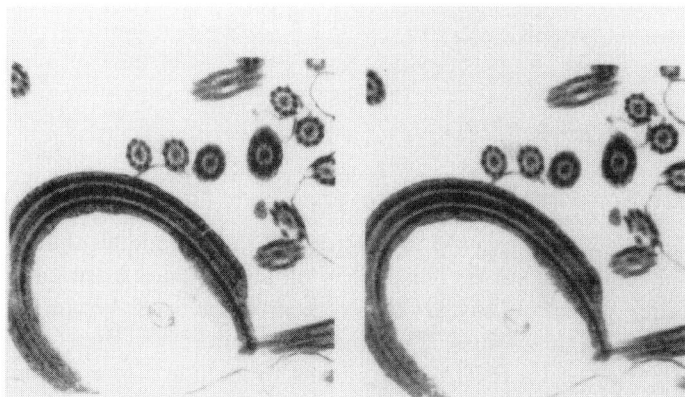

Fig. 71.6: A remarkable example of the high compression rate attainable by transform coding. *Left:* Original image of a thin section of flagella about 200 nm in diameter; 512 × 512 pixels, 8 bits/pixel. *Right:* The same image after Karhunen–Loève transform coding; same pixel size, 0.29 bits/pixel on average.

Sampling

The advantages of scanning patterns other than the traditional raster are argued by Sasov *et al.* (1982), Dain *et al.* (1983) and Sasov and Sokolov (1984, 1985). The basic theory was developed by Petersen and Middleton (1962) and is explained in detail by Mersereau (1979), Koenderink and van Doorn (1979) and Bell *et al.* (1989). Many of the texts on mathematical morphology treat hexagonal and rectangular sampling on an equal footing. The potential interest for electron optics is mentioned by Hawkes (1982a).

The importance of choosing the sampling interval correctly, for scanning instruments in particular, has been stressed by Crewe and Ohtsuki (Crewe, 1980b, 1984; Crewe and Ohtsuki, 1980, 1981). An early paper of Prewitt (1965) is also still of interest.

The need for image compression, particularly if the practice of exchanging digitized images becomes widespread (Burge, 1980), will certainly grow as digital processing of electron micrographs becomes more common. A dramatic example has been published by Burge *et al.* (1982) (Fig. 71.6), in which a gain of nearly 30 (8 bits/pixel reduced to 0.29 bits/pixel) is achieved by Karhunen–Loève coding. The cosine transform is used by Burge and Wu (1981). See too Burge and Clark (1981), Savoji and Burge (1982, 1985), Wu and Burge (1982) and Unser (1984). On compression and restoration, see Hall *et al.* (1982). The impact of work on image databases on electron microscopy is discussed by Hewan–Lowe (1992).

72
Enhancement

The term *enhancement* is used to describe all the techniques that improve an image in some way but do not depend on any detailed knowledge of the image-forming process. They are thus remedies that treat symptoms without enquiring into their cause. They may be grouped into three categories, to which we may add a fourth if we choose to include colour display. The three categories are operations on individual pixels and linear and nonlinear operations on groups of pixels, which may indeed extend to cover the whole image. Some of these operations are essentially cosmetic in character and serve to adapt the image to the visual response, others bring out or 'enhance' aspects of the image that might otherwise have gone unnoticed.

Colour display is attractive for specific purposes; if we have more than one signal per pixel, for example, it may be helpful to display each signal in a different colour. If the rather limited range of grey levels that the eye can discriminate with comfort is insufficient, the grey levels can be mapped into colours, for the eye is capable of distinguishing a far greater number. Communication is then difficult, however, as there is no widely recognized convention for such colour representations, and furthermore, not all display devices are capable of the fullest colour range, in which intensity, hue and saturation can all be controlled. We say no more about this here. Arguments in favour of the use of colour in electron image display have occasionally been advanced (Kanaya *et al.*, 1964, 1973; Hayes *et al.*, 1969; Swift and Brown, 1975; Krakow, 1978, 1985; Crewe, 1980a; Isaacson *et al.*, 1980; Pawley and Hayes, 1980; Antonovsky, 1982; Saparin 1990).

72.1 Operations on individual pixels

72.1.1 Elementary operations
At the simplest extreme, we may replace the original set of intensity values $\{j_{(1)}\}$ by some other set $\{j_{(2)}\}$ that gives a visually more pleasing image. The background, for example, can be subtracted or adjusted by adding a (positive or negative) constant to all the values $\{j_{(1)}\}$ or by subtracting a linear variation of intensity from the $\{j_{(1)}\}$; the latter is useful if the illumination used to obtain the image was uneven so that one side is too bright and the other too dark. We then speak of subtracting an inclined

background plane. Sometimes, division is preferable to subtraction and commercial units, such as the Synoptics 'SysTEM', offer both possibilities routinely.

After this, the contrast may be stretched (or squeezed) by operations of the form

$$j_{(2)} = c j_{(1)}^{1/\gamma} \qquad (\gamma\text{-control}) \qquad (72.1)$$

or

$$j_{(2)} = \begin{cases} c_1 j_{(1)} & 0 \le j_{(1)} < a \\ c_2(j_{(1)} - a) + j_{(2)}(a) & a \le j_{(1)} < b \\ c_3(j_{(1)} - b) + j_{(2)}(b) & b \le j_{(1)} \le N \end{cases} \begin{matrix} \text{(contrast} \\ \\ \text{stretching)} \end{matrix} \qquad (72.2)$$

The first has long been a standard image control on scanning electron microscopes; commercial devices (e.g. LogEtronics) have been developed to perform electronic contrast compensation during photographic printing (see for example von Prosch, 1974 and Jakobs and Katterwe, 1976). The second includes thresholding ($a = b$), in which all intensity values below the threshold are set equal to the same value and all those above it to another value. If the dynamic range of the image is very wide, as is often the case for diffraction patterns, it may be useful to take the logarithm:

$$j_{(2)} = c \log_{10}(1 + |j_{(1)}|) \qquad (72.3)$$

in which c is a scaling constant and we have written $|j_{(1)}|$ in case any earlier pre-processing operations have introduced negative values of $\{j\}$; in practice, of course, it would be wiser to prevent this from happening than to introduce the modulus.

72.1.2 Histogram-based enhancement
Images are often very difficult for the eye to appreciate because the grey levels are concentrated in too narrow a range: the image is too black or too white for comfort. Photographers have long known how to improve matters to some extent, notably by 'dodging', but once the image has been digitized new possibilities arise. In particular, we can plot a histogram showing the number of pixels having each of the possible grey levels. If, for example, the image is too dark for easy visual scrutiny, the grey levels will be concentrated in the zone near zero. The image would be much easier to see if the histogram were distributed evenly over the whole grey-level range and various mappings permit us to achieve this, to a good approximation at least. Thus in the case shown in Fig. 72.1, we map the darkest non-zero grey level to zero and the brightest to 255 and distribute the intermediate values over the whole range.

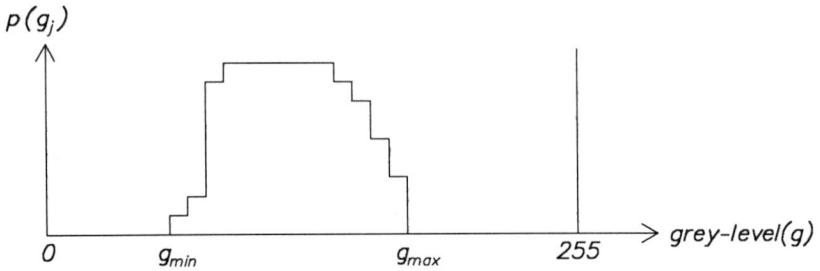

$p(g_j)$ ... grey-level(g)

$0 \quad g_{min} \qquad g_{max} \qquad 255$

Fig. 72.1: Example of a grey-level histogram in which the very light and very dark grey-level values are unoccupied.

The simplest mapping is known as *histogram equalization*. Consider an image with n pixels in all, quantized on G grey levels, which we denote by g_i, $i = 0, 1, \ldots G - 1$. Suppose that the grey level g_i occurs n_i times in the image. We introduce the probability distribution

$$p_1(g_j) = \frac{n_j}{n} \qquad (72.4)$$

This is the *grey-level histogram* of the image. Probability theory tells us that the cumulative probability distribution, defined in terms of $\sum_{j=0}^{k} p_1(g_j)$, would be uniform for a continuous probability distribution and is approximately uniform for discrete distributions. This can be seen by the following argument. Let

$$g'_k := \sum_{j=0}^{k} p_1(g_j) =: T_1(g_k) \qquad (72.5)$$

Then the probability distribution $p_2(g'_k)$ of the new variable g'_k is given by

$$p_2(g'_k) \approx p_1(g_k) \frac{\Delta g_k}{\Delta g'_k} \qquad (72.6)$$

But

$$\frac{\Delta g'_k}{\Delta g_k} = \frac{g'_k - g'_{k-1}}{g_k - g_{k-1}} = \frac{n_k}{n \Delta g_k} = \frac{p_1(g_k)}{\Delta g_k} \qquad (72.7)$$

so that (for $\Delta g_k = 1$)

$$p_2(g'_k) \approx 1 \qquad (72.8)$$

Histogram equalization is thus achieved by establishing the function $T_1(g_k)$ (72.5) and then using this same defining equation to attribute a new grey-level value $\bar{g}_k$ to each of the original values g_k; we note that the discretization may result in several of the original levels being attributed to a single new level.

More generally, the original grey-level histogram may be modified so that it resembles as closely as possible some pre-determined distribution. Suppose that the latter is characterized by the probability $q_1(g_k'')$ and that its cumulative probability distribution is denoted by T_2:

$$g_k' = \sum_{j=1}^{k} q_1(g_k'') =: T_2(g_k'') \qquad (72.9)$$

for which the probability distribution $q_2(g_k')$ is again uniform. We can find the mapping that takes the original grey levels into their new values by equating the cumulative probability distributions, which are (approximately) uniform:

$$g_k'' = T_2^{-1}(g_k') = T_2^{-1}T_1(g_k) \qquad (72.10)$$

Practical difficulties

Histogram equalization or modification is often spectacularly effective (Fig. 72.2). Nevertheless, some precautions must be taken to prevent the creation of spurious detail. The mapping is usually required to spread a few grey levels over the whole black–white range and a grey level in the original may therefore be attributed to any of a range of grey levels: for example, if the grey levels were initially concentrated in the range 10–73 and are redistributed over the full 255 levels ($g_{max} = 73$ in Fig. 72.1), level ten might be attributed any of the values 0–3 after histogram modification, level eleven to the values 4–7 and level 63 to the values 252–255 (this is merely an example, the redistribution will in practice not be as uniform as this). This is aesthetically desirable so that all grey levels are present but can cause problems if the original contained uniform areas, containing no contrast therefore. When the mapping is performed, a test must be introduced to ensure that neighbouring pixels with the same grey level in the original again have the same grey level after transformation.

Refinements

The type of histogram modification described above is the simplest of a growing family of such techniques, which differ in the choice of rule adopted for mapping the grey levels and in the area of the image over which the rule is applied. An example of an unusual choice of rule is the work of Schneider and Craig (1992), who have introduced the notion of fuzzy sets. The uncertainty associated with the presence of several choices is exploited to produce particularly sharp images, suitable for edge detection. It is not essential to study and modify the histogram of the entire image, as this can be disadvantageous: details of interest in small, relatively uniform areas, may well suffer, for example. Local-area forms of histogram equalization have been explored, notably by Ketchum (1976) and by Pizer *et al.* (1987) and Leszczynski and Shalev (1989); more recently, Paranjape *et al.* (1992) and Morrow *et al.* (1992) have proposed a method of choosing the size of the area over which the histogram is modified adaptively, that is, as a function of the contrast variation around each pixel of the image. Gauch (1992) extends this idea by generating a family of histogram-modified images, the different members of the family corresponding to different choices of window-size and of a smoothing parameter. Among the resulting set of enhanced images, some will show certain features more clearly than others and the whole set can be explored to achieve the maximum enhancement. We refer to the original papers for details and tests of these extensions of the basic procedure.

72.2 Linear filtering

The notion of *image filtering* sprang up from the simple argument that all images have a limit of resolution so that any fine detail beyond this limit must be noise. It is therefore reasonable to 'filter out' such detail, and an elementary way of doing this is to truncate the Fourier transform: crudely, we associate coarse detail in the image with low Fourier components and fine detail with higher spatial frequencies. By truncating the transform and hence suppressing all frequency components beyond a certain value, corresponding to the presumed resolution limit, and then performing the inverse transform, we shall in principle have filtered out the noise and retained only meaningful information. In practice the equation between spatial frequency and size of detail, though broadly true, is an over-simplification and application of a sharp cutoff (low-pass filtering) to the transform is a source of image artefacts; nevertheless, the technique does improve images not too close to the limit of resolution, especially if the cutoff is not sharp but a gradual attenuation.

Multiplication of the Fourier transform of an image by some other function, or filter, is equivalent to forming the convolution of the image with the inverse transform of the filter function; this observation encourages us to regard all the many convolution-based ways of enhancing images as different types of filter. We first consider a simple but typical example and then examine filters more generally. Suppose that we are confronted with a noisy image: the measured intensity values at adjoining pixels vary more rapidly than the true values owing to the presence of the rapidly fluctuating noise intensity. This variation can be attenuated by replacing the intensity

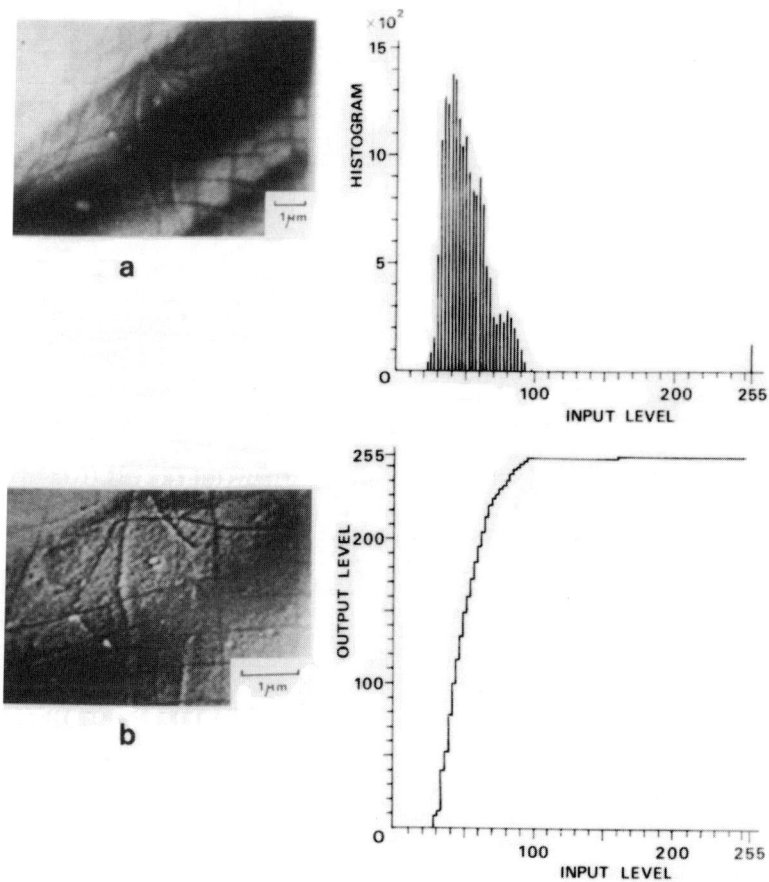

Fig. 72.2: Examples of histogram modification. (a) Image of a dislocation in aluminium obtained at 2 MV and the corresponding grey-level histogram. (b) The same image enhanced by histogram equalization with the Laplacian superimposed on it.

at each pixel by the mean of the intensities at the pixel and its neighbours:

$$g_2(x_i, y_i) = \frac{\sum_{k,l} g_1(x_k, y_l) h(x_{i-k}, y_{j-l})}{\sum_{k,l} h(x_k, y_l)} \qquad (72.11)$$

where g_1 is the original grey level intensity and g_2 is the smoothed intensity.

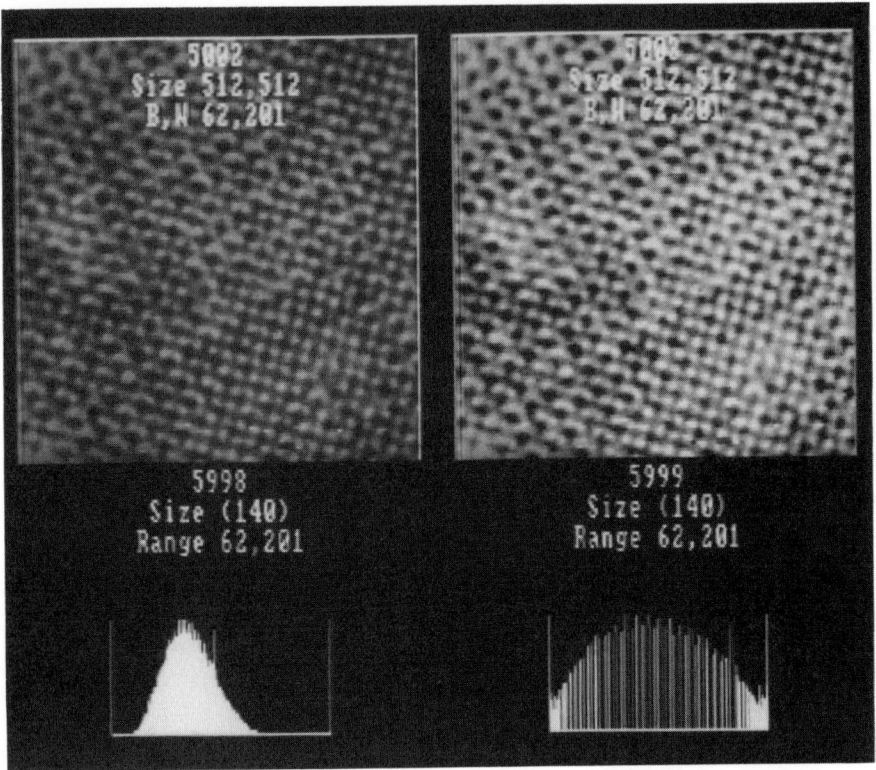

Fig. 72.2: Examples of histogram modification. (c) High-resolution image of a crystalline specimen and its grey-level histogram. (d) The same image enhanced by histogram modification using SysTEM and the SEMPER software from Synoptics.

If the neighbours included in the averaging are those shown in Fig. 72.3, then

$$h = \begin{cases} 1 & \text{for } k = i, i \pm 1 \quad , \quad l = j, j \pm 1 \\ 0 & \text{otherwise} \end{cases} \tag{72.12}$$

or

$$h = \begin{matrix} 1 & 1 & 1 \\ 1 & 1 & 1 \\ 1 & 1 & 1 \end{matrix} \tag{72.13}$$

This example of filtering by convolution of the image with a neighbourhood function h is only one of many and each form of the function h is associated with a particular type of enhancement. The different forms of h of course define which neighbours are to be included and the importance (weight) to be accorded to each of them. There are choices for which

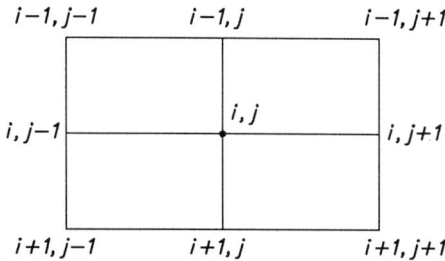

Fig. 72.3: The 8-neighbours of the point (i, j), at which the neighbourhood function h is defined.

$\sum h(x_k, y_l)$ is deliberately made equal to zero and (72.11) is then replaced by

$$g_2(x_i, y_j) = \sum_{k,l} g_1(x_k, y_l) h(x_{i-k}, y_{j-l}) \qquad (72.14)$$

Some useful choices for the array h are as follows.

Low-pass filters

The simplest example is the unweighted average given above (72.13). In practice, it seems intuitively more natural to give greater weight to the pixel in question than to its neighbours and more to the closer neighbours than to those more distant.

$$h = \begin{matrix} 1 & 2 & 1 \\ 2 & 4 & 2 \\ 1 & 2 & 1 \end{matrix} \qquad (72.15)$$

is then used

High-pass filters

Image enhancement may be desirable in order to highlight small differences in contrast. Our instinct is then to differentiate the image in order to render small steps in grey level visible and a number of discrete approximations to differentiation have been developed. These are convolutional filters and we list some examples. If a step in a particular direction is to be enhanced, we can use (Prewitt, 1970)

$$h = \begin{matrix} 1 & 0 & -1 \\ 1 & 0 & -1 \\ 1 & 0 & -1 \end{matrix} \qquad (72.16)$$

(step increasing in the positive x-direction) or

$$h = \begin{matrix} 0 & 1 & 1 \\ -1 & 0 & 1 \\ -1 & -1 & 0 \end{matrix} \qquad (72.17)$$

(step increasing in the direction SW $\to$ NE). These simple expressions are obtained by fitting a quadratic surface over the intensities in a 3×3 zone and evaluating the gradient $(\partial g/\partial x,\ \partial g/\partial y)$ for this surface. Normally, however, the direction of the step is not known and, even more often, we should like to enhance any step, irrespective of its orientation. In the former case (reasonably straight step, orientation arbitrary), the set of eight masks obtained from

$$h = \begin{matrix} 1 & 1 & 1 \\ 1 & -2 & 1 \\ -1 & -1 & -1 \end{matrix} \qquad (72.18)$$

by displacing the succession of 1's and -1's round the pattern one step at a time can be used: the best match gives the largest response (Prewitt, 1970). These are known as 'compass-gradient masks'. Alternative patterns of weights have been proposed by Kirsch (1971) and Sobel*:

$$h = \begin{matrix} 5 & 5 & 5 \\ -3 & 0 & -3 \\ -3 & -3 & -3 \end{matrix} \qquad \text{(Kirsch)} \qquad (72.19)$$

and

$$h = \begin{matrix} 1 & 2 & 1 \\ 0 & 0 & 0 \\ -1 & -2 & -1 \end{matrix} \qquad \text{(Sobel)} \qquad (72.20)$$

with cyclic displacement in each case yielding eight masks. We return to the Kirsch array in Section 72.3.1. The merits of the Sobel array, in its cyclically extended form, are argued by Robinson (1976, 1977); a point of interest is that diagonal directions are weighted more heavily than by the Prewitt arrays (72.16 and 72.18) and by the Kirsch array (72.19).

An interesting way of regarding these cyclic masks has been introduced by Park and Choi (1989). If we write the array in the general form

$$\begin{matrix} h(0) & h(1) & h(2) \\ h(7) & & h(3) \\ h(6) & h(5) & h(4) \end{matrix} \qquad (72.21)$$

---

* This pattern is ascribed to Sobel by Duda and Hart (1973, p. 271, footnote) but with no published reference. Coster and Chermant (1985) give Tennenbaum *et al.* (1969) [sic] as a source but this paper, which is in fact Feldman *et al.* (1969), does not contain a description of the so-called Sobel filter.

and form the 8×8 matrix $\boldsymbol{H}$, the columns of which consist of the elements $h(i)$, $i = 0, 1 \ldots 7$ and the seven cyclic permutations of this, we have

$$H_{ij} = h(i - j \bmod 8) \qquad (72.22)$$

The matrix $\boldsymbol{H}$ thus has the form of a circulant (Bellman, 1970, Section 15 of Chapter 12) and its eigenvectors are hence identical with the columns of the discrete Fourier transform matrix. Multiplication of the intensity values $(\boldsymbol{P})$ at the pixels corresponding to $h(0) \ldots h(7)$ by the matrix $\boldsymbol{H}$ generates an edge-detection vector, $\boldsymbol{E}$ say:

$$\boldsymbol{E} = \boldsymbol{H}\boldsymbol{P} \qquad (72.23)$$

The largest element of $\boldsymbol{E}$ then indicates the orientation of any edge; the above result concerning the form of $\boldsymbol{H}$ shows us that the spectrum (Fourier transform) of $\boldsymbol{E}$ may be regarded as the filtered (weighted) spectrum of $\boldsymbol{P}$, the weights being the eigenvalues of $\boldsymbol{H}$. For the Sobel (72.20), Prewitt (72.16) and Kirsch (72.19) arrays Park and Choi find the following eigenvalues:

$$
\begin{array}{ccc}
\text{Sobel} & \text{Prewitt} & \text{Kirsch} \\
0 & 0 & 0 \\
2(s+1)(1-i) & s(s+1)(1-i) & 4s(s+1)(1-i) \\
0 & 0 & -8i \\
-2(s-1)(1+i) & s(s-1)(1+i) & 4s(s-1)(1+i) \\
0 & 0 & 8 \\
-2(s-1)(1-i) & s(s-1)(1-i) & 4s(s-1)(1-i) \\
0 & 0 & 8i \\
2(s+1)(1+i) & s(s+1)(1+i) & 4s(s+1)(1+i)
\end{array}
\qquad (72.24)
$$

with $s := \sqrt{2}$.

As well as these masks that are deliberately designed to recognize edges and their orientation, there are isotropic masks, which enhance small changes in intensity in any direction. These all use a digital approximation to the Laplacian and we meet ∇^2 in the four-neighbour form

$$
h = \begin{matrix} & -1 & \\ -1 & 4 & -1 \\ & -1 & \end{matrix}
\qquad (72.25a)
$$

and in the 8-neighbour form

$$
h = \begin{matrix} -1 & -1 & -1 \\ -1 & 8 & -1 \\ -1 & -1 & -1 \end{matrix}
\qquad (72.25b)
$$

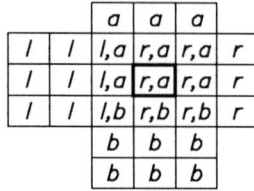

Fig. 72.4: Pixels used in forming the sums of (72.27b).

The difference, $g - k\nabla^2 g$ (k constant), emphasizes higher spatial frequencies for values of k near to unity while having little effect on the low frequency range, which is usually what we should like. A variant on this is the (nonlinear) 'pseudo-Laplacian' filter, introduced by Oho *et al.* (1990) for secondary-electron SEM images; here, each pixel (i, j) and its (5×5) neighbourhood is examined and edge directions are isolated by evaluating $G_{\text{max}} := \max(|G_A|, |G_B|, |G_C|, |G_D|)$ where

$$G_A = g(i+2, j+2) + g(i+1, j+1) - g(i-1, j-1) - g(i-2, j-2)$$
$$G_B = \sqrt{2}\{g(i, j+2) + g(i, j+1) - g(i, j-1) - g(i, j-2)\}$$
$$G_C = g(i-2, j+2) + g(i-1, j+1) - g(i+1, j-1) - g(i+2, j-2)$$
$$G_D = \sqrt{2}\{g(i-2, j) + g(i-1, j) - g(i+1, j) - g(i+2, j)\}$$

$$(72.26)$$

A magnitude E is ascribed to each edge:

$$E := (E_H + E_V)^{\frac{1}{2}} \qquad (72.27a)$$

where

$$E_H = \frac{1}{g}\left(\sum l - \sum r\right) \quad , \quad E_V = \frac{1}{g}\left(\sum a - \sum b\right) \qquad (72.27b)$$

and $\sum l$ denotes the sum of the intensities at all the points to the left of the pixel in question (Fig. 72.4), $\sum r$ all those to the right, $\sum a$ above and $\sum b$ below.

The Laplacian ($\nabla^2 g$) is replaced by the second derivative, modelled by

$$(1 \quad -2 \quad 1) \quad , \text{ for vertical edges}$$

$$\begin{pmatrix} 1 & \cdot & \cdot \\ \cdot & -2 & \cdot \\ \cdot & \cdot & 1 \end{pmatrix} , \text{ for SW – NE edges}$$

and likewise in other directions. The 'constant' k is now allowed to vary as a function of the edge magnitude E. Oho *et al.* (1990) find that the result of processing secondary-electron images in this way is very satisfactory.

This piecemeal presentation of linear neighbourhood operators obscures their underlying unity, studied in detail by Koenderink and van Doorn (1992).

Hexagonal sampling

It has been found that a form of differential gradient operator based on hexagonal tessellation is more efficient than the usual Sobel operator based on square pixels (Staunton, 1989a,b). The idea has been reconsidered by Davies (1991b), who derives suitable operators for the discrete horizontal and vertical derivatives. Davies reminds us of the many advantages of hexagonal sampling, which are offset only by instrumental limitations.

Generalized convolution

The whole notion of convolutional filtering has been enriched by the introduction of a much wider class of convolutions than the usual operation discussed above. This new structure is based on the properties of group algebras, together with an appropriate value function. We refer to the work of Eberly and Wenzel (1991) for a full account of this theory and its application to image processing. A formulation in terms of image algebra is also included in that paper.

Periodic specimens

The diffraction pattern of a specimen that naturally forms a regular pattern or has been prepared in such a way that such a pattern is created consists of isolated spots. The image of such a specimen should therefore be improved by forming its diffraction pattern, masking out any background between the spots and inverting the transform. The resulting image will be rigorously periodic and its detailed structure will be determined by the choice of spots. Such a procedure was used very early in electron image processing, originally in analogue form: an optical bench was used to form the transform and the mask consisted of holes cut in an opaque screen. This was subsequently replaced, or at least complemented, by digital masking, which permitted automatic selection of the spots and soft-edged masking to prevent the creation of spurious effects (Gibbs phenomena). The real-space equivalent of such masking in Fourier space is averaging identical motifs of the structure and the technique is hence often known as lattice averaging. It has various ramifications: the lattices corresponding to the front and back surfaces of a specimen can sometimes be separated, if their orientation is different for example; and the very poor images obtained from fragile periodic specimens in low-dose conditions can be rendered interpretable. For early examples see Klug and Berger (1964), Klug and de Rosier (1966), Berger et al. (1966), Glaeser et al. (1971), Unwin and Henderson (1975) and Unwin (1975) and for extended accounts, Klug (1971), Horne and Markham (1972), Gibbs and Rowe (1976), Donelli and Paoletti (1977) and Chapter 4 of Misell (1978b), who gives many examples. Very full lists of references are included in Hawkes (1978a, 1982c, 1992a).

By combining the ideas of lattice averaging with other processing techniques, many of which are discussed elsewhere in this section, very spectacular improvements have been achieved. Thus Morgan et al. (1992) combine correlation averaging and multivariate statistical analysis (discussed in Section 75.3.3) to extract reliable information from images of tilted and untilted crystals of variable thickness. Fryer et al. (1991) and Dong et al. (1992) obtain structure information at a resolution better than that of the phase data of electron diffraction intensities by invoking entropy maximization (Section 74.5). These are only two of many possible examples, chosen to indicate the way in which processing of this type is developing.

72.3 Nonlinear filters

72.3.1 Nonlinear exploitation of linear filtering

The term 'filter' is conventionally used even though the operations applied to the image no longer have the form of a convolution and can hence not be regarded as filters in Fourier space.

We first mention a few named heuristic filters, then describe median and other nonlinear filters before turning to morphological filters.

The Roberts, Kirsch and Sobel operators are properly speaking nonlinear uses of linear operations. In each case an array or set of arrays such as those we have already met (h in 72.11 or 72.14) is convolved with an image, generating a vector at each point. The elements of this vector are then combined in some simple way and the original pixel value is replaced by the resulting value. Thus the *Roberts operator*, for sharpening edges is defined by

$$g_2(j,k) = \left[\{g_1(j,k) - g_1(j+1,k+1)\}^2 + \{g_1(j,k+1) - g_1(j+1,k)\}^2 \right]^{\frac{1}{2}}$$
(72.28a)

or, in simplified form,

$$g_2(j,k) = |g_1(j,k) - g_1(j+1,k+1)| + |g_1(j,k+1) - g_1(j+1,k)| \quad (72.28b)$$

(Roberts, 1965) and we see that the arrays

$$h = \begin{matrix} 1 & -1 \\ 0 & 0 \end{matrix} \quad \text{and} \quad h = \begin{matrix} 0 & 1 \\ 1 & 0 \end{matrix} \tag{72.29}$$

are used to generate measures of the presence of an edge, which are then combined into a single measure by means of (72.28a or b).

The *Sobel operator* uses

$$g_2(j,k) = (g_{20}^2 + g_{02}^2)^{\frac{1}{2}} \tag{72.30}$$

$$g_{20} = g_1 * h_{20} \qquad g_{02} = g_1 * h_{02}$$

$$h_{20} = \begin{matrix} -1 & 0 & 1 \\ -2 & 0 & 2 \\ -1 & 0 & 1 \end{matrix} \qquad h_{02} = \begin{matrix} 1 & 2 & 1 \\ 0 & 0 & 0 \\ -1 & -2 & -1 \end{matrix} \tag{72.31}$$

(already introduced in 72.20).

The *Kirsch operator* seeks the largest compass-gradient magnitude (but with the current pixel value neglected, the centre value of h in (72.21) is zero):

$$g_2(j,k) = \max\{1, \ \max_{i=1}^{7}(|5g_{20}^i - 3g_{02}^i|)\}$$

$$g_{20}^i = g_i + g_{i+1} + g_{i+2} \tag{72.32}$$

$$g_{02}^i = g_{i+3} + g_{i+4} + g_{i+5} + g_{i+6} + g_{i+7}$$

in which the subscripts are all modulo 8 and

$$\begin{matrix} g_0 = g(j-1,k-1) & g_1 = g(j,k-1) & g_2 = g(j+1,k-1) \\ g_3 = g(j+1,k) & g_4 = g(j+1,k+1) & g_5 = g(j,k+1) \\ g_6 = g(j-1,k+1) & g_7 = g(j-1,k) \end{matrix} \tag{72.33}$$

(cf. 72.19).

A final example is the nonlinear filter tested by Oho *et al.* (1990), already described at the end of Section 72.2 owing to its close relation to Laplace filtering.

In another group of filters, the modulus of the Fourier transform of the image is altered. If the discrete Fourier transform of the image is written as $G(j,k) =: |G(j,k)| \exp i\varphi(j,k)$, then $|G(j,k)|$ is replaced by $|G(j,k)|^\alpha$, $0 \le \alpha < 1$ (root filtering) or by $\log |G(j,k)|$ (homomorphic filtering); in the latter case, the inverse transform of $\log |G(j,k)| \exp i\varphi(j,k)$ is known as the *cepstrum*. For a general study of these extended filtering operations, see Pratt (1978) or Oppenheim *et al.* (1968). Among those that have been found potentially useful for images are the *E-filter* of Moore and

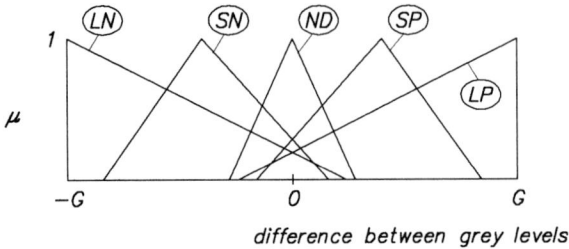

Fig. 72.5: The membership function μ for each of the sets.

Parker (1973) and its generalization by Saito and Cunningham (1990). In its original form (in one dimension), this was defined by

$$g_2(x) := \int_0^x \left\{ 1 + \left(\frac{dg_1}{d\xi} \right)^2 \right\}^{\frac{1}{2}} d\xi \qquad (72.34)$$

and possessed the interesting property that high values of the gradient, $dg_1/d\xi$, were mapped to the lower frequency range and hence survived subsequent low-pass filtering. The generalization consists in weighting the term $(dg_1/d\xi)^2$ suitably.

One weakness of many of the procedures for sharpening images arises from the limited number of deterministic choices of the new value ascribed to each pixel value; indeed, in most cases, only one choice is allowed. A way of escaping from this has been tested by Russo and Ramponi (1992), who express their method in the language of fuzzy set theory. We describe their idea briefly as it seems likely to be of wider interest that their particular application.

Consider a window W centred on a pixel P of an image, the grey level of which may take any value between 0 and G. For each pixel in the window, the difference between its grey level and that of P is calculated. Degrees of membership (μ) are now associated with this difference for each of a number of fuzzy sets, in the present case five. We call these sets 'large negative' (LN), 'small negative' (SN), 'no difference' (ND), 'small positive' (SP) and 'large positive' (LP) and choose the values of μ shown in Fig. 72.5.

We are now ready to assign multiple-choice rules to the enhancement operator. If we write

$$\Delta(k, l) := p(i + k, j + l) - p(i, j)$$

where $p(i, j)$ denotes the grey-level value at (i, j), then we can define rules of the following form: if $\Delta(-1, -1)$ belongs to LP and $\Delta(-1, 0)$ to ND

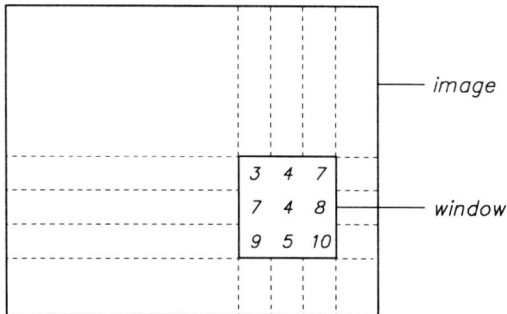

Fig. 72.6: Median filtering. Each pixel value is replaced by the median of the values within the window centred on it. Here, the value 4 at the centre of the window is replaced by 7 (the median of 3 4 4 5 7 7 8 9 10).

and $\Delta(-1, 1)$ to SP, then Δ' must belong to SP; otherwise, $\Delta' = 0$. Here Δ' is the difference between the grey level of pixel (i, j) before and after enhancement. From the figure, it is clear than for many values of Δ, the latter belongs to more than one set and the decision is taken on the basis of the value of the degree of membership, μ.

72.3.2 Median and rank-order filtering

Median filtering is a nonlinear procedure that is at once simple and does not seriously blur sharp edges. It consists in replacing the grey-level value at any point, $g_1(i, j)$ by the *median* of the values at the point and its neighbours. For a 3×3 region, for example, the nine grey-level values are ordered in increasing value and the value at the central point is replaced by the fifth in the list, this being the median. An example is shown in Fig. 72.6. If the window contains n pixels, the median is the $\frac{1}{2}(n + 1)$-th value (n odd) or (arbitrarily) the $\frac{1}{2}n$-th value (n even). Thin curves or sharp corners may be eroded by this filter, however.

A natural extension of the median filter is the *rank-order* or *order-statistic* filter in which the values within a window (n pixels) are again ordered in increasing value and the r-th value (not necessarily the median) is selected. They have been studied extensively (Justusson, 1981; Nodes and Gallagher, 1982; Fitch *et al.*, 1985a,b) and their relation to the morphological operators has been established and exploited by Maragos (1987; see also Maragos and Schafer, 1987, 1990). Rank-order filters (of which the median filter is of course a special case) all belong to the class of *stack filters*, a very important feature from the point of view of efficient VLSI implementation (Wendt *et al.*, 1986a,b). To explain this, we return to the

```
                          ┌─────────────────┐
                          │ median filter,  │
                          │ 3 pixels wide   │
                          └─────────────────┘
                                   ↓
    4  0  2  0  4  1  3  4  1  2   ⟹   4  2  0  2  1  3  3  3  2  2

  ⇓ threshold at 1,2,3,4   ⇓                        ⇧ add

    1  0  0  0  1  0  0  1  0  0   ⟹   1  0  0  0  0  0  0  0  0  0
                                 ↑

    1  0  0  0  1  0  1  1  0  0   ⟹   1  0  0  0  0  1  1  1  0  0
                                 ↑

    1  0  1  0  1  0  1  1  0  1   ⟹   1  1  0  1  0  1  1  1  1  1
                                 ↑

    1  0  1  0  1  1  1  1  1  1   ⟹   1  1  0  1  1  1  1  1  1  1
                                 ↑
                          ┌───────────────────────┐
                          │ median binary filter, │
                          │ 3 pixels wide         │
                          └───────────────────────┘
```

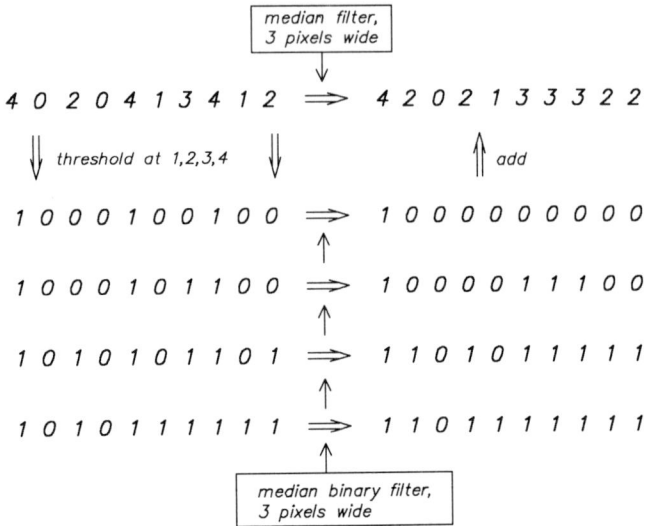

Fig. 72.7: The stacking property of a median filter. A (one-dimensional) median filtering of the original sequence (top left) gives the sequence shown top right. The same result is obtained by forming binary sequences by thresholding at each level in turn, applying a binary median filter to each and adding the 'stacks' of ones.

median filter. The operation of finding the median can be replaced by the following, computationally efficient sequence of steps (Fitch *et al.*, 1984, 1985a,b):

(i)   threshold the input at each level (levels 1–255 for eight-bit quantization);

(ii) apply a *binary* median filter to each output with the same window size;

(iii) add the binary outputs (Fig. 72.7).

The addition is easy because each column of the sum consists of a certain number of zeros stacked on top of ones. It is for this reason that the median and rank-order filters are said to be *stack filters*. For further details we refer to the papers of Wendt *et al.* (1986a,b). There are only twenty possible stack filters with windows three pixels wide but the number increases very rapidly with window size (2^{35} for 7 pixels!); the number for the practically important case of 9 pixels is not known.

Another ingenious extension of the median filtering idea is the 'nearest-neighbour median filtering' of Itoh *et al.* (1988). The nearness of the neighbours here does not signify geometrical proximity but the closeness of their grey levels to that of the pixel in question. Thus suppose that a pixel P is situated within a window of n pixels, centred on P. We first order the grey-level values of these n pixels in ascending order and note the position of the grey level of pixel P in the list. The k nearest neighbours $(k < n)$ of P are then the k pixels with grey levels closest to that of P. The filtering operation then consists in replacing the grey-level value of P by the median of the grey-level values of the k nearest neighbours, defined in this way. The performance depends on the choice of k but generally speaking, detail is well-preserved.

Median or stack filters have a large literature and new suggestions continue to appear (e.g. Sinha and Hong, 1990). For the origins of the subject, see Justusson (1981) and Tyan (1981), and for its development, see Frieden (1976), Gallagher and Wise (1981), Ataman et al. (1981), Narendra (1981), Arce and Gallagher (1982), Nodes and Gallagher (1982, 1983, 1984), Bovik et al. (1983, 1987), Fitch et al. (1984, 1985a,b, 1986), Brownrigg (1984), Arce et al. (1986), Wendt et al. (1986a,b), Ochoa et al. (1987), Coyle (1988), Coyle and Lin (1988), Yli-Harja (1989), Davies (1989, 1991a, 1992b), Wendt (1990a,b) and Yli-Harja et al. (1991), Haavisto et al. (1991), Zeng et al. (1991), Gabbouj et al. (1992a,b), Barner et al. (1992), Butz (1992), Prasad and Lee (1992), Salembier and Kunt (1992), Pitas and Venetsanopoulos (1992), Yin et al. (1993) and Gil and Werman (1993).

72.3.3 Morphological filters

72.3.3.1 Introduction

Measurement has always occupied an important place in image analysis: what is the mean surface area of the particles in the image? How many particles have perimeters longer than a given value? What are the statistical properties of some feature? At the same time as such questions as these were being answered in piecemeal fashion as the need arose, a systematic approach was being developed, by G. Matheron, J. Serra and others in the Ecole des Mines at Fontainebleau, which has come to be called 'mathematical morphology'. In the early days, the theory was limited to binary images (black and white images or those with only two grey levels), which could of course be thresholded grey-level images, but was subsequently extended to include grey-level images. In addition, it was realized that its ideas are of interest far beyond the applications for which it was originally devised, and in particular, 'morphological filtering' is of interest for image enhancement. We return to the question of measurement in Chapter 76, on image analysis. The basic ideas of mathematical morphology are summarized here, however, since they are needed to understand morphological filtering.

At the heart of mathematical morphology are two operations known as *erosion* and *dilation**. As their names indicate, they either remove part of an image or add to it and, even before defining them, we begin to see why such operations might be useful for enhancement: unwanted spikes of noise could be flattened and gaps or crevices in the image, caused by some imperfection in the image-forming process, could be grouted by such operations.

Dilation or erosion is performed with the aid of *structuring elements*, the shape of which may be chosen freely. The choice of shape enables us to control the effect of the operation, just as the shape of woodcarver's tool allows him to produce a variety of surface effects. A structuring element is a cluster of points (pixels) in a plane, defining a square or circle or a cross, for example (or any other shape whatsoever), which it is convenient to regard as a set.

Although binary (black-and-white) images are of little interest in electron image processing, we introduce the notions of dilation and erosion with the aid of such images, for which the definition are easily grasped. We then turn to the case of images with multiple grey-levels, for without the binary background material, the apparent arbitrariness of the definitions can be bewildering.

Binary image morphology

We consider two-dimensional arrays of image intensities, which are allowed to take only the values 0 (white) and 1 (black). The image to be processed is denoted by I and the structuring element that will effect the processing by S. The latter is usually very much smaller than I and may be pictured for simplicity as a bar or circle or cross; nevertheless, it is in practice of interest to consider more complicated, not necessarily connected shapes. We choose to define the sets I and S to be the black pixels, so that for $I = \{(0,1),(0,2),(1,1),(1,3),(2,3),(3,0),(4,2)\}$ and $S = \{(0,0),(0,1)\}$ the image and structuring element are as shown in Fig. 72.8.

The *dilation* of I by S is denoted by $I \oplus S$ and defined by

$$I \oplus S := \{d | d = i + s, i \in I, s \in S\} \qquad (72.35)$$

Figure 72.9 shows $I \oplus S$ for the example of Fig. 72.8.

Dilation possesses a number of familiar properties: it is commutative $(I \oplus S = S \oplus I)$, associative $((I \oplus S_1) \oplus S_2 = I \oplus (S_1 \oplus S_2))$ and increasing $(I_1 \subseteq I_2$ implies $I_1 \oplus S \subseteq I_2 \oplus S)$; the symbol $\subseteq$ signifies 'is contained

---

* This form (inspired by the misleading resemblance between 'dilate' and 'calculate') is invariably used rather than the more awkward 'dilatation', formerly the only correct noun from the verb 'dilate'. *Dilatation* remains the preferred usage in French and German.

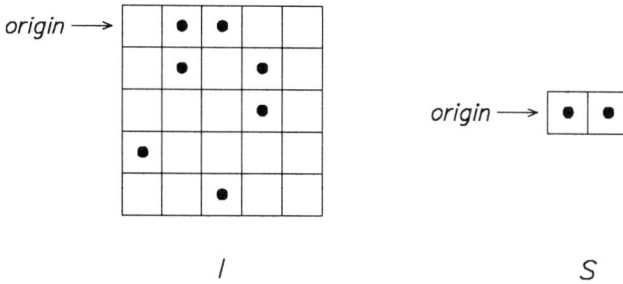

Fig. 72.8: The binary image I and the binary structuring element S. Image $I = \{(0,1),(0,2),(1,1),(1,3),(2,3),(3,0),(4,2)\}$; Structuring element $S = \{(0,0),(0,1)\}$. The ordering is the same as that used for matrix suffixes.

within or the same as.' Another very helpful definition of dilation, easy to picture, involves *translation* of a set. The translation of I by x is denoted by $(I)_x$ and is defined by

$$(I)_x = \{d | d = i + x \text{ for } i \in I\} \tag{72.36}$$

which is of course the everyday meaning of the word 'translation'. If we translate I by each of the elements of the structuring element in turn and form the union of the results we again obtain the dilation of I by S:

$$I \oplus S = \bigcup_{s \in S} (I)_s \tag{72.37}$$

This is commonly referred to as *Minkowski addition*.

For the more usual kinds of structuring element, dilation is *extensive* in the sense that the result of dilation contains the undilated image I. If, however, S does not contain the origin in the sense that $(0,0)$ does not belong to the set S (for example, $S = \{(0,\pm k),(\pm k,0) | k \neq 0\}$), dilation may well not be extensive.

Erosion

Just as dilation typically fills in the original image in a way characterized by the shape of the structuring element, so erosion removes parts of it. It is denoted by $\ominus$ and the erosion of I by S is defined by

$$I \ominus S := \{e | e + s \in I \text{ for all } s \in S\} \tag{72.38}$$

which is equivalent to

$$I \ominus S := \{e | \text{ for every } s \in S, \text{ there is an element } i \in I \text{ such that } e = i - s\} \tag{72.39}$$

(a)

I S $I \oplus S$

(b)

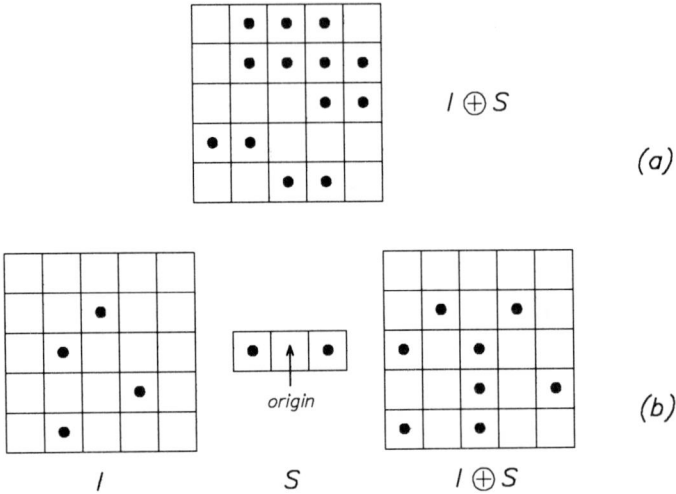

Fig. 72.9: (a) Dilation of the image I by the structuring element S of Fig. 72.8. Note that the result is the union of the translates of each point of I by S. Thus for the first row, for example, the result is a black spot at $(0,1)+(0,2)$ and $(0,1)+(0,1)$ and $(2,0)+(0,0)$ and $(0,2)+(1,0)$.
(b) Dilation by a more elaborate structuring element of which the origin is not a member. The resemblance between I and $I \oplus S$ is slight.

Once again, consideration of translation this time of the structuring element S, gives a simple picture of erosion:

$$I \ominus S = \{e|(S)_e \subseteq I\} \qquad\qquad (72.40)$$

Thus we picture the structuring element being moved over the image, point by point: we retain all points for which the structuring element is wholly within the image and discard the others. The erosion can in consequence be written as the intersection of all translations of I by the elements s of S, with a sign change:

$$I \ominus S = \bigcap_{s \in S} (I)_{-s} \qquad\qquad (72.41)$$

Figure 72.10 illustrates various ways of picturing erosion. Erosion, like dilation, is increasing ($I_1 \subseteq I_2$ implies $I_1 \ominus S \subseteq I_2 \ominus S$). If S contains the origin (that is, $(0,0)$ is 'black'), then erosion is anti-extensive in the sense that $I \ominus S$ is contained within I; this is not necessarily true if S does not contain the origin.

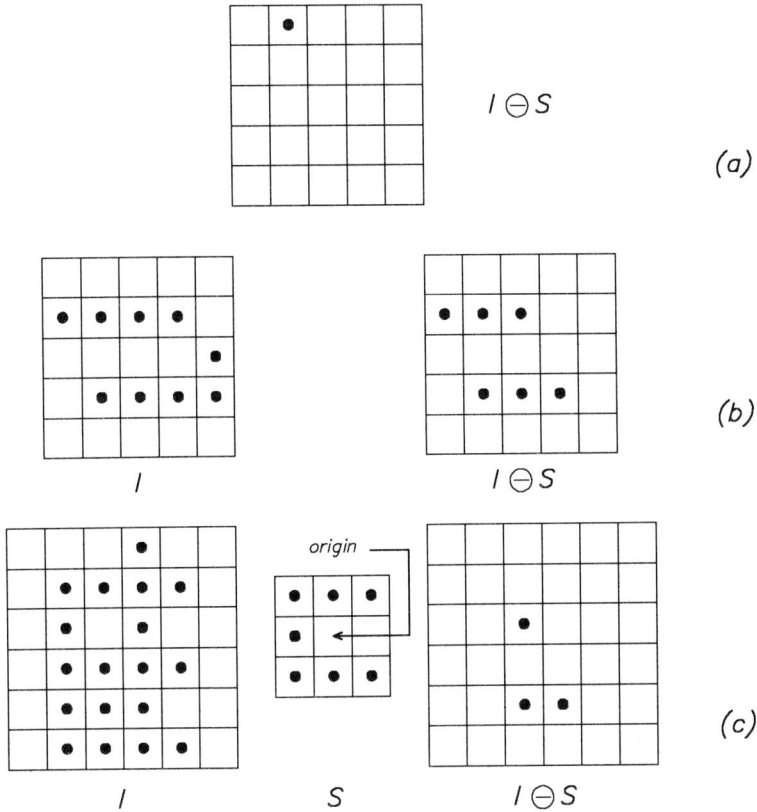

Fig. 72.10: (a) Erosion of the image I by the structuring element S of Fig. 72.8. Note that the eroded image may be obtained by placing the origin of the structuring element at each pixel of I and retaining the pixel if and only if both points of S coincide with points of I. (b) Another example using the same structuring element. (c) Erosion by a more elaborate structuring element, of which the origin is not a member.

Interpretation in terms of convolution

The reader who is meeting dilation and erosion for the first time may well sense that they are not too different from convolution products, though of course they cannot be quite the same since they are not linear: the dilation of a sum is not the same as the sum of dilations. In fact, both dilation and erosion can be written as a convolution followed by a thresholding operation (Huang *et al.*, 1989, Section 4.3, where the proof is given explicitly; Mazille, 1989, which is wholly devoted to 'mathematical morphology and convolutions'; de Bougrenet de la Tocnaye and Hillion, 1992).

Combinations of dilation and erosion

Repeated dilation or repeated erosion of a given image continues to alter the latter. Operations that have no further effect after one application are said to be *idempotent* and it is of interest to enquire what combinations of dilations and erosions exhibit this property, if any. The answer is simple: both dilation followed by erosion and erosion followed by dilation are idempotent. They have hence acquired names: *opening* of an image I by a structuring element S is denoted by $I \circ S$:

$$I \circ S := (I \ominus S) \oplus S \qquad \text{opening} \qquad (72.42)$$

Closure of an image I by a structuring element S is denoted by $\bullet$:

$$I \bullet S := (I \oplus S) \ominus S \qquad \text{closure} \qquad (72.43)$$

Idempotency implies that

$$(I \circ S) \circ S = I \circ S \qquad (72.44)$$

and

$$(I \bullet S) \bullet S = I \bullet S \qquad (72.45)$$

Many extremely interesting and illuminating properties can be found by considering the *complement* of a set, I^c for the image I, which is just the set of white pixels or zeros, and the *reflection* of a set, $\check{S}$ for the structuring element S, which is the reflection of the elements of S about the origin:

$$I^c = \{i | i \not\subset I\}$$
$$\check{S} = \{\check{s}| \text{ for } s \in S, \check{s} = -s\} \qquad (72.46)$$

For example, erosion and dilation are duals in the sense that

$$(I \ominus S)^c = I^c \oplus \check{S} \qquad (72.47)$$

as are opening and closure:

$$(I \bullet S)^c = I^c \circ \check{S} \qquad (72.48)$$

Note

The early work on mathematical morphology is conveniently described in Matheron (1975), Serra's treatises (1982, 1988) and in tutorial papers by him and colleagues (Serra, 1986, 1987; Jeulin, 1983, 1987). Independently, Sternberg had been developing similar ideas, which are described in Sternberg (1980, 1982, 1986). The importance of the early work of the Fontainebleau school was not widely appreciated and it is only relatively recently that it has become better known. This has revealed an inconsistency in notation between the various authors and we therefore state that the notation employed here is that adopted by Haralick, Sternberg and Zhuang (Haralick *et al.*, 1987). This is slightly different from that employed in the books of Coster and Chermant (1985) and Giardina and Dougherty (1988) and those of Serra (1982, 1988) but is the same as that of the important papers of Heijmans and Ronse (1990; Ronse and Heijmans, 1991). A very clear and careful comparison, in which the sources of confusion are identified, is to be found in Maragos and Schafer (1990).

72.3.3.2 Grey-level image morphology

All the results of binary morphology can be extended to images with more than two grey levels reasonably straightforwardly. One way of accomplishing this is to introduce the notions of the *top* of a set and its *umbra*. The top has its everyday meaning: if we picture the grey-level image as a three-dimensional surface, the height of which at any point (or rather, over any pixel) is equal to the grey level there, then the *top* is just the set of surface heights: if $T[I](x, y)$ denotes the top of the image I at (x, y), then

$$T[I](x, y) := \max\{g|(x, y, z) \in I\} \tag{72.49}$$

where $\{g\}$ denotes the set of possible grey levels (e.g. 0–255). The *umbra* U of a surface σ is a set containing the surface itself and everything beneath it:

$$U[\sigma] := \{x, y, g_e | g_e \le \sigma(x, y)\} \tag{72.50}$$

in which $\{g_e\}$ is the set $\{g\}$ plus negative values extending indefinitely, since it is sometimes necessary to regard the umbra as continuing to $-\infty$.

The *dilation* of a function I by a function S is again denoted by $I \oplus S$ and is defined in terms of the umbras:

$$I \oplus S := T[U[I] \oplus U[S]] \tag{72.51}$$

An equivalent definition has a distant relation to convolution:

$$(I \oplus S)(x, y) = \max_{\substack{g \in \{g\} \\ x-g, y-g \in I}} \{I(x - g, y - g) + S(g)\} \tag{72.52}$$

in which $\mathcal{I}$ is the region of the plane over which I is defined. Figure 72.11 shows (in one dimension) how I and S are defined, their umbras and tops and their dilation.

Erosion is defined in similar way:

$$
\begin{aligned}
I \ominus S : &= T[U[I] \ominus U[S]] \\
&= \min_{g \in \{g\}} \{I(x+g, y+g-S(g)\}
\end{aligned}
\tag{72.53}
$$

Again see Fig. 72.11.

The reason why we have been able to extend binary morphology to grey-level images so relatively easily becomes clear once we have seen the 'Umbra Homomorphism', which states that

$$
\begin{aligned}
U[I \oplus S] &= U[I] \oplus U[S] \\
U[I \ominus S] &= U[I] \ominus U[S]
\end{aligned}
\tag{72.54}
$$

It is then easy to show that dilation remains commutative

$$
I \oplus S = S \oplus I
\tag{72.55}
$$

and associative

$$
I \oplus (S_1 \oplus S_2) = (I \oplus S_1) \oplus S_2
\tag{72.56}
$$

and that

$$
(I \ominus S_1) \ominus S_2 = I \ominus (S_1 \oplus S_2)
\tag{72.57}
$$

Opening and closure are defined as in (72.42) and (72.43) and are idempotent. Thus

$$
\begin{aligned}
I \circ S &= (I \ominus S) \oplus S \qquad \text{opening} \tag{72.58} \\
I \bullet S &= (I \oplus S) \ominus S \qquad \text{closure} \tag{72.59}
\end{aligned}
$$

and it can be shown that

$$
I \circ S = T\left[\bigcup_{g|U[S]_g \subseteq U[I]} U[S]_g \right]
\tag{72.60}
$$

This last expression is the formal equivalent of a very convenient pictorial representation of opening. Suppose that the image I is regarded as a surface, the height of which at any point is equal to the grey level at that pixel; the structuring element is then likewise regarded as a smaller block, the top surface of which is likewise defined by the grey level of the element.

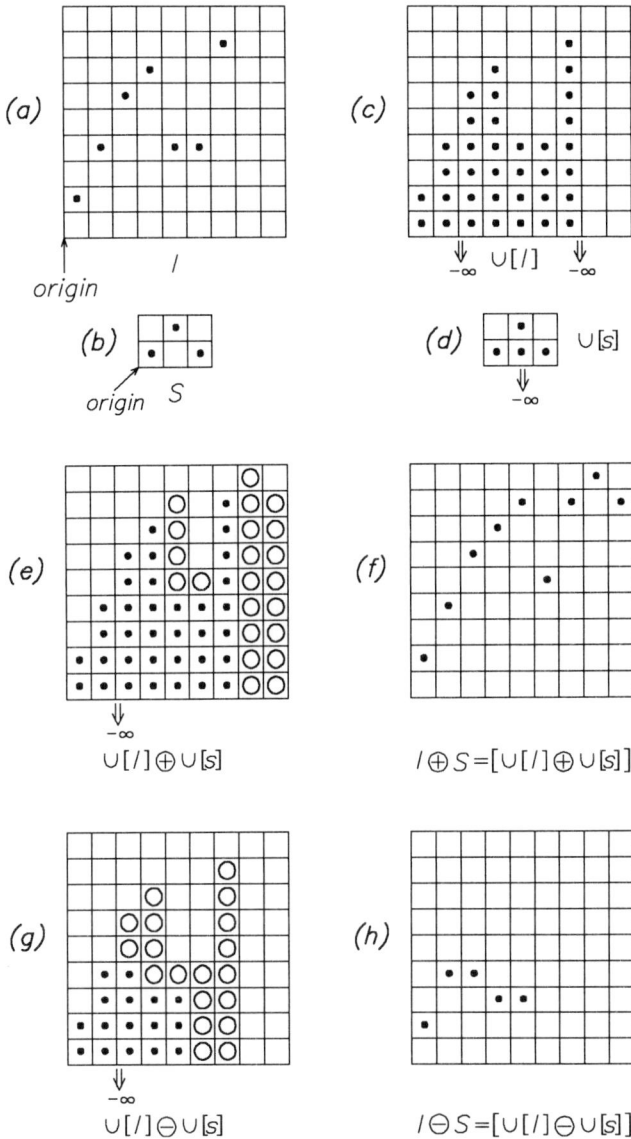

Fig. 72.11: The image I is here one-dimensional and the height above the origin represents the grey level. The top of the set I is thus just the set of the black points in I. The umbra of I is these black points together with all points below them, extending to minus infinity. The top and umbra of the structuring element S are defined similarly. (a) I; (b) S; (c) $U[I]$; (d) $U[S]$; (e) $U[I] \oplus U[S]$ (new points are shown as small circles); (f) $I \oplus S$; (g) $U[I] \ominus U[S]$ (suppressed points are shown as small circles); (h) $I \ominus S$.

The structuring element is then rubbed over the inside of the surface of I and the locus of the highest points reached by S is the opening of I by S. Closure is likewise the locus of the lowest points of S when the latter is inverted and rubbed over the outer surface of I. This is an expression of grey scale dilation–erosion duality and its extension to opening and closure: For $x \in (I \oplus S) \bigcap (I \ominus \check{S})$, we have

$$-(I \oplus S)(x) = ((-I) \ominus \check{S})(x) \qquad (72.61)$$

and

$$-(I \circ S) = (-I) \bullet \check{S} \qquad (72.62)$$

where the hachek ($\check{}$) again denotes reflection:

$$\check{S}(x) := S(-x) \qquad (72.63)$$

A recent complementary approach that appears to have many attractive features is the fuzzy morphology of Sinha and Dougherty (1992). Here, the transition from binary morphology to the morphology of images with more than two grey levels is accomplished by replacing the two grey levels 0 and 1 of binary images by the range $[0, 1]$. All the morphological operations are then defined with the aid of the membership function and the index function. The *membership function* $\mu_A(x)$ of a (fuzzy) set A is a number between zero and one that measures the degree to which x 'belongs' to A or its 'whiteness' or 'blackness'. It is the fuzzy analogue of the characteristic function of a crisp (not fuzzy) set, denoted by χ in (70.18):

$$\chi_A(x) := \begin{cases} 1 & \text{if } x \in A \\ 0 & \text{otherwise} \end{cases}$$

The generalized index function $\mathcal{I}$ proposed by Sinha and Dougherty, based on a number of desirable properties that we do not repeat here, is defined by

$$\mathcal{I}(A, B) = \inf_{x \in U} \mu_{A^c \Delta B}(x)$$

in which U is either the Euclidean plane $\mathbb{R} \times \mathbb{R}$ or the grid $\mathbb{Z} \times \mathbb{Z}$; A^c denotes the complement of A, for which

$$\mu_{A^c}(x) = 1 - \mu_A(x)$$

and the symbol Δ denotes 'bold union', defined by

$$\mu_{A \Delta B}(x) = \min[1, \mu_A(x) + \mu_B(x)]$$

Then erosion and dilation are defined in terms of the membership function and the index function as follows:

$$\mu_{A\ominus B}(x) = \mathcal{I}(\mathcal{T}(B;x), A)$$
$$\mu_{A\oplus B}(x) = \mu_{A^c\ominus B^c}(x) = \mathcal{I}^c(\mathcal{T}(-B;x), A^c)$$

in which $\mathcal{T}$ denotes translation,

$$\mu_{\mathcal{T}(A;v)}(x) = \mu_A(x - v)$$

We shall not pursue this further here. Many practical examples are to be found in the paper cited.

Note

Grey-level morphology is a new subject and many questions remain open. In particular, the above definitions of dilation and erosion, and hence of opening and closure, are certainly acceptable provided that the structuring element is flat (as it tends to be in examples) but can give rise to difficulties when it is not, as Heijmans and Ronse (1990) and Ronse (1990) point out. For discussion of the various approaches, see the work of Serra and Sternberg already cited; an early contribution was the paper of Nakagawa and Rosenfeld (1978). For a clear introductory account, see Dougherty (1992a) and for mathematical developments, Heijmans (1991), Heijmans and Toet (1991), Dougherty (1992b, 1993) and Maragos (1989)

Practical applications

The word filter is used with a particular meaning in many works on mathematical morphology, notably in those of Serra and colleagues; an operation is said to constitute a *morphological filter* if it possesses four properties: it must commute with translation, be continuous, be idempotent and preserve the inclusion relations $\supseteq$ and $\subseteq$. Explicitly, an operator O applied to a grey-level function I is a filter if and only if

(i)

$$O(kI) = kO(I) \qquad k \text{ constant} \tag{72.64}$$

(ii)

$$\text{If } I \text{ is continuous, then so is } O(I) \tag{72.65}$$

(iii)

$$OO(I) = O(I) \tag{72.66}$$

Repeated application of O has no further effect on its argument after the first.

(iv) For two functions I_1 and I_2, $I_1 \subseteq I_2$ implies $O(I_1) \subseteq O(I_2)$. This may be written

$$O(I \vee I_2) \supseteq O(I_1) \vee O(I_2)$$
$$O(I \wedge I_2) \subseteq O(I_1) \wedge O(I_2)$$

(72.67)

where $\vee$ denotes sup and $\wedge$ denotes inf.

We have already mentioned that dilation and erosion are not idempotent and hence cannot be filters in the above sense. Opening and closure, on the other hand, do satisfy all four conditions. These and many simple morphological operations based on them are used for image enhancement. Thus opening an image by a simple structuring element eliminates sharp isolated peaks and hence smooths the image in the way that sandpaper smooths the rough surface of wood. Closing an image fills in narrow crevices.

A number of useful operations on images that can be expressed in morphological terms are not filters in the sense defined in (72.64–67) and some authors (Maragos and Schafer, 1987, 1990) prefer to use the term in a less restrictive sense. Thus median filters and rank-order filters can be constructed from the basic morphological elements. Peaks and valleys can be emphasized rather than eliminated by subtracting the opened or closed image from the original. Structures of gradually increasing size can be enhanced by means of 'alternating sequential filters', in which the size of the structuring element is progressively altered. The structuring element can be *adaptive*, by which we mean that its shape or size can be governed by parameters that are allowed to vary as the structuring element explores the image, the variation being directed by the local properties of the image. Morphological gradients can be defined, in terms of the difference between the dilation and erosion of an image by the same structuring element.

The use of morphological operators for edge detection is explored in detail by Lee *et al.* (1987), who not only devise edge detectors that are insensitive to noise but also compare them with the traditional detectors described earlier. The simplest kind of morphological edge detectors are based on the difference between the original image and its erosion or dilation. In the case of erosion of an image I, we form the edge strength

$$G_e = I - E$$

(72.68)

where

$$E = \min_{(i,j) \in S_0} I(x+i, y+j)$$

(72.69)

and the structuring element S_0 is a flat disc,

$$S_0 = \{(0,-1),(0,1),(-1,0),(1,0)\}$$

(72.70)

It can be shown that

$$G_e = \max_{(i,j)\in N_4} \{I(x,y) - I(i,j)\} \qquad (72.71)$$

where N_4 are the 4-neighbours of (x,y), so that the *erosion residue edge detector* is just the maximum of the four first differences around (x,y) in the directions north, south, east and west. A dilation residue edge detector can be defined similarly but despite various refinements, including extension of the neighbourhood (or change in the shape of S), such detectors are excessively noise-sensitive and may fail to detect sharp edges. For these reasons, Lee *et al.* define modified forms of these detectors, which prove to be free of these defects.

The improved dilation and erosion residue detectors are defined in terms of several structuring elements, namely:

$$S_1 := \{(-1,0),(0,1)\} \qquad S_2 := \{(0,-1),(1,0)\}$$
$$S_3 := \{(-1,0),(0,-1)\} \qquad S_4 := \{(0,1),(1,0)\} \qquad (72.72)$$
$$S := \{(-1,-1),(-1,1),(1,-1),(1,1)\}$$

The improved dilation residue detector is then defined by

$$G'_d := \min\{I \oplus S_0 - I, I \oplus S - I, g_d\} \qquad (72.73a)$$

where

$$g_d := \max\{|(I \oplus S_1 - I) - (I \oplus S_2 - I)|, |(I \oplus S_3 - I) - (I \oplus S_4 - I)|\} \qquad (72.73b)$$

The improved erosion residue detector is defined likewise:

$$G'_e := \min\{I - I \ominus S_0, I - I \ominus S, g_e\} \qquad (72.74a)$$

where

$$g_e := \max\{|(I - I \ominus S_1) - (I - I \ominus S_2)|, |(I - I \ominus S_3) - (I - I \ominus S_4)|\} \qquad (72.74b)$$

Each of these detectors is superior to the detectors G_e and G_d and the sum of the two detects sharp edges and ignores isolated noise points. The smallness of the neighbourhood considered does, however, leave all three sensitive to extended noise. In order to overcome this undesirable sensitivity, Lee *et al.* introduce a 'blur-and-minimum' operator. Here a local blurring is deliberately introduced before calculating the detector, G_b:

$$G_b := \min\{I_b - I_b \ominus S_b, I_b \oplus S - I_b\} \qquad (72.75a)$$

in which I_b is the blurred input image and

$$S_b = (-1, 0, 1) \qquad\qquad (72.75b)$$

We refer to the original article for tests and additional explanations.

The present account is confined to the more elementary parts of the subject. Morphological filters have been the subject of intensive study. For a practical application (to scanning tunnelling microscope images), see Weisman *et al.* (1992) and for additional theoretical studies, see Dougherty and Kraus (1991), Svalbe and Jones (1992), Jones and Svalbe (1992, 1994), Dougherty (1992c,d), Loce and Dougherty (1992a,b), Koskinen and Astola (1992), Dougherty and Haralick (1992), Dougherty and Loce (1993), Song and Delp (1990), Serra and Vincent (1992) and Ronse (1993).

Linearity

The absence of linearity in morphology is usually stressed and is vividly illustrated by recognizing that morphology, with its emphasis on max and min (or sup and inf) operations, is well-adapted to images in which an opaque foreground object obscures some feature in the background: no amount of 'deconvolution' can now retrieve the lost detail, unlike the situation in which the background is blurred by additive noise and can be restored to some extent, by Wiener filtering for example.

In fact this attitude is an over-simplification and a weak form of linearity does exist in mathematical morphology and is of the highest practical importance (Maragos and Ziff, 1990). We merely state the result here, without proof: certain grey-level morphological operations, notably erosion, dilation, opening and closure *by a binary (i.e. flat) structuring element* obey additive superposition in the sense that if the input image is expressed as the sum of its binary thresholds, then the result of applying one of these operations is the sum of the result of applying the same operator to each of the binary threshold images. Maragos and Ziff demonstrate that this weak or *threshold-linear superposition* is applicable to several of the most widely used classes of derived operators, namely differences between an image and its opening (for highlighting spikes or fissures) or its closure (for emphasizing edges); grey-level skeletonization (see Section 76.3.2); and establishment of differential size distributions by using a sequence of structuring elements of different dimensions. This result eases the human interpretation and perhaps the hardware implementation of these operations (Hereford and Rhodes, 1988; Ochoa *et al.*, 1987). Mathematically inclined readers will be interested to know that the class of systems that obey the threshold-linear superposition forms a vector space over the field of real numbers under (suitably defined) vector addition and scalar multiplication (Theorem A of Maragos and Ziff, 1990).

Linear and morphological filters can be represented as different sequences of projections and cross-sections through a *floating stack array*. The latter is obtained from a two-dimensional (or N-dimensional) grey-level image by mapping the latter into a 5-dimensional (or $N + 3$-dimensional) binary space. The new dimensions are stack decomposition, a neighbourhood window and stack decomposition of a related kernel. For extensive analysis of such arrays, see Wilson (1992b).

72.4 Image algebraic representation of enhancement

Formation of the histogram

This is an example of the use of parametrized templates. Let a be a (real-valued) image and let $G = (0, 1 \ldots g)$ be the range of grey-level values, typically $g = 255$. The image a is defined on X. We define a template t_a such that

$$t_{a,x}(y) = \begin{cases} 1 & \text{if } y = a(x) \\ 0 & \text{otherwise} \end{cases} \tag{72.76}$$

The *histogram* h of the image a is then given by

$$h = t_a \oplus 1 \tag{72.77}$$

where 1 is an image with grey-level unity everywhere. The function $h(k) = \sum_x t_{a,x}(k)$ is equal to the number of pixels of a with grey level k.

Convolutional filters

The simplest of these consists in replacing the grey-level value at every pixel by the average of that value and those of its neighbours (four closest neighbours or eight for a 3×3 neighbourhood). For the 3×3 case, we define

$$t = \begin{matrix} 1 & 1 & 1 \\ 1 & ① & 1 \\ 1 & 1 & 1 \end{matrix} \tag{72.78}$$

whereupon local averaging of a is achieved by forming $\frac{1}{9}a \oplus t$.

A more complicated example is Sobel edge detection, in which, as explained in Section 72.3, we first form $a_1 \oplus s_1$ and $a_1 \oplus s_2$, where

$$s_1 = \begin{matrix} -1 & 0 & 1 \\ -2 & & 2 \\ -1 & 0 & 1 \end{matrix} \qquad s_2 = \begin{matrix} -1 & -2 & -1 \\ 0 & & 0 \\ 1 & 2 & 1 \end{matrix} \tag{72.79}$$

and then evaluate

$$a_2 = \{(a_1 \oplus s_1)^2 + (a_1 \oplus s_2)^2\}^{\frac{1}{2}} \qquad (72.80)$$

where a_1 is the original image and a_2 is the edge-enhanced version.

The Kirsch edge detector (Section 72.3) is likewise very simply expressed; here, each pixel of a is replaced by

$$c = \max\{1, \max\{|5(a_i + a_{i+1} + a_{i+2}) - 3(a_{i+3} + \dots a_{i+7}| \,|\, i = 1, 2 \dots 8\}\}$$

in which subscript addition is modulo 8 and the labels are defined as follows:

$$\begin{array}{ccc} a_4 & a_3 & a_2 \\ a_5 & & a_1 \\ a_6 & a_7 & a_8 \end{array} \qquad (72.81)$$

This is encapsulated in the formula

$$1 \vee (\overset{8}{\underset{i=1}{\vee}} |a \oplus t_i|) \qquad (72.82)$$

where t_i is a parametrized template characterized by

$$t_{i,y} = \left\{ x, t_{i,y}(x) \,\middle|\, t_{i,y} \begin{array}{ll} = 5 & \text{if } x = x_i, x_{i+1}, x_{i+2} \\ = 3 & \text{if } x = x_{i+3} \dots x_{i+7} \\ = 0 & \text{otherwise} \end{array} \right\} \qquad (72.83)$$

The neighbourhood labelling is the same as for a.

72.5 Enhancement in electron microscopy

The shortcomings of high-voltage electron micrographs have been compensated by various techniques, notably edge-enhancement and histogram equalization, by Suzuki et al. (1977), Kawata et al. (1978/79), Matsuda et al. (1980). For examples in other branches of electron microscopy, see Heeke (1980), Russ and Russ (1984), Simon (1969, 1970), Braggins et al. (1971), Heinrich et al. (1970), Bahr et al. (1972), Taylor and Gopinath (1973), Vicario and Pitaval (1973), Yew and Pease (1974), Baggett and Glassman (1974), Smith et al. (1977a), Goldfarb and Frank (1978), Vicario et al. (1979), Spivak et al. (1980), Okagaki et al. (1980), Oho et al. (1984, 1985, 1986a,b, 1987, 1990), Osumi et al. (1984), Desai and Reimer (1985), Sasaki et al. (1986a,b), Sasov (1986), Baba et al. (1986b), Hashimoto et al. (1986), Newbury et al. (1988), Russ (1990), Bonnet et al. (1992a,b) and Michel et al. (1993).

73
Linear Restoration

73.1 Introduction

The aim of image restoration is to render the interpretation of the image more reliable or even, in extreme cases, to make it possible. The unreliability and other types of difficulty may arise in different ways. In very high resolution microscopy, for example, detail close in size to the instrumental limit of resolution may not be represented faithfully in the image. We recall that the phase and amplitude transfer functions of the TEM are oscillatory and that even for strongly scattering thin objects, the contrast can be written as the sum of the result of the linear theory and additional quadratic terms. The study of very fragile specimen structures offers another example of the need for restoration; here, the image will have been recorded in low-dose conditions, where the number of electrons per unit area is so low that the unaided eye can discern little or no meaningful contrast. Again, structural information may be available in the form of a large set of images, from which the three-dimensional architecture can perhaps be calculated by computer but never by the brain. As a final example, we note that access to certain information requires a knowledge of the amplitude and phase of the wavefunction whereas recording media respond only to the amplitude.

The methods grouped in these chapters on restoration have an important point in common: they exploit all available information about the image-forming process and indeed, the recording of particular images designed to yield specific numerical information about this process is an important part of some techniques (see Section 77.2). In the remainder of this chapter, we examine the linear methods that have been used to compensate for the adverse effects of the oscillatory nature of the electron microscope transfer functions. In Chapter 74, the nonlinear methods used to solve the 'phase problem' are examined. Finally, three-dimensional reconstruction is the subject of Chapter 75.

73.2 Extended Wiener filters

The transmission electron microscope conveys information from the exit
surface of the specimen to the recording plane in a selective manner, which
is characterized by its transfer function (65.16). If the specimen can be
regarded as a weak scatterer, the corresponding phase and amplitude variations of the object wavefunction will be linearly related to the *intensity*
variation at the image, to a first approximation at least. This is explained
in detail in Chapter 65, where we saw that the Fourier transform of the
image is a superposition of the *filtered* transforms of the specimen phase
and amplitude variations:

$$S_c = \mathcal{F}^-(C) = K_p \tilde{\eta} + K_a \tilde{\sigma} \tag{73.1}$$

Owing to the oscillatory nature of the filters, K_p and K_a, some information about the specimen is lost. This can be remedied by recording a set
of images, obtained in different imaging conditions, and combining these
suitably to recreate a faithful image. Examination of the transfer function
shows that we can create the set of images by altering the aperture function
(T_A) or by varying the defocus (Δ); it is hardly realistic to contemplate
using different C_s values. Even alteration of the aperture function, though
possible in principle by using complementary sector-shaped apertures for
example, has a host of practical disadvantages and variation of the defocus
is really the only feasible possibility.

The problem may be stated as follows: given a set of M images with
measured contrast spectra $S_c^{(j)}$, $j = 1, 2 \ldots M$, what choice of filters, w_j,
will ensure that the estimate

$$\hat{S} = \sum_{j=1}^{M} w_j S_c^{(j)} \tag{73.2}$$

is best in some well-defined sense? It is usual to choose the best estimate in
the *least-squares* or *minimum mean-square error* sense, whereby the mean
of the (modulus)^2 of the difference between the estimate and the true image
is minimized. This is intuitively plausible and has the added attraction that
the problem then *has* an explicit solution. It is not always realized that
this approach has the drawback that 'good' and 'bad' parts of the image
are treated alike. There is a real danger that in improving the latter, we
degrade the former; the unfortunate consequences of 'levelling down' are
notorious.

The use of linear filters of the form (73.2) in electron microscopy was
first proposed by Schiske, who gave the solution for a complex specimen in

the absence of noise (Schiske, 1968) and subsequently analysed the more realistic case including noise (Schiske, 1973). We now examine the latter situation, following Schiske's reasoning in all essentials. We shall see in Section 73.3 that the process can be enriched by incorporating constraints but this extended type of filter has not yet been employed in electron microscopy in the form presented there. The introduction of such constraints is, however, a special case of the general technique of *regularization*, whereby the ill-posed nature of some restoration problems can be remedied, and we shall meet it again in the chapters on the phase problem and three-dimensional reconstruction.

We set out from the equations

$$S_c^{(j)} = K_p^{(j)} \tilde{\eta} + K_a^{(j)} \tilde{\sigma} + N^{(j)} \tag{73.3}$$

in which

$$K_p^{(j)} = \cos\left\{\pi\left(\frac{1}{2}C_s\lambda^3 q^4 - \Delta^{(j)}\lambda q^2\right)\right\}$$
$$K_a^{(j)} = -\sin\left\{\pi\left(\frac{1}{2}C_s\lambda^3 q^4 - \Delta^{(j)}\lambda q^2\right)\right\} \tag{73.4}$$

(65.30 and 66.16) and the $N^{(j)}$ denote the noise spectra. It is convenient to reorganize (73.3) in the form

$$S_c^{(j)} = A^{(j)}c + A^{(j)*}c_-^* + N^{(j)} \tag{73.5}$$

in which

$$A := \frac{1}{2}\exp(-i\chi)$$
$$c := \tilde{\sigma}(\mathbf{q}) - i\tilde{\eta}(\mathbf{q}) \tag{73.6}$$
$$c_- := \tilde{\sigma}(-\mathbf{q}) - i\tilde{\eta}(-\mathbf{q})$$

and seek an estimate, denoted by $\hat{S}$, of the true contrast spectrum, S_c, such that

$$\mathcal{E}|\hat{S} - S_c|^2 \to \min \tag{73.7}$$

As in Section 71.4, $\mathcal{E}$ denotes the mathematical expectation, or mean value. The estimate $\hat{S}$ is chosen to be a linear weighted combination of the recorded spectra, $S_c^{(j)}$ (73.2):

$$\hat{S} = \sum_{j=1}^{M} w_j S_c^{(j)} \tag{73.8}$$

and our task is to find the weights w_j for which (73.7) is satisfied. An ingenious way of simplifying the associated algebra has been found by Schiske (1973), whose procedure we follow in essentials.

The quantity to be minimized (73.7) is

$$\mathcal{E}(\sum_j w_j S^{(j)} - c)(\sum_k w_k^* S^{(k)*} - c^*) \tag{73.9}$$

Introducing

$$u := \sum_j w_j A^{(j)} - 1 \qquad v := \sum_j w_j A^{(j)*} \tag{73.10}$$

and substituting for $S^{(j)}$ from (73.5), we have

$$uu^* \mathcal{E}(cc^*) + uv^* \mathcal{E}(cc_-) + u^* v \mathcal{E}(c^* c_-^*)$$
$$+ vv^* \mathcal{E}(c_-^* c_-) + \sum_j w_j w_j^* \mathcal{E}(N^{(j)} N^{(j)*}) \to \min \tag{73.11}$$

On setting the derivative of this with respect to w_j^* equal to zero, recalling that $\partial u^* / \partial w_j^* = A^{(j)*}$ and $\partial v^* / \partial w_j^* = A^{(j)}$, we find

$$w_j = p A^{(j)*} + q A^{(j)} \tag{73.12}$$

where

$$p := -\frac{1}{\eta} \{ u \mathcal{E}(cc^*) + v \mathcal{E}(c^* c_-^*) \}$$
$$q := -\frac{1}{\eta} \{ u \mathcal{E}(cc_-) + v \mathcal{E}(c_- c_-^*) \} \tag{73.13}$$

and

$$\eta := \mathcal{E}(N^{(j)} N^{(j)*}) \tag{73.14}$$

We now substitute expression (73.12) for w_j back into u and v (73.10), giving

$$u = p a_1 + q a_2 - 1$$
$$v = p a_2^* + q a_1 \tag{73.15}$$

in which

$$a_1 := \sum_j A^{(j)} A^{(j)*} \qquad a_2 := \sum_j A^{(j)} A^{(j)} \tag{73.16}$$

These expressions are substituted into p and q (73.13) and, finally, we obtain a pair of simultaneous equations for p and q, from which the w_j have been eliminated by this iterative sequence.

Writing

$$e_1 = \mathcal{E}(cc^*) \qquad\qquad e_2 = \mathcal{E}(cc_-)$$
$$e_3 = \mathcal{E}(c^* c_-^*) = e_2^* \qquad e_4 = \mathcal{E}(c_- c_-^*) \tag{73.17}$$

we find

$$p\Delta = (e_1 e_4 - e_2^* e_2)a_1 + e_1 \eta$$
$$q\Delta = -(e_1 e_4 - e_2 e_2^*)a_2^* + e_2 \eta \qquad (73.18)$$

where

$$\Delta = (e_1 e_4 - e_2^* e_2)(a_1^2 - a_2 a_2^*) + (a_1 e_1 + a_2 e_2 + a_2^* e_2^* + a_1 e_4)\eta + \eta^2 \quad (73.19)$$

Finally, therefore, from (73.12)

$$w_j = \frac{e_1 e_4 - e_2^* e_2}{\Delta}(a_1 A^{(j)*} - a_2^* A^{(j)}) + \frac{\eta}{\Delta}(e_1 A^{(j)*} + e_2 A^{(j)}) \qquad (73.20)$$

This is a generalization of the Wiener filter (see for example Rosenfeld and Kak, 1982, Section 7.3) to which it reduces in the case of a single image ($j = 1$) with a single component (e.g. $\tilde{\sigma} = 0$ in 73.3). After some calculation, we then find

$$w = \frac{A - A^*}{(A - A^*)^2 + \eta/cc^*} \qquad (73.21)$$

A slightly different form of this minimum mean-square error filter has been proposed and tested by Kirkland (Kirkland, 1979; Kirkland et al., 1980, 1981; Kirkland and Siegel, 1979, 1981). If noise is neglected in (73.3), the minimum least-squares error filter is the so-called Moore–Penrose generalized inverse or pseudo-inverse (Moore, 1920; Penrose, 1955) of the matrix $\boldsymbol{A}$, with elements $A_{1j} = A^{(j)}$ and $A_{2j} = A^{(j)*}$. This pseudo-inverse is given by $(\boldsymbol{A}^\dagger \boldsymbol{A})^{-1} \boldsymbol{A}^\dagger$, in which the dagger (†) denotes the Hermitian adjoint and of course neither $\boldsymbol{A}$ nor its adjoint has an inverse. We denote* the pseudo-inverse by $\boldsymbol{A}^+$:

$$\boldsymbol{A}^+ = (\boldsymbol{A}^\dagger \boldsymbol{A})^{-1} \boldsymbol{A}^\dagger \qquad (73.22)$$

It is not difficult to express the pseudo-inverse as a series in terms of the rows of the singular-value decomposition of $\boldsymbol{A}$, that is, the eigenvectors of $\boldsymbol{A}\boldsymbol{A}^T$ and $\boldsymbol{A}^T\boldsymbol{A}$; for details, see Andrews and Hunt (1977) or Pratt (1978, 1991). Kirkland et al. (1980) obtain the minimum mean-square error solution for the estimate $\hat{S}$ such that

$$\hat{S} = \sum_{j=1}^{M} w A_j^+ S^{(j)} \qquad (73.23)$$

---

* There is no standard notation for the pseudo-inverse: Pratt (1978) uses $\boldsymbol{A}^-$ whereas Andrews and Hunt (1977) employ $[\boldsymbol{A}]^+$; in the mathematical literature, $\boldsymbol{A}^+$ is common (Albert, 1972; Rao and Mitra, 1971; Boullion and Odell, 1971).

This differs from Schiske's solution in that the j-dependence of the weights is predetermined and the unknown weight w, to be found, is a mere scaling factor, independent of j. Kirkland finds

$$w = \frac{1}{1 + \eta \operatorname{Tr}(\boldsymbol{A}^\dagger \boldsymbol{A})^{-1}} \qquad (73.24)$$

where Tr denotes the trace (or spur) of a matrix and

$$\eta := \frac{<N^{(j)}>^2}{M <S^{(j)}>^2} \qquad (73.25)$$

η is thus the mean noise-to-signal ratio.

A technique that may be thought of a continuous form of the inverse filter has been introduced and tested by Ikuta, Taniguchi and colleagues (Ikuta, 1989; Taniguchi *et al.*, 1990a–c, 1992a,b). We recall that the image contrast spectrum may be written

$$S_c(\boldsymbol{q}) = K_p(\boldsymbol{q}; \Delta)\tilde{\eta}(\boldsymbol{q}) + K_a(\boldsymbol{q}; \Delta)\tilde{\sigma}(\boldsymbol{q})$$

in which we show the defocus Δ explicitly as an argument of the transfer functions K_p and K_a. A pair of weighting functions, w_p and w_a, are then sought such that

$$\hat{S}_c(\boldsymbol{q}) = \int\limits_{-\infty}^{\infty} w_p(\Delta)K_p(\boldsymbol{q}; \Delta)\tilde{\eta}(\boldsymbol{q})\,d\Delta + \int\limits_{-\infty}^{\infty} w_a(\Delta)K_a(\boldsymbol{q}; \Delta)\tilde{\sigma}(\boldsymbol{q})\,d\Delta$$

is an improvement on $S_c(\boldsymbol{q})$. We re-write this relation in the form

$$\hat{S}_c(\boldsymbol{q}) =: \hat{K}_p(\boldsymbol{q})\tilde{\eta}(\boldsymbol{q}) + \hat{K}_a(\boldsymbol{q})\tilde{\sigma}(\boldsymbol{q})$$

where clearly

$$\hat{K}_p(\boldsymbol{q}) := \int\limits_{-\infty}^{\infty} w_p(\Delta)K_p(\boldsymbol{q}; \Delta)\,d\Delta$$

$$\hat{K}_a(\boldsymbol{q}) := \int\limits_{-\infty}^{\infty} w_a(\Delta)K_a(\boldsymbol{q}; \Delta)\,d\Delta$$

The sinusoidal form of K_p and K_a makes it easy to solve for w_p and w_a in terms of $\hat{K}_p$ and $\hat{K}_a$:

$$w_p(\Delta) = \int\limits_{0}^{\infty} \hat{K}_p(\boldsymbol{q}) \cos\left\{ \frac{2\pi}{\lambda}\left(\frac{1}{4}C_s\lambda^4 q^4 - \frac{1}{2}\Delta\lambda^2 q^2 \right) \right\} d\left(\frac{1}{2}\lambda q^2 \right)$$

$$w_a(\Delta) = \int\limits_{0}^{\infty} \hat{K}_a(\boldsymbol{q}) \sin\left\{ \frac{2\pi}{\lambda}\left(\frac{1}{4}C_s\lambda^4 q^4 - \frac{1}{2}\Delta\lambda^2 q^2 \right) \right\} d\left(\frac{1}{2}\lambda q^2 \right)$$

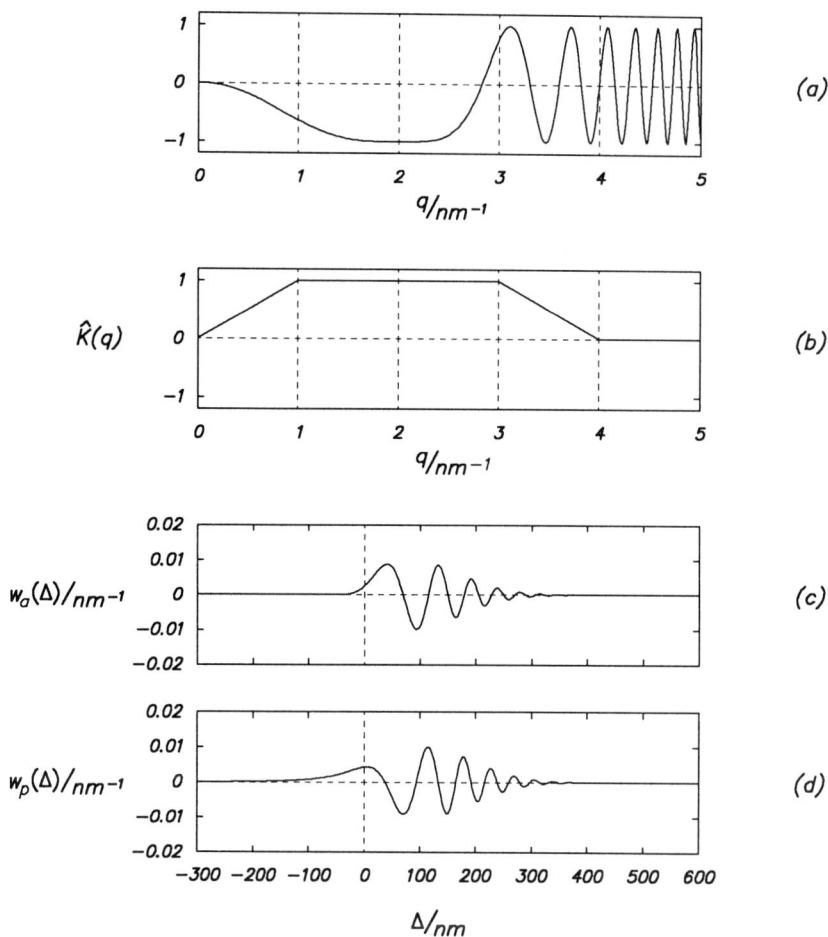

Fig. 73.1: (a) Phase-contrast transfer function for $C_s = 0.85$ mm and $U = 80$ kV at Scherzer focus. (b) Effective transfer function, $\hat{K}(q) = \hat{K}_a(q) = \hat{K}_p(q)$. (c) Amplitude weighting function $w_a(\Delta)$. (d) Phase weighting function $w_p(\Delta)$.

The desired transfer functions $\hat{K}_p$ and $\hat{K}_a$ will be chosen in such a way as to improve the image over the spatial frequency range within which the signal strength is sufficient. Ikuta (1989) and Taniguchi *et al.* (1992a,b) choose a trapezoidal curve for both functions (Fig. 73.1b). The corresponding weighting functions for $C_s = 0.85$ mm and an accelerating voltage of 80 kV ($\lambda = 4.2$ pm) are shown in Figs. 73.1c,d. For further details and discussion of the effect of inaccuracies and partial coherence, we refer to the original papers.

The Wiener filter has been extended to fit various experimental situations, notably those in which the statistical assumptions made in deriving the form of the filter are not satisfied. One such extension has acquired the name 'kriging' in the geostatistical literature and has been introduced into electron microscopy by Prasad *et al.* (1991), who applied it to energy-filtered images and electron energy-loss spectra (EELS). A clear general account is given by Jeulin and Renard (1992). With the notation of Section 70.3, an estimate $\hat{f}$ of the image f is obtained from the measured data g in the neighbourhood of each pixel; in one dimension, we write

$$\hat{f}(j) = \sum_{i \in W} \lambda_i g(i)$$

where the λ_i are weights, to be determined, and W is the window defining the neighbourhood adopted. The weights are obtained by solving the equations

$$\sum_k \lambda_k \gamma(k - i) + \mu = \gamma'(j - i)$$

$$\sum_k \lambda_k = 1$$

in which γ is the *variogram* of g and γ' is the *cross-variogram* of f and g:

$$\gamma(j) := \frac{1}{2} \mathcal{E}\{g(i + j) - g(i)\}^2$$

$$\gamma'(j) := \frac{1}{2} \mathcal{E}[\{f(i + j) - f(i)\}\{g(i + j) - g(i)\}]$$

Prasad *et al.* applied the technique to sets of energy-filtered images and found it to give a significant improvement in signal-to-noise ratio without perceptible loss of spatial resolution.

An intrinsic source of error in inverse and Wiener filters is the finite nature of the calculation. We illustrate this for the simple case of an object with a single component

$$g_{ij} = \sum \sum h_{kl} f_{i-k,j-l} + n_{ij}$$

(70.40 with a small change in notation, $\kappa \to h$ and additive noise n_{ij}). We neglect the noise term, which is not relevant here. The discrete Fourier transform of this equation may be written

$$G_{uv} = H_{uv} F_{uv} + E_{uv}$$

in which $G = \mathcal{F}(g)$, $H = \mathcal{F}(h)$ and $F = \mathcal{F}(f)$. The term in E is an *error* term, which is due to the lack of periodicity of the image

$$
E_{uv} = \sum_k \sum_l h_{kl} W^{ku+lv} \Bigg\{ \sum_{m=1}^{l} (G^{(1)}_{u,-m} - G^{(1)}_{u,N-m}) W^{-mv}
$$
$$
+ \sum_{n=1}^{k} (G^{(2)}_{-n,v} - G^{(2)}_{N-n,v}) W^{-nu}
$$
$$
+ \sum_{m=1}^{l} \sum_{n=1}^{k} (f_{-n,-m} + f_{N-n,N-m} - f_{N-n,-m} - f_{-n,N-m}) W^{-nu-mv} \Bigg\}
$$

This error is called the *edge error* by Lim *et al.* (1991) and Tan *et al.* (1991), who have examined it carefully. To remedy the ill effects of E, it is suggested that the image be divided into regions, typically nine (3 × 3); four square corner regions, four rectangular side regions and a large square central region (for a square image). A window function is associated with each region and chosen so that the resulting error E is as small as possible. The full calculation is given in Tan *et al.* (1991), who find that the appropriate windows are all sums over h_{kl} for different ranges of the indexes k and l apart from the central window, which is set equal to unity.

For further discussion of inverse and Wiener-like filters, see Ansley (1972), Frank (1972a), Herrmann *et al.* (1978), Lee (1983), Lenz (1975), Liu and Gallagher (1974), Misell and Childs (1972), Saxton (1986a), Uyeda *et al.* (1978) and Zemlin (1978). This subject was extensively studied in the 1970s, when many attempts to 'deblur holographically' were made. The abundant literature is listed in Section 2.2 of Hawkes (1978a) and in Hawkes (1982c).

73.3 Filtering with constraints

From the foregoing discussion of filters that are generalizations of the simple Wiener filter, a major disadvantage is apparent: the power spectra of the random fields to which picture and noise are assumed to belong must be known or estimated. Knowledge, or at least a plausible guess, of the noise spectrum is a less demanding requirement and methods of restoring images degraded by noise and by the effect of a transfer function have been devised when only the noise variance is known. This is much closer to the situation in electron microscopy and we therefore present the method here, though it has not yet found much practical application.

Introduced into image processing by Hunt (1972, 1973, 1975), it was extended by Hawkes (1974b) to include complex specimen transparency

($\tilde{\sigma} \neq 0$ in 73.1). The one-dimensional situation had been analysed earlier by Phillips (1962) and Twomey (1963).

We again set out from the basic filter relation in the form (73.3)

$$S_c^{(j)} = A^{(j)}c + A^{(j)*}c_-^* + N^{(j)} \qquad (73.26)$$

Hitherto, we have sought an estimate $\hat{S}$ of S in the form

$$\hat{S} = \sum_j w_j S_c^{(j)} \qquad (73.27)$$

subject to the constraint

$$\mathcal{E}|\hat{S} - S_c|^2 \to \min \qquad (73.28)$$

We now generalize the constraint and require that

$$U = \sum_{p,q} \hat{S}_{pq} C_{pq} C_{pq}^* \hat{S}_{pq}^* \to \text{extremum} \qquad (73.29)$$

subject to

$$V^{(j)} = 0 \qquad \forall j \qquad (73.30)$$

where

$$V^{(j)} := \sum_{p,q} \{(S_{pq}^{(j)} - A_{pq}^{(j)} c_{pq} - A_{N-p,N-q}^{(j)*} c_{N-p,N-q}^*)$$
$$\times (S_{pq}^{(j)*} - A_{pq}^{(j)*} c_{pq}^* - A_{N-p,N-q}^{(j)} c_{N-p,N-q}) - N_{pq} N_{pq}^*\} = 0$$
$$(73.31)$$

Condition (73.29) is satisfied by requiring that

$$\frac{\partial W}{\partial c_{rs}^*} = 0 \qquad (73.32)$$

where

$$W = U + \sum_j \lambda_j V^{(j)} \qquad (73.33)$$

and the λ_j are unknown Lagrange multipliers, which must be determined in the course of the calculation. After a certain amount of straightforward manipulation, given in full in Hawkes (1974b), the following expressions for the estimates $\hat{\eta}$ and $\hat{\sigma}$ of $\tilde{\eta}$ and $\tilde{\sigma}$ respectively are found

$$\hat{\sigma}_{rs} = \sum_j Q_{rs}^{\sigma,j} S_{rs}^{(j)}$$
$$\hat{\eta}_{rs} = \sum_j Q_{rs}^{\eta,j} S_{rs}^{(j)} \qquad (73.34)$$

in which

$$\Delta_{rs}Q_{rs}^{\sigma,j} = -\lambda_j\{(\overline{\mu}_{rs} - \overline{\nu}_{rs})A_{rs}^{(j)*} - (\mu_{rs} - \nu_{rs})A_{N-r,N-s}^{(j)}\}$$
$$\Delta_{rs}Q_{rs}^{\eta,j} = -i\lambda_\alpha\{(\overline{\mu}_{rs} + \overline{\nu}_{rs})A_{rs}^{(j)*} - (\mu_{rs} + \nu_{rs})A_{N-r,N-s}^{(j)}\}$$

$$(73.35)$$

where

$$\Delta_{rs} = \mu_{rs}\overline{\nu}_{rs} - \overline{\mu}_{rs}\nu_{rs} \tag{73.36}$$

and

$$\mu_{rs} = C_{rs}C_{rs}^* + 2\sum_j \lambda_j A_{rs}^{(j)} A_{rs}^{(j)*}$$

$$\nu_{rs} = 2\sum_j \lambda_j A_{rs}^{(j)*} A_{N-r,N-s}^{(j)*}$$

$$(73.37)$$

$$\overline{\mu}_{rs} = 2\sum_j \lambda_j A_{rs}^{(j)} A_{N-r,N-s}^{(j)}$$

$$\overline{\nu}_{rs} = C_{N-r,N-s}C_{N-r,N-s}^* + 2\sum_j \lambda_\alpha A_{N-r,N-s}^{(j)} A_{N-r,N-s}^{(j)*}$$

These expressions can be cast into a form resembling the Schiske filter (73.20) by a suitable choice of notation. For choices of C that impose constraints of practical interest on the solution (smoothness in particular), see the papers by Hunt (1973), Phillips (1962) and Twomey (1963) already mentioned. Very compact matrix expressions for the Wiener, Schiske and constrained filters have been given by Schiske (1984), which greatly simplify their manipulation and comparison.

Another way of introducing constraints into filtering for noise-reduction has been found useful for improving STEM images (Unser, 1990). The idea here is to form a weighted sum of a noisy image, filtered in some way, and the original unfiltered image, thereby restoring detail that might have been smoothed out by the filter. The weights are determined by local properties of the image. The following account is based on Unser's work, which is a development of an earlier proposal of Lee (1980, 1981), also examined by Mastin (1985) and Kuan *et al.* (1985).

The recorded image g is assumed to be the sum of the true image f and zero-mean noise n:

$$g_{ij} = f_{ij} + n_{ij}$$

The variance $\sigma^2 = \mathcal{E}\{n_{ij}^2\}$ is known or estimated. In a first step, a convolutional filter h is applied to g, giving a first estimate of f, denoted by $\hat{f}^{(1)}$:

$$\hat{f}^{(1)} := g * h =: f' + n' \tag{73.38}$$

The filter h might typically be a moving-average, in which each pixel value is replaced by the mean of the pixel values in its immediate neighbourhood. (To ensure that this first estimate is unbiased when f is constant, the array h is normalized so that the sum of the elements is unity.) The residual noise coefficient, p,

$$p := \frac{1}{\sigma^2} \mathcal{E}\{n_{ij} n'_{ij}\} \tag{73.39}$$

can be estimated.

The object of the calculation is to find coefficients a_{ij}, b_{ij} such that the second estimate of f, denoted $\hat{f}^{(2)}$:

$$\hat{f}^{(2)}_{ij} := a_{ij} f_{ij} + b_{ij} \hat{f}^{(1)}_{ij} \tag{73.40}$$

is in some well-defined sense superior to f and $\hat{f}^{(1)}$. The coefficients a and b are taken to be constant over a region $\mathcal{A}$ within the image, though they may vary from region to region. As usual, we seek the best estimate in the least-squares sense:

$$\frac{1}{N_{\mathcal{A}}} \sum_{k,l \in \mathcal{A}} (\hat{f}^{(2)} - f)^2 \rightarrow \min$$

in which $N_{\mathcal{A}}$ is the number of pixels in the region $\mathcal{A}$. It is easily shown that

$$\begin{pmatrix} a \\ b \end{pmatrix} = \begin{pmatrix} S_{00} & S_{01} \\ S_{10} & S_{11} \end{pmatrix}^{-1} \begin{pmatrix} S_{00'} \\ S_{10'} \end{pmatrix} \tag{73.41a}$$

where

$$S_{00} = \frac{1}{N_{\mathcal{A}}} \sum g_{ij} g_{ij}$$
$$S_{01} = \frac{1}{N_{\mathcal{A}}} \sum g_{ij} \hat{f}^{(1)}_{ij}$$
$$S_{11} = \frac{1}{N_{\mathcal{A}}} \sum \hat{f}^{(1)}_{ij} \hat{f}^{(1)}_{ij}$$

$$S_{00'} = \frac{1}{N_{\mathcal{A}}} \sum g_{ij} f'_{ij}$$
$$S_{10'} = \frac{1}{N_{\mathcal{A}}} \sum \hat{f}^{(1)}_{ij} f'_{ij}$$
$$\tag{73.41b}$$

and the sums are taken over $\mathcal{A}$. Unser shows that $S_{00'}$ and $S_{10'}$ can be plausibly replaced by the following estimates:

$$\hat{S}_{00'} = S_{00} - \sigma^2$$
$$\hat{S}_{10'} = S_{01} - p\sigma^2$$
$$\tag{73.42}$$

It is reasonable to require that $a + b = 1$, since this ensures that the mean of $\hat{f}^{(2)}$ is the same as that of $\hat{f}^{(1)}$; stability in the sense that $\hat{f}^{(2)}$ falls between g and $\hat{f}^{(1)}$ is ensured by requiring that in addition, a lies between zero and one.

Unser obtains explicit expressions for a_{ij} and b_{ij}, with and without applying the constraints. For a region $\mathcal{A}$ centred on the current pixel, dimensions $(2r+1) \times (2s+1)$:

$$a_{ij} = \frac{S_{00}S_{11} - S_{01}^2 - \sigma^2 S_{11} + p\sigma^2 S_{01}}{S_{00}S_{11} - S_{01}^2}$$

$$b_{ij} = \frac{\sigma^2 S_{01} + p\sigma^2 S_{00}}{S_{00}S_{11} - S_{01}^2}$$

(73.43)

for the unconstrained case; the suffixes (i, j) have been omitted on the right-hand sides (S_{00} denotes $S_{00}(i, j)$, for example). These define the Adaptive Least-Squares Filter (ALSF). With constraints, (73.43) are replaced by

$$a_{ij} = 1 - \frac{(1-p)\sigma^2}{P(i, j)}$$

(73.44a)

$$b_{ij} = 1 - a_{ij}$$

with

$$P(i, j) := S_{00} + S_{11} - 2S_{01}$$

(73.44b)

giving the Adaptive Constrained Least-Squares Filter (ACLSF).

Unser tests these procedures on real STEM images and compares them with earlier proposals. The constrained form is particularly easy to implement and gives good results.

73.4 Hoenders' procedure

The original feature of Hoenders' approach (Hoenders 1972a,b; Hoenders and Ferwerda, 1973a,b,c) is the use of a new sampling scheme (or discretization) of the image associated with a diffraction-plane wavefunction having finite support (that is, vanishing outside a certain known region, corresponding to the objective aperture). This limitation has wide-ranging repercussions, as we shall see in Chapter 74. Since the Fourier transform of the image wavefunction is confined to a band of values, as indeed is that of the image intensity, such functions are commonly said to be *bandlimited*.

We set out from the image contrast, $C(x_i, y_i)$, now defined by

$$C(x_i, y_i) = \frac{|\psi(x_i, y_i)|^2 - B}{B}$$

$$B := \left| \iint_{\mathcal{D}_0} K(x_i, y_i; x_o, y_o) \, dx_o \, dy_o \right|^2$$

(73.45)

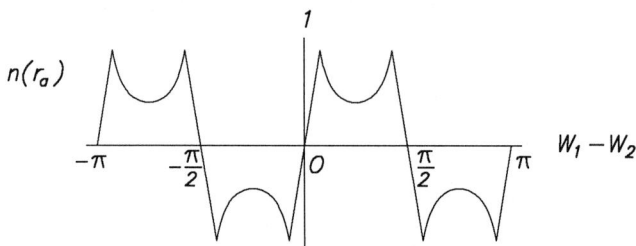

Fig. 73.2: The function $n(r_a)$ consisting of a cosecant curve, replaced by straight segments around multiples of $\pi/2$.

in which the integral of K is taken over the object plane, $\mathcal{D}_0$; Hoenders examines two situations, the unrealistic case of a rectangular aperture and the familiar circular aperture. We shall follow him in considering both, since his proposal is easier to grasp for the rectangular case. Hoenders introduces into the familiar imaging equation for a weakly scattering object the real and imaginary parts of the integral of K:

$$U(x_i, y_i) + \mathrm{i}V(x_i, y_i) := \iint_{\mathcal{D}_0} K(x_i, y_i; x_o, y_o)\, dx_o\, dy_o \qquad (73.46)$$

The contrast then consists of two parts:

$$BC(x_i, y_i) := U C_u + V C_v \qquad (73.47)$$

where

$$C_u := \iint_{\mathcal{D}_0} \iint_{\mathcal{D}_a} \exp \frac{2\pi \mathrm{i}}{\lambda f} \left\{ \left(x_o - \frac{x_i}{M} \right) x_a + \left(y_o - \frac{y_i}{M} \right) y_a \right\}$$
$$\times \left\{ \sin W(r_a)\tilde{\eta}(x_o, y_o) + \cos W(r_a)\tilde{\sigma}(x_o, y_o) \right\} dx_o\, dy_o\, dx_a\, dy_a$$
$$(73.48a)$$

$$C_v := \iint_{\mathcal{D}_0} \iint_{\mathcal{D}_a} \exp \frac{2\pi \mathrm{i}}{\lambda f} \{\cdot\} (\tilde{\eta} \cos W + \tilde{\sigma} \sin W)\, dx_o\, dy_o\, dx_a\, dy_a$$

$$(73.48b)$$

Hoenders argues that $|V| \ll |U|$ and considers two images with different defocus. This enables him to establish a pair of linear equations for $\iint K\tilde{\eta}\, dx_o\, dy_o$ and $\iint K\tilde{\sigma}\, dx_o\, dy_o$, which can be solved explicitly. Hoenders evaluates the resulting expressions as a sum of the form $\sum_{j=0}^{N} d_j \Delta^j$, the coefficients d_j being those occurring in the series expansion of a function $n(r_a)$, chosen to mimic $1/\sin(W_1 - W_2)$:

$$n(r_a) = \mathrm{cosec}\{W_1(r_a) - W_2(r_a)\}$$
$$\text{for } (n-1)\pi + \delta \leq W_1(r_a) - W_2(r_a) \leq n\pi - \delta \qquad (73.49)$$

and $n(r_a)$ is some continuous curve (typically a straight line) joining $(n\pi - \delta, \mathrm{cosec}(n\pi - \delta))$ and $(n\pi + \delta, \mathrm{cosec}(n\pi + \delta))$, as shown in Fig. 73.2; δ is an arbitrarily small positive quantity and $n = 0, \pm 1, \pm 2 \ldots$. Thus Hoenders' approach is all too likely to suffer from the same drawbacks as the simple inverse filter. The reconstruction of $\tilde{\eta}$ and $\tilde{\sigma}$ is completed by Fourier transforming the results of the foregoing calculations.

A much more detailed account of the process, with considerable discussion of the uncertainties involved, is to be found in Hoenders' work. An interesting extension is to be found in Ferwerda and Hoenders (1974), which leans to some extent on Hoenders and Ferwerda (1973a). It is shown there that (in one dimension) the problem of reconstructing the wavefunction for weakly scattering objects can be solved by measuring the recorded intensity at a particular set of points, which do not necessarily coincide with the regular pattern associated with the Whittaker–Shannon sampling theorem (cf. Section 71.2). This set is in turn determined by the kernel of the integral relating image and diffraction pattern. A further development is described by Ferwerda and Hoenders (1975).

73.5 Recursive filtering

The transform-based methods described in Sections 73.2–3 involve operations on large matrices. Moreover, the whole image must be available before the processing can begin. Recursive methods set out to improve the image as it is being acquired and measured and work therefore with a limited data set: the most recent image density measurement and its predecessors. We shall see that the recursive procedures typically answer the question: what linear combination of the current measurement and a subset of the earlier measurements (usually the measurements at the nearest neighbours to the current point) gives the best estimate of the true value at the current point? Before attempting to find answers, we analyse critically the spirit of these methods.

Recursive correction was originally devised for temporal signals and, in particular, to counter the effect of additive noise. The idea of in some way extrapolating past information to correct the current signal value then seems reasonable since past, present and future have unambiguous meanings, and the last of these is obviously inaccessible. For spatial signals,

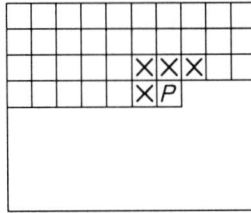

Fig. 73.3: Neighbours employed in recursive filtering (scanning from left to right and from top the bottom). P : current point; × : points already measured. The region covered with squares is $\mathcal{A}$. The pixels labelled × form $\mathcal{D}$.

images, these distinctions are far less natural; even in the case of a scanning microscope image, 'past' and 'future' are accidental, for the specimen could be rotated through 180° and the 'future' would then be scanned before the 'past'. Furthermore, the immediate predecessors of a temporal signal can reasonably be expected to be loosely correlated with the current value and hence useful for extrapolation; for a scanned image, however, the nearest neighbours will be situated on neighbouring scan lines as well as on the current line (Fig. 73.3). If we accept that nearby points on the preceding line are to be used in the extrapolation, we might as well wait until the subsequent line has also been scanned and hence use all the nearest neighbours of the current point and not just its predecessors. The intrinsically asymmetric extrapolation from the past to the present would then be replaced by a symmetric correction involving both future and past. If we waited a little longer, we could use the whole image, which would bring us back to the transform approach. For images, then, the attraction of recursive methods is that only a small portion of the large data set representing the image is employed at any given time, typically parts of three lines of the image matrix if the eight nearest neighbours of each (non-edge) point are employed (Fig. 73.4).

If the image contrast is related to the phase and amplitude components of the specimen transparency as in (73.1), recursive methods become even less attractive. Two images are now required so that for one at least, all the image intensity values are available. Furthermore, the full calculation shows that inaccessible cross-correlations appear in the recursion relations even if we limit these to nearest neighbours.

For all these reasons we give only a very superficial account of recursion; for more details, see Rosenfeld and Kak (1982), Panda and Kak (1976), Jain (1989) and, for discussion of the use of the method in electron microscopy, Hawkes (1981). The historical development of the subject may be traced through the papers of Wong (1968, 1978), Habibi

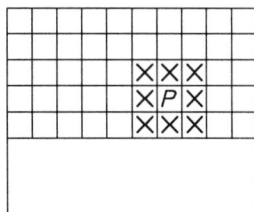

Fig. 73.4: Symmetric neighbourhood of the current point P involving data from three lines of the image, some of which are acquired after P.

(1972), Nahi (1972), Nahi and Franco (1973), Nahi and Habibi (1975), Jain and Angel (1974), Aboutalib and Silverman (1975), Strintzis (1976, 1977, 1978), Aboutalib *et al.* (1977), Jain (1977, 1981), Schoute *et al.* (1977), Woods and Radewan (1977), Murphy and Silverman (1978), Jain and Jain (1978), Chan (1980), O'Connor and Huang (1981), Rajala and de Figueiredo (1981), Ramamoorthy and Bruton (1981), Woods (1978, 1981, 1984), Dikshit (1982), Biemond (1986), Liu and Caelli (1988), Jain and Ranganath (1986) and Sections 8.10–12 of Jain (1989). Jain's very clear account shows explicitly how this technique is related to Kalman filtering.

We set out from the discrete convolution relating the image contrast g, to some specimen transparency function, f:

$$g(i,j) = \sum_{p,q} h(i - p, j - q) f(p,q) + n(i,j) \qquad (73.50)$$

in which h is a suitable point-spread function and n is noise. We now seek an estimate $\hat{f}$ of the true value of f in the form of a weighted linear combination of the contrast values already measured:

$$\hat{f}(i,j) = \sum_{k,l} d_{i,j}(k,l) g(k,l) \qquad (73.51)$$

The values of k and l correspond to measured points only. In Fig. 73.3, which represents an image being scanned from left to right and from top to bottom, for example, k, l must correspond to points in the region $\mathcal{A}(i,j)$. The estimate $\hat{f}$ is to be chosen such that $\mathcal{E}\{|f - \hat{f}|^2\}$ is minimized; the symbol $\mathcal{E}$ (estimation) again denotes averaging, which would ideally be taken over an ensemble of images but in practice will probably be taken over the image itself. (These averages will only be the same if the ergodic hypothesis is satisfied; as we have already observed, this is by no means likely!)

The weights $d_{i,j}(k,l)$ can be found by differentiating $\mathcal{E}(|f - \hat{f}|^2)$ with respect to each of them in turn, giving

$$\mathcal{E}[\{f(i,j) - \hat{f}(i,j)\}g(k,l)] = 0 \qquad (73.52)$$

In practice it is unrealistic to require that (73.52) be satisfied for all values of k,l lying in $\mathcal{A}(i,j)$, for points far from the current point cannot be expected to have much in common with the latter. Furthermore, the calculation becomes so unwieldy as to put its worth in doubt if more than a few neighbours of the current point are retained. We therefore limit the points to be included in the summation (73.51) to the domain $\mathcal{D}$ (Fig. 73.3); these points form the set $\{(i - k, j - l)|(k,l) = (1,0),(0,1),(1,1),(-1,1)\}$. Returning to (73.51), it seems preferable to discard the measured values at each point as soon as the improved estimate is generated and hence use these estimates in the recursion; instead of (73.51), then, we use

$$\hat{f}(i,j) = \sum_{k,l \in \mathcal{D}} d_{i,j}(k,l)\hat{f}(i-k,j-l) + d_{i,j}^{(0)}g(i,j)$$

Substitution of this into $\mathcal{E}\{|f - \hat{f}|^2\}$ and differentiation with respect to the weights generates a set of simultaneous equations. These can be solved straightforwardly to give the set of four weights $\{d_{i,j}(k,l)|k,l \in \mathcal{D}\}$ and $d_{i,j}^{(0)}$.

73.6 Other approaches

Nonlinear methods of solving the linear problem that is the subject of this chapter have also been employed. Thus Geman and Reynolds (1992) have tested a form of constrained restoration that belongs to the general class of regularization techniques some of which are examined in Chapter 74. Sezan and Trussell (1991) and Kuo and Mammone (1992) have employed the set-theoretic technique known as POCS, which we describe in the context of three-dimensional reconstruction (Section 75.2.4).

74

Nonlinear Restoration

74.1 Introduction

The conditions in which the image contrast is linearly related to the specimen transparency as in (73.1) are extremely restrictive and, all too often, unrealistic: at 100 kV, a phase shift of $\pi/2$ corresponds to about 20 nm of carbon and only 7 or 8 nm of tungsten (Misell, 1976, 1978a; Grinton and Cowley, 1971). A large body of work has therefore been devoted to the difficult nonlinear problem of deducing the amplitude and phase of the electron wave at the image, and hence at the exit surface of the specimen, from intensity records. As it is the phase that is difficult to recover, this is known generically as the *phase problem* and is encountered in many fields, not just in electron microscopy. We shall examine the various approaches to this problem in turn but first, make two general comments. It is assumed that information about the phase is present in the recorded images and can be extracted by exploiting knowledge of the image-forming process; this is of course exactly the situation in holography, where the reconstruction step is designed to decode the information stored in the hologram. We have seen in Chapter 63 that electron holography is indeed offering a solution, essentially thanks to the use of digital reconstruction, to the phase problem. The other point concerns the measure of experimental success achieved with these procedures: despite the numerous model and real attempts to test them, the results are disappointing. A comment by Saxton (1987c), who has been deeply involved in the subject since its inception, discourages any over-optimism: "In spite of many attempts during the last fifteen years to extend their range of validity, the only truly viable image restoration schemes are still those relying largely on the weak scattering approximation .... Perhaps what has been most effective in frustrating the various attempts made to extend the thickness range over which [linear] restoration schemes are valid is the fact that it is not simply one approximation that breaks down as the thickness is increased but as many as three quite different ones .... "

The extensive material to be covered is organized as follows. After a formal statement of the problem, we consider the extended linear approach. The use of multiple representations of the image with a standard (round) aperture is then analysed, after which we explain how analyticity can be invoked to provide a solution. We conclude with an account of some other proposals that have been made. We have to emphasize that entire books or long review articles have been devoted to many of these topics and our aim is not to rival these here; we hope merely to provide enough background knowledge to enable the reader to follow them and bring a critical spirit to new proposals. The bibliographic material should make up for any inadequacies.

Formal statement of the problem

The wavefunction in the image plane is a complex quantity; with suitable scaling,

$$
\begin{aligned}
\psi_i &= 1 + \Re(\psi_i - 1) + i\Im(\psi_i) \\
&=: 1 + \psi^{(r)} + i\psi^{(i)}
\end{aligned}
\tag{74.1}
$$

The (scaled) image intensity is therefore

$$
I_i := \psi_i \psi_i^* = 1 + 2\psi^{(r)} + \psi^{(r)2} + \psi^{(i)2}
\tag{74.2}
$$

In the weak-scattering approximation, the quadratic terms are neglected and the relation between image intensity and the real part of the wavefunction is linear. The latter is in turn related to the real and imaginary parts of the specimen transparency via the phase and amplitude transfer functions. The complex specimen transparency can hence be recovered, in principle at least, from two (or more) images of the same object taken at different defocus values (Section 73.2).

In the present chapter, we no longer assume that the specimen scatters weakly; the quadratic terms in (74.2) are retained. We examine in turn the numerous techniques that have been proposed to obtain $\psi^{(i)}$ and $\psi^{(r)}$ in these conditions. None has proved entirely satisfactory.

74.2 Extended linear approximation

If the weak-scattering approximation cannot be defended but is nevertheless not grossly inadequate, an extension of the linear theory may be useful. The idea is simple: estimates of the real and imaginary parts of the specimen transparency (η and σ) are obtained using the methods of Section 73.2; these estimates are then used to recalculate the wavefunction in the defocused image planes, including the nonlinear terms in (74.2), thereby yielding an improved estimate of η and σ. The whole process can be iterated until no further change in η and σ is perceptible. Partial coherence can be included with only modest extra effort.

Perhaps because the class of 'semi-weak' specimens is somewhat vague and difficult to define, the method has hardly ever been used in practice in normal microscope operating conditions, though we shall meet it again in Section 74.4.3 in connection with half-plane apertures. Model studies of a version of the technique have, however, been made by van Toorn *et al.* (1978) and we now summarize these briefly. We follow van Toorn *et al.* in confining the discussion to one-dimensional specimens; their comment that extension to two dimensions should not be difficult in anisoplanatic conditions seems reasonable.

We set out from the expression for the wavefunction in the exit pupil in the absence of aberrations and neglecting constants (65.15a),

$$K(\xi) = \int \psi_o(x_o) \exp\left(-\frac{2\pi i}{\lambda}\xi x_o\right) dx_o \qquad (74.3)$$

where $\xi = x_a/f$ and write

$$Q(\xi) = K(\xi)T_L(\xi) \qquad (74.4)$$

where (65.26)

$$T_L(\xi) = \exp\left\{2\pi i\left(-\frac{1}{4}C_s\lambda^3\xi^4 + \frac{1}{2}\lambda\Delta\xi^2\right)\right\} \qquad (74.5)$$
$$= \exp\{-i\chi(\xi)\}$$

The image intensity is proportional to $I(x_i) := |\psi_i|^2$,

$$\psi_i(x_i) = \int_{-a}^{a} Q(\xi)\exp(2\pi i x_i\xi)\,d\xi \qquad (74.6)$$

in which the limits correspond to the open part of the aperture (where $T_A = 1$ (65.36)).

It is convenient to introduce the *shifted transform*, $i(\xi)$ of $I(x_i)$:

$$i(\xi) := \int_{-\infty}^{\infty} I(x_i)\exp\{2\pi i(a + \xi)x_i\}\,dx_i \qquad (74.7)$$

which can be written, using (74.6), in the form

$$i(\xi) = \int_{\xi}^{a} Q(\xi')Q^*(\xi' + \xi + a)\,d\xi' \qquad (74.8)$$

The function $Q(\xi)$ may be written (71.1')

$$Q(\xi) = \sum_{n=-\infty}^{\infty} Q_n \operatorname{sinc}(\xi/h - n) \tag{74.9}$$

in which h is the sampling interval and $Q_n := Q(nh)$. At the sampling points,

$$i_k := i(kh) = \int_{kh}^{a} \sum_{m,n} Q_m Q_n^* C_k(m,n) \tag{74.10a}$$

$$C_k(m,n) := \int \operatorname{sinc}\left(\frac{\xi - mh}{h}\right) \operatorname{sinc}\left(\frac{\xi + kh + a - nh}{h}\right) d\xi$$

$$\approx h\delta_{m,-k-N/2-n} \tag{74.10b}$$

so that

$$i_k \approx h \sum_{m=h}^{\frac{1}{2}N-1} Q_m Q_{n+k+\frac{1}{2}N}^* \tag{74.11a}$$

$$= h \sum K_m K_{m+k+\frac{1}{2}N}^* \chi(m, m-k)$$

where

$$\chi(m,n) := \exp \mathrm{i}\{\chi(mh) - \chi((n - N/2)h)\} \tag{74.11b}$$

N, taken to be even, denotes $2a/h$.

We are now ready to derive an iteration relation from which ψ_o can be estimated. Suppose that the object is illuminated by a (tilted) plane wave, of the form $\exp(2\pi i \alpha x_o)$ and that the direct beam passes through the aperture; for convenience, we assume that the direction α corresponds to some sampling point in the diffraction plane, $\alpha = lh$ say. We shall need the (reasonable) assumption that this diffraction spot is brighter than its surroundings, and we express this by writing

$$K(\xi) = K_l \operatorname{sinc} \frac{\xi - lh}{h} + \sum_{n=-\infty}^{\infty} k_n \sin \frac{\xi - nh}{h} \tag{74.12}$$

in which $k_n = K_n$ unless $n = l$, in which case $k_l = 0$ and $K_l \gg k_n$ for all n. Given two images with different defocus and hence two values of Δ in (74.5), we write

$$i_k^{(j)} = \sum_{n=k}^{\frac{1}{2}N-1} K_n K_{n+k+N/2}^* \chi^{(j)}(m, n-k) \qquad j = 1,2 \tag{74.13}$$

and distinguish three ranges of k:

$$-\tfrac{1}{2}N + 1 \le k \le -l$$

$$\frac{i_k^{(j)}}{h} = Q_l q_{l-k-\frac{1}{2}N}^* K^{(j)}(l, l-k)$$
$$+ Q_l^* q_{l+k+\frac{1}{2}N} K^{(j)}(l+k+N/2, l+N/2)$$
$$+ \sum_{n=k}^{\frac{1}{2}N-1} q_n q_{n-k-\frac{1}{2}N}^* K^{(j)}(n, n-k) \qquad (a)$$

$$-l < k \le l$$

$$\frac{i_k^{(j)}}{h} = Q_l q_{l-k-\frac{1}{2}N}^* K^{(j)}(l, l-k)$$
$$+ \sum_{n=k}^{\frac{1}{2}N-1} q_n q_{n-k-\frac{1}{2}N}^* K^{(j)}(n, n-k) \qquad (b)$$

$$k > l$$

$$\frac{i_k^{(j)}}{h} = \sum_{n=k}^{\frac{1}{2}N-1} q_n q_{n-k-\frac{1}{2}N}^* K^{(j)}(n, n-k) \qquad (c)$$

$$(74.14)$$

Setting $k = -\tfrac{1}{2}N$ in (74.13), we obtain

$$\frac{i_{-N/2}^{(j)}}{h} = \sum_{n=-\frac{1}{2}N}^{\frac{1}{2}N-1} Q_n Q_n^* K^{(j)}(n, n+N/2)$$

$$= |Q_l|^2 K^{(j)}(l, l+N/2) + \sum_{n=-\frac{1}{2}N}^{\frac{1}{2}N-1} q_n q_n^* K^{(j)}(n, n+N/2)$$

$$(74.15)$$

and neglecting the quadratic terms in k_n, we can obtain a first estimate of $|Q_l|$, $Q_l^{(1)}$:

$$Q_l^{(1)} = \left\{ \frac{i_{-\frac{1}{2}N}^{(j)}}{hK^{(j)}(l, l+\frac{1}{2}N)} \right\}^{\frac{1}{2}} \qquad (74.16)$$

A first estimate of the terms involving q can then be obtained from (74.14a). Substituting then back into (74.15) gives a better approximation for Q_l and continuing in this way we have an iterative sequence for all the terms.

Van Toorn *et al.* (1978) go on to describe an improved procedure in which the diffraction intensities (q_n) are used to constrain the solution.

This renders the method much more tolerant to noise and speeds up convergence. We note that Saxton (1980a) casts doubt on the correctness of their numerical test of the simpler procedure described here, since their findings seem unduly pessimistic.

The extended linear approximation is likely to remain of extremely limited interest for, even if the quadratic terms in (74.2) can be included correctly, there are other factors that cannot. As the specimen thickness is increased, the model that we are using becomes inadequate owing to the growing number of multiple scattering events: a significant proportion of the beam electrons are deflected by several specimen atoms as they traverse the object. Inelastic scattering likewise becomes more important, though this problem could in principle be solved by filtering out electrons that have lost energy.

74.3 Multiple recordings (circular symmetry)

74.3.1 Introduction
The phase problem was first solved in electron microscopy by the use of two or more signals from the same specimen area. In the earliest proposal (Gerchberg and Saxton, 1972, 1973a), these signals were the image and diffraction pattern intensities from the same region of the specimen and, in a subsequent suggestion (Misell, 1973a,b), two images taken at different defocus. At about the same time (Frank, 1973), the fact that the difference between the bright- and dark-field images of the same specimen area contains enough information to yield the object wavefunction was recognized and this technique is still being intermittently investigated. In 1975, Schiske showed that the wavefunction can be determined from the positions of the spots in the diffraction pattern of a crystalline specimen and three (or exceptionally, two) images taken at different and suitably chosen defocus values.

In the following sections, we present the different procedures and draw attention to the extensive literature in which the uniqueness of the results has been discussed and gradually elucidated.

74.3.2 Gerchberg–Saxton algorithm
Given the image and diffraction pattern intensities of the same specimen area, how can the phase of the wavefunction be found? More formally, given

$$I(x_i) = |\psi_i|^2 \qquad (74.17)$$

and

$$D(x_d) = |\psi_d|^2 \qquad (74.18)$$

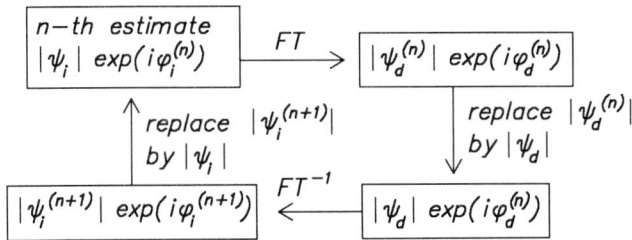

Fig. 74.1: The iterative cycle of the Gerchberg–Saxton algorithm. $|\psi_i|$ and $|\psi_d|$ are known, φ_i and φ_d are to be found.

where

$$\psi_d = \mathcal{F}(\psi_i) \qquad (74.19)$$

apart from unimportant constants and scale factors, how can ψ_i and ψ_d be determined? The algorithm proposed by Gerchberg and Saxton (1972, 1973a) and tested soon after by Gerchberg (1972), Toms $et\ al.$ (1972) and Chapman (1975) iterates between image and diffraction pattern, using the known moduli of ψ_i and ψ_d as constraints. Initially, the phase of ψ_i (say) is chosen arbitrarily and $|\psi_i| \exp i\varphi_i^{(0)}$ is then Fourier transformed to give $\psi_d^{(0)}$ and hence $D^{(0)}(x_d)$. This will not be the same as the measured function $D(x_d)$. The modulus of $\psi_d^{(0)}$ is then replaced by the correct values, giving $|\psi_d| \exp i\varphi_d^{(0)}$. The Fourier transform of this yields a new expression for ψ_i, which we denote $\psi_i^{(1)}$ and hence for $I^{(1)}(x_i)$. Once again the modulus is corrected, giving $|\psi_i| \exp i\varphi_i^{(1)}$. This process is repeated (Fig. 74.1), the modulus being corrected at each step, until no further improvement, within some chosen error range, is observed. The process has then converged and the error has diminished (or remained constant) at each step. Nevertheless, the result need not necessarily be correct if the procedure happens to have become trapped in a local minimum; a way of avoiding this was later suggested by Gerchberg (1986).

That the error never increases may be seen as follows. A convenient measure of the error at any stage is the 'sum-squared error', E, defined by

$$E = \frac{\sum (|\tilde{\psi}_i^{(k)}| - |\psi_d|)^2}{\sum |\psi_d|^2} \qquad (74.20)$$

in which the summation includes all the sample values in the diffraction plane and $\tilde{\psi}_i^{(k)}$ denotes the Fourier transform of the k-th estimate of the image wavefunction. An analogous definition could of course be employed using $|\tilde{\psi}_d^{(k)}|$ and $|\psi_i|$.

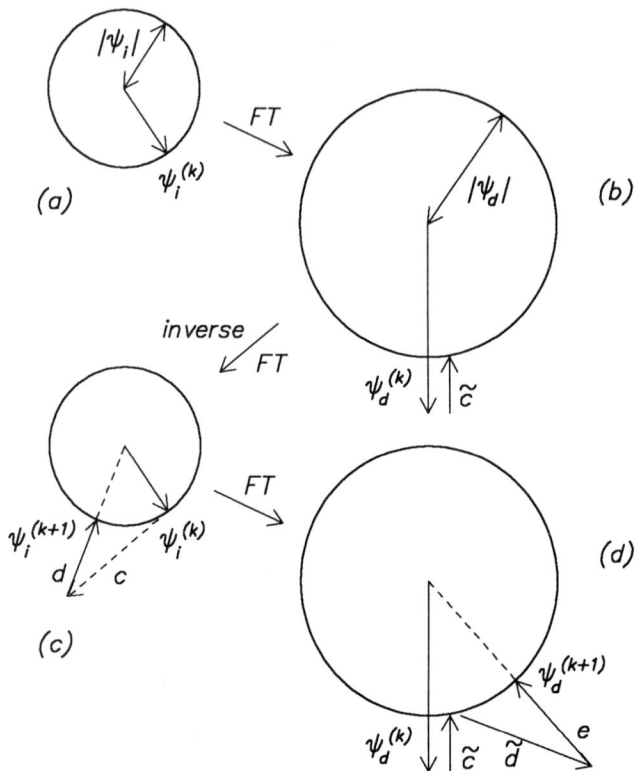

Fig. 74.2: Convergence of the Gerchberg–Saxton iterative algorithm (after Saxton, 1978).

Figure 74.2 shows successive iterations in the form of Argand diagrams corresponding to the same point. Figure 74.2b shows that a correction $\tilde{c}$ has been added to the modulus to alter this to the correct value. In Fig. 74.2c, we see that a correction d must be made in the image plane and in Fig. 74.2d, a further correction e is necessary.

For Fig. 74.2b, the error E_a is simply

$$E_a = \sum |\tilde{c}|^2$$

(setting the common normalization factor equal to unity). For the final stage,

$$E_d = \sum |e|^2$$

But from Fig. 74.2d, we see (using the triangle inequality) that

$$|e| \leq |\tilde{d}|$$

for any point, and so

$$E_d \leq \sum |\tilde{d}|^2$$

Parseval's theorem tells us that

$$N \sum |d|^2 = \sum |\tilde{d}|^2$$

giving

$$E_d \leq N \sum |d|^2$$

and Fig. 74.2c shows that

$$|d| \leq |c|$$

and so

$$E_d \leq N \sum |c|^2$$

Again invoking Parseval's theorem, we have

$$N \sum |c|^2 = \sum |\tilde{c}|^2 = E_a$$

and so finally

$$E_d \leq E_a \tag{74.21}$$

In fact it is not difficult to show that the error diminishes (or remains constant) at each step, not just every two steps as shown above.

In principle, therefore, the Gerchberg–Saxton algorithm seems to offer a satisfactory way of solving the phase problem but in practice, difficulties arise that limit its use to periodic specimens. We have seen that the image and diffraction pattern must be generated by the same object area. For a periodic specimen, illumination of several unit cells of the structure will generate a diffraction pattern corresponding to a single cell, which can then be selected in the image: the data collected in two planes will indeed be in correspondence. If the specimen is not periodic, however, the area illuminated must be as small as that subsequently used in the computation and the relative alignment, particularly in azimuth, of the diffraction pattern and image will be extremely difficult. For non-periodic specimens, noise may also be a problem, especially if the object structure is fragile and must hence be imaged in low-dose conditions. For all these reasons it is not likely that the algorithm will be successful for any but periodic specimens, a conclusion that is borne out by the few practical attempts to use it (Toms *et al.*, 1972; Chapman, 1975; Gerchberg, 1972). The dynamic range of diffraction patterns, commonly very wide, has also been invoked as a problem in the past but the use of CCD cameras, which accept such wide dynamic ranges, presumably removes this particular difficulty. No such technological advance seems likely to eliminate the others.

Very full discussion of the Gerchberg–Saxton algorithm supported by numerous model calculations is to be found in the monograph of Saxton (1978). The account in a review by Misell (1978a) is also clear and objective. Subsequent work on this and related algorithms has been largely concerned with the reliability of the results in the presence of errors of measurement, which render the original assumption that image and diffraction plane wave function form a Fourier transform pair untrue. We return to this 'noisy phase problem' in a later Section (74.4.5).

74.3.3 The multiple-image algorithm

Images of the same specimen area obtained with different amounts of defocus also form a set (usually a pair) of signals from which the phase can in principle be extracted. The signals are now related not by a Fourier transform but by convolution with a quadratic phase shift. A series of iterations exactly analogous to that described in Section 74.3.2 should yield the complex wavefunction (Misell, 1973a,b; Gerchberg and Saxton, 1973b; Misell, 1978). Explicitly, we see from (63.16) that the spectra S_j of the wavefunctions $\psi_i^{(j)}$, $j = 1, 2$ for a pair of images corresponding to defocus values Δz apart, have the form

$$S_j = T_A T_L^{(j)} S_o \qquad (74.22)$$

Thus

$$S_2 = S_1 \frac{T_L^{(2)}}{T_L^{(1)}} = t_L S_1 \qquad (74.23)$$

where

$$t_L := \frac{T_L^{(2)}}{T_L^{(1)}} = \exp\left\{-i(\chi_2 - \chi_1)\right\} = \exp\left(-\pi i \Delta z \frac{\lambda}{f^2}\right) \qquad (74.24)$$

or

$$\psi_i^{(2)} = \psi_i^{(1)} * \mathcal{F}^-(t_L) \qquad (74.25)$$

Alignment should be easier here than in the case of image and diffraction pattern but Misell (1973b, 1978a) observes that a relatively small contribution to the image from inelastic scattering may well have a serious effect on the phase solution. Rather surprisingly, there have been no practical attempts to use the two-image algorithm, apart from the tests recorded by Misell (1973b, 1978a) and Saxton (1978). The fact that convergence is proved by the same reasoning as for the Gerchberg–Saxton algorithm was pointed out by Gerchberg and Saxton (1973b).

For further discussion of the two-defocus and image–diffraction algorithms, see van Toorn and Ferwerda (1976a,b), Drenth *et al.* (1975a,b), Huiser and Ferwerda (1976a,b), Huiser (1979, 1980), Boucher (1980), Pulvermacher (1981) and Huang *et al.* (1988).

A very different way of extracting phase and amplitude information from sets of images of the same specimen has been examined by Kirkland (1982, 1984, 1987; Kirkland *et al.*, 1982) and reconsidered in greater detail by Gribelyuk and Hutchinson (1991, 1992). In this 'multiple-input maximum *a posteriori*' (MIMAP) procedure, we seek to minimize the mean-square difference between the true wavefunction ψ_o emerging from the specimen and the estimated wavefunction $\hat{\psi}_0$, $\int |\psi_o - \hat{\psi}_0|^2 \, d\boldsymbol{x} \to$ min, subject to a constraint of the form

$$\sum_k \int |I_k - I_k^{(0)} - C|^2 \, d\boldsymbol{x} + m\eta(\psi_o - \hat{\psi}_o) \to \text{min}$$

In this last equation, the I_k are the recorded images, $I_k^{(0)}$ is the corresponding restored image, m is the number of images recorded ($k = 1, 2 \ldots m$) and C is a fog-level, which might be introduced during photographic processing for example. The parameter η (unknown initially) is a Lagrange multiplier and the purpose of the term in which it appears is to prevent noise from accumulating. Suitable iterative sequences are given by Gribelyuk and Hutchinson, who found that in realistic conditions—noisy images, absence of exact information about the defocus values used—the method will not function satisfactorily. Furthermore, if the weak-phase object approximation is used as starting-point and hence for the estimation of ψ, the noise-control term is ineffective and it is better to omit it unless some better estimate can be found.

74.3.4 Bright-field / dark-field subtraction

Equation (74.2) shows that the difference between the bright-field intensity and the dark-field intensity gives the real part of the image wavefunction directly[*] (Frank, 1973). The two intensities must be recorded in identical conditions, except that the undiffracted beam, those electrons unaffected by their passage through the specimen, is present in one case (bright-field) and absent in the other (dark-field). The success of the method is therefore governed more by the technical possibility of removing the central beam cleanly, with the minimum perturbation to the remaining electrons, than

---

[*] The paper by Ansley (1973), occasionally cited in this context, is not relevant here.

by the more indirect questions of alignment and the like that bedevilled the image–diffraction and two-image algorithms. For this reason, it continues to attract sporadic attention, whereas the other two algorithms have been "more honoured in the breach than the observance". For details see Frank (1974a), Burge (1976), Saxton (1980c), Saxton and Stobbs (1984).

74.3.5 Direct methods

We have chosen to discuss first the iterative methods based on two (or more) signals but in fact the first solution (Gerchberg and Saxton, 1971) was a direct method and Saxton devotes considerable space to these in his monograph (1978); apart from some work on the uniqueness of the solution obtained, however, they have attracted almost no other attention and we therefore give an extremely brief account of them here. For simplicity, consider N (even) samples of a one-dimensional image wavefunction, ψ_k and the Fourier transform $\tilde{\psi}_k$, $-N/2 \leq k \leq N/2$. We assume that $|\psi_k|$ and $|\tilde{\psi}_k|$ are known (measured intensities in image and diffraction pattern). The Fourier transform of the measured intensity $|\psi_k|^2$ gives the autocorrelation function

$$\tilde{I}_j = \sum_k \tilde{\psi}_k \tilde{\psi}^*_{k-j} \qquad (74.26)$$

Only one term contributes to the outermost point:

$$\tilde{I}_{2N} = \tilde{\psi}_N \tilde{\psi}^*_{-N} \qquad (74.27)$$

Attributing an arbitrary phase to $\tilde{\psi}_N$, we can obtain $\tilde{\psi}^*_{-N}$ immediately. The next term in $\tilde{I}$, $\tilde{I}_{2N-1}$, involves two terms of the summation:

$$\tilde{I}_{2N-1} = \tilde{\psi}_N \tilde{\psi}^*_{-(N-1)} + \tilde{\psi}_{N-1} \tilde{\psi}^*_{-N} \qquad (74.28)$$

There are thus two solutions for each of the new terms and, as we proceed inwards ($\tilde{I}_{2N-2}, \ldots$), a choice of two solutions is introduced for each new term encountered. When $\tilde{I}_N$ is reached, possible solutions for all the ψ_k will have been established and the other N values of $\tilde{I}_N$ must be employed to choose between all the possible solutions corresponding to the many branches of the 'solution tree'.

This method has two immense drawbacks: it is highly sensitive to noise since errors are amplified as we proceed inwards and the outermost values of $\tilde{I}$ are likely to be the least accurate. In addition, exploration of all the branches in the search for the correct solution is prohibitively time-consuming. Saxton (1978) describes ways of accelerating the search process and we refer the interested reader to his account.

Several authors have examined the direct solution of the phase problem; for a proposal that seems likely to be unstable, see Dallas (1975, 1976), and for extensive analysis, see the papers from Ferwerda's laboratory by Drenth *et al.* (1975a,b), Huiser and Ferwerda (1976a,b) and van Toorn and Ferwerda (1976a,b). See too Gu and Yang (1981).

Although we have preferred to open with an account of the use of direct methods for electron images, a proposal inspired by sampling theory had in fact been made earlier, by Arsenault and Lowenthal (1969), who found an answer to the following question: given the intensity distribution in a diffraction pattern, $D = \psi_d \psi_d^*$, what was the object distribution ψ_o, $\psi_d = \mathcal{F}(\psi_o)$? The object support was assumed finite: $\psi_o(x) = 0$ for $|x| > \frac{1}{2}$. The sampling theorem tells us that

$$\psi_d(p) = \sum_{j=-\infty}^{\infty} \psi_d(j)\,\mathrm{sinc}(p - j) \tag{74.29}$$

and hence

$$D(p) = \sum_j \sum_k |\psi_d(j)||\psi_d(k)|\exp\mathrm{i}\{\varphi_d(j) - \varphi_d(k)\}\,\mathrm{sinc}(p - j)\,\mathrm{sinc}(p - k) \tag{74.30}$$

in which φ_d is the phase of ψ_d. We know, however, that the support of the autocorrelation function of ψ_o is $|x| < 1$ (double that of ψ_o) and an alternative sampling expansion for $D(p)$ is therefore

$$D(p) = \sum_j D(j/2)\,\mathrm{sinc}(2p - j) \qquad D(j) = |\psi_d(j)|^2 \tag{74.31j}$$

After some reorganization, it can be shown that

$$D\left(\frac{2m+1}{2}\right) = \frac{4}{\pi^2}\sum_j \sum_k \frac{(-1)^{j+k}|\psi_d(j)||\psi_d(k)|\exp\mathrm{i}\{\varphi_d(j) - \varphi_d(k)\}}{(2j - 2m - 1)(2k - 2m - 1)} \tag{74.32}$$

Solution of these equations for the phases $\varphi_d(m)$ is in principle possible but the number of possible solutions, in the absence of any further constraints, is of course very high.

The method was later perfected by Arsenault and Chalasinska-Macukow (1983) and Chalasinska-Macukow and Arsenault (1983, 1985); their procedure worked satisfactorily for starting phase values reasonably close to the true solution (see too Ferwerda and Frieden, 1990). Algorithms based on ingenious use of the sampling theorem have also been explored in detail by Bates and colleagues (Bates, 1982; Garden and Bates, 1982; Fright and Bates, 1982) and by Robaux (1970) and Robaux and Roizen-Dossier (1970).

74.3.6 Modulation of the incident beam

We remarked in the introduction that the various ways of obtaining phase from intensity records exploit deliberate or inevitable modifications of the wavefunction as it propagates along the instrument. Such a modification could be introduced upstream from the specimen and the resulting product of the incident beam structure and the specimen structure exploited to yield phase information. We shall describe the resulting procedure only briefly, for it is not practical to impress more than the simplest types of variation on an electron beam between source and specimen. The beam can be tilted or converted into a hollow cone but any additional modulations, of amplitude or phase, are beset with difficulties so severe as to have discouraged all but the most sporadic or intrepid interest.

The first suggestion that the incident electron beam could be modulated in a beneficial way was made by Hoppe (Hoppe, 1969a; Hoppe and Strube, 1969). Reasonably encouraging results were obtained from a computer simulation by Hegerl and Hoppe (1972). We summarize their procedure, which Hoppe styled 'ptychography', in one dimension only. Consider a specimen transparency $t(x)$ and an illuminating (source) wave $\psi_Q(x)$; the object wave ψ_o emerging from the specimen, is given by

$$\psi_o = \psi_Q t \tag{74.33}$$

For illumination of finite extent, $\psi_Q(x) = 0$ for $|x| > a/2$, say, the diffraction pattern may be represented by suitably spaced samples (see 71.1'). Ignoring constant factors, we have

$$\psi_d = \mathcal{F}(\psi_o) = \tilde{\psi}_Q * \tilde{t} = \sum_j \tilde{\psi}_Q(p - j/a)t(j/a) \tag{74.34}$$

Hoppe reduces the complexity of the situation by recording two diffraction patterns, for one of which the illumination is simply truncated but otherwise uniform:

$$\psi_Q^{(1)} = \begin{cases} \text{const} & |x| \leq a/2 \\ 0 & \text{otherwise} \end{cases}$$
$$= \text{rect}(x/a) \tag{74.35}$$

For the other, the illumination is again truncated outside $|x| = a/2$ but is no longer uniform:

$$\psi_Q^{(2)} = (1 - e^{-2\pi i x/a})\text{rect}(x/a) \tag{74.36}$$

It is difficult to see how such a distribution could be created and monitored with the necessary precision.

A further suggestion along similar lines has been made by Berndt and Doll (1976, 1978, 1983; cf. Kunath, 1978b). See too Paxman *et al.* (1987).

74.3.7 One image and its derivative with respect to defocus
We have mentioned that our ignorance of the phase information when only an intensity record is available may be pictured by saying that we know where the electrons arrive but not their direction of motion when they arrive. This makes the method based on two images taken at different defocus easy to accept. An extreme case of the method has been proposed by van Dyck and Coene (1987a,b, 1988), who show that the phase of the wavefunction can be obtained from its amplitude and the derivative of the latter in the z-direction. The two images must be taken at very close defocus values and van Dyck and Coene estimate that for a resolution of 1.4 Å at 400 kV, the defocus must be much less than 4 Å, say of the order of 1 Å. This is clearly beyond the performance limits of the present generation of electron microscopes.

The algorithm proposed by van Dyck and Coene proceeds as follows. The propagation in free space of the wavefunction is described (van Dyck, 1985) by

$$\frac{\partial^2 \psi_0}{\partial x^2} + \frac{\partial^2 \psi_0}{\partial y^2} + 2ik\frac{\partial \psi_0}{\partial z} = 0 \tag{74.37}$$

with

$$\psi = \psi_0 \exp(i\boldsymbol{k} \cdot \boldsymbol{r})$$

and writing

$$\psi_0 = a \exp i\theta \tag{74.38}$$

we find that

$$\frac{\partial a}{\partial z} = -\frac{1}{2k}(2\,\mathrm{grad}\,a \cdot \mathrm{grad}\,\theta + a\nabla^2\theta) \tag{74.39}$$

Van Dyck and Coene observe that this equation can be solved for θ, given a and $\partial a/\partial z$, either directly or after Fourier transformation (for further examination of the propagation equation, see Ichikawa *et al.*, 1988 and Streibl, 1984).

A more practical version of this procedure in which the defocus step is bigger has subsequently been tested by van Dyck and Op de Beeck (1990, 1991). The wavefunction in some image plane is written in the form

$$\psi(\boldsymbol{x},0) = C + \int_{\boldsymbol{q}\neq0} \hat{\psi}(\boldsymbol{q})\exp(2\pi i\boldsymbol{q} \cdot \boldsymbol{x})\,d\boldsymbol{q} \tag{74.40}$$

in which the second argument of ψ indicates the relative defocus, set equal to zero in the plane in which the solution is to be obtained. Propagation to a plane with relative defocus Δ gives

$$\psi(\boldsymbol{x}, \Delta) = C + \int_{\boldsymbol{q} \neq 0} \hat{\psi}(\boldsymbol{q}) \exp(2\pi i \boldsymbol{q} \cdot \boldsymbol{x}) \exp(-i\pi \lambda q^2 \Delta) \, dq \qquad (74.41)$$

The Fourier transform with respect to $\boldsymbol{x}$ and Δ of the corresponding image intensity, $I(\boldsymbol{x}, \Delta) = \psi(\boldsymbol{x}, \Delta)\psi^*(\boldsymbol{x}, \Delta)$, has the form

$$
\begin{aligned}
\tilde{I}(\boldsymbol{q}, d) &:= \mathcal{F}\{I(\boldsymbol{x}, \Delta)\} \\
&= |C|^2 \delta(\boldsymbol{q}) + C^* \hat{\psi}(\boldsymbol{q}) \delta(d - \lambda q^2/2) \qquad\qquad\qquad\ \text{(I)} \\
&\quad + C \hat{\psi}^*(-\boldsymbol{q}) \delta(d + \lambda q^2/2) \qquad\qquad\qquad\qquad\ \text{(II)} \\
&\quad + \int_{\substack{q' \neq 0 \\ q+q' \neq 0}} \hat{\psi}^*(\boldsymbol{q})\psi(\boldsymbol{q} + \boldsymbol{q}') \delta(d - \lambda\{(\boldsymbol{q} + \boldsymbol{q}')^2 - q'^2\}/2) \, dq' \quad \text{(III)}
\end{aligned}
$$

$$(74.42)$$

The terms I and II are sharply peaked on the paraboloids in reciprocal space corresponding to the δ-functions, whereas term III makes a more diffuse contribution. In the original proposal, images were to be taken at such close defocus values that the Fourier transform with respect to both the transverse coordinates and the defocus could be calculated and the values on the paraboloid extracted. In view of the difficulty of obtaining such closely spaced images, van Dyck and Op de Beeck offered an alternative solution. Several images are again recorded but at more reasonable focus steps, of the order of 3 nm. The two-dimensional transform, with respect to $\boldsymbol{x}$ but not to Δ, of each intensity $I(\boldsymbol{x}, \Delta)$ is obtained; we denote this by $\tilde{I}_2(\boldsymbol{q}, \Delta)$. The sum $\sum_n \tilde{I}_2(\boldsymbol{q}, \Delta_n) \exp(-i\pi \lambda q^2 \Delta_n)$ is then formed, where Δ_n denote the various defocus values, for positive and for negative values of d. An estimate of the function $\hat{\psi}(\boldsymbol{q})$ can be extracted from the results and the effect of term III in (74.42) can be eliminated by iteration. In this version, however, there is nothing fundamentally new and the method can be traced back to Schiske's early (linear) restoration scheme of 1968 (Saxton, 1994). See too Op de Beeck and van Dyck (1991), van Dyck (1991), Op de Beeck et al. (1992), Coene et al. (1992) and van Dyck et al. (1993). Beeching and Spargo (1992, 1993) and de Ruijter et al. (1992) have also tested this method.

For an analysis of the statistics of this and some of the other methods described here, see Miedema and Buist (1992).

74.3.8 Related problems
We have presented the Gerchberg–Saxton algorithm (Section 74.3.2) in isolation in order to underline its originality and far-reaching effect. It has

$$\text{input} \xrightarrow{\quad FT \quad}$$
$$\psi_k^{(in)}$$

$$\text{(MR)} \uparrow \begin{array}{l} \textit{modifications} \\ \textit{in real space} \end{array} \qquad \text{(MF)} \Big| \begin{array}{l} \textit{modifications in} \\ \textit{Fourier space} \end{array}$$

$$\text{output} \longleftarrow$$
$$\psi_k^{(out)} \qquad FT^{-1}$$

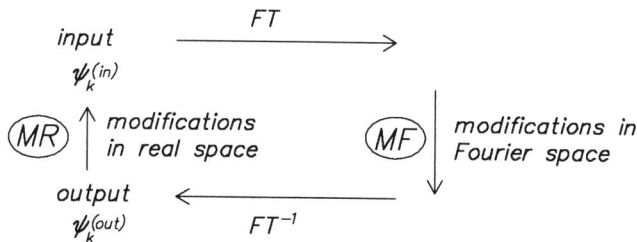

Fig. 74.3: Generalization of the Gerchberg–Saxton algorithm due to Fienup; the modifications in real and Fourier space may use other information than the moduli of the wavefunctions.

suggested ways of solving a family of related problems, which have the following general form: given incomplete information about a complex function and its Fourier transform, find the function. In the Gerchberg–Saxton case, the imperfect information consists of the moduli of the function and its transform. In another common situation, the modulus of the function and the support of the transform, that is, the region occupied by the latter in the Fourier transform plane, are known. Positivity is another powerful clue. This whole class of problems may in principle be solved by an iterative sequence of the general form shown in Fig. 74.3.

In the Gerchberg–Saxton algorithm, the modification in Fourier space (MF) consists in replacing the estimated modulus values by the correct ones and the modification in real space (MR) is the same. The other possibilities, essentially due to Fienup (1978, see 1982, 1984) use different forms of MR; they are referred to as *Fienup's algorithms* except by Fienup himself, who modestly prefers input–output algorithms. The principal choices are:

Basic input–output algorithm

$$\psi_{k+1}^{(in)} = \begin{cases} \psi_k^{(in)} & \text{for points where constraints are satisfied} \\ \psi_k^{(in)} - \beta\psi_k^{(out)} & \text{elsewhere} \end{cases}$$

$$(74.43a)$$

Output–input algorithm

$$\psi_{k+1}^{(in)} = \begin{cases} \psi_k^{(out)} \\ \psi_k^{(out)}(1 - \beta) \end{cases} \qquad (74.43b)$$

Hybrid input–output algorithm

$$\psi_{k+1}^{(in)} = \begin{cases} \psi_k^{(out)} \\ \psi_k^{(in)} - \beta\psi_k^{(out)} \end{cases} \qquad (74.43c)$$

in which β is a constant, the value of which strongly affects the convergence of the various algorithms.

These algorithms differ from those of the Gerchberg–Saxton type in that the constraints in the object plane are not satisfied throughout the iterative process. In this respect, they may be compared with over-relaxation techniques for solving partial differential equations (see Section 11.4), in which successive estimates of the true solution are over-corrected. Over- or under-correction does not, however, seem to be beneficial in the image–diffraction pattern case. Gassmann (1977a,b, 1979) has suggested that in the stage MF, the estimated modulus values should be replaced by a weighted version of the correct values, the weight approaching unity as the process converges. This has not been tested, however, and has the disquieting feature that the current is not conserved.

The ease with which the Gerchberg–Saxton and Fienup routines can be expressed in terms of image algebra has been pointed out by Hawkes (1990a, 1991b,c).

74.4 Analyticity

74.4.1 Introduction

Image formation is accomplished by the propagation of a wavefunction, onto which information about the specimen has been imprinted, and the subsequent recording of some characteristic of this function, typically the square of its modulus. In the course of propagation, the wavefunction may be modified, deliberately or by factors difficult or impossible to control, and these modifications will in all likelihood have repercussions on the image. Our aim is to supplement the recorded modulus of the wavefunction with a knowledge of its phase and, to this end, numerous methods involving deliberate modification of the wavefunction upstream from the recording plane or exploiting natural features of the image-forming chain have been devised. In particular, the beam-limiting effect of the finite dimensions of the lenses or of the deliberate insertion of a diaphragm into the path of the beam, typically in the back focal plane of the objective lens, proves to be a rich source of ways of obtaining phase information: a restriction on the extent of the wavefunction in one plane affects its Fourier transform in a potentially fruitful way.

Before pursuing these ideas in detail, we must forewarn the reader that, despite the apparent strength of some techniques, difficulties remain to be overcome. Above all, the effect of noise is still poorly understood; much of the theory applies strictly to measurements free of error. Although experiment and, to some extent, intuition suggest that a solution obtained in the presence of modest measurement error will be close to the correct solution, the theoretical support for this belief is fragile and is still a subject of debate.

The earlier literature of the subject is relegated to the historical notes in Section 74.4.6 for the following reason. For many years, the passage from one-dimensional analysis to a rigorous two-dimensional treatment was hampered by the fact that the theory of functions of more than one complex variable rarely forms part of the physicist's basic mathematical knowledge. Progress had to wait until the image processing community became aware of the large body of work on these functions in the literature of pure mathematics, after which the situation became very much clearer. (The texts of Fuks, 1983, Levin, 1964 and Ronkin, 1974 are particularly relevant and those of Grauert and Fritzsche, 1976, Gunning and Rossi, 1965, Hörmander, 1966 and Cartan, 1961 contain useful background material.)

We mentioned in Section 74.1 that much of the extensive literature devoted to the phase problem is concerned with the retrieval of phase from a single intensity record together with information about the support of its Fourier transform (the region within which the latter may be non-zero), a situation that arises routinely in some fields, such as astronomy. The fact that the electron microscope is in principle capable of furnishing more than one record of a given object, in different imaging conditions (notably image and diffraction pattern, or images at different defocus) is distinctly enviable: "Other phase-retrieval problems, such as in electron microscopy in which one has squared-modulus measurements in each of two domains ... had been solved; but those earlier successes depended on much greater object-domain constraints than just non-negativity and support" (Seldin and Fienup, 1990a). We include some account of this work in which the raw data consist only of one record and a constraint, for the methods of analysis and of practical exploitation are very instructive.

74.4.2 Analytic continuation of wavefunctions
It is usual to consider first the academic and physically uninteresting case of one-dimensional signals. We adopt this plan here, for the mathematics required for the higher-dimensional case is relatively unfamiliar and may be found easier to follow after examination of the simpler situation. We do so with some reluctance, however, for it will emerge that the one-dimensional case is misleadingly pessimistic and that apparent weaknesses of the analytic approach vanish when we go beyond one dimension*.

---

* "Optical scientists have also paid much attention to one-dimensional phase problems ... presumably to avoid what may have seemed to them at the time as the unnecessary complications of multidimensional analysis, although recent events may make some of them wish they had been more enterprising " (Bates and Fright, 1984).

We set out from the equation that states that the image wavefunction is the Fourier transform of that in the diffraction plane, neglecting constants and scale changes:

$$\psi_i(x) = \int\limits_{-a}^{a} \psi_d(q) \exp(2\pi i x q)\, dq \qquad (74.44)$$

in which the limits of integration, $\pm a$ correspond to the size of the objective aperture. If we regard the variable x as the real part of a complex variable $s = x + iy$, and set

$$\psi_i(s) := \int\limits_{-a}^{a} \psi_d(q) \exp(2\pi i s q)\, dq \qquad (74.45)$$

then it is immediately obvious that the real and imaginary parts of $\psi_i(s)$ satisfy the Cauchy–Riemann relations: with $\psi_i(s) := \psi_R(s) + i\psi_I(s)$, then clearly

$$\frac{\partial \psi_R}{\partial x} = \frac{\partial \psi_I}{\partial y} = -2\pi \int\limits_{-a}^{a} q \exp(-2\pi y q)\Im\{\psi_d(q)\exp(2\pi i x q)\}\, dq \qquad (74.46)$$

$$\frac{\partial \psi_R}{\partial y} = -\frac{\partial \psi_I}{\partial x} = 2\pi \int\limits_{-a}^{a} q \exp(-2\pi y q)\Re\{\psi_d(q)\exp(2\pi i x q)\}\, dq \qquad (74.47)$$

The fact that $\psi_i(s)$ is analytic implies that this function extends to infinity in both the positive and negative x directions and that exact knowledge of it over an arbitrarily small zone is sufficient to reconstruct it everywhere. The second feature is of little practical interest since a very small imprecision in the 'exact' local knowledge of ψ_i can cause an enormous change in the remainder. How does $\psi_i(s)$ behave as $|s| \to \infty$? From (74.45), we see that

$$\psi_i(s)\exp(-2\pi a|s|) = \int\limits_{-a}^{a} \psi_d(q) \exp 2\pi(isq - a|s|)\, dq \qquad (74.48)$$

and the Schwarz inequality then tell us that

$$|\psi_i(s)\exp(-2\pi a|s|)|^2 \le \int\limits_{-a}^{a} |\psi_d(q)|^2\, dq \int\limits_{-a}^{a} |\exp 2\pi(isq - a|s|)|^2\, dq \qquad (74.49)$$

As $|s| \to \infty$, the second integral on the right-hand side must vanish, except at $q = \pm a$:

$$\lim_{|s|\to\infty} |\psi_i(s)| \exp(-2\pi a|s|) = 0 \tag{74.50}$$

Functions that behave in this way are known as *integral* (Titchmarsh, 1939) or *entire* functions (Boas, 1954) and the behaviour described by (74.50) identifies $\psi_i(s)$ as an entire function 'of order unity'. Titchmarsh (1939, §8.2) tells us that "Functions of finite order are, after polynomials, the simplest integral [entire] functions. A polynomial is of order zero; some of the properties of functions of small order are similar to those of polynomials", a helpful observation when we attempt to exploit these properties. In particular, the positions of the zeros of $\psi_i(s)$ will prove to be of great importance; this is not surprising if we think of entire functions of finite order as close relatives of polynomials, which are of course completely characterized by a knowledge of their roots. We now establish the analogous expansion for $\psi(s)$. (The following account, though based on the standard texts on analysis, notably Titchmarsh (1939) and Boas (1954), is also strongly influenced by Saxton (1978).)

The number $n(r)$ of zeros $(s_1, s_2 \ldots)$ of ψ lying within a circle of radius r can be established with the aid of a formula derived by Jensen:

$$\int_0^r \frac{n(x)}{x} \, dx = \frac{1}{2\pi} \int_0^{2\pi} \log \left| \frac{\psi(re^{i\theta})}{\psi(0)} \right| \, d\theta \tag{74.51}$$

Using (74.50), we can show that

$$\int_0^r \frac{n(x)}{x} \, dx < 2\pi a r \quad \text{as } r \to \infty \tag{74.52}$$

and deduce that

$$n(r) \le 2\pi e a r \tag{74.53}$$

The comparison test then shows that the series $\sum_j 1/|s_j|^2$ converges absolutely (Section 8.22 of Titchmarsh, 1939, with $\alpha = 2$ and $\varrho = 1$). We are now ready to invoke Hadamard's factorization theorem (Titchmarsh, Section 8.24), which tells us that for entire functions of order unity,

$$\psi(s) = c' \exp(cs) P(s) \tag{74.54}$$

where $P(s)$ is a so-called 'canonical product', which here takes the form

$$P(s) = \prod_{j=1}^{\infty} \left(1 - \frac{s}{s_j}\right) \exp\left(\frac{s}{s_j}\right) \tag{74.55}$$

Clearly $c' = \psi(0)$ and $c = \psi'(0)/\psi(0)$. Finally, therefore, we have

$$\psi(s) = \psi(0)\exp(cs)\prod_{j=1}^{\infty}\left(1 - \frac{s}{s_j}\right)\exp\left(\frac{s}{s_j}\right)$$

$$c = \frac{\psi'(0)}{\psi(0)}$$

(74.56)

It is this expression for the image-plane wavefunction when the diffraction-plane wavefunction is confined to the region $|x| \leq a$ that is at the heart of the various phase-retrieval techniques based on analyticity. The importance of the positions of the zeros is immediately obvious, and symmetry arguments show that these zeros may be related. If $\psi(x)$ is real or imaginary, then the zeros will occur in pairs reflected in the real axis $(a \pm ib)$. If $\psi(x)$ is symmetric or antisymmetric about the origin, then the zeros again occur in pairs but now reflected in the origin $(a + ib, -a - ib)$. If $\psi(x)$ is conjugate symmetric or antisymmetric, the pairs of zeros are reflected in the imaginary axis $(\pm a + ib)$. If ψ_d is real and positive, ψ_i can have no imaginary zeros and if ψ_d is in addition confined to the interval $(-a, a)$, then the zeros of ψ_i must be at least $1/4a$ away from the imaginary axis.

Suppose now that we consider not the wavefunction ψ_i but the image intensity $|\psi_i|^2$. The transform of this has finite support $(-2a, 2a)$ when the transform of ψ_i is confined to $(-a, a)$. Analytic continuation now enables us to write $|\psi_i|^2$ as $\psi(s)\psi^*(s^*)$, which automatically has pairs of zeros symmetrically placed about the real axis $(a \pm ib)$, as we have already observed. Given any wavefunction consistent with the image intensity, others can therefore be formed by shifting one or more of its zeros to the complex conjugate position ($a + ib$ replaced by $a - ib$). Walther (1963) showed that all other possible solutions can be obtained by this process of 'zero flipping'. Returning to (74.56), we see that this is equivalent to multiplication by $(1 - s/s_j^*)(1 - s/s_j)^{-1}$, the modulus of which tends to unity as $|s| \to \infty$; the limiting behaviour of the image wavefunction is unchanged and the support of ψ_d is likewise not altered.

74.4.3 Use of half-plane apertures
The image wavefunction is even more severely limited if the diffraction plane is confined not merely to an interval $(-a, a)$ symmetric about the origin, or optic axis, but to part of the positive (or negative) half-plane: $(0, a)$ or $(-a, 0)$. This one-sided problem, closely associated with causality when the axis measures time, is encountered in many situations, of which coherence theory is an interesting example (see Born and Wolf, 1959 Section 10.2 for many references).

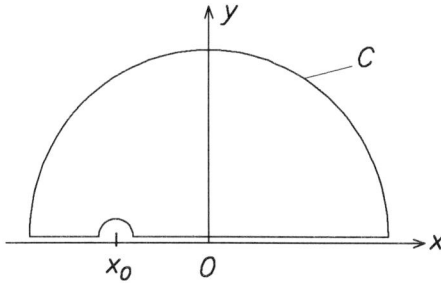

Fig. 74.4: Path of integration C employed in (74.58).

We now have

$$\psi_i(x) = \int_0^a \psi_d(q)\exp(2\pi\mathrm{i}xq)\,dq \qquad (74.57)$$

and provided that ψ_d is square-integrable, which physically of course it must be, $\psi_i(s)$ again vanishes as $|x| \to \infty$ for any value of y ($s = x + \mathrm{i}y$) and vanishes exponentially as $y \to \infty$. Since zero-flipping does not affect the limiting behaviour of ψ_i, ψ_d remains one-sided.

The real and imaginary parts of $\psi_i(s)$ are related by the Hilbert transform. This may be seen by considering the contour integral

$$\int_C \frac{\psi_i(s)}{s - x_0}\,ds \qquad (74.58)$$

where C is the semi-circular path shown in Fig. 74.4, excluding the point $s = x_0$ on the real axis. There will be no contribution from the semicircle and so

$$\mathcal{P}\int_{-\infty}^{\infty} \frac{\psi_i(s)}{x - x_0}\,ds = \pi\mathrm{i}\psi_i(x_0) \qquad (74.59)$$

in which the symbol $\mathcal{P}$ indicates the (Cauchy) principal part. Writing

$$\psi_i = \psi_R + \mathrm{i}\psi_I \qquad (74.60)$$

(74.45) becomes

$$\psi_I(x_0) = -\frac{1}{\pi}\mathcal{P}\int_{-\infty}^{\infty} \frac{\psi_R(x)}{x - x_0}\,dx \qquad (74.61a)$$

$$\psi_R(x_0) = \frac{1}{\pi}\mathcal{P}\int_{-\infty}^{\infty} \frac{\psi_I(x)}{x - x_0}\,dx \qquad (74.61b)$$

The real and imaginary parts of the wavefunction thus form a Hilbert transform pair.

This result is of immediate interest in electron microscopy, since we can arrange for the Fourier transform of the image wavefunction to be one-sided by inserting a semicircular aperture in the back-focal plane of the objective lens. A small notch must be cut in the aperture around the axis to allow the unscattered beam to pass and the straight edge of the diaphragm must be expected to accumulate charge but the experimental conditions otherwise match the mathematical model closely. From (74.2), we see that for weak scattering conditions, the real part of the image wavefunction can be measured directly and the imaginary part is then deduced from (74.61a).

In fact, we could have obtained this result for weakly scattering specimens very much more simply by using (66.16) and defining T_A appropriately. Setting $T_A = 0$ for $p < 0$ and $T_A = 1$ for $0 \le p < a$ we find

$$\begin{aligned} K_a &= \exp\{-i\,\mathrm{sgn}(p)\gamma(p)\} \\ K_p &= i\,\mathrm{sgn}(p)\exp\{-i\,\mathrm{sgn}(p)\gamma(p)\} \end{aligned} \qquad (74.62)$$

The general relation (74.61) may also be established more physically. As before (74.60), we write

$$\psi_i = \psi_R + i\psi_I$$

so that

$$\psi_d = \tilde{\psi}_R + i\tilde{\psi}_I \qquad (74.63)$$

and hence

$$\text{or} \quad \begin{aligned} \tilde{\psi}_R &= -i\tilde{\psi}_I \\ \tilde{\psi}_I &= i\tilde{\psi}_R \end{aligned} \quad \text{where} \quad \psi_d = 0 \qquad (74.64)$$

Taking the complex conjugate of this and recalling that for a real function $f(x)$, $\tilde{f}^*(-x) = \tilde{f}(x)$, we see that

$$\tilde{\psi}_I(p) = -i\,\mathrm{sgn}(p)\tilde{\psi}_R(p) \qquad (74.65)$$

in which the signum function $\mathrm{sgn}(p)$ is defined by

$$\begin{aligned} \mathrm{sgn}(p) &= 1 & p > 1 \\ \mathrm{sgn}(p) &= -1 & p < 0 \\ \mathrm{sgn}(0) &= 0 \end{aligned} \qquad (74.66)$$

The Fourier transform of such a step function* may be regarded as $1/x$, so that (74.65) is equivalent to (74.61).

---

* The Fourier integral for the step function fails to converge but is summable in the Cesàro sense $(C1)$; see Section 8.43 of Whittaker and Watson (1927).

A slightly different form of these expressions is useful when the object is periodic and hence the diffraction pattern consists of individual points (or more realistically, spots). Returning to (74.45), we can now write

$$\psi_i(x) = \sum_{j=-N}^{N} \psi_{d_j} \exp(2\pi i j x) \tag{74.67}$$

which is of course analytic for all s when we continue x to $s = x + iy$. With $w := \exp(2\pi i s)$, we have

$$\psi(s) = \sum_{j=-N}^{N} \psi_{d_j} w^j = w^{-N} \sum_{j=0}^{2N} \tilde{h}_j w^j =: h(w) \tag{74.68}$$

where

$$\tilde{h}_j := \psi_{d,j-N} \tag{74.69}$$

The sum in (74.68) is a polynomial of degree $2N$ and all the $2N$ zeros are at finite distances from the origin. Assuming that $h(1) \neq 0$, we can write

$$h(w) = h(1)w^{-n} \prod_{j=0}^{2N} \frac{w - w_j}{1 - w_j} \tag{74.70}$$

What changes can be made to the zeros, w_j, without affecting $|\psi_i|^2$ on the real axis? Such changes must likewise leave $|h(w)|^2$ unaltered on the unit circle. From the definition (74.68), the continuation of $|h(w)|^2$ away from the unit circle is $h(w)h^*(1/w^*)$, with pairs of zeros at w_j and $1/w_j^*$. Since one member of each of these pairs of zeros must occur in any such function, the only permissible changes to the zeros consist in replacing an arbitrary number of them by $1/w_j^*$. For the half-plane situation, the negative powers in (74.68) are absent, so that

$$h(w) = \sum_{j=1}^{N} \tilde{h}_j w^j \qquad \tilde{h}_j := \psi_{d,j} \tag{74.71}$$

The contour integral

$$I = \frac{1}{2\pi} \mathcal{P} \int h(w) \frac{w + w'}{w - w'} \frac{dw}{w} \tag{74.72}$$

taken around the unit circle gives

$$I = ih(w') \tag{74.73}$$

Replacing h by ψ_d and w by $\exp(2\pi i s)$, we find

$$i\psi_i(x) = \mathcal{P} \int_0^1 \psi(x') \cot \pi(x' - x) \, dx' \qquad (74.74)$$

so that

$$\psi_I(x) = -\mathcal{P} \int_0^1 \psi_R(x') \cot \pi(x' - x) \, dx'$$
$$\qquad (74.75)$$
$$\psi_R(x) = \mathcal{P} \int_0^1 \psi_I(x') \cot \pi(x' - x) \, dx'$$

If we apply the earlier arguments to periodic specimens, we find that

$$\tilde{\psi}_{I,j} = -i \operatorname{sgn}(j)\tilde{\psi}_{R,j} \qquad (74.76)$$

with

$$\psi_i = \sum_{j=1}^{N} \psi_{d,j} \exp(2\pi i j x) \qquad (74.77a)$$

and

$$\psi_d = \tilde{\psi}_R + i\tilde{\psi}_I \qquad (74.77b)$$

Like the Fourier transform of $\operatorname{sgn}(x)$, the series

$$-\sum_{j \neq 0} i \operatorname{sgn}(j) \exp(2\pi i j x) \qquad (74.78)$$

does not converge but does sum($C1$) to $\cot(\pi x)$.

The literature on the use of half-plane apertures falls into two categories: (i) the numerous papers on reconstruction of weakly scattering objects in which transfer theory is used to show that there are no transfer gaps when the aperture is semi-circular; and (ii) papers in which the analytic properties discussed in this section are invoked to solve the phase problem for strongly scattering specimens. This distinction, which is not always made clear, is to be borne in mind when reading through the literature, as is the fact that the distinction is not a sharp one: the weak-scattering approximation is used as a first step in a strong-scattering calculation.

The suggestion that the use of half-plane apertures would solve the problem of 'transfer gaps', that is, the fact that spatial frequencies close to the zeros of the transfer function are lost, was first made by Hanszen and Morgenstern (1965) and Hanszen (1969, 1970) and then by Hoppe *et al.* (1970). Experimental tests in the microscope were reported by Thon (1968, 1971; see Thon and Willasch, 1970) and further contributions followed by Andersen (1972), Cullis and Maher (1974, 1975), Downing and Siegel (1973a,b, 1974, 1975), Sieber (1974), Spence (1974), Harris and Kerr (1976), Nakahara *et al.* (1976), Downing (1975, 1979), Nakahara and Cullis (1977). Further experimental work has been performed recently by Krakow (1991, 1992a,b) and Hohenstein (1991, 1992). The use of a displaced (excentric) aperture to accentuate phase contrast as in light microscopy is relevant here; see Haydon and colleagues (Haydon *et al.*, 1971; Haydon and Lemons, 1972; Haydon; 1974). The use of complementary half-plane apertures to obtain the phase of strongly-scattering objects was proposed by Misell *et al.* (1974a,b) and discussed in terms of Hilbert transforms shortly afterwards by Burge *et al.* (1974), Saxton (1974a) and Misell (1974). Test calculations were performed by Misell and Greenaway (1974). A detailed survey is to be found in Misell (1978a), where the related optical and holographic work is compared. The idea was extended by Lannes (1976), who pointed out that exact complementarity of the two half-planes is not necessary and that two unsymmetric views of the specimen should be sufficient; a similar proposal was made by Burge and Fiddy (1979).

74.4.4 Logarithmic Hilbert transform pairs
One feature of the foregoing discussion was the representation of the complex quantities that occured in terms of their real and imaginary parts. An alternative choice is of course their modulus and phase and since the natural logarithm of a complex number, s, separates the latter into the sum

$$\ln s = \ln |s| + i \arg s \qquad (74.79)$$

we may anticipate that working with $\ln \psi_i$ rather than ψ_i will shed light on $|\psi_i|$ and $\arg \psi_i$ rather than on $\Re(\psi_i)$ and $\Im(\psi_i)$. Since the image wavefunction in bright-field conditions has the form of contrast against a background, it is natural to write

$$\psi_i =: 1 + t \qquad (74.80)$$

(The foregoing sentence is not to be regarded as a justification for adopting this convention but is simply intended to help the reader to accept what at first sight seems an arbitrary, if minor, complication. It is easy to go astray if the separation (74.80) is not made, however – see Peřina, 1972, Burge *et al.*, 1974 and Saxton, 1978, Section 3.6.)

We now apply the reasoning employed to establish (74.61) to the function $\ln(1 + t)$, subject only to the assumption that the background (the term 1 in 74.80) is stronger than the signal (t): $|t(x)| < 1$. The function $\ln(1 + t(s))$ is analytic for all finite values of s with the exception of (logarithmic) branch points at the zeros of $1+t(s)$. Since $t(s)$ vanishes at infinity in the upper half-plane and $|t| < 1$ on the real axis, the maximum-modulus principle (Jeffreys and Jeffreys, 1956, Section 11.161; Titchmarsh, 1939, Chapter 5) tells us that $|t(s)| < 1$ everywhere in the upper half-plane so that $1 + t(s)$ can have no zeros there; $\ln(1 + t)$ must be analytic in the same region. We can therefore integrate $\{\ln(1 + t)\}/(s - x)$ over the contour of Fig. 74.4 to give

$$\pi i \ln\{1 + t(x)\} = \mathcal{P} \int_{-\infty}^{\infty} \frac{\ln\{1 + t(x')\}}{x' - x} \, dx' \qquad (74.81)$$

Hence

$$\arg\{1 + t(x)\} = -\frac{1}{\pi}\mathcal{P} \int_{-\infty}^{\infty} \frac{\ln|1 + t(x')|}{x' - x} \, dx' \qquad (74.82a)$$

$$\ln|1 + t(x)| = \frac{1}{\pi}\mathcal{P} \int_{-\infty}^{\infty} \frac{\arg\{1 + t(x')\}}{x' - x} \, dx' \qquad (74.82b)$$

These relations can be established in other ways. A modification of the method introduced into coherence theory by Peřina (1972) has been explored in detail by Burge et al. (1974); their formulation prevents the expressions given by Peřina, which involve not t but ψ_i, from failing to give a solution. An ingenious argument by Saxton (1978) shows that the Fourier transform of $\ln(1 + t)$ is one-sided if that of ψ_i is, which yields (74.82) directly. The proof is based on the observation that the convolution of two one-sided functions is itself one-sided so that if $\psi(x)$ has a one-sided transform, so do all integral powers of ψ. The same is true of series of such powers

$$\phi(x) = \sum_{0}^{\infty} a_n \psi^n(x)$$

provided that the series is uniformly convergent, a very interesting finding. In particular, it is true of $\ln(1 + t)$ provided that $|t| < 1$. We see at last why it was so important to work with $1 + t$ instead of ψ_i, even though they represent the same physical quantity.

    Equations (74.82) have the obvious attraction that the integrand of (74.82a) contains the measured quantity $|1+t|$ and not the real or imaginary part. In principle, therefore, the argument of $1 + t$ can be obtained without iteration.

The work of Burge *et al.* (1974, 1976a,b) on the logarithmic Hilbert transform method inspired a considerable literature on a process akin to apodization. In an attempt to control the positions of the zeros of the entire function ψ, Ross *et al.* (1977, 1978) sought ways of driving complex zeros away from the axis and hence weakening their effect. They showed that this could be achieved in principle by the use of spatially partially coherent illumination. The suggestion was taken up and explored in great detail by Nakajima and Asakura (1982a,b,c, 1983a,b). Further proposals for using the logarithmic form of the Hilbert transform relation are to be found in Nakajima and Asakura (1985, 1986) and Nakajima (1987, 1988a,b). See also Wood *et al.* (1981). Other suggestions involving apodization (deliberate modification of the wavefunction in the diffraction pattern plane) or controlled modification of the illuminating beam are mentioned in Section 74.3.6.

74.4.5 Uniqueness in one and two dimensions

The passage from signals in one dimension to images in two (or more) dimensions is not trivial. Arguments that shed light on the uniqueness of the results provided by the various methods of phase retrieval in one dimension do not extend easily to images in two dimensions and were indeed in the early days misleading: it seemed that the lack of uniqueness in one dimension could only be aggravated in two dimensions whereas, in fact the presence of 'adjoining' information has proved to be beneficial. The chance of finding a wrong result becomes small and would become negligibly so if the conditions assumed in the mathematical analysis were rigorously satisfied, but in practice they never are.

Naturally enough, a full understanding of the higher dimensional case was only acquired gradually and major contributions were made by several research groups. As early as 1978, Saxton considered briefly functions of two complex variables and concluded that "instead of squaring the degree of ambiguity, the transition from one to two dimensions ... may leave it unaltered" (Saxton, 1978, Section 3.4). Bruck and Sodin (1979, 1983) argued persuasively that a form of the phase problem should have a unique solution in two dimensions and Bates and colleagues advanced further arguments soon after (Bates, 1982, 1983, 1984; Garden and Bates, 1982; Fright and Bates, 1982; Bates and Fright, 1983). A major contribution was made by Sanz and Huang and we shall follow their approach closely since it gives great insight into the analysis, thereby enabling the reader to relate the mathematical abstractions to the physical realities (Sanz *et al.*, 1983, 1984; Sanz and Huang, 1984a–c, 1985; Sanz, 1985).

The discrete and continuous forms of the phase problem require somewhat different treatments and, for later convenience, we restate them formally here.

Continuous form

Let $f(\boldsymbol{x})$, $\boldsymbol{x} = (x_1, x_2 \ldots x_n)$ be a multidimensional signal (for an image, $n = 2$) of finite support, $\mathcal{A}$, with Fourier transform $F(\boldsymbol{q})$:

$$F(\boldsymbol{q}) = \int_{\mathcal{A}} f(\boldsymbol{x}) \exp(2\pi i \boldsymbol{q} \cdot \boldsymbol{x}) \, d\boldsymbol{x} \qquad (74.83)$$

Given $|F(\boldsymbol{q})|$ for all values of the (real) spatial frequency variable $\boldsymbol{q}$, we have to find $f(\boldsymbol{x})$ for all values of $\boldsymbol{x}$ lying in $\mathcal{A}$.

Discrete form*

Let $a(\boldsymbol{j})$ be a multi-dimensional sequence, in which the vector index $\boldsymbol{j} = (j_1, j_2 \ldots j_n)$ runs over N values: $0 \leq j_i \leq N - 1$ for all i. The discrete Fourier series of $a(\boldsymbol{j})$, denoted by $A(\boldsymbol{q})$, is defined by

$$A(\boldsymbol{q}) := \sum_{j_1=0}^{N-1} \cdots \sum_{j_n=0}^{N-1} a(\boldsymbol{j}) \exp(2\pi i \boldsymbol{q} \cdot \boldsymbol{j}) \qquad (74.84)$$

Given

$$|A(\boldsymbol{q})| \quad , \quad 0 \leq q_i < 1 \quad , \quad i = 1, 2 \ldots n \qquad (74.85a)$$

we have to find

$$a(\boldsymbol{j}) \quad , \quad j_i = 0, 1 \ldots N - 1 \quad , \quad i = 1, 2 \ldots n \qquad (74.85b)$$

In both cases, there are of course inevitable ambiguities of little importance. Thus if $f(\boldsymbol{x})$ is a solution of the continuous form of the problem, then so are cf ($|c| = 1$), $f^*(-\boldsymbol{x})$ and, in many cases, $f(\boldsymbol{x}+\boldsymbol{c})$. These 'trivial associates' of a given solution we disregard.

One-dimensional case

For both continuous and discrete functions, it is known that the problem has no unique solution and the reasons for this are well understood (continuous case: Akutowicz, 1956, 1957; Hofstetter, 1964; Walther, 1963; discrete

---

* We follow Sanz in employing a different notation for discrete sequences to distinguish between the two forms of the problem in what follows.

case: Hayes, 1982). As we have already seen, these sets of solutions can be obtained in the continuous case by flipping any number of the complex zeros of the transform of $F(s)$, $s = x + iy$, over the real axis. In the discrete case, we form the z-transform of $a(j)$

$$\tilde{A}(z) = \sum_{j=1}^{N-1} a(j)z^j \qquad (74.86)$$

This polynomial in z is always reducible* provided that $N > 2$ and hence multiple solutions again exist.

Two- or higher-dimensional case. Discrete form

The work of Hayes (1982) extends the one-dimensional theory to more than one dimension. Consider now the discrete autocorrelation of the sequence $a(j_1, j_2 \ldots j_n)$, where j_i and i have the ranges specified earlier (74.85b) and, for the two-dimensional case, $a(\boldsymbol{j})$ reduces to $a(j_1, j_2)$; we write

$$b(j_1, j_2 \ldots j_n) := (a * a_-)(j_1, j_2 \ldots j_n) \qquad (74.87a)$$

where

$$a_-(j_1, j_2 \ldots j_n) := a(-j_1, -j_2 \ldots - j_n) \qquad (74.87b)$$

and

$$(a_1 * a_2)(j_1, j_2 \ldots j_n) := \sum_{\text{all } k_i} a_1(j_1 - k_1, j_2 - k_2 \ldots j_n - k_n)a_2(k_1, k_2 \ldots k_n)$$

$$(74.87c)$$

Then if we denote z-transforms by capital letters with a tilde, so that for example

$$\tilde{A}(z_1, z_2 \ldots z_n) = \sum_{j_1=0}^{N-1} \ldots \sum_{j_n=0}^{N-1} a(j_1, \ldots j_n)z_1^{j_1} \ldots z_N^{j_N} \qquad (z_1 \in \mathbf{C} \, \forall i)$$

$$(74.88)$$

we know that

$$\tilde{B}(z_1, z_2 \ldots z_n) = \tilde{A}(z_1, z_2 \ldots z_n)\tilde{A}(z_1^{-1}, z_2^{-1} \ldots z_n^{-1}) \qquad (74.89)$$

This will be recognized as the convolution theorem for a suitable choice of the $\{z_i\}$. Hayes (1982) showed that if the z-transform $\tilde{A}$ of a, regarded as

---

* A polynomial p is said to be irreducible (over a given field) if its only factors are of the form kp, $k\epsilon$, where k is an element of the same field and ϵ is unity; otherwise p is reducible. Irreducibility of a polynomial corresponds to primeness of an integer.

a polynomial in n variables, is irreducible, then a is uniquely determined by the modulus of the Fourier series, $|\tilde{a}|$, always excepting trivial associates of course. The number of possible solutions is 2^{r-1}, where r denotes the number of non-symmetric irreducible factors of the z-transform $\tilde{A}$; if $n \geq 2$, therefore, a unique solution exists for (almost) all sequences $\boldsymbol{a}$ because (Hayes and McClellan, 1982) the set of reducible polynomials is of zero measure in $\mathbb{R}^N \times \ldots \mathbb{R}^N$. Sanz (1985) and Sanz and Huang (1985) went on to establish even more stringent limits on possible solutions. We follow these papers closely, but state without proof the various lemmas and theorems upon which their findings repose. (The reader who wishes to pursue this problem further is strongly recommended to consult the original papers, which contain much helpful comment and are written in very accessible language.)

Sanz and Huang consider a slightly generalized form of the problem, which brings it closer to the conditions encountered experimentally, where signals are corrupted by noise and often by some linear transformation (for example, a transfer function). Instead of (74.85), they set out to solve the following problem:

Given

$$M_j := |Ha(j)| \quad , \quad j = 0, 1 \ldots n - 1$$

where H ($\mathbb{C}^m \to \mathbb{C}^n$, $n \geq m$) is a linear transform operator and a is real-valued discrete signal of finite support, find a.

Clearly, real solutions of

$$|Ha(j)|^2 = M_j^2 \quad , \quad j = 0, \ldots n - 1 \tag{74.90}$$

are also solutions of the polynomials

$$\sum_k \sum_l h_{jl} h_{jk}^* z_l z_k = M_j^2 \tag{74.91}$$

and likewise real solutions of (74.91) satisfy (74.90); this is not necessarily true of complex solutions of (74.91), of course.

Sanz and Huang then prove two lemmas, which we quote directly.

Lemma 1: Let $P|\mathbb{R}^m \to \mathbb{R}^n$ be a polynomial

$$P(x_1 \ldots x_m) = (P_1(x_1 \ldots x_m), P_2(x_1 \ldots x_m) \ldots P_n(x_1 \ldots x_m))$$

and $m < n$. There exists a polynomial Q, not identically zero, such that the range of P is contained in the set of zeros of Q, that is to say

$$P(\mathbb{R}^m) \subseteq \{x \in \mathbb{R}^n | Q(x) = 0\}$$

Lemma 2: Let $P|\mathbf{C}^m \to \mathbf{C}^n$ be a polynomial mapping with $m \le n$ and such that $P = (P_1, \ldots P_n)$ satisfies $\deg(P_i) = \deg(P_j)$ for all i, j, P_i is homogeneous and $P(z) = 0$ has only the trivial solution $z = 0$. Then the set $P(\mathbf{C}^m)$ is algebraic in the sense that there are polynomials $Q_1 \ldots Q_r$ such that $P(\mathbf{C}^m) = \{z \in \mathbf{C}^n | Q_1(z) = 0 \ldots Q_r(z) = 0\}$. Moreover, the dimension of $P(\mathbf{C}^m)$ is m.

We now apply these general statements to the following polynomial mapping $P_j|\mathbf{C}^m \to \mathbf{C}^n$, $n \ge m$:

$$P_j(z_0 \ldots z_{m-1}) = \sum_{k=0}^{m-1} \sum_{l=0}^{m-1} h_{jk} h_{jl}^* z_k z_l \qquad j = 0 \ldots n-1 \qquad (74.92)$$

which enables us to establish the following theorem: for the over-determined $(n > m)$ phase-retrieval problem (74.90), there exists a real polynomial $Q(\not\equiv 0)$, defined over $\mathbb{R}^n$, such that any admissible data vector b_j, $j = 0 \ldots n-1$, satisfies

$$Q(b_0^2, b_1^2 \ldots b_{n-1}^2) = 0 \qquad (74.93)$$

A further theorem can be established in view of the fact that P_j is homogeneous and $\deg(P_j) = \deg(P_i)$ for all i, j. Provided that the matrix H with elements h_{ij} is such that $P_j(z) = 0$ has only the trivial solution for all $j = 0, 1, \ldots n-1$, then there exist polynomials $Q_1, \ldots Q_r$ such that any admissible data vector $\{b_j\}$ satisfies

$$Q_k(b_0^2, b_1^2 \ldots b_{n-1}^2) = 0 \quad , \qquad k = 1, 2 \ldots r \qquad (74.94)$$

and the set of zeros of these polynomials is an algebraic set of dimension m.

This tells us that there exist one or more algebraic relations between the coordinates of a vector $b = (b_0, b_1 \ldots b_{n-1})$ which form necessary and sufficient conditions for the system of polynomial equations

$$P_0(x) = b_0^2, \qquad P_1(x) = b_1^2, \qquad \ldots, \qquad P_{n-1}(x) = b_{n-1}^2 \qquad (74.95)$$

to have a solution.

A disturbing consequence is that over-determined phase problems (those for which $m < n$) are locally ill-conditioned: if (in two dimensions) $b = (b_0, b_1)$ is an admissible data vector, then a slightly perturbed form of b, $b' = (b_0 + \epsilon_0, b_1 + \epsilon_1)$ will not be admissible. If we consider the specific case of (74.90) in which H corresponds to the discrete Fourier transform, we find that in one dimension the system of equations (74.91) is not over-determined (the number of unknowns is equal to the number of equations for a sequence of real unknown values), but in two (or more) dimensions it is over-determined, by an amount that is a function of the sequence length. This observation is at the heart of the difference between the one-dimensional case, for which no unique solution exists, and the two- or higher-dimensional situation, for which there is a unique solution.

Unlike the local ill-conditioning mentioned above, the irreducibility condition for phase retrieval stated by Hayes (1982) is stable in the sense that, given some irreducible sequence, there exists a small region surrounding this sequence which again generates an irreducible z-transform.

The fundamental results governing the uniqueness of the solution of the phase problem in two or more dimensions have thus been established. Important though they are, there unfortunately remains a gap between them and the physical reality owing to the effects of measurement error in any real situation and to the fact that the 'zero probability' regarded as reassuring above may not be sufficient in practice. Seldin and Fienup (1990a) voice the doubts very graphically: "By this ... method, it is possible to make up an uncountably infinite number of ambiguous cases even though the theory indicates that the ambiguity is rare (of zero probability) in two dimensions. Consider that it is also true that any randomly chosen real number has probability zero of being a rational number (almost all are irrational numbers). Yet any real number, even if irrational, can be approximated arbitrarily well by a rational number. Thus the fact that the probability of any given object's being ambiguous (the Fourier transform being factorable) is zero is not necessarily comforting."

In one dimension, the phase problem in its extreme form—image intensity and width of diffraction pattern—has multiple solutions owing to the fact that all functions of the form (74.56) are potential solutions. In two dimensions, the pairs of zeros of the one-dimensional solution become zero-sheets and the continuity and smoothness of these sheets impose severe restrictions on the potential multiplicity of the solution. If the solution is regarded as a polynomial $\psi(u, v)$ of the two complex variables u, v the solution may fail to be unique if the function ψ can be written as a product of prime factors $P_j(u, v)$:

$$\psi_i(u, v) = \prod_{j=1}^{N} P_j(u, v) P_j^*(u^*, v^*) \qquad (74.96)$$

This is in theory a very rare occurrence and the effect of noise should not be unduly serious. This can be seen immediately from the very telling example imagined by Bruck and Sodin (1979): in one plane, we have a set of points, like a diffraction pattern, on a square grid of spacing d, so that

$$\psi_i(x, y) = \sum_{j,k} w_{jk} \delta(x - jd, y - jd) \tag{74.97}$$

where w_{jk} is the strength of the spot at the point (jd, kd). The diffraction pattern wavefunction is then (with appropriate scaling)

$$\begin{aligned}
\psi_d &= \sum_{j,k} w_{jk} \exp(ij\,dx) \exp(ik\,dy) \\
&= \sum_{j,k} w_{jk} z_x^j z_y^k =: P(z_1, z_2)
\end{aligned} \tag{74.98}$$

where

$$z_x = \exp(i\,dx) \quad , \quad z_y = \exp(i\,dy) \tag{74.99}$$

The diffraction intensity is $|\psi_d|^2$,

$$|\psi_d|^2 = P(z_x, z_y) P^*(z_x^{-1}, z_y^{-1}) \tag{74.100}$$

If the polynomial P cannot be factorized then a unique solution will be obtained, apart from uncertainty between P and P^*. Polynomials in more than one variable usually cannot be factorized, though it is presumably not unlikely that some factors can be found for the fuzzy set of polynomials within the error bounds of the measurements. In practice, there is reason to believe that the correct solution is obtained, corrupted by error or noise to an extent that depends on the measurement errors afflicting the raw data.

74.4.6 Summary and list of further reading

The battle against the phase problem has been waged on many fronts with weapons of varying degrees of mathematical sophistication and the results have been presented in language that ranges from the extreme sobriety of the pure mathematician to the exuberance of the 'Canterbury School'. The outcome of all this is that, in two (or more) dimensions, and excluding pathological cases and minor ambiguities, it is in principle possible to find a unique solution to the various members of the family of phase problems if the data are free of error. Much more important, there is good evidence that the same is true of the realistic situation in which the measurements are imperfect: the methods are robust in the sense that they are tolerant of a certain amount of error. The theory does not yet permit us to set an upper limit on the acceptable error, which explains why this last statement is only qualitative.

The intuitive feeling that the more data are available and the greater the difference between the conditions in which they were obtained, the more likely are the methods to give the right result is justified: in the case of the Gerchberg–Saxton algorithm, for example, *three* sets of measured data (two images at different defocus and one diffraction pattern) cause rapid convergence to the correct solution in test cases, even with noisy measurements.

The literature of the subject is very extensive. Clear presentations of the problem and ways of analysing it are to be found in the papers of Fienup (1982, 1984, 1987a) and more recently, Seldin and Fienup (1990a), Stefanescu (1985), Sanz (1985) and Bates and Mnyama (1986) and in several chapters of Stark (1987), notably those by Hayes, Dainty and Fienup, Levi and Stark, and Fiddy.

The development of thinking about the phase problem in general may be traced through the following papers: Wolf (1962), Walther (1963), O'Neill and Walther (1963), Roman and Marathay (1963), Dialetis (1967), Dialetis and Wolf (1967), Mehta (1965, 1968), Mehta *et al.* (1966), Kohler and Mandel (1973), Schiske (1974), Burge *et al.* (1976a), Ferwerda *et al.* (1977), Baltes (1978, 1980), Lannes (1978), Devaney and Chidlaw (1978), Bates (1978), Ross *et al.* (1978, 1980a,b, 1981), Nieto-Vesperinas and Hignette (1979), Kiedron (1979, 1980, 1982, 1983), Nieto-Vesperinas *et al.* (1980, 1981a,b), Hayes and Oppenheim (1980), Fienup (1981, 1982, 1984, 1987a,b, 1990), Oppenheim and Lim (1981), Devaney (1981), Taylor (1981), Hayes (1982, 1984), Oppenheim *et al.* (1982, 1983), Wingham (1982), Byrne and Wells (1983), Canterakis (1983), Mammone (1983, 1987), Moses and Prosser (1983), Fiddy (1983), van Hove *et al.* (1983), Hayes and Quatieri (1983), Sanz and Huang (1983a–d, 1984a,b, 1985), Bates and Fright (1984), Rosenblatt (1984), Bertero *et al.* (1985), Fan *et al.* (1985), Nieto-Vesperinas (1986), Gonsalves (1985, 1987), Izraelevitz and Lim (1987), Byrne and Fiddy (1987), Lannes *et al.* (1987a,b), Ferwerda (1988), Nakajima (1988a,b), Bates *et al.* (1990), Kim and Hayes (1990) and Perez-Ilzarbe *et al.* (1990).

Zero location is considered especially by Bates *et al.* (1976), McKinnon *et al.*(1976), Greenaway (1977), Ross *et al.* (1977, 1978, 1979, 1980a,b), Fiddy and Greenaway (1979a), Fiddy and Ross (1979), Walker (1981), Fiddy *et al.* (1981a,b,c, 1982), Ross (1982), Wood *et al.* (1983), Atkin and Ross (1984), Nieto-Vesperinas (1984a,b), Scivier *et al.* (1984), Nieto-Vesperinas and Dainty (1985), Scivier and Fiddy (1985a,b), Curtis and Oppenheim (1987) and Root (1987).

The foregoing list is somewhat arbitrarily separated from the following papers which are concerned with the uniqueness of solutions of the phase problem, especially in two dimensions: Hoenders (1975a), Green-

away and Huiser (1976), Huiser *et al.* (1976, 1977), van Toorn and Ferwerda (1977), Robinson (1978), Feldkamp and Fienup (1980), Huiser and
van Toorn (1980), Burge and Fiddy (1981), Fiddy *et al.* (1981a), Foley
and Butts (1981), Tom *et al.* (1981), Bates (1982, 1984), Garden and
Bates (1982), Fright and Bates (1982), Nieto-Vesperinas and Ross (1982),
Darling *et al.* (1983a,b), Fiddy *et al.* (1983), Barakat and Newsam (1984),
Nieto-Vesperinas and Dainty (1984), Sanz and Huang (1984a,b, 1985), Sanz
et al. (1983, 1984), Sault (1984a,b), Berenyi *et al.* (1985), Deighton *et al.*
(1985a,b), Sanz (1985), Won *et al.* (1985), Fiddy (1985), Berenyi and Fiddy
(1986), Lane and Bates (1987), Lane *et al.* (1987), Bates and Lane (1987a),
Fienup and Kowalczyk (1990).

A few papers are specifically aimed at the mathematical techniques
involved: Akutowicz (1956, 1957), Toll (1956), Hofstetter (1964), Nussenzveig (1967, 1981), Requicha (1980), Nieto-Vesperinas (1980, 1984a,b),
Manolitsakis (1982), Byrne *et al.* (1983), Çetin (1989), Nieto-Vesperinas
and Mendez (1986), Lawton and Morrison (1987) and Byrne and Jones
(1990).

The size, shape and topology of the support of whichever function is
bandlimited are of course important for most of the theory but shape and
connectedness are specifically discussed by Crimmins and Fienup (1981),
Fienup *et al.* (1982), Brames (1986, 1987a,b), Byrne and Fiddy (1987),
Crimmins (1987), Crimmins *et al.* (1990) and Hohenstein (1991).

A slightly different approach is required for blind deconvolution, in
which only the convolution of two unknown functions is known, together
with information about their supports; recent papers on this form of the
problem are those of Bates and Lane (1987a,b), Lane and Bates (1987),
Ayers and Dainty (1988), Davey *et al.* (1989), Seldin and Fienup (1990b)
and Millane (1990).

It is again somewhat artificial to detach the iterative algorithms associated with Fienup from the above; the following are, however, concerned particularly with the iterative routines that have evolved from the Gerchberg–
Saxton sequence: Fienup (1978, 1979, 1980, 1981, 1982, 1983, 1984, 1986,
1987a,b), Fienup and Wackerman (1986), Fienup and Kowalczyk (1990),
Taylor (1980), van Toorn *et al.* (1984), Bates (1983), Bates and Fright
(1983), Bates and Tan (1985) and Wackerman and Yagle (1991).

74.5 Maximum entropy and related probabilistic methods

In 1972, Frieden proposed a method of improving image quality based on
the maximization of a quantity that has the form of an entropy. Later in the
same decade, Hunt (1977a) introduced the maximum *a posteriori* (MAP)
procedure, which also exploits knowledge or assumptions about the image
statistics. We describe these only very briefly, for neither has so far been
very useful in electron microscopy for images at least, though they may be
of interest for spectra; even so, interest in them is growing. The reason why
they are not immediately attractive is that, as Frieden explains (1987a),
"Each [method] has its intended domain of use and non-use. [Maximum
entropy] prefers point and line sources ... and MAP prefers dark high
contrast regions." Crystallography (Bryan, 1987, 1988a, 1989, 1990) and
astronomy (Gull and Daniell, 1978, 1979) thus offer more suitable images.
See Skilling (1989a,b, 1990) for recent developments, Rosenfeld and Kak
(1982, Section 7.7) for a very readable introduction and Shore and Johnson
(1980, 1981) for discussion of the fundamentals of this approach.

Consider now an image $g(p, q)$, related as usual to the object via a
point-response function h and degraded by additive noise, n:

$$g(p,q) = \sum_j \sum_k h(p - j, q - k) f(j, k) + n(p, q) \qquad (74.101)$$

We shall need to introduce the natural logarithms of the signal and noise
and we therefore lower the origin of n by writing

$$n'(p, q) := n(p, q) + n_0 \qquad (74.102)$$

where $|n_0|$ is trivially greater than the largest negative value of n. Hence
$n' > 0$; n_0 is typically set equal to twice the standard deviation of the noise.
We form the quantity

$$H_f = -\sum_p \sum_q f(p, q) \ln f(p, q) \qquad (74.103)$$

which takes its greatest value for a uniform image, $f(p, q) = $ const for all
p, q, and seek the maximum of

$$H := H_f + w H_n \qquad (74.104)$$

in which w is a weight and

$$H_n := -\sum \sum n' \ln n' \qquad (74.105)$$

subject to two constraints. The first requires g and f be related by (74.101) while the second merely normalizes the image strength:

$$\sum_p \sum_q f(p,q) =: f_0 \qquad (74.106)$$

where f_0 is a constant. The result of maximizing (74.104) will be an image that is biased towards the (noisy) recorded image but has been subjected to some smoothing. The quantity to be maximized is the functional U:

$$U := \quad H_f \quad + \quad wH_n \quad + \quad \sum_p \sum_q \lambda_{pq} G(p,q) \quad + \quad \lambda_0 F_0$$

$$\begin{array}{cccc} \text{signal} & \text{noise} & \text{first} & \text{second} \\ \text{entropy} & \text{entropy} & \text{constraint} & \text{constraint} \end{array}$$

$$(74.107)$$

in which

$$G(p,q) := g(p,q) - \sum_j \sum_k h(p-j, q-k) f(j,k) - n'(p,q) + n_0$$

$$F_0 = \sum_j \sum_k f(j,k) - f_0$$

$$(74.108)$$

The quantities λ_{pq} and λ_0 are Lagrange multipliers, to be determined. Their presence renders the equations to be solved highly nonlinear.

For further comment, the papers of Frieden (1977), Kikuchi and Soffer (1977), Fiddy and Greenaway (1978, 1979b), Kreinovic and Kosheleva (1979), Dainty et al. (1979), Burch et al. (1983), Skilling and Bryan (1984), Livesey and Skilling (1985), Bryan (1988a,b) and Gonsalves et al. (1990) may be consulted. Lawrence et al. (1989) discuss maximum entropy in connection with electron tomography.

For recent attempts to apply the maximum entropy restoration procedure to electron images, see Farrow and Ottensmeyer (1987, 1988, 1989a,b), Barth et al. (1987, 1989) and Martin et al. (1991).

The MAP technique, which belongs to the field of Bayesian estimation, requires us to find the maximum value of a conditional probability density function, $p(f|g)$, the probability of f conditional on a knowledge of g. The result of such a search will be the most probable form of the estimated image f consistent with the observed image g. The method has been applied to focal series, for which several observed images g_j are available so that $p(f|g) \to$ max is replaced by $\prod_i p(f|g_j) \to$ max, by Kirkland (1984, 1987; Kirkland et al., 1985). Kirkland shows that the MAP equation has the same form as a constrained least-squares problem (Section 73.3) and tests the method in realistic conditions on a high resolution image. See also Coene (1992).

For further information about the MAP approach see Hunt (1977a,b), Andrews and Hunt (1977), Lo and Sawchuk (1979), Trussell and Hunt (1979), Trussell (1980), Geman and Geman (1984), Besag (1986), Kasturi and Walkup (1986), Hanson (1987), Zhuang *et al.* (1987), Frieden (1987a,b) and Skilling (1989a,b, 1990). The proceedings of the MaxEnt workshops provide a record of developments and applications of many kinds (Smith and Grandy, 1985; Smith and Erickson, 1987; Justice, 1986; Erickson and Smith, 1988; Skilling, 1989a; Fougère, 1990; Grandy and Schick, 1991; Smith *et al.*, 1992).

For a non-iterative method based on a weighted sum of shifted records, see Frieden and Oh (1992).

75

Three-dimensional Reconstruction

75.1 Introduction

The two-dimensional image provided by the transmission electron microscope represents some kind of a projection through the specimen, the intensity at each image point corresponding to the cumulative effect on the
electrons of all the structure in their path within the specimen. The ultimate aim of the microscopist is often, however, to establish the three-
dimensional structure of the object and methods have been developed for
deriving a structure in space from projections through the specimen in different directions. These methods are being used increasingly widely, in
molecular biology especially, and have reached a high degree of perfection.
This chapter is of necessity confined to the principles of the subject, of
which several full accounts are available (Frank and Radermacher, 1986;
Moody, 1990; Frank, 1992a), See too earlier surveys by Huxley and Klug
(1971), Hoppe (1970a,b, 1979, 1981), DeRosier (1971), Vainshtein (1978),
Hoppe and Typke (1979), Hoppe and Hegerl (1980), Kiselev (1978, 1980,
1984a,b, 1986), Mellema (1980a,b) and for a historical survey of the whole
subject, Hoppe (1983). The books of Deans (1983), Herman (1979a, 1980b),
Herman and Natterer (1981), Natterer (1986) and Kak and Slaney (1988)
contain invaluable background material.

In this introductory section, we attempt to put the subject in context and impose a pattern on its variety. In the later sections, specific
reconstruction procedures are described briefly, with references to specialized texts for details. This Chapter is concerned only with reconstruction
from projections in different directions; for methods of obtaining three-
dimensional information from *serial sections*, see the survey by Lockhausen
et al. (1987).

We have observed that a three-dimensional reconstruction involves
recording the intensity distributions representing projections through the
specimen in different directions and then suitably redistributing this mass
of data. These projections are acquired in different ways, depending on
the nature of the specimen. At one extreme are crystalline specimens with
reasonably high symmetry; a small number of views may well be sufficient
for good three-dimensional reconstruction because the crystal symmetries

effectively multiply the number of views actually recorded and digitized. Specimens that form regular arrays, either naturally or thanks to ingenious preparation procedures, are also very satisfactory for reconstruction purposes because the regularity can be exploited to create an average image of the unit cell for each projection; the data on which the subsequent calculations are based can therefore be expected to have a reasonably high signal-to-noise ratio. For regular specimens of this kind, it is usual to speak of *electron crystallography** (Glaeser, 1985; Moras *et al.*, 1987; Downing, 1991a,b; Dorset, 1991).

Images of groups of particles that adopt only a small number of orientations and can hence be classified and then averaged are likewise well suited for reconstruction, though the pre-processing tasks are now quite long and difficult. Finally, we have the most difficult case of all, individual particles with little or no simplifying symmetry and no possibility of averaging over identical particles. In these cases, we speak of *electron tomography* (Frank, 1992a).

The steps preceding the actual reconstruction are different in these various cases but their purpose is always the same: to collect a set of views through the specimen as complete as possible.

The reconstruction step is based on a number of assumptions about the image-forming process. In particular, the recorded density must be related in some known and unambiguous way to the projected potential, for it is this potential distribution that governs the image contrast. For most of the specimens that are currently studied, the thickness is such that the image contrast is scattering contrast (electrons scattered through large angles by the heavier atoms in the structure are intercepted by the objective aperture and hence absent from the image). It is well-known that the contrast at the image is then an exponential function of the mass–thickness[†] of the specimen at the corresponding object point (see for example Reimer, 1993, Section 6.1 or Hawkes, 1992c). One condition for successful reconstruction is thus satisfied. At very high resolution, where phase contrast is created by the imaging properties of the microscope, characterized by its contrasttransfer function, the situation is more complex, but once again it can be shown that the image contrast does indeed provide a projection through the specimen. This point is discussed in detail in Hawkes (1992c) and a related experiment is described by Liu (1991b). It must, however, be said that the mechanism of three-dimensional reconstruction is distinctly harder to understand for the phase objects that are examined in electron

---

[*] The same term is used to describe the study of crystalline material by means of convergent-beam electron diffraction (e.g. Steeds, 1984; Steeds and Carlino, 1992).

[†] The quantities needed in scattering theory are commonly expressed in terms of *mass–thickness*, defined as the product of density and geometrical thickness.

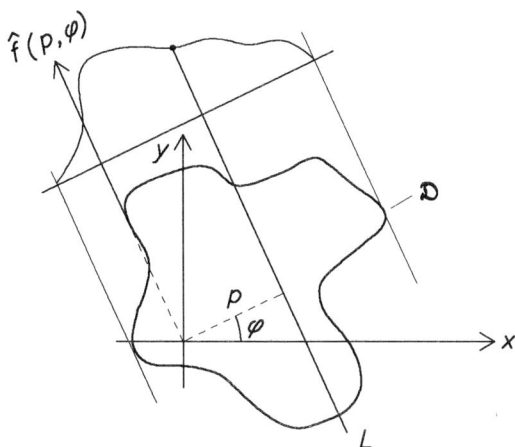

Fig. 75.1: Quantities used in the definition of the Radon transform.

microscopy than for the amplitude objects of medical tomography and some aspects remain unclear. Bright-field imaging of specimens thin enough for single-scattering to be predominant should, however, provide data suitable for reconstruction. Even so, special precautions are necessary owing to the defocus gradient between opposite sides of a steeply tilted specimen. We present a remedy in Section 75.3.

In Section 75.2, the various ways of performing the reconstruction are considered, while in Section 75.3, we return to the types of pre-processing associated with the different kinds of specimen. First, however, we establish certain general results; the derivations are easier to follow if we regard all the functions involved as continuous. The effects of finite sampling and hence discrete data are examined later.

For ease of presentation, we proceed as though our aim were reconstruction of a two-dimensional body from projections but the extension to three dimensions is a formality. Figure 75.1 shows a domain $\mathcal{D}$ and a line L in two dimensions. The line is characterized by two parameters, p and φ, in terms of which we have

$$x \cos \varphi + y \sin \varphi = p \qquad (75.1)$$

With

$$\boldsymbol{a} := (\cos \varphi, \sin \varphi)$$
$$\boldsymbol{x} := (x, y) \qquad (75.2)$$

this may be written

$$\boldsymbol{x} \cdot \boldsymbol{a} = p \qquad (75.3)$$

The *projection* along L of some function f defined within $\mathcal{D}$ is given by

$$\hat{f} := \int_L f(x,y)\,ds \qquad (75.4)$$

and we write

$$\hat{f}(p,\varphi) =: \mathcal{R}f = \int_L f(x,y)\,ds \qquad (75.5)$$

in which the integration is taken along L. This is known as the *Radon transform* of f (Radon, 1917). Provided that f is continuous and has compact support, integration along the set of lines parallel to L defines $\mathcal{R}f$ uniquely. Another form of (75.5) is

$$\begin{aligned}
\hat{f}(p,\boldsymbol{a}) &= \iint f(\boldsymbol{x})\delta(p - \boldsymbol{a}\cdot\boldsymbol{x})\,dx\,dy \\
&= \int f(\boldsymbol{x})\delta(p - \boldsymbol{a}\cdot\boldsymbol{x})\,d\boldsymbol{x}
\end{aligned} \qquad (75.6)$$

and this is convenient because, with only a minor change in interpretation, it is valid in three dimensions also: we have only to regard $\boldsymbol{x}$ as (x,y,z), the angular parameter $\boldsymbol{a}$ likewise has three components and the equation $\boldsymbol{x}\cdot\boldsymbol{a} = p$ defines not a line but a plane.

There is an important relation between the Fourier transform of $f(x,y)$ and its Radon transform. As usual, we set

$$F(\boldsymbol{q}) = \mathcal{F}^-(f) = \int f(\boldsymbol{x})\exp(-2\pi i\boldsymbol{q}\cdot\boldsymbol{x})\,d\boldsymbol{x} \qquad (75.7)$$

But

$$\exp(-2\pi i\boldsymbol{q}\cdot\boldsymbol{x}) \equiv \int \exp(-2\pi iy)\delta(y - \boldsymbol{q}\cdot\boldsymbol{x})\,dy$$

so that if we write

$$\boldsymbol{q} = \boldsymbol{a}s \quad , \quad y = ps \qquad (75.8)$$

we find

$$\begin{aligned}
F(\boldsymbol{a}s) &= |s|\iint f(\boldsymbol{x})e^{-2\pi isp}\delta(sp - s\boldsymbol{a}\cdot\boldsymbol{x})\,dp\,d\boldsymbol{x} \\
&= \int dp\,e^{-2\pi isp}\int d\boldsymbol{x}\,f(\boldsymbol{x})\delta(p - \boldsymbol{a}\cdot\boldsymbol{x}) \\
&= \int \hat{f}(p,\boldsymbol{a})\exp(-2\pi isp)\,dp
\end{aligned} \qquad (75.9)$$

with inverse

$$\hat{f}(p, \boldsymbol{a}) = \int F(\boldsymbol{a}s) \exp(2\pi i s p) \, ds \qquad (75.10)$$

These relations are an expression of the important *projection theorem,* which states that the (two-dimensional) Fourier transform of $f(x, y)$ is the same as the (one-dimensional) Fourier transform of the Radon transform of f.

In three dimensions, a similar relation is found; with $F(\boldsymbol{q}) = \mathcal{F}^- f(\boldsymbol{x})$, $\boldsymbol{x} = (x, y, z)$, and

$$\sigma(x, y) = \int f(x, y, z) \, dz \qquad (75.11)$$

we see that

$$
\begin{aligned}
F(q_x, q_y, 0) &= \iiint f(x, y, z) \exp\{-2\pi i (x q_x + y q_y)\} \, dx \, dy \, dz \\
&= \iint \sigma(x, y) \exp\{-2\pi i (x q_x + y q_y)\} \, dx \, dy \qquad (75.12) \\
&= \mathcal{F}^-(\sigma)
\end{aligned}
$$

Returning to (75.10) and recalling that $F(\boldsymbol{q}) = \mathcal{F}^-(f)$, it is immediately seen that

$$
\begin{aligned}
f &= \mathcal{F}_{(2)} F(\boldsymbol{q}) \\
&= \mathcal{F}_{(2)} \{ \mathcal{F}_{(1)}^- \{\hat{f}\} \} \qquad (75.13) \\
&= \mathcal{F}_{(2)} \{ \mathcal{F}_{(1)}^- \{\mathcal{R}f\} \}
\end{aligned}
$$

We now introduce the notion of *back-projection,* which is a first step towards inversion of the Radon operator, and a vital one in many reconstruction procedures. The back-projection operator, $\mathcal{B}$, of some arbitrary function $g(p, \varphi)$ is defined by

$$b(x, y) := \mathcal{B}g := \int_0^\pi g(\boldsymbol{a} \cdot \boldsymbol{x}, \varphi) \, d\varphi \qquad (75.14)$$

where as before, $p = \boldsymbol{a} \cdot \boldsymbol{x}$, $\boldsymbol{a} = (\cos \varphi, \sin \varphi)$. In polar coordinates, we write

$$b(x, y) := b_p(r, \theta) = \int_0^\pi g(r \cos(\varphi - \theta), \varphi) \, d\varphi \qquad (75.15)$$

The effect of $\mathcal{B}$ is easily seen: the ray-sums of the lines passing through a given point (x, y) are added. Suppose for example that we have N projections and that N lines pass through (x, y). We write g in the form

$$g = \sum_{j=1}^{N} g_j(p)\delta(\varphi - \varphi_j) \tag{75.16}$$

whereupon we see immediately that

$$b_p(r, \theta) = \sum g_j(p_j) \tag{75.17}$$

$(p_j := r\cos(\varphi_j - \theta))$.

Suppose now that the function g is the Radon transform of some function $f(x, y)$,

$$g = \mathcal{R}f$$

Then

$$b(x, y) = \mathcal{B}(\mathcal{R}f) = \int_0^\pi \hat{f}(p, \varphi)\, d\varphi \tag{75.18}$$

proves to be the same as the convolution of f and $1/r$, $r^2 = x^2 + y^2$. This extremely important result can be derived with the aid of (75.13). We have

$$\mathcal{B}\hat{f} = \mathcal{B}\mathcal{F}_{(1)}\{\mathcal{F}_{(2)}^{-}f\} \tag{75.19a}$$

or

$$b_p(r, \theta) = \mathcal{B}\mathcal{F}_{(1)}\tilde{F}(q, \varphi) \tag{75.19b}$$

in which $\tilde{F}(q, \varphi) = F(q_x\cos\varphi, q_y\sin\varphi)$ and $\tan\varphi := q_x/q_y$. The one-dimensional transform acts on q. Thus

$$b_p(r, \theta) = \mathcal{B}\int_{-\infty}^{\infty} \tilde{F}(q, \varphi)\exp\{2\pi iqr\cos(\theta - \varphi)\}\, dq$$

$$= \int_0^\pi \int_{-\infty}^{\infty} \tilde{F}(q, \varphi)\exp\{2\pi iqr\cos(\theta - \varphi)\}\, d\varphi\, dq \tag{75.20}$$

$$= \int_0^{2\pi} \int_0^\infty \frac{\tilde{F}(q, \varphi)}{q}\exp\{2\pi iqr\cos(\theta - \varphi)\}q\, dq\, d\varphi$$

$$= \mathcal{F}_{(2)}\left\{\frac{\tilde{F}(q, \varphi)}{q}\right\}$$

and finally

$$b_p(r, \theta) = f(r, \theta) * \frac{1}{r} \tag{75.21}$$

from which f can in principle be extracted by taking the Fourier transform of both sides, dividing by $\mathcal{F}(1/r)$ and inverse Fourier transforming the result.

We thus have two ways of extracting $f(x, y)$ from its Radon transform, $\hat{f}(p, \varphi)$. Examination of the support of the various functions involved shows that neither is ideal but a remarkable feature of the back-projection operator yields a distinctly better procedure. It will emerge that, instead of filtering the back-projection of $\hat{f}$, as we must if we use (75.21) directly, we can filter the projections first and then apply back-projection. This is shown as follows.

We have

$$f(x, y) = \int\int\limits_{-\infty}^{\infty} F(\mathbf{q}) \exp(2\pi i \mathbf{q} \cdot \mathbf{x}) \, d\mathbf{q}$$

$$= \int\limits_{0}^{2\pi} \int\limits_{0}^{\infty} F_p(q, \varphi) \exp\{2\pi i q (x \cos \varphi + y \sin \varphi)\} q \, dq \, d\varphi \tag{75.22}$$

where as usual, $F_p(q, \varphi) := F(q \cos \varphi, q \sin \varphi)$ and $p = x \cos \varphi + y \sin \varphi$. Reorganising (75.22) gives

$$f(x, y) = \int\limits_{0}^{\pi} \int\limits_{-\infty}^{\infty} |q| F_p(q, \varphi) \exp(2\pi i p q) \, dq \, d\varphi \tag{75.23}$$

but we know (75.10) that $F_p(q, \varphi)$ is the same as the one-dimensional transform of $\hat{f}(p, \varphi)$ with respect to p, which we denote $\hat{F}(q, \varphi)$. Hence

$$f(x, y) = \int\limits_{0}^{\pi} \int\limits_{-\infty}^{\infty} |q| \hat{F}(q, \varphi) \exp(2\pi i p q) \, dq \, d\varphi$$

$$= \int\limits_{0}^{\pi} \tilde{f}(p, \varphi) \, d\varphi \tag{75.24}$$

$$= \mathcal{B} \tilde{f}$$

with

$$\tilde{f}(p,\varphi) = \int\limits_{-\infty}^{\infty} |q|\hat{F}(q,\varphi)\exp(2\pi ipq)\,dq$$

$$= \mathcal{F}_{(1)}\{|q|\hat{F}(q,\varphi)\}$$

$$= \mathcal{F}_{(1)}(|q|) * \hat{f}(p,\varphi)$$

(75.25)

Finally

$$f = \mathcal{B}(h(p) * \hat{f})$$ (75.26)

in which $h(p)$ is the *filter function* $\mathcal{F}_{(1)}(|q|)$ in (75.25). We have introduced a new symbol h because, in practice, not only must a discrete approximation be employed but it may also be preferable to tune the filter to the back-projection in question, as we see below.

We close this introduction with a reference to a change of coordinates introduced by Edholm and Herman (1987), which replaces the relation (75.1) between p, x, y and $\sin\varphi, \cos\varphi$ by a linear relation, not involving the circular functions. These new coordinates have two forms:

$$p = \frac{u}{(1+v^2)^{\frac{1}{2}}} \quad , \quad \varphi = \arctan v$$ (75.27a)

or

$$p = \frac{v}{(1+u^2)^{\frac{1}{2}}} \quad , \quad \varphi = -\operatorname{arccot} u$$ (75.27b)

(Fig. 75.2). Edholm and Herman show that the back-projection operator $\mathcal{B}$ can be expressed in a convenient form and derive a new reconstruction algorithm. For further discussion of this approach, see Edholm *et al.* (1988). More recently, Herman *et al.* (1992) have developed an efficient algorithm for exploiting data organized in this way.

75.2 Methods

The Radon transform has an inverse, given $\mathcal{R}f$ for all lines L (and likewise in higher dimensions): $f(x,y)$ can apparently be reconstructed from $\mathcal{R}f$. But in practice, we never do know $\mathcal{R}f$ for all L but only for a finite number of sample values and furthermore, Smith *et al.* (1977b) have shown that a function of compact support in $\mathbb{R}^2$ is uniquely determined by *any* infinite set, but *no* finite set, of its projections. Arbitrarily good approximations can, however, be found by increasing the number of projections (Hamaker *et al.*, 1980) so, as Deans (1983) puts it graphically, "even though you can't win, you must not give up." In practice, reconstruction from projections

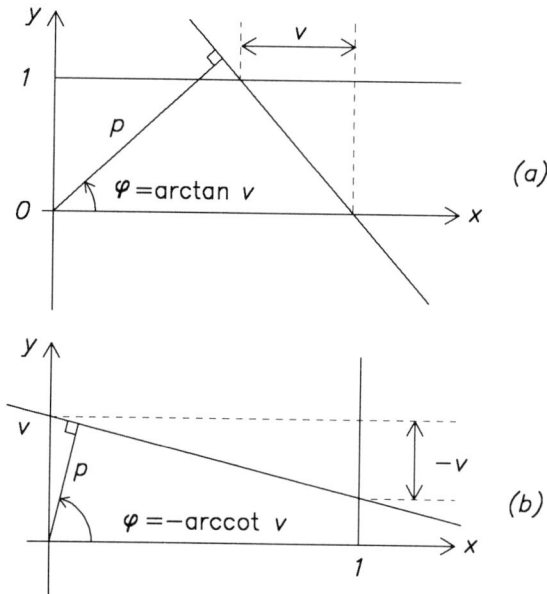

Fig. 75.2: The two definitions of (u, v) introduced by Edholm and Herman. They are both used, the choice depending on the values of the variable.

rarely leads to wrong structures but the closeness of the sampling directions effectively determines the resolution attainable.

The reconstruction algorithms fall into two broad classes: direct methods and iterative methods. The details vary according to the way in which the projections have been collected and we therefore mention that two procedures are in use: single-axis tilting and conical tilting. For special cases, highly symmetrical structures for example, the procedure will usually be simpler.

In single-axis tilting, the specimen is tilted in the microscope by small increments about a single axis over as wide a range of angles as possible, commonly $\pm 60°$. A wedge of information is hence lost (between $\pm 60°$ and $\pm 90°$), which can generate artefacts in the reconstruction and causes the resolution in the direction of the optic axis (z) to be lower than in the lateral directions (x, y).

In conical tilting, the object is tilted through a fixed angle τ and then rotated about the optic axis in small angular steps. There is now a conical gap in the Fourier transform of the structure, of half-angle $(\pi - \tau)$. If the maximum tilt of the single-axis scheme is τ, then the z-resolution will be better with conical tilting and artefacts will be less troublesome but more images must be recorded in order to achieve the same $x - y$-resolution.

75.2.1 Direct methods

In the simplest case, (75.9) is used and back-projection does not appear explicitly. The Radon transform is measured for a range of directions and $f(x, y, z)$ is calculated from (75.13), namely $f = \mathcal{F}_{(2)}\{\mathcal{F}_{(1)}^{-}(\mathcal{R}f)\}$ where as before $\mathcal{F}_{(j)}$ denotes a j-dimensional transform. This approach, which has many ramifications and refinements, was used for the earliest reconstructions made by the Cambridge group (see Crowther *et al.*, 1970a for a full discussion; de Rosier and Klug, 1968; DeRosier and Moore, 1970; Crowther *et al.*, 1970b). For other examples, see Frank (1992a) and the bibliographies of Hawkes (1978a, 1982c, 1992a).

Interpolation is required to redistribute the measured data onto a regular cartesian grid but this can be avoided by suitable choice of coordinates. Thus for single-axis tilting, the measurements can be used directly if cylindrical polar coordinates are adopted and the Fourier–Bessel transform introduced (Crowther *et al.*, 1970a,b). The reconstruction step can be simplified still further by expanding the measured projections as a series in the Chebyshev polynomials and the object as a similar series but in terms of the Zernike polynomials, whereupon the expansion coefficients can be transferred directly from one series to the other (Cormack, 1963, 1964); this idea was introduced independently into electron microscopy by Smith *et al.* (1973) and Zeitler (1974).

An idea of the resolution attainable has been given by de Rosier and Klug (1968) and Crowther *et al.* (1970b), for single-axis tilting:

$$d = \pi D / N \qquad (75.28)$$

where d is the resolution, D is a measure of the size of the object being reconstructed and N is the number of (regularly spaced) views. For further early justification of this formula, see Hoppe (1969b), Hoppe and Hegerl (1980), Klug and Crowther (1972). It has been re-examined by van Heel and Harauz (1986), who show that, for a small number of projections, it must be replaced by

$$d = 2D \sin(\pi / 2N) \qquad (75.29)$$

which of course reduces to (75.28) for large N, for which $\sin(\pi/2N) \approx \pi/2N$. Van Heel and Harauz also derive a formula for resolution in direct three-dimensional reconstruction; we refer to their paper for careful discussion of this expression.

For conical tilting, the following expressions have been proposed (Radermacher and Hoppe, 1980; see Radermacher, 1988):

$$d = \frac{2\pi D}{N} \sin \tau \qquad (N \text{ even})$$

$$d = \frac{2\pi D \sin \tau \cos \tau}{N\{\cos^2 \tau + (\pi/2N)^2 \sin^2 \tau\}^{\frac{1}{2}}} \qquad (N \text{ odd}) \qquad (75.30)$$

with $\tau < \pi/2 - \Delta\varphi/4$, $\Delta\varphi$ being the increment in azimuthal angle (around the optic axis). These resolution measures refer to the x-direction (single-axis) or the $x - y$-plane (conical). In the z-direction, the resolution (d_z) is given by

$$d_z = d\left(\frac{\tau + \sin\tau\cos\tau}{\tau - \sin\tau\cos\tau}\right)^{\frac{1}{2}} \qquad \text{(single-axis)}$$

$$d_z = d\left(\frac{3 - \sin^2\tau}{2\sin^2\tau}\right)^{\frac{1}{2}} \qquad \text{(conical)}$$

(75.31)

In practice, filtered back-projection has supplanted the earlier methods. Equation (75.26) is now the starting point and the methods differ, minor details apart, only in the space in which the filter operation is performed, direct or reciprocal. In both cases, the distribution $f(x, y)$ is obtained from $f = \mathcal{B}(h * \hat{f})$ (75.26) in which h is a one-dimensional filter with frequency response $|q|$. In the direct-space approach, the convolution of $\hat{f}$ and h is first evaluated, giving $h * \hat{f}$, and the result is back-projected. In the transform method, the direct product of $H(\mathbf{q})$, the transform of $h(p)$, equal to $|q|$ in the ideal continuous situation, and $\hat{F}(q, \varphi)$ is formed; the result is inverse Fourier transformed and back-projected.

These ideas have many interesting ramifications, both practical and theoretical; for a good introduction to the latter, the role of the Gegenbauer polynomials in particular, see Deans (1983).

Discrete data

It seems that a good choice of ways of reconstructing two- or three-dimensional density distributions from projections is available but, in practice, difficulties of many kinds arise. These are principally due to the fact that the measured data are discrete, not continuous, and do not fill the angular range over which the various functions have to be integrated. The problem may be exacerbated by the fact that the discrete data may not even be distributed regularly over the space they do cover. Thus in the case of the back-projection algorithms, considerable attention has been paid to the best choice of the filter $h(p)$ to be used in the discrete convolution; a list of candidates is given by Jain (1989, Table 10.3). Harauz and van Heel (1986a) have examined these critically and propose a set of exact filter functions, each member of the set corresponding to a particular projection direction. We refer to their paper for details.

For further examples of reconstructions using these techniques, and guidance on their development, we draw attention to the following papers, in addition to those already cited: Vainshtein (1970, 1971, 1973, 1978) and Kiselev (1980, 1986) where many reference to the Russian school are to be found; Ramachandran (1971), Ramanchandran and Lakshminarayanan (1971a,b); Radermacher and Hoppe (1978); Lewitt and Bates (1978); Lewitt (1983, 1990).

The important special case of helical structures is dealt with in great detail by Stewart (1988) and by Moody (1990).

75.2.2 Iterative methods

Simple back-projection is the starting point of the iterative or 'algebraic' methods in which a first estimate of the structure is used to calculate a new set of projections. The difference between these and the measured projections is then used to correct the first estimate and the process is repeated until convergence is achieved. Thus, suppose that we write $f(x,y)$ approximately in the form of a finite series

$$f(x,y) \approx \sum_{i=1}^{I} \sum_{j=1}^{J} a_{ij} m_{ij}(x,y) \tag{75.32}$$

in which $\{m_{ij}(x,y)\}$ is some set of basis functions. The Radon transform of f written in this form is then

$$\hat{f}(p,\varphi) = \mathcal{R}(f) = \sum\sum a_{ij}\mathcal{R}m_{ij} =: \sum\sum a_{ij}M_{ij}(p,\varphi) \tag{75.33}$$

in which the set of functions $\{M_{ij}\}$ is given by the Radon transforms of $\{m_{ij}\}$. In discrete form, this gives

$$\hat{f}(p_k,\varphi_l) = \sum\sum a_{ij}M_{ij}(p_k,\varphi_l) \tag{75.34}$$

and in practice a noise term must be included, giving us the familiar linear restoration equation, which we have already discussed in great detail. The coefficients a_{ij} can be obtained by Wiener filtering, maximum entropy or calculation of the pseudo-inverse, for example, depending on the nature of the data.

Another way of solving (75.34) has come to be known as the algebraic reconstruction technique, which takes various forms. We rewrite (75.34) in the matrix form

$$\hat{f} = Ma \tag{75.35}$$

and construct the iterative sequence (Kaczmarz, 1937)

$$\overline{a}^{(k+1)} = a^{(k)} + \frac{\hat{f}_{k+1} - M_{k+1}a^{(k)}}{|M_{k+1}|^2} M_{k+1} \tag{75.36}$$

in which M_k is the k-th row of M and $\hat{f}_k$ is the k-th element of $\hat{f}$. The denominator is the scalar product of the vector M_{k+1} with itself. Constraints may be included, as we have indicated by writing $\bar{a}$ rather than a on the left-hand side of (75.36). If $a = \bar{a}$, then the iterative sequence is unconstrained; alternatively, a may be forced to be positive, or again, may be confined within a prescribed interval. The routine (75.36) is not the only choice but it is typical of those employed.

Several schemes for algebraic reconstruction were developed and tested in the early 1970s: the Algebraic Reconstruction Technique, ART (Bender et al., 1970; Gordon et al., 1970; Herman and Rowland, 1971; Frieder and Herman, 1971; Gordon and Bender, 1971; Gordon and Herman, 1971; Crowther and Klug, 1971; Lake, 1971; Herman, 1972; Bellman et al., 1971; Gilbert, 1972a,b; Herman et al., 1973; Barbieri, 1974; Gordon, 1974, Herman and Lent, 1976a,b); the Simultaneous Iterative Reconstruction Technique, SIRT (Gilbert, 1972a; Lakshminarayanan and Lent, 1979); and the Iterative Least-Squares Technique, ILST (Goitein, 1972).

More recent developments are charted by Herman (1979a,b, 1980a,b), Lewitt (1983), Trussell (1984), Herman et al. (1984) and Censor and Herman (1987), which contains a representative bibliography. The book by Kak and Slaney (1988) is also recommended.

75.2.3 Reconstruction from a single view of an oblique section

An oblique section through a crystalline layer provides a series of projections through the structure from which, if certain conditions are satisfied, the three-dimensional architecture can be reconstructed (Crowther, 1984; Crowther and Luther, 1984; Crowther et al., 1990). The relation between the image and the structure can be understood from Fig. 75.3, which shows the oblique section cutting through the crystal layer, thickness t, divided into slices (rows) of thickness ϱ and columns. For each column, the projected density σ_i is given by

$$\sigma_i = \int_{z_i-t/2}^{z_i+t/2} \varrho(z)\,dz = \int_{-\infty}^{\infty} w(z - z_i)\varrho(z)\,dz \qquad (75.37)$$

where

$$w(z) = \begin{cases} 1 & \text{if } |z| < t/2 \\ 0 & \text{if } |z| \geq t/2 \end{cases} \qquad (75.38)$$

This convolution relation may be solved, in principle at least, by a direct least-squares method or by taking the Fourier transform of (75.37) and employing a Wiener filter; an inverse filter cannot be used because the

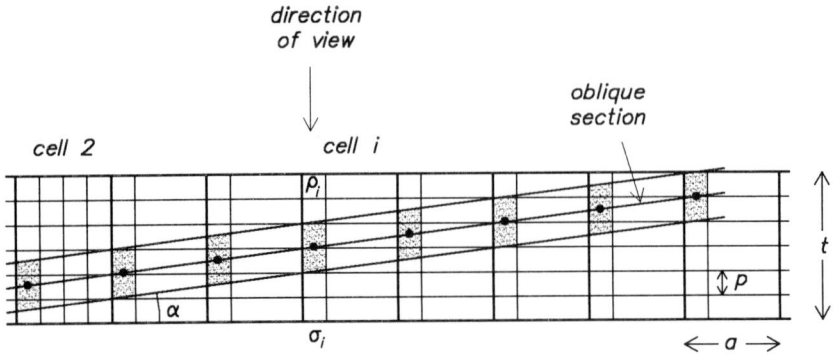

Fig. 75.3: Oblique section through a single layer of a two-dimensional crystal, the plane of which is normal to that of the diagram. As the section rises through the layer, different densities from the first column of the unit cell are selected. Each cell is divided into pixels of height p, density ϱ_j and the projected density in the image from unit cell i corresponding to the first column is denoted by σ_i. The dots represent points at which the convolution of the continuous density (in the first column) and the 'window' defined by the oblique section is sampled (after Crowther and Luther, 1984).

transform of $w(z)$ is oscillatory, passing through zero. For a general account, see Crowther *et al.* (1990) and for examples, Luther and Crowther (1984) and Taylor and Crowther (1986, 1991, 1992).

75.2.4 The missing cone

The data available for three-dimensional reconstruction are neither continuous nor complete. The fact that only a finite number of discrete views is available rather than a quasi-continuous sequence of views limits the resolution that can be obtained but the absence of any views beyond the steepest possible tilting angle raises a much more thorny question: *extrapolation*. We have avoided this huge subject in earlier chapters, for the nature of the electron image-forming process is such that it has little relevance for extending the resolution limit of the microscope. In three-dimensional reconstruction, however, it cannot be dismissed and we therefore discuss it very briefly here.

The various approaches that have been explored can in many cases be subsumed within the very general method that has come to be known as *Projection onto Convex Sets* (POCS) and it is this method that we present here. The idea is simple. When we attempt to reconstruct a three-dimensional structure, information of various and very different kinds will be available. Typically, we might have a finite set of views through the specimen, some knowledge of its size and perhaps shape and some estimate of the extent of its Fourier transform; in addition, constraints such as positivity and the range within which the density values in the specimen must lie will be known. All these pieces of information and any others applicable to special cases can be expressed mathematically as 'projections' onto convex sets and the method consists in seeking the solution that lies in the intersection of all these sets. The word 'projection' is used in a wider sense here than in earlier sections of this chapter, where we used it to mean the projected density in a particular direction. Here, the word has a specific meaning for each of the sets in turn. For example, when we specify the physical limits of the specimen structure, we define a projection P such that

$$Pf = \begin{cases} f & \text{if } x, y, z \text{ lies within the permitted limits} \\ 0 & \text{otherwise} \end{cases}$$

and the associated set is the set of all functions that vanish outside the permitted limits.

Once the sets and projection operators corresponding to a particular situation have been established, the structure f is calculated iteratively:

$$f^{(k+1)} = P_n P_{n-1} \cdots P_1 f^{(k)}$$

in which P_i ($i = 1, \ldots n$) are the projection operators corresponding to the sets C_i. The convergence properties of such iterative routines are known, and the sequence can be expected to converge to a solution compatible with the input information provided that each set is *closed* and *convex*. A set C is *convex* if, for any two members of C, the quantity $af_1 + (1-a)f_2$ is also a member of C for $0 \leq a \leq 1$. More generally, a *relaxed projection operator* p_i is associated with each P_i:

$$p_i := 1 + \lambda_i(P_i - 1) \quad , \quad 0 < \lambda_i < 2$$

and again, the iteration

$$f^{(k+1)} = p_n p_{n-1} \cdots p_1 f^{(k)}$$

converges weakly to a solution compatible with the data.

The method has been tested by Carazo and Carrascosa (1987), who followed Sezan and Stark (1982, 1983, 1984) in using five sets and hence five projection operators P_i ($i = 1, \ldots, 5$):

- Set C_1 is the set of all functions in the Hilbert space H of square-integrable functions $f(x, y, z)$ that vanish outside a given region R:

$$P_1 f = \begin{cases} f & (x, y, z) \in R \\ 0 & \text{otherwise} \end{cases}$$

- Set C_2 is the set of all functions in H for which the Fourier transform takes a given value Φ in a given region r of reciprocal space:

$$P_2 f = \begin{cases} \Phi & \text{if the argument of } \Phi \in r \\ F & \text{otherwise} \end{cases}$$

where F is the Fourier transform of f.

- Set C_3 is the set of all real non-negative functions in H for which $\iiint |f|^2 \, dx \, dy \, dz \leq E$, where E is a given constant:

$$P_3 f = \begin{cases} 0 & \text{if } f_r \leq 0 \\ f_r^+ & \text{if } E_r^+ \leq E \\ f_r^+ (E/E_r^+)^{\frac{1}{2}} & \text{if } E_r^+ > E \end{cases}$$

Here f_r is the real part of f and f_r^+ is equal to f_r whenever $f_r > 0$ but is equal to zero where f_r is zero or negative. The partial 'energy' E_r^+ is equal to $\int |f_r^+|^2 \, dx dy dz$.

- Set C_4 is the set of all functions in H, the amplitudes of which lie within a given interval $[a, b]$, $a, b \geq 0$, $a < b$ in a closed region R. Then

$$P_4 f = \begin{cases} a & \text{if } f < a \\ f & \text{if } a \leq f \leq b \\ b & \text{if } f > b \end{cases}$$

for $(x, y, z) \in R$. Otherwise, $P_4 f = f$.

- Set C_5 is the set of all real functions in H, the Fourier transforms of which over the region outside the 'missing cone' are within a given distance σ from the (noisy measured) data.

$$P_5 f = \begin{cases} F_r & \text{if } \| \chi F_r - G \| < \sigma \\ G + \sigma \frac{\chi F_r - G}{\|\chi F_r - G\|} + (1 - \chi) F_r & \text{if } \| \chi F_r - G \| \geq \sigma \end{cases}$$

in which $f_r = \Re(f)$, F_r is the Fourier transform of f_r and G is the Fourier data set. χ is unity for points inside the measured region and zero in the missing cone. The norm $\| \cdot \|$ is as usual given by

$$\| f \| = \iiint |f|^2 \, dx \, dy \, dz$$

The results obtained are very encouraging; we refer to the papers cited for additional information.

For details of the method and formal discussion of convergence and the choice of sets to be included, see the early papers of Youla (1978) and Youla and Webb (1982) and of Sezan and Stark (1982, 1983, 1984). Very clear surveys have been written by Youla (1987) and Sezan and Stark (1987). A form of the iteration suited to parallel computers is described by Crombez (1990, 1991) and by Carazo *et al.* (1992). See too Sezan (1990, 1992) and Sanjurjo *et al.* (1992).

75.2.5 Final remarks; reconstruction quality

The foregoing sections explain the principal methods that have found a place in electron microscope tomography. The fact that the measured data will inevitably be noisy and hence slightly incompatible is, however, all too likely to hinder the reconstruction process and, to accommodate such imperfect measurements, the maximum-entropy and POCS approaches have been examined. For details, see Colsher (1977), Wernecke (1977), Wernecke and d'Addario (1977) and Minerbo (1979), and for a practical test, see Baba *et al.* (1981); subsequent work is recorded in the papers of Herman (1982, 1985), Censor and Herman (1987) and Mohammed-Djafari and Demoment (1989).

This question of reconstruction from data that are inevitably somewhat noisy and hence formally incompatible raises the problem of quality measures: how can we decide, by studying the results of applying various algorithms to test data, which is the most reliable? Several measures have been employed and a number of these are critically compared by Harauz and van Heel (1986a), namely the cross-correlation coefficient, various distance measures, the similarity ratio, Fourier shell correlation functions and radial power spectra. The cross-correlation coefficient

$$C_{12} := \frac{\sum \{f_1(\boldsymbol{r}) - \overline{f}_1\}\{f_2(\boldsymbol{r}) - \overline{f}_2\}}{[\sum \{f_1(\boldsymbol{r}) - \overline{f}_1\}^2 \sum \{f_2(\boldsymbol{r}) - \overline{f}_2\}^2]^{\frac{1}{2}}} \qquad (75.39)$$

gave much the same information as the distance measures(defined in Herman, 1980b, Section 5.1) and the similarity ratio. The Fourier shell correlation function gives an indication of the similarity between original and reconstruction as a function of spatial frequency; it is defined by

$$C_{FS}^{(q)} := \frac{\sum_r F_1(q)F_2^*(q)}{(\sum |F_1|^2 \sum |F_2|^2)^{\frac{1}{2}}} \qquad (75.40)$$

in which F_1 and F_2 are the (three-dimensional) Fourier transforms of the original and reconstruction evaluated over the spherical shell of radius q in Fourier space. A final measure used by Harauz and van Heel is the averaged radial power spectrum,

$$C_{RPS}(q) = \frac{\sum |F(q)|}{N(q)} \qquad (75.41)$$

in which $N(q)$ is the number of volume-elements (voxels) in the shell of radius q. Their careful analysis shows that C_{FS} and C_{RPS} are both useful measures and that C_{FS} is particularly valuable.

For other work on the statistical properties and reliability of reconstruction, see the general studies in Chapter 8 of Rosenfeld and Kak (1982) and the analysis of back-projection by Liu (1991a).

75.3 Pre-processing

75.3.1 Background
Before the methods of the last section can be applied, the projection data must be collected, a task that varies enormously in difficulty and complexity, depending on the nature of the specimen. In many cases, considerable effort is needed at the specimen-preparation stage to make reconstruction possible at all but, even then, the measurements may not be ready for immediate use. The projection images of fragile specimens, for example, have to be recorded with very low electron doses, sometimes as small as a few electrons per pixel; such images, which are said to be *under-sampled*, are extremely uncontrasty and heavily corrupted by noise. There are specimens for which the projection directions do not form a regular pattern or are even unknown. For isolated particles, we may suspect that more than one type of particle is present or that the same particle is present in more than one orientation or configuration; the individual particle projection images must therefore be classified into groups, each corresponding to different views through identical particles. To make matters worse, these are often low-dose images and hence, as we have mentioned, extremely noisy: to the unaided eye, the differences between the individual images may well be much more striking than the similarities!

Very specific methods have been developed to perform these various tasks, which are designed to put the raw measurements into a form suitable for reconstruction. These may include allowance for the effects of the microscope transfer function, which we have already discussed in Chapter 73. In Sections 75.3.2–4 we shall concentrate on the problems of averaging and organizing the measurements. The final section, 75.3.5 is concerned with a more mundane problem that commonly arises with periodic specimens. The regular nature of these may easily be perturbed during preparation, and this obviously affects techniques based on the assumption that the specimen structure has a perfect pattern-repeat. Correlation averaging can be reasonably tolerant towards such distortions but methods of 'unbending' or 'unwarping' have recently come into use, designed to undistort the image before further processing. We describe these briefly and more developments can be anticipated.

75.3.2 Alignment

We first consider the problem of aligning several images of the same structure or closely similar structures. For this, correlation techniques are extremely powerful. In the worst case, the individual images are rotated through arbitrary angles with respect to one another and their relative origins are unknown. Ignorance of the origin can be made harmless by first calculating the auto-correlation function (ACF) of each individual image since this is always centro-symmetric. These ACFs are then resampled on a polar grid and the resulting arrays of values are cross-correlated. The cross-correlation functions (CCFs) will have peaks at points corresponding to the angular distance between the ACFs and the original particle images can hence be rotated in such a way that they all have the same orientation. Calculation of the cross-correlation between these images will then yield their relative displacement and they can hence be shifted so that they are all in register. They are now ready to be superposed in order to accentuate common features at the expense of minor variations.

The ACF and CCF are defined in Section 76.3.2 where the use of the latter to establish the displacement between two images is likewise explained. See Frank (1972b), Frank et al. (1978) and Steinkilberg and Schramm (1980); many further references are to be found in the reviews already cited. Correlation is also necessary to establish the exact relation between different members of a series of projections through a structure, tilted at successively increasing angles to the electron beam. This too is achieved by seeking the maximum of the CCF between successive projections, for these will almost always have features in common. In practice, the projections must be stretched by a factor $\sec \theta$, where θ is the tilt angle, to normalize the projections, assuming that these are being obtained by rotation around an axis (like a chicken on a spit, see Fig. 16 in Hoppe, 1983). For detailed discussion of this point, see Guckenberger (1982) and the surveys of Frank (1980a, 1982) and Frank and Radermacher (1986); these build upon the earlier work of Hunsmann et al. (1972), Hoppe (1969b, 1974a,b), Hoppe et al. (1974, 1978) and Hoppe and Grill (1977). See too Kunath et al. (1986). An ingenious way of reducing the effect of dynamic range by cross-correlating a function of the images instead of the images themselves has been introduced by van Heel et al. (1991, 1992).

A very different approach to image registration, due to Venot et al. (1984), has been introduced into electron optics by Bonnet and Liehn (1988). This involves not merely alignment but adjustment of brightness ranges, prior to subtraction for example. It is intended for use with both spatial sequences, as encountered in three-dimensional reconstruction, energy-loss series, and even time sequences such as those proposed for 'trace analysis' (Hoppe, 1978). We set out from two similar images, $I^{(1)}$ and $I^{(2)}$, and form a transformed version $I_T^{(2)}$ of $I^{(2)}$ by translation and rotation (T_g) and linear mapping of the grey levels:

$$I_T^{(2)} =: T(I^{(2)}) =: T_g(aI^{(2)} + b)$$

The difference between $I^{(1)}$ and $I_T^{(2)}$ is then formed and values of the parameters a and b and those in T_g are sought such that the number of sign-changes in $I^{(1)} - I_T^{(2)}$ is as great as possible: if the images were perfectly matched and only differed in noise content, these sign-changes would be due to noise and hence very numerous (assuming zero mean). The method is tested by Bonnet and Liehn and its expected robustness is confirmed: regions that match badly may be expected to contribute few sign-changes and hence do not affect the method adversely.

75.3.3 Classification by correspondence analysis

Correspondence analysis, which belongs to the field of multivariate statistical analysis, was introduced into electron image processing by van Heel and Frank (van Heel and Frank, 1980a,b, 1981; Frank and van Heel, 1980, 1982a,b) to solve the following problem: given a gallery of images of individual particles, how can we decide which are in principle the same, and may hence be superposed to improve the ratio of signal to noise? We assume that the individual images have been aligned in orientation and position by the methods of Section 75.3.2. The measured image intensities of the set of I images are now organized into a large matrix X as shown in Fig. 75.4. The j-th row of this matrix contains all the intensity values of the j-th image, following the rule

$$X_{jk} := i_{pq}^{(j)} \quad , \quad k = (p-1)n + q \tag{75.42}$$

in which $i^{(j)}$ is the j-th image and all these individual images are $n \times n$. We have already seen that useful statistical properties of a large array can be deduced by considering the eigenvalues and eigenvectors of the square matrix obtained by multiplying the array by its transpose (Section 71.4). Our object here is to reveal affinities and small differences between the images and pixel values on the rows and columns respectively of X. It is known that this can be achieved by examining a matrix closely related to $X^T X$ (van Heel and Frank, 1981; van Heel, 1984a,b, 1989; Frank and Radermacher, 1986; Frank, 1989a, 1990). We introduce the diagonal matrices N and M:

$$N_{ij} := \begin{cases} \sum_k X_{ik} & \text{if } i = j \\ 0 & \text{if } i \neq j \end{cases} \tag{75.43}$$

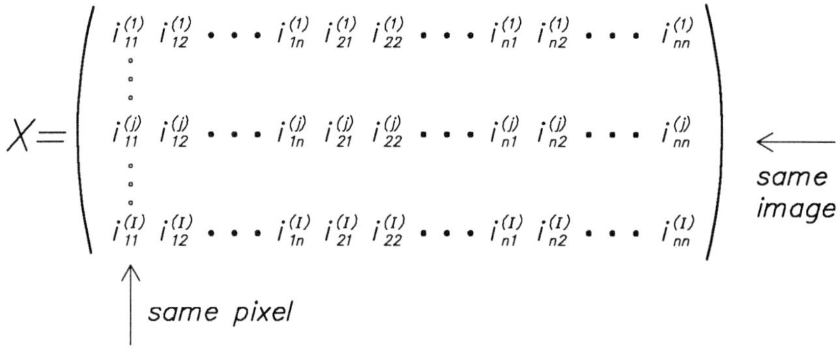

$$X = \begin{pmatrix} i^{(1)}_{11} & i^{(1)}_{12} & \cdots & i^{(1)}_{1n} & i^{(1)}_{21} & i^{(1)}_{22} & \cdots & i^{(1)}_{n1} & i^{(1)}_{n2} & \cdots & i^{(1)}_{nn} \\[2pt] & & & & & & & & & & \\ i^{(j)}_{11} & i^{(j)}_{12} & \cdots & i^{(j)}_{1n} & i^{(j)}_{21} & i^{(j)}_{22} & \cdots & i^{(j)}_{n1} & i^{(j)}_{n2} & \cdots & i^{(j)}_{nn} \\[2pt] & & & & & & & & & & \\ i^{(I)}_{11} & i^{(I)}_{12} & \cdots & i^{(I)}_{1n} & i^{(I)}_{21} & i^{(I)}_{22} & \cdots & i^{(I)}_{n1} & i^{(I)}_{n2} & \cdots & i^{(I)}_{nn} \end{pmatrix} \begin{array}{l} \\ \leftarrow \\ \textit{same} \\ \textit{image} \end{array}$$

$\uparrow$ *same pixel*

Fig. 75.4: The k-th column of X contains the intensity at the same pixel for each of the I images, where as before $k = (p-1)n + q$ for the pixels $i^{(j)}_{pq}$. The matrix X thus has I rows and n^2 columns.

and

$$M_{ij} := \begin{cases} \sum_k X_{kj} & \text{if } i = j \\ 0 & \text{if } i \neq j \end{cases} \tag{75.44}$$

N is $n^2 \times n^2$ and M is $I \times I$.

We now form the square matrix P,

$$P := X^T N X M$$

The presence of the arrays N and M allows us to accord different weights to different images and pixels. It is the eigenvalues and eigenvectors of P that will enable us to classify the images into groups containing images with a high degree of resemblance. We therefore solve the equation

$$P\boldsymbol{u} = \lambda \boldsymbol{u}$$

and project the original data matrix onto the planes defined by the directions of the eigenvectors corresponding to the largest few eigenvalues. The individual images are then classified systematically, typically by 'hierarchical ascendent classification'. This can be understood from Fig. 75.5. The images at the bottom of the figure are merged, that is, said to belong to the same category, if some measure of their difference is smaller than a given value. If the images are thought to belong randomly to many classes, then it is reasonable to anticipate that each class will contain about the same number of members and hence to choose a grouping that yields this result (classification based on class size). If, on the other hand, it is likely that only a few distinct families of images are present, a small number of views

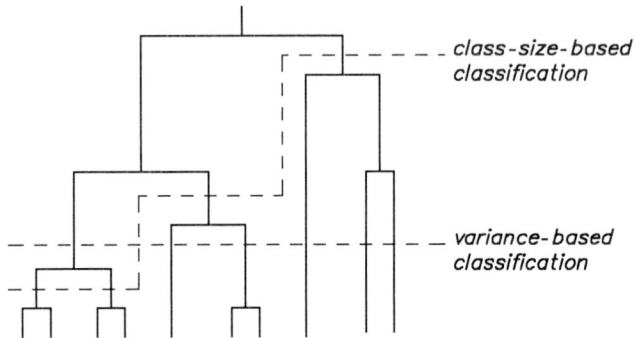

Fig. 75.5: The principle of hierarchical classification.

of the same object, for example, then a classification in which the inter-group variance is as large as possible will probably be preferable. For a very detailed discussion of this procedure, see the work of van Heel (1984a, 1989), where examples are to be found. In the second of these papers several procedures for forming clusters of closely related views are described and compared. A form of hierarchical ascendant classification that allows migration from one class to a neighbouring one appears to be particularly well-suited to the electron microscope situation.

For further comment, see van Heel *et al.* (1982), van Heel (1986, 1987), Unser *et al.* (1986, 1989a,b), Harauz and van Heel (1986b), Harauz *et al.* (1988), Trus*et al.* (1988), Harauz and Fong-Lochovsky (1989), van Heel (1989), Tichelaar and van Heel (1990), Schatz *et al.* (1990) and Schatz and van Heel (1990, 1992).

Angular calibration using factor space is analysed by Pujari and Frank (1990) and Frank *et al.* (1989).

75.3.4 Random tilt series

The electron dose needed to obtain reliable measurements of the projected density for single particles may be so high that the particle is severely damaged long before the full tilt series has been recorded. Efforts have therefore been made to devise methods of reconstructing the structure of particles sprinkled in random orientations on the specimen grid. A method that appears to be successful for particles for which a good starting estimate of the projection direction is available has been devised by Kam (1980) and pursued by Vogel *et al.* (1986), Vogel and Provencher (1988) and Provencher and Vogel (1988, cf. 1983); internal symmetry is a great help here. If the particles are elongated and hence tend to settle in positions that correspond to rotation about their long axis, the projections will form a single-axis tilt series and only the angles will need to be deduced; reconstructions of such specimens have been attempted by van Heel (1983) and by Verschoor *et al.* (1983, 1984). Finally, particles may settle on essentially the same facet on the grid but their relative orientations will inevitably be arbitrary. This case has attracted considerable attention for, by tilting the grid plane and recording a single projection through all the particles, a conical tilt series will be generated with random, or at least unknown, azimuthal tilts in the sense described in Section 75.2. These tilts can, however, be deduced with high accuracy as Radermacher *et al.* (1986a,b,c, 198a,b), and Zhang *et al.* (1987) have shown. Modified versions of the weighted back-projection method have been developed by Radermacher *et al.* (1986a, 1987a) and by Harauz and van Heel (1986a) for the reconstruction step. For applications, see Typke *et al.* (1990), Hegerl *et al.* (1990, 1992), Radermacher *et al.* (1990), Frank (1991, 1992b) and Frank *et al.* (1992).

75.3.5 Removal of distortion
In the preceding sections, we have essentially been discussing classification but this is not the only kind of pre-processing that may be needed. It may easily happen that the physical processes of specimen preparation distort or warp the specimen, so that a pattern of squares, for example, becomes a network with curved meshes (Fig. 75.6). Procedures for removing such perturbations or 'unwarping' have been developed, and these form the subject of this section.

The first attempts to overcome this problem were based on cross-correlation. Since the aim was to superpose units as nearly identical as possible to form an improved averaged image, the process came to be known as *correlation averaging*. A repeat unit is identified in the image by eye and the cross-correlation function (CCF) of this (small) motif and the whole image is calculated. Peaks will be seen in the CCF whenever the motif matches the image. The sharpness of these peaks is governed by the exactness of the match and a certain amount of mismatch can be tolerated. If the peaks form a distorted pattern instead of the expected regular one, the stress and strain in the image can in principle be deduced. Only displacement will be detected, not rotation, but as the latter will usually be small it will merely broaden the CCF peak. At the cost of considerable computing effort, both displacement and rotation can be estimated.

This approach has been thoroughly studied and the relation between electron dose, motif-size, resolution and contrast for both bright-field and dark-field images is well understood (Saxton and Frank, 1977). For extensive discussion and many examples, see Frank (1975a,c, 1976b, 1980a,b, 1982), Frank *et al.* (1978, 1981a, 1988b), Saxton (1980b), Frank and Goldfarb (1980), Saxton and Baumeister (1981, 1982), Crepeau and Fram

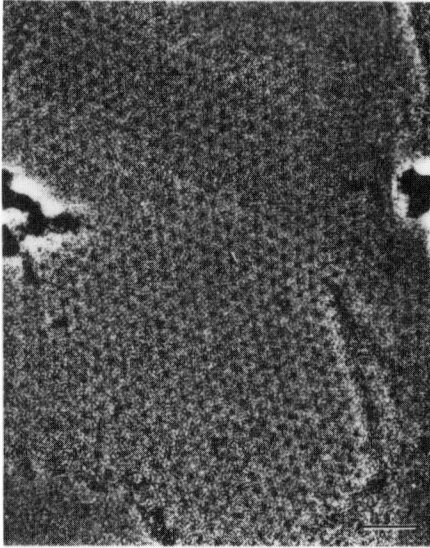

(a)

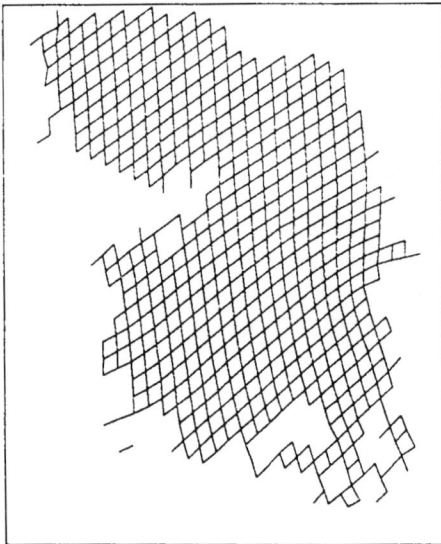

(b)

Fig. 75.6: Crystal warping and its correction.

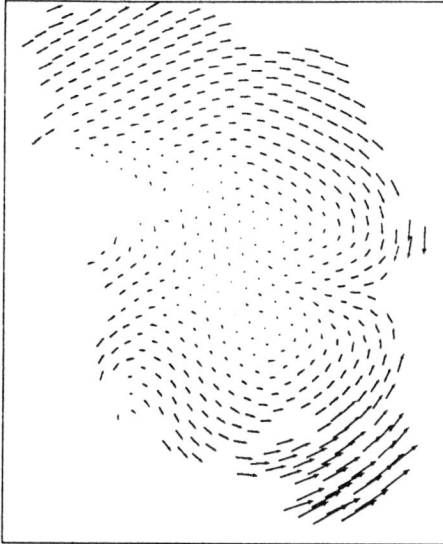

(c)

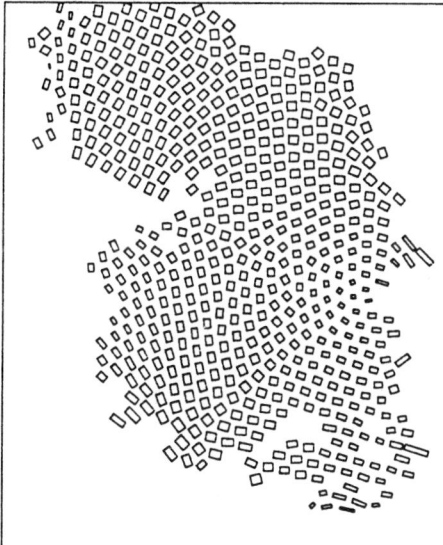

(d)

Fig. 75.6: Crystal warping and its correction.

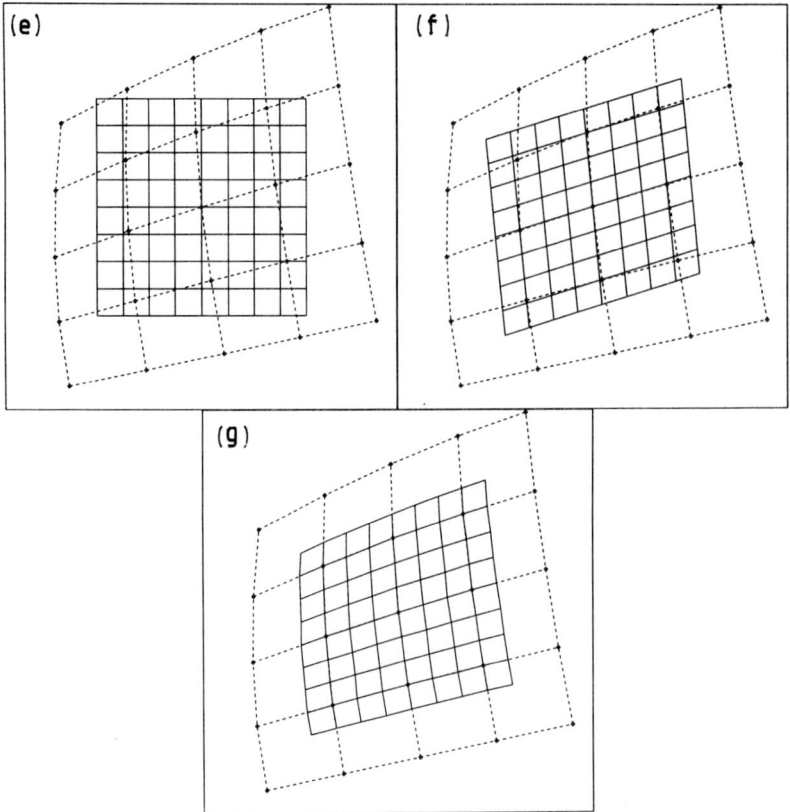

Fig. **75.6**: Crystal warping and its correction. (a) Micrograph of a badly distorted crystal from *Pyrobaculum islandicum*. (b) The corresponding distorted lattice. (c) Displacement vectors for the warped crystal. (d) Distorted squares representing elongation and magnification (×5 for visual convenience). (e–g) Resampling grids used in extracting regions for averaging. (e) Displacement only, as in standard correlation averaging. (f) Constant distortion. (g) Locally varying distortion.

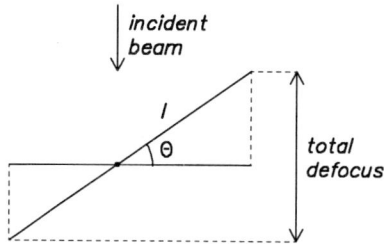

Fig. 75.7: Variation of defocus across a tilted specimen.

(1981), Baumeister (1982), Amos *et al.* (1982), Saxton *et al.* (1984), Frank and Wagenknecht (1984), Glaeser (1985), Trus *et al.* (1986), Hegerl and Baumeister (1988), Lembcke *et al.* (1991).

Although correlation averaging is still widely practised, it is gradually being improved by prior removal of the distortions of the image of the lattice. Early attempts go back to the work of Crowther and Sleytr (1977) and major studies have been made by Henderson *et al.* (1986, 1990) and Hegerl and Baumeister (1988). A procedure based on minimization of the deformation energy has been devised by Dürr *et al.* (1989) and a mapping that is adapted to subsequent filtering is tested by Lembcke *et al.* (1991). The strategy to be followed is discussed by Saxton and Dürr (1990); recent developments are described in Saxton (1991), Baumeister *et al.* (1991), Dürr *et al.* (1991) and Saxton *et al.* (1992). A procedure that requires the recognition of fiduciary points is described by Unser *et al.* (1988), based on earlier work by Steven *et al.* (1986), Eden *et al.* (1986) and Unser *et al.* (1987, 1989c). Yet another approached, based on moiré pattern theory, is employed by Bierwolf *et al.* (1992, 1993). Differences in magnification are considered by Aldroubi *et al.* (1992) and distortions in diffraction patterns by Crowther and Wischik (1985) and by Bellon and Lanzavecchia (1992).

75.3.6 Defocus gradient

When a specimen is imaged in a tilted position, the defocus will vary across the object by $2l \sin \theta$, where l is the appropriate specimen dimension and θ the tilt angle (Fig 75.7).

The phase contrast transfer function is a rapidly varying function of defocus, as we saw in Chapter 66, with the result that the imaging conditions are not constant across the specimen (Schiske, 1978). Figure 75.8 shows a practical example of the magnitude of the effect, for a microscope with $C_s = 1.35$ mm, illumination aperture 0.2 mrad, operating at 100 kV ($\lambda = 3.7$ pm).

This variation raises problems of several kinds: the form of the transfer

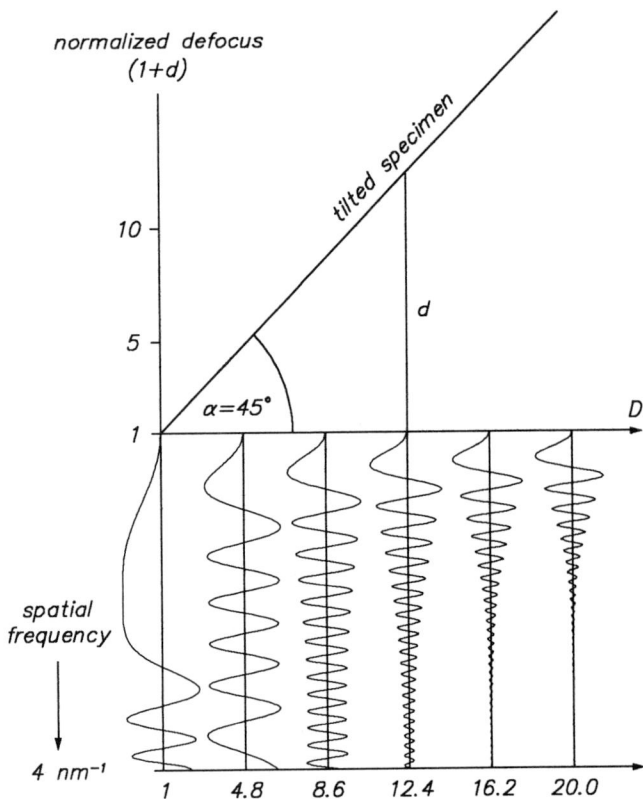

Fig. 75.8: Change of the phase contrast transfer function across a specimen tilted at 45°.

function not only modulates the specimen function in a complicated way but also cuts off much valuable information; simple averaging of elements of periodic specimens cannot now be used to improve the signal-to-noise ratio; the phase shift between corresponding areas of periodic objects causes diffraction spots to be split into two (Hayward and Stroud, 1981; Schiske, 1982; Henderson et al., 1986; Henderson and Baldwin, 1986). These various effects have been examined by Zemlin (1988, 1989), who shows that a technique in which the specimen is explored by a scanning probe and the defocus is varied as the probe climbs the specimen slope can give significant improvement. Zemlin shows that, in practice, a probe diameter of about 100 nm offers a good compromise between the gain due to focal variation and the loss of spatial coherence (increase of illumination angle with reduction in probe size).

This technique is appropriate when the specimen is liable to be damaged by the beam, and *small-spot scanning* has indeed been used to protect the specimen (e.g. Henderson and Glaeser, 1985; Bullough and Henderson, 1987; Downing, 1988, 1991a,b, 1992; Brink et al., 1992). If, however, radiation damage is not a major problem, the beam can be scanned beyond the specimen and the selected-area aperture used to image only a small area of the specimen. The specimen is then illuminated with the full beam but the defocus is varied as the imaging beam is swept over the selecting aperture.

75.4 Concluding remarks

Three-dimensional reconstruction is probably the largest single activity in electron image processing, in transmission microscopy at least, and new ways of collecting and exploiting the projection data can be expected to emerge. The present account is confined to the essentials and a number of interesting topics have had to be excluded — we draw attention in particular to projects for microscopes designed specifically for three-dimensional microscopy (Hoppe, 1972; Typke and Hoppe, 1972; Typke et al., 1976, 1980, 1991, 1992; Plies and Typke, 1978; Typke, 1981; Plies, 1981; Dierksen et al., 1992, 1993).

We note that correspondence analysis is coming to be used for analysing large bodies of image measurements for purposes other than three-dimensional reconstruction; for discussion and examples, see Bonnet and Hannequin (1988a,b, 1989), Hannequin and Bonnet (1988), Bonnet et al. (1990, 1991, 1992a) and Quintana and Ollacarizqueta (1989).

We shall not devote a separate section to image algebraic representation of the various algorithms for reconstruction; see Hawkes (1991a,b).

Further reading

The references cited in this chapter are representative but very far from complete. In the following list, we indicate papers that should provide useful starting points for a more thorough exploration of the subject. Crowther (1971), Crowther and Amos (1971), Crowther et al. (1972), Zwick and Zeitler (1973), Amos (1974, 1975, 1976), Crowther and Klug (1974, 1975), Gordon and Herman (1974), Mellema and Schepman (1976), Herman and Lent (1976b,c), Baba and Murata (1977, 1981), Klug (1978/79), Baba et al. (1979, 1984), Herman and Lewitt (1979), Herman et al. (1979, 1980), Herman (1980a), Mellema (1980a,b), Buoncuore et al. (1981), Amos et al. (1982), Frank et al. (1982, 1986, 1988a, 1991), Hayner and Jenkins (1984), Lepault and Pitt (1984), Ximen and Shao (1985), Ximen and Kapp (1985, 1986), Kapp and Ximen (1985, 1986), Verschoor et al. (1985, 1986,

1989), Baker and Winkelmann (1986), Olins (1986), Steven (1986), Ximen and Li (1986), Ximen (1986, 1990), Censor and Herman (1987), Medoff (1987), Wagenknecht *et al.* (1988a,b), Frank (1989a,b), Lawrence *et al.* (1989), Elsner *et al.* (1991), Harauz and Chiu (1991), Jap *et al.* (1991), Stewart (1991), Zemlin (1991), Zhang (1991), Schiske (1993) and the series by Bates and colleagues (Lewitt and Bates, 1978; Lewitt *et al.*, 1978; Bates and Heffernan, 1980; Heffernan and Bates, 1982; Garden and Bates, 1984; Tan *et al.*, 1986). The many papers by Hoppe and colleagues are conveniently classified and cited in Hoppe (1983); see in particular Hoppe (1969b, 1970a,b, 1971, 1978/79, 1980), Hoppe and Hegerl (1981), Hoppe and Köstler (1976), Hoppe *et al.* (1973, 1975, 1976a,b, 1977, 1978), Knauer and Hoppe (1980). Numerous applications are listed in the bibliographies of Hawkes (1978a, 1982c, 1992a). Clack (1992) is a general study of the mathematics of filtered back-projection.

For the use of special-purpose computers see Carazo *et al.* (1990, 1992) and Zapata *et al.* (1992) who employ a hypercube.

76

Image Analysis

76.1 Introduction

The purpose of image analysis is not to enhance the visual quality of the image or to restore it but to extract features of interest, characterize them, measure them, match them against patterns or describe them. With the exception of measurement, these are all qualitative and to some extent subjective activities, with the result that much of the effort that is being put into image analysis is concerned with translating them into mathematical terms.

Only a few aspects of this vast topic, which ranges from forensic science (fingerprint analysis, for example) to robot vision, are of direct interest in electron optics. In transmission microscopy, sophisticated procedures may need to be employed to classify sub-images before applying the methods of enhancement or three-dimensional reconstruction already described. In scanning microscopy, however, image analysis is relatively common and the scanning microscope and an image analysis unit nowadays tend to be designed as a single entity. The SEM image can then be segmented into regions possessing different properties; particles can be counted and characterized by area, ellipticity, texture or chemical composition, for example; features of interest can be described in ways that are insensitive to orientation or to a certain degree of distortion.

As in so much of image processing, the techniques in use in image analysis tend to have been developed piecemeal and the texts in this field remind one of nothing so much as recipe books. Some order has been achieved by the use of the morphological formalism but much remains to be done in this direction.

It is hardly appropriate to reproduce lists of recipes in a book of principles, nor would it be useful without according a disproportionate amount of space to them. Instead, we just draw attention to the methods used in the various branches of image analysis, leaving the reader to seek detailed guidance elsewhere.

Before discussing segmentation and feature extraction (Section 76.3), classification (Section 76.4) and description (Section 76.5), we give a brief account of the notions of discrete geometry (Section 76.2). Although we

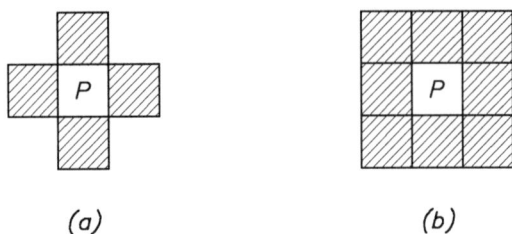

Fig. 76.1: Neighbours. (a) 4-neighbours of the point P. (b) 8-neighbours of P.

shall not use this extensively, a knowledge of the geometry appropriate to discrete data is essential for understanding much of the literature of the subject, where it is nowadays often taken for granted, and we shall need it occasionally in later sections.

76.2 Discrete geometry

Familiar notions such as connectedness or even distance need to be re-defined when we are confronted with discrete data rather than continuous curves or surfaces. What meaning can we ascribe to curvature or gradient, for example, for a sampled image represented by a matrix of grey levels? The solution is to be found in *discrete* geometry and we present the more relevant ideas in this section: neighbourhood, distance, connectedness, simplicity and the derived notion of border. We conclude these introductory remarks with an important warning. The form of discrete geometry presented here is known to be imperfect and an approach that does not suffer from its weaknesses can be based on the theory of finite topological spaces and in particular of abstract cell complexes (Kovalesvsky, 1989 and especially 1992; Herman, 1990); the ideas presented below are those currently in use but we anticipate that they will give way in time to an improved theory. The account by Lee *et al.* (1993) is an indispensable introduction to algebraic topology in two and three dimensions of rare clarity.

Neighbours

The pixels adjoining some pixel P in the vertical and horizontal directions are known as the 4-*neighbours* of P (Fig. 76.1a) and are said to be 4-*adjacent* to P. These, together with the four diagonally neighbouring pixels (Fig. 76.1b), form the 8-*neighbours* of P and are said to be 8-*adjacent* to P.

Suppose that A and B are regions of a digital image I; regions A and

```
4  3  2  3  4              2  2  2  2  2
3  2  1  2  3              2  1  1  1  2
2  1  P  1  2              2  1  P  1  2
3  2  1  2  3              2  1  1  1  2
4  3  2  3  4              2  2  2  2  2
```

        (a)   *(b)*

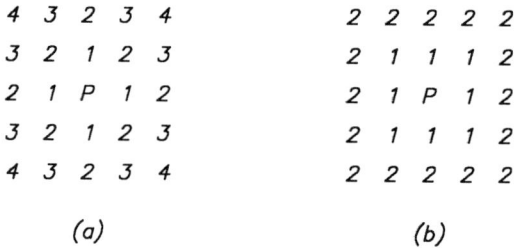

Fig. 76.2: Discrete distance measures: (a) d_4-distances and (b) d_8-distances from P to nearby pixels.

B are said to be 4- (or 8-) adjacent if at least one pixel of A is 4- (or 8-) adjacent to one or more pixels of B.

Distance

A number of discrete distance measures are in use, of which the most common are the *city-block distance* d_4 and the *chessboard distance* d_8. For two points $P(i, j)$ and $Q(k, l)$,

$$d_4(P, Q) := |k - i| + |l - j|$$
$$d_8(P, Q) := \max(|k - i|, |l - j|) \tag{76.1}$$

These both possess the expected properties of a distance, namely, they are always non-negative and vanish only if P and Q coincide; the distance from P to Q is the same as that from Q to P; and the triangle inequality is satisfied in that the direct route is never longer than any other. Figure 76.2a shows d_4-distances from P to neighbouring pixels and Fig. 76.2b shows the corresponding d_8-distances; note that in the first case, the points at unit distance from P are its 4-neighbours while in the second case, they are its 8-neighbours.

For a distance measure in which the city-block and chessboard distances are used alternately, giving an 'octagonal' distance, see Rosenfeld and Pfalz (1968) and Das (1990) and for distances based on the knight's move in chess (unlike the chessboard distance which is based on the castle's or rook's move), see Das and Chatterji (1988) and Das and Mukherjee (1990).

We shall see in Section 76.3.2 that a form of Euclidean distance is advantageous in connection with the calculation of skeletons.

Connectedness

A path π of length n from P_0 to P_n in a digital image I is a sequence of pixels $P_0, P_1 \ldots, P_i, \ldots P_n$ such that P_i and P_{i+1} are neighbours for all

i along the path. Depending on the definition of neighbour, π may be a 4-*path* or an 8-*path*.

Suppose now that A is a subset of the pixels of I and that P_0 and P_n both lie in A. P_0 and P_n are *connected* in A if a path can be found from P_0 to P_n, all the pixels of which lie in A. If we consider an arbitrary pixel P in A, the remaining pixels will be either connected to P in A or not connected; the set of pixels that are connected to P is called a *connected component* of A and if it happens that A has exactly one component, then it is called a *connected set*.

Border*

The border of the subset A of the digital image I is the set of pixels of A that are adjacent to its complement $(\overline{A})$. The pixels that do not lie on the border form the 'interior'. We now consider two components, C_A in A and $C_{\overline{A}}$ in $\overline{A}$; $C_{\overline{A}}$ is chosen to be adjacent to C_A. The set of pixels of C_A that are adjacent to $C_{\overline{A}}$ form the $C_{\overline{A}}$-*border* of C_A while the set of pixels of $C_{\overline{A}}$ that are adjacent to C_A form the C_A-border of $C_{\overline{A}}$. Provided that we use 4-connectedness for A and 8-connectedness for $\overline{A}$, or *vice versa*, these borders will be *closed curves*.

Simplicity

If a pixel P lies on the border of A, then we say that it is 8-*simple* if the set of 8-neighbours of P that lie within A has exactly one component that is 8-adjacent to P; it is 4-*simple* if this same set of 8-neighbours has exactly one component that is 4-adjacent to P. The examples of Fig. 76.3 should make this rather obscure definition, which will be needed when we examine thinning, easier to understand.

76.3 Segmentation and feature extraction

The first task in image analysis is to identify the structures or 'features' of the image to be analysed. In low-dose microscopy, for example, the image might consist of many views of similar objects, distributed over the grid, and it must therefore be divided up into sub-images, each containing a single view of an object, prior to alignment and superposition. Segmentation of the image and feature extraction in such a case are thus intimately connected, since identification of an individual object is tantamount to

---

* This definition of border is an example of the imperfections of the discrete geometry described here; Kovalevsky asserts that the problem vanishes when abstract cell complexes are introduced.

```
0 | 1 | 1
0 | P | 0          P is 4–simple but              (a)
1   0   0          not 8–simple
```

```
0 | 1 | 1
0 | P | 1          P is 4–simple but              (b)
1   0   0          not 8–simple
```

```
0   1   1
0   P   0          P is neither 4–simple          (c)
0   1   0          nor 8–simple
```

```
0 | 1 | 0
0 | P | 1          P is not 4–simple              (d)
0   0   0          but is 8–simple
```

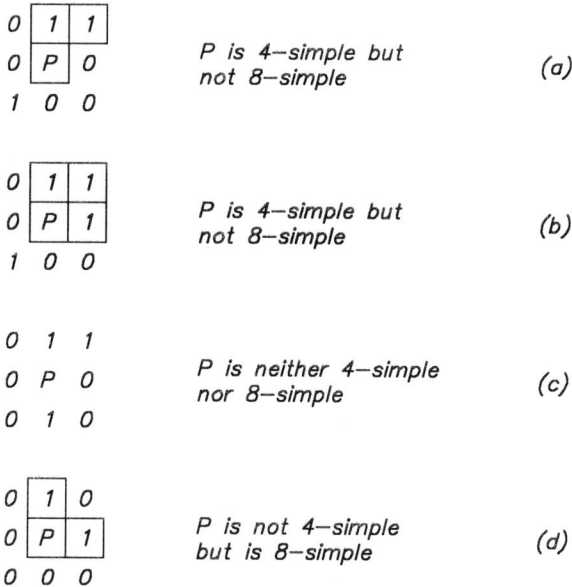

Fig. 76.3: Simplicity. The difference between 4-simple and 8-simple points.

segmentation. There are other situations in which the image is segmented blindly, however, and the contents of the various segments are analysed afterwards: the SEM image of Fig. 76.4, for example, clearly consists of zones and it is natural to segment it along the zone boundaries and then examine the contents of each zone in turn. The roles of segmentation and feature extraction are thus strongly dependent on the problem in question and although we separate them in the following account for clarity, they are really different ways of considering the same activity.

76.3.1 Segmentation
The simplest way of segmenting an image is to threshold it (Section 72.1); for a single threshold, at grey level g_t say, all grey levels above g_t are set equal to a maximum value, g_{max} and the remainder are set equal to a minimum value, g_{min}. More thresholds may be allowed, in which case the original grey-level image will be replaced by a new grey-level image, in which the number of grey levels present is one greater than the number of thresholds.

Typically, two thresholds will produce an image with g_{min}, $g_{intermed}$ and g_{max}. Guidance on suitable values of the threshold will usually be obtained from the histogram (Section 72.1); if this suggests that the grey levels fall into classes, then the thresholds will be chosen accordingly (Fig. 76.5).

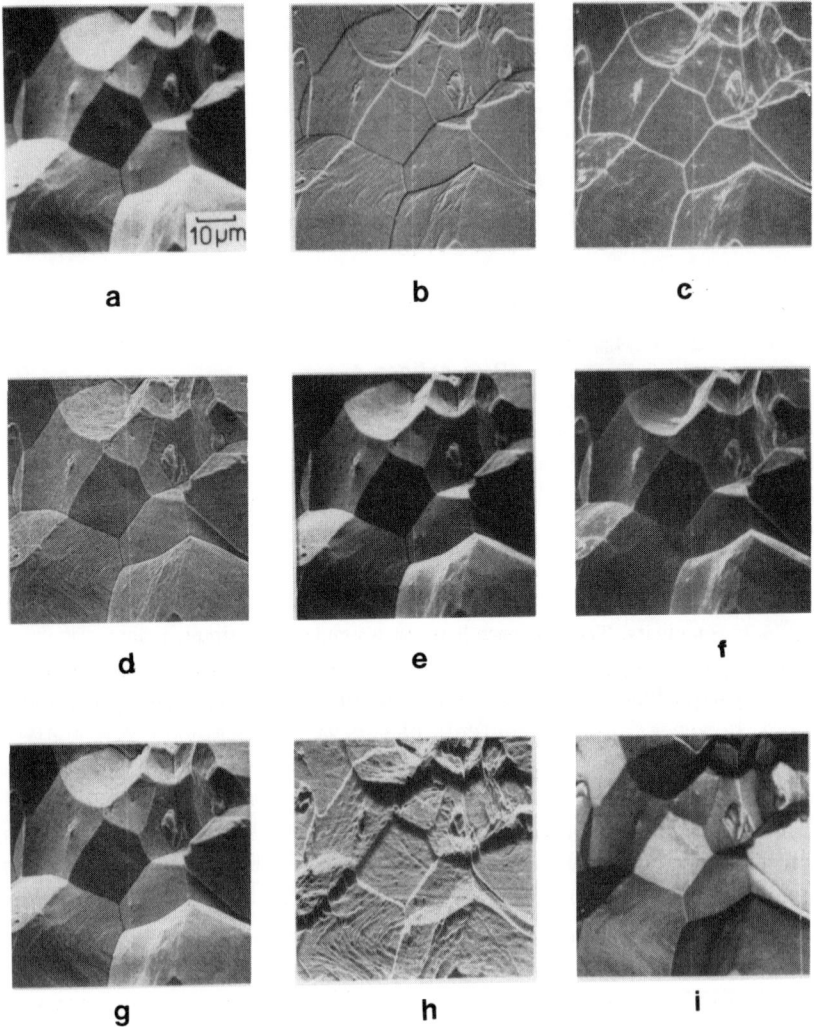

Fig. 76.4: A scanning electron micrograph suitable for segmentation together with various simple transformations. (a) Secondary electron image of fracture surface of pure iron. No image processing applied. (b) First time derivative. Note flattening and apparently oblique illumination. Scanned orthogonally (i.e. horizontal and vertical scans are superimposed). (c) Absolute value of first time derivative. Note strong outlining of edges. Scanned orthogonally. (d) Second time derivative. Note impression of vertical illumination and less apparent flattening. Scanned orthogonally. (e) Same as (b) but mixed in equal proportions with direct signals scanned orthogonally. Note crispening of image. (f) Same as (c) but mixed in equal proportions with direct signal. Scanned orthogonally. (g) Same as (d) but mixed in equal proportions with direct signal. Scanned orthogonally. Note crispening of image as well as apparent vertical illumination. (h) Y-modulation image. Note distortion of fracture facets. (i) Inverted γ-processed image with $\gamma = 4$.

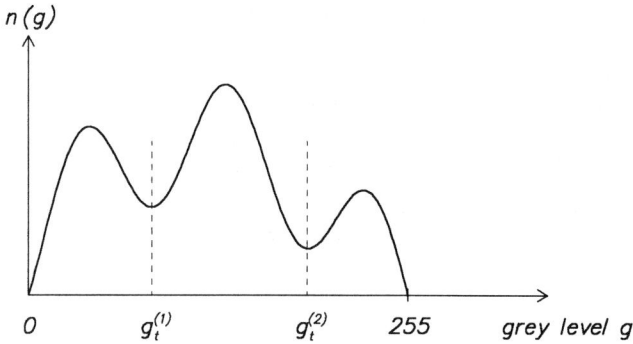

Fig. 76.5: A grey-level histogram with several peaks. The bottoms of the valleys are natural positions for thresholds, $g_t^{(1)}$ and $g_t^{(2)}$.

Thresholding is most suitable for images in which structures stand out against a background (recognition of printed or written characters, for example, or chromosome arrays for karyotyping) and transmission electron micrographs are rarely of this nature; for some scanning micrographs, however, and for diffraction patterns, it may be very suitable.

For images that naturally divide into zones (Fig. 76.4), techniques designed to detect boundaries are very effective. We have already met many of these in the chapter on enhancement (72) and we recall that they are usually based on the response to convolution of the image with a simple mask or set of masks. The boundaries may not, however, be continuous, the detection method having failed to recognize some boundary points; it is frequently necessary to fill in the gaps, and the method used for this will depend strongly on the nature of the image. For the rather straight boundaries of Fig. 76.4, for example, gaps could be filled by simple linear interpolation or even manually, by eye. For curved but still relatively uncomplicated boundaries, curve-fitting will often be sufficient. For more complicated situations, mathematical morphology will probably offer a better solution. For recent developments in edge-detection, see Davies (1992a), Rosenfeld and Banerjee (1992) and the relevant chapter of Haralick and Shapiro (1992).

As an alternative to detecting the boundaries directly, we can find them by examining the *texture* of the image and identifying the lines or strips across which the texture changes. Texture is commonly characterized by statistical properties and in particular, by the co-occurrence matrix, which helps us to appreciate how often the numerous possible pairs of grey levels occur in given relative positions. Consider a displacement d, characterized as usual by $(k - i, l - j)$; we construct a matrix, M_d, the element (p, q)

1	1	2	3	0
1	2	2	3	3
0	2	3	3	3
1	1	2	2	2
0	0	0	1	1

00	01	02	03
10	11	12	13
20	21	22	23
30	31	32	33

2	1	1	0
0	3	3	0
0	0	3	3
1	0	0	3

image   *combinations* M_d , $d = (1,0)$
                             to be counted

Fig. 76.6: Example of a co-occurrence matrix. Examination of the grey levels of the image (left) shows that 16 combinations of grey levels may occur. The co-occurrence matrix M_d for $d = (1,0)$ has the form shown (right). Note that the number of horizontal transitions $(d = (1,0))$ in an image of size $N \times N$ is $N(N-1)$ and that this must be equal to the sum of the elements of M_d. Similar rules can be established for other forms of d.

of which is equal to the number of times that a pixel with grey level g_p is found at a displacement d from a pixel with grey level g_q. If the number of grey levels in the image is G, then $0 \leq p, q \leq G-1$. A very simple example is shown in Fig. 76.6.

It is convenient to normalize the elements of M_d with respect to the number of point pairs in the image separated by d. In the example of Fig. 76.6, where $d = (1,0)$, the number of point pairs N_d is $N(N-1)$ and this will become smaller as d represents progressively larger separations. The matrix P_d, obtained by dividing each element of M_d by N_d,

$$P_d(i,j) := \frac{M_d(i,j)}{N_d} \tag{76.2}$$

is known as the *grey-level co-occurrence matrix* of the image; its (p,q)-th element is an estimate of the joint probability that pixels separated by d will have grey levels (g_p, g_q).

Although there is little point in studying P_d for separations d of more than a few pixels, it is clearly necessary to characterize the P_d that are retained in some compact way if they are to be useful diagnostic tools. Several proposals have been made, and the earlier suggestions are summarized and illustrated in a survey by Haralick (1979). These have such forms as $\sum_p \sum_q (p-q)^2 P_d(p,q)$ (or other powers of $p-q$ and of P_d), which is a

Fig. 76.7: Structuring element for texture analysis.

kind of contrast measure; the inverse difference moment,

$$\sum_p \sum_q P_d(p,q)\{1 + (p-q)^2\}^{-1}$$

in the modified form given by Rosenfeld and Kak (1982, vol. 2, p.297), where again, other powers may be tried; and an entropy,

$$-\sum_p \sum_q P_d(p,q)\ln P_d(p,q)$$

The uses of the co-occurrence matrix have been extended and the matrix itself generalized by Davis *et al.* (1979, 1981), Dyer *et al.* (1980) and Werman *et al.* (1985). A way of detecting directionality is described by Davis (1981). The large literature of texture may be traced through the annual surveys of Rosenfeld (1969 onwards). The ability of various texture measures to discriminate between electron micrographs has been investigated by Zinzindohoué (Zinzindohoué *et al.*, 1988; Zinzindohoué, 1991, 1992). For a full and careful account of work on texture analysis and modelling with particular reference to electron microscopy, see Burge and Ali (1987).

Finally, we mention that the methods of mathematical morphology lend themselves well to the analysis of texture. The basic idea is to erode the image with a series of structuring elements, typically consisting of two pixels or groups of pixels; the members of the series correspond to increasing distances between the pixels (Fig. 76.7). The result of the erosion is characterized by some suitable parameter, the behaviour of which is used to describe the texture of the image or sub-image analysed.

For extensive discussion, see Serra (1982, 1988), Coster and Chermant (1985) and, for a simple introduction, Haralick (1979).

76.3.2 Feature extraction

The features to be extracted from an image may be described in many different ways; we may know their exact shape, for example, or we may know little more than that they possess pronounced directionality, or again we may know more about their spatial spectrum than about their appearance in direct space. Each of the numerous approaches to feature extraction is often best suited to a particular way of characterizing features.

Alternatively, we may not be seeking a feature possessing specific properties but rather, any repeated structure or family of objects that bear a strong resemblance to one another. It will probably be useful to break the image down to very simple constituent parts in this case, which will reveal the presence of the desired features.

Let us first suppose that we wish to find all the structures in the image closely resembling a given structure. We return to problems of orientation below and assume for the moment that the structures to be found all have the same orientation as that of the pattern, or template. For an image i and template t, we wish to find the positions of t relative to i for which some measure of the mismatch is as small as possible. Although we could measure the mismatch in terms of $\max|i - t|$ or $\sum \sum |i - t|$ over the region within which $t \neq 0$ (this will be much smaller than i), it is more convenient to use $\sum \sum (i - t)^2$. For given i and t, we can then adopt $\sum \sum it$ as a measure of the mismatch. Thus by moving thetemplate over the image and calculating $\sum \sum t(p, q)i(p + p', q + q')$ for all possible shifts (p', q'), we can identify the points at which the template matches the image. This quantity is just the cross-correlation between i and t and in practice must be normalized with respect to $\sum \sum i^2(p + p', q + q')$. Maxima of the resulting normalized cross-correlation function correspond to positions of the template at which t and i match best. (For a very full account of correlation techniques see Frank, 1980a.) In practice, various precautions are necessary: normalization of the grey scale of picture and template; padding of the template array if the correlation is performed in Fourier space to take advantage of the convolution theorem; multiplication of the spatial spectra by a damping function to reduce artifical edge effects.

The basic procedure has many ramifications. In particular, the template may itself be derived from the image, particularly if the latter is approximately periodic, by identifying a unit cell with the aid of the Fourier transform. The template may be used iteratively, for recognition of common structures in low-dose images, for example, where an averaged template, obtained by superposing the features extracted on a first sweep using an imperfect template, is used for a second sweep.

The initial assumption concerning the common orientation of the template and image feature is very unlikely to be true, except in the case of periodic structures. Cross-correlation can be used to align sub-images in both position and orientation and, indeed in magnification if necessary. In this case, the auto-correlation functions of the images to be aligned are first calculated, since these are centrally symmetric; each of these functions is then resampled around concentric circles and the resulting measurements disposed on a rectangular array as the elements of a matrix. The cross-correlation function of these new matrices will have a peak, the position of which indicates the relative orientation of the auto-correlation functions, and hence of the original images, and also any difference between the magnifications. For an analysis in terms of the Mellin transform, see Hawkes (1976) and for an experimental test, Saxton (1974b).

If the features to be detected are characterized by the possession of qualitative properties, methods based on the histogram or texture or on the spatial spectrum of the image may be appropriate. In the first two cases, the image is explored by a window and the histogram or some textural measure is computed for each window position. The histogram will be characterized by one or more derived quantities (organized as a 'feature vector'), such as its

moments

$$m_j = \sum_{i=0}^{G-1} g_i^j p(g_i)$$

(76.3)

(where g_i, $i = 0, \ldots G - 1$ are the grey levels and $p(g_i)$ their probability distribution described by the histogram),

absolute moments

$$\hat{m}_j = \sum_i |g_i|^j p(g_i) \qquad \text{if } g_i \text{ is allowed to have negative values}$$

central moments

$$\mu_j = \sum_i (g_i - m_1)^j p(g_i)$$

absolute central moments

$$\hat{\mu}_j = \sum_i |g_i - m_1|^j p(g_i)$$

entropy

$$H = -\sum_i p(g_i) \log_2 p(g_i)$$

(76.4)

and especially the dispersion ($\hat{\mu}_1$), the mean (m_1), the variance (μ_2), skewness (μ_3) and kurtosis ($\mu_4 - 3$), the median and the most probable value (known as the 'mode'). For suitable textural measures, see the discussion of texture in Section 76.3.1 or the Rosenfeld surveys.

The use of image transforms for feature extraction is limited, though it is possible that more extensive exploration of transforms other than the Fourier transform would be fruitful. The Fourier transform is of course a powerful tool for recognizing the unit cell of a periodic structure, which can then be used for straightening or unwarping distorted lattices (see

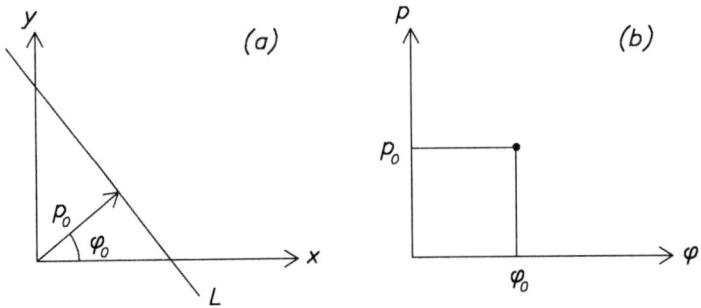

Fig. 76.8: The mapping of a line into a point is the basis for the Hough transform.

Section 75.5). Highly directional aspects of the image can be identified by seeking privileged directions in the transform. This has been exploited in great detail in scanning microscopy by Tovey and colleagues (Tovey, 1971a,b, 1980; Tovey and Wong, 1978; Tovey and Smart, 1986; Smart and Tovey, 1982, 1988; Tovey et al., 1989, 1991; Unitt, 1975; Unitt and Smith, 1976; Smith et al., 1977a). See too Bamberger and Smith (1992).

Lines and, less easily, curves of given shape can be extracted with the aid of a transform closely related to the Radon transform introduced in Chapter 75, which is known as the *Hough transform*. The idea behind this is to find a coordinate transform that maps all points lying on a particular kind of curve into a single point. The simplest example is the straight line. We recall that a line L (Fig. 76.8a) in a plane may be characterized by $x \cos \varphi + y \sin \varphi = p$. If we transfer to the p–φ plane (Fig. 76.8b), all points on L map to a point, which is the point of intersection of the Radon transforms of the points on the line. How could this be used to find lines in an image? Any line may be written as $y = mx + c$ and we scan the image and draw the line $c = -xm + y$ weighted by the grey level at the point in question (x, y) for each point. In practice, the raw image will be thresholded and only the surviving points retained. Points that lay on a straight line in the original (or thresholded) image will generate lines intersecting at a single point. Thus peaks in the c–m plane indicate the presence of lines in the original.

The basic idea has been extended to enable more complicated curves to be extracted (Leavers, 1992; Princen et al., 1992; Jeng and Tsai, 1991; Davis, 1982; Ballard, 1981; Sklansky, 1978; Shapiro, 1978) and has recently been employed to detect structure in diffraction patterns (Russ et al., 1989a,b; Russ, 1989a,b, 1990) and back-scattering patterns (Dorkel et al., 1993).

If little is known about the features to be detected, a rather different approach is indicated. The structure of the image may be simplified by tracing boundaries and edges, using the various edge-detectors already described. Alternatively, the image may be thinned down to its *medial axes* or to a *skeleton*, suitably defined. We present these two thinning operations first in the language of digital geometry and then in that of mathematical morphology.

The medial axis of a region may be pictured as the quench point of a fire lit around its periphery: we suppose that such a fire propagates uniformly across the region and that it will go out whenever the fire reaches the same point from different directions, for want of combustibles; such points will form lines that reflect the shape of the region but only imperfectly for they are likely to consist of isolated segments, not joined together as a continuous figure. More formally, consider a region A of an image I and a set of upright squares of side n pixels (odd). We centre such a square at each point of A and note the size and centre of the largest square that is contained in A; squares wholly contained within others are discarded and the surviving centres then form the *medial axis*. The set of centres and associated sizes are said to form the medial axis transform (or MAT) of A. It is not difficult to reconcile this definition with the idea of the quench points of a fire by noting that the set of centres coincide with the points of A, the d_8-distances of which from $\overline{A}$, the complement of A, or equivalently, from its border are local maxima. Likewise, if we had defined the medial axis in terms of diamonds or lozenges instead of squares, we should have found that the medial axis was the set of points, the d_4-distances of which from $\overline{A}$ were local maxima.

Although a knowledge of the medial axis transform of a region is sufficient for us to reproduce that region, the medial axis is not very useful for our present purposes, owing to its disconnected nature. This inconvenience disappears when we consider true *skeletons*. We first consider binary pictures and then extend the discussion to grey-level images.

A typical set of rules for thinning a region down to a skeleton is as follows: we consider each of the border points in turn, first all those encountered when we approach from above (north), then from the right (east), south and west and begin the cycle again from the north. For each direction of approach, border points are removed provided that they are simple and that they have more than one neighbour. The first condition preserves connectedness and the second prevents the ends of branches from being cut off.

This is not the only way of thinning a binary picture down to a skeleton; see Arcelli *et al.* (1981), Davies and Plummer (1981), Toriwaki and Yokoi (1981), Hilditch (1983), Arcelli and Sanniti di Baja (1985) and Lam *et al.* (1992) for critical discussion and Klein and Kübler (1987) for a more recent examination of the problem and a procedure based on a Euclidean distance (rather than d_4 or d_8). Haralick (1992) proposes a way of characterizing the performance of such algorithms.

Both the medial axis and the skeleton are well-adapted to morphology: the medial axis is generated by $\bigcup(A \ominus nS)/(A \ominus nS)_S$, in which $A \ominus nS$ represents n iterations of the form $(A \ominus S) \ominus S \ldots \ominus S$ and the union is taken over values of n from 1 to that at which the eroded set is empty; $(\cdot)_S$ denotes the opening of $(\cdot)$ by S and the oblique stroke $(/)$ indicates set difference. Thinning is represented by $A/\{(A \ominus S_1)/(A \oplus S_2)\}$, in which S_1 and S_2 are pixels in the structuring element S that are required to belong to the interior and exterior of A, respectively. See Maragos and Schafer (1986) and Maragos and Ziff (1988).

A thinning procedure capable of generating a skeleton from a *grey-level* image has been developed (Dyer and Rosenfeld, 1979). In order to explain this, we define the *strength* of a path from P to Q as the smallest grey-level value of the pixels along the path and the *degree of connectedness* of P and Q as the greatest value of the strengths of all paths connecting P and Q. A point P is then said to be *simple* if the degree of connectedness of any two points in its 8-neighbourhood is not decreased by replacing the grey level of P by the smallest grey level among its neighbours. *Thinning* is then defined as a local-minimum operation in which the grey-level value of a point is replaced by the minimum grey-level value of its neighbours, provided that the point is simple and has more than one neighbour with a higher grey level. As before, thinning is applied first to all points encountered on approaching from the north, then to all points encountered from each of the other directions, in sequence.

A large chapter of morphology is concerned with thinning grey-level images and, in particular, with finding local maxima and minima and 'watersheds' in such images, regarded as surfaces the height of which represents the grey level at the underlying pixel. We shall not go into this here. We simply refer to the papers of Beucher and Meyer, some of whose work is concerned with SEM images (Meyer and Beucher, 1990,1991; Beucher, 1990, 1991; Meyer, 1989, 1990), to the many papers cited by Jeulin (1987) and to the work of Bleau *et al.* (1992).

76.3.3 Measurement

Once features have been identified or regions of the image recognized, measurements of area or perimeter or other geometrical or topological characteristics may be needed. This is largely a matter of definition, though of course precautions must be taken to ensure that the regions or features are clearly delimited and not artificially broken or modified by noise. A detailed study of measurements on images with multiple grey levels has been made by Eberly and Lancaster (1991) and Eberly *et al.* (1991), to which we refer for extensive discussion. Much fundamental work on digital topology is due to Rosenfeld and colleagues (e.g. Kong and Rosenfeld, 1989; Lee *et al.*, 1991) and can be traced through the annual literature surveys of Rosenfeld. For applications in electron microscopy, see Russ (1990).

There is an extensive literature on morphological methods of characterizing granulometry, which has found application in SEM. In addition to the work already cited, we list that of Dougherty (Dougherty, 1990; Dougherty and Pelz, 1991; Dougherty *et al.*, 1992).

The development of parallel acquisition in electron energy-loss spectroscopy (EELS) has resulted in the notion of the (three-dimensional) *spectrum–image* (Jeanguillaume and Colliex, 1989) in which the x and y axes (conventionally horizontal) represent image coordinates and the E (energy) axis is vertical. A horizontal section through the image thus represents the energy-loss image or energy-filtered image for the corresponding value of E. A vertical section shows the energy loss spectrum for each member of the row of pixels selected.

Such spectrum-images can be accumulated in various ways, using a TEM or a STEM (Hunt and Williams, 1991; Balossier *et al.*, 1991; Williams and Hunt, 1992; Lavergne *et al.*, 1992). The subsequent task of data-handling, into which extensive checks and corrections must be incorporated, has been studied in great detail by Hunt and Williams and software has been developed for extracting practical information from these three-dimensional arrays (Hunt and Williams, 1991).

76.4 Classification

Once features of interest in the image have been detected, they will generally be combined in some way, to produce an enhanced image (by superposition to improve the signal-to-noise ratio, for example) or to create a three-dimensional image from a set of projections. It may happen that an intermediate step is needed, in which the sub-images detected are classified in some way. In transmission electron microscopy, this is most likely to involve sorting the sub-images into groups that correspond to the various structures present in the image: the classic example is the haemocyanin half-molecule of *Limulus*, studied by van Heel and Frank (1980a,b, 1981), which could rest on the grid in either of two positions, like a chair with one leg shorter than the other three. It was the problem of classifying arrays of low-dose images that led van Heel and Frank to introduce the notions of correspondence analysis into electron image analysis. We have, however, preferred to present this in the context of reconstruction from projections

(Section 75.3.3) and we discuss it no more here. Three-dimensional reconstruction is not, however, the only branch of electron optics in which multivariate statistical analysis is useful. It could conceivably be useful for studying the dynamic behaviour of certain specimens (Frank, 1982), as dreamed of by Hoppe in the form of 'trace analysis' (Hoppe, 1975, 1978, 1982). It is already being used to condense the vast amount of data in sequences of energy-filtered images (Hannequin and Bonnet, 1988; Bonnet and Hannequin 1988a,b, 1989; Bonnet et al., 1990, 1992a; Bonnet and Trebbia, 1991). These are images of the same area obtained in a microscope equipped with an imaging filter (see Section 53.4.6, Colliex, 1984 or Reimer, 1993, Section 9.2.1) and we consider this case in more detail.

We assume that M images have been recorded, corresponding to different energy windows, and organize the measured pixel values into a large matrix, thus:

$$
\boldsymbol{m} := \begin{pmatrix}
x & x & x & \dots & x & x \\
x & x & x & \dots & x & x \\
x & x & x & \dots & x & x \\
\vdots & \vdots & \vdots & \dots & \vdots & \vdots \\
\vdots & \vdots & \vdots & \dots & \vdots & \vdots \\
x & x & x & \dots & x & x \\
x & x & x & \dots & x & x
\end{pmatrix}
\quad \longleftarrow \quad \begin{array}{l} \text{same image,} \\ \text{same energy window} \end{array}
$$

$\uparrow$

same pixel in each different image

As explained in Section 75.3.3, the rows of the original image matrices $(n \times n)$ are rearranged into a single row of the new matrix, which hence contains n^2 elements per row; the new matrix is $n^2 \times M$. Each of the columns represents the grey-level value of the same pixel in the energy series. We denote the j-th column by $Y(j)$, and suppose, for the sake of example, that we wish to identify a particular chemical element, A. Then pixel j may or may not contain it and we write

$$
\boldsymbol{Y}(j) = A(j)\boldsymbol{f}_A + B(j)\boldsymbol{f}_B + \boldsymbol{N} \tag{76.5}
$$

in which $\boldsymbol{f}_A$ is a vector, the elements of which describe the change in grey-level of pixels that contain A while $\boldsymbol{f}_B$ is a vector describing the same quantity for those that do not. $A(j)$ and $B(j)$ are weights and $\boldsymbol{N}$ is a noise vector. The aim is to calculate the elements of $\boldsymbol{f}_A$ and $\boldsymbol{f}_B$ and the values of the weights $A(j)$ and $B(j)$. This is achieved by calculating the principal components of the 'variance-covariance' matrix $(\boldsymbol{m}\boldsymbol{m}^T)$, which are the eigenvectors corresponding to the largest eigenvalues. If we assume that f_A and f_B are associated with the first two principal components, C_1 and C_2 say, then we need to know how they are related. They will not in general be related identically, because C_1 and C_2 are orthogonal, which f_A and f_B need not be; furthermore the weights $a(j)$ and $b(j)$ in

$$Y(j) =: a(j)C_1 + b(j)C_2 + N \qquad (76.6)$$

may not be non-negative, whereas $A(j)$ and $B(j)$ cannot be negative. The relation between f and C and between $A(j)$, $a(j)$, $b(j)$ and $B(j)$ is established by 'oblique' analysis: oblique (non-orthogonal) directions are sought in the C_1–C_2 plane such that the elements of f_A, f_B, $A(j)$ and $B(j)$ are all positive.

For further details and tests, see the papers by Bonnet *et al.* cited above, and for an application in cell biology, see Quintana and Ollacarizqueta (1989).

76.5 Description

Picture description is the branch of image analysis in which the details of grey-level values are replaced by a set of descriptors that somehow encapsulate the useful information in the image in an abstract or general way. It has so far found no application in electron microscopy even though, as we have mentioned (Hawkes, 1990b), it can be expected to attract interest as fast microscope–computer transfer becomes more usual. For example, if we could describe all the different structures of interest in a particular application in a nearly context-independent way, we could compare these template-descriptions with descriptions of structures seen in the microscope and isolate those of interest automatically. In this way, fragile specimens would not be exposed to a dose greater than that needed to acquire an image for subsequent processing; specimens such as those commonly studied in cell biology could be explored rapidly and the grid region containing the cell component of interest presented to the microscopist for further study; patterns that the eye finds difficult or tiring to study, such as certain dislocation images, could be surveyed automatically in a preliminary step. It is easy to imagine other examples. In view of such possible developments as these, we give a very brief introduction to *description*.

We have already encountered a number of rather crude image descriptors in other contexts: moments (76.3–4) and the associated invariants; texture measures (76.2); (local) histograms, autocorrelation functions, transform spectra. We shall say nothing about the graph-theoretical methods, which do not seem to lend themselves to busy grey-level images but may find a place in scanning electron micrograph studies. The only approach that we consider here is that based on *grammatical* or *syntactic* models.

We limit this account to an explanation of the idea in qualitative terms. A feature of an image can be broken down into a number of relatively simple curves and straight lines, joined in various ways. If we give a symbol (letter) to each of the curves and lines and another to each of the ways of connecting them, the feature will be described by a sequence or string of symbols and we can compare this with strings extracted from the image. Clearly this description can be made insensitive to size and orientation and may be tolerant to quite major distortions, provided that the symbols are defined with some latitude. Furthermore, the comparison between the strings can be programmed to allow some degree of mismatch. The classic example is the 'chromosome grammar' of Ledley (Ledley, 1964; Ledley *et al.*, 1965), which distinguishes between submedian and telocentric chromosomes (Fig. 76.9) and its extension by Lee and Fu (1972) to distinguish between median, submedian and acrocentric chromosomes.

Figure 76.9 shows suggested *pattern primitives* forming a *terminal set* and a submedian and a telocentric chromosome labelled in terms of these. The gradual decomposition from an intact submedian chromosome to its pattern primitives is shown in Fig. 76.10. If we attribute symbols to the intermediate (non-terminal) structures, shown in this figure, then we can write down rules for the decomposition. Thus with

$$
\begin{array}{llll}
A & \rightarrow & \text{arm-pair} & D & \rightarrow & \text{arm} \\
B & \rightarrow & \text{bottom} & E & \rightarrow & \text{right-part} \\
C & \rightarrow & \text{side} & F & \rightarrow & \text{left-part}
\end{array}
$$

the following are all plausible:

$$
\begin{array}{llll}
A \rightarrow CA & A \rightarrow DE & A \rightarrow AC & A \rightarrow FD \\
B \rightarrow bB & B \rightarrow Bb & B \rightarrow e & \\
C \rightarrow Cb & C \rightarrow bC & C \rightarrow d & C \rightarrow b \\
D \rightarrow Db & D \rightarrow bD & D \rightarrow a & \\
E \rightarrow cD & & F \rightarrow Dc &
\end{array}
$$

If S_1 is a submedian chromosome, then the description in terms of pattern primitive starts from $S_1 \rightarrow AA$; in the telocentric case, S_2, from $S_2 \rightarrow BA$.

We see that several kinds of quantities are involved:

(i) terminals (a, b, c, d and e);

(ii) non-terminals (S_1, S_2, A, B, C, D, E, F and S), where S is a starting symbol and includes S_1 and S_2: to the list above we must add $S \rightarrow S_1$ and $S \rightarrow S_2$ as possibilities,

(iii) the set of rules listed above, which are known as *productions*.

(a)

babcbabdacad

(b)

ebabcbab

(c)

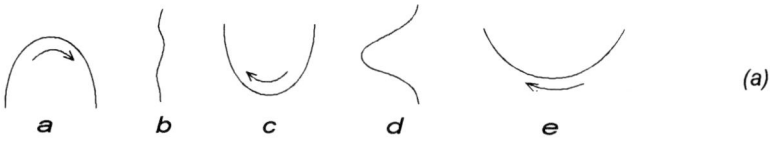

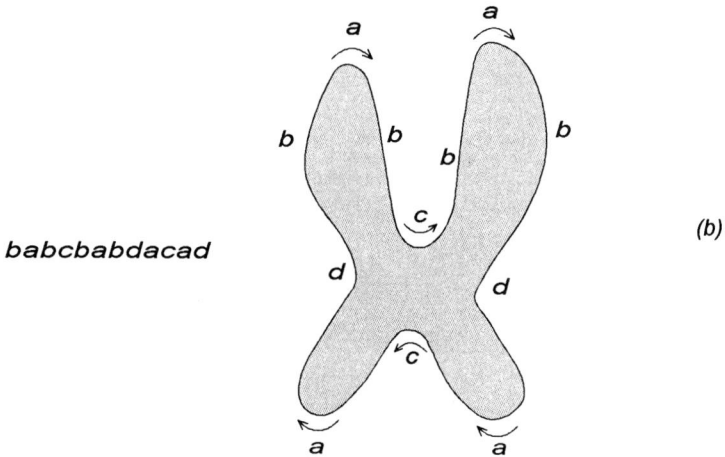

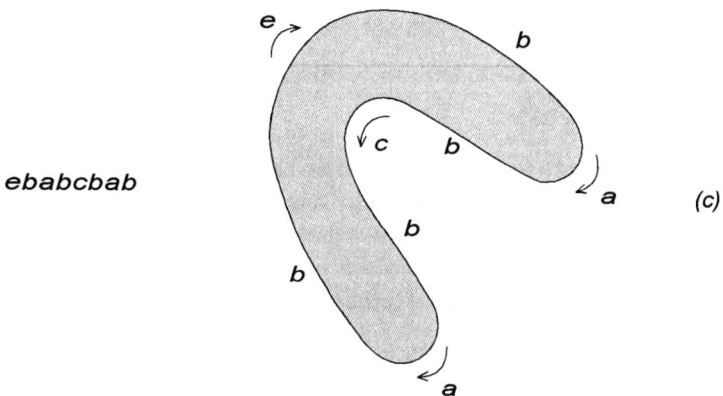

Fig. 76.9: Ledley's 'Chromosome grammar'. (a) Primitive pattern elements, each labelled with a symbol. (b) Submedian and (c) telocentric chromosomes broken down into these elements.

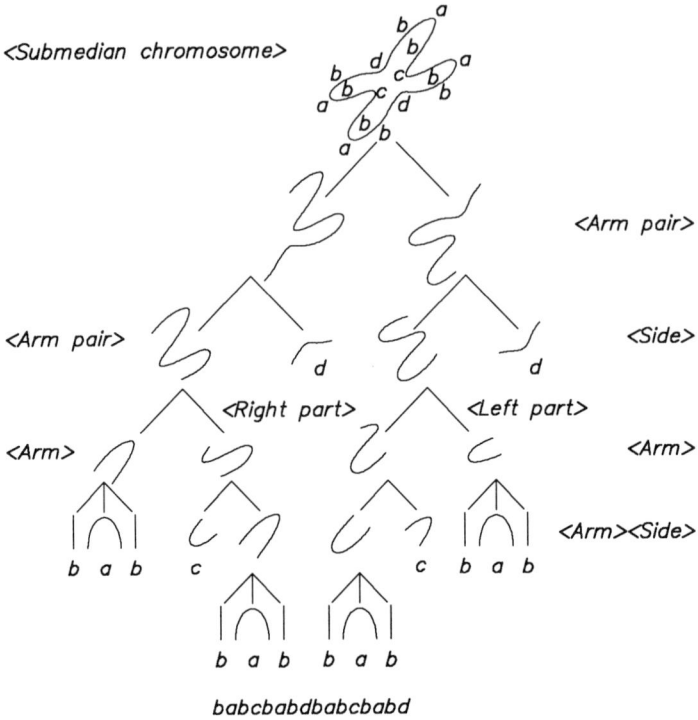

Fig. 76.10: Hierachical decomposition of a submedian chromosome into a character sequence.

A *grammar* G is defined by four quantities:

$$G = (\text{non-terminals, terminals, productions, starting-symbol})$$
$$= (N, \Sigma, P, S) \qquad (S \in N)$$

This notion proves to be extremely rich and such a context-free string grammar is at the simplest extremity of the many that have been studied. Tree grammars, which are suited to structures of the form shown in Fig. 76.11 would be more suitable for dislocation patterns; web grammars and shape grammars are each suited to patterns of particular kinds.

These few words are intended to do no more than draw attention to the existence of a whole way of describing pictures that seems well adapted to some tasks in electron microscopy. For details, see the introductory treatises of Gonzalez and Thomason (1978) and especially Fu (1982) and for more abstract thinking on these matters, Pavel (1989) and Grenander (1975–1981). Some possible uses in electron microscopy are suggested in Hawkes (1990b).

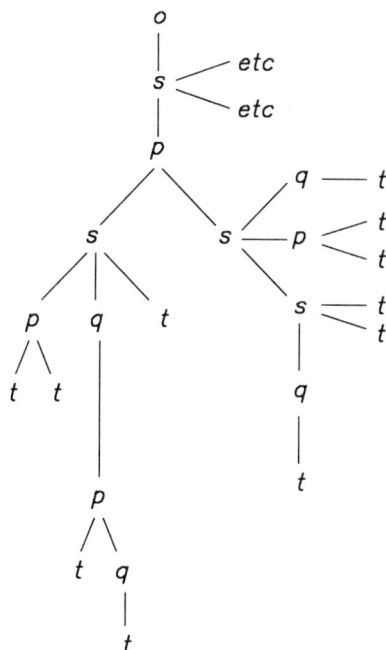

Fig. 76.11: Tree grammars are well suited to structures of this form and hence to dislocation networks.

76.6 Further reading

For general articles on image analysis and its applications, frequently in connection with the scanning electron microscope, see the papers of White *et al.* (1970, 1971, 1972a), Heinrich *et al.* (1970), Newbury and Joy (1973), Fiori *et al.* (1974), Newbury (1975), Krisch *et al.* (1975), Dinger and White (1976), Montoto *et al.* (1978), Stott and Chatfield (1979), Dixon and Taylor (1979), Moza *et al.* (1979), Lee and Kelly (1980), Hillman *et al.* (1980), Kelly *et al.* (1980), Huggins *et al.* (1980), Bauer *et al.* (1983), Lee *et al.* (1985), Smith *et al.* (1986), Mott (1987), Russ and Russ (1987a,b), Bright (1987a,b), Lee and Russ (1989). Many more references, especially short conference abstracts, are listed in Hawkes (1992a).

For recognition techniques applicable to diffraction patterns, see Bright and Steel (1985, 1987, 1988), Goehner *et al.* (1976), Rao and Goehner (1975), Goehner and Rao (1977), Schwarzer and Weiland (1984), Schwarzer (1985), Weiland and Schwarzer (1985, 1986).

For early work on recognition of preferred orientations, see Finlay and Brown (1971), on contour extraction, Matson *et al.* (1970) and on thresholding, McMillan *et al.* (1969). On segmentation,feature discrimination and extraction, see Phillips *et al.* (1983), Russ and Russ (1987a,b), Hood and Howitt (1989); on boundary following, see Selfridge (1986) and on skeleton calculation, Russ (1984). Measurements on particle distribution are described by Niedermeyer and Wilke (1982), Sergeev *et al.* (1984), Raeymaekers *et al.* (1984) and Kovbasa *et al.* (1988).

Many references to the use of mathematical morphology in electron microscopy are to be found in the papers of Beucher (1991), Grillon (1991), Jeulin (1991) and Prod'homme *et al.* (1991); see also Meyer and Beucher (1990) and Froehling (1991).

Morphology is being extended to three dimensions, notably by Gratin and Meyer (1992) and Meyer (1992) and to multi-valued images by Serra (1993).

77

Instrument Control and Instrumental Image Manipulation

77.1 Introduction

The electron image contains information not only about the specimen but also about the operating conditions of the electron microscope. In conventional transmission microscopy, the defocus and spherical aberration coefficient can be deduced from the spatial spectrum of a reasonably amorphous region of the specimen and, with a suitable test specimen, the source size and energy spread of the illumination can also be estimated. The astigmatism can likewise be measured and, if the spatial spectrum is available on line, it can be used to adjust the stigmator correctly. New techniques are being devised for aligning the microscope accurately by exploiting information provided by the image.

The information thus acquired may be exploited manually by the microscope user or fed to a control unit, the role of which is to adjust the microscope, automatically or in response to simple high-level commands. Such control units are gradually coming into use and we can anticipate a time when they will be designed to perform much more complicated tasks of pattern recognition, seeking the structures of interest to the microscopist and relieving him of the task of searching the grid, for example.

In the case of the scanning transmission microscope, the situation is rather different; since each pixel may contribute to several images but each image is generated sequentially, it has long been natural to couple such microscopes to computer memory and hence to process them digitally on-line or, at least, shortly after acquisition. Some useful specimen information can be revealed by skilful detector design, or by mimicking this in the computer memory; this aspect of STEM imagery has already been mentioned in Chapter 67. We make no attempt to describe the various microscope–computer links that have been designed, however, partly because these go beyond the scope of the book but mainly because the associated technology is evolving so rapidly that review articles and conference proceedings are a safer source.

The scanning electron microscope (SEM) is entering a new phase of its existence, with the introduction of field-emission guns and the proliferation of ancillary image-analysis equipment. For many years regarded as limited to resolutions modest compared with that of a good TEM, not much higher than 5–10 nm at best and usually operated at very much lower values, it has recently entered the high resolution domain and the types of image processing required are no longer confined to image analysis. We shall say no more about this here, except to comment that many of the aspects of image processing that have hitherto been of interest exclusively for TEM images will no doubt be finding their way into scanning microscopy.

77.2 Measurement of microscope operating parameters

77.2.1 TEM

We have seen that the image of an amorphous, weakly scattering specimen, traditionally a thin layer of carbon, is modulated by the sinusoidal variation of intensity characteristic of the phase contrast-transfer function. The positions of the maxima and minima of the modulation can be used to calculate the spherical aberration coefficient C_s and the defocus Δ; if sufficient care is taken with the recording, information about the source size and the energy spread of the illuminating beam can be extracted from the values of the intensity of the maxima.

The most obvious way of calculating C_s and Δ is to find the best least-squares fit of the sinusoidal contrast-transfer function to the measured curve (Frank et al., 1970a,b). An ingenious way of obtaining C_s and Δ from the diffractogram was proposed by Krivanek (1976), who noticed that the condition for $\sin\gamma$ to reach a maximum (n an odd integer) or a zero (n an even integer)

$$\frac{1}{4}C_s\lambda^3 p^4 - \frac{1}{2}\Delta\lambda p^2 = \frac{1}{4}n$$

can be rewritten

$$C_s\lambda^3 p^2 - 2\Delta\lambda = \frac{n}{p^2}$$

On plotting n/p^2 against p^2, therefore, for different values of n, we obtain a family of curves (Fig. 77.1) onto which the values of (p, n) given by the spatial spectrum of an amorphous specimen are marked; they lie on a straight line, the slope of which is $C_s\lambda^3$ and the intercept with the ordinate axis $2\lambda\Delta$. Krivanek examines a diffractogram obtained with a JEOL 100 B instrument. By establishing Δ in the directions corresponding to the major and minor axes of the astigmatism ellipse, the difference in defocus between the astigmatic line foci can likewise be estimated.

This technique has been considerably improved by Coene and Denteneer (1991), in various ways: several diffractograms corresponding to dif-

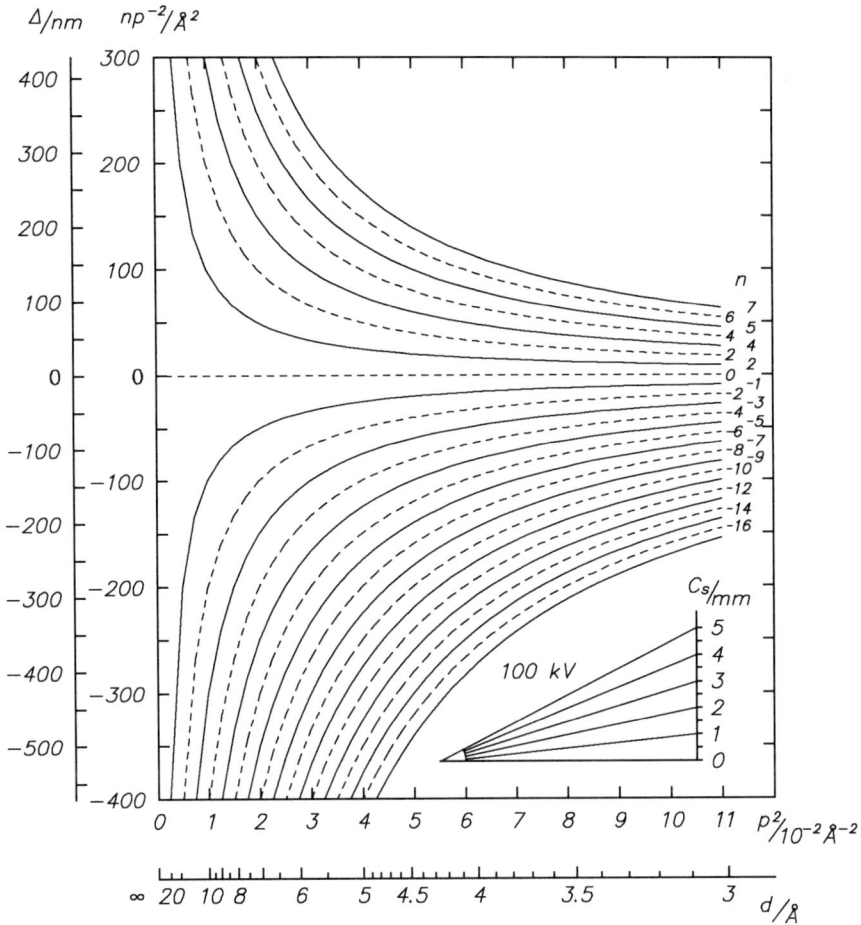

Fig. 77.1: Plot of n/p^2 (in Å^2) against $100p^2$ (in Å^{-2}), for a wide range of values of n. The lower abscissa scale shows $d(=1/p)$ in Å. The left ordinate scale shows defocus Δ in nm for 100 kV operation. The pencil of lines inset shows the gradient corresponding to various values of C_s, again for 100 kV.

ferent values of defocus are used to provide a least-squares fit and a conic (a parabola) of the form $C_s\lambda^3 x^2 - 2\Delta\lambda x = n$, $x := p^2$, is used in the fitting instead of the straight line $y = C_s\lambda^3 x - c$, $y := n/p^2$, $x := p^2$, $c := 2\Delta\lambda$. Coene and Denteneer analyse the accuracy to be expected and employ the method to determine C_s for the Philips SuperTWIN and UltraTWIN lenses.

If the effects of non-vanishing source-size and energy spread of the illu-

minating beam can be represented by envelope functions, which modulate
the phase contrast transfer function $\sin \chi$, the parameters characterizing
these envelopes can be estimated from the rate of attenuation of the latter
(Frank, 1975c; Saxton, 1977a,b). In reduced coordinates, we have

$$\hat{K}_p(Q) = \sin\{\pi(DQ^2 - \frac{1}{2}Q^4)\}$$

$$t_k(q) \to \hat{t}_k(Q) = \exp\{-\pi^2\alpha^2 Q^2(Q^2 - D)^2\}$$

$$t_v(q) \to \hat{t}_v(Q) = \exp(-\frac{\pi^2}{4}B^2 Q^4)$$

in which

$$\alpha := \alpha_c(C_s/\lambda)^{\frac{1}{4}} \quad , \quad B := v_m C'_c/(C_s\lambda)^{\frac{1}{2}}$$

(66.83). The product of $\hat{t}_v$ and the structure factor of the film used
to generate the diffractogram can in principle be found by measuring the
diffractogram at points at which $-\sin \chi$ and t_k are both unity, namely,
$Q = n^{\frac{1}{4}}$ and $D = Q^2 = n^{\frac{1}{2}}$, where n is an odd integer. The angular
spread can then be obtained from measurements of the overall attenuation
of $\hat{K}_p(Q)$. Figure 77.2 shows the result for an amorphous silicon film about
10 nm thick and a Siemens 102 microscope.

It has also been suggested that the point beyond which the attenuation
due to t_k and t_v effectively cuts off the function $\sin \chi$ could be used to
estimate the parameters characterizing the illumination. We then write

$$\pi^2\alpha^2 Q_c^2(Q_c^2 - D)^2 + \frac{\pi^2}{4}B^2 Q_c^4 = \tau$$

where τ is some threshold value. If a set of values of Q_c can be obtained for
different values of the defocus, then $(Q_c^3 - DQ_c)/Q_c^4$ can be plotted against
$1/Q_c^4$ to give a straight line; the gradient of the latter will be $\tau/\pi^2\alpha^2$ and the
intercept, $B^2/4\alpha^2$. Both α and B can hence be estimated if a reasonably
satisfactory value of τ is available, but this is clearly a weak point in the
calculation; if, however, a reliable value of α is already known, B can be
obtained without using the inevitably subjective estimate of τ.

Another sensitive method of measuring C_s, based on the image displacement caused by tilting the beam incident on the specimen, has been
devised by Koster and de Jong (1991). We defer discussion of this to the
next section, where tilt-induced displacement is discussed in the context of
control.

A technique for establishing α and the specimen thickness that relies
on comparison of the recorded image with simulated images has been developed by Thust and Urban (1992) for use with periodic specimens. Here,

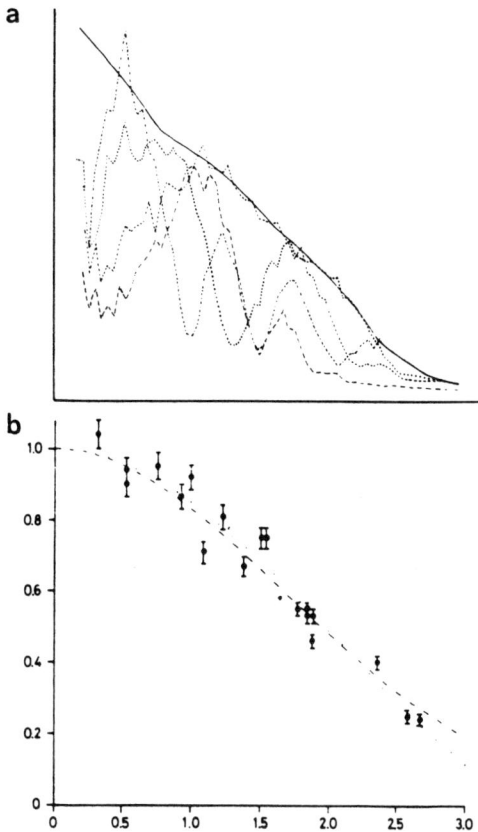

Fig. 77.2: (a) Smoothed radial sections through diffractograms of an amorphous silicon film. The continuous curve represents the combined effect of a narrow energy spread and the structure factor. The dotted curves correspond to various members of the focal series. (b) The experimentally determined envelope due to finite source-size. The fitted curve corresponds to a Gaussian source; the curve for a disc-shaped source is very similar. The abscissa is $|Q(Q^2 - D)|$.

members of a focal series are compared with a large number of simulated images, corresponding to different values of defocus and specimen thickness. By minimizing a suitable distance measure, essentially cross-correlation in Fourier space, the two free parameters can be found. See too King and Campbell (1993).

An elaborate procedure for determining the parameters of the envelope associated with energy spread has been developed by Möbus *et al.* (1992, 1993), a full discussion of which is to be found in Möbus and Rühle (1993).

Among the merits of their method is the separation of the two envelope functions from each other and from the coherent transfer function without using the weak-scattering approximation; amorphous layers of heavy elements with large scattering factors at high frequencies can thus be used. We refer to their papers for details and experimental tests.

77.2.2 STEM

The methods used to establish the defocus and spherical aberration coefficient for TEM bright-field imagery can also be employed in STEM. The earlier attempt by Hammel et al. (1990) to extract these parameters from the transfer function has been repeated with improvements by Wong et al. (1991, 1992), who show that the method is satisfactory. For earlier attempts to establish the STEM parameters in other ways, see Carpenter et al. (1982), Browne et al. (1981) and Lin and Cowley (1986a,b).

77.3 Control

The user of the TEM is required to align the instrument with a high degree of accuracy for high-resolution work, to correct the astigmatism by adjusting the stigmator and to choose the degree of defocus. The ease and accuracy with which these tasks can be performed improved immeasurably in the 1980s with the arrival of fast on-line analysis of certain image parameters; the methods may be said to have evolved as a function of the power of the analytic tools available, but it must be emphasized that the principles had been clearly laid down earlier, notably by Zemlin et al. (1978) and Zemlin (1979). The earlier proposals for autotuning the TEM relied on a modest amount of information, skilfully chosen (Koops and Walter, 1980; Le Poole and Groot, 1980; van der Mast, 1984; Nomura et al., 1986; Nomura and Isakozawa, 1987). Subsequent efforts use a much larger volume of diagnostic information, namely the form of the contrast-transfer function obtained in various operating conditions. Two ways of exploiting these data have emerged and commercial systems based on them have been developed*. These two approaches are based on the form of the auto-correlation function (ACF) and on the image dispacement caused by (deliberate) beam tilt.

The use of the ACF goes back to the work of Frank (1975b), Frank and Al-Ali (1975a,b) and Al-Ali and Frank (1980); Frank and Al-Ali (1975a,b) examined the focus-dependence of the image variance, defined by $\int |I|^2 \, d\boldsymbol{p}$ in which I is the product of three spatial-frequency-dependent factors: the

---

* E.g. TEMDIPS from Tietz (Video and Image Processing Systems, Gauting) and SysTEM from Synoptics, Cambridge.

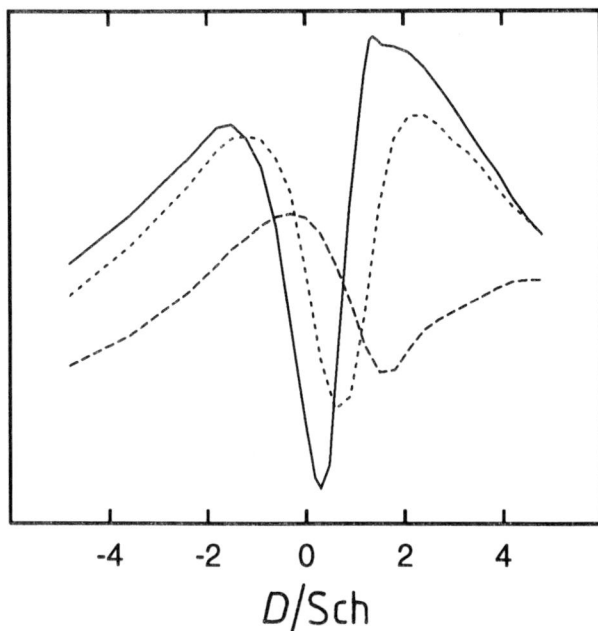

Fig. 77.3: Dependence of image variance (arbitrary units) on reduced defocus (in scherzers) for beam tilts of zero (continuous curve), 0.5 (short dashes) and 1 (long dashes) Gl/Sch.

structure factor of the (amorphous carbon) specimen, the phase contrast-transfer function and the modulation-transfer function of the recording unit. This quantity passes through a minimum close to Gaussian focus (Fig. 77.3) (less sharp in Frank and Al-Ali's work, where the effect of the third factor, the modulation-transfer of the recording unit, was not considered) and through maxima at Scherzer focus ($D = 1$ Sch) and at the subsequent preferred values of D ($D = \sqrt{n}$, n an odd integer). Figure 77.3 also shows that beam tilt shifts the minimum and changes the form of the curve considerably (cf. Erasmus and Smith, 1982; Saxton *et al.*, 1982). Figure 77.4 shows variance as a function of beam tilt for various values of defocus. The next degree of complexity involves the simultaneous presence of astigmatism and beam tilt; this is illustrated in Figs 77.5 and 77.6, which show variance as a function of astigmatism with and without tilt and variance as a function of tilt with and without astigmatism.

   These curves show that the variance exhibits a number of useful properties: Gaussian focus, correct alignment and cancellation of astigmatism coincide with a global minimum of the variance; the extremum of the vari-

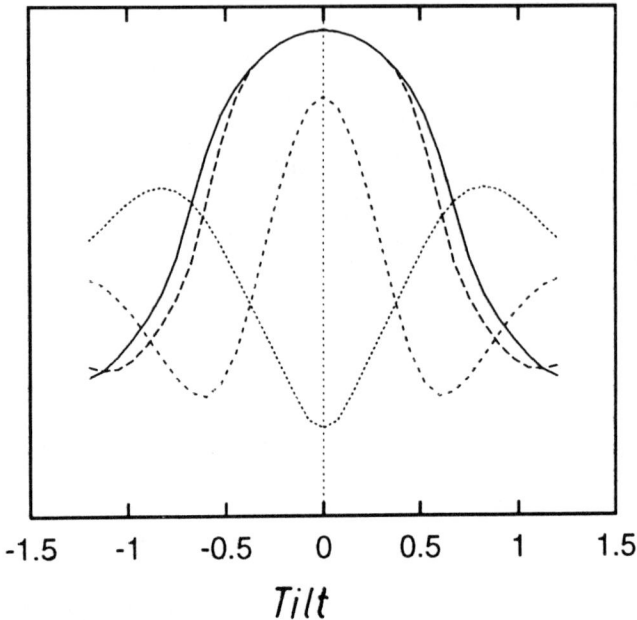

Fig. 77.4: Dependence of image variance on reduced beam tilt, measured in Gl/Sch for four values of reduced defocus: 2 (continuous curve); $\sqrt{3}$ (long dashes); 1 (short dashes); 0 (dots).

ance with respect to tilt is unaffected by the presence of astigmatism but the converse is not true, so that tilt must be corrected first.

The focus dependence of the ACF (the Fourier transform of the diffractogram or power spectrum of the image) was first explored by Al-Ali and Frank (1980), in support of Frank's earlier suggestion (1975b) that the ACF be used to assist astigmatism correction and focusing. It has the disadvantage that tilts that are very small, though large enough to affect the image ($\sim$ 1 mrad), have little or no perceptible effect on it (Saxton and O'Keefe, 1981; Smith et al., 1983); furthermore, the power spectrum for a tilted beam can be the same as that for a beam with a particular defocus and astigmatism (Zemlin et al., 1978).

In the light of all this, Saxton et al. (1983) concluded that "Both diffraction and ACF approaches ... present further problems of data reduction before they can lead to simple prescriptions for supply current adjustments. Perhaps it is the elimination of any requirement for such further data reduction that makes the last approach—simple contrast minimization/maximization on the basis of computed contrast figures, as described

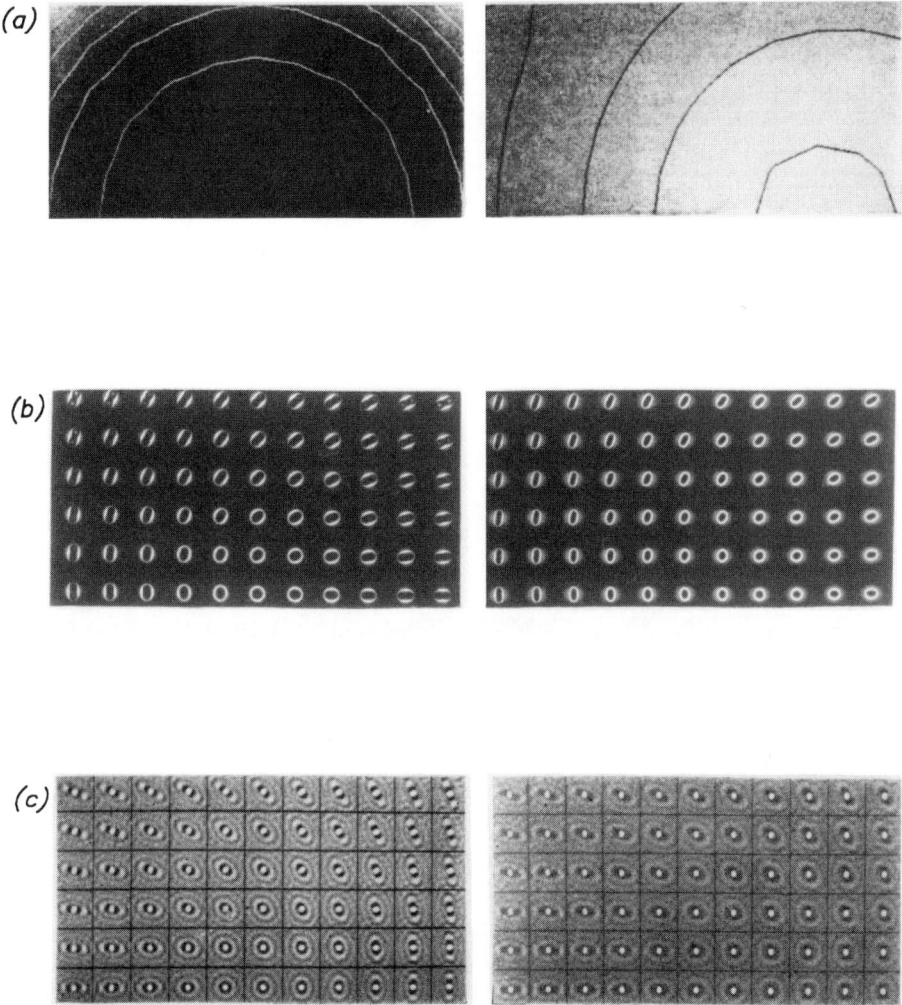

Fig. 77.5: (a) Dependence of image variance on astigmatism. Dark areas correspond to low variance, bright areas to high variance. The abscissa measures x-stigmator current and the ordinate, the y-current. The maximum astigmatism in each direction is one scherzer. (b) Diffractograms at Gaussian focus for the same range of stigmator currents. (c) The corresponding auto-correlation functions (Fourier transforms of (b)). The left-hand member of each pair corresponds to zero tilt, the right-hand member to a tilt of 0.5 Gl/Sch.

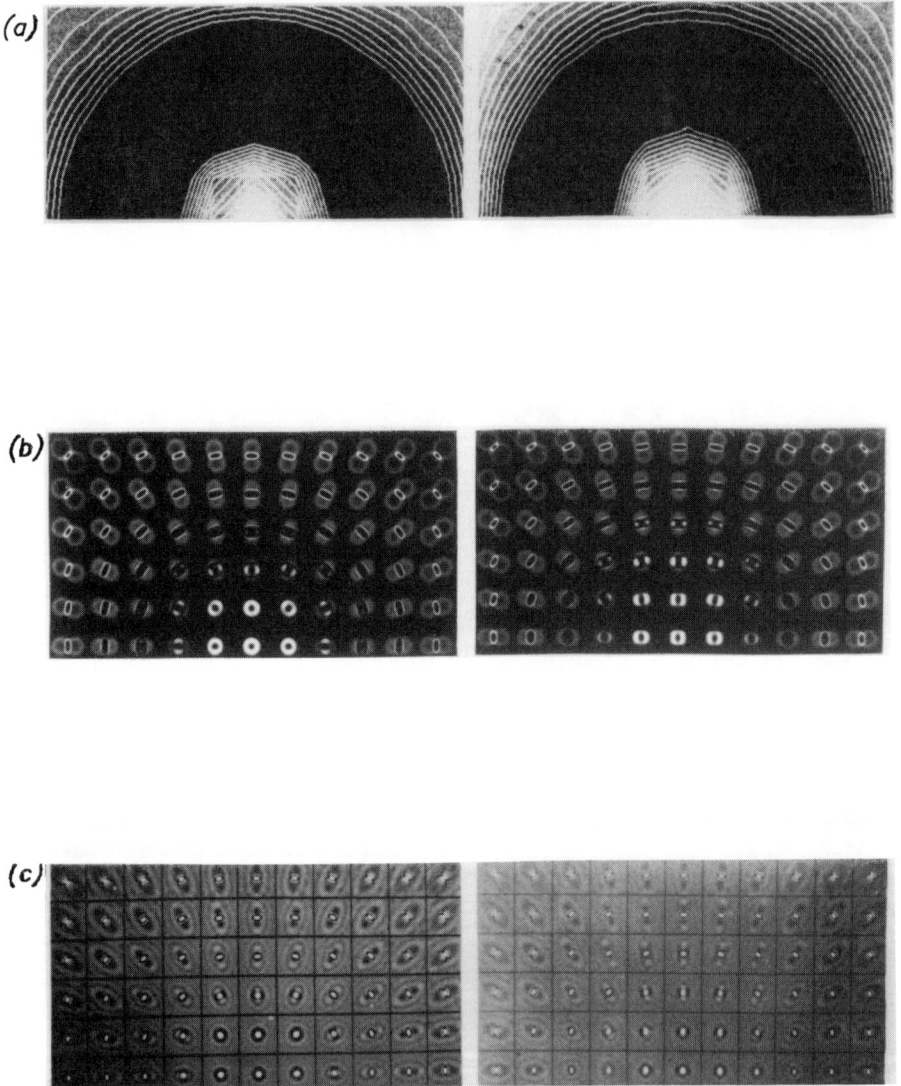

Fig. 77.6: (a) Dependence of image variance on beam tilt. Representation as in Fig. 77.5a. (b) Diffractograms at Scherzer focus. (c) Autocorrelation functions at Scherzer focus.

by Erasmus and Smith (1982)—so successful."

A major disadvantage of the variance method is that the radiation dose imposed on the specimen is high. This was one of the reasons for concentrating on the effect of beam tilt in the image, in the Delft auto-tuning project, which we now describe (Koster et al., 1986, 1987a,b, 1989, 1992a,b; Koster and Ruijter, 1988; van der Mast and Koster, 1988). This account is limited to the principles; an extremely detailed study is available (Koster et al., 1989). See too Tietz (1992).

A careful analysis of the form of the phase contrast-transfer function shows that a set of three images, a direct view and a pair of images obtained with the incident beam tilted by equal and opposite angles about the first direction, is capable of furnishing instructions for aligning the microscope on the coma-free axis. Two more tilt pairs, again through equal and opposite angles about the axis, in perpendicular directions, then give estimates of the defocus and astigmatism. Koster et al. discuss with great care the accuracy of the method and possible sources of error. They likewise consider the electron dose needed for the method to be reliable and, in a later paper (Koster and de Ruijter, 1992), they present practical tests of the procedure.

Koster and de Jong also pointed out (1991) that by plotting image displacement against incident beam tilt, a cubic is obtained characterized by $C_s\lambda^3$ and $\Delta_0\lambda$. Provided that the necessary precautions are taken, in particular concerning the alignment of the microscope, very accurate values of $C_s\lambda^3$ and $\Delta_0\lambda$ can be found, and hence of C_s and Δ_0 provided that λ is known accurately. See also de Jong and Koster (1991), de Jong (1992) and de Jong et al. (1992).

This account is limited to the major recent contributions to direct control of the electron microscope based on image information acquired on-line. For other contributions to the theme, see Baba et al. (1986b, 1987), Ogasawara et al. (1988), Henderson et al. (1986), Kunath et al. (1986, 1987), Koops (1976), Nomura and Isakozawa (1987), Nomura et al. (1986), Saxton (1987b) and Scheerschmidt et al. (1992). For the role of automatic control in the acquisition of data for electron tomography, see Koster et al. (1992a,b), Typke et al. (1992a,b) and Dierksen et al. (1993). Control and the STEM is discussed by Colling and von Harrach (1992). Stage control is examined by Russ (1993).

It has only recently been realised that the threefold axial astigmatism (65.29) becomes important when very high resolution is required (Krivanek and Leber, 1993). Crossed sextupoles are needed to correct it and the aid of a computer is probably indispensable for adjusting them correctly.

Part XVI

Coherence, Brightness and Spectral Functions

78
Coherence and the Brightness Functions

78.1 Introduction

All real sources have areas that may be very small but are never vanishingly small and emit radiation that may have a narrow wavelength or energy spread but is never monochromatic. We have shown in Chapter 66 that the resulting effects can usually be represented by envelope functions that modulate the ideal transfer functions, ideal in the sense that source-size and energy spread are neglected. This finding was, however, based on a phenomenological argument: contributions from different source points and at different wavelengths were simply added. Since these contributions were currents, we were assuming that they were all independent and thus that the source was *incoherent*. The formal theory of partial coherence is designed to explore the relations between emissions from different source points more thoroughly and we therefore begin with a recapitulation of this theory, based closely on that to be found in textbooks of light optics, notably Born and Wolf (1980).

Source properties are also characterized by various forms of brightness, as we explained at some length in Chapter 47. The brightness function (47.1b) and the generalized brightness functions (47.2) characterize the electron current distribution in both position and direction and allow us to calculate various radiometric quantities. The functions that measure coherence likewise have two pairs of arguments and the radiometric quantities can be derived from them. What, if any, is the relation between the functions of coherence theory and the brightness? What brightness function should be employed when the source is partially coherent? These questions have been explored in great detail in light optics and we now give some account of these quite recent developments.

In Section 78.2, we recapitulate the main points of classical coherence theory. In the following short section, the vocabulary and formalism of radiometry are introduced and linked whenever possible to the brightness theory of Chapter 47. We are then ready to present the new ideas that bring these two themes together. Section 78.4 contains an account of the fundamentally new relation between brightness and coherence adumbrated by

Walther in 1968 and developed rigorously and in great detail by Marchand and Wolf (1972a,b, 1974a,b). The next two sections are concerned with the consequences of these new ideas for the van Cittert–Zernike theorem and with a new eigenfunction expansion of the coherence functions discovered by Wolf (1982, 1984). It is not until Section 78.7 that we introduce the extremely important source model known as the quasihomogeneous source, which is a generalization of the Schell source, also defined there. This leads to Section 78.8, in which recent thinking on the subject of brightness is presented.

In the penultimate section, we give a brief account of the relation between temporal and spatial coherence, which was formally established by Wolf et al. (1981) and Wolf and Devaney (1981), inspired by a very original paper of Sudarshan (1969). We conclude with some complementary bibliographic information, which includes topics to which little attention is given here, notably the use of the Wigner and ambiguity functions.

It will be clear from the few references cited above that the present chapter leans very heavily on the work of Wolf and his school, of which no connected account is available. In that work, the frequency variable mostly employed is the circular frequency ω, whereas the frequency ν is preferable here because the definition of the Fourier transform that we use consistently has 2π in the exponents. The frequency ν is, however, employed in the work of Foley and Wolf (1985, 1991), which contains some of the most significant findings. In the following account, the frequency ν is used consistently throughout, with the result that many of the functions employed differ trivially from those encountered in the cited papers. For the most part, we use the same notation as in the latter and minor discrepancies, notably in the powers of 2π, are due to this choice of frequency variable. The symbols used for the many arguments are usually those adopted in the most recent publications.

78.2 Coherence

78.2.1 Definitions
We have seen in Chapter 57 that an ideal monochromatic point source would produce a wave field in which the amplitude at any point remained constant while the phase varied linearly. If we consider a more realistic source in which some wavelength spread is present and the emissive area is not vanishingly small, both phase and amplitude will fluctuate; intuition suggests that the fluctuations at neighbouring points will be similar. Situations such as this are conveniently characterized by a cross-correlation function, which is a measure of the similarity between the wavefunctions

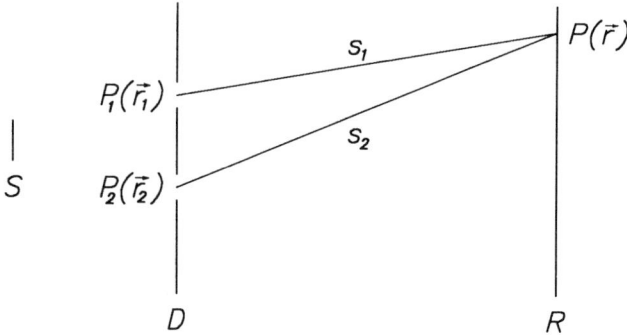

Fig. 78.1: A source, S, a screen D containing two openings P_1 at r_1 and P_2 at r_2 and an arbitrary point P at r in a recording plane R.

at pairs of points. The definitions of the various coherence functions are inspired by these ideas.

Consider now a simple optical system consisting of a source S, an opaque screen D containing two small openings P_1 and P_2 and a recording plane R (Fig. 78.1). If the wavefunction at P_1 is $\Psi(r_1,t)$ and that at P_2 is $\Psi(r_2,t)$, then the wavefunction $\Psi(r,t)$ at some point P in the recording plane will be of the form

$$\Psi(r,t) = c_1\Psi(r_1, t-t_1) + c_2\Psi(r_2, t-t_2)$$

where t_j denotes the time of flight from P_j to $P(j=1,2)$. The instantaneous intensity $I(r,t)$ at P will be given by $\langle\Psi(r,t)\Psi^*(r,t)\rangle$:

$$\begin{aligned}
I(r,t) :&= \langle\Psi(r,t)\Psi^*(r,t)\rangle \\
&= c_1 c_1^* \langle\Psi(r_1, t-t_1)\Psi^*(r_1, t-t_1)\rangle \\
&\quad + c_2 c_2^* \langle\Psi(r_2, t-t_2)\Psi^*(r_2, t-t_2)\rangle \\
&\quad + c_1 c_2^* \langle\Psi(r_1, t-t_1)\Psi^*(r_2, t-t_2)\rangle \\
&\quad + c_1^* c_2 \langle\Psi^*(r_1, t-t_1)\Psi(r_2, t-t_2)\rangle
\end{aligned} \tag{78.1}$$

For convenience, we use the term 'intensity' for $I(r,t)$ here. It is of course related to the current density by a scaling factor.

It is reasonable to assume that the field is (wide-sense) stationary, by which we mean that the time-average of $\Psi(r,t)$ at any point is independent of time; this implies that $\langle\Psi(r_j, t+t_1)\Psi^*(r_k, t+t_2)\rangle$ depends only on the time difference, $t_2 - t_1 =: \tau$. If we introduce the notation

$$I_j := \langle\Psi(r_j,t)\,\Psi^*(r_j,t)\rangle \quad j=1,2 \tag{78.2}$$

(78.1) becomes

$$\langle I(\mathbf{r},t) \rangle = c_1 c_1^* I_1 + c_2 c_2^* I_2 + 2|c_1 c_2| \Re \Gamma_{12}(\tau) \tag{78.3}$$

where

$$\Gamma_{12}(\tau) := \Gamma(\mathbf{r}_1, \mathbf{r}_2; \tau) := \langle \Psi(\mathbf{r}_1, t+\tau) \Psi^*(\mathbf{r}_2, t) \rangle \tag{78.4}$$

The time τ is obviously equal to the difference between the distances s_1 and s_2 (Fig. 78.1) divided by the velocity, in field-free space. We note that $c_j c_j^* I_j$ is the intensity that would be found at P if only the j-th opening were present and it is hence convenient to write

$$I^{(j)}(\mathbf{r},t) := c_j c_j^* I(\mathbf{r}_j, t) \quad j = 1, 2 \tag{78.5}$$

giving

$$I = I^{(1)} + I^{(2)} + 2|c_1 c_2| \Re \Gamma_{12}(\tau) \tag{78.6}$$

The function $\Gamma_{12}(\tau)$ is just the cross-correlation function of the individual wavefunctions $\Psi(\mathbf{r}_1, t)$ and $\Psi(\mathbf{r}_2, t)$. In coherence theory, it is known as the *mutual coherence function*. When the points P_1 and P_2 coincide, $\Gamma_{12}(\tau)$ collapses to the autocorrelation function of $\Psi(\mathbf{r}_1, t)$ (or $\Psi(\mathbf{r}_2, t)$), which is called the *self-coherence* at P_1:

$$\Gamma_{11}(\tau) = \langle \Psi(\mathbf{r}_1, t+\tau) \Psi^*(\mathbf{r}_1, t) \rangle \tag{78.7}$$

When the time difference τ also vanishes, we recover the intensity

$$\Gamma_{11}(0) = \Gamma(\mathbf{r}_1, \mathbf{r}_1; 0) = \langle \Psi(\mathbf{r}_1, t) \Psi^*(\mathbf{r}_1^*, t) \rangle$$
$$= I^{(1)} \tag{78.8}$$

A normalized form of $\Gamma_{12}(\tau)$ is in practice more convenient. We write

$$\gamma_{12}(\tau) := \gamma(\mathbf{r}_1, \mathbf{r}_2; \tau) := \frac{\Gamma_{12}(\tau)}{\{\Gamma_{11}(0)\Gamma_{22}(0)\}^{\frac{1}{2}}} = \frac{\Gamma_{12}(\tau)}{(I^{(1)} I^{(2)})^{\frac{1}{2}}} \tag{78.9}$$

and it is not difficult to show that

$$|\Gamma_{12}(\tau)|^2 \leq \Gamma_{11}(0)\Gamma_{22}(0)$$

and hence that

$$|\gamma_{12}(\tau)| \leq 1 \tag{78.10}$$

The function $\gamma_{12}(\tau)$ is known as the *complex degree of coherence*. It may be related directly to the sharpness of certain interference fringes, as we now show.

We have seen that the intensity at a point P in the recording plane R is given by (78.6). The complex function $\gamma_{12}(\tau)$ may be written

$$\gamma_{12}(\tau) = |\gamma_{12}(\tau)| \exp[2\pi \mathrm{i}\{\gamma_{12}(\tau) - \nu_0 \tau\}] \qquad (78.11)$$

in which ν_0 is the mean frequency of the source and we assume that the latter is quasi-monochromatic (we return to this point below). Then from (78.6) and (78.9) we see that

$$
\begin{aligned}
I(\mathbf{r}) :&= \langle I(\mathbf{r}, t) \rangle \\
&= I^{(1)}(\mathbf{r}) + I^{(2)}(\mathbf{r}) + 2(I^{(1)} I^{(2)})^{\frac{1}{2}} |\gamma_{12}(\tau)| \cos 2\pi\{\gamma_{12}(\tau) - \alpha_0\}
\end{aligned}
$$
$$(78.12)$$

where $\alpha_0 := \nu_0 \tau$. The intensities $I^{(j)}(\mathbf{r})$ and the functions $|\gamma_{12}|$ and α_{12} vary slowly over a region for which $s_1 - s_2 \ll \lambda_0^2/\Delta\lambda$ and the interference fringes that are seen on R will hence be determined by the term in α_0. The modulus and argument of $\gamma_{12}(\tau)$ may hence be obtained by measuring the visibility of the interference fringes, defined as $(I_{max} - I_{min})/(I_{max} + I_{min})$, and the positions of their maxima or minima.

The extreme values of $|\gamma_{12}(\tau)|$, namely zero and one, correspond to familiar physical conditions, as we can see immediately from (78.12). When $|\gamma_{12}(\tau)| = 0$, we have

$$I(\mathbf{r}) = I^{(1)}(\mathbf{r}) + I^{(2)}(\mathbf{r}) \qquad (78.13)$$

The intensities generated by the two openings add and there are no interference effects: the wavefunctions add *incoherently* at $P(\mathbf{r})$. If, at the other extreme, $|\gamma_{12}(\tau)| = 1$, then

$$I(\mathbf{r}) = |(\hat{\Psi}_1 + \hat{\Psi}_2)|^2 \qquad (78.14)$$

where $|\hat{\Psi}_j|^2 = I^{(j)}(\mathbf{r})$ and $\arg \hat{\Psi}_2 - \arg \hat{\Psi}_1 = \arg \gamma_{12}(\tau)$ so that the intensity at $P(\mathbf{r})$ is the same as that created by two sources $P(\mathbf{r}_1)$ and $P(\mathbf{r}_2)$ radiating with a *fixed* phase difference and hence *coherent*.

Before proceeding further with the theory, we pause to examine the concept of a quasi-monochromatic source, used above without formal definition. It will be recalled that the chromatic aberration of electron lenses is so severe that the energy spread in the illuminating beam of a microscope has to be kept extremely small, so that typically $\Delta\lambda/\lambda_0 \lesssim 10^{-5}$, where λ_0 is the mean wavelength and $\Delta\lambda$ is some convenient measure of departures from λ_0. The situation in the illuminating beam or in the crossover is not, however, the same as at the source itself, the cathode; here the electrons have low energies and the ratio of the energy (or wavelength) spread to the mean value may well be of the order of unity, far greater indeed than the corresponding ratio of wavelengths in an everyday incandescent light source. How can such a highly polychromatic source become transformed into a quasi-monochromatic source in the space between filament and crossover? In the language of geometrical optics, the answer is so obvious that the question does not need to be asked. But in terms of coherence theory, we need a result of Wolf and Carter (Wolf and Carter, 1975; Carter and Wolf, 1975), which is derived in Section 78.7. Radiation for which the condition

$$\frac{\Delta\lambda}{\lambda_0} \ll 1 \tag{78.15}$$

is satisfied is said to be *quasi-monochromatic*. In practice, the time differences τ of interest are also small and in this case, simpler formulae than those introduced above are adequate. We define the spectrum of $\gamma_{12}(\tau)$, $\tilde{\gamma}_{12} = \mathcal{F}(\gamma_{12})$ or

$$\gamma_{12}(\tau) =: \int_{-\infty}^{\infty} \tilde{\gamma}_{12}(\nu) \exp(-2\pi i \nu \tau)\, d\nu \tag{78.16}$$

From (78.11), we have $|\gamma_{12}(\tau)| \exp\{2\pi i \alpha_{12}(\tau)\} = \gamma_{12}(\tau)\exp(2\pi i \nu_0 \tau)$. Hence

$$|\gamma_{12}(\tau)| \exp\{2\pi i \alpha_{12}(\tau)\} = \int_{-\infty}^{\infty} \tilde{\gamma}_{12}(\nu) \exp\{-2\pi i (\nu - \nu_0)\tau\}\, d\nu \tag{78.17}$$

For values of τ and $\nu - \nu_0$ such that

$$|(\nu - \nu_0)\tau| \ll 1 \tag{78.18}$$

the integral in (78.17) may be replaced by $\int_{-\infty}^{\infty} \tilde{\gamma}_{12}(\nu)\, d\nu$, which is equal to $|\gamma_{12}(0)| \exp\{2\pi i \alpha_{12}(0)\}$. Thus if and only if (78.18) is true, we may write

$$\Gamma_{12}(\tau) \approx \Gamma_{12}(0) \exp(-2\pi i \nu_0 \tau)$$
$$\gamma_{12}(\tau) \approx \gamma_{12}(0) \exp(-2\pi i \nu_0 \tau) \tag{78.19}$$

The quantity $\Gamma_{12}(0)$ is known as the *mutual intensity* and both $\gamma_{12}(\tau)$ and $\gamma_{12}(0)$ as the *complex degree of coherence* or *degree of spatial coherence*. The reason for this last name will become clear below when we have encountered the van Cittert–Zernike theorem. For electron sources, condition (78.18) implies that

$$|\tau \Delta\nu| = \left|\left(\frac{\lambda\Delta\phi}{2\pi v}\right)\left(\frac{v\Delta\lambda}{\lambda^2}\right)\right| = \left|\frac{\Delta\phi\Delta\lambda}{2\pi\lambda}\right| = \left|\frac{\Delta\phi}{4\pi}\frac{\Delta U}{U}\right| \ll 1$$

where U is the accelerating voltage and $\Delta\phi$ is the phase difference corresponding to the time lag τ. For $\Delta U/U = 10^{-5}$, therefore, we have $|\Delta\phi| \ll 10^6$ rad.

Before turning to the various spatial and temporal spectral functions in terms of which the next stage of the theory is expressed, we comment briefly on the relations between spatial and temporal coherence and the physical properties of a source, its size and the associated wavelength spread. We have seen in Chapter 66 that the effects of non-vanishing source size and of energy spread can usually be separated and it is physically plausible to associate them with partial spatial and temporal coherence respectively. This can be understood, qualitatively at least, in terms of the mutual coherence function $\Gamma_{12}(\tau)$ or the complex degree of coherence $\gamma_{12}(\tau)$. Consider first a beam divided into two, by a biprism for example, then recombined. The intensity in the recording plane will be given by $\langle\Psi(\boldsymbol{r},t+\tau)\Psi^*(\boldsymbol{r},t)\rangle$, where τ is a measure of the difference in path-length: this quantity is just the self-coherence $\Gamma(\tau)$. The narrower the wavelength spread in the initial beam, the greater the path-difference and hence τ can be before the visibility of the interference fringes becomes impaired. *Coherence time* and *effective spectral width* are hence defined in terms of $\Gamma(\tau)$ or $\gamma(\tau)$ and we associate these quantities with temporal coherence and the effect of wavelength spread in an illuminating beam.

A similar argument allows us to associate $\Gamma_{12}(0)$ or $\gamma_{12}(0)$ with non-vanishing source size. Consider the fringes produced on the recording screen by two pinholes illuminated by a small but not vanishingly small source. For sufficiently closely spaced pinholes, some fringes will be visible around the central point and their visibility there will be $\Gamma_{12}(0)$. It must not, however, be concluded from these qualitative arguments that the temporal and spatial effects can be separated, though it does seem likely that they can be treated independently to a good approximation. Strictly speaking, they can be separated only if Γ_{12} or γ_{12} can be written as the product of a spatial part and a temporal part, that is, if these functions have the form

$$\Gamma_{12}(\tau) = \Gamma_{12}(0)\Gamma(\tau)$$
$$\gamma_{12}(\tau) = \gamma_{12}(0)\gamma(\tau) \tag{78.20}$$

Such sources are said to be cross-spectrally pure (Mandel, 1961; Mandel and Wolf, 1976).

78.2.2 Spectral functions

The mutual coherence function $\Gamma(\boldsymbol{r}_1,\boldsymbol{r}_2;\tau)$ is a cross-correlation function. The underlying stochastic process has a cross-power spectrum $W(r_1,r_2;\nu)$, which is the Fourier transform*of $\Gamma(\mathbf{r}_1, \mathbf{r}_2; \tau)$:

$$\Gamma(\mathbf{r}_1, \mathbf{r}_2; \tau) =: \int_0^\infty W(\mathbf{r}_1, \mathbf{r}_2; \nu) \exp(-2\pi i \nu \tau) \, d\nu \qquad (78.21a)$$

$$W(\mathbf{r}_1, \mathbf{r}_2; \nu) := \int_{-\infty}^\infty \Gamma(r_1, r_2; \tau) \exp(2\pi i \nu \tau) \, d\tau \qquad (78.21b)$$

or

$$\Gamma(\mathbf{r}_1, \mathbf{r}_2; \tau) = \mathcal{F}_\nu^- W(\mathbf{r}_1, \mathbf{r}_2; \nu)$$
$$W(\mathbf{r}_1, \mathbf{r}_2; \nu) = \mathcal{F}_\tau W(\mathbf{r}_1, \mathbf{r}_2; \tau) \qquad (78.22)$$

Note that the lower limit of integration in the definition of W (78.21a) is zero because the electron frequency is given by $\nu = E/h$ and hence the wavefunction $\Psi(\mathbf{r}, t)$ is one-sided:

$$\Psi(\mathbf{r}, t) = \int_0^\infty \tilde{\Psi}(\mathbf{r}, \nu) \exp(-2\pi i \nu t) \, d\nu \qquad (78.23)$$

From this we see that

$$\Gamma(\mathbf{r}_1, \mathbf{r}_2; \tau) = \int_0^\infty \int_0^\infty \langle \tilde{\Psi}(\mathbf{r}_1, \nu) \tilde{\Psi}^*(\mathbf{r}_2, \nu') \rangle e^{-2\pi i (\nu - \nu') t} e^{-2\pi i \nu \tau} \, d\nu \, d\nu'$$

But the average $\langle \cdot \rangle$ must be of the form

$$\langle \tilde{\Psi}(\mathbf{r}_1, \nu) \tilde{\Psi}^*(\mathbf{r}_2, \nu') \rangle =: W(\mathbf{r}_1, \mathbf{r}_2; \nu) \delta(\nu - \nu')$$

and hence

$$\Gamma(\mathbf{r}_1, \mathbf{r}_2; \tau) = \int_0^\infty W(\mathbf{r}_1, \mathbf{r}_2; \nu) \exp(-2\pi i \nu \tau) \, d\nu \qquad (78.24)$$

as stated above.

Thus the *cross-spectral density* W is essentially the *temporal* spectrum of $\Gamma_{12}(\tau)$. The *spatial* spectrum of Γ_{12} is also of great importance. This is defined with the aid of the angular correlation function $\mathcal{A}$:

$$W(\mathbf{r}_1, \mathbf{r}_2; \nu) =: \int_{-\infty}^\infty \mathcal{A}(\mathbf{s}_1, \mathbf{s}_2; \nu)$$

$$\times \exp\left\{ \frac{2\pi i}{\lambda}(\mathbf{s}_1 \cdot \mathbf{x}_1 + \sigma_1 z_1 - \mathbf{s}_2 \cdot \mathbf{x}_2 - \sigma_2 z_2) \right\} \, d\mathbf{s}_1 \, d\mathbf{s}_2$$
$$(78.25)$$

---

* The variable of integration is indicated by a subscript in the $\mathcal{F}$-notation for Fourier transforms whenever the function being transformed has several arguments.

so that

$$\Gamma(\boldsymbol{r}_1, \boldsymbol{r}_2; \tau) = \int_0^\infty d\nu e^{-2\pi i \nu \tau}$$

$$\times \int A(\boldsymbol{s}_1, \boldsymbol{s}_2; \nu) e^{\{2\pi i (\boldsymbol{s}_1 \cdot \boldsymbol{x}_1 + \sigma_1 z_1 - \boldsymbol{s}_2 \cdot \boldsymbol{x}_2 - \sigma_2 z_2)/\lambda\}} \, d\boldsymbol{s}_1 \, d\boldsymbol{s}_2$$

$$(78.26)$$

This definition may be understood by considering the plane-wave expansion of $\tilde{\Psi}(\boldsymbol{r}, \nu.)$, in terms of which we define the quantities $\boldsymbol{s}$ and σ undefined in (78.25-26). We have

$$\tilde{\Psi}(\boldsymbol{r}, \nu) =: \int_{-\infty}^\infty a(\boldsymbol{s}, \nu) \exp\left\{ \frac{2\pi i}{\lambda} (\boldsymbol{s} \cdot \boldsymbol{x} + \sigma z) \right\} \, d\boldsymbol{s} \qquad (78.27)$$

in which the coordinates of the point $\boldsymbol{r}$ are $(\boldsymbol{x}, z)$ and $\sigma := (1 - \boldsymbol{s} \cdot \boldsymbol{s})^{1/2}$. Then

$$W(\boldsymbol{r}_1, \boldsymbol{r}_2; \nu) \delta(\nu - \nu') = \langle \tilde{\Psi}(\boldsymbol{r}_1, \nu) \tilde{\Psi}^*(\boldsymbol{r}_2, \nu') \rangle$$

$$= \int \langle a(\boldsymbol{s}_1, \nu) a^*(\boldsymbol{s}_2, \nu') \rangle \exp\left\{ \frac{2\pi i}{\lambda} (\boldsymbol{s}_1 \cdot \boldsymbol{x}_1 + \sigma_1 z_1) \right\}$$

$$\times \exp\left\{ -\frac{2\pi i}{\lambda'} (\boldsymbol{s}_2 \cdot \boldsymbol{x}_2 + \sigma_2 z_2) \right\} d\boldsymbol{s}_1 \, d\boldsymbol{s}_2$$

which implies that averaged term must be of the form

$$\langle a(\boldsymbol{s}_1, \nu) a^*(\boldsymbol{s}_2, \nu') \rangle = A(\boldsymbol{s}_1, \boldsymbol{s}_2; \nu) \delta(\nu - \nu')$$

which gives (78.25) immediately.

How is the *angular correlation function* A related to the *spatial* Fourier transform of W? The latter is defined by

$$W(\boldsymbol{x}_1, z_1; \boldsymbol{x}_2, z_2; \nu) =: \int \hat{W}(\boldsymbol{s}_1, z_1; \boldsymbol{s}_2, z_2; \nu)$$

$$\times \exp\{2\pi i (\boldsymbol{s}_1 \cdot \boldsymbol{x}_1 + \boldsymbol{s}_2 \cdot \boldsymbol{x}_2)\} d\boldsymbol{s}_1 \, d\boldsymbol{s}_2 \quad (78.28)$$

relative to the transverse coordinates $\boldsymbol{x}_1, \boldsymbol{x}_2$. Comparison of (78.25) with (78.28) shows that

$$A(\boldsymbol{s}_1, \boldsymbol{s}_2; \nu) = \frac{1}{\lambda^4} \hat{W}(\boldsymbol{s}_1/\lambda, z_1; -\boldsymbol{s}_2/\lambda, z_2; \nu) \exp\left\{ \frac{2\pi i}{\lambda} (\sigma_2 z_2 - \sigma_1 z_1) \right\}$$

$$= \lambda^{-4} \hat{W}(\boldsymbol{s}_1/\lambda, 0; -\boldsymbol{s}_2/\lambda, 0; \nu) \qquad (78.29)$$

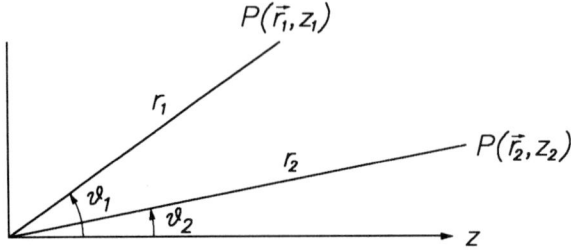

Fig. 78.2: Notation used in deriving far-zone behaviour.

The far-zone behaviour of these various functions will be needed in connection with the brightness. We now obtain expressions for W and Γ as the normalized distances $2\pi(x_1^2 + y_1^2 + z_1^2)^{1/2}/\lambda$ and $2\pi(x_2^2 + y_2^2 + z_2^2)^{1/2}/\lambda$ to the points $P_1(\boldsymbol{r}_1)$ and $P_2(\boldsymbol{r}_2)$ tend to infinity in the directions characterized by $(\boldsymbol{s}_1, \sigma_1) = (x_1/r_1, y_1/r_1, z_1/r_1)$ and $(\boldsymbol{s}_2, \sigma_2) = (x_2/r_2, y_2/r_2, z_2/r_2)$ respectively. We give the results without proof (see Marchand and Wolf, 1972b for further details):

$$W(\boldsymbol{r}_1, \boldsymbol{r}_2; \nu) \sim \lambda^2 \cos\theta_1 \cos\theta_2 \mathcal{A}(\boldsymbol{s}_1, \boldsymbol{s}_2; \nu)$$
$$\times \frac{\exp\{2\pi i(r_1 - r_2)/\lambda\}}{r_1 r_2} \qquad (78.30a)$$

$$\Gamma(\boldsymbol{r}_1, \boldsymbol{r}_2; \tau) \sim v^2 \frac{\cos\theta_1 \cos\theta_2}{r_1 r_2}$$
$$\times \int_{-\infty}^{\infty} \mathcal{A}(\boldsymbol{s}_1, \boldsymbol{s}_2; \nu) \exp\left\{-2\pi i\nu\left(\tau - \frac{r_1 - r_2}{v}\right)\right\} \frac{d\nu}{v^2} \qquad (78.30b)$$

where

$$\sigma_1 = z_1/r_1 = \cos\theta_1 \qquad \sigma_2 = z_2/r_2 = \cos\theta_2 \qquad (78.31)$$

(Fig. 78.2). The symbol $\sim$ signifies 'tends asymptotically to'. The far-zone form of W may also be expressed in terms of $\hat{W}$:

$$W(\boldsymbol{r}_1, \boldsymbol{r}_2; \nu) \sim \frac{1}{\lambda^2} \frac{\cos\theta_1 \cos\theta_2}{r_1 r_2} \hat{W}(\boldsymbol{s}_1/\lambda, 0; -\boldsymbol{s}_2/\lambda, 0; \nu) e^{\{2\pi i(r_1 - r_2)/\lambda\}} \qquad (78.32)$$

If the points $P(\boldsymbol{r}_1)$ and $P(\boldsymbol{r}_2)$ coincide, $W(\boldsymbol{r}_1, \boldsymbol{r}_2; \nu)$ collapses to $W(\boldsymbol{r}, \boldsymbol{r}; \nu)$, which is just the spectrum of the radiation in the far zone. The mutual intensity $\Gamma(\boldsymbol{r}_1, \boldsymbol{r}_2; \tau)$ collapses to the self-coherence and this in turn reduces to the total intensity in the corresponding direction when we set $\tau = 0$:

$$I(\boldsymbol{r}) = \Gamma(\boldsymbol{r}; \boldsymbol{r}; 0) \sim v^2 \frac{\cos^2 \theta}{r^2} \int \mathcal{A}(\boldsymbol{s}_1, \boldsymbol{s}_2; \nu) \frac{d\nu}{\nu^2} \qquad (78.33)$$

A normalized form of the cross-spectral density is also of interest. We write

$$\mu(\boldsymbol{r}_1, \boldsymbol{r}_2; \nu) := \mu_{12}(\nu) := \frac{W(\boldsymbol{r}_1, \boldsymbol{r}_2; \nu)}{\{W(\boldsymbol{r}_1, \boldsymbol{r}_1; \nu) W(\boldsymbol{r}_2, \boldsymbol{r}_2; \nu)\}^{\frac{1}{2}}} \qquad (78.34)$$

and can show easily that

$$0 \leq |\mu(\boldsymbol{r}_1, \boldsymbol{r}_2; \nu)| \leq 1 \qquad (78.35)$$

(Suitable precautions must be taken if $W(\boldsymbol{r}, \boldsymbol{r}; \nu)$ vanishes.) The function $\mu(\boldsymbol{r}_1, \boldsymbol{r}_2; \nu)$ is known as the *(complex) degree of spectral coherence* or occasionally as the *degree of spatial coherence*. A straightforward calculation shows that μ is equal to a normalized form of the Fourier transform of $\gamma(\boldsymbol{r}, \boldsymbol{r}_2; \tau)$. If we again write (78.16)

$$\tilde{\gamma}(\boldsymbol{r}_1, \boldsymbol{r}_2; \nu) = \mathcal{F}_\tau \gamma(\boldsymbol{r}_1, \boldsymbol{r}_2; \tau)$$

then

$$\mu(\boldsymbol{r}_1, \boldsymbol{r}_2; \nu) = \frac{\tilde{\gamma}(\boldsymbol{r}_1, \boldsymbol{r}_2; \nu)}{\{\tilde{\gamma}(\boldsymbol{r}_1, \boldsymbol{r}_1; \nu) \tilde{\gamma}(\boldsymbol{r}_2, \boldsymbol{r}_2; \nu)\}^{\frac{1}{2}}} \qquad (78.36)$$

Surprisingly, the functions $\gamma_{12}(\tau)$ and $\mu_{12}(\nu)$ do not contain exactly equivalent information, as Carter (1992) has shown. In particular, a field that is perfectly coherent in the sense that $|\gamma_{12}(\tau)| = 1$ for all point pairs in a given domain and all values of τ has a mutual coherence $\Gamma_{12}(\tau)$ of the form

$$\Gamma_{12}(\tau) = U(\boldsymbol{r}_1, \nu) U^*(\boldsymbol{r}_2, \nu) \exp(-2\pi i \nu_0 \tau)$$

It is then true that the cross-spectral density may be written

$$W(\boldsymbol{r}_1, \boldsymbol{r}_2; \nu) = U'(\boldsymbol{r}_1, \nu) U'^*(\boldsymbol{r}_2, \nu)$$

and $|\mu(\boldsymbol{r}_1, \boldsymbol{r}_2; \nu)| = 1$. The converse is not true, however, and Carter concludes that coherence in the space–time space is somehow more fundamental than coherence in the space–frequency space.

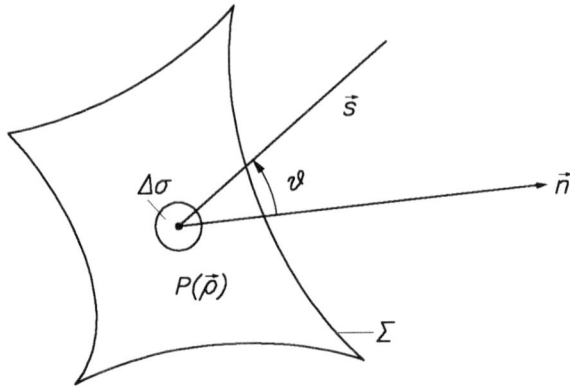

Fig. 78.3: Notation used in connection with radiometry.

78.3 Radiometry

The quantities used to describe the flow of luminous energy for light or of current density for electrons are defined in terms of flux, area and direction. Consider a surface emitting electrons or a surface through which a beam of electrons passes, Σ (Fig. 78.3). The current flowing through an element of area $\Delta\sigma$ around a point P lying on the surface in the direction characterized by s inclined at an angle θ to the normal to Σ at P may be written

$$\Delta I = B(\rho, s; \nu) \cos\theta \Delta\sigma \Delta\Omega \tag{78.37}$$

Here, $\Delta\Omega$ denotes the element of solid angle through which the current is flowing and $B(\rho, s; \nu)$ is a form of brightness. The argument ρ indicates the position of P on the surface Σ in some suitable coordinate frame. Comparison with (47.2) shows that $B(\rho, s; \nu)$ is essentially the same as the generalized brightness function introduced by Lenz (1957); the factor $\cos\theta$ in (78.37) merely reflects the fact that the element of area Δa in (47.2) is normal to the direction s.

It is usual to write

$$\Delta I = (B\cos\theta\Delta\sigma)\Delta\Omega$$
$$= (B\cos\theta\Delta\Omega)\Delta\sigma \tag{78.38}$$

and define

$$J(s, \nu) = \int B\cos\theta d\sigma \tag{78.39}$$

$$E(\boldsymbol{\rho}, \nu) = \int B \cos \theta d\Omega \qquad (78.40)$$

In light optics, E is known as the *radiant emittance* and J as the *radiant intensity*. The total current is then given by integration,

$$I = \int_{2\pi} J(\boldsymbol{s}, \nu)\, d\Omega = \int E(\boldsymbol{\rho}, \boldsymbol{\nu})\, d\sigma \qquad (78.41)$$

If the function $B(\boldsymbol{\rho}, \boldsymbol{s}; \nu)$ is independent of the direction, $\boldsymbol{s}$, then $J(\boldsymbol{s}, \nu)$ reduces to

$$J(\boldsymbol{s}, \nu) \to J_0 \cos \theta, \quad J_0 = \int B(\boldsymbol{\rho}, \nu)\, d\sigma \qquad (78.42)$$

The radiant intensity in a given direction then varies as the cosine of the angle between the normal and the direction in question: the source is said to obey Lambert's (cosine) Law (44.17). The calculations in Chapter 44 show that the models used to describe electron microscope sources implicitly make the assumption that $B(\boldsymbol{\rho}, \boldsymbol{s}; \nu)$ is independent of direction.

A serious weakness of the standard radiometric theory is that individual contributions to total currents are simply added: any possible correlation between the current elements is ignored. What function should we be using to calculate currents, current densities and the like when the source is partially coherent? This is the subject of the next section.

78.4 The brightness of partially coherent sources

We saw in Section 78.2 that the intensity at a point $\boldsymbol{r}$ in the far zone may be expressed in terms of the angular spectrum $\mathcal{A}(\boldsymbol{s}, \boldsymbol{s}; \nu)$:

$$I(\boldsymbol{r}) \sim \int \left(\frac{\lambda}{r}\right)^2 \cos^2 \theta \mathcal{A}(\boldsymbol{s}, \boldsymbol{s}; \nu)\, d\nu \qquad (78.43)$$

The current flowing through a surface element $d\sigma$ at the far-zone point $P(\boldsymbol{r})$, normal to the line joining the source to P, is then

$$dI = \int I(\boldsymbol{r}) r^2\, d\Omega\, d\nu =: \int dI_\nu\, d\nu \qquad (78.44a)$$

where

$$dI_\nu = \lambda^2 \cos^2 \theta \mathcal{A}(\boldsymbol{s}, \boldsymbol{s}; \nu)\, d\Omega \qquad (78.44b)$$

From (78.29), we know that

$$\mathcal{A}(s,s;\nu) = \lambda^{-4} \iint_{z=0} W(\boldsymbol{\rho}_1,\boldsymbol{\rho}_2;\nu)\exp\left\{-\frac{2\pi i}{\lambda}(\boldsymbol{\rho}_1 - \boldsymbol{\rho}_2)\cdot s\right\} d\boldsymbol{\rho}_1\,d\boldsymbol{\rho}_2$$

(78.45)

in which the points* $\boldsymbol{\rho}_1$ and $\boldsymbol{\rho}_2$ lie in the source plane $z = 0$. Let us now define

$$\boldsymbol{\rho}' := \boldsymbol{\rho}_1 - \boldsymbol{\rho}_2$$
$$\boldsymbol{\rho} := \frac{1}{2}(\boldsymbol{\rho}_1 + \boldsymbol{\rho}_2)$$

(78.46)

whereupon $\mathcal{A}(s,s;\nu)$ becomes

$$\mathcal{A}(s,s;\nu) =: \lambda^{-2}\sec\theta \int_{z=0} b(\boldsymbol{\rho},s;\nu)\,d\boldsymbol{\rho}$$

(78.47)

where

$$b(\boldsymbol{\rho},s;\nu) := \lambda^{-2}\cos\theta \int_{z'=0} W(\boldsymbol{\rho}+\boldsymbol{\rho}'/2, \boldsymbol{\rho}-\boldsymbol{\rho}'/2;\nu)\exp\left(-\frac{2\pi i}{\lambda}\boldsymbol{\rho}'\cdot s\right) d\boldsymbol{\rho}'$$

(78.48)

Substituting these expressions into (78.44b), we find

$$dI_\nu \to \cos\theta\,d\Omega \int_{z=0} b(\boldsymbol{\rho},s;\nu)\,d\boldsymbol{\rho}$$

(78.49)

and so

$$I \to \int_{2\pi} d\Omega\cos\theta \int_{z=0} b(\boldsymbol{\rho},s;\nu)\,d\boldsymbol{\rho}$$

(78.50)

This equation has exactly the same form as (78.41) and hence represents the current produced by a source in any state of coherence. The function $b(\boldsymbol{\rho},s;\nu)$ replaces the 'incoherent' brightness function $B(\boldsymbol{\rho},s;\nu)$ and is known as the *generalized radiance*; it was introduced "tentatively" by Walther (1968) and justified, criticized and investigated in very great detail in the following decades. The first thorough study was made by Marchand and Wolf (1974b), building on their work of 1972; we chart the subsequent literature in Section 78.10. We shall usually refer to it as the (generalized) *brightness*.

---

* We use the notation ρ, ρ_1, etc. consistently when the point in question is in the source plane.

The form of (78.49) leads us to define generalized forms of the radiant emittance (78.40) and radiant intensity (78.39); we write

$$e(\boldsymbol{r}, \nu) = \int_{2\pi} b(\boldsymbol{r}, \boldsymbol{s}; \nu) \cos \theta \, d\Omega \qquad (78.51)$$

$$j(\boldsymbol{s}, \nu) = \int_{z=0} b(\boldsymbol{r}, \boldsymbol{s}; \nu) \, d\boldsymbol{r} \qquad (78.52)$$

We now examine these generalized quantities in more detail and, in particular, draw attention to their weaknesses.

Two other expressions for $b(\boldsymbol{r}, \boldsymbol{s}; \nu)$ are useful. It is not difficult to show that

$$b(\boldsymbol{r}, \boldsymbol{s}; \nu) = \frac{\cos \theta}{\lambda^2} \int_{\boldsymbol{f}} \hat{W}(\boldsymbol{s}/\lambda + \boldsymbol{f}/2, -\boldsymbol{s}/\lambda + \boldsymbol{f}/2; \nu) \exp(2\pi i \boldsymbol{f} \cdot \boldsymbol{r}) \, d\boldsymbol{f} \quad (78.53)$$

with

$$\hat{W}(\boldsymbol{f}_1, \boldsymbol{f}_2; \nu) = \int \int_{z=0} W(\boldsymbol{r}_1, \boldsymbol{r}_2; \nu) \exp\{-2\pi i(\boldsymbol{f}_1 \cdot \boldsymbol{r}_1 + \boldsymbol{f}_2 \cdot \boldsymbol{r}_2)\} \, d\boldsymbol{r}_1 \, d\boldsymbol{r}_2$$

$$(78.54)$$

and that

$$b(\boldsymbol{r}, \boldsymbol{s}; \nu) = \cos \theta \int \int_{-\infty}^{\infty} \mathcal{A}(\boldsymbol{s} + \boldsymbol{s}'/2, \boldsymbol{s} - \boldsymbol{s}'/2; \nu) \exp(ik\boldsymbol{s}' \cdot \boldsymbol{r}) \, d\boldsymbol{s}' \quad (78.55)$$

The generalized radiance is real but not necessarily positive. That it is real can be seen by forming the complex conjugate of b:

$$b^*(\boldsymbol{r}, \boldsymbol{s}; \nu) = (1/\lambda)^2 \cos \theta \int_{z'=0} W^*(\boldsymbol{r} + \boldsymbol{r}'/2, \boldsymbol{r} - \boldsymbol{r}'/2; \nu) \exp(2\pi i \boldsymbol{s} \cdot \boldsymbol{r}'/\lambda) \, d\boldsymbol{r}'$$

(from the definition 78.48) and using the fact that

$$W^*(\boldsymbol{r}_1, \boldsymbol{r}_2; \nu) = W(\boldsymbol{r}_2, \boldsymbol{r}_1; \nu) \qquad (78.56)$$

Hence

$$b^*(\boldsymbol{r}, \boldsymbol{s}; \nu) = (1/\lambda)^2 \cos \theta \int_{z'=0} W(\boldsymbol{r} - \boldsymbol{r}'/2, \boldsymbol{r} + \boldsymbol{r}'/2; \nu) \exp(2\pi i \boldsymbol{s} \cdot \boldsymbol{r}'/\lambda) \, d\boldsymbol{r}'$$

$$= (1/\lambda)^2 \cos \theta \int_{z'=0} W(\boldsymbol{r} + \boldsymbol{r}'/2, \boldsymbol{r} - \boldsymbol{r}'/2; \nu)$$

$$\times \exp(-2\pi i \boldsymbol{s} \cdot \boldsymbol{r}'/\lambda) \, d\boldsymbol{r}' \qquad (78.57)$$

which is identical with the expression for $b(\mathbf{r}, \mathbf{s}; \nu)$. Thus $b = b^*$ and so b is real. The fact that b is not necessarily positive has been shown by considering the special case of coherent sources; we refer to Marchand and Wolf (1974) for details, where it is also shown that the generalized radiant emittance can also become negative. We shall have little occasion to use it below and simply quote without proof a convenient expression for it:

$$e(\mathbf{r}, \nu) = \frac{k^2}{2(2\pi)^{1/2}} \int\limits_{z'=0} W(\mathbf{r} + \mathbf{r}'/2, \mathbf{r} - \mathbf{r}'/2; \nu) \frac{J_{3/2}(kr')}{(kr')^{3/2}} \, d\mathbf{r}' \qquad (78.58)$$

The definition (78.48) expresses $b(\mathbf{r}, \mathbf{s}; \nu)$ as a *linear* combination of different values of the cross-spectral density. We have seen that it is not entirely satisfactory. Is there any other linear combination that possesses all the properties that we expect of a brightness? This question was posed by Friberg (1979) who showed that there is not, an important negative result. Specifically, Friberg stated that any radiance or brightness function $B(\mathbf{r}, \mathbf{s}; \nu)$ associated with a finite, planar, partially coherent source to which a physical meaning can be attributed in the sense of traditional radiometry must satisfy the following four conditions:

$$B(\mathbf{r}, \mathbf{s}; \nu) = \mathsf{L}\{W(\mathbf{r}, \mathbf{r}_2; \nu)\}$$

$$B(\mathbf{r}, \mathbf{s}; \nu) \geq 0$$

$$B(\mathbf{r}, \mathbf{s}; \nu) = 0 \quad \text{when } \mathbf{r} \text{ lies outside the source area} \qquad (78.59)$$

$$\int B(\mathbf{r}, \mathbf{s}; \nu) \, d\mathbf{r} = \frac{\cos\theta}{\lambda^2} \hat{W}(\mathbf{s}/\lambda, -\mathbf{s}/\lambda; \nu)$$

L is any linear operator, that is, an operator for which

$$\mathsf{L}\{a_1 F_1(\mathbf{r}_1, \mathbf{r}_2; \nu) + a_2 F_2(\mathbf{r}_1, \mathbf{r}_2; \nu)\} = a_1 \mathsf{L}\{F_1(\mathbf{r}_1, \mathbf{r}_2; \nu)\} + a_2 \mathsf{L}\{F_2(\mathbf{r}_1, \mathbf{r}_2; \nu)\}$$
$$(78.60)$$

He then goes on to prove that there is no linear operator for which all four conditions are in general satisfied. It is perhaps worth stressing that there is no reason why the four requirements should not be satisfied in particular circumstances, and notably for particular forms of the cross-spectral density W. We shall see that the quasihomogeneous source is such a case. The importance of Friberg's result is that there is clearly no point in exploring other linear combinations of W in the hope of finding a universally satisfactory expression.

The generalized radiant intensity will prove to have the same physical interpretation as the ordinary radiant intensity (78.39). We have

$$j(\mathbf{s}, \nu) = \cos\theta \int\limits_{z=0} b(\mathbf{r}, \mathbf{s}; \nu) \, d\mathbf{r} \qquad (78.61)$$

But

$$\int b(\boldsymbol{r},\boldsymbol{s};\nu)\,d\boldsymbol{r} = \frac{\cos\theta}{\lambda^2}\int \hat{W}(\boldsymbol{s}/\lambda + \boldsymbol{f}/2, -\boldsymbol{s}/\lambda + \boldsymbol{f}/2; \nu)\exp(2\pi\mathrm{i}\boldsymbol{f}\cdot\boldsymbol{r})\,d\boldsymbol{f}\,d\boldsymbol{r}$$

$$= \frac{\cos\theta}{\lambda^2}\hat{W}(\boldsymbol{s}/\lambda, -\boldsymbol{s}/\lambda; \nu) \tag{78.62}$$

and so

$$j(\boldsymbol{s},\nu) = \hat{W}(\boldsymbol{s}/\lambda, -\boldsymbol{s}/\lambda; \nu)\frac{\cos^2\theta}{\lambda^2} \sim R^2 W(R\boldsymbol{s}, R\boldsymbol{s}; \nu) \quad R/\lambda \gg 1 \tag{78.63}$$

This last expression demonstrates that $j(\boldsymbol{s},\nu)$ is never negative. We also have

$$j(\boldsymbol{s},\nu) = \lambda^2 \cos^2\theta \mathcal{A}(\boldsymbol{s},\boldsymbol{s};\nu) \tag{78.64}$$

Comparison with (78.39) shows that j measures exactly the same quantity as the radiant intensity $J(\boldsymbol{s},\nu)$ and we need not distinguish between the two quantities.

We shall see in Section 78.8 that the brightness (78.48) is in fact appropriate for many practical situations but, meanwhile, we must introduce the eigenfunction expansion for W and its consequences and discuss the notion of the quasihomogeneous source.

78.5 Consequences for the van Cittert–Zernike theorem

An expression for the mutual intensity and complex degree of coherence in a plane illuminated by an extended incoherent quasi-monochromatic source was first obtained by van Cittert (1934) and later by Zernike (1938) and is known as the van Cittert–Zernike theorem. A straightforward generalization gives an expression for the same quantities when the source and screen are separated by lenses and situated in regions of different refractive index (Hopkins, 1951). A further generalization that extends the theorem to partially coherent sources was made by Wolf and Carter (1976). We merely list the results here, referring to their work for derivations.

The far-field cross-spectral density can be shown to be related to the generalized radiance b by the formula

$$W(R_1\boldsymbol{s}_1, R_2\boldsymbol{s}_2; \nu) = \frac{\cos\theta_1\cos\theta_2}{\cos\theta_{12}}\frac{\exp\{\mathrm{i}k(R_1 - R_2)\}}{R_1 R_2}$$

$$\times \int_{z=0} b\left(\boldsymbol{\rho}, \frac{\boldsymbol{s}_1 + \boldsymbol{s}_2}{2};\nu\right)\exp\{-\mathrm{i}k(\boldsymbol{s}_1 - \boldsymbol{s}_2)\cdot\boldsymbol{\rho}\}\,d\boldsymbol{\rho}$$

$$\tag{78.65}$$

in which

$$\cos\theta = \sigma = \sqrt{1 - \boldsymbol{s}\cdot\boldsymbol{s}} \quad \text{and} \quad \boldsymbol{s} = (s_x, s_y)$$

$$\cos\theta_{12} = \left\{1 - \frac{1}{4}(\boldsymbol{s}_1 + \boldsymbol{s}_2)^2\right\}^{\frac{1}{2}}$$

The other quantities are shown in Fig. 78.2.

The complex degree of coherence (78.34) is then given by

$$\begin{aligned}
\mu(R_1\boldsymbol{s}_1, R_2\boldsymbol{s}_2; \nu) &= \frac{W(R_1\boldsymbol{s}_1, R_2\boldsymbol{s}_2; \nu)}{\{W(R_1\boldsymbol{s}_1, R_1\boldsymbol{s}_1; \nu)W(R_2\boldsymbol{s}_2, R_2\boldsymbol{s}_2; \nu)\}^{\frac{1}{2}}} \\
&= \frac{\cos\theta_1 \cos\theta_2}{\cos_{12}} \frac{\exp\{ik(R_1 - R_2)\}}{\{j(\boldsymbol{s}_1; \nu)j(\boldsymbol{s}_2; \nu)\}^{\frac{1}{2}}} \\
&\quad \times \int_{z=0} b\left(\boldsymbol{\rho}, \frac{\boldsymbol{s}_1 + \boldsymbol{s}_2}{2}; \nu\right) \exp\{-ik(\boldsymbol{s}_1 - \boldsymbol{s}_2)\cdot\boldsymbol{\rho}\} d\boldsymbol{\rho}
\end{aligned}$$

$$\tag{78.66}$$

where j is the radiant intensity (78.61).

If we consider the case of an incoherent source, the usual form of the van Cittert–Zernike theorem is recovered.

A further analogue of this theorem has been established by Wolf *et al.* (1990), who find that the degree of spatial coherence of statistically homogeneous fields is related in a simple way to the complex amplitude in an associated monochromatic field.

78.6 Eigenfunction expansions of the coherence functions

78.6.1 The expansions

We set out from the relation (78.21b) between the cross-spectral density $W(\boldsymbol{r}_1, \boldsymbol{r}_2; \nu)$ and the mutual coherence $\Gamma(\boldsymbol{r}_1, \boldsymbol{r}_2; \tau)$ of a primary or secondary source distribution $\Psi(\boldsymbol{r}, t)$:

$$W(\boldsymbol{r}_1, \boldsymbol{r}_2; \nu) = \int_{-\infty}^{\infty} \Gamma(\boldsymbol{r}_1, \boldsymbol{r}_2; \tau) \exp(2\pi i\nu\tau) d\tau \tag{78.67}$$

where $\Gamma(\boldsymbol{r}_1, \boldsymbol{r}_2; \tau) = \langle \Psi^*(\boldsymbol{r}_1, t)\Psi(\boldsymbol{r}_2, t + \tau) \rangle$ and it is assumed that $|\Gamma|$ falls off sufficiently rapidly with τ to ensure that $\int_{-\infty}^{\infty} |\Gamma| d\tau < \infty$ for all points $\boldsymbol{r}_1$ and $\boldsymbol{r}_2$ in the source domain. Then the spectrum of $\Psi(\boldsymbol{r}, t)$ at the point $\boldsymbol{r}$ is given by the 'diagonal element' of W :

$$S(\boldsymbol{r}, \nu) = W(\boldsymbol{r}, \boldsymbol{r}; \nu) \tag{78.68}$$

The foregoing summary is incomplete in the sense that considerable care is necessary whenever Fourier transforms are used in connection with functions that are random variables. The appropriate analysis was developed in the 1930s by Wiener and Khintchine and is extremely clearly recapitulated in Wolf (1982). A discussion of this point is beyond our present concerns but it must be remembered that the spectra introduced here are usually Wiener–Khintchine spectra, for which it is the autocorrelation function of a random process that is Fourier transformed and not the random process itself.

The fact that $\int |\Gamma| \, d\tau < \infty$ here implies that $\int |\Gamma|^2 \, d\tau < \infty$ and Plancherel's theorem tells us that $\int |W|^2 \, d\nu < \infty$ so that (78.24) can be inverted, to give

$$\Gamma(\boldsymbol{r}_1, \boldsymbol{r}_2; \tau) = \int W(\boldsymbol{r}_1, \boldsymbol{r}_2; \nu) \exp(-2\pi i \nu \tau) \, d\nu \qquad (78.69)$$

The cross-spectral density possesses three properties that are essential for the subsequent analysis. Provided that Γ is a continuous function of $\boldsymbol{r}_1$ and $\boldsymbol{r}_2$ and is uniformly bounded throughout the source domain $\mathcal{D}$ by a function (of τ) that is everywhere integrable, $W(\boldsymbol{r}_1, \boldsymbol{r}_2; \nu)$ will likewise be a continuous function of $\boldsymbol{r}_1$ and $\boldsymbol{r}_2$, bounded in the same domain:

$$\iint_{\mathcal{D}} |W(\boldsymbol{r}_1, \boldsymbol{r}_2; \nu)|^2 \, d\boldsymbol{r}_1 \, d\boldsymbol{r}_2 < \infty \qquad (78.70)$$

Furthermore,

$$W(\boldsymbol{r}_2, \boldsymbol{r}_1; \nu) = W^*(\boldsymbol{r}_1, \boldsymbol{r}_2; \nu) \qquad (78.71)$$

and, much less obvious,

$$\iint_{\mathcal{D}} W(\boldsymbol{r}_1, \boldsymbol{r}_2; \nu) f^*(\boldsymbol{r}_1) f(\boldsymbol{r}_2) \, d\boldsymbol{r}_1 \, d\boldsymbol{r}_2 \geq 0 \qquad (78.72)$$

for *any* square-integrable function f. A proof of this last property is to be found in Wolf (1982).

These three conditions (78.70–72) tell us that the cross-spectral density function W is a *Hermitian, non-negative Hilbert–Schmidt kernel.* Such kernels can be expanded in terms of the eigenfunctions of W, that is, in terms of the solutions of the integral equation

$$\int_{\mathcal{D}} W(\boldsymbol{r}_1, \boldsymbol{r}_2; \nu) \phi_k(\boldsymbol{r}_1; \nu) \, d\boldsymbol{r}_1 = \lambda_k(\nu) \phi_k(\boldsymbol{r}_2; \nu) \qquad (78.73)$$

as follows:

$$W(\mathbf{r}_1, \mathbf{r}_2; \nu) = \sum_k \lambda_k(\nu) \phi_k^*(\mathbf{r}_1; \nu) \phi_k(\mathbf{r}_2, \nu) \qquad (78.74)$$

Furthermore, all the eigenvalues are real and non-negative (78.71–72):

$$\lambda_k(\nu) \geq 0 \quad \forall k \qquad (78.75)$$

The eigenfunctions are square-integrable and orthogonal (with the usual precautions if some eigenvalues are equal):

$$\int_{\mathcal{D}} \phi_j^*(\mathbf{r}; \nu) \phi_k(\mathbf{r}; \nu) \, d\mathbf{r} = \delta_{jk} \qquad (78.76)$$

Numerous interesting results can be derived from the expansion (78.74). These are discussed at length in the seminal paper of Wolf (1982) and we present them only briefly here.

The series expansion (78.74) may be re-written in the more suggestive form

$$W(\mathbf{r}_1, \mathbf{r}_2; \nu) =: \sum_k W^{(k)}(\mathbf{r}_1, \mathbf{r}_2; \nu) \qquad (78.77)$$

where

$$W^{(k)}(\mathbf{r}_1, \mathbf{r}_2; \nu) = \lambda_k(\nu) \phi_k^*(\mathbf{r}_1, \nu) \phi_k(\mathbf{r}_2, \nu) \qquad (78.78)$$

Each component has the form of a cross-spectral density and W may thus be pictured as a sum of individual cross-spectral densities $W^{(k)}$, each belonging to a 'natural mode' of the source. With each mode we associate its degree of spatial coherence μ (78.34):

$$\begin{aligned}
\mu^{(k)}(\mathbf{r}_1, \mathbf{r}_2; \nu) :&= \frac{W^{(k)}(\mathbf{r}_1, \mathbf{r}_2; \nu)}{\{W^{(k)}(\mathbf{r}_1, \mathbf{r}_1; \nu) W^{(k)}(\mathbf{r}_2, \mathbf{r}_2; \nu)\}^{\frac{1}{2}}} \\
&= \frac{\phi_k^*(\mathbf{r}_1, \nu) \phi_k(\mathbf{r}_2, \nu)}{|\phi_k(\mathbf{r}_1, \nu)||\phi_k(\mathbf{r}_2, \nu)|}
\end{aligned} \qquad (78.79)$$

Furthermore, $|\mu^{(k)}(\mathbf{r}_1, \mathbf{r}_1; \nu)|$ is obviously equal to unity for all values of its arguments, which indicates that each individual mode arose from a *coherent* source element. An immediate consequence is that a source described by only one mode must be completely coherent.

The spectrum $S(\mathbf{r}, \nu)$ may also be written straightforwardly in terms of the eigenfunctions ϕ_k and their eigenvalues, since it forms the diagonal elements of W (78.68):

$$\begin{aligned}
S(\mathbf{r}, \nu) :&= W(\mathbf{r}, \mathbf{r}; \nu) \\
&= \sum_k \lambda_k(\nu) |\phi_k(\mathbf{r}, \nu)|^2 =: \sum_k S^{(k)}(\mathbf{r}, \nu)
\end{aligned} \qquad (78.80)$$

Integrating over the source area $\mathcal{D}$, we find

$$\int S(\boldsymbol{r}, \nu)\, d\boldsymbol{r} = \int \sum_k S^{(k)}(\boldsymbol{r}, \nu)\, d\boldsymbol{r}$$

$$= \sum_k \lambda_k(\nu) \tag{78.81}$$

The individual cross-spectral densities are orthogonal, in the sense that

$$\iint_{\mathcal{D}} W^{(k)*} W^{(\ell)}\, d\boldsymbol{r}_1\, d\boldsymbol{r}_2 = \lambda_k^2(\nu)\delta_{k\ell} \tag{78.82}$$

and so

$$\iint_{\mathcal{D}} |W|^2\, d\boldsymbol{r}_1\, d\boldsymbol{r}_2 = \sum_k \lambda_k^2(\nu) \tag{78.83}$$

Related results may be established for the mutual coherence function $\Gamma(\boldsymbol{r}_1, \boldsymbol{r}_2; \tau)$ and its normalized counterpart, $\gamma(\boldsymbol{r}_1, \boldsymbol{r}_2; \tau)$:

$$\Gamma(\boldsymbol{r}_1, \boldsymbol{r}_2; \tau) = \sum_k \Gamma^{(k)}(\boldsymbol{r}_1, \boldsymbol{r}_2; \tau) \tag{78.84a}$$

where

$$\Gamma^{(k)}(\boldsymbol{r}_1, \boldsymbol{r}_2; \tau) = \mathcal{F}_\nu^- W^{(k)}(\boldsymbol{r}_1, \boldsymbol{r}_2; \nu)$$

$$= \int_{-\infty}^{\infty} \lambda_k(\nu)\phi_k^*(\boldsymbol{r}_1, \nu)\phi_k(\boldsymbol{r}_1, \nu)e^{-2\pi i \nu \tau}\, d\nu \tag{78.84b}$$

The modes $\Gamma^{(k)}$ are orthogonal in the sense that

$$\iiint \Gamma^{(k)*}(\boldsymbol{r}_1, \boldsymbol{r}_2; \tau)\Gamma^{(\ell)}(\boldsymbol{r}_1, \boldsymbol{r}_2; \tau)\, d\boldsymbol{r}_1\, d\boldsymbol{r}_2\, d\tau = \int \lambda_k^2(\nu)\, d\nu \cdot \delta_{k\ell} \tag{78.85}$$

It can be shown that

$$\int_{\mathcal{D}} \Gamma^{(k)}(\boldsymbol{r}, \boldsymbol{r}; 0)\, d\boldsymbol{r} = \int \lambda_k(\nu)\, d\nu$$

The definition of $\gamma^{(k)}(\boldsymbol{r}_1, \boldsymbol{r}_2; \tau)$ in terms of $\Gamma^{(k)}(\boldsymbol{r}_1, \boldsymbol{r}_2; \tau)$, analogous to (78.9), shows that, unlike $|\mu^{(k)}|$, $|\gamma^{(k)}|$ is not in general equal to unity so that the contribution of a given mode of the cross-spectral density of W to the mutual coherence function Γ is not associated with spatial or temporal coherence.

A further development of considerable theoretical importance concerns the possibility of expressing the cross-spectral density as an ensemble average over a set of monochromatic emitters. Wolf (1982) shows that this is possible and draws attention to the repercussions of this proof. We mention only one point, which emerged later (Foley and Wolf, 1985). The cross-spectral density can be written

$$W(\boldsymbol{r}_1, \boldsymbol{r}_2; \tau) = \langle U^*(\boldsymbol{r}_1, \nu) U(\boldsymbol{r}_2, \nu) \rangle \qquad (78.86)$$

where $\{U(\boldsymbol{r}, \nu) \exp(-2\pi i \nu t)\}$ is a set of monochromatic wavefields, each of which can be expanded in the form of an angular spectrum of plane waves:

$$U(\boldsymbol{r}, \nu) = \int a(\boldsymbol{s}, \nu) \exp(ik\boldsymbol{r} \cdot \boldsymbol{s}) \, d\boldsymbol{s}$$

It is not difficult to show that a complex quantity of which the real part is identical with the brightness $b(\boldsymbol{r}, \boldsymbol{s}; \nu)$ defined by (78.48) can be expressed compactly in terms of U and a. We write

$$\hat{b}(\boldsymbol{r}, \boldsymbol{s}; \nu) := \sigma \langle U^*(\boldsymbol{r}, \nu) a(\boldsymbol{s}, \nu) \rangle \exp(ik\boldsymbol{r} \cdot \boldsymbol{s}) \qquad (78.87)$$

and we recall (78.47) that

$$\frac{1}{\lambda^2 \cos \theta} \int b(\boldsymbol{r}, \boldsymbol{s}; \nu) \, d\boldsymbol{r} = \mathcal{A}(\boldsymbol{s}, \boldsymbol{s}; \nu)$$

Substituting the angular spectral expression for $U^*(\boldsymbol{r}, \nu)$ in $\hat{b}(\boldsymbol{r}, \boldsymbol{s}; \nu)$, we obtain

$$\hat{b}(\boldsymbol{r}, \boldsymbol{s}; \nu) = \sigma \int \langle a^*(\boldsymbol{s}', \nu) a(\boldsymbol{s}, \nu) \rangle \exp\{ik\boldsymbol{r} \cdot (\boldsymbol{s} - \boldsymbol{s}')\} \, d\boldsymbol{s}'$$

$$= \sigma \int \mathcal{A}(\boldsymbol{s}, \boldsymbol{s}'; \nu) \exp\{ik\boldsymbol{r} \cdot (\boldsymbol{s} - \boldsymbol{s}')\} \, d\boldsymbol{s}'$$

and so

$$\int \hat{b}(\boldsymbol{r}, \boldsymbol{s}; \nu) \, d\boldsymbol{r} = \lambda^2 \sigma \int \mathcal{A}(\boldsymbol{s}, \boldsymbol{s}'; \nu) \delta(\boldsymbol{s} - \boldsymbol{s}') \, d\boldsymbol{s}'$$

$$= \lambda^2 \sigma \mathcal{A}(\boldsymbol{s}, \boldsymbol{s}; \nu)$$

in agreement with (78.47), since $\sigma = \cos\theta$.

78.6.2 A new set of brightness formulae

The form of the expansion (78.74) led Martínez-Herrero and Mejías (1984) to re-examine the definition of generalized brightness and to propose an expression for the latter that satisfies the physical requirements and does not conflict with the finding of Friberg (1979) discussed in Section 78.4. A more general study of new definitions inspired by (78.74) of which the Martínez-Herrero–Mejías formula is a special case was made by Foley and Nieto-Vesperinas (1985) and we follow their analysis closely.

The object of the analysis is to find an expression for the cross- spectral density of separated form, $W(\boldsymbol{r}_1, \boldsymbol{r}_2; \nu) = \int_{\mathcal{D}} G^*(\boldsymbol{r}_1, \boldsymbol{r}; \nu) G(\boldsymbol{r}_2, \boldsymbol{r}; \nu)\, d\boldsymbol{r}$ in Martínez-Herrero and Mejías (1984), so that the expression for brightness takes the form of an integral over a positive definite integrand and hence cannot become negative.

If we consider a generating function $G(\boldsymbol{r}', \boldsymbol{r}; \nu)$ of the form

$$G(\boldsymbol{r}', \boldsymbol{r}; \nu) := \chi_{\mathcal{D}}(\boldsymbol{r}) \sum_k \sum_\ell a_{k\ell}(\nu) \phi_\ell(\boldsymbol{r}', \nu) \phi_k^*(\boldsymbol{r}, \nu) \tag{78.88a}$$

with

$$\sum_k \sum_\ell |a_{k\ell}(\nu)|^2 < \infty \tag{78.88b}$$

then

$$\int_{\mathcal{D}} G^*(\boldsymbol{r}_1, \boldsymbol{r}; \nu) G(\boldsymbol{r}_2, \boldsymbol{r}; \nu)\, d\boldsymbol{r} = \sum_k \sum_\ell \sum_m a_{k\ell}^*(\nu) a_{km}(\nu) \phi_\ell^*(\boldsymbol{r}_1, \nu) \phi_m(\boldsymbol{r}_2, \nu) \tag{78.89}$$

where the function $\chi_{\mathcal{D}}(\boldsymbol{r})$ essentially characterizes the source area and ensures that G vanishes outside it:

$$\chi_{\mathcal{D}}(\boldsymbol{r}) = \begin{cases} 1 & \text{if } \boldsymbol{r} \in \mathcal{D} \\ 0 & \text{otherwise} \end{cases}$$

Since

$$W(\boldsymbol{r}_1, \boldsymbol{r}_2; \nu) = \sum_k \lambda_k(\nu) \phi_k^*(\boldsymbol{r}_1, \nu) \phi_k(\boldsymbol{r}_2, \nu)$$

$$\equiv \sum_\ell \sum_m \lambda_\ell(\nu) \delta_{\ell m} \phi_\ell^*(\boldsymbol{r}_1, \nu) \phi_m(\boldsymbol{r}_2, \nu) \tag{78.91}$$

the quantity in (78.89) can be identical with W if and only if

$$\sum_k a_{k\ell}^* a_{km} = \lambda_\ell \delta_{\ell m} \tag{78.92}$$

We then have

$$W(\boldsymbol{r}_1, \boldsymbol{r}_2; \nu) = \int G^*(\boldsymbol{r}_1, \boldsymbol{r}; \nu) G(\boldsymbol{r}_2, \boldsymbol{r}; \nu) \, d\boldsymbol{r} \tag{78.93}$$

From (78.48), however, we expect the brightness β to be such that

$$\int \beta(\boldsymbol{r}, \boldsymbol{s}; \nu) \, d\boldsymbol{r} = \frac{\sigma}{\lambda^2} \iint_{\mathcal{D}} W(\boldsymbol{r}_1, \boldsymbol{r}_2; \nu) \exp\left\{\frac{2\pi i}{\lambda} \boldsymbol{s} \cdot (\boldsymbol{r}_2 - \boldsymbol{r}_1)\right\} \, d\boldsymbol{r}_1 \, d\boldsymbol{r}_2$$

and with W given by (78.93), we find

$$\beta(\boldsymbol{r}, \boldsymbol{s}; \nu) = \frac{\sigma}{\lambda^2} \int G^*(\boldsymbol{r}_1, \boldsymbol{r}; \nu) \exp\left(-\frac{2\pi i}{\lambda} \boldsymbol{s} \cdot \boldsymbol{r}_1\right) \, d\boldsymbol{r}_1$$

$$\times \int G(\boldsymbol{r}_2, \boldsymbol{r}; \nu) \exp\left(\frac{2\pi i}{\lambda} \boldsymbol{s} \cdot \boldsymbol{r}_2\right) \, d\boldsymbol{r}_2$$

$$= \frac{\sigma}{\lambda^2} \left| \iint G(\boldsymbol{r}', \boldsymbol{r}; \nu) \exp\left(\frac{2\pi i}{\lambda} \boldsymbol{s} \cdot \boldsymbol{r}'\right) \, d\boldsymbol{r}' \right|^2 \tag{78.94a}$$

or

$$\beta(\boldsymbol{r}, \lambda\boldsymbol{s}; \nu) = \frac{\sigma}{\lambda^2} |\mathcal{F}_{\boldsymbol{r}'} G(\boldsymbol{r}', \boldsymbol{r}; \nu)|^2 \tag{78.94b}$$

A simple choice of the coefficients $a_{\ell m}$ that satisfies (78.92) is

$$a_{\ell m}(\nu) = \lambda_\ell^{\frac{1}{2}}(\nu) \delta_{\ell m} \tag{78.95}$$

which was the value originally suggested by Martínez-Herrero and Mejías (1984). Functions of the form (78.93) thus satisfy the four requirements for a brightness function and in particular, are non-negative and vanish outside the source area. It appears, however, that they do not propagate according to the appropriate law (Foley and Wolf, 1985).

78.7 The quasihomogeneous source

Many of the difficulties in relating coherence and the radiometric quantities can be resolved for a family of source-models to which real sources often belong. A number of early results, not described here (e.g. Wolf and Carter, 1975; Baltes *et al.*, 1976) were concerned with sources that were statistically homogeneous and hence large—in theory, of unlimited extent. A much more realistic situation consists in allowing the intensity distribution over the source to vary 'slowly'. Such sources are said to be *quasihomogeneous* and were first introduced and investigated by Carter and Wolf (1977). In a major later paper, the same authors (Carter and Wolf, 1981) extended this work and, in particular, generalized the planar sources studied in the 1977 paper to three-dimensional source distributions.

In order to understand the reasons underlying the definition of a quasi-homogeneous source, we first consider the truly homogeneous case, for which W depends on r_1 and r_2 only as $r_1 - r_2$:

$$W(r_1, r_2; \nu) =: w(r_1 - r_2; \nu) \qquad (78.96)$$

Then the intensity at a point r is just $W(r, r; \nu)$ which is equal to $w(0; \nu)$ and hence constant, independent of r. Furthermore, the complex degree of spatial coherence μ (78.34) is of the form

$$\mu(r_1, r_2; \nu) = w(r_1 - r_2; \nu)/w(0; \nu) \qquad (78.97)$$

We recall that in general, μ is related to W by

$$W(r_1, r_2; \nu) = \{I(r_1, \nu)I(r_2, \nu)\}^{1/2} \mu(r_1, r_2; \nu) \qquad (78.98)$$

so that if $I(r, \nu)$ is independent of r and μ is a function of $r_1 - r_2$, then the source must be homogeneous and, in principle at least, must occupy the entire source plane. If we relax these requirements, we obtain a more physically realistic situation.

We now consider the situation in which $I(r, \nu)$ is allowed to vary and not only does the function $\mu(r_1, r_2; \nu)$ have the form

$$\mu(r_1, r_2; \nu) =: m(r_1 - r_2; \nu) =: m(r'; \nu), \quad r' := r_1 - r_2 \qquad (78.99)$$

but also $m(r'; \nu)$ is appreciably different from zero only within a domain that is much smaller than the source area, of maximum width l, say. It is assumed that $I(r, \nu)$ vanishes outside the source area and varies slowly enough to be regarded as constant over distances of the order of l. It can be argued that many real sources satisfy these requirements, and much subsequent coherence theory is specifically applied to this quasihomogeneous case.

In (78.98), $\{I(\boldsymbol{r}_1,\nu)I(\boldsymbol{r}_2,\nu)\}^{1/2}$ can be replaced by $I\left(\frac{\boldsymbol{r}_1+\boldsymbol{r}_2}{2},\nu\right)$ with little error and *for quasihomogeneous sources*, therefore, we have

$$W(\boldsymbol{r}_1,\boldsymbol{r}_2;\nu) = I\left(\frac{\boldsymbol{r}_1+\boldsymbol{r}_2}{2},\nu\right)m(\boldsymbol{r}_1-\boldsymbol{r}_2;\nu) \qquad (78.100)$$

The function I is said to be a *slow* function of its spatial argument while m is said to be a *fast* function of $\boldsymbol{r}'$. Edge effects, where the requirements on I and m will not be exactly satisfied, are assumed to be negligible.

For this source model, simpler expressions can be derived for the various coherence functions. We list these here, referring to the paper of Carter and Wolf (1977) for proofs.

The cross-spectral density in the far zone becomes

$$W(R_1\boldsymbol{s}_1,R_2\boldsymbol{s}_2;\nu) \sim \{I^{(\infty)}(R_1\boldsymbol{s}_1,\nu)I^\infty(R_2\boldsymbol{s}_2,\nu)\}^{\frac{1}{2}}m^\infty(R_1\boldsymbol{s}_1,R_2\boldsymbol{s}_2;\nu) \qquad (78.101)$$

in which the intensity $I^{(\infty)}$ is given by

$$I^{(\infty)}(R\boldsymbol{s},\nu) = \widetilde{I}^{(0)}\widetilde{m}^{(0)}(\boldsymbol{s}/\lambda;\nu)\frac{\cos^2\theta}{R^2\lambda^2} \qquad (78.102)$$

and the complex degree of spatial coherence by

$$m^{(\infty)}(R_1\boldsymbol{s}_1,R_2\boldsymbol{s}_2;\nu) \approx \frac{\widetilde{I}^{(0)}\left(\frac{\boldsymbol{s}_1-\boldsymbol{s}_2}{\lambda},\nu\right)}{\widetilde{I}^{(0)}(0,\nu)}\exp\{ik(R_1-R_2)\} \qquad (78.103)$$

The term $\widetilde{m}^{(0)}\left(\frac{\boldsymbol{s}_1+\boldsymbol{s}_2}{2\lambda};\nu\right)/\{\widetilde{m}^{(0)}(\boldsymbol{s}_1/\lambda;\nu)\widetilde{m}^{(0)}(\boldsymbol{s}_2/\lambda;\nu)\}^{1/2}$ has been set equal to unity here. The transforms $\widetilde{I}$ and $\widetilde{m}$ are defined by

$$\widetilde{I}^{(0)}(\boldsymbol{p},\nu) := \int I^{(0)}(\boldsymbol{r};\nu)\exp(-2\pi i\boldsymbol{p}\cdot\boldsymbol{r})\,d\boldsymbol{r}$$

$$\qquad (78.104)$$

$$\widetilde{m}^{(0)}(\boldsymbol{p};\nu) := \int m^{(0)}(\boldsymbol{r}';\nu)\exp(-2\pi i\boldsymbol{p}\cdot\boldsymbol{r}')\,d\boldsymbol{r}'$$

The brightness, radiant intensity and radiant emittance may now be obtained.

$$b(\boldsymbol{r},\boldsymbol{s};\nu) = \frac{\cos\theta}{\lambda^2}\int W(\boldsymbol{r}+\boldsymbol{r}'/2,\boldsymbol{r}-\boldsymbol{r}'/2;\nu)\exp\left(-\frac{2\pi i}{\lambda}\boldsymbol{r}'\cdot\boldsymbol{s}\right)d\boldsymbol{r}' \qquad (78.105)$$

But

$$W^{(0)}(\boldsymbol{\rho}_1, \boldsymbol{\rho}_2; \nu) = I\left(\frac{\boldsymbol{\rho}_1 + \boldsymbol{\rho}_2}{2}, \nu\right) m(\boldsymbol{\rho}_1 - \boldsymbol{\rho}_2; \nu) \tag{78.106}$$

or

$$W^{(0)}\left(\boldsymbol{\rho} + \frac{1}{2}\boldsymbol{\rho}', \boldsymbol{\rho} - \frac{1}{2}\boldsymbol{\rho}'; \nu\right) = I^{(0)}(\boldsymbol{\rho}, \nu) m^{(0)}(\boldsymbol{\rho}'; \nu) \tag{78.107}$$

and so

$$b(\boldsymbol{\rho}, \boldsymbol{s}; \nu) = \frac{\cos\theta}{\lambda^2} \int I^{(0)}(\boldsymbol{\rho}, \nu) m^{(0)}(\boldsymbol{\rho}'; \nu) \exp\left(-\frac{2\pi i \boldsymbol{\rho}' \cdot \boldsymbol{s}}{\lambda}\right) d\boldsymbol{\rho}'$$

$$= \frac{\cos\theta}{\lambda^2} I^{(0)}(\boldsymbol{\rho}, \nu) \widetilde{m}^{(0)}(\boldsymbol{s}/\lambda; \nu) \tag{78.108}$$

Thus for quasihomogeneous sources, the brightness defined by (78.105) separates into two factors, one a function of position only, the other a function of direction only. Furthermore, *the brightness is now always non-negative*: I can never be negative and it is easy to see that the same is true of $\widetilde{m}$. The latter is the Fourier transform of a non-negative definite function and is hence likewise non-negative for all values of its argument (Bochner's theorem):

$$b(\boldsymbol{\rho}, \boldsymbol{s}; \nu) \geq 0 \tag{78.109}$$

The radiant intensity $J(\boldsymbol{s}, \nu)$ is given by

$$J(\boldsymbol{s}, \nu) = r^2 I^\infty(r\boldsymbol{s}, \nu) \quad kr \to \infty$$

$$= \cos\theta \int b(\boldsymbol{r}, \boldsymbol{s}; \nu) \, d\boldsymbol{r}$$

$$= \frac{\cos^2\theta}{\lambda^2} \widetilde{m}^{(0)}(\boldsymbol{s}/\lambda; \nu) \int I^{(0)}(\boldsymbol{r}, \nu) \, d\boldsymbol{r}$$

$$= \frac{\cos^2\theta}{\lambda^2} \widetilde{m}^{(0)}(\boldsymbol{s}/\lambda; \nu) I^{(0)}(0, \nu) \tag{78.110}$$

The radiant emittance is obtained by integration over solid angle,

$$e(\boldsymbol{r}, \nu) = \int b(\boldsymbol{r}, \boldsymbol{s}; \nu) \cos\theta \, d\Omega =: I^{(0)}(\boldsymbol{r}) C_g \tag{78.111a}$$

where

$$C_g = \frac{1}{\lambda^2} \int_{2\pi} \widetilde{m}^{(0)}(\boldsymbol{s}/\lambda; \nu) \cos^2\theta \, d\Omega$$

$$= \left(\frac{\pi}{2}\right)^{\frac{1}{2}} \frac{1}{\lambda^2} \int m^{(0)}(\boldsymbol{r}'; \nu) \frac{J_{3/2}(kr')}{(kr')^{3/2}} \, d\boldsymbol{r}' \tag{78.111b}$$

Finally, the total radiant flux is obtained by further integration, either over solid angle or over the source area:

$$I = \int_{z=0} e(\boldsymbol{r}, \nu)\, d\boldsymbol{r} = \int_{2\pi} J(\boldsymbol{s}, \nu)\, d\Omega$$

$$= C_g \tilde{I}^{(0)}(0, \nu) \tag{78.112}$$

A straightforward calculation (Wolf and Carter, 1984) shows that the angular correlation function $\mathcal{A}(\boldsymbol{s}_1, \boldsymbol{s}_2; \nu)$ defined by (78.25) is given by

$$\mathcal{A}(\boldsymbol{s}_1, \boldsymbol{s}_2; \nu) = \frac{1}{\lambda^4} \tilde{I}^{(0)}\left(\frac{\boldsymbol{s}_1 - \boldsymbol{s}_2}{\lambda}, \nu\right) \tilde{m}^{(0)}\left(\frac{\boldsymbol{s}_1 + \boldsymbol{s}_2}{2\lambda}; \nu\right) \tag{78.113}$$

The cross-spectral density at arbitrary points $\boldsymbol{r}_1, \boldsymbol{r}_2$ is then given by

$$W(\boldsymbol{r}_1, \boldsymbol{r}_2; \nu) = \frac{1}{\lambda^4} \int \tilde{I}^{(0)}\left(\frac{\boldsymbol{s}_1 - \boldsymbol{s}_2}{\lambda}, \nu\right) \tilde{m}^{(0)}\left(\frac{\boldsymbol{s}_1 + \boldsymbol{s}_2}{2\lambda}; \nu\right)$$

$$\times \exp\left\{\frac{2\pi i}{\lambda}(\boldsymbol{s}_1 \cdot \boldsymbol{r}_1 - \boldsymbol{s}_2^* \cdot \boldsymbol{r}_2)\right\} d\boldsymbol{s}_1\, d\boldsymbol{s}_2 \tag{78.114}$$

and so

$$\widehat{W}(\boldsymbol{s}_1/\lambda, z_1, \boldsymbol{s}_2/\lambda, z_2; \nu)$$

$$= \tilde{I}^{(0)}\left(\frac{\boldsymbol{s}_1 + \boldsymbol{s}_2}{\lambda}, \nu\right) \tilde{m}^{(0)}\left(\frac{\boldsymbol{s}_1 - \boldsymbol{s}_2}{2\lambda}; \nu\right) \exp\left\{\frac{2\pi i}{\lambda}(\sigma_1 z_1 - \sigma_2 z_2)\right\} \tag{78.115}$$

This suggests that the intensity and degree of spectral coherence of the source can be determined by measuring the cross-spectral density far from the source (Carter and Wolf, 1985). We refer to their paper for details and for extended study of the particular case of sources for which both $I^{(0)}(\boldsymbol{r}, \nu)$ and $m^{(0)}(\boldsymbol{r}'; \nu)$ are Gaussian.

The spectroscopy of partially coherent sources is investigated much more fully in a paper by Wolf (1992b), which is the culmination of numerous studies on the propagation of spectra in free space, notably Wolf (1986b, 1987a,b,c), Gamliel and Wolf (1988) and Foley (1990). We shall not describe this in detail but just repeat the conclusions, which may well be relevant for studies on the extremely small sources of the future even if they do not seem to be of direct interest for the traditional types of source. For a planar, uniform, secondary, quasihomogeneous source, we know that the far-zone spectrum is given by

$$S^{(\infty)}(r\boldsymbol{s}, \nu) = \left(\frac{\sigma}{r\lambda}\right)^2 \tilde{I}^{(0)}(0) s^{(0)}(\nu) \tilde{m}^{(0)}(\boldsymbol{s}/\lambda; \nu) \tag{78.116}$$

where the normalized source spectrum $s^{(0)}(\nu)$ is just

$$s^{(0)}(\nu) = \frac{S^{(0)}(\nu)}{\int S^{(0)}(\nu)\,d\nu} \tag{78.117}$$

Wolf shows that the normalized spectrum $s^{(0)}(\nu)$ and the degree of spectral coherence $m^{(0)}(\boldsymbol{\rho'};\nu)$ of the source can be determined from a knowledge of the far-zone spectrum $S^{(\infty)}(r\boldsymbol{s},\nu)$ over the large hemisphere centred on the origin in the source region. The reconstructed expressions are given by

$$s^{(0)}(\nu) = \frac{\int(1/\sigma)S^{\infty}(r\boldsymbol{s},\nu)\,d\boldsymbol{s}}{\iint(1/\sigma)S^{\infty}(r\boldsymbol{s},\nu)\,d\boldsymbol{s}\,d\nu}$$

$$m^{(0)}(\boldsymbol{\rho'};\nu) = \frac{\int(1/\sigma)S^{\infty}(r\boldsymbol{s},\nu)\exp(\mathrm{i}k\boldsymbol{s}\cdot\boldsymbol{\rho'})\,d\boldsymbol{s}}{\int(1/\sigma)S^{\infty}(r\boldsymbol{s},\nu)\,d\boldsymbol{s}} \tag{78.118}$$

If thenormalized spectrum is independent of the direction of $\boldsymbol{s}$ in the far-zone, then

$$s^{(\infty)}(r\boldsymbol{s},\nu) := \frac{S^{(\infty)}(r\boldsymbol{s},\nu)}{\int S^{(\infty)}(r\boldsymbol{s},\nu)\,d\nu} \tag{78.119}$$

collapses to a function of ν only, $s^{(\infty)}(\nu)$ and the reconstructed spectrum $S^{(0)}(\nu)$ is found to be equal to $s^{(\infty)}(\nu)$. The reconstructed degree of spectral coherence is then a function of $k\rho' = k(\boldsymbol{\rho_2} - \boldsymbol{\rho_1})$ only ($k = 2\pi/\lambda$ as usual) and is said to obey the *scaling law*(introduced in Wolf, 1986).

Quasihomogeneous sources form a special case of a more general source model, the Schell model. In this, the degree of spectral coherence $\mu(\boldsymbol{r}_1,\boldsymbol{r}_2;\nu)$ is again of the form $m(\boldsymbol{r}_1 - \boldsymbol{r}_2;\nu)$ but the relative variation of the intensity and degree of spectral coherence with their arguments does not necessarily satisfy the conditions discussed at the beginning of this section. They are of considerable interest for light sources and exhibit a property that is of importance here: the set of eigenfunctions $\{\phi_k(\boldsymbol{r},\nu)\}$ of (78.73) associated with non-zero eigenvalues of any finite Schell-model source is complete in a Hilbert space of square-integrable functions (Wolf, 1984). Since quasihomogeneous sources are special cases of Schell sources, the same is true of them also.

We shall not discuss Schell-model sources further here. See Wolf (1984) for references to early papers, including those of Schell (1961, 1967).

78.8 Brightness, coherence and quasihomogeneity

In this final passage in the saga that began with the realization that a radiometry suitable for partially coherent sources based on the traditional, essentially incoherent, theory contained unacceptable weaknesses, the definition of brightness adumbrated by Walther and studied in detail by Marchand and Wolf is reconsidered in the light of a result derived in Section 78.7 and an observation of Foley and Wolf (1985), pursued in greater detail in Foley and Wolf (1991). "In my beginning is my end...In my end is my beginning." The important result is that the brightness of quasihomogeneous sources is non-negative when that definition is adopted; the advance made by Foley and Wolf emerged from an examination of the short-wavelength limit, $k = 2\pi/\lambda \to \infty$.

We have seen that it is not possible to define a brightness function that depends linearly on the cross-spectral density and satisfies the physical requirements of radiometry when the source is partially coherent (Friberg, 1979, discussed in Section 78.4). Furthermore, some of the proposed definitions of brightness can become negative and the definitions of Martínez-Herrero and Mejías (1984) and Foley and Nieto-Vesperinas (1985) seem to have another weakness (Foley and Wolf, 1985). These problems were reconsidered by Foley and Wolf (1985) who showed that the formulae of traditional radiometry are obtained for quasihomogeneous sources when the brightness (78.48) is adopted in the short-wavelength limit.

It is convenient to set out from the complex form of the brightness, which we mentioned in Section 78.4, $\hat{b}(\boldsymbol{r}, \boldsymbol{s}; \nu)$. We know that the cross-spectral density can be written as an average over monochromatic wavefields (78.86), $U(\boldsymbol{r}, \nu) \exp(-2\pi i \nu t)$:

$$W(\boldsymbol{r}_1, \boldsymbol{r}_2; \nu) = \langle U^*(\boldsymbol{r}_1, \nu) U(\boldsymbol{r}_2, \nu) \rangle$$

and that the plane-wave expansion of U is

$$U(\boldsymbol{r}, \nu) = \int a(\boldsymbol{s}, \nu) \exp(ik\boldsymbol{r} \cdot \boldsymbol{s}) \, d\boldsymbol{s}$$

The complex brightness may then be written

$$\hat{b}(\boldsymbol{r}, \boldsymbol{s}; \nu) = \sigma \langle U^*(\boldsymbol{r}, \nu) a(\boldsymbol{s}, \nu) \rangle \exp(ik\boldsymbol{s} \cdot \boldsymbol{r})$$

This in turn may be expressed in terms of the complex brightness in the source plane by using the relation

$$U(\boldsymbol{r}, \nu) = \int G(\boldsymbol{R}, \nu) U^{(0)}(\boldsymbol{\rho}, \nu) \, d\boldsymbol{\rho}$$

$$\boldsymbol{R} := \boldsymbol{r} - \boldsymbol{\rho}, \quad G(\boldsymbol{r}, \nu) = -\frac{1}{2\pi} \frac{\partial}{\partial z} \left\{ \frac{\exp(ikR)}{R} \right\}$$

(78.120)

whereupon we find

$$\hat{b}(\boldsymbol{r}, \boldsymbol{s}; \nu) = \exp(\mathrm{i}k\boldsymbol{s} \cdot \boldsymbol{r}) \int G^*(\boldsymbol{r}, \nu)\hat{b}^{(0)}(\boldsymbol{\rho}, \boldsymbol{s}; \nu) \exp(-\mathrm{i}k\boldsymbol{s} \cdot \boldsymbol{\rho}) \, d\boldsymbol{\rho} \quad (78.121)$$

The complex brightness in the source plane is given by

$$\hat{b}^{(0)}(\boldsymbol{\rho}, \boldsymbol{s}; \nu) = \sigma\langle U^{(0)*}(\boldsymbol{\rho}, \nu)a(\boldsymbol{s}, \nu)\rangle \exp(\mathrm{i}k\boldsymbol{s} \cdot \boldsymbol{\rho}) \quad (78.122)$$

For a quasihomogeneous source, the various approximations permit us to write

$$\hat{b}^{(0)}(\boldsymbol{\rho}, \boldsymbol{s}; \nu) \approx \frac{\sigma}{\lambda^2} I^{(0)}(\boldsymbol{\rho}, \nu)\widetilde{m}^{(0)}(\boldsymbol{s}/\lambda; \nu) \quad (78.123)$$

(cf. 78.100). Thus

$$\hat{b}(\boldsymbol{\rho}, \boldsymbol{s}; \nu) = \frac{\sigma}{\lambda^2}\widetilde{m}^{(0)}(\boldsymbol{s}/\lambda; \nu)C^*(\boldsymbol{r}, \boldsymbol{s}; \nu) \exp(\mathrm{i}k\boldsymbol{s} \cdot \boldsymbol{r}) \quad (78.124a)$$

with

$$C(\boldsymbol{r}, \boldsymbol{s}; \nu) = \int G(\boldsymbol{R}, \nu)I^{(0)}(\boldsymbol{\rho}, \nu)\exp(\mathrm{i}k\boldsymbol{s} \cdot \boldsymbol{\rho}) \, d\boldsymbol{\rho} \quad (78.124b)$$

A careful examination of the behaviour of C as the wavenumber k tends to infinity, given in detail in Foley and Wolf (1985, 1991), shows that when the point S_0, for which $\boldsymbol{\rho} = \boldsymbol{r}_t - (z/\sigma)\boldsymbol{s}$, $(\boldsymbol{r}_t = (x, y, 0))$, lies within the source,

$$C(\boldsymbol{r}, \boldsymbol{s}; \nu) \sim I^{(0)}(\boldsymbol{r}_t - (z/\sigma)\boldsymbol{s}, \nu)\exp(\mathrm{i}k\boldsymbol{s} \cdot \boldsymbol{r}) \quad (78.125)$$

When S_0 lies outside the source, $C \sim 0$. Thus

$$\hat{b}(\boldsymbol{r}, \boldsymbol{s}; \nu) \sim b(\boldsymbol{r}, \boldsymbol{s}; \nu) := \frac{\sigma}{\lambda^2} I^{(0)}(\boldsymbol{r}_t - z\boldsymbol{s}/\sigma, \nu)\widetilde{m}^{(0)}(\boldsymbol{s}/\lambda; \nu) \quad (78.126)$$

The rate at which energy passes through an element of area da per unit solid angle around the direction $\boldsymbol{s}$ is given by $b(\boldsymbol{r}, \boldsymbol{s}; \nu)\boldsymbol{s} \cdot \boldsymbol{n} \, da$ in which $\boldsymbol{n}$ is a unit vector normal to da. The radiant intensity through a plane z is then

$$j(\boldsymbol{s}, \nu) = \sigma \int_z b(\boldsymbol{r}, \boldsymbol{s}; \nu) \, d\boldsymbol{r}_t \quad (78.127)$$

and with the expression (78.126) for b, we find

$$j(\boldsymbol{s}, \nu) = \left(\frac{\sigma}{\lambda}\right)^2 \widetilde{I}^{(0)}(0, \nu)\widetilde{m}^{(0)}(\boldsymbol{s}/\lambda; \nu) \quad (78.128)$$

with

$$\widetilde{I}^{(0)}(0, \nu) = \int I^{(0)}(\rho, \nu) \, d\rho \tag{78.129}$$

Equation (78.128) agrees exactly with (78.110), when we recall that $\sigma = \cos \theta$.

We already know that $\widetilde{m}^{(0)}(s/\lambda; \nu)$ is non-negative and hence so is the brightness defined by (78.126). Furthermore, the latter vanishes outside the source area in the source plane and it can be shown easily that, in free space, $b(r, s; \nu)$ is constant along straight lines from a source point to a point at r, which defines the direction s.

When the brightness $\hat{b}(r, s; \nu)$ defined by (78.126) is substituted into the usual radiometric formulae, therefore, the correct results will be obtained for quasihomogeneous sources in the limit $k \to \infty$.

This result has been applied to several types of source of practical importance (Foley and Wolf, 1991): sources obeying Lambert's cosine law and Gaussian-correlated sources. We recapitulate their findings briefly.

(a) *Uniform Lambertian source*

The source is taken to be circular, with a radius a much greater than the wavelength, and uniform in the sense that the spectral density is the same for all points of the source:

$$S^{(0)}(\rho, \nu) = \begin{cases} S^{(0)}(\nu) & \rho \le a \\ 0 & \rho > a \end{cases} \tag{78.130}$$

For Lambertian sources,

$$m^{(0)}(\rho_2 - \rho_1; \nu) = \frac{\sin(k|\rho_2 - \rho_1|)}{k|\rho_2 - \rho_1|} \tag{78.131}$$

and

$$\widetilde{m}^{(0)}(s/\lambda; \nu) = \frac{\lambda^2}{\sigma} \tag{78.132}$$

and so

$$b^{(0)}(\rho, s; \nu) = \begin{cases} S^{(0)}(\nu)/2\pi & \text{for } \rho \le a \\ 0 & \text{for } \rho > a \end{cases} \tag{78.133a}$$

Hence

$$b(r, s; \nu) = \begin{cases} S^{(0)}(\nu)/2\pi & \text{for } s \in \Omega \\ 0 & \text{for } s \notin \Omega \end{cases} \tag{78.133b}$$

in which Ω is the solid angle generated by the lines connecting the edge of the source to the point r.

(b) *Lambertian source with Gaussian intensity profile*

The spectral density is now defined by

$$S^{(0)}(\boldsymbol{\rho}, \nu) = S^{(0)}(\nu) \exp\left(-\frac{\rho^2}{2\sigma_0^2(\nu)}\right) \text{circ}\left(\frac{\boldsymbol{\rho}}{a}\right) \tag{78.134}$$

where

$$\text{circ}(\boldsymbol{r}) = \begin{cases} 1 & |r| \leq 1 \\ 0 & |r| > 1 \end{cases}$$

and now

$$b^{(0)}(\boldsymbol{\rho}, \boldsymbol{s}; \nu) = S^{(0)}(\nu) \exp\{-\rho^2/2\sigma_0^2(\nu)\} \text{ circ}(\boldsymbol{\rho}/a)$$

$$b(\boldsymbol{r}, \boldsymbol{s}; \nu) = \begin{cases} S^{(0)}(\nu) \exp\left\{-|\boldsymbol{r} - z\boldsymbol{s}/\sigma|^2/2\sigma_0^2(\nu)\right\} & \boldsymbol{s} \in \Omega \\ 0 & \boldsymbol{s} \notin \Omega \end{cases} \tag{78.135}$$

(c) *Uniform Gaussian-correlated source*

The function $m^{(0)}$ is now Gaussian:

$$m^{(0)}(\boldsymbol{\rho}'; \nu) = \exp\{-\rho'^2/2\sigma_0^2(\nu)\} \tag{78.136}$$

but $S^{(0)}$ is again constant (uniform intensity):

$$S^{(0)}(\boldsymbol{\rho}, \nu) = S^{(0)}(\nu)\text{circ}(\boldsymbol{\rho}/a) \tag{78.137}$$

Foley and Wolf (1991) find

$$b^{(0)}(\boldsymbol{\rho}, \boldsymbol{s}; \nu) = A(\nu)\sigma \exp\left\{-\frac{1}{2}k^2\sigma_0^2(\nu)s^2\right\} \text{ circ}(\boldsymbol{\rho}/a)$$

$$b(\boldsymbol{r}, \boldsymbol{s}; \nu) = \begin{cases} A(\nu)\sigma \exp\left\{-\frac{1}{2}k^2\sigma_0^2(\nu)s^2\right\} & \boldsymbol{s} \in \Omega \\ 0 & \boldsymbol{s} \notin \Omega \end{cases} \tag{78.138}$$

where

$$A(\nu) = \frac{\{k\sigma_0(\nu)\}^2}{2\pi} S^{(0)}(\nu) \tag{78.139}$$

(d) *Gaussian-correlated source with Gaussian intensity*

$$S^{(0)}(\boldsymbol{\rho}, \nu) = S^{(0)}(\nu) \exp\left\{-\frac{\rho^2}{2\sigma_s^2(\nu)}\right\} \text{circ}(\boldsymbol{\rho}/a)$$

$$m^{(0)}(\boldsymbol{\rho}'; \nu) = \exp\left\{-\frac{\rho'^2}{2\sigma_g^2(\nu)}\right\} \tag{78.140}$$

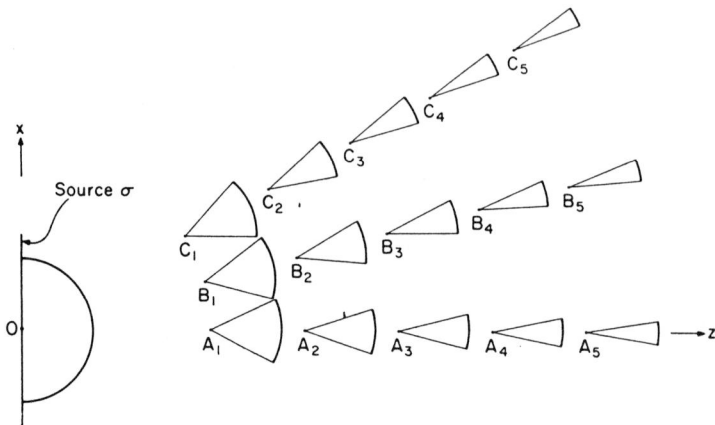

Fig. 78.4: Polar diagram of the brightness produced by a uniform circular Lambertian source.

which lead to

$$
b^{(0)}(\boldsymbol{\rho}, \boldsymbol{s}; \nu) = A(\nu)\sigma \exp\left\{-\frac{\rho^2}{2\sigma_s^2(\nu)}\right\} \exp\left(-\frac{k^2\sigma_s^2 s^2}{2}\right) \operatorname{circ}\left(\frac{\boldsymbol{\rho}}{a}\right)
$$

$$
b(\boldsymbol{r}, \boldsymbol{s}; \nu) = \begin{cases} A(\nu)\sigma \exp\left\{-|r_t - z\boldsymbol{s}/\sigma|^2/2\sigma_s^2(\nu)\right\} \\ \quad \times \exp\left\{-\frac{1}{2}k^2\sigma_g^2(\nu)\right\} & \boldsymbol{s} \in \Omega \\ 0 & \text{otherwise} \end{cases}
$$

$$(78.141)$$

The polar diagrams for each of these types of source give a helpful physical picture of the way in which their brightness varies with distance from the source in different directions. Figures 78.4–7 show these polar diagrams for the four cases discussed above.

78.9 Temporal and spatial coherence

We have noticed on several occasions that the temporal and spatial coherence properties of sources can rarely be regarded completely separately, though a separate treatment of each may well be satisfactory in practice. In Chapter 66, for example, we saw that the effect of modest departures from perfect coherence (point source, monochromatic emission) can usually be represented to an excellent approximation by two separate envelope functions, one describing spatial partial coherence, the other its temporal

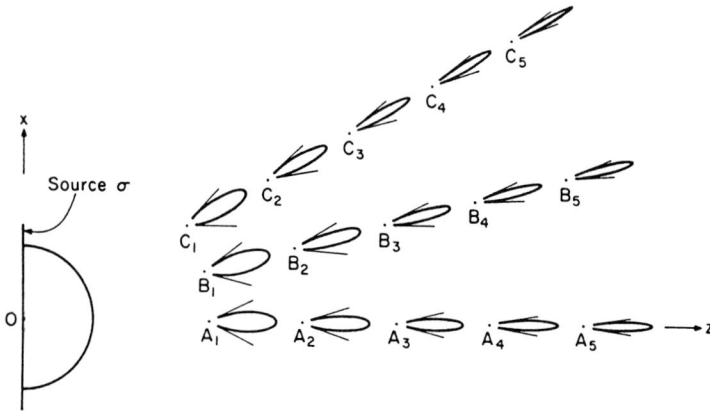

Fig. 78.5: Polar diagram of the brightness produced by a Gaussian Lambertian source.

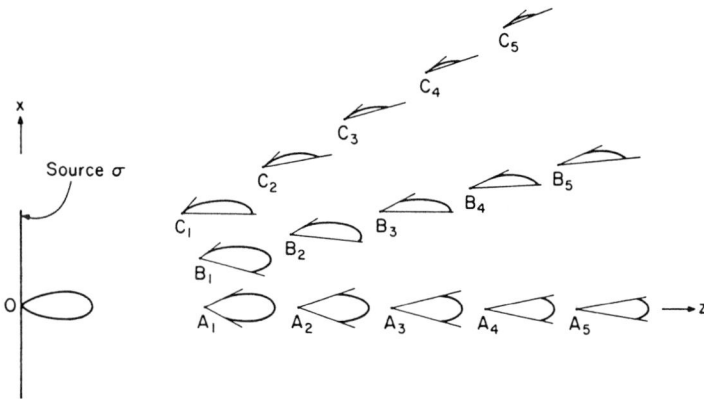

Fig. 78.6: Polar diagram of the brightness produced by a uniform Gaussian-correlated source.

counterpart. It therefore comes as a surprise* to learn that the two are strictly speaking related and that a knowledge of one gives considerable information about the other.

The first suggestion that such a relation exists is to be found in a "remarkable but little known paper" (Wolf *et al.*, 1981) by Sudarshan (1969).

---

* In their paper of 1981, Wolf and Devaney wrote ". . . Sudarshan made a number of predictions regarding some previously unsuspected relationships between spatial coherence properties and temporal coherence properties of fluctuating free fields."

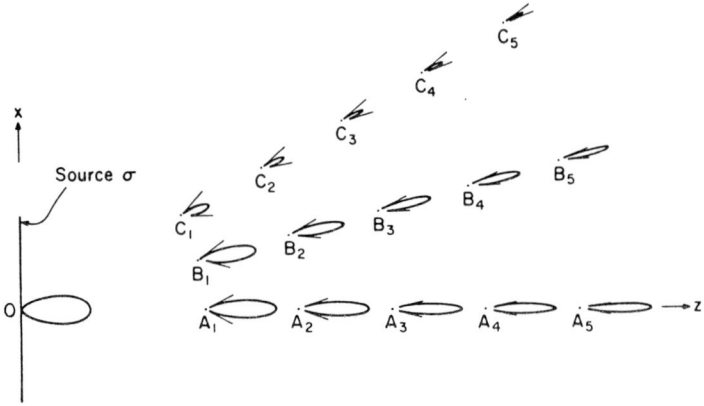

Fig. 78.7: Polar diagram of the brightness produced by a Gaussian, Gaussian-correlated source.

The importance of this was recognized by E. Wolf, A.J. Devaney and J.T. Foley who presented their first analysis at the ICO Conference in 1980 (Wolf et al., 1981) and a more detailed study soon followed (Wolf and Devaney, 1981).

We set out from the plane-wave expansion for the wavefunction, which we here denote U rather than Ψ since the original discussion was concerned with waves satisfying the wave equation $\nabla^2 U = (1/c^2)\partial^2 U/\partial t^2$. Nevertheless, there seems to be no reason why the theory should be fundamentally different for solutions of Schrödinger's equation; this has not yet been studied, however. Thus

$$U(\boldsymbol{r},t) = \int a(\boldsymbol{k}) \exp\{\mathrm{i}(\boldsymbol{k}\cdot\boldsymbol{r} - kct)\}\,d\boldsymbol{k} \qquad (78.142)$$

and hence

$$\frac{\partial U}{\partial t} = -\mathrm{i}c \int ka(\boldsymbol{k}) \exp\{\mathrm{i}(\boldsymbol{k}\cdot\boldsymbol{r} - kct)\}\,d\boldsymbol{k} \qquad (78.143)$$

We now define an operator D as follows:

$$\mathrm{D}\hat{f}(\boldsymbol{r}) := \int k f(\boldsymbol{k}) \exp(\mathrm{i}\boldsymbol{k}\cdot\boldsymbol{r})\,d\boldsymbol{k} \qquad (78.144)$$

in which

$$\hat{f}(\boldsymbol{r}) = \mathcal{F}(f(\boldsymbol{k}))$$

Then (78.143) may be written

$$DU = (i/c)\partial U/\partial t \qquad (78.145)$$

We now consider the ensemble average of U, which will of course reduce to the mutual coherence function when we consider stationary fields:

$$\Gamma(\boldsymbol{r}_1, t_1; \boldsymbol{r}_2, t_2) := \langle U^*(\boldsymbol{r}_1, t_1) U(\boldsymbol{r}_2, t_2) \rangle$$

Then

$$\begin{aligned} D_1\Gamma &= -(i/c)\partial\Gamma/\partial t_1 \\ D_2\Gamma &= (i/c)\partial\Gamma/\partial t_2 \end{aligned} \qquad (78.146)$$

where the suffix 1, 2 attached to D implies that the operator is to be applied to $\boldsymbol{r}_1$, $\boldsymbol{r}_2$ respectively. These are *Sudarshan's equations* in their general form. If we specialize to the stationary case in which the time dependence of Γ is of the form $t_2 - t_1 =: \tau$, they reduce to

$$\begin{aligned} D_1\Gamma(\boldsymbol{r}_1, \boldsymbol{r}_2; \tau) &= (i/c)\partial\Gamma(\boldsymbol{r}_1, \boldsymbol{r}_2; \tau)\partial\tau \\ D_2\Gamma(\boldsymbol{r}_1, \boldsymbol{r}_2; \tau) &= (i/c)\partial\Gamma(\boldsymbol{r}_1, \boldsymbol{r}_2; \tau)\partial\tau \end{aligned} \qquad (78.147)$$

After a certain amount of analysis, given explicitly by Wolf *et al.* (1981), it is found that

$$\Gamma(\boldsymbol{r}_1, \boldsymbol{r}_2; \tau) = \int \Gamma(\boldsymbol{r}_1, \boldsymbol{r}_2'; \tau') G(\boldsymbol{r}_2 - \boldsymbol{r}_2'; \tau - \tau')\, d\boldsymbol{r}_2' \qquad (78.148)$$

in which $G(\boldsymbol{r}, \tau)$ is defined by

$$G(\boldsymbol{r}, \tau) := -\frac{1}{2\pi r}\frac{\partial}{\partial r}\left\{ \delta^{(+)}(r - c\tau) + \delta^{(-)}(r + c\tau) \right\} \qquad (78.149)$$

and

$$\delta^{(+)}(x) = \frac{1}{2\pi}\int_0^\infty \exp(ikx)\, dk\ , \quad \delta^{(-)}(x) = \frac{1}{2\pi}\int_0^\infty \exp(-ikx)\, dk \quad (78.150a)$$

or

$$\delta^{(\pm)}(x) = \pm\frac{i}{2\pi}\mathcal{P}(1/x) + \frac{1}{2}\delta(x) \qquad (78.150b)$$

and $\mathcal{P}$ denotes the Cauchy principal value.

On setting $\tau' = 0$ in (78.148), Γ reduces to the mutual intensity $J(r_1, r_2)$ and we find that

$$\Gamma(r_1, r_2; \tau) = \int J(r_1, r_2) G(r_2 - r_2'; \tau) \, dr_2' \qquad (78.151)$$

A more interesting result is obtained by bringing the points r_1 and r_2 together, $r_1 = r_2 =: r$. Then

$$\Gamma(r, r; \tau) = \int J(r, r') G(r - r'; \tau) \, dr' \qquad (78.152)$$

If we identify $\Gamma(r, r; \tau)$ with the temporal coherence and $J(r, r')$ with the spatial coherence, we see that a knowledge of J should enable us to establish $\Gamma(r, r; \tau)$.

The cross-spectral density can now be expressed in terms of J and hence so too can the ordinary spectrum of the radiation, given by $W(r, r; \nu)$, as is shown by Wolf and Devaney (1981). We have

$$W(r_1, r_2; \nu) = \int J(r_1, r_2') \tilde{G}(r_2 - r_2'; \nu) \, dr_2'$$

in which

$$\tilde{G}(r; \nu) = \int_{-\infty}^{\infty} G(r, \tau) \exp(2\pi i \nu \tau) \, d\tau$$

$$= \frac{k^2}{\pi c} \frac{\sin kr}{kr}$$

and so*

$$W(r_1, r_2; \nu) = \frac{k^2}{2\pi^2 c} \int J(r_1, r_2') \frac{\sin k|r_2 - r_2'|}{k|r_2 - r_2'|} \, dr_2' \qquad (78.153)$$

The spectrum $S(r, \nu)$ is finally obtained by setting $r_1 = r_2$ in W, giving

$$S(r, \nu) = W(r, r; \nu) = \frac{k^2}{2\pi^2 c} \int J(r, r') \frac{\sin k|r - r'|}{k|r - r'|} \, dr' \qquad (78.154)$$

---

* A form of this expression that is symmetric in r_1 and r_2 can be obtained by using the relation $W(r_1, r_2; \nu) = W^*(r_2, r_1; \nu)$.

This equation expresses the spectrum of the radiation at some point in terms of the mutual intensity. Some consequences of this revolutionary relation are examined in Wolf *et al.* (1983).

78.10 Related work

For ease of presentation, we have confined the foregoing account to studies expressed in the traditional language of coherence and radiometry and have even neglected some of the more peripheral features of the theory altogether. One way of expressing some aspects at least of these recent developments employs *Wigner functions* and the various approaches often differ in little more than vocabulary and notation. In Section 78.10.2 we chart some of the main stages in the use of these functions. First, however, in Section 78.10.1, we draw attention to an important paper by Agarwal *et al.* (1987), which could have been incorporated in Section 78.7 but which is treated here separately as its implications for electron optics are still unclear.

78.10.1 Operator formalism

The starting point here is the familiar quantum mechanical practice of associating an operator with certain variables. In particular, we associate the operators $\hat{x}$ and $\hat{y}$ with the coordinates x and y, $\boldsymbol{\rho} = (x, y)$ in some source plane. The eigenstates of the operators are as usual the solutions of the equations

$$\hat{x}\,|x> \,=\, x\,|x> \qquad \hat{y}\,|y> \,=\, y\,|y> \qquad (78.155a)$$

with the Dirac bra and ket notation, and so

$$\hat{\boldsymbol{\rho}}\,|\boldsymbol{\rho}> \,=\, \boldsymbol{\rho}\,|\boldsymbol{\rho}> \qquad (78.155b)$$

We have written $\hat{\boldsymbol{\rho}} := (\hat{x}, \hat{y})$. We have already encountered the angular spectrum representation (78.27) of a (monochromatic) wavefield $U(\boldsymbol{r}, \nu)$:

$$U(\boldsymbol{\rho}, \nu) =: \int a(\boldsymbol{s}, \nu) \exp(\mathrm{i}k\boldsymbol{s} \cdot \boldsymbol{\rho})\,d\boldsymbol{s}$$

Operating on this with the transverse Laplacian, $\nabla_t := (\partial/\partial x, \partial/\partial y)$, we find that

$$-\mathrm{i}\lambda\nabla_t U(\boldsymbol{\rho}, \nu) = \int \boldsymbol{s}a(\boldsymbol{s}, \nu) \exp(\mathrm{i}k\boldsymbol{s} \cdot \boldsymbol{\rho})\,d\boldsymbol{s} \qquad (78.156)$$

with

$$\lambda := \lambda/2\pi = 1/k \tag{78.157}$$

The form of (78.156) suggests that the operator associated with s should be defined by

$$\hat{s} := -i\lambda\nabla_t \tag{78.158}$$

It is not difficult to show that the commutation relations between $\hat{s}$ and $\hat{\rho}$ are very similar to those between position and momentum:

$$[\hat{x}, \hat{s}_x] = i\lambda, \quad [\hat{y}, \hat{s}_y] = i\lambda \tag{78.159}$$

The brightness defined by (78.55) may be written

$$b(\rho, s; \nu) := \sigma F(\rho, s; \nu)/\lambda^2$$

$$F(\rho, s; \nu) := \int W(\rho + \rho'/2, \rho - \rho'/2; \nu) \exp(-iks \cdot \rho') d\rho' \tag{78.160}$$

and in $F(\rho, s; \nu)$ we recognize the 'Wigner representative' of an operator $\hat{G} := G(\hat{\rho}, \hat{s})$:

$$F(\rho, s; \nu) = \int <\rho + \rho'/2| \, G(\hat{\rho}, \hat{s}) \, |\rho - \rho'/2> \exp(-iks \cdot \rho') d\rho' \tag{78.161}$$

where

$$<\rho_1| \, G(\hat{\rho}, \hat{s}) \, |\rho_2>:= W(\rho_1, \rho_2; \nu) \tag{78.162}$$

This formalism was first developed by Agarwal *et al.* (1987), who stress the classical (non-quantum) basis of the theory, despite its formal resemblance to the phase-space presentation of quantum mechanics. It is this point that has led us to describe this approach separately, for the effect of the transition from electromagnetic waves to electrons on this observation has not yet been thoroughly investigated. Multiplication of the commutation relations by $\hbar k$ and use of $g = \hbar k s$ obviously renders them identical with the quantum mechanical expressions, as we should expect.

Agarwal *et al.* then demonstrate that a large class of generalized radiance or brightness functions can be created by expressing $\hat{G}$ in the different ways permitted by the commutation relations and recover a number of known results concerning the brightness functions considered by various authors. These different phase-space representatives F of $\hat{G}$ and hence the corresponding brightness functions "are associated with different rules of ordering of products involving the noncommuting operators $\hat{\rho}$ and $\hat{s}$." But in the limit as $\lambda \to 0$, the operators *do* commute and all the different brightnesses will become identical. Since we know that the brightness $b(\boldsymbol{\rho}, \boldsymbol{s}; \nu)$ of (78.55) possesses all the properties that are associated with a radiance function in traditional radiometry for quasihomogeneous sources, we can conclude that all linearly related brightness functions will collapse to the same form in the short-wavelength limit and that this form can be used safely in radiometry.

78.10.2 Use of Wigner and ambiguity functions

The need to find a way of conveying information about a signal and its spectrum or about an image and its spatial frequency spectrum has generated some work on the use of two representative signals, the Wigner distribution function and the ambiguity function. We have met the former briefly in the preceding section and the latter is essentially its Fourier transform. We define them here and point out why they are of interest since the reader is likely to encounter them in publications dealing with the subject of this chapter; they have also been mentioned briefly in connection with visual representation of the microscope transfer functions (Chapter 66). We limit this account to functions of a single variable but the extension to two or more dimensions is trivial.

The Wigner distribution function of a signal $\psi(x)$ is defined by

$$W(x,p) = \int \psi(x + x'/2)\psi^*(x - x'/2)\exp(2\pi i p x')\,dx' \qquad (78.163a)$$

or in terms of the Fourier transform of ψ:

$$W(x,p) = \int \tilde{\psi}(p + p'/2)\tilde{\psi}^*(p - p'/2)\exp(-2\pi i p' x)\,dp' \qquad (78.163b)$$

with

$$\psi(x) = \int \tilde{\psi}(p)\exp(-2\pi i p x)\,dx = \mathcal{F}^-(\tilde{\psi})$$

The ambiguity function is the Fourier transform of W with respect to both its arguments:

$$A(y,q) := \iint W(x,p)\exp\{2\pi i(xy - pq)\}\,dx\,dp \qquad (78.164)$$

and clearly

$$A(y,q) = \int \psi(x + q/2)\psi^*(x - q/2)\exp(2\pi i x y)\,dx \qquad (78.165a)$$

or again

$$A(y,q) = \int \tilde{\psi}(p + y/2)\tilde{\psi}^*(p - y/2)\exp(-2\pi ipq)\, dp \qquad (78.165b)$$

The close similarity in form between these functions and the cross-spectral density is obvious. For discussion of the use of these functions in optics, see Bastiaans (1978, 1979a-c) and in particular his surveys (Bastiaans, 1981a,b), Papoulis (1974), Guigay (1978), Bartelt et $al.$ (1980), Brenner et $al.$ (1983) and Brenner and Ojeda-Castañeda (1984). Their use in electron optics is considered by Castaño (1988, 1989), Castaño et $al.$ (1991) and Polo et $al.$ (1992).

78.10.3 Further reading

Among the voluminous literature on the themes of this chapter, we draw attention to general papers by Bastiaans (1977) and Hawkes (1994); to further studies of the quasihomogeneous case by Friberg (1981) and Pedersen (1982); to a note on the van Cittert–Zernike theorem in connection with Lambertian sources by Winthrop (1972); to a study of the Gaussian Schell-model case by Sudol and Friberg (1984); to later papers by Walther (1973; see Marchand and Wolf, 1974a,b and Walther 1974, 1978a,b); to complementary work by Wolf, Carter and others (Wolf and Carter, 1975, 1978; Carter and Wolf, 1975, 1978, 1981, 1985, 1987; Wolf, 1976, 1978a,b, 1984; Carter, 1978, 1988, 1990; Wolf and Collett, 1978; Collett and Wolf, 1979, 1980; Mandel and Wolf, 1981; Friberg and Wolf, 1983; Wolf et $al.$ 1983; Friberg 1986a,b; Foley et $al.$, 1986; Winston and Ning, 1986; Kim and Wolf, 1987; Cairns and Wolf, 1987; Gori et $al.$, 1988; Foley and Wolf, 1988; Foley et $al.$, 1988; Foley, 1991; James and Wolf, 1989; Winston and Welford, 1990, 1991).

Pedersen (1982) cites some Russian work roughly contemporary with Walther's first paper of 1968. For further details of this, see the review article of Apresyan and Kravtsov (1984).

The evolution of the various brightness functions along rays is examined by Littlejohn and Winston (1993), who also propose a new expression that is exactly conserved as in incoherent radiometry.

79

Instrumental Aspects of Coherence

79.1 Introduction

Coherence has been discussed phenomenologically in Chapter 66 and more formally in Chapter 78. We are now ready to harmonize these two approaches by considering the *propagation* of the mutual coherence and cross-spectral density in an optical instrument, in the transmission electron microscope in particular (Section 79.2). Some other practical aspects of coherence will then be examined briefly, notably illuminating systems and the role of coherence in STEM imagery.

79.2 The propagation of coherence functions

79.2.1 Propagation of the mutual intensity in free space
A straightforward calculation shows that the mutual intensity $\Gamma_{12}(0)$ between two points P_1, P_2 in an arbitrary plane distant R from a plane source S (Fig. 79.1) is given by

$$\Gamma_{12}(0) = \int I(\rho)\frac{\exp\{ik(R_1 - R_2)\}}{R_1 R_2}\, d\rho \qquad (79.1)$$

and as usual

$$\gamma_{12}(0) = \frac{\Gamma_{12}(0)}{\{\Gamma_{11}(0)\Gamma_{22}(0)\}^{\frac{1}{2}}}$$

This is the basic form of the van Cittert–Zernike theorem, already encountered in Section 78.5. It can be seen immediately that the complex degree of coherence in the recording plane "is the same as the normalized complex amplitude in a certain diffraction pattern centred on P_2. This pattern would be obtained on replacing the source by a diffracting aperture of the same size and shape as the source and on filling it with a spherical wave converging on P_2, the amplitude distribution over the wavefront in the aperture being proportional to the intensity distribution across the source" (Born and Wolf, 1959, 1980, Section 10.4.2). For small sources, and points

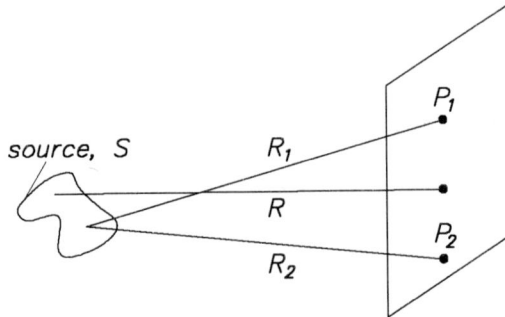

Fig. 79.1: Notation relating to the van Cittert–Zernike theorem.

that are close together relative to R_1 or R_2, a familiar calculation tells us that

$$\gamma_{12}(0) = \frac{\int I(\rho)\exp(-ik\boldsymbol{s}\cdot\rho)\,d\rho}{\int I(\rho)\,d\rho}\exp(i\psi) \qquad (79.2)$$

where $\boldsymbol{s} = (\boldsymbol{r}_1 - \boldsymbol{r}_2)/R$ and $R \approx R_1 \approx R_2$. The angle ψ is a measure of the phase difference between waves arriving at P_1 and P_2 from the centre of the source. For a uniform circular source, for which $I(\rho) = \text{const}$, $|\rho| \leq a$ and zero otherwise, we have

$$\gamma_{12}(0) = \frac{2J_1(\theta)\exp(i\psi)}{\theta}, \qquad \theta := \frac{2\pi}{\lambda}\frac{a}{R}|\boldsymbol{r}_1 - \boldsymbol{r}_2| \qquad (79.3)$$

Thus $|\gamma_{12}(0)| = 1$ at $\theta = 0$, when P_1 and P_2 coincide. At $\theta = 1$, we find $|\gamma_{12}(0)| = 0.88$ and this value is regarded as defining a convenient estimate of the coherently illuminated area. The distance d_{12} between P_1 and P_2 is then given by

$$d_{12} := |\boldsymbol{r}_1 - \boldsymbol{r}_2| = \frac{0.16R\lambda}{a} \qquad (79.4)$$

For $\lambda = 4$ pm, for example, we see that $d_{12} = 100$ nm for $a/R \approx 6$ μrad.

79.2.2 Propagation of the mutual intensity through a lens system

We now set out from the expression given in Chapter 65 for the wavefunction $\psi_i(\boldsymbol{u}_i)$ in an image plane in terms of the corresponding function in the conjugate object plane, $\psi_o(\boldsymbol{u}_o)$:

$$\psi_i(\boldsymbol{u}_i) = \iint G(\boldsymbol{u}_i - \boldsymbol{u}_o)\psi_o(\boldsymbol{u}_o)\,d\boldsymbol{u}_o$$

(65.13), in which, we recall, numerous scaling operations are implicit. Hence

$$\Gamma(\boldsymbol{u}_{i1}, \boldsymbol{u}_{i2}; \nu) = \langle \psi_i(\boldsymbol{u}_{i1}; \nu) \psi_i^*(\boldsymbol{u}_{i2}; \nu) \rangle$$

$$= \iint G(\boldsymbol{u}_{i1} - \boldsymbol{u}_{o1}; \nu) G^*(\boldsymbol{u}_{i2} - \boldsymbol{u}_{o2}; \nu) \langle \psi_o(\boldsymbol{u}_{o1}; \nu) \psi_o(\boldsymbol{u}_{o2}; \nu) \rangle \, d\boldsymbol{u}_{o1} \, d\boldsymbol{u}_{o2}$$

$$= \iint \Gamma(\boldsymbol{u}_{o1}, \boldsymbol{u}_{o2}; \nu) G(\boldsymbol{u}_{i1} - \boldsymbol{u}_{o1}; \nu) G^*(\boldsymbol{u}_{i2} - \boldsymbol{u}_{o2}; \nu) \, d\boldsymbol{u}_{o1} \, d\boldsymbol{u}_{o2}$$

(79.5)

This has the form of a double convolution and the Fourier transforms of the various members are hence related by a direct product. With

$$\tilde{\Gamma}(\boldsymbol{q}_1, \boldsymbol{q}_2; \nu) = \int \Gamma(\boldsymbol{u}_1, \boldsymbol{u}_2; \nu) \exp\{2\pi i(\boldsymbol{q}_1 \cdot \boldsymbol{u}_1 + \boldsymbol{q}_2 \boldsymbol{u}_2)\} \, d\boldsymbol{u}_1 \, d\boldsymbol{u}_2 \qquad (79.6)$$

we have

$$\tilde{\Gamma}_i(\boldsymbol{q}_1, \boldsymbol{q}_2; \nu) = \tilde{\Gamma}_o(\boldsymbol{q}_1, \boldsymbol{q}_2; \nu) \tilde{g}(\boldsymbol{q}_1, \boldsymbol{q}_2; \nu) \qquad (79.7)$$

where

$$\tilde{g}(\boldsymbol{q}_1, \boldsymbol{q}_2; \nu) := \int G(\boldsymbol{u}_1; \nu) G^*(\boldsymbol{u}_2; \nu) \exp\{2\pi i(\boldsymbol{u}_1 \cdot \boldsymbol{q}_1 + \boldsymbol{u}_2 \cdot \boldsymbol{q}_2)\} \, d\boldsymbol{u}_1 \, d\boldsymbol{u}_2$$

(79.8)

The mutual intensity is transferred *linearly* from object to image and the filter that relates the two is *scalar*: each pair of spatial frequencies $(\boldsymbol{q}_1, \boldsymbol{q}_2)$ at the object contributes only to the same pair at the image—there is no intermixing or 'cross-talk.' The filter $\tilde{g}(\boldsymbol{q}_1, \boldsymbol{q}_2; \nu)$ may be written explicitly in terms of $T(\boldsymbol{q}) = \mathcal{F}^-(G)$ (65.15c) and hence in terms of $T_A(\boldsymbol{q})$ and $T_L(\boldsymbol{q}) = \exp\{-i\chi(\boldsymbol{q})\}$ (65.30). Neglecting the argument ν, we have

$$\tilde{g}(\boldsymbol{q}_1, \boldsymbol{q}_2) = T(\boldsymbol{q}_1) T^*(-\boldsymbol{q}_2)$$
$$= T_A(\boldsymbol{q}_1) T_A^*(-\boldsymbol{q}_2) \exp[-i\{\chi(\boldsymbol{q}_1) - \chi(\boldsymbol{q}_2)\}] \qquad (79.9)$$

The terms in T_A will thus *truncate* the spectrum wherever q_1^2 or q_2^2 exceeds the cutoff value $a^2/\lambda^2 f^2$, a denoting the radius of the objective aperture. The exponential term acts as a *spatial frequency filter*.

79.2.3 Propagation of the cross-spectral density and brightness through a lens system
The propagation of the cross-spectral density and of the various radiometric functions defined in terms of it has been investigated in detail in two papers by Hawkes (1977a,b). In the the first, general relations are established

valid for any system and, in the second, the corresponding expressions for isoplanatic systems are derived. This work is not reproduced here. We simply state the important result that in isoplanatic conditions, the optical system acts as a scalar filter on the spectrum of the brightness function $b(\boldsymbol{r}, \boldsymbol{s}; \nu)$. By the spectrum, we understand the Fourier transform with respect to the first argument, $\boldsymbol{r}$. The elements of the filter are determined by $T_A T_L$.

79.2.4 Introduction of a specimen

As usual, we assume that the specimen can be represented by a two-dimensional complex transmission function, $O(\boldsymbol{u}_o)$. If the incident mutual intensity (on the source side) is $\Gamma_s(\boldsymbol{u}_{o1}, \boldsymbol{u}_{o2})$, neglecting the frequency ν, the emergent mutual intensity will be given by As usual, we assume that the specimen can be represented by a two-dimensional complex transmission function, $O(\boldsymbol{u}_o)$. If the incident mutual intensity (on the source side) is $\Gamma_s(\boldsymbol{u}_{o1}, \boldsymbol{u}_{o2})$, neglecting the frequency ν, the emergent mutual intensity will be given by

$$\Gamma_o(\boldsymbol{u}_{o1}, \boldsymbol{u}_{o2}) = O(\boldsymbol{u}_{o1})O^*(\boldsymbol{u}_{o2})\Gamma_s(\boldsymbol{u}_{o1}, \boldsymbol{u}_{o2}) \qquad (79.10)$$

The current density $j(\boldsymbol{u}_i)$ at the image is then obtained by substituting (78.10) in (79.5) and setting $\boldsymbol{u}_{i1} = \boldsymbol{u}_{i2}$. We obtain

$$j(\boldsymbol{u}_i): = \Gamma(\boldsymbol{u}_i, \boldsymbol{u}_i)$$

$$= \int O(\boldsymbol{u}_{o1})O^*(\boldsymbol{u}_{o2})\Gamma_s(\boldsymbol{u}_{o1} - \boldsymbol{u}_{o2})G(\boldsymbol{u}_{i1} - \boldsymbol{u}_{o1})$$

$$\times G^*(\boldsymbol{u}_{i2} - \boldsymbol{u}_{o2}) \, d\boldsymbol{u}_{o1} \, d\boldsymbol{u}_{o2} \qquad (79.11)$$

On introducing Fourier transforms throughout, we obtain a relation between the image *spectrum* $S_i = \mathcal{F}(j)$ and the object *spectrum* $S_o = \mathcal{F}(O)$ of the form

$$S_i(\boldsymbol{q}) = \int S_o(\boldsymbol{q}' + \boldsymbol{q})S_o^*(\boldsymbol{q}')\tilde{T}(\boldsymbol{q}' + \boldsymbol{q}, \boldsymbol{q}) \, d\boldsymbol{q}' \qquad (79.12)$$

in which

$$\tilde{T}(\boldsymbol{q}, \boldsymbol{q}') = \int \tilde{\Gamma}_s(\boldsymbol{v})\tilde{G}(\boldsymbol{q} + \boldsymbol{v})\tilde{G}^*(\boldsymbol{q}' + \boldsymbol{v}) \, d\boldsymbol{v} \qquad (79.13)$$

The function $\tilde{T}$ is known as the *transmission cross-coefficient*. Equation (79.12) shows that the instrument is now acting as a *vector* filter: all the spatial frequency components of the object spectrum may contribute to a given spatial frequency component of the image spectrum.

The situation becomes distinctly simpler when we specialize to weakly scattering objects, writing

$$O(\boldsymbol{u}_o) \approx 1 + i\eta_o - \sigma_o, \quad \eta_o, \sigma_o \ll 1$$

and neglect quadratic terms in η_o and σ_o. After a certain amount of straightforward calculation, given in detail in Hawkes (1978), the image contrast spectrum takes the form

$$S_c(\boldsymbol{q}) = T_a(\boldsymbol{q})\tilde{\sigma}(\boldsymbol{q}) + T_p(\boldsymbol{q})\tilde{\eta}(\boldsymbol{q})$$

which has exactly the same form as (66.17c) but instead of $K_a = \cos\chi$ and $K_p = \sin\chi$ we now have

$$\begin{aligned} T_a &= -\{\tilde{T}(\boldsymbol{q},0) + \tilde{T}(0,-\boldsymbol{q})\} \\ T_p &= i\{\tilde{T}(\boldsymbol{q},0) - \tilde{T}(0,-\boldsymbol{q})\} \end{aligned} \tag{79.14}$$

These in turn may be written in terms of $T_A T_L$; for a symmetric source we have

$$T_a(\boldsymbol{q}) = -\int \tilde{\Gamma}_s(\boldsymbol{v}) T_A(\boldsymbol{q}+\boldsymbol{v}) T_A(\boldsymbol{v}) \cos\{\chi(\boldsymbol{q}+\boldsymbol{v}) - \chi(\boldsymbol{v})\}\, d\boldsymbol{v}$$

$$\tag{79.15}$$

$$T_p(\boldsymbol{q}) = \int \tilde{\Gamma}_s(\boldsymbol{v}) T_A(\boldsymbol{q}+\boldsymbol{v}) T_A(\boldsymbol{v}) \sin\{\chi(\boldsymbol{q}+\boldsymbol{v}) - \chi(\boldsymbol{v})\}\, d\boldsymbol{v}$$

Frank (1973) has shown that when, as is often the case, the effect of T_A can be neglected and values of $|\boldsymbol{q}|$ greater than the source aperture are primarily of interest, further simplifications are permissible. In particular, we may set

$$\chi(\boldsymbol{q}+\boldsymbol{v}) \approx \chi(\boldsymbol{q}) + \boldsymbol{v}\cdot\operatorname{grad}\chi(\boldsymbol{q}) \tag{79.16}$$

and obtain

$$\begin{aligned} T_a(\boldsymbol{q}) &= -\cos\chi(\boldsymbol{q})E(\boldsymbol{q}) \\ T_p(\boldsymbol{q}) &= \sin\chi(\boldsymbol{q})E(\boldsymbol{q}) \end{aligned} \tag{79.17}$$

with

$$E(\boldsymbol{q}) = \int \tilde{\Gamma}_s(\boldsymbol{v}) \cos\{\boldsymbol{v}\cdot\operatorname{grad}\chi(\boldsymbol{q})\}\, d\boldsymbol{v}$$

$$\tag{79.18}$$

$$= \Gamma_s\left(\frac{\operatorname{grad}\chi(\boldsymbol{q})}{2\pi}\right)$$

which was stated without proof at the end of Section 66.2.

For a disc-shaped source,

$$s(\boldsymbol{v}) = \widetilde{\Gamma}_s(\boldsymbol{v}) = \begin{cases} 1/\pi v_0^2 & |\boldsymbol{v}| \le v \\ 0 & \text{otherwise} \end{cases}$$

we have

$$E(\boldsymbol{q}) = 2\frac{J_1(v_0|\operatorname{grad}\chi(\boldsymbol{q})|)}{v_0|\operatorname{grad}\chi(\boldsymbol{q})|} \tag{79.20}$$
$$= 2\operatorname{besinc}(v_0|\operatorname{grad}\chi(\boldsymbol{q})|)$$

For a Gaussian source, for which

$$s(\boldsymbol{v}) = \frac{1}{\pi v_0^2}\exp\left(-\frac{v^2}{v_0^2}\right) \tag{79.21}$$

we find

$$E(\boldsymbol{q}) = \exp\left\{-\frac{v_0^2|\operatorname{grad}\chi(\boldsymbol{q})|^2}{4}\right\} \tag{79.22}$$

A slight improvement can be achieved by retaining the next higher term in the expansion for $\chi(\boldsymbol{q}+\boldsymbol{v})$. For details, see Frank (1973). In the case of a Gaussian source, it is found that

$$\begin{aligned} T_a(\boldsymbol{q}) &= -E(\boldsymbol{q})\cos\{\chi - e(\boldsymbol{q})\cos\chi\} \\ &\approx -E(\boldsymbol{q})\cos(\chi+\varepsilon) \\ T_p(\boldsymbol{q}) &= E(\boldsymbol{q})\sin\{\chi + e(\boldsymbol{q})\cos\chi\} \\ &\approx -E(\boldsymbol{q})\sin(\chi+\varepsilon) \end{aligned} \tag{79.23}$$

for small values of $e(\boldsymbol{q})$, where

$$e(\boldsymbol{q}): = \widehat{E}(\boldsymbol{q})/E(\boldsymbol{q})$$

$$\widehat{E}(\boldsymbol{q}) = \int \pi C_s \lambda^3 \{(\boldsymbol{q}\cdot\boldsymbol{q})(\boldsymbol{v}\cdot\boldsymbol{v}) + 2(\boldsymbol{q}\cdot\boldsymbol{v})^2\}\widehat{\Gamma}(\boldsymbol{v})\cos\{\boldsymbol{v}\cdot\operatorname{grad}\chi(\boldsymbol{q})\}\,d\boldsymbol{v}$$

$$\tag{79.24}$$

The angular shift ε is given by

$$\varepsilon = \arcsin e(\boldsymbol{q}) \tag{79.25}$$

This form of the envelope representation of the effect of spatial coherence is therefore capable of representing not just an attenuation but also the small shifts in the zeros of the coherent transfer function caused by the departure from perfect coherence.

79.3 Coherence and illumination

There are two modes of illumination in common use, *critical illumination* and *Köhler illumination* (see Chapter 64). In the former, the specimen plane is conjugate to the effective source plane (crossover plane) whereas in the Köhler case, the effective source plane is conjugate to the image focal plane of the final condenser lens, which may in fact be the prefield of the objective; rays from each source point then form a parallel beam at the specimen and the angular spread of the beam will be determined by the source-size and focal length f of the total condenser system. If α is the semi-angular spread and a the source radius, then $\alpha = a/f$.

We have already seen that $\lambda/2\pi\alpha$ is a useful estimate of the size of the coherently illuminated patch at the object. It can be shown that the degree of coherence at the specimen is the same for both modes of illumination. Another useful measure of the coherence at the specimen is the *coherence ratio* or *parameter*, m, defined by

$$m = \alpha/\alpha_0 \qquad (79.26)$$

where α_0 is the greatest angle at which electrons leaving the specimen can reach the image; it is thus determined by the size of the objective aperture. For coherent image formation, the coherently illuminated disc at the object must be at least as large as the Airy disc, which implies that

$$\frac{\lambda}{2\pi\alpha} \gtrsim \frac{0.6\lambda}{\alpha_0} \qquad (79.27a)$$

and hence that

$$m = \frac{\alpha}{\alpha_0} \lesssim 0.26 \qquad (79.27b)$$

For the scanning transmission electron microscope, arguments based on reciprocity or straightforward calculation show that we should attach the idea of coherence to the detector. We shall not pursue this in detail here, referring instead to the standard study by Burge and Dainty (1976) and to the discussion in Hawkes (1978) for an extended account of the earlier work and to the texts of Spence (1988) and Spence and Zuo (1992), where the practical implications of partial coherence at the STEM probe are discussed at length.

APPENDIX

Additional major references

Dorset, D.L. (1995) *Structural Electron Crystallography* (Plenum, New York and London)

Frank, J. (1996). *Three-dimensional Electron Microscopy of Macromolecular Assemblies* (Academic Press, Orlando and London).

Hawkes, P.W., ed. (1994). *Selected Papers in Electron Optics* (Spie Optical Engineering Press, Bellingham) SPIE Milestone Series Vol. **MS94**.

Mandel, L. and Wolf, E. (1995). *Optical Coherence and Quantum Optics* (Cambridge University Press, Cambridge and New York).

Munro, E. and Freund, H.P., eds (1995). *Electron-beam Sources and Charged-particle Optics*. Proc SPIE **2522**.

Ritter, G. and J.N. Wilson (1996). *Handbook of Computer Vision Algorithms in Image Algebra* (CRC Press, Boca Raton).

Thompson, W.B., Sato, M. and Crewe, A.V., eds (1993). *Charged-particle Optics*. Proc SPIE **2014**.

Tonomura, A., Allard, L.F., Pozzi, G., Joy, D.C. and Ono, Y.A. (1995). *Electron Holography* (Elsevier, Amsterdam, Lausanne, New York, Oxford, Shannon and Tokyo).

Wang, Z.L. (1995). *Elastic and Inelastic Scattering in Electron Diffraction and Imaging* (Plenum, New York and London).

Notes and References

Notes and References

The lists of references follow the main divisions of the book in this volume. In order to avoid repetition, standard abbreviations have been adopted for several series of conference proceedings, namely, the European and International conferences on electron microscopy, which have alternated every two years since 1954 (prior to that date, they were not quite so regular); the occasional conferences on high-voltage electron microscopy; the annual meetings of the Electron Microscopy Society of America, now the Microscopy Society of America; and the biennial meetings of the Electron Microscopy and Analysis Group of the (British) Institute of Physics and of the Società Italiana di Microscopia Elettronica. The European and International Congresses are identified by date and place, the high-voltage conferences by HVEM and place, the American meetings by date followed by EMSA or MSA and meeting number, the British meetings by date followed by EMAG and the Italian meetings by date followed by SIME. Full bibliographic details of all these conference volumes are to be found at the end of this Section, after the references for Part XVI.

Some of the lists that follow contain papers that are not cited in the text. All these additional references are, however, cited in the notes that precede the lists. The list for the Preface and the introductory Chapter 54 also includes publishing details of a number of relevant books, not cited explicitly; some of the other lists also include uncited books. In all cases, these are labelled with an asterisk.

Despite the length of these lists, we make no claim to completeness. Where good bibliographies are available, we have referred to these rather than repeating their contents. Nevertheless, for most of the topics discussed here, the literature is very scattered and we have attempted to give rather thorough coverage. We shall be most grateful to have any errors and serious omissions drawn to our attention.

Preface and Chapter 54

Agarwal, G.S., Foley, J.T. and Wolf, E. (1987). Opt. Commun. **62**, 67–72.

Aharonov, Y. and Bohm, D. (1959). Phys. Rev. **115**, 485–491.

*Amelinckx, S., Gevers, R., Remaut, G. and Landuyt. J. van, eds (1970). *Modern Diffraction and Imaging Techniques in Material Science* (North-Holland, Amsterdam and London).

*Baumeister, W. and Vogell, W., eds (1980). *Electron Microscopy at Molecular Dimensions. State of the art and strategies for the future* (Springer, Berlin and New York).

*Bethge, H. and Heydenreich, J., eds (1982). *Elektronenmikroskopie in der Festkörperphysik* (Deutscher Verlag der Wissenschaften, Berlin and Springer, Berlin and New York).

*Bethge, W. and Heydenreich, J., eds (1987). *Electron Microscopy in Solid State Physics* (Elsevier, Amsterdam, Oxford, New York and Tokyo). This is a translation with revisions of Bethge and Heidenreich (1982).

Blackman, M. (1978). In *Electron Diffraction 1927–1977* (Dobson, P.J., Pendry, J.B. and Humphreys, C.J., eds) p. v (Institute of Physics, Bristol). Conference Series No. 41.

Boersch, H. (1936). Ann. Physik **26**, 631–644 and **27**, 75–80.

Boersch, H. (1940). Naturwissenschaften **28**, 709–711.

Boersch, H. (1943). Phys. Z. **44**, 32–38 and 202–211.

Boersch, H. (1946). Monatshefte Chem. **76**, 86–92 and 163–167.

Boersch, H. (1947a). Publ. Inst. Rech. Sci. Tettnang No 3–4, 37–42.

Boersch, H. (1947b). Z. Naturforsch. **2a**, 615–633.

Boersch, H. (1948). Monatshefte Chem. **78**, 163–171.

Bonhomme, P. and Beorchia, A. (1980). The Hague, vol. 1, pp. 134–135.

Bonnet, N., Troyon, M. and Gallion, P. (1978). Toronto, vol. 1, pp. 222–223.

Born, M. and Wolf, E. (1959). *Principles of Optics* (Pergamon, Oxford and New York).

*Borries, B. von (1949). *Die Übermikroskopie. Einführung, Untersuchung ihrer Grenzen und Abriss ihrer Ergebnisse* (Editio Cantor, Aulendorf/Württ).

Bristol (1984). *Convergent Beam Electron Diffraction of Alloy Phases* by the Bristol Group under the direction of John Steeds, compiled by J. Mansfield (Adam Hilger, Bristol and Boston).

Broglie, L. de (1923). C. R. Acad. Sci. Paris **177**, 507–510, 548–550 and 630–632.

Broglie, L. de (1925). Ann. Physique **3**, 22–128 ; republished in Ann. Fond. Louis de Broglie **17** (1992) 1–109.

Broglie, L. de (1950). *Optique Electronique et Corpusculaire* (Hermann, Paris).

*Buseck, P.R., Cowley, J.M. and Eyring, L., eds (1988). *High-resolution Electron Microscopy and Associated Techniques* (Oxford University Press, New York and Oxford).

Carter, W.H. and Wolf, E. (1977). J. Opt. Soc. Am. **67**, 785–794.

*Chapman, J.N. and Craven, A.J., eds (1984). *Quantitative Electron Microscopy* (Scottish Universities Summer School in Physics, Edinburgh).

Coene, W. and Dyck, D. van (1984). Ultramicroscopy **15**, 41–50 and 287–299.

Cowley, J.M. (1975). *Diffraction Physics* (North-Holland, Amsterdam, New York and Oxford); 2nd ed. (1981).

Cowley, J.M. (1981). In Goodman (1981) pp. 271–275.

*Cowley, J.M., ed. (1992, 1993). *Electron Diffraction Techniques*, 2 vols (Oxford University Press, Oxford and New York).

Cowley, J.M. and Moodie A.F. (1957a). Proc. Phys. Soc. (London) **B70**, 486–496, 497–504 and 505–513.

Cowley, J.M. and Moodie, A.F. (1957b). Acta Cryst. **10**, 609–619.

Cowley, J.M. and Moodie, A.F. (1958). Proc. Phys. Soc. (London) **71**, 533–545.

Cowley, J.M. and Moodie, A.F. (1959). Acta Cryst. **12**, 353–359 and 360–367.

Cowley, J.M. and Moodie, A.F. (1960). Proc. Phys. Soc. (London) **76**, 378–384.

Crewe, A.V., Wall, J. and Welter, L.M. (1968). J. Appl. Phys. **39**, 5861–5868.

Croce, P. (1956). Rev. Opt. **35**, 569–589 and 642–656.

Crowther, R.A. (1971). Phil. Trans. Roy. Soc. (London) **B261**, 221–230.

Crowther, R.A., Amos, L.A., Finch, J.T., Rosier, D.J. de and Klug, A. (1970a). Nature **226**, 421–425.

Crowther, R.A., DeRosier, D.J. and Klug, A. (1970b). Proc. Roy. Soc. (London) **A317**, 319–340.

Czapski, S. and Eppenstein, O. (1924). *Grundzüge der Theorie der Optischen Instrumente nach Abbe* (Barth, Leipzig).

Davisson, C.J. and Germer, L.H. (1927). Phys. Rev. **30**, 705–740.

Dekkers, N.H and Lang, H. de (1974). Optik **41**, 452–456.

Duffieux, [P.] M. (1940a). Bull. Soc. Sci. Bretagne **17**, 107–114.

Duffieux, P.M. (1940b). Ann. Physique **14**, 302–338.

Duffieux, P.M. (1942). Ann. Physique **17**, 209–236.

Duffieux, P.-M. (1946). *L'Intégrale de Fourier et ses Applications à l'Optique* (privately printed by S.A. des Imprimeries Oberthur, Rennes). A second edition appeared in 1970 (Masson, Paris), translated into English as *The Fourier Transform and its Applications to Optics* (Wiley, New York and Chichester, 1983).

Duffieux, P.M. (1970). Comment j'ai pris contact avec la transformation de Fourier. In *Applications de l'Holographie* (Viénot, J.-C., Bulabois, J. and Pasteur, J., eds) pp. xviii–xx (Université de Besançon, Besançon).

Dyck, D. van (1980). J. Microscopy **119**, 141–152.

Dyck, D. van and Coene, W. (1984). Ultramicroscopy **15**, 29–40.

Ehrenberg, W. and Siday, R.E. (1949). Proc. Phys. Soc. (London) **B62**, 8–21.

Elsasser, W. (1925). Naturwissenschaften **13**, 711.

Frank, J. (1973). Optik **38**, 519–536.

*Frank, J., ed. (1992). *Electron Tomography. Three-dimensional imaging with the transmission electron microscope* (Plenum, New York and London).

Frank, J. and Heel, M. van (1980). The Hague, vol. 2, pp. 690–691.

Frisch, R. and Stern, O. (1933). Beugung von Materiestrahlen. In *Handbuch der Physik* (Geiger, H. and Scheel, K., eds) vol. 24, Pt. 1, pp. 313–354 (Springer, Berlin).

*Fryer, J.R. (1979). *The Chemical Applications of Transmission Electron Microscopy* (Academic Press, London and New York).

*Fryer, J.R. and Dorset, D.L., eds (1991). *Electron Crystallography of Organic Molecules* (Kluwer, Dordrecht, Boston MA and London).

*Gabor, D. (n.d., 1945). *The Electron Microscope* (Hulton, London).

Gabor, D. (1948). Nature **161**, 777–778.

Gabor, D. (1968). Preface to Marton (1968).

Gallion, P., Troyon, M. and Beorchia, A. (1975). Opt. Acta **22**, 731–743.

Gerchberg, R.W. and Saxton, W.O. (1972). Optik **35**, 237–246.

Gerchberg, R.W. and Saxton, W.O. (1973). In *Image Processing and Computer-aided Design in Electron Optics* (Hawkes, P.W., ed.) pp. 66–81 (Academic Press, London and New York).

Glaser, W. (1943). Z. Physik **121**, 647–666; additional note *ibid.* **125** (1949) 541.

Glaser, W. (1949). Acta Phys. Austriaca **3**, 38–51.

Glaser, W. (1950a). Sitzungsber. Öst. Akad. Wiss., Mathem.-Naturw. Klasse, Abt. IIa, **159**, 297–360.

Glaser, W. (1950b). Paris, vol 1, pp. 63–72 and 164.

Glaser, W. (1952). *Grundlagen der Elektronenoptik* (Springer, Vienna).

Glaser, W. (1956). Elektronen- und Ionenoptik. In *Handbuch der Physik* (Flügge, S., ed.) vol. 33, pp. 123–395 (Springer, Berlin).

Glaser, W. and Braun, G. (1954). Acta Phys. Austriaca **9**, 41–74.

Glaser, W. and Braun, G. (1955). Acta Phys. Austriaca **9**, 267–296.

Glaser, W. and Schiske, P. (1953). Ann. Physik **12**, 240–266 and 267–280.

Goodman, P., ed. (1981). *Fifty Years of Electron Diffraction* (Reidel, Dordrecht, Boston MA and London).

Haine, M.E. and Mulvey, T. (1952). J. Opt. Soc. Am. **42**, 763–773.

*Hall, C.E. (1953). *Introduction to Electron Microscopy* (McGraw–Hill, New York and London) ; 2nd ed. (1966).

Hanszen, K.-J. (1966). Z. angew. Phys. **20**, 427–435.

Hanszen, K.-J. (1971). Adv. Opt. Electron Microsc. **4**, 1–84.

Hanszen, K.-J. (1973). In *Image Processing and Computer-aided Design in Electron Optics* (Hawkes, P.W., ed.) pp. 16–53 (Academic Press, London and New York).

Hanszen, K.-J. (1982). Adv. Electron. Electron Phys. **59**, 1–77.

Hanszen, K.-J. and Morgenstern, B. (1965). Z. angew. Phys. **19**, 215–227.

Hanszen, K.-J. and Trepte, L. (1971). Optik **32**, 519–538.

Hart, R.G. (1968). Science **159**, 1464–1467.

Hashimoto, H. (1954). J. Phys. Soc. Japan **9**, 150–161.

Hashimoto, H., Howie, A. and Whelan, M.J. (1960). Phil. Mag. **5**, 967–974.

Hashimoto, H., Mannami, M. and Naiki, T. (1961). Phil. Trans. Roy. Soc. (London) **253**, 459–489 and 490–516.

Hashimoto, H., Howie, A. and Whelan, M.J. (1962). Proc. Roy. Soc. (London) **A269**, 80–103.

*Hawkes, P.W., ed. (1973). *Image Processing and Computer-aided Design in Electron Optics* (Academic Press, London and New York).

*Hawkes, P.W., ed. (1980). *Computer Processing of Electron Microscope Images* (Springer, Berlin and New York).

*Hawkes, P.W. and Valdrè, U., eds (1990). *Biophysical Electron Microscopy* (Academic Press, London and New York).

*Hayman, P. (1974). *Théorie Dynamique de la Microscopie et Diffraction Electroniques* (Presses Universitaires de France, Paris).

Head, A.K., Humble, P., Clarebrough, L.M., Morton, A.J. and Forwood, C.T. (1973). *Computed Electron Micrographs and Defect Identification* (North-Holland, Amsterdam and London; American Elsevier, New York). Defects in Crystalline Solids (Amelinckx, S., Gevers, R. and Nihoul, J., eds) vol. 7.

Heel, M. van and Frank, J. (1980). The Hague, vol. 2, pp. 692–693.

Heel, M. van and Frank, J. (1981). Ultramicroscopy **6**, 187–194.

Heidenreich, R.D. (1949). J. Appl. Phys. **20**, 993–1010.

Heidenreich, R.D. (1951). Bell System Tech. J. **30**, 867–887.

*Heidenreich, R.D. (1964). *Fundamentals of Transmission Electron Microscopy* (Wiley–Interscience, New York and London).

Hibi, T. (1956). J. Electronmicrosc. **4**, 10–15.

Hibi, T. (1985). In *The Beginnings of Electron Microscopy* (Hawkes, P.W., ed.) pp. 297–315. Adv. Electron. Electron Phys. Suppl. 16.

Hillier, J. (1940). Phys. Rev. **58**, 842.

Hillier, J. (1941). Phys. Rev. **60**, 743–745.

Hirsch, P.B. (1980). Proc. Roy. Soc. (London) A**371**, 160–164.

Hirsch, P.B. (1986). Mat. Sci. Eng. **84**, 1–10.

Hirsch, P.B., Horne, R.W. and Whelan, M.J. (1956). Phil. Mag. **1**, 677–684.

Hirsch, P.B., Howie, A. and Whelan, M.J. (1960). Phil. Trans. Roy. Soc. (London) A**252**, 499–529.

Hirsch, P.B., Howie, A., Nicholson, R.B., Pashley, D.W. and Whelan, M.J. (1965). *Electron Microscopy of Thin Crystals* (Butterworths, London) ; reprinted with an additional chapter by Krieger, New York (1977).

*Holt, D.B. and Joy, D.C., eds (1989). *SEM Microcharacterization of Semiconductors* (Academic Press, London and San Diego).

*Holt, D.B., Muir, M.D., Grant, P.R. and Boswarva, I.M., eds (1974). *Quantitative Scanning Electron Microscopy* (Academic Press, London and New York).

Hoppe, W. (1961). Naturwissenschaften **48**, 736–737.

Hoppe, W. (1983). Angew. Chemie **95**, 465–494; Int. Ed. in English **22**, 456–485.

Hoppe, W., Langer, R., Knesch, G. and Poppe, C. (1968). Naturwissenschaften **55**, 333–336.

Hoppe, W., Gassmann, J., Hunsmann, N., Schramm, H.J. and Sturm, M. (1974). Hoppe–Seyler's Z. Physiol. Chem. **355**, 1483–1487.

Hoppe, W., Schramm, H.J., Sturm, M., Hunsmann, N. and Gassmann, J. (1976). Z. Naturforsch. **31a**, 645–655, 1370–1379 and 1380–1390.

*Howie, A. and Valdrè, U., eds (1988). *Surface and Interface Characterization by Electron Optical Methods* (Plenum, New York and London). Vol. 191 of the NATO ASI Series B.

Howie, A. and Whelan, M.J. (1960). Delft, vol.1, pp. 181–185 and 194–198.

Howie, A. and Whelan, M.J. (1961). Proc. Roy. Soc. (London) A**263**, 217–237.

Howie, A. and Whelan, M.J. (1962). Proc. Roy. Soc. (London) A**267**, 206–230.

Ishizuka, K. and Uyeda, N. (1977). Acta Cryst. A**33**, 740–749.

Kikuchi, S. (1928). Proc. Imp. Acad. Japan **4**, 271–274, 275–278, 354–356 and 471–474.

*Koehler, J.K., ed. (1973, 1978, 1986). *Advanced Techniques in Biological Electron Microscopy* I, II, III (Springer, Berlin and New York).

Kossel, W. (1941). Ann. Physik **40**, 17–38.

Kossel, W. and Möllenstedt, G. (1938). Naturwissenschaften **26**, 660–661.

Kossel, W. and Möllenstedt, G. (1939). Ann. Physik **36**, 113–140.

Kossel, W. and Möllenstedt, G. (1942). Ann. Physik **42**, 287–293.

*Krakow, W. and O'Keefe, M., eds (1989). *Computer Simulation of Electron Microscope Diffraction and Images* (Minerals, Metals and Materials Society, Warrendale PA).

*Krakow, W., Smith, D.A. and Hobbs, L.W., eds (1984). *Electron Microscopy of Materials* (North-Holland, New York, Amsterdam and Oxford).

*Lannes, A. and Pérez, J.-P. (1983). *Optique de Fourier en Microscopie Electronique* (Masson, Paris and New York).

*Larsen, P.K. and Dobson, P.J., eds (1988). *Reflection High-energy Electron Diffraction and Reflection Electron Imaging of Surfaces* (Plenum, New York and London). Vol. 188 of the NATO ASI Series B.

Leith, E.N. and Upatnieks, J. (1962). J. Opt. Soc. Am. **52**, 1123–30.

Leith, E.N. and Upatnieks, J. (1963). J. Opt. Soc. Am. **53**, 1377–1381.

Lenz, F. (1953). Naturwissenschaften **39**, 265.

Lenz, F. (1954). Z. Naturforsch. **9a**, 185–204.

Lynch, D.F. and O'Keefe M.A. (1972). Acta Cryst. **A28**, 536–548.

MacDonald, N.C. (1968). EMSA **26**, pp. 362–363.

MacDonald, N.C. (1969). Scanning Electron Microsc. 431–437.

MacGillavry, C.H. (1940a). Nature **145**, 189–190.

MacGillavry, C.H. (1940b). Physica **7**, 329–343.

Marchand, E.W. and Wolf, E. (1972a). Opt. Commun. **6**, 305–308.

Marchand, E.W. and Wolf, E. (1972b). J. Opt. Soc. Am. **62**, 379–385.

Marchand, E.W. and Wolf, E. (1974). J. Opt. Soc. Am. **64**, 1219–1226.

Maréchal, A. and Croce, P. (1953). C.R. Acad. Sci. Paris **237**, 607–609.

Maréchal, A. and Françon, M. (1960). *Diffraction. Structure des Images* (Editions de la Revue d'Optique Théorique et Instrumentale, Paris).

Marton, L. (1934a). Bull. Acad. Roy. Belgique (Classe Sci.) **20**, 439–446.

Marton, L. (1934b). Ann. Bull. Soc. Roy. Sci. Med. Nat. Bruxelles No. 5–6, 92–106.

Marton, L. (1936). Physica **3**, 959–967.

Marton, L. (1952). Phys. Rev. **85**, 1057–1058.

Marton, L. (1968). *Early History of the Electron Microscope* (San Francisco Press, San Francisco).

Marton, L. and Schiff, L.I. (1941). J. Appl. Phys. **12**, 759–765.

Marton, L., Simpson, J.A. and Suddeth, J.A. (1953). Phys. Rev. **90**, 490–491.

Menter, J.W. (1956). Proc. Roy. Soc. (London) **A236**, 119–135.

*Merli, P.G. and Antisari, M.V., eds (1992). *Electron Microscopy in Materials Science* (World Scientific, Singapore, River Edge NJ and London).

*Misell, D.L. and Brown, E.B. (1987). *Electron Diffraction: an Introduction for Biologists* (Elsevier, Amsterdam, New York and Oxford). Practical Methods in Electron Microscopy (Glauert, A., ed.) vol. 12.

*Misell, D.L. (1978). *Image Analysis, Enhancement and Interpretation* (North-Holland, Amsterdam, New York and Oxford). Practical Methods in Electron Microscopy (Glauert, A., ed.) vol. 7.

Möllenstedt, G. (1941). Ann. Physik **40**, 39–65.

Möllenstedt, G. (1989). Phys. Stat. Sol (a) **116**, 13–22.

Möllenstedt, G. (1991). The invention of the electron Fresnel interference biprism. Adv. Opt. Electron Microsc. **12**, 1–23.

Möllenstedt, G. and Düker, H. (1955). Naturwissenschaften **42**, 41.

Möllenstedt, G. and Düker, H. (1956). Z. Physik **145**, 377–397.

Möllenstedt, G. and Wahl, H. (1968). Naturwissenschaften **55**, 340–341.

Munch, J. (1975). Optik **43**, 79–99.

Nishikawa, S. and Kikuchi, S. (1928). Proc. Imp. Acad. Japan **4**, 475–477.

O'Keefe, M.A. (1973). Acta Cryst. A**29**, 389–401.

Peshkin, M. and Tonomura, A. (1989). *The Aharonov–Bohm Effect* (Springer, Berlin, New York and London). Lecture Notes in Physics, vol. 340.

Picht, J. (1939). *Einführung in die Theorie der Elektronenoptik* (Barth, Leipzig); 2nd ed. (1957); 3rd ed. (1963).

*Picht, J. (1945). *Was Wir über Elektronen Wissen* (Carl Winter, Universitätsverlag, Heidelberg).

*Picht, J. and Heydenreich, J. (1966). *Einführung in die Elektronenmikroskopie* (Verlag Technik, Berlin).

Pinsker, Z.G. (1949). *Diffraktsia Elektronov* (Izd. Akad. Nauk SSSR, Moscow and Leningrad); English translation : *Electron Diffraction* (Butterworths, London 1953).

Rayleigh, Lord (1881). Phil. Mag. **11**, 196–205

*Reimer, L. (1984). *Transmission Electron Microscopy: Physics of image formation and microanalysis* (Springer, Berlin and New York); 2nd ed. (1989); 3rd ed. (1993).

*Reimer, L. (1985). *Scanning Electron Microscopy: Physics of image formation and microanalysis* (Springer, Berlin and New York).

*Reimer, L. (1993). *Image formation in Low-voltage Scanning Electron Microscopy* (SPIE Optical Engineering Press, Bellingham).

*Reimer, L., ed. (1994). *Energy-filtering Transmission Electron Microscopy* (Springer, Berlin and New York).

*Reimer, L. and Pfefferkorn, G. (1973). *Raster-Elektronenmikroskopie* (Springer, Berlin and New York); 2nd ed. (1977).

Rose, H. (1974). Optik **39**, 416–436.

Rosier, D.J. de and Klug, A. (1968). Nature **217**, 130–134.'

*Ruedl, E. and Valdrè, U., eds (1976). *Electron Microscopy in Materials Science*, Parts III and IV (Commission of the European Communities, Luxemburg)

Ruska, E. (1934). Z. Physik **87**, 580–602.

Ruska, E. (1979). Die Frühe Entwicklung der Elektronenlinsen und der Elektronen-mikroskopie. Acta Hist. Leopoldina, Nr.12. English translation (1980): *The Early Development of Electron Lenses and Electron Microscopy* (Hirzel, Stuttgart); also published as Supplement 5 to Microsc. Acta.

Sanders, J.V. and Goodman, P. (1981). In Goodman (1981), pp. 281–283.

*Saxton, W.O. (1978). *Computer Techniques for Image Processing in Electron Microscopy* (Academic Press, New York and London). Adv. Electron. Electron Phys. Supplement 10.

Scherzer, O. (1949). J. Appl. Phys. **20**, 20–29.

Schiff, L.I. (1942). Phys. Rev. **61**, 391 and 721–722.

Schiske, P. (1968). Rome, vol. 1, pp 145–146.

Schiske, P. (1973). In *Image Processing and Computer-aided Design in Electron Optics* (Hawkes, P.W., ed.) pp. 82–90 (Academic Press, London and New York).

*Siegel, B.M. and Beaman, D.R., eds (1975). *Physical Aspects of Electron Microscopy and Microbeam Analysis* (Wiley, New York and London).

*Spence, J.C.H. (1980). *Experimental High-resolution Electron Microscopy* (Oxford University Press, New York and Oxford); 2nd. ed. (1988).

*Spence, J.C.H. and Zuo, J.M. (1992). *Electron Microdiffraction* (Plenum, New York and London).

Süsskind, C. (1985). L.L. Marton, 1901–1979. In *The Beginnings of Electron Microscopy* (Hawkes, P.W., ed.) pp. 501–523. Adv. Electron. Electron Phys. Suppl. 16.

*Synge, J.L. (1954). *Geometrical Mechanics and de Broglie Waves* (Cambridge University Press, Cambridge).

Talbot, H.F. (1836). Phil. Mag. **9**, 401–407.

Tanaka, M. and Terauchi, M. (1985). *Convergent-beam Electron Diffraction* (JEOL, Tokyo).

Tanaka, M., Terauchi, M. and Kaneyama, T. (1988). *Convergent-beam Electron Diffraction II* (JEOL, Tokyo).

Thomson, G.P. (1927). Nature **120,** 802.

Thomson, G.P. (1928). Proc. Roy. Soc. (London) **A117**, 600–609.

Thomson, G.P. and Reid, A. (1927). Nature **119**, 890.

Thon, F. (1966a). Z. Naturforsch. **21a**, 476–478.

Thon, F. (1966b). Kyoto, vol. 1, pp. 23–24.

Tomita, H., Matsuda, T. and Komoda, T. (1970a). Grenoble, vol. 1, pp. 151–152.

Tomita, H., Matsuda, T. and Komoda, T. (1970b). Japan. J. Appl. Phys. **9**, 719.

Tomita, H., Matsuda, T. and Komoda, T. (1972). Japan. J. Appl. Phys. **11**, 143–149.

Tonomura, A. (1969). J. Electronmicrosc. **18**, 77–78.

Tonomura, A. and Watanabe, H. (1968). Nihon Butsuri Gakkai-shi [Proc. Phys. Soc. Japan] **23**, 683–684.

Tonomura, A., Fukuhara, A., Watanabe, H. and Komoda, T. (1968). Japan. J. Appl. Phys. **7**, 295.

Tonomura, A., Osakabe, N., Matsuda, T., Kawasaki, T., Endo, J., Yano, S. and Yamada, H. (1986). Phys. Rev. Lett. **56**, 792–795.

Troyon, M., Gallion, P. and Laberrigue, A. (1976). Jerusalem, vol.1, pp. 344–345.

Vainshtein, B.K. (1956). *Strukturnaya Elektronografiya* (Izd. Akad. Nauk SSSR, Moscow and Leningrad); English translation : *Structure Analysis by Electron Diffraction* (Pergamon, Oxford and New York 1964).

*Valdrè, U., ed. (1971). *Electron Microscopy in Material Science* (Academic Press, New York and London).

*Valdrè, U. and Ruedl, E., eds (1976). *Electron Microscopy in Materials Science*, Parts I and II (Commission of the European Communities, Luxembourg).

Wade, R.H. and Frank, J. (1977). Optik **49**, 81–92.

Wahl, H. (1974). Optik **39**, 585–588

Wahl, H. (1975). *Bildebenenholographie mit Elektronen.* Habilitationsschrift, Tübingen.

Walther, A. (1968). J. Opt. Soc. Am. **58**, 1256–1259.

Watanabe, H. and Tonomura, A. (1969). Nihon Kessho Gakkai-shi [J. Crystallograph. Soc. Japan] **11**, 23–25.

White, E. W., McKinstry, H. A. and Johnson, G. G. (1968). Scanning Electron Microsc. 95–103.

Weisel, H. (1910). Ann. Physik **33**, 995–1031.

Wolfke, M. (1913). Ann. Physik **40**, 194–200

Zeitler, E. and Thomson, M.G.R. (1970). Optik **31**, 258–280 and 359–366.

*Zworykin, V.K., Morton, G.A., Ramberg, E.G., Hillier, J. and Vance, E.E. (1945). *Electron Optics and the Electron Microscope* (Wiley, New York and Chapman and Hall, London).

Part XI, Chapters 55–60

Other studies on wave propagation and diffraction have been made by Bremmer (1951a,b) and by Komrska, whose work is recapitulated in a long review (Komrska, 1971).

We recall here that the paper by Ehrenberg and Siday in which the 'Aharonov–Bohm' effect was first noticed was primarily concerned with the definition of the refractive index in electron optics. Their observation provoked a response by Glaser (1951a,b) and further comment by Ehrenberg and Siday (1951).

Abramowitz, M. and Stegun, I.A., eds (1965). *Handbook of Mathematical Functions* (Dover, New York).

Aharonov, Y. and Bohm, D. (1959). Phys. Rev. **115**, 485–491.

Aharonov, Y. and Bohm, D. (1961). Phys. Rev. **123**, 1511–1524.

Boersch, H. (1951). Z. Physik **131**, 78–81.

Boersch, H. (1968). Z. Physik **215**, 28–33.

Boersch, H., Hamisch, H., Grohmann, K. and Wohlleben, D. (1961). Z. Physik **165**, 79–93.

Boersch, H., Hamisch, H. and Grohmann, K. (1962). Z. Physik **169**, 263–272.

Boersch, H., Grohmann, K., Hamisch, H., Lischke, B. and Wohlleben, D. (1981). Lett. Nuovo Cim. **30**, 257–258.

Born, M. and Wolf, E. (1959). *Principles of Optics* (Pergamon, Oxford and New York) ; 6th ed. (1980).

Bremmer, H. (1951a). Physica **17**, 63–70.

Bremmer, H. (1951b). Washington, pp. 145–158.

Chambers, R.G. (1960). Phys. Rev. Lett. **5**, 3–5.

Ditchburn, R.W. (1963). *Light,* 2nd ed. (Blackie, London). See Section 6.46.

Durand, E. (1953). C. R. Acad. Sci. Paris **236**, 1337–1339.

Ehrenberg, W. and Siday, R.E. (1949). Proc. Phys. Soc. (London) **B62**, 8–21.

Ehrenberg, W. and Siday, R.E. (1951). Proc. Phys. Soc. (London) **B64**, 1088–1089.

Ferwerda, H.A., Hoenders, B.J. and Slump, C.H. (1986). Opt. Acta **33**, 145–157 and 159–183.

Fowler, H.A., Marton, L., Simpson, J.A. and Suddeth, J.A. (1961). J. Appl. Phys. **32**, 1153–1155.

Glaser, W. (1943). Z. Physik **121**, 647–666.

Glaser, W. (1950a). Sitzungsber. Öst. Akad. Wiss., Math.-Naturwiss. Kl., Abt. II a, **159**, 297–360.

Glaser, W. (1950b). Paris, vol. 1, pp. 63–72 and 164.

Glaser, W. (1951a). Proc. Phys. Soc. (London) **B64**, 114–118.

Glaser, W. (1951b). Proc. Phys. Soc. (London) **B64**, 1089.

Glaser, W. (1951c). Washington, pp. 111–126.

Glaser, W. (1952). *Grundlagen der Elektronenoptik* (Springer, Vienna).

Glaser, W. (1953). Öst. Ing.-Arch. **7**, 144–152.

Glaser, W. (1954). Optik **11**, 101–117.

Glaser, W. (1956). Elektronen- und Ionenoptik. *Handbuch der Physik* **33**, 123–395.

Glaser, W. and Braun, G. (1954). Acta Phys. Austriaca **9**, 41–74.

Glaser, W. and Braun, G. (1955). Acta Phys. Austriaca **9**, 267–296.

Glaser, W. and Schiske, P. (1953). Ann. Physik **12**, 240–266 and 267–280.

Grivet, P. (1965). *Electron Optics* (Pergamon, Oxford and New York); 2nd ed. (1972).

Hoppe, W. (1961). Naturwissenschaften **48**, 736–737.

Hoppe, W. (1963). Optik **20**, 599–606.

Hosemann, R. and Joerchel, D. (1954). Z. Physik **138**, 209–221.

Jagannathan, R. (1990). Phys. Rev. **A42**, 6674–6689; corrigendum, *ibid.* **A44** (1991) 7856.

Jagannathan, R., Simon, R., Sudarshan, E.C.G. and Mukunda, N. (1989). Phys. Lett. **A134**, 457–464.

Kamke, E. (1977). *Differentialgleichungen, Lösungsmethoden und Lösungen* (Teubner, Stuttgart).

Kasper, E. (1973). Z. Naturforsch. **28a**, 216–221.

Komrska, J. (1971). Scalar diffraction theory in electron optics. Adv. Electron. Electron Phys. **30**, 139–234.

Krimmel, E. (1960). Z. Physik **158**, 35–38.

Krimmel, E. (1961). Z. Physik **163**, 339–355.

Lenz, F. (1961). Z. Physik **164**, 425–427.

Lenz, F. (1962). Naturwissenschaften **49**, 82.

Lenz, F. (1963). Z. Physik **172**, 498–502.

Lenz, F. (1964). Optik **21**, 489–493.

Lenz, F. (1965). Lab. Invest. **14**, 808–818.

Lenz, F. and Krimmel, E. (1961). Z. Physik **163**, 356–362.

Lenz, F. and Krimmel, E. (1963). Z. Physik **175**, 235–241.

Lenz, F. and Wilska, A.P. (1966/67). Optik **24**, 383–396.

Lipson, H. and Walkley, K. (1968). Opt. Acta **15**, 83–91.

Möllenstedt, G. and Bayh, W. (1962). Naturwissenschaften **49**, 81–82.

Möllenstedt, G., Grote, K.H. von and Jönsson, C. (1963). In *X-ray Optics and X-ray Microanalysis* (Pattee, H.H., Cosslett, V.E. and Engström, A., eds) pp. 73–79 (Academic Press, New York and London).

Phan–Van–Loc (1953). C.R. Acad. Sci. Paris **237**, 649–651.

Phan–Van–Loc (1954). C.R. Acad. Sci. Paris **238**, 2494–2496.

Phan–Van–Loc (1955). Ann. Fac. Sci. Univ. Toulouse **18**, 178–192.

Phan–Van–Loc (1958a). Interprétation physique de l'expression mathématique du principe de Huygens en théorie de l'électron de Dirac. Cahiers de Physique No. 97, **12**, 327–340.

Phan–Van–Loc (1958b). C.R. Acad. Sci. Paris **246**, 388–390.

Phan–Van–Loc (1960). *Principes de Huygens en Théorie de l'Electron de Dirac*. Thèse, Toulouse.

Rang, O. (1964). Optik **21**, 59–65.

Rang, O. (1977). Ultramicroscopy **2**, 149–151.

Reimer, L. (1985). *Scanning Electron Microscopy* (Springer, Berlin and New York).

Rubinowicz, A. (1934). Acta Phys. Polon. **3**, 143–163.

Rubinowicz, A. (1957). *Die Beugungswelle in der Kirchhoffschen Theorie der Beugung* (Państwowe Wydawnictwo Naukowe, Warsaw); (1966) 2nd ed. (PWN, Warsaw and Springer, Berlin and New York).

Rubinowicz, A. (1963). Acta Phys. Polon. **23**, 727–744.

Rubinowicz, A. (1965). The Miyamoto–Wolf diffraction wave. Prog. Opt. **4**, 199–240. See Part II, Section 3: Diffraction wave in the Kirchhoff theory of Dirac-electron waves.

Saxton, W.O., O'Keefe, M.A., Cockayne, D.J. and Wilkens, M. (1983). Ultramicroscopy **12**, 75–77 ; corrigenda *ibid.* **13**, 349–350.

Sommerfeld, A. (1954). *Optics* (Academic Press, New York).

Spence, J.C.H. (1988). *Experimental High-resolution Electron Microscopy*, 2nd ed. (Oxford University Press, New York and Oxford).

Part XII, Chapters 61–63

Ade, G. (1973). *Der Einfluss der Bildfehler dritter Ordnung auf die elektronenmikros-kopische Abbildung und die Korrektur dieser Fehler durch holographische Rekon-struktion*. Dissertation, Braunschweig and PTB–Bericht APh–3.

Ade, G. (1980). The Hague, vol. 1, pp. 138–139.

Ade, G. (1982). Hamburg, vol. 1, pp. 425–426.

Ade, G. (1986). Kyoto, vol. 1, pp. 687–688.

Ade, G. (1988). York, vol. 1, pp. 201–202.

Ade, G. (1992). Granada, vol. 1, pp. 489–490.

Ade, G. (1994). Digital techniques in electron off-axis holography. Adv. Electron. Electron Phys. **89**.

Ade, G. and Lauer, R. (1988). York, vol. 1, pp. 203–204.

Ade, G. and Lauer, R. (1990). Seattle, vol. 1, pp. 232–233.

Ade, G. and Lauer, R. (1991). Optik **88**, 103–108.

Ade, G. and Lauer, R. (1992a). PTB-Mitt. **102**, 181–187.

Ade, G. and Lauer, R. (1992b). Optik **91**, 5–10.

Ade, G., Hanszen, K.-J. and Lauer, R. (1984). Budapest, vol. 1, pp. 283–284.

Aharonov, Y. and Bohm, D. (1959). Phys . Rev. **115**, 485–491.

Aharonov, Y. and Bohm, D. (1961). Phys . Rev. **123**, 1511–1524.

Allard, L.F., Nolan, T.A., Joy, D.C. and Hashimoto, H. (1992). EMSA **50**, pp. 944–945.

Anaskin, I.F. and Stoyanova, I.G. (1967). Dokl. Akad. Nauk SSSR **174**, 56–59 ; Sov. Phys. Dokl. **12**, 447–450.

Anaskin, I. and Stoyanova, I. (1968a). Rome, vol. 1, pp. 149–150.

Anaskin, I. and Stoyanova, I.G. (1968b). Zh. Eksp. Theor. Fiz. **54**, 1687–1689; J. Exp. Theor. Phys. **27**, 904–905.

Anaskin, I.F. and Stoyanova, I.G. (1968c). Radiotekh. Elektron. **13**, 913–920 and 1031–1040 ; Radio Eng. Electron Phys. **13**, 789–794 and 895–902.

Anaskin, I.F. and Stoyanova, I.G. (1972). Manchester, pp. 636–637.

Anaskin, I.F., Stoyanova, I.G. and Chyapas, A.F. (1966). Izv. Akad. Nauk SSSR (Ser. Fiz.) **30**, 766–768; Bull. Acad. Sci. USSR (Phys. Ser.) **30**, 793–796.

Anaskin, I.F., Stoyanova, I.G. and Shpagina, M.D. (1968). Izv. Akad. Nauk SSSR **32**, 1016–1021 ; Bull. Acad. Sci. USSR **32**, 941–946..

Arii, T., Mihama, K., Matsuda, T. and Tonomura, A (1981). J. Electron Microsc. **30**, 121–127.

Banzhof, H. and Herrmann, K.-H. (1993). Ultramicroscopy **48**, 475–481.

Banzhof, H., Herrmann, K.-H. and Lichte, H. (1988). York, vol. 1, pp. 263–264.

Banzhof, H., Herrmann, K.-H. and Lichte, H. (1992). Microsc. Res. Tech. **20**, 450–456.

Bartell, L.S. (1972). Trans. Am. Cryst. Assoc. **8**, 37–57.

Bartell, L.S. (1975). Optik **43**, 373–390 and 403–418.

Bartell, L.S. and Johnson, R.D. (1977). Nature **268**, 707–708.

Bartell, L.S. and Ritz, C.L. (1974). Science **185**, 1163–1165.

Bates, R.H. and Lewitt, R.M. (1975). Optik **44**, 1–16.

Bayh, W. (1962). Z. Physik **169**, 492–510.

Boersch, H. (1967). Phys. Blätt. **23**, 393–404.

Boersch, H. and Lischke, B. (1970a). Z. Physik **237**, 449–468.

Boersch, H. and Lischke, B. (1970b). Grenoble, vol. 1, pp. 69–70.

Boersch, H., Hamisch, H.J., Wohlleben, D. and Grohmann, K. (1960). Z. Physik **159**, 397–404.

Boersch, H., Hamisch, H., Grohmann, K. and Wohlleben, D. (1961). Z. Physik **165**, 79–93.

Boersch, H., Hamisch, H.J., Grohmann, K. and Wohlleben, D. (1962a). Z. Physik **167**, 72–82.

Boersch, H., Hamisch, H. and Grohmann, K. (1962b). Z. Physik **169**, 263–272.

Boersch, H., Lischke, B. and Söllig, H. (1974). Phys. Stat. Sol. (b) **61**, 215–222.

Bonhomme, P. and Beorchia, A. (1980). The Hague, vol. 1, pp. 134–135.

Bonnet, N., Troyon, M. and Gallion, P. (1978). Toronto, vol. 1, pp. 222–223.

Borzjak, P.G., Kuljupin, Ju. A., Nepijko, S.A., Ostranitsa, A.P. and Shamonja, V.G. (1977). Dopovidi Akad. Nauk. Ukr. RSR, Ser. A, No. 10, 920–922.

Boseck, S., Block, H., Schmidt, B. and Reuber, E. (1986). Kyoto, vol. 1, pp. 683–684.

Braun, K.-J. (1972). *Untersuchung der Kohärenzeigenschaften von Elektronenwellen mit dem Elektroneninterferometer.* Diplomarbeit, Tübingen.

Brünger, W. (1968). Naturwissenschaften **55**, 295.

Brünger, W. (1972). Z. Physik **250**, 263–272.

Brünger, W. and Klein, M. (1977). Surface Sci. **62**, 317–320.

Buhl, R. (1958). Berlin, vol. 1, pp. 233–234.

Buhl, R. (1959). Z. Physik **155**, 395–412.

Buhl, R. (1961a). Z. angew. Phys. **13**, 232–235.

Buhl, R. (1961b). Naturwissenschaften **48**, 298–299.

Chambers, R.G. (1960). Phys. Rev. Lett. **5**, 3–5.

Chen, J.W., Matteucci, G., Missiroli, G.F. and Pozzi, G. (1987). SIME, pp. 259–260.

Chen, J., Hirayama, T., Lai, G., Tanji, T., Ishizuka, K. and Tonomura, A. (1993). Opt. Lett. **18**, 1887–1889.

Costa, A., Matteucci, G. and Patuelli, S. (1989). SIME, pp. 267–268.

Cowley, J.M. (1990). Ultramicroscopy **34**, 293–297.

Cowley, J.M. (1991). EMSA **49**, pp. 650–651.

Cowley, J.M. (1992). Ultramicroscopy **41**, 335–348.

Cowley, J.M. and Walker, D.J. (1981). Ultramicroscopy **6**, 71–75.

Crewe, A.V. and Saxon, J. (1971). EMSA **29**, 12–13.

Daberkow, I., Dreher, W., Fu, Q., Lichte, H. and Völkl, E. (1988). York, vol. 1, pp. 193–194.

Donati, O., Missiroli, G.F. and Pozzi, G. (1973). Am. J. Phys. **41**, 639–644.

Drahoš, V. and Delong, A. (1963). Čs. Čas. Fys. **13**, 278–286.

Drahoš, V. and Delong, A. (1964). Opt. Acta **11**, 173–181.

Düker, H. (1955). Z. Naturforsch. **10a**, 256.

Durand, M., Faget, J., Ferré, J. and Fert, C. (1958). C. R. Acad. Sci. Paris **247**, 590–593.

Dyson, J. (1950). Paris, vol. 1, pp. 126–128.

Ehrenberg, W. and Siday, R.E. (1949). Proc. Phys. Soc. (London) **B62**, 8–21.

Endo, J. and Tonomura, A. (1990). Mat. Trans. JIM **31**, 551–560.

Endo, J., Matsuda, T. and Tonomura, A. (1979). Japan. J. Appl. Phys. **18**, 2291–2294.

Endo, J., Osakabe, N., Matsuda, T., Yano,, S., Yamada, H., Kawasaki, T. and Tonomura, A. (1986). EMSA **44**, pp. 614–615.

Endo, J., Kawasaki, T., Masuda, T. and Tonomura, A. (1989). EMSA **47**, pp. 104–105.

Estrada, F.F.M., Misiroli, G.F. and Nichelatti, E. (1991). SIME, pp. 411–412.

Fabbri, R., Matteucci, G. and Pozzi, G. (1987). Phys. Stat. Sol. (a) **102**, K127–K129.

Faget, J. (1961). Rev. Opt. **40**, 347–381.

Faget, J. and Fert, C. (1956). C.R. Acad. Sci. Paris **243**, 2028–2029.

Faget, J. and Fert, C. (1957). Cahiers de Physique **11**, 285–296.

Faget, J., Ferré, J. and Fert, C. (1958). C.R. Acad. Sci. Paris **246**, 1404–1407.

Faget, J., Fagot, M. and Fert, C. (1960). Delft, vol. 1, pp. 18–24.

Fagot, M., Ferré, J. and Fert, C. (1961). C.R. Acad. Sci. Paris **252**, 3766–3768.

Feltynowski, A. (1963). Z. angew. Phys. **15**, 312–315.

Fert, C. (1961). In *Traité de Microscopie Electronique* (Magnan, C., ed.) vol. 1, pp. 333–390 (Hermann, Paris).

Fert, C. (1962). J. Electronmicrosc. **11**, 1–9.

Fert, C. and Faget, J. (1958). Berlin, vol. 1, pp. 234–239.

Fert, C., Faget, J., Fagot, M. and Ferré, J. (1962a). J. Microscopie **1**, 1–12.

Fert, C., Faget, J., Fagot, M. and Ferré, J. (1962b). J. Phys. Soc. Japan **17**, Suppl. B–II, 186–190.

Fink, H.-W. and Kreuzer, H.J. (1992). Phys. Rev. Lett. **68**, 3257.

Fink, H.-W., Schmid, H., Kreuzer, H.J. and Wierbicki, A. (1991). Phys. Rev. Lett. **67**, 1543–1546. Cf. Spence *et al.* (1992).

Fischer, D. and Lischke, B. (1967). Z. Physik **205**, 458–464.

Fowler, H.A., Marton, L., Simpson, J.A. and Suddeth, J.A. (1961). J. Appl. Phys. **32**, 1153–1155.

Frabboni, S., Matteucci, G., Pozzi, G. and Vanzi, M. (1985). Phys. Rev. Lett. **55**, 2196–2199.

Frabboni, S., Matteucci, G., Missiroli, G.F., Pozzi, G., Pizzochero, G. and Vanzi, M. (1986). Kyoto, vol. 1, pp. 685–686.

Frabboni, S., Matteucci, G. and Pozzi, G. (1987). Ultramicroscopy **23**, 29–38.

Franke, F.J., Herrmann, K.-H. and Lichte, H. (1986). Kyoto, vol. 1, pp. 677–678.

Franke, F.J., Hermann, K.-H. and Lichte, H. (1987). Niagara Falls, pp. 59–67.

Frost, B. and Lichte, H. (1988). York, vol. 2, pp. 267–268.

Fu, S., Chen, J.W., Wang, Z. and Cao, H. (1987). Optik **76**, 45–47.

Fu, Q., Lichte, H. and Völkl, E. (1991). Phys. Rev. Lett. **67**, 2319–2322.

Fukuhara, A., Shinagawa, K., Tonomura, A. and Fujiwara, H. (1983). Phys. Rev. **B27**, 1839–1843.

Gabor, D. (1948). Nature **161**, 777–778.

Gabor, D. (1949a). Delft, pp. 55–59.

Gabor, D. (1949b). Proc. Roy. Soc. (London) **A197**, 454–487.

Gabor, D. (1950). Paris, vol. 1, pp. 129–137.

Gabor, D. (1951a). Proc. Phys. Soc. (London) **B64**, 449–469.

Gabor, D. (1951b). Washington, pp. 237–245.

Gabor, D. (1956). Rev. Mod. Phys. **28**, 260–276.

Gabor, D. (1968/69). Optik **28**, 437–441.

Gabor, D. (1980). Israel J. Technol. **18**, 209–213.

Gabor, D., Stroke, G.W., Restrick, R., Funkhouser, A. and Brumm, D. (1965). Phys. Lett. **18**, 116–118.

Gajdardziska–Josifovska, M., McCartney, M.R., Ruijter, W.J. de, Smith, D.J., Weiss, J.K. and Zuo, J.M. (1993). Ultramicroscopy **50**, 285–299.

Garcia, N. (1989). SIME, pp. 255–256.

Greenaway, A.H. and Huiser, A.M.J. (1976). Optik **45**, 295–300.

Gribelyuk, M.A. and Cowley, J.M. (1991). EMSA **49**, pp. 684–685.

Gribelyuk, M.A. and Cowley, J.M. (1992a) Granada, vol. 1, pp. 649–650.

Gribelyuk, M.A. and Cowley, J.M. (1992b). Ultramicroscopy **45**, 103–113.

Gribelyuk, M.A. and Cowley, J.M. (1993). Ultramicroscopy **50**, 29–40.

Haine, M.E. and Dyson, J. (1950). Nature **166**, 315–316.

Haine, M.E. and Mulvey, T. (1950). Paris, vol. 1, pp. 120–125.

Haine, M.E. and Mulvey, T. (1951). Washington, pp. 247–250.

Haine, M.E. and Mulvey, T. (1952). J. Opt. Soc. Am. **42**, 763–773.

Hanszen, K.-J. (1969). Z. Naturforsch. **24a**, 1849.

Hanszen, K.-J. (1970a). Grenoble, vol. 1, pp. 21–22.

Hanszen, K.-J. (1970b). Optik **32**, 74–90.

Hanszen, K.-J. (1971). Adv. Opt. Electron Microsc. **4**, 1–84.

Hanszen, K.-J. (1972a). Optik **35**, 431–444.

Hanszen, K.-J. (1972b). Optik **36**, 41–54.

Hanszen, K.-J. (1973a). In *Image Processing and Computer-aided Design in Electron Optics* (Hawkes, P.W., ed.) pp. 16–53 (Academic Press, London and New York).

Hanszen, K.-J. (1973b). PTB–Bericht APh–4, 30 pp.

Hanszen, K.-J. (1974). Optik **39**, 520–542.

Hanszen, K.-J. (1976). Jerusalem, vol. 1, pp. 95–96.

Hanszen, K.-J. (1980). The Hague, vol. 1, pp. 136–137.

Hanszen, K.-J. (1982a). Holography in electron microscopy. Adv. Electron. Electron Phys. **59**, 1–77.

Hanszen, K.-J. (1982b). Ultramicroscopy **9**, 159–166.

Hanszen, K.-J. (1982c). PTB–Bericht APh–17, 34 pp.

Hanszen, K.-J. (1982d). Hamburg, vol. 1, pp. 423–424.

Hanszen, K.-J. (1983). Optik **65**, 153–177.

Hanszen, K.-J. (1984). Budapest, vol. 1, pp. 279–280.

Hanszen, K.-J. (1985). Optik **71**, 155–162.

Hanszen, K.-J. (1986a). J. Phys. D: Appl. Phys. **19**, 373–395.

Hanszen, K.-J. (1986b). Beijing, pp. 66–69.

Hanszen, K.-J. (1987). Optik **77**, 57–61.

Hanszen, K.-J. (1990a). 40 Jahre elektronenoptische Forschung in der Physikalisch-Technischen Bundsanstalt. PTB–Bericht APh–33 (2 parts).

Hanszen, K.-J. (1990b). PTB–Mitt. **100**, 363–368.

Hanszen, K.-J. and Ade, G. (1974). PTB–Bericht APh–5, 68 pp.

Hanszen, K.-J. and Ade, G. (1976a). Jerusalem, vol. 1, pp. 446–447.

Hanszen, K.-J. and Ade, G. (1976b). PTB–Bericht APh–10, 38 pp.

Hanszen, K.-J. and Ade, G. (1977). A consistent Fourier optical representation of image formation in the conventional fixed beam electron microscope, in the scanning transmission electron microscope and of holographic reconstruction. PTB–Bericht APh–11, 31pp.

Hanszen, K.-J. and Ade, G. (1983). Optik **63**, 247–264.

Hanszen, K.-J. and Ade, G.(1984). Optik **68**, 81–95.

Hanszen, K.-J. and Lauer, R. (1980). The Hague, vol. 1, pp. 140–141.

Hanszen, K.-J., Ade, G. and Lauer, R. (1972a). Optik **35**, 567–590.

Hanszen, K.-J., Lauer, R. and Ade, G. (1972b). Optik **36**, 156–159.

Hanszen, K.-J., Lauer, R. and Ade, G. (1980). PTB–Bericht APh–15, 38 pp.

Hanszen, K.-J., Lauer, R. and Ade, G. (1981). PTB–Bericht APh–16, 35 pp.

Hanszen, K.-J., Lauer, R. and Ade, G. (1982). PTB–Bericht APh–19, 37 pp.

Hanszen, K.-J., Lauer, R. and Ade, G. (1983a). Optik **63**, 285–303.

Hanszen, K.-J., Lauer, R. and Ade, G. (1983b). PTB–Bericht APh–20, 35 pp.

Hanszen, K.-J., Ade, G. and Lauer, R. (1984). PTB–Bericht APh–23, 31 pp.

Hanszen, K.-J., Lauer, R. and Ade, G. (1985). PTB–Bericht APh–25, 56 pp.

Hanszen, K.-J., Lauer, R. and Ade, G. (1986). PTB–Bericht APh–30, 49 pp.

Harada, K. and Shimizu, R. (1991). J. Electron Microsc. **40**, 92–96.

Harada, K., Endoh, H. and Shimizu, R. (1988). J. Electron Microsc. **37**, 199–201.

Harada, K., Ogai, K. and Shimizu, R. (1990a). J. Electron Microsc. **39**, 465–469.

Harada, K., Ogai, K. and Shimizu, R. (1990b). J. Electron Microsc. **39**, 470–476.

Harada, K., Matsuda, T., Bonevich, J., Igarashi, M., Kondo, S., Pozzi, G., Kawabe, U. and Tonomura, A. (1992). Nature **360**, 51–53.

Hasegawa, S., Kawasaki, T., Endo, J., Tonomura, A., Honda, Y., Futamoto, M., Yoshida, K., Kugiya, F. and Koizumi, M. (1989). J. Appl. Phys. **65**, 2000–2004.

Hasselbach, F. (1988). Z. Physik **B71**, 443–449.

Hasselbach, F. (1992). Recent contributions of electron interferometry to wave–particle duality. In *Wave-Particle Duality* (Selleri, F., ed.) pp. 109–125 (Plenum, New York and London).

Hasselbach, F. and Nicklaus, M. (1986). Kyoto, vol. 1, pp. 691–692.

Hasselbach, F. and Nicklaus, M. (1988). Physica **B151**, 230–234.

Hasselbach, F. and Nicklaus, M. (1990). Seattle, vol. 1, pp. 212–213.

Hasselbach, F. and Nicklaus, M. (1993). Phys. Rev. **A48**, 143–151.

Hawkes, P.W. (1978). Adv. Opt. Electron Microsc. **7**, 101–184.

Hawkes, P.W. (1980). In *Computer Processing of Electron Microscope Images* (Hawkes, P.W., ed.) pp. v–vi (Springer, Berlin and New York).

Hawkes, P.W. (1992). Ultramicroscopy **41**, 441.

Herman, R.M. (1992). Found. Phys. **22**, 713–725.

Herring, R.A. and Tanji, T. (1993). MSA **51**, pp. 1086–1087.

Herring, R.A., Tanji, T. and Tonomura, A. (1992). EMSA **50**, pp. 990–991.

Herring, R.A., Pozzi, G., Tanji, T. and Tonomura, A. (1993a). Ultramicroscopy **50**, 94–100.

Herring, R.A., Pozzi, G., Tanji, T. and Tonomura, A. (1993b). MSA **51**, pp. 1056–1057.

Herrmann, K.-H. and Lichte, H. (1986). Kyoto, vol. 1, pp. 679–680.

Herrmann, K.-H., Reuber, E. and Schiske, P. (1978). Toronto, vol. 1, pp. 226–227.

Hibi, T. (1956). J. Electronmicrosc. **4**, 10–15.

Hibi, T. and Takahashi, S. (1963). J. Electronmicrosc. **12**, 129–133.

Hibi, T. and Takahashi, S. (1969). Z. angew. Phys. **27**, 132–138.

Hibi, T. and Yada, K. (1976). Electron interference microscope. In *Principles and Techniques of Electron Microscopy* (Hayat, M.A., ed.) vol. 6, pp. 312–343 (Van Nostrand Reinhold, New York and London).

Hoffmann, H. and Jönsson, C. (1965). Z. Physik **182**, 360–365.

Home, D. and Selleri, F. (1992). The Aharonov–Bohm effect from the point of view of local realism. In *Wave–Particle Duality* (Selleri, F., ed.) pp. 127–137 (Plenum, New York and London).

Ishizuka, K. (1993). Ultramicroscopy **52**, 1–5.

Ishizuka, K. (1994). Ultramicroscopy

Ishizuka, K., Tanji, T. and Tonomura, A. (1991). Cambridge, pp. 423–432.

Jönsson, C. (1961). Z. Physik **161**, 454–474.

Jönsson, C., Hoffmann, H. and Möllenstedt, G. (1965). Phys. kond. Mat. **3**, 193–199.

Jönsson, C., Brandt, D. and Hirschi, S. (1974). Am. J. Phys. **42**, 4–11.

Joy, D.C., Zhang, Y.-S., Zhang, X., Hashimoto, T., Bunn, R.D., Allard, L. and Nolan, T.A. (1993). Ultramicroscopy **51**, 1–14.

Kamefuchi, S., Ezawa, H., Murayama, Y., Namiki, M., Nomura, S., Ohnuki, Y. and Yajima, T., eds (1984). *Proc. Int. Symp. Foundations of Quantum Mechanics in the Light of New Technology* , Tokyo 1983 (Physical Society of Japan, Tokyo).

Kasper, E. (1992). Optik **92**, 45–47.

Kawasaki, T. and Rodenburg, J.M. (1993). Ultramicroscopy **52**, 248–252.

Kawasaki, T., Endo, J., Matsuda, T. and Tonomura, A. (1990a). Seattle, vol. 1, pp. 222–223.

Kawasaki, T., Matsuda, T., Endo, J. and Tonomura, A. (1990b). Japan. J. Appl. Phys. **29**, L508–L510.

Kawasaki, T., Ru, Q.X., Matsuda, T., Bando, Y. and Tonomura, A. (1991a). Japan. J. Appl. Phys. **30**, L1830–L1832.

Kawasaki, T., Ru, Q.X., Matsuda , T. and Tonomura, A. (1991b). EMAG, pp. 483–486.

Kawasaki, T., Pozzi, G. and Tonomura, A. (1992a). Granada, vol. 1, pp. 651–652.

Kawasaki, T., Ru, Q.X. and Tonomura, A. (1992b). Granada, vol. 1, pp. 653–654.

Keller, M. (1958). Berlin, vol. 1, pp. 230–232.

Keller, M. (1961). Z. Physik **164**, 274–291 and 292–294.

Keller, A. (1962). Optik **19**, 117–121.

Kerschbaumer, E. (1967). Z. Physik **201**, 200–208.

Kobayashi, S.-i., Ezawa, H., Murayama, Y. and Nomura, S., eds (1990). *Proc. 3rd. Int. Symp. Foundations of Quantum Mechanics in the Light of New Technology* (Physical Society of Japan, Tokyo).

Kobe, D.H., Aguilera-Navarro, V.C. and Ricotta, R.M. (1992). Phys. Rev. A **45**, 6192–6197.

Komrska, J. (1971). Scalar diffraction theory in electron optics. Adv. Electron. Electron Phys. **30**, 139–234.

Komrska, J. (1975). Čs. Čas. Fyz. **A25**, 1–13.

Komrska, J. and Lenc, M. (1970). Grenoble, vol. 1, pp. 67–68.

Komrska, J. and Vlachová, B. (1973). Opt. Acta **20**, 207–215.

Komrska, J., Drahoš, V. and Delong, A. (1964a). Czech. J. Phys. **B14**, 753–756.

Komrska, J., Drahoš, V. and Delong, A. (1964b). Opt. Acta **11**, 145–157.

Komrska, J., Drahoš, V. and Delong, A. (1967). Opt. Acta **14**, 147–167.

Konnert, J. and d'Antonio, P. (1992). Ultramicroscopy **45**, 281–290.

Krimmel, E. (1960). Z. Physik **158**, 35–38.

Krimmel, E. (1961). Z. Physik **163**, 339–355.

Krimmel, E., Möllenstedt, G. and Rothemund, W. (1964). Appl. Phys. Lett. **5**, 209–210.

Kulick, J.H., Ball, M.D., Benton, S.A., Gallerneault, C., Krantz, E., Wilson, L. and Whitmarsh, S. (1987). EMAG, pp. 213–216.

Kulyupin, Yu. A., Nepijko, S.A., Sedov, N.N. and Shamonya, V.G. (1978/79). Optik **52**, 101–109.

Kuper, C. G. (1980). Phys. Lett. **79A**, 413–416.

Laberrigue, A., Balossier, G., Beorchia, A., Bonhomme, P., Bonnet, N. and Troyon, M. (1980). J. Microsc. Spectrosc. Electron. **5**, 655–664.

Langbein, W. (1958). Naturwissenschaften **45**, 510–511.

Lannes, A. (1978). Toronto, vol. 1, pp. 228–229.

Lannes, A. (1980). J. Math. Anal. Applics **74**, 530–559.

Lannes, A. (1982). J. Optics (Paris) **13**, 27–39.

Lau, B. and Pozzi, G. (1978). Optik **51**, 287–296.

Lauer, R. (1982). Hamburg, vol. 1, pp. 427–428.

Lauer, R. (1984a). Optik **66**, 159–174.

Lauer, R. (1984b). Optik **67**, 291–293.

Lauer, R. and Ade, G. (1990). Seattle, vol. 1, pp. 230–231.

Lauer, R. and Ade, G. (1992). Granada, vol. 1. pp. 523–524.

Lauer, R. and Hanszen, K.-J. (1986). Kyoto, vol. 1, pp. 681–682.

Lauer, R. and Lickfeld, K. (1988). York, vol. 3, pp. 351–352.

Leith, E.N. and Upatnieks, J. (1962). J. Opt. Soc. Am. **52**, 1123–1130.

Leith, E.N. and Upatnieks, J. (1963). J. Opt. Soc. Am. **53**, 1377–1381.

Leith, E.N. and Upatnieks, J. (1967). Prog. Opt. **6**, 1–52.

Lenz, F. (1962). Phys. Blätt. **18**, 305–307.

Lenz, F. (1965). Optik **22**, 270–288.

Lenz, F. (1972). Z. Physik **249**, 462–464.

Lenz, F. (1987). In *Proc. 2nd Int. Symp. Foundations of Quantum Mechanics*, Tokyo 1986 (Namiki, M., Ohnuki, Y., Murayama, Y. and Nomura, S., eds) pp. 112–116 (Physical Society of Japan, Tokyo).

Lenz, F. (1988). Optik **79**, 13–14.

Lenz, F. and Völkl, E. (1990). Seattle, vol. 1, pp. 228–229.

Lenz, F. and Wohland, G. (1984). Optik **67**, 315–329.

Leuthner, T., Lichte, H. and Herrmann, K.-H. (1988). York, vol. 1, pp. 177–178.

Leuthner, T., Lichte, H. and Herrmann, K.-H. (1989). Phys. Stat. Sol. (a) **116**, 113–132.

Leuthner, T., Lichte, H., Herrmann, K.-H. and Sum, J. (1990). Seattle, vol. 1, pp. 224–225.

Lichte, H. (1979). PTB–Mitt. 229–236.

Lichte, H. (1980a). Optik **57**, 35–67.

Lichte, H. (1980b). The Hague, vol. 1, pp. 30–31.

Lichte, H. (1982). Hamburg, vol. 1, pp. 411–418.

Lichte, H. (1984a). Budapest, vol. 1, pp. 281–282.

Lichte, H. (1984b). In *Proc. Int. Symp. Foundations of Quantum Mechanics in the Light of New Technology*, Tokyo 1983 (Kamefuchi, S., Ezawa, H., Murayama, Y., Namiki, M., Nomura, S., Ohnuki, Y. and Yajima, T., eds) pp. 29–38 (Physical Society of Japan, Tokyo).

Lichte, H. (1985). Optik **70**, 176–177.

Lichte, H. (1986a). Ultramicroscopy **20**, 293–304.

Lichte, H. (1986b). Kyoto, vol. 1, pp. 675–676.

Lichte, H. (1986c). Beijing, pp. 150–153.

Lichte, H. (1986d). Ann. New York Acad. Sci. **480**, 175–189.

Lichte, H. (1988). Physica **B151**, 214–222.

Lichte, H. (1989). EMSA **47**, 2–3.

Lichte, H. (1990). Seattle, vol. 1, pp. 208–209.

Lichte, H. (1991a). Electron image plane off-axis holography of atomic structures. Adv. Opt. Electron Microsc. **12**, 25–91.

Lichte, H. (1991b). Ultramicroscopy **38**, 13–22.

Lichte, H. (1991c). EMAG, pp. 465–471.

Lichte, H. (1991d). EMSA **49**, pp. 494–495.

Lichte, H. (1991e). Cambridge, pp. 433–440.

Lichte, H. (1992a). Granada, vol. 1, pp. 637–641.

Lichte, H. (1992b). Beijing, vol. 1, pp. 234–237.

Lichte, H. (1992c). Ultramicroscopy **47**, 223–230.

Lichte, H. (1993). Ultramicroscopy **51**, 15–20.

Lichte, H. and Hornstein, R. (1982). Hamburg, vol. 1, pp. 431–432.

Lichte, H. and Möllenstedt, G. (1977). In *Eighth Int. Cong. X-ray Optics and Microanalysis* (Beaman, D.R., Ogilvie, R.E. and Wittry, D.B., eds) pp. 211–212 (Pendell, Midland MI).

Lichte, H. and Möllenstedt, G. (1979). J. Phys. E: Sci. Instrum. **12**, 941–944.

Lichte, H. and Völkl, E. (1988). York, vol. 1, pp. 191–192.

Lichte, H. and Völkl, E. (1991). EMSA **49**, pp. 670–671.

Lichte, H. and Weierstall, U. (1988). York, vol. 3, pp. 325–326.

Lichte, H., Möllenstedt, G. and Wahl, H. (1972). Z. Physik **249**, 456–461.

Lichte, H., Herrmann, K.-H. and Lenz, F. (1987). Optik **77**, 135–140.

Lichte, H., Herrmann, K.-H. and Lenz, F. (1988). York, vol. 1, pp. 189–190.

Lichte, H. , Völkl, E. and Scheerschmidt, K. (1992). Ultramicroscopy **47**, 231–240.

Lin, J.A. and Cowley, J.M. (1986). Ultramicroscopy **19**, 179–190.

Lischke, B. (1969). Phys. Rev. Lett. **22**, 1366–1368.

Lischke, B. (1970). Z. Physik **237**, 469–474 and **239**, 360–378.

Mankos, M., Wang, S.-Y., Weiss, J.K. and Cowley, J.M. (1992). EMSA **50**, pp. 102–103.

Marton, L. (1952). Phys. Rev. **85**, 1057–1058.

Marton, L. (1954). London, pp. 272–279.

Marton, L., Simpson, J.A. and Suddeth, J.A. (1953). Phys. Rev. **90**, 490–491.

Marton, L., Simpson, J.A. and Suddeth, J.A. (1954). Rev. Sci. Instrum. **25**, 1099–1104.

Matsuda, T., Tonomura, A. and Komoda, T. (1978). Japan. J. Appl. Phys. **17**, 2073–2074.

Matsuda, T., Tonomura, A., Suzuki, R., Endo, J., Osakabe, N., Umezaki, H., Tanabe, H., Sugita, Y. and Fujiwara, H. (1982). J. Appl. Phys. **53**, 5444–5446.

Matsuda, T., Hasegawa, S., Igarashi, M., Kobayashi, T., Naito, M., Kajiyama, H., Endo, J., Osakabe, N. and Tonomura, A. (1989). Phys. Rev. Lett. **62**, 2519–2522.

Matsuda, T., Hasegawa, S., Endo, J., Osakabe, N., Tonomura, A. and Aoki, R. (1990). Seattle, vol. 1, pp. 210–211.

Matsuda, T., Fukuhara, A., Yoshida, T., Hasegawa, S., Tonomura, A. and Ru, Q. (1991). Phys. Rev. Lett. **66**, 457–460.

Matsumoto, T., Osakabe, N., Endo, J., Matsuda, T. and Tonomura, A. (1991). EMSA **49**, pp. 686–687.

Matteucci, G. (1978). J. Microsc. Spectrosc. Electron. **3**, 69–71.

Matteucci, G. and Muccini, M. (1992). Granada, vol. 1, pp. 657–658.

Matteucci, G. and Pozzi, G. (1980). Ultramicroscopy **5**, 219–222.

Matteucci, G., Missiroli, G.F., Pozzi, G., Merli, P.G. and Vecchi, I. (1979). Thin Solid Films **62**, 5–17.

Matteucci, G., Missiroli, G.F. and Pozzi, G. (1981). Ultramicroscopy **6**, 109–114.

Matteucci, G., Missiroli, G.F. and Pozzi, G. (1982a). Ultramicroscopy **8**, 403–408.

Matteucci, G., Missiroli, G.F. and Pozzi, G. (1982b). Ultramicroscopy **10**, 247–252.

Matteucci, G., Missiroli, G.F. and Pozzi, G. (1982c). Hamburg, vol. 1, pp. 429–430.

Matteucci, G., Missiroli, G.F. and Pozzi, G. (1984). IEEE Trans. **MAG**–20, 1870–1875.

Matteucci, G., Missiroli, G.F., Nichelatti, E. Chen, J.W. and Pozzi, G. (1987). SIME, pp. 257–258.

Matteucci, G., Missiroli, G.F., Chen, J.W. and Pozzi, G. (1988a). Appl. Phys. Lett. **52**, 176–178.

Matteucci, G., Migliori, A., Pozzi, G. and Vanzi, M. (1988b). York, vol. 1, pp. 195–196.

Matteucci, G., Missiroli, G.F., Nichelatti, E. and Pozzi, G. (1988c). York, vol. 2, pp. 265–266.

Matteucci, G., Migliori, A., Missiroli, G. F., Nichelatti, E., Pozzi, G. and Vanzi, M. (1989). SIME, pp. 259–260.

Matteucci, G., Medina, F.F. and Pozzi, G. (1992a). Ultramicroscopy **41**, 255–268.

Matteucci, G., Missiroli, G.F., Muccini, M. and Pozzi, G. (1992b). Ultramicroscopy **45**, 77–83.

Medina, F.F. and Pozzi, G. (1990). J. Opt. Soc. Am. **A7**, 1027–1033.

Menu, C. and Evrard, D. (1971). C.R. Acad. Sci. Paris **B273**, 309–312.

Menzel, E., Mirandé, W. and Weingärtner, I. (1973). *Fourier-Optik und Holographie* (Springer, Vienna and New York) especially Section 10.9-a.

Merli, P.G. and Pozzi, G. (1978). Optik**51**, 39–48.

Merli, P.G., Missiroli, G.F. and Pozzi, G. (1974). J. Phys. E: Sci. Instrum. **7**, 729–732.

Merli, P.G., Missiroli, G.F. and Pozzi, G. (1976a). Am. J. Phys. **44**, 306–307.

Merli, P.G., Missiroli, G.F. and Pozzi, G. (1976b). Jerusalem, vol. 1, pp. 478–479.

Migliori, A. and Pozzi, G. (1991). SIME, pp. 407–408.

Migliori, A. and Pozzi, G. (1992). Ultramicroscopy **41**, 169–179.

Migliori, A., Pozzi, G. and Tonomura, A. (1993). Ultramicroscopy **49**, 87–94.

Missiroli, G.F., Pozzi, G. and Valdrè, U. (1981). J. Phys. E: Sci. Instrum. **14**, 649–671.

Missiroli, G.F., Muccini, M. and Pozzi, G. (1991). SIME, pp. 409–410

Möllenstedt, G. (1960). Delft, vol. 1, pp. 1–17.

Möllenstedt, G. (1962). Philadelphia, post-deadline paper.

Möllenstedt, G. (1987). Denshikenbikyo **21**, 190–200.

Möllenstedt, G. (1988). Physica **B151**, 201–205.

Möllenstedt, G. (1991). The invention of the electron Fresnel interference biprism. Adv. Opt. Electron. Microsc. **12**, 1–23.

Möllenstedt, G. and Bayh, W. (1961a). Phys. Verhandl. **12**, 142.

Möllenstedt, G. and Bayh, W. (1961b). Naturwissenschaften **48**, 400.

Möllenstedt, G. and Bayh, W. (1962a). Naturwissenschaften **49**, 81–82.

Möllenstedt, G. and Bayh, W. (1962b). Phys. Blätt. **18**, 299–305.

Möllenstedt, G. and Buhl, R. (1957). Phys. Blätt. **13**, 357–360.

Möllenstedt, G. and Düker, H. (1955). Naturwissenschaften **42**, 41.

Möllenstedt, G. and Düker, H. (1956). Z. Physik **145**, 377–397.

Möllenstedt, G. and Jönsson, C. (1959). Z. Physik **155**, 472–474.

Möllenstedt, G. and Keller, M. (1957). Z. Physik **148**, 34–37.

Möllenstedt, G. and Krimmel, E. (1964). Mikroskopie **19**, 29.

Möllenstedt, G. and Lenz, F. (1962). J. Phys. Soc. Japan **17**, Suppl. B–II, 183–186.

Möllenstedt, G. and Lichte, H. (1978a). Optik **51**, 423–428.

Möllenstedt, G. and Lichte, H. (1978b). Toronto, vol. 1, pp. 178–179.

Möllenstedt, G. and Lichte, H. (1979). Electron interferometry. In *Neutron Interferometry* (Bonse, U. and Rauch, H. eds) pp. 363–388 (Clarendon, Oxford).

Möllenstedt, G. and Lichte, H. (1989). Nova Acta Leopoldina **64**, No. 274, 57–62.

Möllenstedt, G. and Wahl, H. (1968). Naturwissenschaften **55**, 340–341.

Möllenstedt, G. and Wohland, G. (1980). The Hague, vol. 1, pp. 28–29.

Munch, J. (1975). Optik **43**, 79–99.

Munch, J. and Zeitler, E. (1974). EMSA **32**, pp. 386–387.

Namiki, M., Ohnuki, Y., Murayama, Y. and Nomura, S., eds (1987). *Proceedings of the 2nd International Symposium Foundations of Quantum Mechanics in the Light of New Technology* , Tokyo 1986 (Physical Society of Japan, Tokyo).

Nicklaus, M. (1989). *Ein Sagnac-Experiment mit Elektronenwellen.* Dissertation, Tübingen.

Nicklaus, M. and Hasselbach, F. (1993). Phys. Rev. A**48**, 152–160.

Ogai, K., Kimura, Y., Shimizu, R., Ishibashi, K., Aoyagi, Y. and Namba, S. (1991). Japan. J. Appl. Phys. **30**, 3272–3276.

Ohshita, A., Minamide, H., Saito, Y. and Tomita, H. (1990). Seattle, vol. 1, pp. 220–221.

Ohtsuki, M. and Zeitler, E. (1977). Ultramicroscopy **2**, 147–148.

Olivei, A. (1969). Optik **30**, 27–43.

O'Raifeartaigh, L., Straumann, N. and Wipf, A. (1991). Comments Nucl. Particle Phys. **20**, 15–22.

Osakabe, N. (1992). Microsc. Res. Tech. **20**, 457–462.

Osakabe, N., Yoshida, K., Horiuchi, Y., Matsuda, T., Tanabe, H., Okuwaki, T., Endo, J., Fujiwara, H. and Tonomura, A. (1983). Appl. Phys. Lett. **42**, 746–748.

Osakabe, N., Matsuda, T., Kawasaki, T., Endo, J., Tonomura, A., Yano, S. and Yamada, H. (1986). Phys. Rev. **A34**, 815–822.

Osakabe, N., Matsuda, T., Endo, J. and Tonomura, A. (1988). Japan. J. Appl. Phys. **27**, L1772–L1774.

Osakabe, N., Endo, J., Matsuda, T., Tonomura, A. and Fukuhara, A. (1989). Phys. Rev. Lett. **62**, 2969–2972.

Osakabe, N., Matsuda, T., Endo, J. and Tonomura, A. (1993). Ultramicroscopy **48**, 483–488.

Peshkin, M. and Tonomura, A. (1989). *The Aharonov–Bohm Effect* (Springer, Berlin and New York). Lecture Notes in Physics, vol. 340.

Plass, R. and Marks, L. D. (1992). EMSA **50**, pp. 984–985.

Pozzi, G. (1975). Optik **42**, 97–102.

Pozzi, G. (1977). Optik **47**, 105–107.

Pozzi, G. (1980a). The Hague, vol. 1, pp. 32–33.

Pozzi, G. (1980b). Optik **56**, 243–250.

Pozzi, G. (1983a). Optik **63**, 227–238.

Pozzi, G. (1983b). Optik **66**, 91–100.

Pozzi, G. (1992). In *Electron Microscopy in Materials Science* (Merli, P.G. and Antisari, M.V., eds) pp. 269–278 (World Scientific, Singapore, River Edge NJ and London).

Pozzi, G. (1993). SIME, pp. 3–6.

Pozzi, G. and Missiroli, G.F. (1973). J. Microscopie **18**, 103–108.

Pozzi, G. and Prola, R. (1987). SIME, pp. 255–256.

Pozzi, G. and Vanzi, M. (1982). Optik **60**, 175–180.

Qian, W., Scheinfein, M.R. and Spence, J.C.H. (1993). Appl. Phys. Lett. **62**, 315–317.

Rang, O. (1953). Z. Physik **136**, 464–479.

Rau, W.D., Lichte, H., Völkl, E. and Weierstall, U. (1991a). EMSA **49**, pp. 680–681.

Rau, W.D., Lichte, H., Völkl, E. and Weierstall, U. (1991b). J. Computer-assist. Microsc. **3**, 51–63.

Rogers, G.L. (1950a). Nature **166**, 237.

Rogers, G.L. (1950b). Nature **166**, 1027.

Rogers, G.L. (1952). Proc. Roy. Soc. Edinburgh **A63**, 193–221.

Rogers, G.L. (1970). Opt. Acta **17**, 527–538.

Rogers, J. (1978). In *Optica Hoy y Mañana* (Bescos, J., Hidalgo, A., Plaza, L. and Santamaria, J., eds) pp. 235–238 (Sociedad Española de Optica, Madrid).

Rogers, J. (1980). Electron holography. In *Imaging Processes and Coherence in Physics* (Schlenker, M., Fink, M. Goedgebuer, J.P., Malgrange, C., Viénot, J.C. and Wade, R.H., eds) pp. 365–370 (Springer, Berlin and New York).

Ru, Q., Endo, J., Tanji, T. and Tonomura, A. (1991a). Appl. Phys. Lett. **59**, 2372–2374.

Ru, Q., Matsuda, T., Fukuhara, A. and Tonomura, A. (1991b) J. Opt. Soc. Am. **A8**, 1739–1745.

Ru, Q., Harayama, T., Endo, J. and Tonomura, A. (1992a). Japan. J. Appl. Phys. **31**, 1919–1921.

Ru, Q., Hirayama, T. Endo, J. and Tonomura, A. (1992b). Granada, vol. 1, pp. 661–662.

Ruijter, W.J. de and Weiss, J.K. (1993). Ultramicroscopy **50**, 269–283.

Saldin, D. K. (1991). Adv. Mat. **3**, 159–161.

Saxon, G. (1972a). Optik **35**, 195–210.

Saxon, G. (1972b). Optik **35**, 359–375.

Schaal, G. (1971). Z. Physik **241**, 65–81.

Schaal, G., Jönsson, C. and Krimmel, E.F. (1966/67). Optik **24**, 529–538.

Scheinfein, M.R., Qian, W. and Spence, J.C.H. (1993). J. Appl. Phys. **73**, 2057–2068.

Schmid, H. (1984). Budapest, vol. 1, pp. 285–286.

Selleri, F., ed. (1992). *Wave–Particle Duality* (Plenum, New York and London).

Septier, A. (1959). C.R. Acad. Sci. Paris **249**, 662–664.

Simpson, J.A. (1954). Rev. Sci. Instrum. **25**, 1105–1109.

Simpson, J.A. (1956). Rev. Mod. Phys. **28**, 254–260.

Sonier, F. (1968). C.R. Acad. Sci. Paris **B267**, 187–190.

Sonier, F. (1970). C.R. Acad. Sci. Paris **B270**, 1536–1539.

Sonier, F. (1971). J. Microscopie **12**, 17–32.

Speidel, R. and Kurz, D. (1977). Optik **49**, 173–185.

Spence, J.C.H. and Qian, W. (1992). Phys Rev. **B45**, 10271–10279.

Spence, J.C.H., Zuo, J.M. and Qian, W. (1992). Phys. Rev. Lett. **68**, 3256. See Fink *et al.* (1991).

Spence, J.C.H., Cowley, J.M. and Zuo, J.M. (1993). Appl. Phys. Lett. **62**, 2446–2447. See Zhang *et al.* (1993a).

Stoyanova, I. and Anaskin, I. (1968). Rome, vol. 1, p. 161.

Stoyanova, I. and Anaskin, I. (1972). *Fizicheskie Osnovy Metodov Prosvechivayushchei Elektronnoi Mikroskopii* (Nauka, Moscow)..

Stroke, G.W. (1967). In *Rec. IEEE 9th Ann. Symp. Electron, Ion and Laser Beam Technology* (Pease, R.F.W., ed.) pp. 287–294 (San Francisco Press, San Francisco).

Stroke, G.W. and Halioua, M. (1972). Optik **35**, 50–65 and 489–508.

Stroke, G.W. and Halioua, M. (1973). Optik **37**, 192–203 and 249–264.

Stroke, G.W., Halioua, M., Saffir, A.J. and Evins, D.J. (1971a). Scanning Electron Microsc. (I), pp. 57–64.

Stroke, G.W., Halioua, M., Saffir, A.J. and Evins, D.J. (1971b). EMSA **29**, pp. 92–93.

Stroke, G.W., Halioua, M. and Saffir, A.J. (1973). Phys. Lett. **44A**, 115–117.

Stroke, G.W., Halioua, M., Thon, F. and Willasch, D. (1974). Optik **41**, 319–343.

Stroke, G.W., Halioua, M., Thon, F. and Willasch, D.H. (1977). Proc. IEEE **65**, 39–62.

Stumpp, H. (1984). Optik **68**, 193–207.

Stumpp, H., Lichte, H. and Möllenstedt, G. (1984). Optik **68**, 147–152.

Subbarao, A. (1991). Optik **89**, 91.

Takeda, M. and Ru, Q.-s. (1985). Appl. Opt. **24**, 3068–3071.

Takeguchi, M., Harada, K. and Shimizu, R. (1990). J. Electron Microsc. **39**, 269–272.

Takeguchi, M., Hanqing, C., Shibata, K. and Shimizu, R. (1992). Beijing, vol. 1, pp. 238–239.

Tanji, T., Urata, K. and Ishizuka, K. (1991) EMSA **49,** 672–673.

Tomita, H. and Savelli, M. (1968). C. R. Acad. Sci. Paris **B267**, 580–583.

Tomita, H., Matsuda, T. and Komoda, T. (1970a). Japan. J. Appl. Phys. **9**, 719.

Tomita, H., Matsuda, T. and Komoda, T. (1970b). Grenoble, vol. 1, pp. 151–152.

Tomita, H., Matsuda, T. and Komoda, T. (1972). Japan. J. Appl. Phys. **11**, 143–149.

Tong, S.Y., Li, H. and Huang, H. (1991). Phys. Rev. Lett. **67**, 3102–3105.

Tonomura, A. (1969). J. Electron Microsc. **18**, 77–78.

Tonomura, A. (1972). Japan. J. Appl. Phys. **11**, 493–502.

Tonomura, A. (1983). J. Magnetism Magnet. Mat. **31–34**, 963–969.

Tonomura, A. (1984). J. Electron Microsc. **33**, 101–115.

Tonomura, A. (1986a). Electron holography. Prog. Opt. **23**, 183–220.

Tonomura, A. (1986b). Kyoto, vol. 1, pp. 9–14.

Tonomura, A. (1987a). Applications of electron holography. Rev. Mod. Phys. **59**, 639–669.

Tonomura, A. (1987b). J. Appl. Phys. **61**, 4297–4302.

Tonomura, A. (1987c). SIME, pp. 151–154.

Tonomura, A. (1987d). SIME, pp. 321–324.

Tonomura, A. (1989). J. Electron Microsc. **38**, S43–S50.

Tonomura, A. (1990). Phys. Today **43**(4), 22–29.

Tonomura, A. (1991a). EMSA **49**, pp. 676–677.

Tonomura, A. (1991b). SIME, pp. 403–406.

Tonomura, A. (1992a). Adv. Phys. **41**, 59–103.

Tonomura, A. (1992b). Ultramicroscopy **47**, 419–424.

Tonomura, A. (1992c). Experiments on the Aharonov–Bohm effect. In *Wave–Particle Duality* (Selleri, F., ed.) pp. 291–299 (Plenum, New York and London).

Tonomura, A. (1992d). Beijing, vol. 1, pp. 230–233.

Tonomura, A. (1993). *Electron Holography* (Springer, Berlin and New York); Springer Series in Optical Sciences, vol. 70.

Tonomura, A. and Matsuda, T. (1980). The Hague, vol. 1, pp. 24–25.

Tonomura, A. and Watanabe, H. (1968). Nihon Butsuri Gakkai-shi [Proc. Phys. Soc. Japan] **23**, 683–684.

Tonomura, A., Fukuhara, A., Watanabe, H. and Komoda, T. (1968a). Japan. J. Appl. Phys. **7**, 295.

Tonomura, A., Fukuhara, A., Watanabe, H. and Komoda, T. (1968b). Rome, vol. 1, pp. 277–278.

Tonomura, A., Matsuda, T. and Komoda, T. (1978a). Toronto, vol. 1, pp. 224–225.

Tonomura, A., Matsuda, T. and Komoda, T. (1978b). Toronto, vol. 1, p. 670.

Tonomura, A., Matsuda, T. and Komoda, T. (1978c). Japan. J. Appl. Phys. **17**, 1137–1138.

Tonomura, A., Matsuda, T. and Endo, J. (1979a). Japan J. Appl. Phys. **18**, 9–14.

Tonomura, A., Matsuda, T. and Endo, J. (1979b). Japan J. Appl. Phys. **18**, 1373–1377.

Tonomura, A., Matsuda, T. and Endo, J. (1979c). Denshikenbikyo **14**, 47–52.

Tonomura, A., Endo, J. and Matsuda, T. (1979d). Optik **53**, 143–146.

Tonomura, A., Matsuda, T., Endo, J., Arii, T. and Mihama, K. (1980a). Phys. Rev. Lett. **44**, 1430–1433.

Tonomura, A., Matsuda, T., Endo, J., Arii, T. and Mihama, K. (1980b). The Hague, vol. 1, pp. 22–23.

Tonomura, A., Matsuda, T., Suzuki, R., Fukuhara, A., Osakabe, N., Umezaki, H., Endo, J., Shinagawa, K., Sugita, Y. and Fujiwara, H. (1982a). Phys. Rev. Lett. **48**, 1443–1446.

Tonomura, A., Matsuda, T, Tanabe, H., Osakabe, N., Endo, J., Fukuhara, A., Shinagawa, K. and Fujiwara, H. (1982b). Phys. Rev. **B25**, 6799–6804.

Tonomura, A., Matsuda, T., Tanabe, H., Osakabe, N. and Fujiwara, H. (1982c). Hamburg, vol. 1, pp. 419–420.

Tonomura, A., Matsuda, T., Endo, J., Suzuki, R., Osakabe, N., Umezaki, H., Sugita, Y. and Fujiwara, H. (1982d). Hamburg, vol. 1, pp. 421–422.

Tonomura, A., Umezaki, H., Matsuda, T., Osakabe, N., Endo, J. and Sugita, Y. (1984). In *Proc. Int. Symp. Foundations of Quantum Mechanics in the Light of New Technology* , Tokyo 1983 (Kamefuchi, S., Ezawa, H., Murayama, Y., Namiki, M., Nomura, S., Ohnuki, Y. and Yajima, T., eds) pp. 20–28 (Physical Society of Japan, Tokyo).

Tonomura, A., Matsuda, T., Kawasaki, T., Endo, J. and Osakabe, N. (1985). Phys. Rev. Lett. **54**, 60–62.

Tonomura, A., Osakabe, N., Matsuda, T., Kawasaki, T., Endo, J., Yano, S. and Yamada, H. (1986). Phys. Rev. Lett. **56**, 792–795.

Tonomura, A., Yano, S., Osakabe, N., Matsuda, T., Yamada, H. Kawasaki, T. and Endo, J. (1987). In *Proc. 2nd Int. Symp. Foundations of Quantum Mechanics* , Tokyo 1986 (Namiki, M., Ohnuki, Y., Murayama, Y. and Nomura, S., eds). pp. 97–105 (Physical Society of Japan, Tokyo).

Tonomura, A., Endo, J., Matsuda, T., Kawasaki, T. and Ezawa, H. (1989). Am. J. Phys. **57**, 117–120.

Tonomura, A., Matsuda, T., Hasegawa, S., Igarashi, M., Kobayashi, T., Naito, M., Kajiyama, M., Endo, J., Osakabe, N. and Aoki, R. (1990). Electron-interferometric observation of magnetic flux quanta using the Aharonov–Bohm effect. In *Proc. 3rd Int. Symp. Foundations of Quantum Mechanics*, Tokyo 1989 (Kobayashi,S.-i., Ezawa, H., Murayama, Y. and Nomura, S., eds) pp. 15–24 (Physical Society of Japan, Tokyo).

Tonomura, A., Umezaki, H., Matsuda, T., Osakabe, N., Endo, J. and Sugita, Y. (1983). Phys. Rev. Lett. **51**, 331–334.

Valdrè, U. (1974). In *Atti del Convegno del Gruppo di Strumentazione e Tecniche Non-biologiche della Società Italiana di Microscopia Elettronica* (Merli, P.G., Armigliato, A. and Pedulli, L., eds.) pp. 37–56 (CLUE, Bologna).

Valdrè, U. (1979). J. Microscopy **117**, 55–75.

Veneklasen, L.H. (1975). Optik **44**, 447–468.

Völkl, E. (1991). *Höchstauflösende Elektronenholographie.* Dissertation, Tübingen.

Völkl, E. and Lichte, H. (1990a). Seattle, vol. 1, pp. 226–227.

Völkl, E. and Lichte, H. (1990b). Ultramicroscopy **32**, 177–180.

Voronin, Yu.M., Demenchenok, I.P., Mokhnatkin, A.V. and Khaitlina, R.Yu. (1972). Izv. Akad. Nauk SSSR (Ser. Fiz.) **36**, 1293–1296 ; Bull. Acad. Sci. USSR (Phys. Ser.) **36**, 1154–1156.

Wade, R.H. (1974). Optik **40**, 201–216.

Wade, R.H. (1975). EMAG, pp. 197–200.

Wade, R.H. (1980). Holographic methods in electron microscopy. In *Computer Processing of Electron Microscope Images* (Hawkes, P.W., ed.) pp. 223–255.

Wahl, H. (1968/69). Optik **28**, 417–420.

Wahl, H. (1970a). Ber. Bunsen-Ges. Phys. Chem. **74**, 1142–1148.

Wahl, H. (1970b). Optik **30**, 508–520 and 577–589.

Wahl, H. (1974). Optik **39**, 585–588.

Wahl, H. (1975). *Bildebenenholographie mit Elektronen.* Habilitationsschrift, Tübingen, 63 pp.

Wahl, H. and Lau, B. (1979). Optik **54**, 27–36.

Wang, S.-y. and Cowley, J.M. (1991). EMSA **49**, pp. 682–683.

Watanabe, H. and Tonomura, A. (1969). Nihon Kessho Gakkai-shi [J. Crystallographic Soc. Japan] **11**, 23–25.

Weingärtner, I., Mirandé, W. and Menzel, E. (1969a). Optik **29**, 87–104.

Weingärtner, I., Mirandé, W. and Menzel, E. (1969b). Optik **30**, 318–322.

Weingärtner, I., Mirandé, W. and Menzel, E. (1970). Optik **31**, 335–353.

Weingärtner, I., Mirandé, W. and Menzel, E. (1971). Ann. Physik **26**, 289–301.

Weiss, J.K., Ruijter, W.J. de, Gajdardziska-Josifovska, M., Smith, D.J., Völkl, E. and Lichte, H. (1991). EMSA **49**, pp. 674–675.

Weiss, J.K., Ruijter, W.J. de, Gajdardziska-Josifovska, M., McCartney, M.R. and Smith, D.J. (1993). Ultramicroscopy **50**, 301–311.

Werner, F.G. and Brill, D.R. (1960). Phys. Rev. Lett. **4**, 344–346.

Yada, K., Shibata, K. and Hibi, T. (1973). J. Electron Microsc. **22**, 223–230.

Yatagai, T., Ohmura, K., Iwasaki, S., Hasegawa, S., Endo, J. and Tonomura, A. (1987). Appl. Opt. **26**, 377–382.

Yoshida, K., Okuwaki, T., Osakabe, N., Tanabe, H., Horiuchi, Y., Matsuda, T., Shinagawa, K., Tonomura, A. and Fujiwara, H. (1983). IEEE Trans. **MAG–19**, 1600–1604.

Zeitler, E. (1979). EMSA **37**, pp. 376–379.

Zhang, X. and Joy, D. (1991). EMSA **49**, pp. 678–679.

Zhang, X., Hashimoto, T. and Joy, D.C. (1992). Appl. Phys. ·Lett. **60**, 784–786.

Zhang, X., Hashimoto, T. and Joy, D.C. (1993a). Appl. Phys. Lett. **62**, 2447. Cf. Spence *et al.* (1993).

Zhang, X., Joy, D.C., Zhang, Y., Hashimoto, T., Allard, L. and Nolan, T.A. (1993b). Ultramicroscopy **51**, 21–30.

Part XIII, Chapters 64–67

The literature on the detection and detectability of single atoms is not examined here but we do list the early comment of Eisenhandler and Siegel (1966a) and of Zeitler and Thomson (1969), forerunners of much later work. A paper by van Dorsten *et al.* (1968) considers phase and amplitude. Incoherent transfer is studied by van Heel (1978).

Ade, G. (1977a). Optik **49**, 113–116.

Ade, G. (1977b). Probleme der Kontrastübertragung bei der Verarbeitung von Differenz-signalen. PTB–Bericht APh–12, 21 pp.

Ade, G. (1978). Optik **50**, 143–162.

Agar, A.W., Revell, R.S.M. and Scott, R.A. (1949). Delft, pp. 52–54.

Anaskin, I.F. and Ageev, E.V. (1974). Izv. Akad. Nauk SSSR (Ser.Fiz.) **38**, 1389–1392; Bull. Acad. Sci. USSR (Phys. Ser.) **38** (7), 26–29.

Badde, H.G. and Reimer, L. (1970). Z. Naturforsch. **25a**, 760–765.

Balossier, G. and Bonnet, N. (1981). Optik **58**, 361–376.

Balossier, G. and Thomas, X. (1984). J. Microsc. Spectrosc. Electron. **9**, 343–350.

Balossier, G., Bonnet, N., Genotel, D. and Laberrigue, A. (1980). The Hague, vol. 1, pp. 26–27.

Banbury, J.R. (1974). Canberra, vol. 1, pp. 44–45.

Banbury, J.R. and Bance, U.R. (1973a). EMSA **31**, pp. 296–297.

Banbury, J.R. and Bance, U.R. (1973b). EMAG, pp. 164–168.

Banbury, J.R., Drummond, I.W. and Ray, I.L.F. (1975). EMSA **33**, pp. 112–113.

Barnett, M.E. (1973). Optik **38**, 585–588.

Barnett, M.E. (1974). J. Microscopy **102**, 1–28.

Bates, R.H.T. and Rodenburg, J.M. (1989). Ultramicroscopy **31**, 303–307.

Beck, V. (1977). Ultramicroscopy **2**, 351–360.

Beck, V. and Crewe, A.V. (1975). Ultramicroscopy **1**, 137–144.

Beer, M., Frank, J., Hanszen, K.-J., Kellenberger, E. and Williams, R.C. (1975). Q. Rev. Biophys. **7**, 211–238.

Benner, G., Bihr, J. and Prinz, M. (1990). Seattle, vol. 1, pp. 138–139.

Benner, G., Bihr, J. and Weimer, E. (1991). EMSA **49**, pp. 1008–1009.

Beorchia, A. and Bonhomme, P. (1974). Optik **39**, 437–442.

Boersch, H. (1947). Z. Naturforsch. **2a**, 615–633.

Bonhomme, P., Beorchia, A. and Bonnet, N. (1973). C. R. Acad. Sci. Paris **B277**, 83–86.

Bonnet, N. and Bonhomme, P. (1980). Optik **56**, 353–362.

Bonnet, N., Beorchia, A. and Bonhomme, P. (1978). J. Microsc. Spectrosc. Electron. **3**, 497–511.

Born, M. and Wolf, E. (1980). *Principles of Optics* 6th ed. (Pergamon, Oxford and New York).

Borries, B. von and Lenz, F. (1956). Stockholm, pp. 60–64.

Bos, A. van den (1991). EMSA **49**, pp. 492–493.

Bos, A. van den (1992). Ultramicroscopy **47**, 298–306.

Boussakta, S. and Holt, A.G.J. (1992). Electron. Lett. **28**, 1683–1684.

Bouwhuis, G. and Dekkers, N.H. (1980). Optik **56**, 233–242.

Bovey, P.E. and Nicholls, A.W. (1987). SIME, pp. 247–250

Brown, L.M. (1981). J. Phys. F: Metal Phys. **11**, 1–26.

Browne, M.T. and Ward, J.F.L. (1982). Ultramicroscopy **7**, 249–262.

Browne, M.T., Lackovic, S. and Burge, R.E. (1975). EMAG, pp. 27–30.

Browning, N.D., Chisholm, M.F. and Pennycook, S.J. (1993). Nature **366**, 143–146.

Burge, R.E. (1977). Scanning transmission electron microscopy at high resolution. In *Analytical and Quantitative Methods in Microscopy* (Meek, G.A. and Elder, H.Y., eds) pp. 171–191 (University Press, Cambridge).

Burge, R.E. and Clark, A.F. (1981). EMAG, pp. 315–320.

Burge, R.E. and Dainty, J.C. (1976). Optik **46**, 229–240.

Burge, R.E. and Derome, M. (1976). Optik **46**, 161–182.

Burge, R.E. and Toorn, P. van (1979). EMAG, pp. 249–252.

Burge, R.E. and Toorn, P. van (1980). Scanning Electron Microsc. (I), 81–91.

Burge, R.E., Browne, M.T., Dainty, J.C., Derome, M. and Lackovic, S. (1976). Jerusalem, vol. 1, pp. 442–443.

Burge, R.E., Browne, M.T., Lackovic, S. and Ward, J.F.L. (1979a). Scanning Electron Microsc. (I), 127–136.

Burge, R.E., Browne, M.T., Dainty, J.C., Lackovic, S., Robinson, D. and Ward, J. (1979b). In *Machine-aided Image Analysis 1978* (Gardner, W.E., ed.) pp. 107–113 (Institute of Physics, Bristol). Conference Series **44**.

Burge, R.E., Browne, M.T., Charalambous, P., Clark, A. and Wu, J.K. (1982). J. Microscopy **127**, 47–60.

Chapman, J.N. (1989). Mater. Sci. Eng. **B3**, 355–358.

Chapman, J.N. and Morrison, G.R. (1983). J. Magnetism Magnet. Mater. **35**, 254–260.

Chapman, J.N., Batson, P.E., Waddell, E.M. and Ferrier, R.P. (1978a). Ultramicroscopy **3**, 203–214.

Chapman, J.N., Batson, P.E., Waddell, E.M. and Ferrier, R.P. (1978b). Toronto, vol. 1, pp. 466–467.

Chapman, J.N., Morrison, G.R., Jakubovics, J.P. and Taylor, R.A. (1983). EMAG, pp. 197–200.

Chapman, J.N., McFadyen, I.R. and McVitie, S. (1990). IEEE Trans. Magnetics 26, 1506–1511.

Christenson, K.K. and Eades, J.A. (1986). Ultramicroscopy 19, 191–194.

Coene, W. and Janssen, A.J.E.M. (1991). Cambridge, pp. 379–403.

Colliex, C. (1985). J. Microsc. Spectrosc. Electron. 10, 313–332.

Colliex, C., Craven, A.J. and Wilson, C.J. (1977). Ultramicroscopy 2, 327–335.

Colliex, C., Jeanguillaume, C. and Mory, C. (1984). J. Ultrastruct. Res. 88, 177–206.

Cowley, J.M. (1969). Appl. Phys. Lett. 15, 58–59.

Cowley, J.M. (1970). J. Appl. Cryst. 3, 49–58.

Cowley, J.M. (1975). J. Phys. D : Appl. Phys. 8, L77–L79.

Cowley, J.M. (1976a). Ultramicroscopy 1, 255–262.

Cowley, J.M. (1976b). Ultramicroscopy 2, 3–16. (Cf. Dekkers and de Lang, 1978b.)

Cowley, J.M. (1976c). EMSA 34, pp. 466–467.

Cowley, J.M. (1978). Electron microdiffraction. Adv. Electron. Electron Phys. 46, 1–53.

Cowley, J.M. (1978/9). Chem. Scripta 14, 33–38.

Cowley, J.M. (1980). Micron 11, 229–233.

Cowley, J.M. (1981). Ultramicroscopy 7, 19–26.

Cowley, J.M. (1983). J. Microscopy 129, 253–261.

Cowley, J.M. (1984). J. Electron Microsc. Tech. 1, 83–94.

Cowley, J.M. (1985). EMSA 43, 134–135.

Cowley, J.M. (1986). J. Electron Microsc. Tech. 3, 25–44.

Cowley, J.M. (1987). Annu. Rev. Phys. Chem. 38, 57–88.

Cowley, J.M. (1993). Ultramicroscopy 49, 4–13.

Cowley, J.M. and Au, A.Y. (1978a). Scanning Electron Microsc. (I), 53–60.

Cowley, J.M. and Au, A.Y. (1978b). Toronto, vol. 1, pp. 172–173.

Cowley, J.M. and Disko, M.M. (1980). Ultramicroscopy 5, 469–477.

Cowley, J.M. and Jap, B.K. (1976a). Scanning Electron Microsc. (I), 377–384 and 344.

Cowley, J.M. and Jap, B.K. (1976b). EMSA 34, 460–461.

Cowley, J.M. and Spence, J.C.H. (1979). Ultramicroscopy 3, 433–438.

Cowley, J.M. and Spence, J.C.H. (1981). Ultramicroscopy 6, 359–366.

Cowley, J.M. and Walker, D.J. (1981). Ultramicroscopy 6, 71–75.

Cowley, J.M., Strahm, M. and Butler, J.H. (1980). Micron **11**, 285–286.

Crewe, A.V. (1966). Science **154**, 729–738.

Crewe, A.V. (1970). Q. Rev. Biophys. **3**, 137–175.

Crewe, A.V. (1971). Phil. Trans. Roy. Soc. (London) **B261**, 61–70.

Crewe, A.V. (1973). Prog. Opt. **11**, 223–246.

Crewe, A.V. (1974). J. Microscopy **100**, 247–259.

Crewe, A.V. (1978/9). Chem. Scr. **14**, 17–20.

Crewe, A.V. (1979). J. Electron Microsc. **28**, S9–S16.

Crewe, A.V. (1980a). Repts Prog. Phys. **43**, 621–639.

Crewe, A.V. (1980b). Ultramicroscopy **5**, 131–138.

Crewe, A.V. (1980c). EMSA **38**, pp. 60–61.

Crewe, A.V. (1983). Science **221**, 325–330

Crewe, A.V. (1985). Scanning Electron Microsc. (II), 467–472.

Crewe, A.V. and Groves, T. (1974). J. Appl. Phys. **45**, 3662–3672.

Crewe, A.V. and Kopf, D. (1980). Optik **55**, 325–327.

Crewe, A.V. and Ohtsuki, M. (1980). The Hague, vol. 2, pp. 588–589.

Crewe, A.V. and Ohtsuki, M. (1981a). EMSA **39**, pp. 236–237.

Crewe, A.V. and Ohtsuki, M. (1981b). Ultramicroscopy **7**, 13–18.

Crewe, A.V. and Salzman, D.B. (1982). Ultramicroscopy **9**, 373–377.

Crewe, A.V. and Wall, J.S. (1970a). J. Mol. Biol. **48**, 375–393.

Crewe, A.V. and Wall, J.S. (1970b). Optik **30**, 461–474.

Crewe, A.V., Isaacson, M. and Johnson, D. (1969). Rev. Sci. Instrum. **40**, 241–246.

Crewe, A.V., Wall, J. and Langmore, J. (1970a). Science **168**, 1338–1340.

Crewe, A.V., Langmore, J., Wall, J., and Beer, M. (1970b). EMSA **28**, pp. 250–251.

Crewe, A.V., Isaacson, M.S. and Zeitler, E. (1979). Adv. Structure Res. Diffraction Methods **7**, 23–48.

Daberkow, I. and Herrmann, K.-H. (1984). Budapest, vol. 1, pp. 115–116.

Daberkow, I. and Herrmann, K.-H. (1988). York, vol. 1, pp. 125–126.

Daberkow, I., Herrmann, K.-H. and Lenz, F. (1993). Ultramicroscopy **50**, 75–82.

Dekkers, N.H. (1979). Optik **53**, 131–142.

Dekkers, N.H. and Lang, H. de (1974). Optik **41**, 452–456.

Dekkers, N.H. and Lang, H. de (1977). Philips Tech. Rev. **37** (1), 1–9.

Dekkers, N.H. and Lang, H. de (1978a). Optik **51**, 83–92.

Dekkers, N.H. and Lang, H. de (1978b). Ultramicroscopy **3**, 101–102.

Dekkers, N.H., Lang, H. de and Mast, K. van der (1976). J. Microsc. Spectrosc. Electron. 1, 511–512.

Desseaux, J., Renault, A. and Bourret, A. (1977). Phil. Mag. 35, 357–372.

Dorsten, A.C. van, Mellema, J.E. and Premsela, H.F. (1968). Rome, vol. 2, pp. 103–104.

Downing, K.H. (1975). Optik 43, 199–203.

Dumontet, P. (1955a). Opt. Acta 2, 53–63.

Dumontet, P. (1955b). Publ. Sci. Univ. Alger B1, 33–44.

Dumontet, P. (1956). Publ. Sci. Univ. Alger B2, 151–179 and 203–294.

Dyck, D. van (1989). SIME, pp. 245–246.

Dyck, D. van (1991). EMSA 49, pp. 448–449 and 496–497.

Dyck, D. van (1992). In Electron Microscopy in Materials Science (Merli, P.G. and Antisari, M.V., eds) pp. 193–268 (World Scientific, Singapore, River Edge NJ and London).

Dyck, D. van and Jong, A. F. de (1992). Ultramicroscopy 47, 266–281.

Eisenhandler, C.B. and Siegel, B.M. (1966a). J. Appl. Phys. 37, 1613–1620.

Eisenhandler, C.B. and Siegel, B.M. (1966b). Appl. Phys. Lett. 8, 258–260.

Endoh, H. and Hashimoto, H. (1977). HVEM Kyoto, pp. 293–296.

Engel, A. (1974). Optik 41, 117–126.

Engel, A., Wiggins, J.W. and Woodruff, D.C. (1974). J. Appl. Phys. 45, 2739–2747.

Erickson, H.P. (1973). The Fourier transform of an electron micrograph–first order and second order theory of image formation. Adv. Opt. Electron Microsc. 5, 163–199.

Erickson, H.P. and Klug, A. (1970a). Ber. Bunsen-Ges. Phys. Chem. 74, 1129–1137.

Erickson, H.P. and Klug, A. (1970b). EMSA 28, pp. 248–249.

Erickson, H.P. and Klug, A. (1971). Phil. Trans. Roy. Soc. (London) B261, 105–118.

Faget, J., Ferré, J. and Fert, C. (1960a). C. R. Acad. Sci. Paris 251, 526–528.

Faget, J., Fagot, M. and Fert, C. (1960b). Delft, vol. 1, pp. 18–24.

Faget, J., Fagot, M., Ferré, J. and Fert, C. (1962). Philadelphia, vol. 1, pp. A–7.

Fejes, P.L. (1977). Acta Cryst. A33, 109–113.

Fertig, J. and Rose, H. (1977). Ultramicroscopy 2, 269–279.

Fertig, J. and Rose, H. (1978a). Optik 51, 213–220.

Fertig, J. and Rose, H. (1978b). Toronto, vol. 1, pp. 238–239.

Fertig, J. and Rose, H. (1979). Optik 54, 165–191.

Fertig, J. and Rose, H. (1981). Optik 59, 407–429.

Frank, J. (1969). Optik 30, 171–180.

Frank, J. (1973). Optik 38, 519–536.

Frank, J. (1975a). Optik 43, 25–34.

Frank, J. (1975b). Optik **43**, 103–109.

Frank, J. (1976a). Optik **44**, 379–391.

Frank, J. (1976b). Jerusalem, vol. 1, pp. 97–98.

Frank, J., Bussler, P., Langer, R. and Hoppe, W. (1970). Ber. Bunsen-Ges. Phys. Chem. **74**, 1105–1115.

Frank, J., McFarlane, S.C. and Downing, K.H. (1978/9). Optik **52**, 49–60.

Franke, G. (1966). Optik **23**, 20–25.

Freeman, L.A., Howie, A. and Treacy, M.M.J. (1977). J. Microscopy **111**, 165–178.

Freeman, L.A., Howie, A. and Mistry, A.B. (1980). J. Microscopy **119**, 3–18.

Frieden, B.R. (1966). J. Opt. Soc. Am. **56**, 1355–1362.

Friedman, S.L., Rodenburg, J.M. and McCallum, B.C. (1991). EMAG, pp. 491–494.

Glaser, W. (1956). Elektronen- und Ionenoptik. In *Handbuch der Physik* **33**, 123–395.

Goldfarb, W., Krakow, W., Ast, D. and Siegel, B.M. (1975). EMSA **33**, 186–187.

Groves, T. (1975). Ultramicroscopy **1**, 15–31 and 170–172.

Guigay, J.P., Wade, R.H. and Delpla, C. (1971). EMAG, pp. 238–239.

Hahn, M. (1973). *Theoretische und experimentelle Untersuchungen zum Nachweis von Einzelatomen mit Durchstrahlungs-Elektronenmikroskopen.* Dissertation, Düsseldorf.

Hahn, M. and Seredynski, J. (1974). Canberra, vol. 1, pp. 234–235.

Haider, M., Boulin, C. and Epstein, A. (1988). York, vol. 1, pp. 123–124.

Hammel, M. and Rose, H. (1993). Ultramicroscopy **49**, 81–86.

Hammel, M., Kohl, H. and Rose, H. (1990). Seattle, vol. 1, pp. 120–121.

Hanszen, K.-J. (1966a). Z. angew. Phys. **20**, 427–435.

Hanszen, K.-J. (1966b). Kyoto, vol. 1, pp. 39–40.

Hanszen, K.-J. (1967). Naturwissenschaften **54**, 125–133.

Hanszen, K.-J. (1969). Z. angew. Phys. **27**, 125–131.

Hanszen, K.-J. (1971). The optical transfer theory of the electron microscope: fundamental principles and applications. Adv. Opt. Electron Microsc. **4**, 1–84.

Hanszen, K.-J. (1974). The relevance of dark field illumination in conventional and scanning transmission electron microscopy. PTB–Bericht APh–7, 26 pp.

Hanszen, K.-J. (1976). Jerusalem, vol. 1, pp. 95–96.

Hanszen, K.-J. (1982). Holography in electron microscopy. Adv. Electron. Electron Phys. **59**, 1–77.

Hanszen, K.-J. (1990). 40 Jahre elektronenoptische Forschung in der Physikalisch-Technischen Bundesanstalt. Teil I: Aus der Geschichte des Laboratoriums für Elektronenoptik; Teil II: Kurzberichte aus den PTB–Mitteilungen; Veröffentlichungs- und Vortragslisten. PTB–Bericht APh–33.

Hanszen, K.-J. and Ade, G. (1974). Canberra, vol. 1, pp. 196–197.

Hanszen, K.-J. and Ade, G. (1976a). Jerusalem, vol.1, pp. 446–447.

Hanszen, K.-J. and Ade, G. (1976b). Optik **44**, 237–249.

Hanszen, K.-J. and Ade, G. (1977). A consistent Fourier optical representation of image formation in the conventional fixed beam electron microscope, in the scanning transmission electron microscope and of holographic reconstruction. PTB–Bericht APh–11, 31 pp.

Hanszen, K.-J. and Ade, G. (1978). Optik **51**, 119–126.

Hanszen, K.-J. and Morgenstern, B. (1965). Z. angew. Phys. **19**, 215–227.

Hanszen, K.-J. and Trepte, L. (1970). Grenoble, vol. 1, pp. 45–46.

Hanszen, K.-J. and Trepte, L. (1971a). Optik **32**, 519–538.

Hanszen, K.-J. and Trepte, L. (1971b). Optik **33**, 166–198.

Hanszen, K.-J., Morgenstern, B. and Rosenbruch, K.-J. (1964). Z. angew. Phys. **16**, 477–486.

Hanszen, K.-J., Rosenbruch, K.-J. and Sunder-Plassmann, F.-A. (1965). Z. angew. Phys. **18**, 345–350.

Harada, Y., Tamura, N. and Goto, T. (1975). EMAG, pp. 15–18.

Hashimoto, H. and Endoh, H. (1978). In *Electron Diffraction 1927–1977* (Dobson, P.J., Pendry, J.B. and Humphreys, C.J., eds) pp. 188–194 (Institute of Physics, Bristol). Conference series **41**.

Hashimoto, H., Endoh, H., Tanji, T., Ono, A. and Watanabe, E. (1977). J. Phys. Soc. Japan **42**, 1073–1074.

Hawkes, P.W. (1971). EMAG, pp. 230–232.

Hawkes, P.W. (1972). Manchester, pp. 398–399.

Hawkes, P.W. (1973a). Optik **37**, 366–375 and 376–384.

Hawkes, P.W. (1973b). Introduction to electron optical transfer theory. In *Image Processing and Computer-aided Design in Electron Optics* (Hawkes, P.W., ed.) pp. 2–14 (Academic Press, London and New York).

Hawkes, P.W. (1974a). Optik **40**, 539–556.

Hawkes, P.W. (1974b). Canberra, vol. 1, pp. 202–203.

Hawkes, P.W. (1975). Optik **42**, 433–438.

Hawkes, P.W. (1978a). Electron image processing: a survey. Comput. Graph. Image Proc. **8**, 406–446.

Hawkes, P.W. (1978b). J. Optics (Paris) **9**, 235–241.

Hawkes, P.W. (1978c). Coherence in electron optics. Adv. Opt. Electron Microsc. **7**, 101–184.

Hawkes, P.W. (1980a). Optik **55**, 207–212.

Hawkes, P.W. (1980b). Ultramicroscopy **5**, 67–70.

Hawkes, P.W. (1980c). Scanning Electron Microsc. (I), 93–98.

Hawkes, P.W. (1982a). Electron image processing: 1978–1980. Comput. Graph. Image Proc. **18**, 58–96.

Hawkes, P.W. (1982b). Ultramicroscopy **9**, 27–30.

Hawkes, P.W. (1985). J. Microsc. Spectrosc. Electron. **10**, 395–398.

Hawkes, P.W. (1992). Electron image processing: 1981–1990. J. Computer-assist. Microsc. **4**, 1–72 and **5** (1993) 255.

Heel, M.G. van (1978). Optik **49**, 389–408.

Herring, R.A. (1991). EMSA **49**, pp. 692–693.

Hirt, A. and Hoppe, W. (1972). Manchester, pp. 12–13.

Hoppe, W. (1961). Naturwissenschaften **48**, 736–737.

Hoppe, W. (1963). Optik **20**, 599–606.

Hoppe, W. (1970). Ber. Bunsen-Ges. Phys. Chem. **74**, 1090–1100.

Hoppe, W. (1971). Phil. Trans. Roy. Soc. (London) **B261**, 71–94.

Hoppe, W. (1974). Naturwissenschaften **61**, 239–249.

Hoppe, W. (1982). Ultramicroscopy **10**, 187–198.

Hoppe, W. and Köstler, D. (1976). Jerusalem, vol. 1, pp. 99–104.

Hoppe, W., Langer, R. and Thon, F. (1970). Optik **30**, 538–545.

Hoppe, W., Köstler, D. and Sieber, P. (1974). Z. Naturforsch. **29a**, 1933–1934.

Hoppe, W., Köstler, D., Typke, D. and Hunsmann, N. (1975). Optik **42**, 43–56.

Howie, A. (1972). Manchester, pp. 408–413.

Howie, A. (1974). In *Quantitative Scanning Electron Microscopy* (Holt, D.B., Muir, M.D., Grant, P.R. and Boswarva, I.M., eds) pp. 183–211 (Academic Press, London and New York).

Howie, A. (1978). J. Non-cryst. Sol. **31**, 41–55.

Howie, A. (1979). J. Microscopy **117**, 11–23.

Howie, A. (1983). J. Microscopy **129**, 239–251.

Howie, A., Krivanek, O. and Rudee, M.L. (1972). Manchester, pp. 450–451.

Howie, A., Krivanek, O.L. and Rudee, M.L. (1973). Phil. Mag. **27**, 235–255.

Hubert, G., Krisch, B. and Willasch, D. (1978). Toronto **1**, 22–23.

Humphreys, C.J. (1981). Ultramicroscopy **7**, 7–12.

Isaacson, M., Utlaut, M. and Kopf, D. (1980). In *Computer Processing of Electron Microscope Images* (Hawkes, P.W., ed.), pp. 257–283 (Springer, Berlin and New York).

Ishizuka, K. (1980). Ultramicroscopy **5**, 55–65.

Ishizuka, K. (1989). In *Computer Simulation of Electron Microscope Diffraction and Images* (Krakow, W. and O'Keefe, M., eds) pp. 43–55 (Minerals, Metals and Materials Society, Warrendale PA).

Ishizuka, K. (1990). Seattle, vol. 1, pp. 60–61.

Izui, K., Furuno, S. and Otsu, H. (1977). J. Electron Microsc. **26**, 129–132.

Izui, K., Furuno, S., Nishida, T., Otsu, H. and Kuwabara, S. (1978). J. Electron Microsc. **27**, 171–179.

Jenkins, W.K. (1979). *Contrast Transfer in Bright Field Electron Microscopy of Amorphous Specimens.* Dissertation, Cambridge.

Jenkins, W.K. and Wade, R.H. (1977). EMAG, pp. 115–118.

Jesson, D.E. and Pennycook, S.J. (1990). Seattle, vol. 1, pp. 74–75.

Jones, A.V. (1988). EMSA **46**, pp. 648–649.

Jones, A.V. and Haider, M. (1989). Scanning Microsc. **3**, 33–42.

Jong, A.F. de and Dyck, D. van (1993). Ultramicroscopy **49**, 66–80.

Kanaya, K. and Kawakatsu, (1958). Berlin, vol. 1, pp. 308–316.

Kanaya, K., Inoue, Y. and Ishikawa, A. (1954). London, pp. 46–60.

Kanaya, K., Kawakatsu, H. and Ishikawa, A. (1957). Bull. Electrotech. Lab. **21**, 825–833.

Kanaya, K., Kawakatsu, H. and Yotsumoto, H. (1958a). J. Electron Microsc. **6**, 1–4.

Kanaya, K., Kawakatsu, H., Ito, K. and Yotsumoto, H. (1958b). J. Appl. Phys. **29**, 1046–1049.

Kanaya, K., Oho, E., Naka, M., Koyanagi, T. and Sasaki, T. (1985). J. Electron Microsc. Tech. **2**, 73–87.

Kanaya, K., Baba, N., Oho, E. and Sasaki, T. (1986). Kyoto, vol. 1, pp. 435–436 and Beijing, pp. 107–109.

Kermisch, D. (1977). J. Opt. Soc. Am. **67**, 1357–1360.

Kiselev, A.G. and Sherman, M.B. (1976). Optik **46**, 55–60.

Kohl, H. (1986). Kyoto, vol. 1, pp. 777–778.

Kohl, H. and Rose, H. (1985). Theory of image formation by inelastically scattered electrons in the electron microscope. Adv. Electron. Electron Phys. **65**, 173–227.

Köhler, A. (1893). Z. Wiss. Mikrosk. **10**, 433–440; English translation Proc. Roy. Microsc. Soc. **28** (1993) 181–185.

Köhler, A. (1899). Z. Wiss. Mikrosk. **16**, 1–28.

Koike, H., Harada, Y., Goto, T., Kokubo, Y., Yamada, K., Someya, T. and Watanabe, M. (1974). Canberra, vol. 1, pp. 42–43.

Krakow, W. (1976a). Ultramicroscopy 1, 203–221.

Krakow, W. (1976b). EMSA 34, 566–567.

Krakow, W. (1977). EMSA35, pp. 72–73.

Krakow, W. (1978). Ultramicroscopy 3, 291–301.

Krakow, W. (1982). Hamburg, vol. 1, pp. 203–204.

Krakow, W. (1984). J. Electron Microsc. Tech. 1, 107–130.

Krakow, W. and Howland, L.A. (1976). Ultramicroscopy 2, 53–67.

Krakow, W. and Siegel, B.M. (1975). Optik 42, 245–268.

Krakow, W., Ast, D.C., Goldfarb, W. and Siegel, B.M. (1976). Phil. Mag. 33, 985–1014.

Krisch, B., Müller, K.-H., Schliepe, R. and Willasch, D. (1976). Siemens Rev. 43, 390–393 ; Siemens Z. 50 (1), 47–50.

Krisch, B., Müller, K.-H., Rauch, M.v., Schliepe, R. and Thieringer, H.M. (1977). Scanning Electron Microsc. (I), 423–430.

Krivanek, O.L. (1975). Optik 43, 361–372.

Krivanek, O.L. (1976). Jerusalem, vol. 1, pp. 263–264.

Krivanek, O.L. (1978). Toronto, vol. 1, pp. 168–169.

Krivanek, O.L. and Howie, A. (1975). J. Appl. Cryst. 8, 213–219.

Krivanek, O.L., Gaskell, P.H. and Howie, A. (1976). Nature 262, 454–457.

Kunath, W. (1976). Jerusalem, vol. 1, pp. 340–341.

Kunath, W. (1979). Ultramicroscopy 4, 3–7.

Kunath, W. and Gross, H. (1985). Ultramicroscopy 16, 349–356.

Kunath, W. and Weiss, K. (1980). The Hague, 1 , 114–115.

Kunath, W., Weiss, K. and Zeitler, E. (1981). EMSA 39, pp. 226–227.

Kunath, W., Zemlin, F. and Weiss, K. (1985). Ultramicroscopy 16, 123–138.

Kunath, W., Gao, P., Schiske, P. and Weiss, K. (1986). Kyoto, vol. 1, pp. 775–776.

Kuypers, W., Thompson, M.N. and Andersen, W.H.J. (1973). Scanning Electron Microsc. 9–16.

Laberrigue, A., Balossier, G., Beorchia, A., Bonhomme, P., Bonnet, N. and Troyon, M. (1980). J. Microsc. Spectrosc. Electron. 5, 655–663.

Lackovic, S., Browne, M.T. and Burge, R.E. (1979). Scanning Electron Microsc. (I), 137–144.

Lang, H. de and Dekkers, N.H. (1979). Optik 53, 353–365.

Langer, R. and Hoppe, W. (1966/7). Optik 24, 470–489.

Langer, R. and Hoppe, W. (1967). Optik 25, 413–428 and 507–522.

Lannes, A., Tanaka, M. and Temple, P. (1981). Optik **60**, 1–28.

Lenz, F. (1963). Z. Physik **172**, 498–502.

Lenz, F. (1964). Optik **21**, 489–493.

Lenz, F. (1965). Lab. Invest. **14**, 808–818.

Lenz, F.A. (1971a). Transfer of image information in the electron microscope. In *Electron Microscopy in Material Science* (Valdrè, U., ed.) pp. 540–569 (Academic Press, New York and London).

Lenz, F. (1971b). EMAG, pp. 224–229.

Lenz, F. and Scheffels, W. (1958). Z. Naturforsch. **13a**, 226–230.

Li, Z.G. and Dorignac, D. (1986). Kyoto, vol. 1, pp. 427–428.

Linfoot, E.H. (1956). J. Opt. Soc. Am. **46**, 740–752.

Linfoot, E.H. (1957). Opt. Acta **4**, 12–16.

Linfoot, E.H. (1960). *Qualitätsbewertung optischer Bilder* (Vieweg, Braunschweig).

Linfoot, E.H. (1964). *Fourier Methods in Optical Image Evaluation* (Focal Press, London and New York).

Locquin, M. (1954). London, pp. 285–289.

Locquin, M. (1955). Z. Wiss. Mikrosk. **62**, 220–223.

Locquin, M. (1956). Stockholm, pp. 78–79.

McCallum, B.C. and Rodenburg, J.M. (1992). Ultramicroscopy **45**, 371–380.

McCallum, B.C. and Rodenburg, J.M. (1993a). J. Opt. Soc. Am. **A10**, 1–9.

McCallum, B.C. and Rodenburg, J.M. (1993b). Ultramicroscopy **52**, 85–99.

McFarlane, S.C. (1975). J. Phys. C: Solid State Phys. **8**, 2819–2836.

McFarlane, S.C. and Cochrane, W. (1975). J. Phys. C: Solid State Phys. **8**, 1311–1321.

Misell, D.L. (1971). J. Phys. A: Gen. Phys. **4**, 782–797 and 798–812.

Misell, D.L. (1973). J. Phys. A: Math. Nucl. Gen. **6**, 62–78 and 205–217.

Misell, D.L. and Atkins, A.J. (1973). J. Phys. A: Math. Nucl. Gen. **6**, 218–235.

Misell, D.L. (1977). J. Phys. D: Appl. Phys. **10**, 1085–1107.

Misell, D.L., Stroke, G.W. and Halioua, M. (1974). J. Phys. D: Appl. Phys. **7**, L113–L117.

Möllenstedt, G. Speidel, R., Hoppe, W., Langer, R., Katerbau, K.-H. and Thon, F. (1968). Rome, vol. 1, pp. 125–126.

Morrison, G.R. and Chapman, J.N. (1981). EMAG, pp. 329–332.

Morrison, G.R. and Chapman, J.N. (1982). Hamburg, vol. 1, pp. 211–212.

Morrison, G.R. and Chapman, J.N. (1983). Optik **64**, 1–12.

Morrison, G.R., Chapman, J.N. and Craven, A.J. (1979). EMAG, pp. 257–260.

Mory, C. and Colliex, C. (1985). J. Microsc. Spectrosc. Electron. **10**, 389–394.

Mory, C., Colliex, C. and Cowley, J.M. (1987a). Ultramicroscopy **21**, 171–177.

Mory, C., Bonnet, N., Colliex, C., Kohl, H. and Tencé, M. (1987b). Niagara Falls, pp. 329–342.

Müller, K.-H. (1971). Optik **33**, 296–311 and 331–343.

Müller, K.-H. (1976). Optik **45**, 73–85.

Müller, K.-H. and Rindfleisch, V. (1971). EMSA **29**, 48–49.

Niehrs, H. (1973). Optik **38**, 44–63.

Nussbaumer, H.J. (1982). *Fast Fourier Transform and Convolution Algorithms* (Springer, Berlin and New York).

Ohtsuki, M. and Crewe, A.V. (1980). EMSA **38**, pp. 62–63.

O'Keefe, M.A. (1979). EMSA **37**, 556–557.

O'Keefe, M.A. (1992). Ultramicroscopy **47**, 282–297.

O'Keefe, M.A. and Spence, J.C.H. (1991). EMSA **49**, pp. 498–499.

O'Neill, E.L. (1963). *Introduction to Statistical Optics* (Addison–Wesley, Reading MA and London).

Parsons, D.F. and Johnson, H.M. (1972). Appl. Opt. **11**, 2840–2843.

Parsons, J.R. and Hoelke, C.W. (1974). Phil. Mag. **30**, 135–143.

Pennycook, S.J. (1981). J. Microscopy **124**, 15–22.

Pennycook, S.J. (1989). Ultramicroscopy **30**, 58–69.

Pennycook, S.J. (1992). Z-contrast transmission electron microscopy: direct atomic imagery of materials. Annu. Rev. Mat. Sci. **22**, 171–195.

Pennycook, S.J. and Boatner, L.A. (1988). Nature **336,** 565–567.

Pennycook, S.J. and Jesson, D.E. (1990). Phys. Rev. Lett. **64**, 938–941.

Pennycook, S.J. and Jesson, D.E. (1992). In *Electron Microscopy in Materials Science* (Merli, P.G. and Antisari, M.V., eds) pp. 333–362 (World Scientific, Singapore, River Edge NJ and London).

Pennycook , S.J., Craven, A.J. and Brown, L.M. (1977). EMAG, pp. 69–72.

Pennycook, S.J., Jesson, D.E., Chisholm, M.F., Ferridge, A.G. and Seddon, M.J. (1991). Cambridge, pp. 233–243

Plamann, T. and Rodenburg, J.M. (1992). Granada, vol. 1, pp. 659–660.

Probst, W., Benner, G., Bihr, J., Bauer, R. and Weimer, E. (1991a). SIME, pp. 369–370.

Probst, W., Bauer, R., Benner, G. and Lehmann, J.L. (1991b). EMSA **49**, pp. 1010–1011.

Ray, I.L.F., Drummond, I.W. and Banbury, J.R. (1975). EMAG, pp. 11–14.

Reed, I.S., Truong, T.K., Kwoh, Y.S. and Hall, E.L. (1977). In *Image Science Mathematics* (Wilde, C.O. and Barrett, E., eds) pp. 229–233 (Western Periodicals, North Hollywood).

Reichelt, R. and Engel, A. (1985). J. Microsc. Spectrosc. Electron. **10**, 491–498.

Reichelt, R. and Engel, A. (1986). Ultramicroscopy **19**, 43–56.

Reimer, L. and Badde, H.G. (1970). Grenoble, vol. 1, pp. 15–16.

Reimer, L. and Hagemann, P. (1977). Optik **47**, 325–336.

Reimer, L., Gentsch, P. and Hagemann, P. (1975). Optik **43**, 431–452.

Robinson, D.W. (1979). EMAG, pp. 253–256.

Rodenburg, J.M. (1989a). Ultramicroscopy **27**, 413–422.

Rodenburg, J.M. (1989b). EMAG, vol. 1, pp. 103–106.

Rodenburg, J.M. and Bates, R.H.T. (1992). Phil. Trans. Roy. Soc. London **A339**, 521–553.

Rodenburg, J.M. and McCallum, B.C. (1991). Cambridge, pp. 223–232.

Rodenburg, J.M. and McCallum, B.C. (1992). Granada, vol. 1, pp. 125–129.

Rodenburg, J.M., McCallum, B.C. and Nellist, P.D. (1993). Ultramicroscopy **48**, 304–314.

Röhler, R. (1967). *Informationstheorie in der Optik* (Wissenschaftliche Verlagsgesellschaft, Stuttgart).

Rose, H. (1974a). Optik **39**, 416–436.

Rose, H. (1974b). Canberra, vol. 1, pp. 212–213.

Rose, H. (1975). Optik **42**, 217–244.

Rose, H. (1977). Ultramicroscopy **2**, 251–267.

Rose, H. (1978). Ann. New York Acad. Sci. **306**, 47–61.

Rose, H. (1984). Ultramicroscopy **15**, 173–191.

Rose, H. and Fertig, J. (1976). Ultramicroscopy **2**, 77–87.

Rose, H. and Fertig, J. (1977). EMSA **35**, pp. 200–201.

Sarikaya, M. and Howe, J. M. (1992). Ultramicroscopy **47**, 145–161.

Saxton, W.O. (1977). EMAG, pp. 111–114.

Saxton, W.O. (1986). Kyoto, post-deadline paper 1.

Saxton, W.O. (1987). EMSA **45**, 10–13.

Saxton, W.O. and Smith, D.J. (1979). EMAG, pp. 265–268.

Saxton, W.O., Howie, A., Mistry, A. and Pitt, A. (1977). EMAG, pp. 119–122.

Saxton, W.O., Jenkins, W.K., Freeman, L.A. and Smith, D.J. (1978). Optik **49**, 505–510.

Scherzer, O. (1949). J. Appl. Phys. **20**, 20–29.

Schiske, P. (1973). In *Image Processing and Computer-aided Design in Electron Optics* (Hawkes, P.W., ed.) pp. 54–65 (Academic Press, London and New York).

Sieber, P. and Tonar, K. (1975). Optik **42**, 375–380.

Sieber, P. and Tonar, K. (1976). Optik **44**, 361–364.

Siegel, B.M., Eisenhandler, C.B. and Coan, M.G. (1966). Kyoto, vol. 1, pp. 41–42.

Smith, K.C.A. and Erasmus, S.J. (1982). J. Microscopy **122**, RP1–RP2.

Someya, T., Goto, T., Harada, Y., Yamada, K., Koike, H., Kokubo, Y. and Watanabe, M. (1974). Optik **41**, 225–244

Spence, J.C.H. (1978). Scanning Electron Microsc. (I), 61–68.

Spence, J.C.H. (1988). *Experimental High-resolution Electron Microscopy*, 2nd. ed. (Oxford University Press, New York and Oxford).

Spence, J.C.H. (1992). Optik **92**, 57–68.

Spence, J.C.H. and Cowley, J.M. (1978). Optik **50**, 129–142.

Spence, J.C.H. and Zuo, J.M. (1992). *Electron Microdiffraction* (Plenum, New York and London).

Steeds, J.W. and Carlino, E. (1992). In *Electron Microscopy in Materials Science* (Merli, P.G. and Antisari, M.V., eds) pp. 279–313 (World Scientific, Singapore, River Edge NJ and London).

Strehl, K. (1902). Z. Instrumentenkunde **22**, 213–217.

Thompson, M.N. (1973a). EMAG, pp. 176–181.

Thompson, M.N. (1973b). EMSA **31**, pp. 154–155.

Thompson, M.N. (1975). EMAG, pp. 31–34.

Thomson, M.G.R. (1973). Optik **39**, 15–38.

Thon, F. (1965). Z. Naturforsch. **20a**, 154–155.

Thon, F. (1966a). Z. Naturforsch. **21a**, 476–478.

Thon, F. (1966b). Kyoto, vol. 1, pp. 23–24.

Thon, F. (1967). Probleme der Hochauflösungs-Elektronenmikroskopie. Phys. Blätt. **23**, 450–458.

Thon, F. (1968a). *Zur Deutung der Bildstrukturen in hochaufgelösten elektronenmikroskopischen Anfnahmen dünner amorpher Objekte*. Dissertation, Tübingen.

Thon, F. (1968b). Rome, vol. 1, pp. 127–128.

Thon, F. (1971). Phase contrast electron microscopy. In *Electron Microscopy in Material Science* (Valdrè, U., ed.). pp. 570–625 (Academic Press, New York and London).

Thon, F. (1974). Canberra, vol. 1, pp. 238–239.

Thon, F. and Willasch, D. (1970). Grenoble, vol. 1, pp. 3–4.

Thon, F. and Willasch, D. (1971). EMSA **29**, pp. 38–39 and 46–47.

Thon, F. and Willasch, D. (1972a). Optik **36**, 55–58.

Thon, F. and Willasch, D. (1972b). Manchester, pp. 650–651.

Tochigi, H., Nakatsuka, H., Fukami, A. and Kanaya, K. (1970). Grenoble, vol. 1, pp. 73–74.

Tochigi, H., Ishikawa, A. and Satake, Y. (1974). Canberra, vol. 1, pp. 50–51.

Toorn, P. van and Robinson, D.W. (1980). Optik **56**, 323–331.

Treacy, M.M.J. (1981). Scanning Electron Microsc. (I), 185–197.

Treacy, M.M.J., Howie, A. and Wilson, C.J. (1978). Phil. Mag. **A38**, 569–585.

Treacy, M.M.J., Howie, A. and Pennycook, S.J. (1979). EMAG, pp. 261–264.

Tsuno, K. (1993). Ultramicroscopy **50**. 245–253.

Typke, D. and Köstler, D. (1976). Optik **45**, 495–498.

Typke, D. and Köstler, D. (1977). Ultramicroscopy **2**, 285–295.

Unser, M., Trus, B.L. and Steven, A.C. (1987a). Ultramicroscopy **23**, 39–52.

Unser, M., Trus, B.L. and Steven, A.C. (1987b). EMSA **45**, 746–747.

Unwin, P.N.T. (1970a). Ber. Bunsen-Ges. Phys. Chem. **74**, 1137–1141.

Unwin, P.N.T. (1970b). Grenoble, vol. 1, pp. 65–66.

Unwin, P.N.T. (1971). Phil. Trans. Roy. Soc. (London) **B261**, 95–104.

Unwin, P.N.T. (1972). Proc. Roy. Soc. (London) **A329**, 327–359.

Unwin, P.N.T. (1974). Z. Naturforsch. **29a**, 158–163.

Valdrè, U. (1974). In *Atti del Convegno del Gruppo di Strumentazione e Tecniche Non-biologiche della Società Italiana di Microscopia Elettronica* (Merli, P.G., Armigliato, A. and Pedulli, L., eds) pp. 37–56 (CLUE, Bologna).

Valdrè, U. (1979). J. Microscopy **117**, 55–75.

Veneklasen, L.H. (1975). Optik **44**, 447–468.

Waddell, E.M. and Chapman, J.N. (1979). Optik **54**, 83–96.

Waddell, E.M., Chapman, J.N. and Ferrier, R.P. (1977). EMAG, pp. 267–270.

Waddell, E.M., Chapman, J.N. and Ferrier, R.P. (1978). Toronto, vol. 1, pp. 176–177.

Wade, R.H. (1976a). Optik **45**, 87–91.

Wade, R.H. (1976b). Phys. Stat. Sol. (a) **37**, 247–256.

Wade, R.H. (1978). Ultramicroscopy **3**, 329–334.

Wade, R.H. and Frank, J. (1977). Optik **49**, 81–92.

Wade, R.H. and Jenkins, W.K. (1978). Optik **50**, 1–17.

Wang, S.-Y., Mankos, M. and Cowley, J.M. (1992). EMSA **50**, pp. 982–983.

Ward, J.F.L., Browne, M.T. and Burge, R.E. (1979). EMAG, pp. 85–88.

Wardell, I.R.M., Morphew, J. and Bovey, P.E. (1973). EMAG, pp. 182–185.

Welford, W.T. (1972). J. Microscopy **96**, 105–107.

Willasch, D. (1973). *Versuche zur Kontrastverbesserung in der Elektronenmikroskopie durch Hellfeldabbildung mittels Phasenplatten und Dunkelfeldabbildung bei hohlkegelförmiger Beleuchtung.* Dissertation, Tübingen.

Willasch, D. (1975a). Optik **44**, 17–36.

Willasch, D. (1975b). EMAG, pp. 185–190.

Willasch, D. (1976). J. Microsc. Spectrosc. Electron. **1**, 505–506.

Zeitler, E. (1975). Scanning Electron Microsc. 671–678.

Zeitler, E. and Thomson, M.G.R. (1969). EMSA **27**, 170–171.

Zeitler, E. and Thomson, M.G.R. (1970). Optik **31**, 258–280 and 359–366.

Zemlin, F. and Schiske, P. (1980). Ultramicroscopy **5**, 139–145.

Zemlin, F. and Weiss, K. (1993). Ultramicroscopy **50**, 123–126.

Zemlin, F., Kunath, W. and Weiss, K. (1982). Hamburg, vol. 1, pp. 199–200.

Part XIV, Chapters 68–69

The bibliography of this Part is very selective and the reader is urged to consult the books and surveys cited for further guidance and Moodie (1981) for a historical account. Of the earlier literature on scattering theory, the following papers are particularly relevant: Schiff (1942), Koppe (1947), Leisegang (1952), Schomaker and Glauber (1952), Glauber and Schomaker (1953), Hoerni and Ibers (1953), Ibers and Hoerni (1954), Ibers (1958), Vainshtein and Aibers [Ibers] (1958), Aibers [Ibers] and Vainshtein (1959), Lippert (1956), Halliday (1960), Halliday and Quinn (1960), Burge and Smith (1962), Smith and Burge (1962, 1963), Clementi (1962, 1964), Karle and Bonham (1964), Strand and Bonham (1964), Cox and Bonham (1967), Reimer (1966), Peacher and Wills (1967), Kimura *et al.* (1967), Doyle and Turner (1968), Radi (1970), Langer (1973) and Goodman and Moodie (1974). The relevant sections of Wilson (1992) contain lengthy lists, with titles.

The paper by Doyle and Turner (1968) contained tables of the relativistic Hartree–Fock X-ray and electron atomic scattering (or form) factors $(\hat{f}(q)$ and $F(q)$ in Section 68.3) for 76 atoms and ions in the range $0 \le (\sin\theta)/\lambda \le 60$ nm^{-1}. The list was completed by Cromer and Waber for their tables in the 1974 edition of the *International Tables for X-ray Crystallography*. Doyle and Turner also fitted the X-ray scattering factors to Gaussians,

$$f(\overline{q}) = \sum_{i=1}^{4} a_i \exp(-b_i \overline{q}^2) + c \quad , \quad \overline{q} := (\sin\theta)/\lambda$$

and tabulated the nine constants $\{a_i, b_i, c\}$ for each element or ion. However, the fit is acceptable only for small angles, $\overline{q} < 20$ nm^{-1}. A much improved fit has been obtained by Fox *et al.* (1989), using

$$\ln f(\overline{q}) = a_0 + a_1 \overline{q} + a_2 \overline{q}^2 + a_3 \overline{q}^3$$

Aibers, Dzh.A. [Ibers, J.A.] and Vainshtein, B.K. (1959). Kristallografiya **4**, 641–645; Sov. Phys. Cryst. **4**, 601–605.

Amelinckx, S. (1992). In *Electron Microscopy in Materials Science* (Merli, P.G. and Antisari, M.V., eds) pp. 39–181 (World Scientific, Singapore, River Edge NJ and London).

Amelinckx, S. and Dyck, D. van (1993). Electron microscope imaging and diffraction contrast. In *Electron Diffraction Techniques* (Cowley, J.M., ed.) vol. 2, pp. 1–222 (Oxford University Press, Oxford and New York).

Amelinckx, S., Gevers, R., Remaut, G. and Landuyt, J. van, eds (1970). *Modern Diffraction and Imaging Techniques in Material Science* (North-Holland, Amsterdam and London).

Barry, J. (1992). Computing for high-resolution images and diffraction patterns. In *Electron Diffraction Techniques* (Cowley, J.M., ed.) vol. 1, pp. 170–211 (Oxford University Press, Oxford and New York).

Bethe, H.A. (1928). Ann. Physik **87**, 55–129.

Bethe, H.A. (1933). Quantenmechanik. In *Handbuch der Physik* (Geiger, H. and Scheel, K., eds) vol. 24, Pt. 1, pp. 273–560 (Springer, Berlin).

Bonham, R.A. and Strand, T.G. (1963). J. Chem. Phys. **39**, 2200–2204.

Bonse, U. and Hart, M. (1965). Appl. Phys. Lett. **6**, 155–156.

Borrmann, G. (1941). Z. Physik **42**, 157–162.

Bothe, W. (1921). Z. Physik **5**, 63–69.

Bothe, W. (1933). Durchgang von Elektronen durch Materie. In *Handbuch der Physik* (Geiger, H. and Scheel, K., eds) vol. 22, Pt. 2, pp. 1–74 (Springer, Berlin).

Burge, R.E. and Smith, G.H. (1962). Proc. Phys. Soc. (London) **79**, 673–690.

Buseck, P.R., Cowley, J.M. and Eyring, L., eds (1988). *High-resolution Transmission Electron Microscopy and Associated Techniques* (Oxford University Press, New York and Oxford).

Byatt, W.J. (1956). Phys. Rev. **104**, 1298–1300.

Clementi, E. (1962). J. Chem. Phys. **38**, 996–1000 and 1001–1008.

Clementi, E. (1964). J. Chem. Phys. **41**, 295–302 and 303–315.

Coene, W. and Dyck, D. van (1984). Ultramicroscopy **15**, 41–50 and 287–299.

Cowley, J.M. (1981). *Diffraction Physics,* 2nd ed. (North-Holland, Amsterdam, New York and Oxford).

Cowley, J.M. (1992). Coherent convergent beam diffraction. In *Electron Diffraction Techniques* (Cowley, J.M., ed.) vol. 1, pp. 439–464 (Oxford University Press, Oxford and New York).

Cowley, J.M. and Moodie, A.F. (1957). Acta Cryst. **10**, 609–619.

Cox, H.L. and Bonham, R.A. (1967). J. Chem. Phys. **47**, 2599–2608.

Darwin, C.G. (1914). Phil. Mag. **27**, 315–333 and 675–690.

Disko, M.M., Ahn, C.C. and Fultz, B (1992). *Transmission Electron Energy Loss Spectrometry in Materials Science* (Minerals, Metals and Materials Society, Warrendale PA).

Doyle, P.A. and Turner, P.S. (1968). Acta Cryst. **A24**, 390–397.

Dyck, D. van (1975). Phys. Stat. Sol. (b) **72**, 321–336.

Dyck, D. van (1980). J. Microscopy **119**, 141–152.

Dyck, D. van (1983). J. Microscopy **132**, 31–42.

Dyck, D. van (1985). Image calculations in high-resolution electron microscopy: problems, progress and prospects. Adv. Electron. Electron Phys. **65**, 295–355.

Dyck, D. van and Coene, W. (1984). Ultramicroscopy **15**, 29–40.

Eades, J.A. (1992). Convergent-beam diffraction. In *Electron Diffraction Techniques* (Cowley, J.M., ed.) vol. 1, pp. 313–359 (Oxford University Press, Oxford and New York).

Egerton, R.F. (1989). *Electron Energy Loss Spectroscopy in the Electron Microscope* (Plenum, New York and London).

Fanidis, C., Dyck, D. van and Landuyt, J. van (1992). Ultramicroscopy **41**, 55–64.

Fanidis, C., Dyck, D. van and Landuyt, J. van (1993). Ultramicroscopy **48**, 133–164.

Ferwerda, H.A. and Visser, F.P.C. (1973). In *Image Processing and Computer-aided Design in Electron Optics* (Hawkes, P.W., ed.) pp. 212–219 (Academic Press, London and New York).

Forwood, C.T. and Clarebrough, L.M. (1991). *Electron Microscopy of Interfaces in Metals and Alloys* (Adam Hilger, Bristol and Philadelphia).

Fox, A.G., O'Keefe, M.A. and Tabbernor, M.A. (1989). Acta Cryst. **A45**, 786–793.

Gevers, R. (1970). Kinematical theory of electron diffraction. In *Modern Diffraction and Imaging Techniques in Material Science* (Amelinckx, S., Gevers, R., Remaut, G. and Landuyt, J. van, eds) pp. 1–33 (North-Holland, Amsterdam and London).

Gjønnes, J. (1993). Disorder and defect scattering. In *Electron Diffraction Techniques* (Cowley, J.M., ed.) vol. 2, pp. 223–259 (Oxford University Press, Oxford and New York).

Glauber, R.J. (1959). High-energy collision theory. In *Lectures in Theoretical Physics* (Brittin, W.E. and Dunham, L.G., eds) pp. 315–414 (Interscience, New York and London).

Glauber, R. and Schomaker, V. (1953). Phys. Rev. **89**, 667–671.

Goodman, P. and Moodie, A.F. (1974). Acta Cryst. **A30**, 280–290.

Goringe, M.J. (1976). Diffraction contrast from imperfect crystals. In *Electron Microscopy in Materials Science* (Valdrè , U. and Ruedl, E., eds) Pt. II, pp. 553–589 (Commission of the European Communities, Luxemburg).

Goudsmit, S. and Saunderson, J.L. (1940). Phys. Rev. **57**, 24–29 and **58**, 36–42.

Haase, J. (1966). Z. Naturforsch. **21a**, 187–192.

Haase, J. (1968). Z. Naturforsch. **23a**, 1000–1019.

Haase, J. (1970). Z. Naturforsch. **25a**, 936–945 and 1219–1235.

Halliday, J.S. (1960). Brit. J. Appl. Phys. **11**, 259–263.

Halliday, J.S. and Quinn, T.F.J. (1960). Brit. J. Appl. Phys. **11**, 486–491.

Hashimoto, H., Howie, A. and Whelan, M.J. (1962). Proc. Roy. Soc. (London) **A269**, 80–103.

Hirsch, P.B., Howie, A. and Whelan, M.J. (1960). Phil. Trans. Roy. Soc. (London) **A252**, 499–529.

Hirsch, P.B., Howie, A., Nicholson, R.B., Pashley, D.W. and Whelan, M.J. (1965). *Electron Microscopy of Thin Crystals* (Butterworths, London); reprinted with an additional chapter by Krieger, New York (1977).

Hoerni, J.A. and Ibers, J.A. (1953). Phys. Rev. **91**, 1182–1185.

Howie, A. (1963). Proc. Roy. Soc. (London) **A271**, 268–287.

Howie, A. (1971). The theory of electron diffraction image contrast. In *Electron Microscopy in Material Science* (Valdrè, U., ed.) pp. 274 – 301 (Academic Press, New York and London).

Howie, A. (1978). The theory of high energy electron diffraction. In *Modern Diffraction and Imaging Techniques in Material Science* (Amelinckx, S., Gevers, R., Remaut, G. and Landuyt, J. van, eds) pp. 295–339 (North-Holland, Amsterdam and London).

Howie, A. (1984). Elastic scattering, image formation and interpretation. In *Quantitative Electron Microscopy* (Chapman, J.N. and Craven, A.J., eds) pp. 1–48 (Scottish Universities Summer School in Physics, Edinburgh).

Howie, A. and Whelan, M.J. (1961). Proc. Roy. Soc. (London) **A263**, 217–237.

Howie, A. and Whelan, M.J. (1962). Proc. Roy. Soc. (London) **A267**, 206–228.

Humphreys, C.J. and Bithell, E.G. (1992). Electron diffraction theory. In *Electron Diffraction Techniques* (Cowley, J.M., ed.) vol. 1, pp. 75–151 (Oxford University Press, Oxford and New York).

Humphreys, C.J. and Hirsch P.B (1968). Phil. Mag. **18**, 115–122.

Ibers, J.A. (1958). Acta Cryst. **11**, 178–183.

Ibers, J.A. and Hoerni, J.A. (1954) Acta Cryst. **7**, 405–408.

Ishizuka, K. and Uyeda, N. (1977). Acta Cryst. **A33**, 740–749.

Jost, K. and Kessler, J. (1963). Z. Physik **176**, 126–142.

Karle, J. and Bonham, R.A. (1964). J. Chem. Phys. **40**, 1396–1401.

*Kästner, G. (1993). *Many-beam Electron Diffraction related to Electron Microscope Diffraction Contrast* (Akademie Verlag, Berlin).

Kimura, M., Konaka, S. and Ogasawara, M. (1967). J. Chem. Phys. **46**, 2599–2603.

Koppe, H. (1947). Z. Physik **124**, 658–664.

Krakow, W. and O'Keefe, M.A., eds (1989). *Computer Simulation of Electron Microscope Diffraction and Images* (Minerals, Metals and Materials Society, Warrendale PA).

Langer, R.E. (1973). Phys. Rev. **51**, 669–676.

Leisegang, S. (1952). Z. Physik **132**, 183–194.

Lenz, F. (1954). Z. Naturforsch. **9a**, 185–204.

Lippert, W. (1956). Optik **13**, 506–515.

Metherell, A.J.F. (1976). Diffraction of electrons by perfect crystals. In *Electron Microscopy in Materials Science* (Valdrè , U. and Ruedl, E., eds) Pt II, pp. 397–552 (Commission of the European Communities, Luxemburg).

Molière, G. (1947). Z. Naturforsch. **2a**, 133–145.

Moodie, A.F. (1981). Notes on the theory of forward elastic scattering. In *Fifty Years of Electron Diffraction* (Goodman, P., ed.) pp. 327–337 (Reidel, Dordrecht, Boston MA and London).

Mott, N.F. and Massey, H.S.W. (1965). *The Theory of Atomic Collisions*, 3rd ed. (Oxford University Press, Oxford).

Niehrs, H. (1969). Optik **30**, 273–293.

Niehrs, H. (1970). Optik **31**, 51–71.

Oswald, R. (1992). *Numerische Untersuchung der elastischen Streuung von Elektronen an Atomen und ihre Rückstreuung an Oberflächen amorpher Substanzen im Energiebereich unter 2000eV*. Dissertation, Tübingen.

Peacher, J.L. and Wills, J.G. (1967). J. Chem. Phys. **46**, 4809–4814.

Pozzi, G. (1989a). Ultramicroscopy **30**, 417–424.

Pozzi, G. (1989b). SIME, pp. 247–248.

Pozzi, G. (1990). Optik **85**, 15–18.

Pozzi, G. (1992). In *Electron Microscopy in Materials Science* (Merli, P.G. and Antisari, M.V., eds) pp.183–191 (World Scientific, Singapore, River Edge NJ and London).

Radi, G. (1970). Acta Cryst. **A26**, 41–56.

Radi, G. (1968). Z. Physik **212**, 146–168.

Raith, H. (1968). Acta Cryst. **A24**, 85–93.

Reimer, L. (1966). Z. Naturforsch. **21a**, 1489–1499.

Reimer, L. (1969). Z. Naturforsch. **24a**, 377–389.

Reimer, L. (1985). *Scanning Electron Microscopy* (Springer, Berlin and New York). Springer Series in Optical Sciences, vol. 45.

Reimer, L. (1993). *Transmission Electron Microscopy*, 3rd ed. (Springer, Berlin and New York). Springer Series in Optical Sciences, vol. 36.

Reimer, L. and Gilde, H. (1973). In *Image Processing and Computer-aided Design in Electron Optics* (Hawkes, P.W., ed.) pp. 138–167 (Academic Press, London and New York).

Reimer, L. and Lödding, B. (1984). Scanning **6**, 128–151.

Reimer, L. and Sommer, K.H. (1968). Z. Naturforsch. **22a**, 1569–1582.

Schiff, L.I. (1942). Phys. Rev. **61**, 391 and 721–722.

Schomaker, V. and Glauber, R. (1952). Nature **170**, 290–291.

Schwertfeger, W. (1974). *Zur Kleinwinkelstreuung von mittelschnellen Elektronen beim Durchgang durch amorphe Festkörperschichten.* Dissertation, Tübingen.

Sebastiano, A. di and Pozzi, G. (1992). Granada, vol. 1, pp. 645–646.

Smith, G.H. and Burge, R.E. (1962). Acta Cryst. **15**, 182–186.

Smith, G.H. and Burge, R.E. (1963). Proc. Phys. Soc. (London) **81**, 612–632.

Spence, J.C.H. (1988). *Experimental High-Resolution Electron Microscopy*, 2nd ed. (Oxford University Press, New York and Oxford).

Spence, J.C. H. (1992). Accurate structure factor amplitude and phase determination. In *Electron Diffraction Techniques* (Cowley, J.M., ed.) vol. 1, pp. 360–438 (Oxford University Press, Oxford and New York).

Spence, J.C.H. and Zuo, J.M. (1992). *Electron Microdiffraction* (Plenum, New York and London).

Steeds, J.W. (1984). Electron crystallography. In *Quantitative Electron Microscopy* (Chapman, J.N. and Craven, A.J., eds) pp. 49–96 (Scottish Universities Summer School in Physics, Edinburgh).

Steeds, J.W. and Carlino, E. (1992). In *Electron Microscopy in Materials Science* (Merli, P.G. and Antisari, M.V., eds) pp. 279–313 (World Scientific, Singapore, River Edge NJ and London).

Strand, T.G. and Bonham, R.A. (1964). J. Chem. Phys. **40**, 1686–1691.

Takagi, S. (1962). Acta Cryst. **15**, 1311–1312.

Thomas, G. and Goringe, M.J. (1979). *Transmission Electron Microscopy of Materials* (Wiley–Interscience, New York and Chichester).

Tochilin, S.B. and Whelan, M.J. (1993). Ultramicroscopy **50**, 313–320.

Typke, D. and Radermacher, M. (1982). Ultramicroscopy **9**, 131–138.

Vainshtein, B.K. and Aibers, Dzh.A. [Ibers, J.A.] (1958). Kristallografiya **3**, 416–419; Sov. Phys. Cryst. **3**, 417–420.

Whelan, M.J. (1970). Dynamical theory of electron diffraction. In *Modern Diffraction and Imaging Techniques in Material Science* (Amelinckx, S., Gevers, R., Remaut, G. and Landuyt, J. van, eds) pp. 35–98 (North-Holland, Amsterdam and London).

Wilson, A.J.C., ed. (1992). *International Tables for Crystallography*, vol. C, Mathematical, Physical and Chemical Tables (Kluwer, Dordrecht, Boston MA and London).

Wu, T.Y. and Ohmura, T. (1962). *Quantum Theory of Scattering* (Prentice–Hall, Englewood Cliffs, NJ and London).

Yates, A.C. (1971). A program for calculating relativistic electron–atom collision data. Comp. Phys. Commun. **2**, 175–179.

Yoshioka, H. (1957). J. Phys. Soc. Japan **12**, 618–628.

Zeitler, E. and Olsen, H. (1964). Phys. Rev. **136**, 1546–1552.

Zeitler, E. and Olsen, H. (1966). Z. Naturforsch. **21a**, 1321–1327.

Zeitler, E. and Olsen, H. (1967). Phys. Rev. **162**, 1439–1447.

Part XV, Chapters 70–77

Chapter 70 Image algebra continues to generate many publications; we note in particular those of Davidson (1992b) and Ritter and Zhu (1992) on a form of matrix product in which the sum of products of matrix elements is replaced by the maximum of their sums. In an extremely readable survey, Davidson and Hummer (1993) point out this is precisely the operation needed in the study of a new class of neural networks, the *morphology neural networks*. The subject is set out with great clarity in a later paper by Davidson (1993).

An operator designed to facilitate the manipulation of multi-valued images, such as those generated by a SEM or STEM, which furnish several signals per pixel, is introduced in Hawkes (1992b) and further analysis of such images in the context of electron microscopy is to be found in Hawkes (1993).

Chapter 71 The special features of sampling for holographic image restoration are discussed by Ishizuka (1993, 1994).

Chapter 72 Other aspects of enhancement are considered by Frei (1977, histogram modification), Crowley and Parker (1980, operator formalism) and Prewitt and Mendelsohn (1966, sharpening). A very interesting discussion of the ways in which generalized rank-order filters can be used for filtering or for sharpening is to be found in Hardie and Boncelet (1993).

Chapter 74 Image restoration or sophisticated enhancement is the subject of the papers by Abbiss *et al.* (1983), Baba *et al.* (1986a,b), Bertero (1989), Frank *et al.* (1970a), Hayes (1986), Hoenders (1975b, 1978, 1979), de Jong *et al.* (1989), Kanaya *et al.* (1981, 1982c, 1983a,b), Kunath (1978, 1979), Kreznar (1977), Li *et al.* (1986), Luttrell (1985), Mammone and Rothacker (1987), Morris *et al.* (1987, 1988), Podilchuk and Mammone (1990), Pulvermacher (1976a,b), Rushforth (1987), Sanz and Huang (1983c), Saxton (1986b, 1987a) and Slump and Ferwerda (1982). Inverse filtering based on the Mellin transform instead of the Fourier transform is described in Hawkes (1975).

Cross-correlation of different members of focal series is examined by Al-Ali (1975) and correlation averaging is further studied by Baumeister *et al.* (1986), Carlsson (1992), Frank (1974b, 1976a, 1978a,b, 1979, 1984), Frank *et al.* (1978), Furcinitti *et al.* (1986, in STEM), Hegerl and Knauer (1982), Hegerl *et al.*(1978, 1986), Miller *et al.* (1991), Orlova (1991) and Rasch *et al.* (1984). An early optical device for revealing periodicities by convolution was built by Elliott *et al.* (1968). Image registration is discussed by Tsai and Huang (1984).

Entropy-based restoration is considered by Bryan (1988b), Ferrige *et al.* (1992), Frieden (1987b), Frieden and Wells (1978), Gilmore *et al.* (1993) and Nahi and Assefi (1972).

The role that the wavelet transform could play in electron microscopy is discussed by Beltrán del Río *et al.* (1991) and Gómez *et al.* (1991).

Chapter 76 The papers of Frei and Chen (1977, boundary detection), Shen and Castan (1972, edge detection), Keller *et al.* (1982) and Lipkin *et al.* (1986), both on recognition of particular details, are relevant in image analysis. Edge detection is surveyed by Petrou (1994).

SEM image processing Further aspects of this subject are the subject of Aristov *et al.*(1986, tomography), Geuens *et al.* (1992, image analysis), Hawkes (1977, 1982b, 1984, 1989, surveys), Holburn and Smith (1979, topography), Hounslow and Tovey (1991, segmentation), Jones and Smith (1978, survey), Kanaya *et al.* (1985, enhancement), Nomura and Ichikawa (1984, Fourier filtering), Oho (1992, enhancement by median filtering and histogram modification), Petigand *et al.* (1990, mathematical morphology), Rautureau *et al.* (1991, enhancement), Smith (1972, 1980, 1985, surveys), Tee *et al.* (1977, automatic focusing and astigmatism correction) and Tricart *et al.* (1991).

Chapter 77 Control hardware is the subject of many more papers than are cited in the text. See Koops (1976, use of correlation to adjust his aberration corrector); Erasmus and Smith (1980, focusing and astigmatism correction); Crewe and Ohtsuki (1982, image processing system for STEM); Oudet *et al.* (1988, use of the FFT on-line); Bonnet and Zinzindohoué (1989, microscope automation); Fan *et al.* (1991, use of CCD camera); Rez *et al.* (1991, comparison of acquisition hardware); Downing *et al.* (1992, control); Kujawa and Krahl (1992, CCD camera); McClean *et al.* (1992, SEM image acquisition); de Ruijter and Weiss (1992, CCD camera); de Ruijter *et al.* (1991, control); Ponce and Hikashi (1991, control); Krivanek and Fan (1991, CCD camera; 1992, autotuning); Scheerschmidt *et al.* (1992, autotuning using crystalline specimens); Krivanek and Mooney (1993, CCD camera); de Ruijter *et al.* (1993, CCD camera); Krivanek *et al.* (1993, TEM and STEM automation); Mooney *et al.* (1993, compensation of camera MTF); Fan and Ellisman (1993, CCD camera).

In addition to the suites mentioned in the main text, specialized software is referred to by Smith (1978, three-dimensionl reconstruction); Llinas *et al.* (1979, computer graphics and STEM); Hegerl (1980) and Hegerl and Altbauer (1982; EM program suite); Trus *et al.* (1991, PIC); Flifla *et al.* (1992, 3-D software); Hegerl (1992, software package for three-dimensional reconstruction); and Schmidt *et al.* (1993, crystal images).

The widely used packages SEMPER, SPIDER, and IMAGIC were first described in Horner (1975), Saxton *et al.* (1979), Frank *et al.*(1981b) and van Heel and Keegstra (1981); all have of course been very extensively developed since those early versions were written.

A "distributed laboratory" for the three-dimensional reconstruction process is proposed by Mercurio *et al.* (1992)

Finally we list some survey articles on a variety of image processing themes: Frieden (1975); Saxton (1975, 1984); Ferwerda (1976, 1978, 1981); Hawkes (1978b, 1980b); Kübler (1980); Kanaya *et al.* (1982a,b); Slump and Ferwerda (1986); and Baba *et al.* (1989).

An early paper by Reed *et al.* (1977) introduced the idea of using transforms other than the Fourier transform in image processing.

Abbiss, J.B., Defrise, M., Mol, C. de and Dhadwal, H.S. (1983). J. Opt. Soc. Am. **73**, 1470–1475.

Aboutalib, A.O. and Silverman, L.M. (1975). IEEE Trans. **CAS–22**, 278–286.

Aboutalib, A.O., Murphy, M.S. and Silverman, L.M. (1977). IEEE Trans. **AC–22**, 294–302.

Ahmed, N. and Rao, K.R. (1975). *Orthogonal Transforms for Digital Signal Processing* (Springer, Berlin and New York).

Ahmed, N., Natarajan, T. and Rao, K.R. (1974). IEEE Trans. **C–23**, 90–93.

Akutowicz, E.J. (1956). Trans. Am. Math. Soc. **83**, 179–192.

Akutowicz, E.J. (1957). Proc. Am. Math. Soc. **8**, 234–238.

Al-Ali, L.S. (1975). EMAG, pp. 225–228.

Al-Ali, L. and Frank, J. (1980). Optik **56**, 31–40.

Albert, A. (1972). *Regression and the Moore–Penrose Pseudoinverse* (Academic Press, New York and London).

Aldroubi, A., Trus, B.L., Unser, M., Booy, F.P. and Steven, A.C. (1992). Ultramicroscopy **46**, 175–188.

Amos, L.A. (1974). J. Microscopy **100**, 143–152.

Amos, L.A. (1975). EMSA **33**, pp. 290–291.

Amos, L.A. (1976). Jerusalem, vol. 1, pp. 14–19.

Amos, L.A., Henderson, R. and Unwin, P.N.T. (1982). Three-dimensional structure determination by electron microscopy of two-dimensional crystals. Prog. Biophys. Mol. Biol. **39**, 183–231.

Andersen, W.H.J. (1972). Manchester, pp. 396–397.

*Andrews, H.C. (1972). *Introduction to Mathematical Techniques in Pattern Recognition* (Wiley–Interscience, New York and London).

Andrews, H.C. and Hunt, B.R. (1977). *Digital Image Restoration* (Prentice–Hall, Englewood Cliffs NJ).

*Andrews, H.C., Pratt, W.K. and Caspari, K. (1970). *Computer Techniques in Image Processing* (Academic Press, New York and London).

Ansley, D.A. (1972). EMSA, pp. 596–597.

Ansley, D.A. (1973). Opt. Commun. **8**, 140–141.

Antonovsky, A. (1982). Hamburg, vol. 1, p. 453.

Arce, G.R. and Gallagher, N.C. (1982). IEEE Trans. **ASSP–30**, 894–902.

Arce, G.R., Gallagher, N.C. and Nodes, T.A. (1986). Median filters: theory for one- and two-dimensional filters. Adv. Comput. Vision Image Proc. **2**, 89–166.

Arcelli, C. and Sanniti di Baja, G. (1985). IEEE Trans. **PAMI–7**, 463–474.

Arcelli, C., Cordella, L.P. and Levialdi, S. (1981). IEEE Trans. **PAMI–3**, 134–143.

Aristov, V.V., Ushakov, N.G. and Zaitsev, S.I. (1986). Kyoto, vol. 1, pp. 475–476.

Arsenault, H.A. and Chalasinska-Macukow, K. (1983). Opt. Commun. **47**, 380–386.

Arsenault, H. and Lowenthal, S. (1969). C.R. Acad. Sci. Paris **B269**, 518–521.

Ataman, E., Aatre, V.K. and Wong, K.M. (1981). IEEE Trans. **ASSP–29**, 1073–1075.

Atkin, G.K. and Ross, G. (1984). Opt. Acta **31**, 7–21.

Ayers, G.R. and Dainty, J.C. (1988). Opt. Lett. **13**, 547–549.

Baba, N. and Murata, K. (1977). J. Opt. Soc. Am. **67**, 662–668.

Baba, N. and Murata, K. (1981). Opt. Commun. **38**, 91–95.

Baba, N., Murata, K., Okada, K. and Fujimoto, Y. (1979). Optik **54**, 97–105.

Baba, N., Murata, K., Okada, K. and Fujimoto, Y. (1981). Optik **58**, 233–239.

Baba, N., Shirakawa, H. and Ose, T. (1984). Optik **68**, 335–340.

Baba, N., Oho, E., Mukai, M. and Kanaya, K. (1986a). Kyoto, vol. 1, pp. 465–466.

Baba, N., Oho, E., Sasaki, T. and Kanaya, K. (1986b). Kyoto, vol. 1, pp. 491–492.

Baba, N., Oho, E. and Kanaya, K. (1987). Scanning Microsc. **1**, 1507–1514.

Baba, N., Kokubo, Y. and Endoh, H. (1989). J. Electron Microsc. **38**, S30–S36.

Baggett, M.C. and Glassman, L.H. (1974). Scanning Electron Microsc. 199–206.

Bahr, H.J., Dunger, B. and Schmidt, D. (1972). BEDO **5**, 337–346.

Baker, T.S. and Winkelmann, D.A. (1986). EMSA **44,** pp. 26–29.

Ballard, D.H. (1981). Pattern Recognition **13**, 111–122.

Balossier, G., Thomas, X., Michel, J., Wagner, D., Bonhomme, P., Puchelle, E., Ploton, D., Bonhomme, A. and Pinon, J.-M. (1991). Microsc. Microanal. Microstruct. **2**, 387–394.

Baltes, H.P., ed. (1978). *Inverse Source Problems in Optics* (Springer, Berlin and New York).

Baltes, H.P., ed. (1980). *Inverse Scattering Problems in Optics* (Springer, Berlin and New York).

Bamberger, R.H. and Smith, M.J.T. (1992). IEEE Trans. Signal Proc. **40**, 882–893.

Barakat, R. and Newsam, G. (1984). J. Math. Phys. **25**, 3190–3193.

Barbieri, M. (1974). J. Theor. Biol. **48**, 451–467.

Barner, K.E., Arce, G.R. and Lin, J. (1992). Circuits Syst. Signal Proc. **11**, 153–169.

Barth, M., Bryan, R.K., Hegerl, R. and Baumeister, W. (1987). Niagara Falls, pp. 277–284.

Barth, M., Bryan, R.K. and Hegerl, R. (1989). Ultramicroscopy **31**, 365–378.

Bates, R.H.T. (1978). Optik **51**, 161–170 and 223–234.

Bates, R.H.T. (1982). Optik **61**, 247–262.

Bates, R.H.T. (1983). Proc. SPIE **413**, 208–211.

Bates, R.H.T. (1984). Comput. Vision Graph. Im. Proc. **25**, 205–217.

Bates, R.H.T. and Fright, W.R. (1983). J. Opt. Soc. Am. **73**, 358–365.

Bates, R.H.T. and Fright, W.R. (1984). Reconstructing images from their Fourier intensities. Adv. Comput. Vision Image Proc. **1**, 227–264.

Bates, R.H.T. and Heffernan, P.B. (1980). Optik **56**, 101–112.

Bates, R.H.T. and Lane, R.G. (1987a). Niagara Falls, pp. 149–156.

Bates, R.H.T. and Lane, R.G. (1987b). Proc. SPIE **828**, 158–164.

*Bates, R.H.T. and McDonnell, M.J. (1986). *Image Restoration and Reconstruction* (Clarendon, Oxford).

Bates, R.H.T. and Mnyama, D. (1986). The status of practical Fourier phase retrieval. Adv. Electron. Electron Phys. **67**, 1–64.

Bates, R.H.T. and Tan, D.G.H. (1985). Proc. SPIE **558**, 54–59.

Bates, R.H.T., Napier, P.J., McKinnon, A.E. and McDonnell, M.J. (1976). Optik **44**, 183–201.

Bates, R.H.T., Quek, B.K. and Parker, C.R. (1990). J. Opt. Soc. Am. **A7**, 468–479.

Bauer, B., Schwarz, H. and Thanh, N. van (1983). J. Microscopy **130**, 325–330.

Baumeister, W. (1982). Ultramicroscopy **9**, 151–158.

Baumeister, W. and Herrmann, K.-H. (1990). High-resolution electron microscopy in biology: sample preparation, image recording and processing. In *Biophysical Electron Microscopy* (Hawkes, P.W. and Valdrè , U., eds) pp. 109–131 (Academic Press, London and San Diego).

Baumeister, W., Rachel, R., Guckenberger, R. and Hegerl, R. (1986). EMSA **44**, pp. 136–139.

Baumeister, W., Lembcke, G., Dürr, R. and Phipps, B. (1991). In *Electron Crystallography of Organic Molecules* (Fryer, J.R. and Dorset, D.L., eds.) pp. 283–296 (Kluwer, Dordrecht, Boston MA and London).

Beeching, M.J. and Spargo, A.E.C. (1992). EMSA **50**, pp. 128–129.

Beeching, M.J. and Spargo, A.E.C. (1993). Ultramicroscopy **52**, 243–247.

Bell, S.B.M., Holroyd, F.C. and Mason, D.C. (1989). Image Vision Comput. **7**, 194–204.

Bellman, R. (1970). *Introduction to Matrix Analysis,* 2nd ed. (McGraw–Hill, New York and London).

Bellman, S.H., Bender, R., Gordon, R. and Rowe, J.E. (1971). J. Theor. Biol. **32**, 205–216.

Bellon, P.L. and Lanzavecchia, S. (1992). J. Microscopy **168**, 33–45.

Beltrán del Río, L., Gómez, A. and José-Yacamán, M. (1991). Ultramicroscopy **38**, 319–324.

Bender, R., Bellman, S.H. and Gordon, R. (1970). J. Theor. Biol. **29**, 483–487.

Bendinelli, M., Consortini, A., Ronchi, L. and Frieden, B.R. (1974). J. Opt. Soc. Am. **64**, 1498–1502.

Berenyi, H.M. and Fiddy, M.A. (1986). Opt. Commun. **59**, 342–344.

Berenyi, H.M., Deighton, H.V. and Fiddy, M.A. (1985). Opt. Acta **32**, 689–701.

Berger, J.E., Zobel, C.R. and Engler, P.E. (1966). Science **153**, 168–170.

Berndt, H. and Doll, R. (1976). Optik **46**, 309–332.

Berndt, H. and Doll, R. (1978). Optik **51**, 93–96.

Berndt, H. and Doll, R. (1983). Optik **64**, 349–366.

Bertero, M. (1989). Linear inverse and ill-posed problems. Adv. Electron. Electron Phys. **75**, 1–120.

Bertero, M., Mol, C. de and Pike, E.R. (1985). In *Inverse Methods in Electromagnetic Imaging* (Boerner, W.-M., Brand, H., Cram, L.A., Gjessing, D.T., Jordan, A.K., Keydel, W., Schwierz, G. and Vogel, M., eds) pp. 319–328 and 329–340 (Reidel, Dordrecht, Boston MA and Lancaster).

Besag, J. (1986). J. Roy. Statist. Soc. **B48**, 259–302.

Beucher, S. (1990). Proc. SPIE **1350**, 70–84.

Beucher, S. (1991). Cambridge, pp. 299–314.

Beucher, S. and Meyer, F. (1993). The morphological approach to segmentation— the watershed transformation. In *Mathematical Morphology in Image Processing* (Dougherty, E. R., ed.) pp. 433–481 (Dekker, New York and Basel).

Biemond, J. (1986). Stochastic linear image restoration. Adv. Comput. Vision Image Proc. **2**, 213–273.

Bierwolf, R., Hohensteins, M.[sic], Phillipp, F., Brandt, O. and Ploog, K. (1992). Granada, vol. 1, pp. 495–456.

Bierwolf, R., Hohenstein, M., Phillipp, F., Brandt, O., Crook, G.E. and Ploog, K. (1993). Ultramicroscopy **49**, 273–285.

Bleau, A., Guise, J. de and LeBlanc, A.-R. (1992). CVGIP: Image Understanding **56**, 178–209 and 210–229.

Boas, R.P. (1954). *Entire Functions* (Academic Press, New York).

*Boerner, W.-M., Brand, H., Cram, L.A., Gjessing, D.T., Jordan, A.K., Keydel, W., Schwierz, G. and Vogel, M., eds (1985). *Inverse Methods in Electromagnetic Imaging*, 2 vols (Reidel, Dordrecht, Boston MA and Lancaster). NATO ASI Series C, vol. 143.

Bonnet, N. and Hannequin, P. (1988a). York, vol. 2, pp. 181–182.

Bonnet, N. and Hannequin, P. (1988b). York, vol. 2, pp. 183–184.

Bonnet, N. and Hannequin, P. (1989). Ultramicroscopy **28**, 248–251..

Bonnet, N. and Liehn, J.-C. (1988). J. Electron Microsc. Tech. **10**, 27–33.

Bonnet, N. and Trebbia, P. (1991). Cambridge, pp. 163–177.

Bonnet, N. and Zinzindohoué, P. (1989). J. Electron Microsc. Tech. **11**, 196–201.

Bonnet, N., Simova, E. and Lebonvallet, S. (1990). Seattle, vol. 1, pp. 460–461.

Bonnet, N., Simova, E. and Thomas, X. (1991). Microsc. Microanal. Microstruct. **2**, 129–142.

Bonnet, N., Simova, E., Lebonvallet, S. and Kaplan, H. (1992a). Ultramicroscopy **40**, 1–11.

Bonnet, N., Michel, J., Wagner, D. and Balossier, G. (1992b). Ultramicroscopy **41**, 105–114

Born, M. and Wolf, E. (1959). *Principles of Optics* (Pergamon, Oxford and New York) ; 6th ed. 1980.

Borus, E. (1975). BEDO **8**, 279–291.

Boucher, R.H. (1980). Proc. SPIE **231**, 130–141.

Bougrenet de la Tocnaye, J.-L. de and Hillion, A. (1992). Signal Proc. **26**, 243–246.

Boullion, T.L. and Odell, P.L. (1971). *Generalized Inverse Matrices* (Wiley–Interscience, New York and London).

Bovik, A.C., Huang, T.S. and Munson, D.C. (1983). IEEE Trans. **ASSP–31**, 1342–1350.

Bovik, A.C., Huang, T.C. and Munson, D.C. (1987). IEEE Trans. **PAMI–9**, 181–194.

Braggins, D.W., Gardner, G.M. and Gibbard, D.W. (1971). Scanning Electron Microsc. 393–400.

Brames, B.J. (1986). Opt. Lett. **11**, 61–63.

Brames, B.J. (1987a). Opt. Commun. **64**, 333–337.

Brames, B.J. (1987b). J. Opt. Soc. Am. **A4**, 135–147.

Bright, D.S. (1987a). J. Microscopy **148**, 51–87.

Bright, D.S. (1987b). In *Microbeam Analysis* 1987 (Geiss, R.H., ed.) pp. 290–292 (San Francisco Press, San Francisco).

Bright, D.S. and Steel, E.B. (1985). In *Microbeam Analysis 1985* (Armstrong, J.T., ed.) pp. 163–166 (San Francisco Press, San Francisco).

Bright, D.S. and Steel, E.B. (1987). J. Microscopy **146**, 191–200.

Bright, D.S. and Steel, E.B. (1988). J. Microscopy **150**, 167–180.

Brink, J., Chiu, W. and Dougherty, M. (1992). Ultramicroscopy **46**, 229–240.

Browne, M.T., Charalambous, P. and Burge, R.E. (1981). EMAG, pp. 43–44.

Brownrigg, D.R.K. (1984). Commun. Assoc. Comput. Mach. **27**, 807–818.

Bruck, Yu.M. and Sodin, L.G. (1979). Opt. Commun. **30**, 304–308.

Bruck, Yu.M. and Sodin, L.G. (1983). Opt. Acta **30**, 995–999.

Bryan, R.K. (1987). Niagara Falls, pp. 99–105.

Bryan, R.K. (1988a). In *Maximum-Entropy and Bayesian Methods in Science and Engineering* (Erickson, G.J. and Smith, C.R., eds) vol. 2, pp. 155–169 (Kluwer, Dordrecht, Boston MA and London).

Bryan, R.K. (1988b). In *Maximum-Entropy and Bayesian Methods in Science and Engineering* (Erickson, G.J. and Smith, C.R., eds) vol. 2, pp. 171–179 (Kluwer, Dordrecht, Boston MA and London).

Bryan, R.K. (1989). Maximum entropy in crystallography. In *Maximum Entropy and Bayesian Methods* (Skilling, J., ed.) pp. 213–224 (Kluwer, Dordrecht, Boston MA and London).

Bryan, R.K. (1990). In *Maximum Entropy and Bayesian Methods* (Fougère, P.F., ed.) pp. 221–232 (Kluwer, Dordrecht, Boston MA and London).

Bullough, P. and Henderson, R. (1987). Ultramicroscopy **21**, 223–230.

Buonocore, M.H., Brody, W.R. and Macovski, A. (1981). IEEE Trans. **BME–28**, 69–78.

Burch, S.F., Gull, S.F. and Skilling, J. (1983). Comput Vision Graph. Im. Proc. **23**, 113–128.

Burge, R.E. (1976). Contrast and image formation of biological specimens. In *Principles and Techniques of Electron Microscopy* (Hayat, M.A., ed.) vol. 6, pp. 85–116 (Van Nostrand Reinhold, New York and London).

Burge, R.E. (1980). Proc. Roy. Microsc. Soc. **15**, 267–269.

Burge, R.E. and Ali, S.M. (1987). Niagara Falls, pp. 191–212.

Burge R.E. and Clark, A.F. (1981). EMAG, pp. 315–320.

Burge, R.E. and Fiddy, M.A. (1979). Optik **54**, 21–26.

Burge, R.E. and Fiddy, M.A. (1981). J. Appl. Cryst. **14**, 455–461.

Burge, R.E. and Wu, J.K. (1981). Ultramicroscopy **7**, 169–180.

Burge, R.E., Fiddy, M.A., Greenaway, A.H. and Ross, G. (1974). J. Phys. D: Appl. Phys. **7**, L65–L68.

Burge, R.E., Fiddy, M.A., Greenaway, A.H. and Ross, G. (1976a). Proc. Roy. Soc. (London) **A350**, 191–212.

Burge, R.E., Fiddy, M.A., Greenaway, A.H. and Ross, G. (1976b). Holography and other methods of phase retrieval for investigating the structure of inhomogeneous media. In *Recent Advances in Optical Physics* (Havelka, B. and Blabla, J., eds) pp. 687–694 (Palacký University, Olomouc and Society of Czechoslovak Mathematicians and Physicists, Prague).

Burge, R.E., Browne, M.T., Charalambous, P., Clark, A. and Wu, J.K. (1982). J. Microscopy **127**, 47–60.

Burmester, C. and Schröder, R.R. (1992). Granada, vol. 1, pp. 95–96.

Butz, A.R. (1992). IEEE Trans. Signal Proc. **40**, 32–43.

Byrne, C.L. and Fiddy, M.A. (1987). J. Opt. Soc. Am. **A4**, 112–117.

Byrne, C. and Jones, L. (1990). Proc. SPIE **1351**, 50–55.

Byrne, C.L. and Wells, D.M. (1983). Opt. Lett. **8**, 526–527.

Byrne, C.L., Fitzgerald, R.M., Fiddy, M.A., Hall, T.J. and Darling, A.M. (1983). J. Opt. Soc. Am. **73**, 1481–1487.

Canterakis, N. (1983). IEEE Trans. **ASSP–31**, 1256–1262.

Carazo, J.M. and Carrascosa, J.L. (1987). J. Microscopy **145**, 23–43 and 159–177.

Carazo, J.M., Benavides, I., Marco, S., Carrascosa, J.L. and Zapata, E.L. (1990). Seattle, vol. 1, pp. 454–455.

Carazo, J.M., Benavides, I., Rivera, F.F. and Zapata, E.L. (1992). Ultramicroscopy **40**, 13–32.

Carlsson, A. (1992). Granada, vol. 1, pp. 497–498.

Carpenter, R.W., Chan, I. and Cowley, J.M. (1982). EMSA **40**, 696–697.

Cartan, H. (1961). *Théorie Elémentaire des Fonctions Analytiques d'une ou plusieurs Variables Complexes* (Hermann, Paris); English translation (1963) : *Elementary Theory of Analytic Functions of one or several Complex Variables* (Hermann, Paris and Addison–Wesley, Reading MA and London).

Censor, Y. and Herman, G.T. (1987). Appl. Numer. Math. **3**, 365–391.

Çetin, A.E. (1989). Signal Proc. **16**, 129–148.

Chalasinska-Macukow, K. and Arsenault, H.H. (1983). J. Opt. Soc. Am. **73**, 1875.

Chalasinska-Macukow, K. and Arsenault, H.H. (1985). J. Opt. Soc. Am. **A2**, 46–50.

Chan, D.S.K. (1980). IEEE Trans. **AC–25**, 663–673.

Chapman, J.N. (1975). Phil. Mag. **32**, 527–540 and 541–552.

Clack, R. (1992). Phys. Med. Biol. **37**, 645–660.

Clarke, R.J. (1985). *Transform Coding of Images* (Academic Press, London and Orlando).

Coene, W. (1992). EMSA **50**, pp. 986–987.

Coene, W.M.J. and Denteneer, T.J.J. (1991). Ultramicroscopy **38**, 225–233.

Coene, W., Janssen, G., Op de Beeck, M. and Dyck, D. van (1992). Phys. Rev. Lett. **69**, 3743–3746.

Colliex, C. (1984). Electron energy loss spectroscopy in the electron microscope. Adv. Opt. Electron Microsc. **9**, 65–177.

Colling, J.A. and Harrach, H.S. von (1992). EMSA **50**, pp. 970–971.

Colsher, J.G. (1977). Comput. Graph. Im. Proc. **6**, 513–537.

Cormack, A.M. (1963). J. Appl. Phys. **34**, 2722–2727.

Cormack, A.M. (1964). J. Appl. Phys. **35**, 2908–2913.

Coster, M. and Chermant, J.-L. (1985). *Précis d'Analyse d'Images* (Editions du CNRS, Paris).

Coyle, E.J. (1988). IEEE Trans. **ASSP–36**, 63–76.

Coyle, E.J. and Lin, J.-h. (1988). IEEE Trans. **ASSP–36**, 1244–1254.

Crepeau, R.H. and Fram, E.K. (1981). Ultramicroscopy **6**, 7–17.

Crewe, A.V. (1980a). Scanning **3**, 176–181.

Crewe, A.V. (1980b). Ultramicroscopy **5**, 131–138.

Crewe, A.V. (1984). J. Ultrastruct. Res. **88**, 94–104.

Crewe, A.V. and Ohtsuki, M. (1980). The Hague, vol. 2, pp. 588–589.

Crewe, A.V. and Ohtsuki, M. (1981). Ultramicroscopy 7, 13–18.

Crewe, A.V. and Ohtsuki, M. (1982). Ultramicroscopy 9, 101–108.

Crimmins, T.R. (1987). J. Opt. Soc. Am. A4, 124–134.

Crimmins, T.R. and Fienup, J.R. (1981). J. Opt. Soc. Am. 71, 1026–1028.

Crimmins, T.R., Fienup, J.R. and Thelen, B.J. (1990). J. Opt. Soc. Am. A7, 3–13.

Crombez, G. (1990). Glasnik Matemat. 25, 87–93.

Crombez, G. (1991). J. Math. Anal. Applics 155, 413–419.

Crowley, J. and Parker, A. (1980). Transfer function analysis of picture processing operators. In *Issues in Digital Image Processing* (Haralick, R.M. and Simon, J.C., eds)
pp. 3–30 (Sijthoff and Noordhoff, Alphen aan den Rijn and Germantown MD).

Crowther, R.A. (1971). Phil. Trans. Roy. Soc. (London) B261, 221–230.

Crowther, R.A. (1984). Ultramicroscopy 13, 295–304.

Crowther, R.A. and Amos, L.A. (1971). J. Mol. Biol. 60, 123–130.

Crowther, R.A. and Klug, A. (1971). J. Theor. Biol. 32, 199–203.

Crowther, R.A. and Klug, A. (1974) Nature 251, 490–492.

Crowther, R.A. and Klug, A. (1975). Structural analysis of macromolecular assemblies
by image reconstruction from electron micrographs. Annu. Rev. Biochem. 44,
161–182.

Crowther, R.A. and Luther, P.K. (1984). Nature 307, 569–570.

Crowther, R.A. and Sleytr, U.B. (1977). J. Ultrastruct. Res. 58, 41–49.

Crowther, R.A. and Wischik, C.M. (1985). EMBO J. 4, 3661–3665.

Crowther, R.A., Amos, L.A., Finch, J.T., DeRosier, D.J. and Klug, A. (1970a). Nature
226, 421–425.

Crowther, R.A., DeRosier, D.J. and Klug, A. (1970b). Proc. Roy. Soc. (London) A317,
319–340.

Crowther, R.A., Amos, L.A. and Klug, A. (1972). Manchester, pp. 593–597.

Crowther, R.A., Luther, P.K. and Taylor, K.A. (1990). Electron Microsc. Rev. 3, 29–42.

Cullis, A.G. and Maher, D.M. (1974). Phil. Mag. 30, 447–451.

Cullis, A.G. and Maher, D.M. (1975). Ultramicroscopy 1, 97–112.

Curtis, S.R. and Oppenheim, A.V. (1987). J. Opt. Soc. Am. A4, 221–231.

Dain, B.N., Rau, E.I., Savin, D.O., Sasov, A.Yu. and Spivak, G.V. (1983). Izv. Akad.
Nauk SSSR (Ser. Fiz.) 47, 1103–1107 ; Bull Acad. Sci. USSR (Phys. Ser.) 47(6),
60–64.

Dainty, J.C. and Fienup, J.R. (1987). Phase retrieval and image reconstruction for
astronomy. In *Image Recovery: Theory and Application* (Stark, H., ed.) pp.
231–275 (Academic Press, Orlando and London).

Dainty, J.C., Fiddy, M.A. and Greenaway, A.H. (1979). On the danger of applying statistical reconstruction methods in the case of missing phase information. In *Image Formation from Coherence Functions in Astronomy* (Schooneveld, C. van, ed.) pp. 95–101 (Reidel, Dordrecht).

Dallas, W.J. (1975). Optik **44**, 45–59.

Dallas, W.J. (1976). Opt. Commun. **18**, 317–320.

Darling, A.M., Hall, T.J. and Fiddy, M.A. (1983a). J. Opt. Soc. Am. **73**, 1466–1469.

Darling, A.M., Deighton, H.V. and Fiddy, M.A. (1983b). Proc. SPIE **413**, 197–201.

Das, P.P. (1990). Patt. Rec. Lett. **11**, 663–667.

Das, P.P. and Chatterji, B.N. (1988). Patt. Rec. Lett. **7**, 215–226.

Das, P.P. and Mukherjee, J. (1990). Patt. Rec. Lett. **11**, 601–604.

Daubechies, I. (1990). IEEE Trans. **IT–36**, 961–1005.

Davey, B.L.K., Lane, R.G. and Bates, R.H.T. (1989). Opt. Commun. **69**, 353–356.

Davidson, J.L. (1992a). Adv. Electron. Electron Phys. **84**, 61–130.

Davidson, J.L. (1992b). J. Math. Imaging Vision **1**, 169–192.

Davidson, J.L. (1993). Classification of lattice transformations in image processing. CVGIP: Image Understanding **57**, 283–306.

Davidson, J.L. and Hummer, F. (1993). Circuits Systems Sig. Proc. **12**, 177–210.

Davies, E.R. (1989). Signal Proc. **16**, 83–96.

Davies, E.R. (1991a). Electron. Lett. **27**, 826–828.

Davies, E.R. (1991b). Electron. Lett. **27**, 1526–1527.

Davies, E.R. (1992a). Signal Proc. **26**, 1–16.

Davies, E.R. (1992b). J. Mod. Opt. **39**, 103–113.

Davies, E.R. and Plummer, A.P.N. (1981). Pattern Recognition **14**, 53–63.

Davis, L.S. (1981). Pattern Recognition **13**, 219–223.

Davis, L.S. (1982). Pattern Recognition **15**, 277–285.

Davis, L.S., Johns, S.A. and Aggarwal, J.K. (1979). IEEE Trans. **PAMI–1**, 251–259.

Davis, L.S., Clearman, M. and Aggarwal, J.K. (1981). IEEE Trans. **PAMI–3**, 214–221.

Deans, S.R. (1983). *The Radon Transform and Some of its Applications* (Wiley, New York and Chichester).

Deighton, H.V., Scivier, M.S. and Fiddy, M.A. (1985a). Opt. Lett. **10**, 250–251.

Deighton, H.V., Scivier, M.S., Berenyi, H.M. and Fiddy, M.A. (1985b). Proc. SPIE **558**, 65–72.

*Delp, E.J., ed. (1990). *Nonlinear Image Processing*. Proc. SPIE **1247**.

DeRosier, D.J.: see also Rosier, D.J. de.

DeRosier, D.J. (1971). Contemp. Phys. **12**, 437–452.

DeRosier, D.J. and Moore, P.B. (1970). J. Mol. Biol. **52**, 355–369.

Desai, V. and Reimer, L. (1985). Scanning **7**, 185–197.

Devaney, A.J. (1981). In *Optics in Four Dimensions-1980* (Machado, M.A. and Narducci, L.M., eds) pp. 613–626 (AIP Conf. Proc. No. 65, American Institute of Physics, New York).

Devaney, A.J. and Chidlaw, R. (1978). J. Opt. Soc. Am. **68**, 1352–1354.

Dialetis, D. (1967). J. Math. Phys. **8**, 1641–1649.

Dialetis, D. and Wolf, E. (1967). Nuovo Cimento **47**, 113–116.

Dierksen, K., Typke, D., Hegerl, R., Koster, A.J. and Baumeister, W. (1992). Ultramicroscopy **40**, 71–87.

Dierksen, K., Typke, D., Hegerl, R. and Baumeister, W. (1993). Ultramicroscopy **49**, 109–120.

Dikshit, S.S. (1982). IEEE Trans. **ASSP–30**, 125–129.

Dinger, D.R. and White, E.W. (1976). Scanning Electron Microsc. (I), 409–416.

Dixon, R.N. and Taylor, C.J. (1979). Scanning Electron Microsc. (II), 361–366 and 490.

Donelli, G. and Paoletti, L. (1977). Adv. Electron. Electron Phys. **43**, 1–42.

Dong, W., Baird, T., Fryer, J.R., Gilmore, C.J., Macnicol, D.D., Bricogne, G., Smith, D.J., O'Keefe, M.A. and Hövmoller, S. (1992). Nature **355**, 605–609.

Dorkel, S., Schuster, D. and Blanc, G. (1993). J. Computer-assist. Microsc. **5**, 151–157.

Dorset, D.L. (1991). Ultramicroscopy **38**, 23–40.

Dougherty, E.R. (1989). J. Imaging Sci. **33**, 136–143 and 144–149.

Dougherty, E.R. (1990). Proc. SPIE **1350**, 129–137.

Dougherty, E.R. (1992a). *An Introduction to Morphological Image Processing* (SPIE Optical Engineering Press, Bellingham).

Dougherty, E.R. (1992b). J. Math. Imaging Vision **1**, 7–21.

Dougherty, E.R. (1992c). J. Math. Imaging Vision **2**, 185–192.

Dougherty, E.R. (1992d). CVGIP: Image Understanding **55**, 36–54 and 55–72.

Dougherty, E.R., ed. (1993). *Mathematical Morphology in Image Processing* (Dekker, New York and Basel).

*Dougherty, E.R. and Giardina, C.R. (1987). *Image Processing–Continuous to Discrete* (Prentice–Hall, Englewood Cliffs NJ); vol. 1, Geometric, transform and statistical methods.

Dougherty, E.R. and Haralick, R.M. (1992). J. Math. Imaging Vision **2**, 173–183.

Dougherty, E.R. and Kraus, E.J. (1991). SIAM J. Appl. Math. **51**, 1764–1781.

Dougherty, E.R. and Loce, R.P. (1993). Efficient design strategies for the optimal binary digital morphological filter: probabilities, constraints and structuring-element libraries. In *Mathematical Morphology in Image Processing* (Dougherty, E.R. , ed.) pp. 43–92 (Dekker, New York and Basel).

Dougherty, E.R. and Pelz, J. (1991). Opt. Eng. **30**, 438–445.

Dougherty, E.R., Pelz, J.B., Sand, F. and Lent, A. (1992). J. Electron. Imaging **1**, 46–60.

Downing, K.H. (1975). BEDO **8**, 299–307.

Downing, K.H. (1979). Ultramicroscopy **4**, 13–31.

Downing, K.H. (1988). Ultramicroscopy **24**, 387–397.

Downing, K.H. (1991a). EMSA **49**, pp. 420–421.

Downing, K.H. (1991b). Cambridge, pp. 43–52.

Downing, K.H. (1992). Ultramicroscopy **46**, 199–206.

Downing, K.H. and Glaeser, R.M. (1986). Ultramicroscopy **20**, 269–278.

Downing, K.H. and Siegel, B.M. (1973a). EMSA **31**, pp. 266–267.

Downing, K.H. and Siegel, B.M. (1973b). Optik **38**, 21–28.

Downing, K.H. and Siegel, B.M. (1974). Canberra, vol. 1, pp. 326–327.

Downing, K.H. and Siegel, B.M. (1975). Optik **42**, 155–175.

Downing, K.H., Koster, A.J. and Typke, D. (1992). Ultramicroscopy **46**, 189–197.

Drenth, A.J.J., Ferwerda, H.A., Hoenders, B.J., Huiser, A.M.J. and Toorn, P. van (1975a). EMAG, pp. 205–208.

Drenth, A.J.J., Huiser, A.M.J. and Ferwerda, H.A. (1975b). Opt. Acta **22**, 615–628.

Duda, R.O. and Hart, P.E. (1973). *Pattern Classification and Scene Analysis* (Wiley–Interscience, New York and London).

Dürr, R., Peterhans, E. and Heydt, R. (1989). Eur. J. Cell. Biol. **48**, Suppl. 25, 85–88.

Dürr, R., Hegerl, R., Volker, S., Santarius, U. and Baumeister, W. (1991). J. Struct. Biol. **106**, 181–190.

Dyck, D. van (1985). Image calculation in high resolution electron microscopy: problems, progress and prospects. Adv. Electron. Electron Phys. **65**, 295–355.

Dyck, D. van (1991). EMSA **49**, pp. 654–655.

Dyck, D. van and Coene, W. (1987a). Optik **77**, 125–128.

Dyck, D. van and Coene, W. (1987b). Niagara Falls, pp. 131–137.

Dyck, D. van and Coene, W. (1988). J. Microsc. Spectrosc. Electron. **13**, 463–477.

Dyck, D. van and Op de Beeck, M. (1990). Seattle, vol. 1, pp. 26–27.

Dyck, D. van and Op de Beeck, M. (1991). Cambridge, pp. 115–120.

Dyck, D. van, Op de Beeck, M. and Coene, W. (1993). Optik **93**, 103–107.

Dyer, C.R. and Rosenfeld, A. (1979). IEEE Trans. **PAMI–1**, 88–89.

Dyer, C.R., Hong, T.-h. and Rosenfeld, A. (1980). IEEE Trans. **SMC–10**, 158–163.

Eberly, D. and Lancaster, J. (1991). CVGIP: Graph. Models Image Proc. **53**, 538–549.

Eberly, D. and Wenzel, D. (1991). CVGIP: Graph. Models Image Proc. **53**, 340–348.

Eberly, D., Lancaster, J. and Alyassin, A. (1991). CVGIP: Graph. Models Image Proc. **53**, 550–562.

Eden, M., Unser, M. and Leonardi, R. (1986). Signal Proc. **10**, 385–393.

Edholm, P.R. and Herman, G.T. (1987). IEEE Trans. **MI–6**, 301–307.

Edholm, P., Herman, G.T. and Roberts, D.A. (1988). IEEE Trans. **MI–7**, 239–246.

Ekelund, S. (1975). EMAG, pp. 167–170.

Ekelund, S. and Werlefors, T. (1976a). Jerusalem, vol. 1, pp. 168–169.

Ekelund, S. and Werlefors, T. (1976b). Scanning Electron Microsc. (I), 417–424.

Ekelund, S., Werlefors, T. and Eskilsson, C. (1977). EMAG, pp. 323–328.

Elliott, A., Lowy, J. and Squire, J.M. (1968). Nature **219**, 1224–1226.

Elsner, L., Koltracht, I. and Lancaster, P. (1991). Numer. Math. **59**, 91–106.

Erasmus, S.J. and Smith, K.C.A. (1980). The Hague, vol. 1, pp. 494–495.

Erasmus, S.J. and Smith, K.C.A. (1981). EMAG, pp. 115–118.

Erasmus, S.J. and Smith, K.C.A. (1982). J. Microscopy **127**, 185–199.

Erasmus, S.J., Holburn, D.M. and Smith, K.C.A. (1979). EMAG, pp. 73–76.

Erasmus, S.J., Holburn, D.M. and Smith, K.C.A. (1980). Scanning **3**, 273–279.

Erasmus, S.J., Smith, K.C.A. and Smith, D.J. (1982). Hamburg, vol. 1, pp. 529–530.

Erickson, G.J. and Smith, C.R., eds. (1988). *Maximum-Entropy and Bayesian Methods in Science and Engineering,* 2 vols (Kluwer, Dordrecht, Boston MA and London). Papers from the 5th, 6th and 7th workshops on maximum entropy and Bayesian methods, 1985, 1986 and 1987.

Fan, G.Y. and Ellisman, M.H. (1993). Ultramicroscopy **52**, 21–29.

Fan, G.Y. and Krivanek, O.L. (1990). Seattle, vol. 1, pp. 532–533.

Fan, H.F., Zhong, Z.Y., Zheng, C.D. and Li, F.H. (1985). Acta Cryst. **A41**, 163–165.

Fan, G.Y., Gubbens, A.J., Krivanek, O.L., Leber, M.L. and Mooney, P.E. (1991). EMSA **49**, pp. 524–525.

Farrow, N.A. and Ottensmeyer, F.P. (1987). Niagara Falls, pp. 75–81.

Farrow, N.A. and Ottensmeyer, F.P. (1988). In *Maximum-Entropy and Bayesian Methods in Science and Engineering* (Erickson, G.J. and Smith, C.R.) vol. 2, pp. 313–322 (Kluwer, Dordrecht, Boston MA and London).

Farrow, N.A. and Ottensmeyer, F.P. (1989a). Ultramicroscopy **31**, 275–284.

Farrow, N.A. and Ottensmeyer, F.P. (1989b). In *Maximum Entropy and Bayesian Methods* (Skilling, J., ed.) pp. 181–189 (Kluwer, Dordrecht, Boston MA and London).

Feldkamp, G.B. and Fienup, J.R. (1980). Proc. SPIE **231**, 84–93.

Feldman, J.A., Feldman, G.M., Falk, G., Grape, G., Pearlman, J., Sobel, I. and Tenebaum, J.M. (1969). In *Proc. Int. Joint Conf. on Artificial Intelligence* (Walker, D.A. and Norton, L.M., eds) pp. 521–526.

Ferrige, A.G., Seddon, M.J., Pennycook, S., Chisholm, M.F. and Robinson, D.R.T. (1992). In *Maximum Entropy and Bayesian Methods*, Seattle 1991 (Smith, C.R., Erickson, G.J. and Neudorfer, P.O., eds) pp. 337–344 (Kluwer, Dordrecht, Boston and London).

Ferwerda, H.A. (1976). Jerusalem, vol. 1, pp. 1–3.

Ferwerda, H.A. (1978). The phase reconstruction problem for wave amplitudes and coherence functions. In *Inverse Source Problems in Optics* (Baltes, H.P. ed.) pp. 13–39 (Springer, Berlin and New York).

Ferwerda, H.A. (1981). In *Optics in Four Dimensions-1980* (Machado, M.A. and Narducci, L.M., eds.) pp. 402–411 (AIP Conf. Proc. No. 65, American Institute of Physics, New York).

Ferwerda, H.A. (1988). Appl. Opt. **27**, 405–408.

Ferwerda, H.A. and Frieden, B.R. (1990). Proc. SPIE **1351**, 161–172.

Ferwerda, H.A. and Hoenders, B.J. (1974). Optik **39**, 317–326.

Ferwerda, H.A. and Hoenders, B.J. (1975). Opt. Acta **22**, 25–34 and 35–36.

Ferwerda, H.A., Hoenders, B.J., Huiser, A.M.J. and Toorn, P. van (1977). Phot. Sci. Eng. **21**, 282–289.

Fiddy, M.A. (1983). Proc. SPIE **413**, 176–181.

Fiddy, M.A. (1985). Object reconstruction from partial information. In *Inverse Methods in Electromagnetic Imaging* (Boerner, W.-M., Brand, H., Cram, L.A., Gjessing, D.T., Jordan, A.K., Keydel, W., Schwierz, G. and Vogel, M., eds) pp. 341–349 (Reidel, Dordrecht, Boston MA and Lancaster).

Fiddy, M.A. (1987). The role of analyticity in image recovery. In *Image Recovery: Theory and Application* (Stark, H., ed.) pp. 499–529 (Academic Press, Orlando and London).

Fiddy, M.A. and Greenaway, A.H. (1978). Nature **276**, 421.

Fiddy, M.A. and Greenaway, A.H. (1979a). Opt. Commun. **29**, 270–272.

Fiddy, M.A. and Greenaway, A.H. (1979b). Nature **281**, 709.

Fiddy, M.A. and Hall, T.J. (1981). J. Opt. Soc. Am. **71**, 1406–1407.

Fiddy, M.A. and Ross, G. (1979). Opt. Acta **26**, 1139–1146.

Fiddy, M.A., Ross, G. and Nieto-Vesperinas, M. (1981a). IEEE Trans. **AP–29**, 406–408.

Fiddy, M.A., Hall, T.J., Nieto-Vesperinas, M., Ross, G. and Wood, J. (1981b). Optik **59**, 381–388.

Fiddy, M.A., Ross, G., Wood, J. and Nieto-Vesperinas, M. (1981c). In *Optics in Four Dimensions-1980* (Machado, M.A. and Narducci, L.M., eds) pp. 658–667 (AIP

Conf. Proc. No. 65, American Institute of Physics, New York).

Fiddy, M.A., Ross, G., Nieto-Vesperinas, M. and Huiser, A.M.J. (1982). Opt. Acta **29**, 23–40.

Fiddy, M.A., Brames, B.J. and Dainty, J.C. (1983). Opt. Lett. **8**, 96–98.

Fienup, J.R. (1978). Opt. Lett. **3**, 27–29.

Fienup, J.R. (1979). Opt. Eng. **18**, 529–534.

Fienup, J.R. (1980). Opt. Eng. **19**, 297–305.

Fienup, J.R. (1981). Proc. SPIE **373**, 147–160.

Fienup, J.R. (1982). Appl. Opt. **21**, 2758–2769.

Fienup, J.R. (1983). J. Opt. Soc. Am. **73**, 1421–1426.

Fienup, J.R. (1984). Comparison of phase retrieval algorithms. Adv. Comput. Vision Image Proc. **1**, 191–225.

Fienup, J.R. (1986). J. Opt. Soc. Am. **A3**, 284–288.

Fienup, J.R. (1987a). Proc. SPIE. **828**, 13–17.

Fienup, J.R. (1987b). J. Opt. Soc. Am. **A4**, 118–123.

Fienup, J.R. (1990). Proc. SPIE **1351**, 652–660.

Fienup, J.R. and Kowalczyk, A.M. (1990). J. Opt. Soc. Am. **A7**, 450–458.

Fienup, J.R. and Wackerman, C.C. (1986). J. Opt. Soc. Am. **A3**, 1897–1907.

Fienup, J.R., Crimmins, T.R. and Holsztynski, W. (1982). J. Opt. Soc. Am. **72**, 610–624.

Finlay, B. and Brown, I.A. (1971). BEDO **4**(1), 47–57.

Fiori, C.E., Yakowitz, H. and Newbury, D.E. (1974). Scanning Electron Microsc. 167–174.

Fitch, J.P., Coyle, E.J. and Gallagher, N.C. (1984). IEEE Trans. **ASSP–32**, 1183–1188.

Fitch, J.P., Coyle, E.J. and Gallagher, N.C. (1985a). IEEE Trans. **ASSP–33**, 230–240.

Fitch, J.P., Coyle, E.J. and Gallagher, N.C. (1985b). IEEE Trans. **CAS–32**, 445–450.

Fitch, J.P., Coyle, E.J. and Gallagher, N.C. (1986). IEEE Trans. **CAS–33**, 94–102.

Flifla, M.J., Garreau, M., Rolland, J.-P., Coatrieux, J.-L. and Thomas, D. (1992). Cabios **8**, 583–586.

Foley, J.T. and Butts, R.R. (1981). J. Opt. Soc. Am. **71**, 1008–1014.

Fougère, P.F., ed. (1990). *Maximum Entropy and Bayesian Methods* (Kluwer, Dordrecht, Boston MA and London). Proceedings of the 9th MaxEnt workshop, Dartmouth NH, 1989.

Frank, J. (1969). Optik **30**, 171–180.

Frank, J. (1972a). Biophys. J. **12**, 484–511.

Frank, J. (1972b). Manchester, pp. 622–623.

Frank, J. (1973). Optik **38**, 582–584.

Frank, J. (1974a). Optik **41**, 90–91.

Frank, J. (1974b). EMSA **32**, pp. 336–337.

Frank, J. (1975a). Ultramicroscopy **1**, 159–162.

Frank, J. (1975b). J. Phys. E: Sci. Instrum. **8**, 582–587.

Frank, J. (1975c). EMSA **33**, pp. 12–13.

Frank, J. (1975d). Optik **43**, 25–34.

Frank, J. (1976a). Jerusalem, vol. 1, pp. 273–274.

Frank, J. (1976b). EMSA **34**, pp. 478–479.

Frank, J. (1976c). Optik **44**, 379–391.

Frank, J. (1978a). Ann. New York Acad. Sci. **306**, 112–120.

Frank, J. (1978b). Toronto, vol. 3, pp. 87–93.

Frank, J. (1979). J. Microscopy **117**, 25–38.

Frank, J. (1980a). The role of correlation techniques in computer image processing. In *Computer Processing of Electron Microscope Images* (Hawkes, P.W. ed.) pp. 187–222 (Springer, Berlin and New York).

Frank, J. (1980b). The Hague, vol. 2, pp. 694–695.

Frank, J. (1982). Optik **63**, 67–89.

Frank, J. (1984). Budapest, vol. 2, pp. 1307–1316.

Frank, J. (1989a). Image analysis of single macromolecules. Electron Microsc. Rev. **2**, 53–74.

Frank, J. (1989b). BioTechniques **7**, 164–173.

Frank, J. (1990). Classification of macromolecular assemblies studied as 'single particles'. Quart. Rev. Biophys. **23**, 281–329.

Frank, J. (1991). EMSA **49**, pp. 422–423.

Frank, J., ed. (1992a). *Electron Tomography* (Plenum, New York and London).

Frank, J. (1992b). Granada, vol. 1, pp. 401–402.

Frank, J. and Al-Ali, L. (1975a). EMAG, pp. 229–231.

Frank, J. and Al-Ali, L. (1975b). Nature **256**, 376–379.

Frank, J. and Goldfarb, W. (1980). In *Electron Microscopy at Molecular Dimensions* (Baumeister, W. and Vogell, W., eds) pp. 261–269 (Springer, Berlin and NewYork).

Frank, J. and Heel, M. van (1980). The Hague, vol. 2, pp. 690–691.

Frank, J. and Heel, M. van (1982a). Hamburg, vol. 1, pp. 107–108.

Frank, J. and Heel, M. van (1982b). J. Mol. Biol. **161**, 134–137 [Appendix to Frank *et al.*, 1982].

Frank, J. and Radermacher, M. (1986). Three-dimensional reconstruction of nonperiodic macromolecular assemblies from electron micrographs. In *Advanced Techniques in Biological Electron Microscopy III* (Koehler, J.K., ed.) pp. 1–72 (Springer, Berlin and New York).

Frank, J. and Radermacher, M. (1992). Ultramicroscopy **46**, 241–262.

Frank, J. and Wagenknecht, T. (1984). Ultramicroscopy **12**, 169–175.

Frank, J., Bussler, P., Langer, R. and Hoppe, W. (1970a). Ber. Bunsen-Ges. Phys. Chem. **74**, 1105–1115.

Frank, J., Bussler, P.H., Langer, R. and Hoppe, W. (1970b). Grenoble, vol. 1, pp. 17–18.

Frank, J., Goldfarb, W., Eisenberg, D. and Baker, T.S. (1978). Ultramicroscopy **3**, 283–290 ; addendum *ibid* **4** (1979) 247.

Frank, J., Verschoor, A. and Boublik, M. (1981a). Science **214**, 1353–1355.

Frank, J., Shimkin, B. and Dowse, H. (1981b). Ultramicroscopy **6**, 343–358.

Frank, J., Verschoor, A. and Boublik, M. (1982). J. Mol. Biol. **161**, 107–137.

Frank, J., McEwen, B.F., Radermacher, M. and Rieder, C.L. (1986). EMSA **44**, pp. 18–21.

Frank, J., Bretaudière, J.-P., Carazo, J.-M., Verschoor, A. and Wagenknecht, T. (1988a). J. Microscopy **150**, 99–115.

Frank, J., Chiu, W. and Degn, L. (1988b). Ultramicroscopy **26**, 345–360.

Frank, J. Carazo, J.-M. and Radermacher, M. (1989). Eur. J. Cell Biol. **48**, 143–146.

Frank, J., Penczek, P. and Liu, W. (1991). Cambridge, pp. 11–22

Frank, J., Liu, W. and Boisset, N. (1992). Granada, vol. 1, pp. 427–429.

Frei, W. (1977). Comput. Graph. Im. Proc. **6**, 286–294.

Frei, W. and Chen, C.-c. (1977). IEEE Trans. **C−26**, 988–998.

Frieden, B.R. (1971). Evaluation, design and extrapolation methods for optical signals, based on the use of the prolate functions. Prog. Opt. **9**, 311–407.

Frieden, B.R. (1972). J. Opt. Soc. Am. **62**, 511–518 and 1202–1210.

Frieden, B.R. (1975). Image enhancement and restoration. In *Picture Processing and Digital Filtering* (Huang, T.S., ed.) pp. 177–248 (Springer, Berlin and New York); 2nd ed. 1979.

Frieden, B.R. (1976). J. Opt. Soc. Am. **66**, 280–282.

Frieden, B.R. (1977). In *Image Science Mathematics* (Wilde, C.O. and Barrett, E., eds) pp. 140–145 (Western Periodicals, North Hollywood).

Frieden, B.R. (1987a). Niagara Falls, pp. 107–111.

Frieden, B.R. (1987b). J. Opt. Soc. Am. **A4**, 232–235.

Frieden, B.R. and Oh, C. (1992). Appl. Opt. **31**, 1103–1108.

Frieden, B.R. and Wells, D.C. (1978). J. Opt. Soc. Am. **68**, 93–102.

Frieder, G. and Herman, G.T. (1971). J. Theor. Biol. **33**, 189–211.

Fright, W.R. and Bates, R.H.T. (1982). Optik **62**, 219–230.

Froehling, P.E. (1991). J. Microscopy **164**, 81–87.

Fryer, J.R., Gilmore, C.J. and Dong, W. (1991). EMAG, pp. 487–490.

Fu, K.S. (1982). *Syntactic Pattern Recognition and Applications* (Prentice–Hall, Englewood Cliffs NJ).

Fuks, B.A. (1983). *Introduction to the Theory of Analytic Functions of Several Complex Variables* (American Mathematical Society Translations, Vol. 8, Providence RI); translated from the 2nd Russian ed., published by Gos. Izd. Fiz–Mat. Lit. (Moscow, 1962).

Furcinitti, P.S., Oostrum, J. van, Wall, J.S. and Burnett, R.M. (1986). EMSA **44**, pp 152–153.

Gabbouj, M., Yu, P.-T. and Coyle, E.J. (1992a). Circuits Syst. Signal Proc. **11**, 171–193.

Gabbouj, M., Coyle, E.J. and Gallagher, N.C. (1992b). Circuits Syst. Signal Proc. **11**, 7–45.

*Gader, P.D., ed. (1990). *Image Algebra and Morphological Image Processing*. Proc. SPIE **1350**.

Gallagher, N.C. and Wise, G.L. (1981). IEEE Trans. **ASSP–29**, 1136–1141.

Garden, K.L. and Bates, R.H.T. (1982). Optik **62**, 131–142.

Garden, K.L. and Bates, R.H.T. (1984). Optik **68**, 161–173.

Gassmann, J. (1977a). Optik **48**, 347–356.

Gassmann, J. (1977b). Acta Cryst. **A33**, 474–479.

Gassmann, J. (1979). In *Unconventional Electron Microscopy for Molecular Structure Determination* (Hoppe, W. and Mason, R., eds) pp. 121–136. Adv. Struct. Res. Diffract. Methods, vol. **7**.

Gauch, J.M. (1992). CVGIP: Graph. Models Image Proc. **54**, 269–280.

Geman, S. and Geman, D. (1984). IEEE Trans. **PAMI–6**, 721–741.

Geman, D. and Reynolds, G. (1992). IEEE Trans. **PAMI–14**, 367–383.

Gerchberg, R.W. (1972). Nature **240**, 404–406.

Gerchberg, R.W. (1986). Optik **74**, 91–93.

Gerchberg, R.W. and Saxton, W.O. (1971). Optik **34**, 275–284.

Gerchberg, R.W. and Saxton, W.O. (1972). Optik **35**, 237–246.

Gerchberg, R.W. and Saxton, W.O. (1973a). Wave phase from image and diffraction plane pictures. In *Image Processing and Computer-aided Design in Electron Optics* (Hawkes, P.W., ed.) pp. 66–81 (Academic Press, London and New York).

Gerchberg, R.W. and Saxton, W.O. (1973b). J. Phys. D: Appl. Phys. **6**, L31–L32.

Gersho, A. and Gray, R.M. (1992). *Vector Quantization and Signal Compression* (Kluwer, Dordrecht).

Geuens, I., Gijbels, R., Jacob, W., Verbeeck, A. and Dekeyzer, R. (1992). J. Imaging Sci. Technol. **36**, 534–539.

Giardina, C.R. and Dougherty, E.R. (1988). *Morphological Methods in Image and Signal Processing* (Prentice–Hall, Englewood Cliffs NJ).

Gibbs, A.J. and Rowe, A.J. (1976). Optical analysis and reconstruction of images. In *Principles and Techniques of Electron Microscopy* (Hayat, M.A., ed.) vol. 7, pp. 202–230 (Van Nostrand Reinhold, New York and London).

Gibson, J.D. and Sayood, K. (1988). Lattice quantization. Adv. Electron. Electron Phys. **72**, 259–330.

Gil, J. and Werman, M. (1993). IEEE Trans. **PAMI–15**, 504–507.

Gilbert, P.F.C. (1972a). Proc. Roy. Soc. (London). **B182**, 89–102.

Gilbert, P.F.C. (1972b). J. Theor. Biol **36**, 105–117.

Gilmore, C.J., Shankland, K. and Fryer, J.R. (1993). Ultramicroscopy **49**, 132–146.

Glaeser, R.M. (1985). Electron crystallography of biological macromolecules. Annu. Rev. Phys. Chem. **36**, 243–275.

Glaeser, R.M., Kuo, I. and Budinger, T.F. (1971). EMSA **29**, pp. 466–467.

Goehner, R.P. and Rao, P. (1977). Metallography **10**, 415–424.

Goehner, R.P., Hatfield, W.T. and Rao, P. (1976). EMSA **34**, pp. 542–543.

Goitein, M. (1972). Nucl. Instrum. Meth. **101**, 509–518.

Golay, M.J.E. (1969). IEEE Trans. **C–18**, 733–740.

Goldfarb, W. and Frank, J. (1978). Toronto, vol. 2, pp. 22–23.

Gómez, A., Beltrán del Río, L., Romeu, D. and Yacamán, M.J. (1991). Cambridge, pp. 153–161.

Gonsalves, R.A. (1985). Proc. SPIE **528**, 202–215.

Gonsalves, R.A. (1987). J. Opt. Soc. Am. **A4**, 166–170.

Gonsalves, R.A., Kennealy, J.P., Korte, R.M. and Price, S.D. (1990). In *Maximum Entropy and Bayesian Methods* (Fougère, P.F., ed.) pp. 369–382 (Kluwer, Dordrecht, Boston MA and London).

Gonzalez, R.C. and Thomason, M.G. (1978). *Syntactic Pattern Recognition, an Introduction* (Addison–Wesley, Reading MA and London).

*Gonzalez, R.C. and Wintz, P. (1977). *Digital Image Processing* (Addison–Wesley, Reading MA and London); 2nd ed., 1987 (Addison–Wesley, Reading MA and Wokingham).

*Gonzalez, R.C. and Woods, R.E. (1992). *Digital Image Processing* (Addison–Wesley, Reading MA and Wokingham); new ed. of Gonzalez and Wintz (1987).

Gordon, R. (1974). IEEE Trans. **NS–21**, 78–93.

Gordon, R. and Bender, R. (1971). EMSA **29**, pp. 82–83.

Gordon, R. and Herman, G. T. (1971). Commun. Assoc. Comput. Mach. **14**, 759–768.

Gordon, R. and Herman, G. T. (1974). Int. Rev. Cytol. **38**, 111–151.

Gordon, R., Bender, R. and Herman, G.T. (1970). J. Theor. Biol. **29**, 471–481.

Gori, F. (1974). J. Opt. Soc. Am. **64**, 1237–1243.

Gori, F. and Guattari, G. (1973). Opt. Commun. **7**, 163–165.

Gori, F. and Guattari, G. (1974). J. Opt. Soc. Am. **64**, 453–458.

Gori, F. and Guattari, G. (1975). Opt. Acta **22**, 93–101.

Gori, F., Paolucci, S. and Ronchi, L. (1975). J. Opt. Soc. Am. **65**, 495–501.

Grandy, W.T. and Schick, L.H., eds (1991). *Maximum Entropy and Bayesian Methods* (Kluwer, Dordrecht, Boston MA and London). Proceedings of the 10th MaxEnt workshop, Laramie WY, 1990.

Gratin, C. and Meyer, F. (1991). Cambridge, pp. 129–135.

Grauert, H. and Fritzsche, K. (1976). *Several Complex Variables* (Springer, New York and Heidelberg).

Greenaway, A.H. (1977). Opt. Lett. **1**, 10–12.

Greenaway, A.H. and Huiser, A.M.J. (1976). Optik **45**, 295–300.

Grenander, U. (1975–1981). *Lectures in Pattern Theory.* I : Pattern Synthesis ; II : Pattern Analysis ; III : Regular Structures (Springer, New York and Berlin).

Gribelyuk, M.A. and Hutchison, J.L. (1991). EMSA **49**, 550–551.

Gribelyuk, M.A. and Hutchison, J.L. (1992). Ultramicroscopy **45**, 127–143.

Grillon, F. (1991). Cambridge, pp. 147–151.

Grinton, G.R. and Cowley, J.M. (1971). Optik **34**, 221–233.

Gu, B. and Yang, G. (1981). Acta Opt. Sin. **1**, 519–522.

Guckenberger, R. (1982). Hamburg, vol. 1, pp. 559–560.

Gull, S. F. and Daniell, G. J. (1978). Nature **272**, 686–690.

Gull, S. F. and Daniell, G. J. (1979). In *Image Formation from Coherence Functions in Astronomy* (Schooneveld, C. van, ed.) pp. 219–225 (Reidel, Dordrecht).

Gunning, R.C. and Rossi, H. (1965). *Analytic Functions of Several Complex Variables* (Prentice–Hall, Englewood Cliffs NJ).

Haavisto, P., Gabbouj, M. and Neuvo, Y. (1991). J. Circuits Syst. Comput. **1**, 125–148.

Habibi, A. (1972). Two-dimensional Bayesian estimate of images. Proc. IEEE **60**, 878–883 ; reply to criticism by Strintzis, M.G. (1976) : Proc. IEEE **64**, 1257.

Hall, T.J., Darling, A.M. and Fiddy, M.A. (1982). Opt. Lett. **7**, 467–468.

Hamaker, C. , Smith, K.T. , Solmon, D.C. and Wagnor, S. L. (1980). Rocky Mountain J. Math. **10**, 253–283.

Hammel, M., Colliex, C., Mory, C., Kohl, H. and Rose, H. (1990). Ultramicroscopy **34**, 257–269.

Hannequin, P. and Bonnet, N. (1988). Optik **81**, 6–11.

Hanson, K. M. (1987). Bayesian and related methods in image reconstruction from incomplete data. In *Image Recovery: Theory and Application* (Stark, H., ed.) pp. 79–125 (Academic Press, Orlando and London).

Hanszen, K.-J. (1969). Z. Naturforsch. **24a**, 1849.

Hanszen, K.-J. (1970). Grenoble, vol. 1, pp. 21–22.

Hanszen, K.-J. and Morgenstern, B. (1965). Z. angew. Phys. **19**, 215–227.

Haralick, R.M. (1979). Proc. IEEE **67**, 786–804.

Haralick, R.M. (1992). Patt. Rec. Lett. **13**, 5–12.

Haralick, R.M. and Shapiro, L.G. (1992). *Computer and Robot Vision*, 2 vols (Addison-Wesley, Reading MA and Wokingham).

Haralick, R.M., Sternberg, S.R. and Zhuang, X.-h. (1987). Image analysis using mathematical morphology. IEEE Trans. **PAMI-9**, 532–550.

Haralick, R.M., Zhuang, X., Lin, C. and Lee, J.S.J. (1989). IEEE Trans. **ASSP-37**, 2067–2090.

Harauz, G. and Chiu, D. K. Y. (1991). Ultramicroscopy **38**, 305–317.

Harauz, G. and Fong-Lochovsky, A. (1989). Ultramicroscopy **31**, 333–344.

Harauz, G. and Heel, M. van (1986a). Optik **73**, 146–156.

Harauz, G. and Heel, M. van (1986b). In *Pattern Recognition in Practice II* (Gelsema, E.S. and Kanal, L.N., eds) pp. 279–288 (North-Holland, Amsterdam, New York and Oxford).

Harauz, G., Boekema, E.J. and Heel, M. van (1988). Methods in Enzymology **164**, 35–49.

Hardie, R.C. and Boncelet, C.G. (1993). IEEE Trans. Signal Processing **41**, 1061–1076.

Harris, J.R. and Kerr, J. (1976). J. Microscopy **108**, 51–59.

Hashimoto, H., Endoh, H., Kuwabara, M. and Yokota, Y. (1986). Kyoto, vol. 2, pp. 945–950.

*Hawkes, P.W., ed. (1973). *Image Processing and Computer-aided Design in Electron Optics* (Academic Press, London and New York).

Hawkes, P.W. (1974a). Optik **40**, 539–556.

Hawkes, P.W. (1974b). Optik **41**, 64–68.

Hawkes, P.W. (1975). Pattern Recognition **7**, 59–60.

Hawkes, P.W. (1976). Optik **45**, 427–435.

Hawkes, P.W. (1977). J. Microsc. Spectrosc. Electron. **2**, 437–449.

Hawkes, P.W. (1978a). Electron image processing: a survey. Comput. Graph. Im. Proc. **8**, 406–446.

Hawkes, P.W. (1978b). Computer processing of electron micrographs. In *Principles and Techniques of Electron Microscopy* (Hayat, M.A., ed.) vol. 8, pp. 262–306 (Van Nostrand Reinhold, New York and London).

*Hawkes, P.W., ed. (1980a). *Computer Processing of Electron Microscope Images* (Springer, Berlin and New York).

Hawkes, P.W. (1980b). Image processing based on the linear theory of image formation. In *Computer Processing of Electron Microscope Images* (Hawkes, P.W., ed.) pp. 1–33 (Springer, Berlin and New York).

Hawkes, P.W. (1981). EMAG, pp. 325–328.

Hawkes, P.W. (1982a). Hamburg, vol. 1, pp. 467–474.

Hawkes, P.W. (1982b). J. Microsc. Spectrosc. Electron. **7**, 57–76.

Hawkes, P.W. (1982c). Electron image processing: 1978–1980. Comput. Graph. Im. Proc. **18**, 58–96.

Hawkes, P.W. (1984). J. Physique **45**, C2: 195–200.

Hawkes, P.W. (1989). J. Microsc. Spectrosc. Electron. **14**, 229–236.

Hawkes, P.W. (1990a). Proc. SPIE **1350**, 193–200.

Hawkes, P.W. (1990b). Scanning **12**, 107–111.

Hawkes, P.W. (1991a). EMAG, pp. 495–498.

Hawkes, P.W. (1991b). Cambridge, pp. 179–184.

Hawkes, P.W. (1991c). J. Opt. (Paris) **22**, 219–222.

Hawkes, P.W. (1992a). Electron image processing, 1981–1990. J. Computer-assist. Microsc. **4**, 1–72 and **5** (1993) 255.

Hawkes, P.W. (1992b). J. Math. Imaging Vision **2**, 83–85.

Hawkes, P.W. (1992c). The electron microscope as a structure projector. In *Electron Tomography* (Frank, J., ed.) pp. 17–38 (Plenum, New York and London).

Hawkes, P.W. (1993). Optik **93**, 149–154.

*Hawkes, P.W. and Valdrè, U., eds. (1990). *Biophysical Electron Microscopy, Basic Concepts and Modern Techniques* (Academic Press, London and San Diego).

*Hawkes, P.W., Ottensmeyer, F.P., Saxton, W.O. and Rosenfeld, A., eds (1988). *Image and Signal Processing in Electron Microscopy* (Scanning Microscopy International, AMF O'Hare/Chicago). Scanning Microsc., Supplement 2.

Hawkes, P.W., Saxton, W.O. and O'Keefe, M.A., eds (1992). *Signal and Image Processing in Microscopy and Microanalysis* (Scanning Microscopy Int., Chicago). Scanning Microsc., Suppl. 6, published 1994.

Hayakawa, S., Ichihara, S., Hoshino, M., Yamaguchi, H., Sakuma, S., Hanaichi, T., Kamiya, Y. and Arii, T. (1987). J. Electron Microsc. **36**, 1–8.

Haydon, G.B. (1974). Optical shadowing. In *Principles and Techniques of Electron Microscopy* (Hayat, M.A., ed.) vol. 4, pp. 1–15 (Van Nostrand Reinhold, New York and London).

Haydon, G.B. and Lemons, R.A. (1972). J. Microscopy **95**, 483–491.

Haydon, G.B., Hill, B.C. and Lemons, R.A. (1971). EMSA **29**, pp. 438–439.

Hayes, M.H. (1982). IEEE Trans. **ASSP–30**, 140–154.

Hayes, M.H. (1984). Signal reconstruction from spectral phase or spectral magnitude. Adv. Comput. Vision Image Proc. **1**, 145–189.

Hayes, M.H. (1987). The unique reconstruction of multidimensional sequences from Fourier transform magnitude or phase. In *Image Recovery: Theory and Application* (Stark, H., ed.) pp. 195–230 (Academic Press, Orlando and London).

Hayes, M.H. and McClellan, J.H. (1982). Proc. IEEE **70**, 197–198.

Hayes, M.H. and Oppenheim, A.V. (1980). IEEE Trans. **ASSP–28**, 672–680.

Hayes, M.H. and Quatieri, T.F. (1983). J. Opt. Soc. Am. **73**, 1427–1433.

Hayes, T. L., Glaeser, R.M. and Pawley, J.B. (1969). EMSA **27**, pp. 410–411.

Hayner, D.A. and Jenkins, W.K. (1984). The missing cone problem in computer tomography. Adv. Comput. Vision Image Proc. **1**, 83–144.

Hayward, S. B. and Stroud, R. (1981). J. Mol. Biol. **151**, 491–517.

Heeke, G. (1980). The Hague, vol. 1, pp. 504–505.

Heel, M. van (1983). Ultramicroscopy **11**, 307–313.

Heel, M. van (1984a). Ultramicroscopy **13**, 165–183.

Heel, M. van (1984b). Budapest, vol. 2, pp. 1317–1325.

Heel, M. van (1986). In *Pattern Recognition in Practice II* (Gelsema, E.S. and Kanal, L.N., eds) pp. 291–299 (North-Holland, Amsterdam, New York and Oxford).

Heel, M. van (1987). Ultramicroscopy **21**, 111–124.

Heel, M. van (1989). Optik **82**, 114–126.

Heel, M. van and Frank, J. (1980a). The Hague, vol. 2, pp. 692–693.

Heel, M. van and Frank, J. (1980b). In *Pattern Recognition in Practice* (Gelsema, E.S and Kanal, L.N., eds) pp. 235–243 (North-Holland, Amsterdam, New York and Oxford).

Heel, M. van and Frank, J. (1981). Ultramicroscopy **6**, 187–194.

Heel, M. van and Harauz, G. (1986). Optik **72**, 85.

Heel, M. van and Keegstra, W. (1981). Ultramicroscopy **6**, 113–130.

Heel, M. van, Bretaudière, J.-P. and Frank, J. (1982). Hamburg, vol. 1, pp. 563–564.

Heel, M. van, Winkler, H.-P., Orlova, E. and Schatz, M. (1991). Cambridge, pp. 23–42.

Heel, M. van, Schatz, M. and Orlova, E. (1992). Ultramicroscopy **46**, 307–316.

Heffernan, P.B. and Bates, R.H.T. (1982). Optik **60**, 129–142.

Hegerl, R. (1980). The Hague, vol. 2, pp. 700–701.

Hegerl, R. (1992). Ultramicroscopy **46**, 417–423.

Hegerl, R. and Altbauer, A. (1982). Ultramicroscopy **9**, 109–116.

Hegerl, R. and Baumeister, W. (1988). J. Electron Microsc. Tech. **9**, 413–419.

Hegerl, R. and Hoppe, W. (1972). Manchester, pp. 628–629.

Hegerl, R. and Knauer, V. (1982). Hamburg, vol. 1, pp. 565–566.

Hegerl, R., Feltynowski, A. and Grill, B. (1978). Toronto, vol. 1, pp. 214–215.

Hegerl, R., Cejka, Z. and Baumeister, W. (1986). EMSA **44**, pp. 148–151.

Hegerl, R., Pfeifer, G., Dahlmann, B. and Baumeister, W. (1990). Seattle, vol. 1, pp. 272–273.

Hegerl, R., Typke, D. and Dierksen, K. (1992). Granada. vol. 1, pp. 403–404.

Heijmans, H.J.A.M. (1991). IEEE Trans. **PAMI-13**, 568–582.

Heijmans, H.J.A.M. and Ronse, C. (1990). Comput. Vision Graph. Im. Proc. **50**, 245–295.

Heijmans, H.J.A.M. and Toet, A. (1991). CVGIP : Image Understanding **54**, 384–400.

Heinrich, K.F.J., Fiori, C. and Yakowitz, H. (1970). Science **167**, 1129–1131.

Henderson, R. and Baldwin, J. M. (1986). EMSA **44**, pp. 6–9.

Henderson R. and Glaeser R.M. (1985). Ultramicroscopy **16**, 139–150.

Henderson, R., Baldwin, J.M., Downing, K.H., Lepault, J. and Zemlin, F. (1986). Ultramicroscopy **19**, 147–178.

Henderson, R., Baldwin, J.M., Ceska, T.A., Zemlin, F., Beckmann, E. and Downing, K.H. (1990). J. Mol. Biol. **213**, 899–929.

Hereford, J.M. and Rhodes, W.T. (1988). Opt. Eng. **27**, 274–279.

Herman, G.T. (1972). Comput. Graph. Im. Proc. **1**, 123–144.

Herman, G.T., ed. (1979a). *Image Reconstruction from Projections* (Springer, Berlin and New York); Topics in Applied Physics, vol. 32.

Herman, G.T. (1979b). Comput. Biol. Med. **9**, 271–276.

Herman, G.T. (1980a). Comput. Graph. Im. Proc. **12**, 271–285.

Herman, G.T. (1980b). *Image Reconstruction from Projections, the Fundamentals of Computerized Tomography* (Academic Press, New York and London).

Herman, G.T. (1982). Math. Programming Study **20**, 96–112.

Herman, G.T. (1985). In *Maximum Entropy and Bayesian Methods in Inverse Problems* (Smith, C.R. and Grandy, W.T., eds) pp. 319–338 (Reidel, Dordrecht).

Herman, G.T. (1990). Comput. Vision Graph. Im. Proc. **52**, 409–415.

Herman, G.T. and Lent, A. (1976a). Comput. Graph. Im. Proc. **5**, 319–332.

Herman, G.T. and Lent, A. (1976b). Information & Control **31**, 364–384.

Herman, G.T. and Lent, A. (1976c). Comput. Biol. Med. **6**, 273–294.

Herman, G.T. and Lewitt, R.M. (1979). Overview of image reconstruction from projections. In *Image Reconstruction from Projections* (Herman, G.T., ed.) pp. 1–8 (Springer, Berlin and New York).

Herman, G.T. and Natterer, F., eds (1981). *Mathematical Aspects of Computerized Tomography* (Springer, Berlin and New York).

Herman, G.T. and Rowland, S. (1971). J. Theor. Biol. **33**, 213–223.

Herman, G.T., Lent, A. and Rowland, S.W. (1973). J. Theor. Biol. **42**, 1–32.

Herman, G.T., Hurwitz, H., Lent, A. and Lung, H.-p. (1979). Information & Control **42**, 60–71.

Herman, G.T., Lent, A. and Hurwitz, H. (1980). J. Inst. Maths Applics **25**, 361–366.

Herman, G.T., Levkowitz, H., Tuy, H.K. and McCormick, S. (1984). Multilevel image reconstruction. In *Multiresolution Image Processing and Analysis* (Rosenfeld, A., ed.) pp. 121–135 (Springer, Berlin and New York).

Herman, G.T., Roberts, D. and Axel, L. (1992). Phys. Med. Biol. **37**, 673–687.

Herrmann, K.-H. (1990). Seattle, vol. 1, pp. 112–113.

Herrmann, K.-H. and Krahl, D. (1984). Electronic image recording in conventional electron microscopy. Adv. Opt. Electron Microsc. **9**, 1–64.

Herrmann, K.-H., Reuber, E. and Schiske, P. (1978). Toronto, vol. 1, pp. 226–227.

Herzog, R.F. and Everhart, T.E. (1973). EMAG, pp. 2–5.

Herzog, R.F., Lewis, B.L. and Everhart, T.E. (1974). Scanning Electron Microsc. 175–182.

Hewan-Lowe, K. (1992). Ultrastruct. Pathol. **16**, 155–163.

Hilditch, C.J. (1983). Image Vision Comput. **1**, 115–132.

Hillman, D.E., Llinas, R., Chujo, M. and Crank, R. (1980). Scanning Electron Microsc. (I), 125–135.

Hinterberger, J., Hillje, G., Köditz, W., Kugler, J. and Wieczorek, R. (1980). The Hague, vol. 1, pp. 534–535.

Hoenders, B.J. (1972a). *Investigations on the Reconstruction Problem in Electron Microscopy*. Proefschrift, Groningen.

Hoenders, B.J. (1972b). Optik **35**, 116–133.

Hoenders, B.J. (1975a). J. Math. Phys. **16**, 1719–1725.

Hoenders, B.J. (1975b). Optik **42**, 335–350.

Hoenders, B.J. (1978). The uniqueness of inverse problems. In *Inverse Source Problems in Optics* (Baltes, H.P., ed.) pp. 41–82 (Springer, Berlin and New York).

Hoenders, B.J. (1979). Opt. Acta **26**, 711–730.

Hoenders, B.J. and Ferwerda, H.A. (1973a). Optik **37**, 542–556.

Hoenders, B.J. and Ferwerda, H.A. (1973b). Optik **38**, 80–94.

Hoenders, B.J. and Ferwerda, H.A. (1973c). Investigation of the reconstruction problem in electron microscopy. In *Image Processing and Computer-aided Design in Electron Optics* (Hawkes, P.W., ed.) pp. 220–228 (Academic Press, London and New York).

Hofstetter, E.M. (1964). IEEE Trans. **IT–10**, 119–126.

Hohenstein, M. (1991). Ultramicroscopy **35**, 119–129.

Hohenstein, M. (1992). Appl. Phys. **A 54**, 485–492.

Holburn, D.M. and Smith, K.C.A. (1979). Scanning Electron Microsc. (II), 47–52 and 46.

Hood, P.J. and Howitt, D.G. (1989). Scanning **11**, 181–189 and 223–229.

Hoppe, W. (1969a). Acta Cryst. **A25**, 495–501 and 508–514.

Hoppe, W. (1969b). Optik **29**, 617–621.

Hoppe, W. (1970a). Acta Cryst. **A26**, 414–426.

Hoppe, W. (1970b). Ber. Bunsen-Ges. Phys. Chem. **74**, 1090–1100.

Hoppe, W. (1971). Z. Naturforsch. **26a**, 1155–1168.

Hoppe, W. (1972). Z. Naturforsch. **27a**, 919–929.

Hoppe, W. (1974a). Naturwissenschaften **61**, 239–249.

Hoppe, W. (1974b). Naturwissenschaften **61**, 534–536.

Hoppe, W. (1975). Z. Naturforsch. **30a**, 1188–1199.

Hoppe, W. (1978). Ann. New York Acad. Sci. **306**, 121–144.

Hoppe, W. (1978/79). Chemica Scripta **14**, 227–243.

Hoppe, W. (1979). Adv. Struct. Res. Diffract. Meth. **7**, 191–220.

Hoppe, W. (1980). In *Electron Microscopy at Molecular Dimensions* (Baumeister, W. and Vogell, W., eds) pp. 278–287 (Springer, Berlin and New York).

Hoppe, W. (1981). Three-dimensional electron microscopy. Annu. Rev. Biophys. Bioeng. **10**, 563–592.

Hoppe, W. (1982). Ultramicroscopy **10**, 187–198.

Hoppe, W. (1983). Angew. Chem. Int. Ed. Engl. **22**, 456–485.

Hoppe, W. (1985). Ultramicroscopy **16**, 139–150.

Hoppe, W. and Grill, B. (1977). Ultramicroscopy **2**, 153–168.

Hoppe, W. and Hegerl, R. (1980). Three-dimensional structure determination by electron microscopy. In *Computer Processing of Electron Microscope Images* (Hawkes, P.W., ed.) pp. 127–185 (Springer, Berlin and New York).

Hoppe, W. and Hegerl, R. (1981). Ultramicroscopy **6**, 205–206.

Hoppe, W. and Köstler, D. (1976). Jerusalem, vol. 1, pp. 99–104.

Hoppe, W. and Strube, G. (1969). Acta Cryst. **A25**, 502–507.

Hoppe, W. and Typke, D. (1979). Adv. Struct. Res. Diffract. Meth. **7**, 137–190.

Hoppe, W., Langer, R. and Thon, F. (1970). Optik **30**, 538–545.

Hoppe, W., Bussler, P., Feltynowski, A., Hunsmann, N. and Hirt, A. (1973). Some experience with computerized image reconstruction methods. In *Image Processing and Computer-aided Design in Electron Optics* (Hawkes, P.W., ed.) pp. 92–126 (Academic Press, London and New York).

Hoppe, W., Gassmann, J., Hunsmann, N., Schramm, H.J. and Sturm, M. (1974). Hoppe–Seyler's Z. Physiol. Chem. **355**, 1483–1487.

Hoppe, W., Gassmann, J., Hunsmann, N., Schramm, H.J. and Sturm, M. (1975). Hoppe–Seyler's Z. Physiol. Chem. **356**, 1317–1320.

Hoppe, W., Schramm, H.J., Sturm, M., Hunsmann, N. and Gassmann, J. (1976a). Z. Naturforsch. **31a**, 645–655, 1370–1379 and 1380–1390.

Hoppe, W., Hunsmann, N., Schramm, H.J., Sturm, M., Grill, B. and Gassmann, J. (1976b). Jerusalem, vol. 1, pp. 8–13.

Hoppe, W., Wenzl, H. and Schramm, H.J. (1977). Hoppe–Seyler's Z. Physiol. Chem. **358**, 1069–1076.

Hoppe, W., Hegerl, R. and Guckenberger, R. (1978). Z. Naturforsch. **33a**, 857–858.

Hörmander, L. (1966). *An Introduction to Complex Analysis in Several Variables* (Van Nostrand, Princeton NJ); 2nd ed. (1973) published by North-Holland (Amsterdam and London) and American Elsevier (New York).

Horne, R.W. and Markham, R. (1972). In *Practical Methods in Electron Microscopy* (Glauert, A.M., ed.) vol. 1, pp. 325–434 (North-Holland, Amsterdam and London).

Horner, M. (1975). EMAG, pp. 209–212.

Hounslow, M.W. and Tovey, N.K. (1991). Cambridge, pp. 245–254.

Hove, P.L. van, Hayes, M.H., Lim, J.S. and Oppenheim, A.V. (1983). IEEE Trans. **ASSP–31**, 1286–1293.

*Huang, T.S., ed. (1975). *Picture Processing and Digital Filtering* (Springer, Berlin and New York) ; 2nd ed., 1979.

*Huang, T.S., ed. (1981). *Two-dimensional Signal Processing* (Springer, Berlin and New York) 2 vols.

Huang, T.T., Sanz, J.L.C. and Blanz, W.-E. (1988). IEEE Trans. **ASSP–36**, 1292–1304.

Huang, K.S., Jenkins, B.K. and Sawchuk, A.A. (1989). Comput. Vision Graph. Im. Proc. **45**, 295–345.

Huffman, D.A. (1962). Proc. IRE **40**, 1098–1101.

Huggins, F.E., Kosmack, D.A., Huffman, G.P. and Lee, R.J. (1980). Scanning Electron Microsc. (I), 531–540.

Huiser, A.M.J. (1979). *Fundamental Problems in the Evaluation of Electron Micrographs.* Proefschrift, Groningen.

Huiser, A.M.J. (1980). Optik **55**, 241–252.

Huiser, A.M.J. and Ferwerda, H.A. (1976a). Optik **46**, 407–420.

Huiser, A.M.J. and Ferwerda, H.A. (1976b). Opt. Acta **23**, 445–456.

Huiser, A.M.J. and Toorn, P. van (1980). Opt. Lett. **5**, 499–501.

Huiser, A.M.J., Drenth, A.J.J. and Ferwerda, H.A. (1976). Optik **45**, 303–316.

Huiser, A.M.J., Toorn, P. van and Ferwerda, H.A. (1977). Optik **47**, 1–8.

Hunsmann, N., Bussler, P. and Hoppe, W. (1972). Manchester, pp. 654–655.

Hunt, B.R. (1972). IEEE Trans. **AC–17**, 703–705.

Hunt, B.R. (1973). IEEE Trans. **C–22**, 805–812.

Hunt, B.R. (1975). Proc. IEEE **63**, 693–708.

Hunt, B.R. (1977a). IEEE Trans. **C–26**, 219–229.

Hunt, B.R. (1977b). In *Image Science Mathematics* (Wilde, C.O. and Barrett, E., eds) pp. 123–137 (Western Periodicals, North Hollywood).

Hunt, B.R. (1983). Adv. Electron. Electron Phys. **60**, 161–221.

Hunt, J.A. and Williams, D.B. (1991). Ultramicroscopy **38**, 47–73.

*Hurt, N.E. (1989). *Phase Retrieval and Zero Crossings* (Kluwer, Dordrecht, Boston MA and London).

Huxley, H.E. and Klug, A., eds (1971). *A Discussion on New Developments in Electron Microscopy with Special Emphasis on their Application in Biology.* Phil. Trans. Roy. Soc. (London) **B261**, 1–230.

Ichikawa, K., Lohmann, A.W. and Takeda, M. (1988). Appl. Opt. **27**, 3433–3436.

Ikuta, T. (1989). J. Electron Microsc. **38**, 415–422.

Isaacson, M., Utlaut, M. and Kopf, D. (1980). Analog computer processing of scanning transmission electron microscope images. In *Computer Processing of Electron Microscope Images* (Hawkes, P.W., ed.) pp. 257–283 (Springer, Berlin and New York).

Ishizuka, K. (1993). Ultramicroscopy **52**, 1–5.

Ishizuka, K. (1994). Ultramicroscopy

Isoda, S., Saitoh, K., Moriguchi, S. and Kobayashi, T. (1990). Seattle, vol. 1, pp. 168–169.

Isoda, S., Saitoh, K., Ogawa, T., Moriguchi, S. and Kobayashi, T. (1992). Ultramicroscopy **41**, 99–104.

Itoh, K., Ichioka, Y. and Minami, T. (1988). Appl. Opt. **27**, 3445–3450.

Izraelevitz, D. and Lim, J.S. (1987). IEEE Trans. **ASSP**–**35**, 511–519.

Jain, A.K. (1977). J. Optim. Theor. Applics **23**, 65–91.

Jain, A.K. (1981). Proc. IEEE **69**, 502–528.

Jain, A.K. (1989). *Fundamentals of Digital Image Processing* (Prentice–Hall, Englewood Cliffs NJ).

Jain, A.K. and Angel, E. (1974). IEEE Trans. **C**–**23**, 470–476.

Jain, A.K. and Jain J.R. (1978). IEEE Trans. **AC**–**23**, 817–834.

Jain, A.K. and Ranganath, S. (1986). Two-dimensional linear prediction models and spectral estimation. Adv. Comput. Vision Image Proc. **2**, 323–372.

Jakobs, R.-H. and Katterwe, H. (1976). Microsc. Acta **78**, 238–243.

Jap, B.K., Walian, P.J. and Gehring, K. (1991). In *Electron Crystallography of Organic Molecules* (Fryer, J.R. and Dorset, D.L., eds) pp. 309–315 (Kluwer, Dordrecht, Boston MA and London).

Jeanguillaume, C. and Colliex, C. (1989). Ultramicroscopy **28**, 252–257.

Jeffreys, H. and Jeffreys, B. (1956). *Methods of Mathematical Physics*, 3rd ed. (University Press, Cambridge).

Jeng, S.-C. and Tsai, W.-H. (1991). Pattern Recognition **24**, 1037–1051.

Jerri, A.J. (1977). The Shannon sampling theorem—its various extensions and applications: a tutorial review. Proc. IEEE **65**, 1565–1596.

Jeulin, D. (1983). J. Microsc. Spectrosc. Electron. **8**, 1–18.

Jeulin, D. (1987). Niagara Falls, pp. 165–183.

Jeulin, D. (1991). Cambridge, pp. 121–128.

Jeulin, D. and Renard, D. (1992). Microsc. Microanal. Microstruct. **3**, 333–361.

Jones, A.V. and Smith, K.C.A. (1978). Scanning Electron Microsc. (I), 13–26.

Jones, R. and Svalbe, I. (1992). Patt. Rec. Lett. **13**, 175–181.

Jones, R. and Svalbe, I.D. (1994). Basis algorithm in mathematical morphology. Adv. Electron. Electron Phys. **89**.

Jones, A.V. and Unitt, B.M. (1980). Scanning Electron Microsc. (I), 113–124.

Jones, A.V. and Unitt, B.M. (1982). J. Microscopy **127**, 61–68.

Jong, A.F. de (1992). Granada, vol. 1, pp. 137–138.

Jong, A.F. de and Koster, A.J. (1991). Cambridge, pp. 95–103.

Jong, A.F. de, Coene, W. and Dyck, D. van (1989). Ultramicroscopy **27**, 53–66.

Jong, A.F. de, Coene, W.M.J. and Koster, A.J. (1992). EMSA **50**, pp. 136–137.

Joy, D.C. and Verney, G.E. (1973). EMAG, pp. 50–55.

Justice, J.H., ed. (1986). *Maximum Entropy and Bayesian Methods in Applied Statistics* (University Press, Cambridge). Proceedings of the 4th maximum entropy workshop, Calgary, 1984.

Justusson, B.I. (1981). Median filtering: statistical properties. In *Two-dimensional Digital Signal Processing, II, Transforms and Median Filters* (Huang, T.S., ed.) pp. 161–196 (Springer, Berlin and New York).

Kaczmarz, M. S. (1937). Bull. Int. Acad. Polon. Sci. Lett., Cl. Sci. Mat. Nat., Ser A, 355–357.

Kak, A.C. and Slaney, M. (1988). *Principles of Computerized Tomographic Imaging* (IEEE Press, New York).

Kam, Z. (1980). In *Electron Microscopy at Molecular Dimensions* (Baumeister, W. and Vogell, W., eds) pp. 270–277 (Springer, Berlin and New York).

Kanaya, K., Kawakatsu, H., Matsui, S. and Yamazaki, H. (1964). Optik **21**, 399–422.

Kanaya, K., Kawakatsu, H., Atoda, N., Yotsumoto, H. and Ono, A. (1973). J. Phys. D: Appl. Phys. **6**, 6–20.

Kanaya, K., Baba, N. and Takamiya, K. (1981). Micron **12**, 353–364.

Kanaya, K., Baba, N., Kai, M., Oho, E. and Muranaka, Y. (1982a). Scanning Electron Microsc., 61–72.

Kanaya, K., Baba, N., Shino, M., Ichijo, T., Osumi, M. and Horne, R.W. (1982b). Scanning Electron Microsc. 1395–1410.

Kanaya, K., Baba, N., Shino, M., Takamiya, K. and Oikawa, T. (1982c). Micron **13**, 205–219.

Kanaya, K., Takamiya, K., Shino, M. and Shinohara, C. (1983a). Micron **14**, 119–134.

Kanaya, K., Baba, N., Shinohara, C. and Osumi, M. (1983b). Micron Microsc. Acta **14**, 233–247.

Kanaya, K., Oho, E., Naka, M., Koyanagi, T. and Sasaki, T. (1985). J. Electron Microsc. Tech. **2**, 73–87.

Kapp, O.H. and Ximen, J.-y. (1985). Optik **70**, 146–151.

Kapp, O.H. and Ximen, J.-y. (1986). Optik **73**, 51–55.

Kasturi, R. and Walkup, J.F. (1986). Nonlinear image restoration in signal-dependent noise. Adv. Comput. Vision Image Proc. **2**, 167–212.

Kawata, S., Ichioka, Y. and Suzuki, T. (1978/9). Optik **52**, 235–246.

Keller, H., Favre, A., Comazzi, A. and Weibel, E.R. (1982). Hamburg, vol. 1, pp. 535–536.

Kelly, J.F., Lee, R.J. and Lentz, S. (1980). Scanning Electron Microsc. (I), 311–322.

Kerzendorf, W. and Hoppe, W. (1980). The Hague, vol. 1, pp. 510–511.

Ketcham, D.J. (1976). Proc. SPIE **74**, 120–125.

Kiedron, P. (1979). Opt. Applic. **9**, 125–127.

Kiedron, P. (1980). Opt. Applic. **10**, 149–154 and 253–265.

Kiedron, P. (1982). Proc. SPIE. **413**, 189–196.

Kiedron, P. (1983). Optik **64**, 25–36.

Kikuchi, R. and Soffer, B.H. (1977). In *Image Science Mathematics* (Wilde, C.O. and Barrett, E., eds) pp. 146–147 (Western Periodicals, North Hollywood).

Kim, W. and Hayes, M.H. (1990). J. Opt. Soc. Am. **A7**, 441–449.

King, W.E. and Campbell, G.H. (1993). Ultramicroscopy **51**, 128–135.

Kirkland, E.J. (1979). EMSA **37**, pp. 558–559.

Kirkland, E.J. (1982). Ultramicroscopy **9, 45–64.

Kirkland, E.J. (1984). Ultramicroscopy **15**, 151–172.

Kirkland, E.J. (1987). Niagara Falls, pp. 139–147.

Kirkland, E.J. and Siegel, B.M. (1979). Optik **53**, 181–196.

Kirkland, E.J. and Siegel, B.M. (1981). Ultramicroscopy **6**, 169–180.

Kirkland, E.J., Siegel, B.M., Uyeda, N. and Fujiyoshi, Y. (1980). Ultramicroscopy **5**, 479–503.

Kirkland, E.J., Siegel, B.M., Uyeda, N. and Fujiyoshi, Y. (1981). EMSA **39**, pp. 234–235.

Kirkland, E.J., Siegel, B.M., Uyeda, N. and Fujiyoshi, Y. (1982). Ultramicroscopy **9**, 65–74.

Kirkland, E.J., Siegel, B.M., Uyeda, N. and Fujiyoshi, Y. (1985). Ultramicroscopy **17**, 87–103.

Kirsch, R.A. (1971). Comput. Biomed. Res. **4**, 315–328.

Kiselev, N.A. (1978). Toronto, vol. 3, pp. 94–106.

Kiselev, N.A. (1980). The Hague, vol. 2, pp. 572–579.

Kiselev, N.A. (1984a). Budapest, vol. 2, pp. 1451–1458.

Kiselev, N.A. (1984b). Vestnik Akad. Nauk, 102–115.

Kiselev, N.A. (1986). Electron microscopy of protein molecules and ribosomes. Sov. Sci. Rev. D: Physicochem. Biol. **6**, 161–199.

Klein, F. and Kübler, O. (1987). Patt. Rec. Lett. **5**, 19–29.

Klug, A. (1971). Phil. Trans. Roy. Soc. (London) B**261**, 173–179.

Klug, A. (1978/79). Chem. Scripta **14**, 245–246.

Klug, A. and Berger, J.E. (1964). J. Mol. Biol. **10**, 565–569.

Klug, A. and Crowther, R.A. (1972). Nature **238**, 435–440.

Klug, A. and Rosier, D.J. de (1966). Nature **212**, 29–32.

Kluge, N. (1970). BEDO **3**, 69–77.

Knauer, V. and Hoppe, W. (1980). The Hague, vol. 2, pp. 702–703.

Koenderink, J.J. and Doorn, A.J. van (1979). Proc. IEEE **67**, 1465–1470.

Koenderink, J.J. and Doorn, A.J. van (1992). IEEE Trans. **PAMI-14,** 597–605.

Kohler, D. and Mandel, L. (1973). J. Opt. Soc. Am. **63**, 126–134.

Kong, T.Y. and Rosenfeld, A. (1989). Comput. Vision Graph. Im. Proc. **48**, 357–393.

Koops, H. (1976). Jerusalem, vol. 1, pp. 369–371.

Koops, H. and Walter, G. (1980). The Hague, vol. 1, pp. 40–41.

Koskinen, L. and Astola, J. (1992). J. Math. Imaging Vision **2**, 117–135.

Koster, A.J. and Jong, A.F. de (1991). Ultramicroscopy **38**, 235–240.

Koster, A.J. and Ruijter, W.J. de (1988). York, vol. 1, pp. 83–84.

Koster, A.J. and Ruijter, W.J. de (1992). Ultramicroscopy **40**, 89–107.

Koster, A.J., Bos, A. van den, Mast, K.D. van der and Kruit, P. (1986). Kyoto, vol. 1, pp. 501–502.

Koster, A.J., Bos, A. van den and Mast, K.D. van der (1987a). Ultramicroscopy **21**, 209–222.

Koster, A.J., Bos, A. van den and Mast, K. van der (1987b). Niagara Falls, pp. 83–92.

Koster, A.J., Ruijter, W.J. de, Bos, A. van den and Mast, K.D. van der (1989). Ultramicroscopy **27**, 251–272.

Koster, A.J., Braunfeld, M.B., Sedat, J.W and Agard, D.A. (1992a). Granada, vol. 1, pp. 119–123.

Koster, A.J., Chen, H., Sedat, J.W. and Agard, D.A. (1992b). Ultramicroscopy **46**, 207–227.

Kovalevsky, V.A. (1989). Comput. Vision Graph. Im. Proc. **46**, 141–161.

Kovalevsky, V.A. (1992). Adv. Electron. Electron Phys. **84,** 197–259.

Kovbasa, S.I., Tolkachev, M.D. and Sokolov, V.N. (1988). Opt. Mekh. Prom. **55** (5), 13–15 ; Sov. J. Opt. Technol. **55**, 271–273.

Krakow, W. (1978). Microsc. Acta **80**, 391–404.

Krakow, W. (1985). J. Electron Microsc. Tech. **2**, 405–424.

Krakow, W. (1991). EMSA **49**, pp. 412–413.

Krakow, W. (1992a). In *Electron Microscopy in Materials Science* (Merli, P.G. and Antisari, M.V., eds) pp. 269–280 (World Scientific, Singapore, River Edge NJ and London).

Krakow, W. (1992b). Ultramicroscopy **45**, 269–280.

Kreinovic, V.Ja. and Kosheleva, O.M. (1979). Nature **281**, 708–709.

Kreznar, J.E. (1977). In *Image Science Mathematics* (Wilde, C.O. and Barrett, E., eds) pp. 153–162 (Western Periodicals, North Hollywood).

Krisch, B., Müller, K.-H., Rindfleisch, V. and Schliepe, R. (1975). BEDO **8**, 325–330.

Krivanek, O.L. (1976). Optik **45**, 97–101.

Krivanek, O.L., and Fan, G.Y. (1991). Cambridge, pp. 105–114.

Krivanek, O.L. and Fan, G.Y. (1992). EMSA **50**, pp. 96–97.

Krivanek, O.L. and Leber, M.L. (1993). MSA **51**, pp. 972–973.

Krivanek, O.L. and Mooney, P.E. (1993). Ultramicroscopy **49**, 95–108.

Krivanek, O.L., Mooney, P.E., Fan, G.Y., Leber, M.L. and Meyer, C.E. (1991). EMAG, pp. 523–526.

Krivanek, O.L., Ruijter, W.J. de, Meyer, C.E., Leber, M.L. and Wilbrink, J. (1993). MSA **51**, pp. 546–547.

Kuan, D.T., Sawchuk, A.A., Strand, T.C. and Chavel, P. (1985). IEEE Trans. **PAMI–7**, 165–177.

Kübler, O. (1980). Image processing in electron microscopy: non-periodic objects. In *Imaging Processes and Coherence in Physics* (Schlenker, M., Fink, M., Goedgebuer, J.P., Malgrange, C., Viénot, J.C. and Wade, R.H., eds) pp. 545–554 (Springer, Berlin and New York).

Kujawa, S. and Krahl, D. (1992). Ultramicroscopy **46**, 395–403.

Kunath, W. (1978). Toronto, vol. 1, pp. 232–233.

Kunath, W. (1979). Ultramicroscopy **4**, 3–7.

Kunath, W., Giersig, M., Sack-Kongehl, H. and Heel, M. van (1986). Kyoto, vol. 1, pp. 489–490.

Kunath, W., Zemlin, F. and Weiss, K. (1987). Optik **76**, 122–131.

Kuo, S.-s. and Mammone, R.J. (1992). IEEE Trans. Signal Proc. **40,** 159–168.

Lake, J.A. (1971). EMSA **29**, pp. 90–91.

Lakshminarayanan, A.V. and Lent, A. (1979). J. Theor. Biol. **76**, 267–295.

Lam, L., Lee, S.-W. and Suen, C.Y. (1992). IEEE Trans. **PAMI–14**, 869–885.

Landau, H.J. and Pollak, H.O. (1961). Bell System Tech. J. **40**, 65–84.

Landau, H.J. and Pollak, H.O. (1962). Bell System Tech. J. **41**, 1295–1336.

Lane, R.G. and Bates, R.H.T. (1987). J. Opt. Soc. Am. **A4**, 180–188.

Lane, R.G., Fright, W.R. and Bates, R.H.T. (1987). IEEE Trans. **ASSP–35**, 520–526.

Lannes, A. (1976). J. Phys. D: Appl. Phys. **9**, 2533–2544.

Lannes, A. (1978). Toronto, vol. 1, pp. 228–229.

Lannes, A., Roques, S. and Casanove, M.J. (1987a). J. Opt. Soc. Am. **A4**, 189–199.

Lannes, A., Roques, S. and Casanove, M.J. (1987b). J. Mod. Opt. **34**, 161–226 and 321–370.

Lavergne, J.-L., Martin, J.-M. and Belin, M. (1992). Microsc. Microanal. Microstruct. **3**, 517–528.

Lawrence, M.C., Jaffer, M.A. and Sewell, B.T. (1989). Ultramicroscopy **31**, 285–302.

Lawton, W. and Morrison, J. (1987). J. Opt. Soc. Am. **A4**, 105–111.

Leavers, V.F. (1992). CVGIP: Image Understanding 56, 381–398.

Ledley, R.S. (1964). Science **146**, 216–223.

Ledley, R.S., Rotolo, L.S., Golab, T.J., Jacobsen, J.D., Ginsberg, M.D. and Wilson, J.B. (1965). In *Optical and Electro-Optical Information Processing* (Tippett, J.T., Berkowitz, D.A., Clapp, L.C., Koester, C.J. and Vanderburgh, A., eds) pp. 591–613 (Massachusetts Institute of Technology Press, Cambridge MA and London).

Lee, J.-S. (1980). IEEE Trans. **PAMI–2**, 165–168.

Lee, J.-S. (1981). Comput. Graph. Im. Proc. **15**, 380–389.

Lee, J.-S. (1983). Comput. Vision Graph. Im. Proc. **24**, 255–269.

Lee, H.C. and Fu, K.S. (1972). IEEE Trans. **C–21**, 660–666.

Lee, R.J. and Kelly, J.F. (1980). Scanning Electron Microsc. (I), 303–310.

Lee, J.-h. and Russ, J.C. (1989). J. Computer–assist. Microsc. **1**, 79–90.

Lee, J.W., Delp, E.J. and Brinkley, L.L. (1985). Comput. Biomed. Res. **18**, 587–604.

Lee, J.S.J., Haralick, R.M. and Shapiro, L.G. (1987). IEEE J. **RA–3**, 142–156.

Lee, C.-N., Poston, T. and Rosenfeld, A. (1991). CVGIP: Graph. Models Image Proc. **53**, 522–537.

Lee, C.-N., Poston, T. and Rosenfeld, A. (1993). CVGIP: Graph. Models Image Proc. **55**, 20–47.

Lembcke, G., Dürr, R., Hegerl, R. and Baumeister, W. (1991). J. Microscopy **161**, 263–278.

Lenz, F. (1975). EMAG, pp. 179–184.

Lepault, J. and Pitt, T. (1984). EMBO J. **3**, 101–105.

Le Poole, J.B. and Groot, L.E.M. de (1980). The Hague, vol. 1, pp. 469–497.

Leszczynski, K.W. and Shalev, S. (1989). Image Vision Comput. **7**, 205–209.

Levi, A. and Stark, H. (1987). Restoration from phase and magnitude by generalized projections. In *Image Recovery: Theory and Application* (Stark, H., ed.) pp. 277–320 (Academic Press, Orlando and London).

Levin, B.Ja. (1964). *Distribution of Zeros of Entire Functions* (American Mathematical Society, Providence RI); Russian ed. published by Gos. Izd. Tekhniko–Teoret. Lit. (Moscow, 1956).

Lewitt, R.M. (1983). Proc. IEEE **71**, 390–408.

Lewitt, R.M. (1990). J. Opt. Soc. Am. **A7**, 1834–1846.

Lewitt, R.M. and Bates, R.H.T (1978). Optik **50**, 19–33 and 189–204.

Lewitt, R.M., Bates, R.H.T. and Peters, T.M. (1978). Optik **50**, 85–109.

Li, F.H., Fan, H.F., Tang, D., Han, F.S., Zhong, Z.Y. and Zheng, C.D. (1986). Kyoto, vol. 1, pp. 515–516.

Lim, H., Tan, K.-C. and Tan, B.T.G. (1991). Comput. Vision Graph. Im. Proc. **53**, 186–195.

Lin, J.-a and Cowley J.M. (1986a). Appl. Opt. **25**, 2245–2246.

Lin, J.A. and Cowley, J.M. (1986b). Ultramicroscopy **19**, 31–42.

Lo, C.M. and Sawchuk, A.A. (1979). Proc. SPIE **207**, 84–95.

Lipkin, L.E., Watt, W.C. and Kirsch, R.A. (1966). Ann. New York Acad. Sci. **128**, 984–1012.

Liu, W. (1991a). EMSA **49**, pp. 542–543

Liu, Y. (1991b). EMSA **49**, pp. 552–553.

Liu, Z.-q. and Caelli, T. (1988). Comput. Vision Graph. Im. Proc. **44**, 332–349.

Liu, B. and Gallagher, N.C. (1974). J. Opt. Soc. Am. **64**, 1227–1236.

Livesey, A.K. and Skilling, J. (1985). Acta Cryst. **A41**, 113–122.

Llinas, R., Spitzer, R., Hillman, D. and Chujo, M. (1979). Scanning Electron Microsc. (II), 367–374.

Lo, C. M. and Sawchuk, A. A. (1979). Proc. SPIE **207**, 84–95.

Loce, R.P. and Dougherty, E.R. (1992a). Proc. SPIE **1769**, 94–105.

Loce, R.P. and Dougherty, E.R. (1992b). Opt. Eng. **31**, 1008–1025.

Lockhausen, J., Kristen, U., Menhardt, W. and Dallas, B. (1987). BEDO **20**, 77–82.

Luczak, E. and Rosenfeld, A. (1976). IEEE Trans. **C–25**, 532–533

Luther, P.K. and Crowther, R.A. (1984). Nature **307**, 566–568.

Luttrell, S.P. (1985). Opt. Acta **32**, 703–716.

MacDonald, N.C. (1968). EMSA **26**, pp. 362–363.

MacDonald, N.C. (1969). Scanning Electron Microsc. 431–437.

Mammone, R.J. (1983). J. Opt. Soc. Am. **73**, 1476–1480.

Mammone, R. (1987). Image restoration using linear programming. In *Image Recovery: Theory and Application* (Stark, H., ed.) pp. 127–156 (Academic Press, Orlando and London).

Mammone, R.J. and Rothacker, R.J. (1987). J. Opt. Soc. Am. **A4**, 208–215.

Manolitsakis, I. (1982). J. Math. Phys. **23**, 2291–2298.

Maragos, P. (1987). Opt. Eng. **26** , 623–632

Maragos, P. (1989). IEEE Trans. **PAMI–11**, 586–599.

Maragos, P. and Schafer, R.W. (1986). IEEE Trans. **ASSP–34**, 1228–1244.

Maragos, P. and Schafer, R.W. (1987). IEEE Trans. **ASSP–35**, 1153–1169 and 1170–1184.

Maragos, P. and Schafer, R.W. (1990). Proc. IEEE **78**, 690–710.

Maragos, P. and Ziff, R.D. (1988). Proc. SPIE **1001**, 106–115.

Maragos, P. and Ziff, R.D. (1990). IEEE Trans. **PAMI–12**, 498–504.

Marks, R.J. (1991). *Introduction to Shannon Sampling and Interpolation Theory* (Springer, New York, Berlin and London).

Marks, R.J., ed. (1993). *Advanced Topics in Shannon Sampling and Interpolation Theory* (Springer, Berlin and New York).

Martin, D.C., Schaffer, K.R. and Thomas, E. L. (1991). In *Electron Crystallography of Organic Molecules* (Fryer, J.R. and Dorset, D.L., eds.) pp. 129–145 (Kluwer, Dordrecht, Boston MA and London).

Mast, K.D. van der (1984). Budapest, vol. 1, pp. 3–10.

Mast, K.D. van der and Koster, A.J. (1988). EMSA **46,** pp. 646–647.

Mastin, G. A. (1985). Comput. Vision Graph. Im. Proc. **31**, 103–121.

Matheron, G. (1975). *Random Sets and Integral Geometry* (Wiley, New York and London)..

Matson, W.L., McKinstry, H.A., Johnson, G.G., White, E.W. and McMillan, R.E. (1970). Pattern Recognition **2**, 303–312.

Matsuda, J.-i., Horiguchi, A. and Ura, K. (1980). Rev. Sci. Instrum. **51**, 1225–1230.

Mazille, J.E. (1989). J. Microscopy **156**, 3–13.

McCallum, B.C. and Rodenburg, J.M. (1992). Granada, vol. 1, pp. 431–435.

McClean, J.A., Dhariwal, R.S. and Milne, N.G. (1992). Scanning **14,** 174–182.

McKinnon, A.E., McDonnell, M.J., Napier, P.J. and Bates, R.H.T. (1976). Optik **44**, 253–272.

McMillan, R.E., Johnson, G.G. and White, E.W. (1969). Scanning Electron Microsc. 439–444.

Medoff, B. P. (1987). Image reconstruction from limited data: theory and applications in computerized tomography. In *Image Recovery: Theory and Application* (Stark, H., ed.) pp. 321–368 (Academic Press, Orlando and London).

Mehta, C.L. (1965). Nuovo Cim. **36**, 202–205.

Mehta, C.L. (1968). J. Opt. Soc. Am. **58**, 1233–1234.

Mehta, C.L., Wolf, E. and Balachandran, A.P. (1966). J. Math. Phys. **7**, 133–138.

Mellema, J.E. (1980a). Computer reconstruction of regular biological objects. In *Computer Processing of Electron Microscope Images* (Hawkes, P.W., ed.) pp. 89–126 (Springer, Berlin and New York).

Mellema, J.E. (1980b). Image processing of regular biological objects. In *Imaging Processes and Coherence in Physics* (Schlenker, M., Fink, M., Goedgebuer, J.P., Malgrange, C., Viénot, J.C. and Wade, R.H., eds.) pp. 523–532 (Springer, Berlin and New York).

Mellema, J.E. and Schepman, A.M.H. (1976). Jerusalem, vol. 1, pp. 4–7.

Mercurio, P.J., Elvins, T.T., Young, S.J., Cohen, P.S., Fall, K.R. and Ellisman, M.H. (1992). Commun. Assoc. Comput. Mach. **35**(6), 55–63.

Mersereau, R.M. (1979). Proc. IEEE **67**, 930–949.

Meyer, F. (1989). Signal Proc. **16**, 335–364.

Meyer, F. (1990). Proc. SPIE **1350**, 85–102.

Meyer, F. (1992). J. Microscopy **165**, 5–28.

Meyer, F. and Beucher, S. (1990). J. Visual Commun. Image Represent. **1**, 21–46.

Michel, J., Bonnet, N., Wagner, D., Balossier, G. and Bonhomme, P. (1993). Ultramicroscopy **48**, 121–132.

Miedema, M.A.O. and Buist, A.H. (1992). Granada, vol. 1, pp. 451–452.

Millane, R.P. (1990). J. Opt. Soc. Am. **A7**, 394–411.

Miller, G., Fryer, J.R., Kunath, W. and Weiss, K. (1991). In *Electron Crystallography of Organic Molecules* (Fryer, J.R. and Dorset, D.L., eds) pp. 343–353 (Kluwer, Dordrecht, Boston MA and London).

Minerbo, G. (1979). Comput. Graph. Im. Proc. **10**, 48–68.

Misell, D.L. (1973a). J. Phys. D: Appl. Phys. **6**, L6–L9.

Misell, D.L. (1973b). J. Phys. D: Appl. Phys. **6**, 2200–2216 and 2217–2225..

Misell, D.L. (1974). J. Phys. D: Appl. Phys. **7**, L69–L71.

Misell, D.L. (1976). J. Phys. D: Appl. Phys. **9**, 1849–1866.

Misell, D.L. (1978a). The phase problem in electron microscopy. Adv. Opt. Electron Microsc. **7**, 185–279.

Misell, D.L. (1978b). *Image Analysis, Enhancement and Interpretation* (North-Holland, Amsterdam, New York and Oxford). Practical Methods in Electron Microscopy (Glauert, A.M., ed.) vol. 7.

Misell, D.L. and Childs, P.A. (1972). J. Phys. D: Appl. Phys. **5**, 1760–1768.

Misell, D.L. and Greenaway, A.H. (1974). J. Phys. D: Appl. Phys. **7**, 832–855 and 1660–1669.

Misell, D.L., Burge, R.E. and Greenaway, A.H. (1974a). Nature **247**, 401–402.

Misell, D.L., Burge, R.E. and Greenaway, A.H. (1974b). J. Phys. D: Appl. Phys. **7**, L27–L30.

Miyokawa, T., Date, N., Niikura, T., Iwashita, K., Kobayashi, T. and Nakano, Y. (1988). York, vol. 1, pp. 155–156.

Möbus, G. and Rühle, M. (1993). Optik **93**, 108–118.

Möbus, G., Hohenstein, M., Phillip, F., Ernst, F., Necker, G. and Rühle, M. (1992). Granada, vol. 1, pp. 531–532.

Möbus, G., Necker, G. and Rühle, M. (1993). Ultramicroscopy **49**, 46–65.

Mohammad–Djafari, A. and Demoment, G. (1989). In *Maximum Entropy and Bayesian Methods* (Skilling, J., ed.) pp. 195–201 (Kluwer, Dordrecht, Boston MA and London).

Montoto, L., Montoto, M. and Bel-Lan, A. (1978). Toronto, vol. 1, pp. 212–213.

Moody, M.F. (1990). Image analysis of electron micrographs. In *Biophysical Electron Microscopy* (Hawkes, P.W. and Valdrè, U., eds.) pp. 145–287 (Academic Press, London and San Diego).

Mooney, P.E., Ruijter, W.J. de and Krivanek, O.L. (1993). MSA **51**, pp. 262–263.

Moore, E. H. (1920). Bull. Am. Math. Soc. **26**, 394–395.

Moore, D.J.H and Parker, D.J. (1973). IEEE Trans. **IT–19**, 415–422.

Moras, D., Drenth, J., Strandberg, B., Suck, D. and Wilson, K., eds (1987). *Crystallography in Molecular Biology* (Plenum, New York and London). NATO ASI Series A, vol. 126.

Morgan, D.G., Grant, R.A., Chiu, W. and Frank, J. (1992). J. Struct. Biol. **108**, 245–256.

Mori, N., Oikawa, T., Katoh, T., Miyahara, J. and Harada, Y. (1988). Ultramicroscopy **25**, 195–202.

Mori, N., Oikawa, T., Harada, Y., Miyahara, J. and Matsuo, T. (1989). EMSA **47**, pp. 100–101.

Morris, C.E., Richards, M.A. and Hayes, M.H. (1987). J. Opt. Soc. Am. **A4**, 200–207.

Morris, C.E., Richards, M.A. and Hayes, M.H. (1988). IEEE Trans. **ASSP–36**, 1017–1025.

Morrow, W.M., Paranjape, R.B., Rangayyan, R.M. and Desautels, J.E.L. (1992). IEEE Trans. **MI–11**, 392–406.

Moses, H.E. and Prosser, R.T. (1983). J. Opt. Soc. Am. **73**, 1451–1454.

Mott, R.B. (1987). In *Microbeam Analysis 1987* (Geiss, R.H., ed.) pp. 287–289 (San Francisco Press, San Francisco).

Moza, A.K., Austin, L.G. and Johnson, G.G. (1979). Scanning Electron Microsc. (I), 473–476 and 472.

Murphy, M.S. and Silverman, L.M (1978). IEEE Trans. **AC–23**, 809–816.

Nahi, N.E. (1972). Proc. IEEE **60**, 872–877.

Nahi, N.E. and Assefi, T. (1972). IEEE Trans. **C–21**, 734–738.

Nahi, N.E. and Franco, C.A. (1973). IEEE Trans. **COM–21**, 305–311.

Nahi, N.E. and Habibi, A. (1975). IEEE Trans. **CAS–22**, 286–293.

Nakagawa, Y. and Rosenfeld, A. (1978). IEEE Trans. **SMC–8**, 632–635.

Nakahara, S. and Cullis, A.G. (1977). EMAG, pp. 263–266.

Nakahara, S., Maher, D.M. and Cullis, A.G. (1976). Jerusalem, vol. 1, pp. 85–90.

Nakajima, N. (1987). J. Opt. Soc. Am. **A4**, 154–158.

Nakajima, N. (1988a). Jpn. J. Appl. Phys. **27**, 244–252.

Nakajima, N. (1988b). J. Opt. Soc. Am. **A5**, 257–262.

Nakajima, N. and Asakura, T. (1982a). Optik **60**, 181–198.

Nakajima, N. and Asakura, T. (1982b). Optik **60**, 289–305.

Nakajima, N. and Asakura, T. (1982c). Opt. Commun. **41**, 89–94.

Nakajima, N. and Asakura, T. (1983a). Optik **63**, 99–108.

Nakajima, N. and Asakura, T. (1983b). Optik **64**, 37–49.

Nakajima, N. and Asakura, T. (1985). Opt. Acta **32**, 647–658.

Nakajima, N. and Asakura, T. (1986). J. Phys. D: Appl. Phys. **19**, 319–331.

Narendra, P.M. (1981). IEEE Trans. **PAMI–3**, 20–29.

Natterer, F. (1986). *The Mathematics of Computerized Tomography* (Wiley, Chichester and New York and Teubner, Stuttgart).

Netravali, A.N. and Haskell, B.G. (1988). *Digital Pictures, Representation and Compression* (Plenum, New York and London).

Newbury, D.E. (1975). Scanning Electron Microsc. 727–736.

Newbury, D.E. and Joy, D.C. (1973). Scanning Electron Microsc. 151–158.

Newbury, D.E., Marinenko, R.B., Bright, D.S. and Myklebust, R.L. (1988). Scanning **10**, 213–225.

Niedermeyer, W. and Wilke, H. (1982). J. Microscopy **126**, 259–273.

Nieto-Vesperinas, M. (1980). Optik **56**, 377–384.

Nieto-Vesperinas, M. (1984a). J. Math. Phys. **25**, 1592–1598.

Nieto-Vesperinas, M. (1984b). J. Math. Phys. **25**, 2109–2115.

Nieto-Vesperinas, M. (1986). Opt. Acta **33**, 713–722.

Nieto-Vesperinas, M. and Dainty, J.C. (1984). Opt. Commun. **52**, 94–98.

Nieto-Vesperinas, M. and Dainty, J.C. (1985). Opt. Commun. **54**, 333–334.

Nieto-Vesperinas, M. and Hignette, O. (1979). Opt. Pura Aplic. **12**, 175–180.

Nieto-Vesperinas, M. and Mendez, J.A. (1986). Opt. Commun. **59**, 249–254.

Nieto-Vesperinas, M. and Ross, G. (1982). Optik **62**, 87–92.

Nieto-Vesperinas, M., Ross, G. and Fiddy, M.A. (1980). Optik **55**, 165–171.

Nieto-Vesperinas, M., Vences, M.A., Ross, G. and Fiddy, M.A. (1981a). Opt. Commun. **36**, 169–174.

Nieto-Vesperinas, M., Gea, M., Ross, G. and Fiddy, M.A. (1981b). In *Optics in Four Dimensions-1980* (Machado, M.A. and Narducci, L.M., eds.) pp. 679–687 (AIP Conf. Proc. No. 65, American Institute of Physics, New York).

Nodes, T.A. and Gallagher, N.C. (1982). IEEE Trans. **ASSP–30**, 739–746.

Nodes, T.A. and Gallagher, N.C. (1983). IEEE Trans. **ASSP-31**, 1350–1365.

Nodes, T.A. and Gallagher, N.C. (1984). IEEE Trans. **COM-32**, 532–541.

Nomura, S. and Ichikawa, M. (1984). Budapest, vol. 1, pp. 301–302.

Nomura, S. and Isakozawa, S. (1987). J. Electron Microsc. **36,** 157–162.

Nomura, S., Isakozawa, S. and Kamimura, S. (1986). Kyoto, vol. 1, pp. 499–500.

Nussbaumer, H.J. (1982). *Fast Fourier Transform and Convolution Algorithms,* 2nd . ed. (Springer, Berlin and New York).

Nussenzveig, H.M. (1967). J. Math. Phys. **8**, 561–572.

Nussenzveig, H.M. (1981). In *Optics in Four Dimensions–1980* (Machado, M.A. and Narducci, L.M., eds.) pp. 9–30 (AIP Conf. Proc. No. 65, American Institute of Physics, New York).

Ochoa, E., Allebach, J.P. and Sweeney, D.W. (1987). Appl. Opt. **26**, 252–260.

O'Connor, B.T. and Huang, T.S. (1981). Stability of general two-dimensional recursive filters. In *Two-dimensional Digital Signal Processing, I, Linear Filters* (Huang, T.S., ed.) pp. 85–154 (Springer, Berlin and New–York).

Ogasawara, M., Baba, N., Oho, E. and Kanaya, K. (1988). York, vol. 1, pp. 199–200.

Oho, E. (1992). Scanning 14, 335–344.

Oho, E., Baba, N., Katoh, M., Nagatani, T., Osumi, M., Amako, K. and Kanaya, K. (1984). J. Electron Microsc. Tech. **1**, 331–340.

Oho, E., Sasaki, T. and Kanaya, K. (1985). J. Electron Microsc. **34,** 427–429.

Oho, E., Sasaki, T., Baba, N, Muranaka, Y., Osumi, M., Amako, K. and Kanaya, K. (1986a). Kyoto, vol. 1, pp. 493–494.

Oho, E., Sasaki, T. and Kanaya, K. (1986b). J. Electron Microsc. Tech. **4**, 157–162.

Oho, E., Sasaki, T., Ogihara, A. and Kanaya, K. (1987). Scanning 9, 173–176.

Oho, E., Ogihara, A. and Kanaya, K. (1990). J. Microscopy **159**, 33–41.

Oikawa, T., Mori, N., Ichikawa, R., Ohnishi, M. and Kawasaki, M. (1989a). EMSA **47**, pp. 102–103.

Oikawa, T., Mori, N., Ichikawa, R. and Ohnishi, M. (1989b). EMAG, vol. 1, pp. 543–546.

Oikawa, T., Mori, N., Osuna, T., Ohnishi, M. and Kawasaki, M. (1990). Seattle, vol. 1, pp. 170–171.

Oikawa, T., Mori, N., Hosokawa, F. and Kawasaki, M. (1991). EMSA **49**, pp 546–547.

Oikawa, T., Shindo, D., Hiraga, K. and Critchell, J. (1992). Granada, vol. 1, pp. 147–148.

Okagaki, T., Clark, B. and Ferro, J. (1980). Scanning Electron Microsc. (I), 137–142.

Olins, A.L. (1986). EMSA **44**, pp. 22–25.

O'Neill, E.L. and Walther, A. (1963). Opt. Acta **10**, 33–40.

Op de Beeck, M. and Dyck, D. van (1991). Micron Microsc. Acta **22**, 279–280.

Op de Beeck, M., Dyck, D. van and Coene, W. (1992). Granada, vol. 1, pp. 149–150.

Oppenheim, A.V. and Lim, J.S. (1981). Proc. IEEE **69**, 529–541.

Oppenheim, A.V., Schafer, R.W. and Stockham, T.G. (1968). Proc. IEEE **56**, 1264–1291.

Oppenheim, A.V., Hayes, M.H. and Lim, J.S. (1982). Opt. Eng. **21**, 122–127.

Oppenheim, A.V., Lim, J.S. and Curtis, S.R. (1983). J. Opt. Soc. Am. **73**, 1413–1420.

Orlova, E.V. (1991). In *Electron Crystallography of Organic Molecules* (Fryer, J.R. and Dorset, D.L., eds) pp. 327–332 (Kluwer, Dordrecht, Boston MA and London).

Oron, M. and Gilbert, D. (1976). Scanning Electron Microsc. (I), 121–128 and 120.

Osumi, M., Yamada, N., Nagano, M., Murakami, S., Baba, N., Oho, E. and Kanaya, K. (1984). Scanning Electron Microsc. (I), 111–119.

Oudet, F., Vejux, A. and Tietz, H.R. (1988). J. Microsc. Spectrosc. Electron. **13**, 405–412.

Owens, R., Venkatesh, S. and Ross, J. (1989). Patt. Recog. Lett. **9**, 233–244.

Panda, D.P. and Kak, A.C. (1976). Image restoration and enhancement. Purdue University Report TR–EE 76–17, 181 pp.

Paranjape, R.B., Morrow, W.M. and Rangayyan, R.M. (1992). CVGIP : Graph. Models Image Proc. **54**, 259–267.

Park, R.-H. and Choi, W.Y. (1989). Comput. Vision Graph. Im. Proc. **47**, 259–265.

Pavel, M. (1989). *Fundamentals of Pattern Recognition* (Dekker, New York and Basel); 2nd ed., 1993

Pawley, J.B. and Hayes, T.L. (1980). Scanning **3**, 1161–164.

Paxman, R.G., Fienup, J.R. and Clinthorne, J.T. (1987). Proc. SPIE **828**, 184–189.

*Pennington, K.S. and Moorhead, R.J., eds. (1990). *Image Processing Algorithms and Techniques*. Proc. SPIE **1244**.

Penrose, R. (1955). Proc. Cambridge Philos. Soc. **51**, 406–413.

Perez-Ilzarbe, M.J., Nieto-Vesperinas, M. and Navarro, R. (1990). J. Opt. Soc. Am. **A7**, 434–440.

Peřina, J. (1972). *Coherence of Light* (Van Nostrand Reinhold, London and New York); *Teorie Koherence* (SNTL, Nakladatelství Technické Literatury, Prague, 1975) contains some of the same material but is not a translation.

Petersen, D. P. and Middleton, D. (1962). Information & Control **5**, 279–323.

Petitgand, H., Benoit, D., Moukassi, M. and Debyser, B. (1990). ISIJ Int. **30**, 546–551.

Petrou, M. (1994). The differentiating filter approach to edge detection. Adv. Electron. Electron Phys. **88**, 297–345.

Phillips, D.L. (1962). J. Assoc. Comput. Mach. **9**, 84–97.

Phillips, T.-Y., Davis, L.S. and Rosenfeld, A. (1983). Pattern Recognition **16**, 385–400.

Pitas, I. and Venetsanopoulos, A.N. (1992). Proc. IEEE **80**, 1893–1921.

Pizer, S.M., Amburn, E.P., Austin, J.D., Cromartie, R., Geselowitz, A., Greer, T., Haar Remeny, B. ter, Zimmerman, J.B. and Zuiderveld, K. (1987). Comput. Vision Graph. Im. Proc. **39**, 355–368.

Plies, E. (1981). Nucl. Instrum. Meth. **187**, 227–235.

Plies, E. and Typke, D. (1978). Z. Naturforsch. **33a**, 1361–1377.

Podilchuk, C.I. and Mammone, R.J. (1990). J. Opt. Soc. Am. **A7**, 517–521.

Ponce, F.A. and Hikashi, H. (1991). Cambridge, pp. 339–345.

Prasad, M.K. and Lee, Y.H. (1992). Circuits Syst. Signal Proc. **11**, 115–136.

Prasad, P.B., Jeulin, D., Daly, C., Mory, C., Tencé, M. and Colliex, C. (1991). Microsc. Microanal. Microstruct. **2**, 107–127.

Pratt, W.K. (1978). *Digital Image Processing* (Wiley, New York and Chichester); 2nd ed., 1991.

Preteux, F. (1993). On a distance-function approach for grey-level mathematical morphology. In *Mathematical Morphology in Image Processing* (Dougherty, E. R., ed.) pp. 323–349 (Dekker, New York and Basel).

Prewitt, J.M.S. (1965). IEEE Trans. **BME–12**, 14–21.

Prewitt, J.M.S. (1970). Object enhancement and extraction. In *Picture Processing and Psychopictorics* (Lipkin, B.S. and Rosenfeld, A., eds) pp. 75–149 (Academic Press, New York and London).

Prewitt, J.M.S. and Mendelsohn, M.L. (1966). Ann. New York Acad. Sci. **128**, 1035–1053.

Princen, J., Illingworth, J. and Kittler, J. (1992). J. Math. Imaging Vision **1**, 153–168.

Prod'homme, M., Coster, M., Chermant, L. and Chermant, J.-L. (1991). Cambridge, pp. 255–268.

Prosch, H. von (1974). Mikroskopie **30**, 321–326.

Provencher, S.W. and Vogel, R.H. (1983). In *Numerical Treatment of Inverse Problems in Differential and Integral Equations* (Deuflhard, P. and Hairer, E., eds) pp. 304–319; vol. 2 of Progress in Scientific Computing (Birkhäuser, Boston MA, Basel and Stuttgart).

Provencher, S.W. and Vogel, R.H. (1988). Ultramicroscopy **25**, 209–222.

Pujari, M. and Frank, J. (1990). Seattle, vol. 1, pp. 462–463.

Pulvermacher, H. (1976a). Optik **44**, 413–426 and **45**, 1–10.

Pulvermacher, H. (1976b). Optik **45**, 111–121.

Pulvermacher, H. (1981). Optik **58**, 259–266.

Quintana, C. and Ollacarizqueta, A. (1989). Ultramicroscopy **28**, 315–319.

Radermacher, M. (1988). J. Electron Microsc. Tech. **9**, 359–394.

Radermacher, M. and Hoppe, W. (1978). Toronto, vol. 1, pp. 218–219.

Radermacher, M. and Hoppe, W. (1980). The Hague, vol. 1, pp. 132-133.

Radermacher, M., Wagenknecht, T., Verschoor, A. and Frank, J. (1986a). EMSA **44**, pp. 40-43.

Radermacher, M., Frank, J. and Mannella, C.A. (1986b). EMSA **44**, pp. 140-143.

Radermacher, M., Wagenknecht, T., Verschoor, A. and Frank, J. (1986c). J. Microscopy **141**, RP1-RP2.

Radermacher, M., Wagenknecht, T., Verschoor, A. and Frank, J. (1987a). J. Microscopy **146**, 113-136.

Radermacher, M., Wagenknecht, T., Verschoor, A. and Frank, J. (1987b). EMBO J. **6**, 1107-1114.

Radermacher, M., Nowotny, V., Grassucci, R. and Frank, J. (1990). Seattle, vol. 1, pp. 284-285.

Radon, J. (1917). Über die Bestimmung von Funktionen durch ihre Integralwerte längs gewisser Mannigfaltigkeiten. Ber. Verhandl. K. Sächs. Ges. Wiss. Leipzig, Math.-Phys. Kl. **69**, 262-277.

Raeymaekers, B., Espen, P. van and Adams, F. (1984). Mikrochim. Acta (II), 437-454.

Rajala, S.A. and Figueiredo, R.J.P. de (1981). IEEE Trans. **ASSP-29**, 1033-1042.

Ramachandran, G.N. (1971). Proc. Ind. Acad. Sci. **A74**, 14-24.

Ramachandran, G.N. and Lakshminarayanan, A.V. (1971a). Proc. Natl. Acad. Sci. USA **68**, 2236-2240.

Ramachandran, G.N. and Lakshminarayanan, A.V. (1971b). Ind. J. Pure Appl. Sci. **9**, 997-1003.

Ramamoorthy, P.A. and Bruton, L.T. (1981). Design of two-dimensional recursive filters. In *Two-dimensional Digital Signal Processing, I, Linear Filters* (Huang, T.S., ed.) pp. 41-83 (Springer, Berlin and New York).

Rao, P. and Goehner, R.P. (1975). EMSA **33**, pp. 222-223.

Rao, C.R. and Mitra, S.K. (1971). *Generalized Inverse of Matrices and its Applications* (Wiley, New York & London).

Rao, K.R. and Yip, P. (1990). *Discrete Cosine Transform* (Academic Press, Boston and London).

Rasch, M., Saxton, W.O. and Baumeister, W. (1984). FEMS Microbiol. Lett. **24**, 285-290.

Rautureau, M., Harba, R. and Jacquet, G. (1991). Cambridge, pp. 293-298.

Reimer, L. (1984). *Transmission Electron Microscopy* (Springer, Berlin, New York and London); 2nd ed., 1989; 3rd ed., 1993.

Requicha, A.A.G. (1980). Proc. IEEE **68**, 308-328.

Rez, P., Weiss, J.K. and Ruijter, W.J. de (1991). Cambridge, pp. 81-94.

Ritter, G.X. (1991). Recent developments in image algebra. Adv. Electron. Electron Phys. **80**, 243–308.

Ritter, G.X. and Gader, P.D. (1987). J. Parallel Distrib. Comput. **4**, 7–44.

Ritter, G.X. and Zhu, H. (1992). J. Math. Imaging Vision **1**, 201–213.

Ritter, G.X., Wilson, J.N. and Davidson, J.L. (1990). Image algebra: an overview. Comput. Vision Graph. Im. Proc. **49**, 297–331.

Robaux, O. (1970). Opt. Acta **17**, 811–822.

Robaux, O. and Roizen-Dossier, B. (1970). Opt. Acta **17**, 733–746.

Roberts, L.G. (1965). Machine perception of three-dimensional solids. In *Optical and Electro-optical Information Processing* (Tippett, J.T., Berkowitz, D.A., Clapp, L.C., Koester, C.J. and Vanderburgh, A., eds) pp. 159–197 (Massachusetts Institute of Technology Press, Cambridge MA and London).

Robinson, G. S. (1976). Proc. SPIE **87**, 117–125.

Robinson, G.S. (1977). Comput. Graph. Im. Proc. **6**, 492–501.

Robinson, S.R. (1978). J. Opt. Soc. Am. **68**, 87–92.

Rohr, H.P. (1975). BEDO **8**, 293–298.

Roman, P. and Marathay, A.S. (1963). Nuovo Cim. **30**, 1452–1464.

Ronkin, L.I. (1974). *Introduction to the Theory of Entire Functions of Several Variables* (American Mathematical Society Translations of Mathematical Monographs, vol. 44, Providence RI ; Russian ed. published by Izd. Nauka, Moscow, 1971).

Ronse, C. (1990). Signal Proc. **21**, 129–154.

Ronse, C. (1993). IEEE Trans. **PAMI-15**, 484–491.

Ronse, C. and Heijmans, H.J.A.M. (1991). CVGIP: Image Understanding **54**, 74–97.

Root, W.L. (1987). J. Opt. Soc. Am. **A4**, 171–179.

Rosenblatt, J. (1984). Commun. Math. Phys. **95**, 317–343.

Rosenfeld, A. (1969). Picture processing by computer. Computing Surveys **1**, 147–176.

Rosenfeld, A. (1972). Picture processing : 1972. Comput. Graph. Im. Proc. **1**, 394–416.

Rosenfeld, A. (1973). Progress in picture processing. Computing Surveys **5**, 81–108.

Rosenfeld, A. (1974). Picture processing: 1973. Comput. Graph. Im. Proc. **3**, 178–194.

Rosenfeld, A. (1975). Picture processing: 1974. Comput. Graph. Im. Proc. **4**, 133–155.

Rosenfeld, A. (1976). Picture processing: 1975. Comput. Graph. Im. Proc. **5**, 215–237.

Rosenfeld, A. (1977). Picture processing: 1976. Comput. Graph. Im. Proc. **6**, 157–183.

Rosenfeld, A. (1978). Picture processing: 1977. Comput. Graph. Im. Proc. **7**, 211–242.

Rosenfeld, A. (1979). Picture processing: 1978. Comput. Graph. Im. Proc. **9**, 354–393.

Rosenfeld, A. (1980). Picture processing: 1979. Comput. Graph. Im. Proc. **13**, 46–79.

Rosenfeld, A. (1981). Picture processing: 1980. Comput. Graph. Im. Proc. **16**, 52–89.

Rosenfeld, A. (1982). Picture processing: 1981. Comput. Graph. Im. Proc. **19**, 35–75.

Rosenfeld, A. (1983). Picture processing: 1982. Comput. Graph. Im. Proc. **22**, 339–387.

Rosenfeld, A. (1984). Picture processing: 1983. Comput. Graph. Im. Proc. **26**, 347–393.

Rosenfeld, A. (1985). Picture processing: 1984. Comput. Graph. Im. Proc. **30**, 189–242.

Rosenfeld, A. (1986). Picture processing: 1985. Comput. Graph. Im. Proc. **34**, 204–251.

Rosenfeld, A. (1987). Picture processing: 1986. Comput. Vision Graph. Im. Proc. **38**, 147–225.

Rosenfeld, A. (1988). Image analysis and computer vision: 1987. Comput. Vision Graph. Im. Proc. **42**, 234–293.

Rosenfeld, A. (1989). Image analysis and computer vision: 1988. Comput. Vision Graph. Im. Proc. **46**, 196–264.

Rosenfeld, A. (1990). Image analysis and computer vision: 1989. Comput. Vision Graph. Im. Proc. **50**, 188–240.

Rosenfeld, A. (1991). Image analysis and computer vision: 1990. CVGIP: Image Understanding **53**, 322–365.

Rosenfeld, A. (1992). Image analysis and computer vision: 1991. CVGIP: Image Understanding **55**, 349–380

Rosenfeld, A. and Banerjee, S. (1992). CVGIP: Image Understanding **55**, 1–13.

Rosenfeld, A. and Kak, A.C. (1976). *Digital Picture Processing* (Academic Press, New York and London); 2nd ed., 1982 (2 vols).

Rosenfeld, A. and Pfalz, J.L. (1968). Pattern Recognition **1**, 33–61.

Rosier, D.J. de: see also DeRosier, D.J.

Rosier, D.J. de and Klug, A. (1968). Nature **217**, 130–134.

Ross, G. (1982). Opt. Acta **29**, 1523–1542.

Ross, G., Fiddy, M.A., Nieto-Vesperinas, M. and Wheeler, M.W.L. (1977). Optik **49**, 71–80.

Ross, G., Fiddy, M.A., Nieto-Vesperinas, M. and Wheeler, M.W.L. (1978). Proc. Roy. Soc. (London) **A360**, 25–45.

Ross, G., Fiddy, M.A., Nieto-Vesperinas, M. and Manolitsakis, I. (1979). Opt. Acta **26**, 229–238.

Ross, G., Fiddy, M.A. and Moezzi, H. (1980a). Opt. Acta **27**, 1433–1444.

Ross, G., Fiddy, M.A. and Nieto-Vesperinas, M. (1980b). The inverse scattering problem in structural determinations. In *Inverse Scattering Problems in Optics* (Baltes, H.P., ed.) pp. 15–71 (Springer, Berlin and New York).

Ross, G., Fiddy, M.A. and Nieto-Vesperinas, M. (1981). In *Optics in Four Dimensions-1980* (Machado, M.A. and Narducci, L.M., eds) pp. 643–651 (AIP Conf. Proc. No. 65, American Institute of Physics, New York).

Ruijter, W.J. de and Weiss, J.K. (1992). Rev. Sci. Instrum. **63**, 4314–4321.

Ruijter, W.J. de, McCartney, M.R., Sharma, R., Smith, D.J. and Weiss, J.K. (1991). Cambridge, pp. 347–359.

Ruijter, W.J. de, McCartney, M.R., Smith, D.J. and Weiss, J.K. (1992). EMSA **50**, pp. 988–989.

Ruijter, W.J. de, Mooney, P.E. and Krivanek, O.L. (1993). MSA **51**, pp. 1062–1063.

Rushforth, C.K. (1987). Signal restoration, functional analysis, and Fredholm integral equations of the first kind. In *Image Recovery: Theory and Application* (Stark, H., ed.) pp. 1–27 (Academic Press, Orlando and London).

Russ, J.C. (1984). J. Microscopy **136**, RP7–RP8.

Russ, J.C. (1989a). J. Computer-assist. Microsc. **1**, 39–77.

Russ, J.C. (1989b). J. Computer-assist. Microsc. **1**, 105–129.

Russ, J.C. (1990). *Computer-assisted Microscopy, the Measurement and Analysis of Images* (Plenum, New York and London).

Russ, J.C. (1992). *The Image Processing Handbook* (CRC Press, London).

Russ, J.C. (1993). J. Computer–assist. Microsc. **5**, 159–163.

Russ, J.C. and Russ, J.C. (1984). J. Microscopy **135**, 89–102.

Russ, J.C. and Russ, J.C. (1987a). J. Microscopy **148**, 263–277.

Russ, J.C. and Russ, J.C. (1987b). In *Microbeam Analysis 1987* (Geiss, R.H., ed.) pp. 277–286 (San Francisco Press, San Francisco).

Russ, J.C., Bright, D.S., Russ, J.C. and Hare, T.M. (1989a). J. Computer–assist. Microsc. **1**, 3–37.

Russ, J.C., Taguchi, T., Peters, P.M., Chatfield, E., Russ, J.C. and Stewart, W.D. (1989b). Adv. X-ray Anal. **32**, 593–600.

Russo, F. and Ramponi, G. (1992). Electron. Lett. **28**, 1715–1716

Sabatier, P. C., ed. (1987). *Inverse Problems: an Interdisciplinary Study*. Adv. Electron. Electron Phys. Suppl. 19.

Saito, N. and Cunninngham, M.A. (1990). IEEE Trans. **PAMI–12**, 814–817.

Saleh, B. (1977). J. Opt. Soc. Am. **67**, 71–76.

Salembier, P. and Kunt, M. (1992). Signal Proc. **27**, 205–241.

Sanjurjo, J., Zapata, E.L., García, I., Roca, J. and Carazo, J.M. (1992). Granada, vol. 1, pp. 455–456.

Sanz, J.L.C. (1985). SIAM J. Appl. Math. **45**, 651–664.

Sanz, J.L.C. and Huang, T.S. (1983a). IEEE Trans. **ASSP–31**, 643–649.

Sanz, J.L.C. and Huang, T.S. (1983b). IEEE Trans. **ASSP–31**, 1276–1285.

Sanz, J.L.C. and Huang, T.S. (1983c). IEEE Trans. **ASSP–31**, 1492–1501.

Sanz, J.L.C. and Huang, T.S. (1983d). J. Opt. Soc. Am. **73**, 1446–1450.

Sanz, J.L.C. and Huang, T.S. (1983e). J. Opt. Soc. Am. **73**, 1455–1465.

Sanz, J.L.C. and Huang, T.S. (1984a). IEEE Trans. **ASSP–32**, 403–409.

Sanz, J.L.C. and Huang, T.S. (1984b). Support-limited signal and image extrapolation. Adv. Comput. Vision Image Proc. **1**, 1–82.

Sanz, J.L.C. and Huang, T.S. (1984c). J. Math. Anal. Applics **104**, 302–308.

Sanz, J.L.C. and Huang, T.S. (1985). IEEE Trans. **ASSP–33**, 997–1004.

Sanz, J.L.C., Huang, T.S. and Cukierman, F. (1983). J. Opt. Soc. Am. **73**, 1442–1445.

Sanz, J.L.C., Huang, T.S. and Wu, T.-f. (1984). IEEE Trans. **ASSP–32**, 1251–1254.

Saparin, G.V. (1990). In *Biophysical Electron Microscopy* (Hawkes, P.W. and Valdrè, U., eds) pp. 451–478 (Academic Press, London and San Diego).

Sasaki, T., Oho, E., Muranaka, Y. and Kanaya, K. (1986a). Kyoto, vol. 1, pp. 447–448.

Sasaki, T., Kobayashi, M., Oho, E. and Kanaya, K. (1986b). Kyoto, vol. 1, pp. 449–450.

Sasov, A.Yu. (1986). Kyoto, vol. 1, pp. 469–470.

Sasov, A.Yu. and Sokolov, V.N. (1984). Budapest, vol. 1, pp. 303–304.

Sasov, A.Yu. and Sokolov, V.N. (1985). Scanning **7**, 244–253.

Sasov, A.Yu., Sokolov, V.N. and Rau, E.I. (1982). Scanning Electron Microsc. 17–27.

Sault, R.J. (1984a). Austral. J. Phys. **37**, 209–229.

Sault, R.J. (1984b). Opt. Lett. **9**, 325–326.

Savoji, M.H. and Burge, R.E. (1982). Hamburg, vol. 1, pp. 509–510.

Savoji, M.H. and Burge, R.E. (1985). Comput. Vision Graph. Im. Proc. **29**, 259–269.

Saxton, W.O. (1974a). J. Phys. **D7**, L63–L64.

Saxton, W.O. (1974b). Canberra, vol. 1, pp. 314–315.

Saxton, W.O. (1975). EMAG, pp. 191–196.

Saxton, W.O. (1977a). Optik **49**, 51–62.

Saxton, W.O. (1977b). EMAG, pp. 111–114.

Saxton, W.O. (1978). *Computer Techniques for Image Processing in Electron Microscopy* (Academic Press, New York and London). Adv. Electron. Electron Phys. Supplement 10.

Saxton, W.O. (1980a). Recovery of specimen information for strongly scattering objects. In *Computer Processing of Electron Microscope Images* (Hawkes, P.W., ed.) pp. 35–87 (Springer, Berlin and New York).

Saxton, W.O. (1980b). The Hague, vol. 1, pp. 486–493 or vol. 2, pp. 682–689.

Saxton, W. O. (1980c). J. Microsc. Spectrosc. Electron. **5**, 661–670.

Saxton, W.O. (1984). Budapest, vol. 2, pp. 1299–1306.

Saxton, W.O. (1986a). Kyoto, Post–deadline paper 1, 4 pp.

Saxton, W.O. (1986b). EMSA **44**, pp. 526–529.

Saxton, W.O. (1987a). EMSA **45**, pp. 10–13.

Saxton, W.O. (1987b). EMSA **45**, pp. 58–61.

Saxton, W.O. (1987c). Niagara Falls, pp. 213–224.

Saxton, W.O. (1991). Cambridge, pp. 53–70.

Saxton, W.O. (1994). Ultramicroscopy.

Saxton, W.O. and Baumeister, W. (1981). EMAG, pp. 333–336.

Saxton, W.O. and Baumeister, W. (1982). J. Microscopy **127**, 127–138.

Saxton, W.O. and Chang, M. (1988). York, vol. 1, pp. 59–64.

Saxton, W.O. and Dürr, R. (1990). Seattle, vol. 1, pp. 96–97.

Saxton, W.O. and Frank, J. (1977). Ultramicroscopy **2**, 219–227.

Saxton, W.O. and O'Keefe, M.A. (1981). EMAG, pp. 343–346.

Saxton, W.O. and Stobbs, W. (1984). Budapest, vol. 1, pp. 287–288.

Saxton, W.O., Pitt, T.J. and Horner, M. (1979). Ultramicroscopy **4**, 343–354.

Saxton, W.O., Smith, D.J., Erasmus, S.J. and Smith, K.C.A (1982). Hamburg, vol. 1, pp. 527–528.

Saxton, W.O., Smith, D.J. and Erasmus, S.J. (1983). J. Microscopy **130**, 187–201.

Saxton, W.O., Baumeister, W. and Hahn, M. (1984). Ultramicroscopy **13**, 57–70.

Saxton, W.O., Dürr, R. and Baumeister, W. (1992). Ultramicroscopy **46**, 287–306.

Schatz, M. and Heel, M. van (1990). Ultramicroscopy **32**, 255–264.

Schatz, M. and Heel, M. van (1992). Ultramicroscopy, **45,** 15–22.

Schatz, M., Jäger, J. and Heel, M. van (1990). Seattle, vol. 1. pp. 450–451.

Scheerschmidt, K., Hillebrand, R., Bierwolf, R. and Hohenstein, M. (1992). Granada, vol. 1, pp. 541–542.

Schiske, P. (1968). Rome, vol. 1, pp. 145–146.

Schiske, P. (1973). Image processing using additional statistical information about the object. In *Image Processing and Computer-aided Design in Electron Optics* (Hawkes, P.W., ed.) pp. 82–90 (Academic Press, London and New York).

Schiske, P. (1974). Optik **40**, 261–275.

Schiske, P. (1975). J. Phys. D: Appl. Phys. **8**, 1372–1386.

Schiske, P. (1978). Toronto, vol. 1, pp. 216–217.

Schiske, P. (1982). Ultramicroscopy **9**, 17–26.

Schiske, P. (1984). Optik **69**, 13–16.

Schiske, P. (1993). Ultramicroscopy **49**, 121–122.

Schmid, M.F., Dargahi, R. and Tam, M.W. (1993). Ultramicroscopy **48,** 251–264.

Schmiesser, H. (1974). BEDO **7**, 205–219.

Schmiesser, H. (1975). BEDO **8**, 267–278.

Schneider, M. and Craig, M. (1992). Fuzzy Sets Syst. **45**, 271–278.

Schoute, F.C., Horst, M.F. ter and Willems, J.C. (1977). IEEE Trans. **CAS**–**24**, 67–78.

Schwarzer, R. (1985). BEDO **18**, 61–68.

Schwarzer, R.A. and Weiland, H. (1984). Budapest, vol. 1, 341–342.

Scivier, M.S. and Fiddy, M.A. (1985a). Opt. Lett. **10**, 369–371.

Scivier, M.S. and Fiddy, M.A. (1985b). J. Opt. Soc. Am. **A2**, 693–697.

Scivier, M.S., Hall, T.J. and Fiddy, M.A. (1984). Opt. Acta **31**, 619–623.

Seldin, J.H. and Fienup, J.R. (1990a). J. Opt. Soc. Am. **A7**, 412–427.

Seldin, J.H. and Fienup, J.R. (1990b). J. Opt. Soc. Am. **A7**, 428–433.

Selfridge, P.G. (1986). Comput. Vision Graph. Im. Proc. **34**, 156–165.

Sergeev, Y.M., Spivak, G.V., Sasov, A.Y., Osipov, V.I., Sokolov, V.N. and Rau, E.I. (1984). J. Microscopy **135**, 1–12 and 13–24.

Serra, J. (1982). *Image Analysis and Mathematical Morphology* (Academic Press, London and New York).

Serra, J. (1986). Comput. Vision Graph. Im. Proc. **35**, 283–305.

Serra, J. (1987). J. Microscopy **145**, 1–22.

Serra, J., ed. (1988). *Image Analysis and Mathematical Morphology.* Vol. 2 : *Theoretical advances* (Academic Press, London and San Diego).

Serra, J. (1993). Anamorphoses and function lattices (multivalued morphology). In *Mathematical Morphology in Image Processing* (Dougherty, E. R., ed.) pp. 483–523 (Dekker, New York and Basel).

Serra, J. and Laÿ, B. (1985). Signal Proc. **9**, 1–13.

Serra, J. and Vincent, L. (1992). An overview of morphological filtering. Circuits Syst. Signal Proc. **11**, 47–108.

Sezan, M.I. (1990). Seattle, vol. 1, pp. 446–447.

Sezan, M.I. (1992). Ultramicroscopy **40**, 55–67.

Sezan, M.I. and Stark, H. (1982). IEEE Trans. **MI–1**, 95–101.

Sezan, M.I. and Stark, H. (1983). Appl. Opt. **22**, 2781–2789.

Sezan, M.I. and Stark, H. (1984). IEEE Trans. **MI–3**, 91–98.

Sezan, M.I. and Stark, H. (1987). Applications of convex projection theory to image recovery in tomography and related areas. In *Image Recovery: Theory and Application* (Stark, H., ed.) pp. 415–462 (Academic Press, Orlando and London).

Sezan, M.I. and Trussell, H.J. (1991). IEEE Trans. Signal Proc. **39**, 2275–2285.

Shapiro, S.D. (1978). Pattern Recognition **10**, 129–143.

Shen, J. and Castan, S. (1992). CVGIP: Graph. Models Image Proc. **54**, 112–133.

Shindo, D., Hiraga, K., Hirabayashi, M., Oikawa, T. and Mori, N. (1990). Seattle, vol. 1, pp. 172–173.

Shindo, D. Hiraga, K. Oku, T. and Oikawa, T. (1991). Ultramicroscopy **39**, 50–57.

Shindo, D., Hiraga, K., Iijima, S., Kudoh, J., Oikawa, T. and Critchell, J. (1992). Granada, vol. 1, pp. 151–152.

Shore, J.E. and Johnson, R.W. (1980). IEEE Trans. **IT–26**, 26–37.

Shore, J.E. and Johnson, R.W. (1981). IEEE Trans. **IT–27**, 472–482.

Sieber, P. (1974). Canberra, vol. 1, pp. 274–275.

Simon, R. (1969). Scanning Electron Microsc. 445–457.

Simon, R. (1970). J. Appl. Phys. **41**, 4632–4641.

Sinha, D. and Dougherty, E.R. (1992). J. Visual Commun. Image Represent. **3**, 286–302.

Sinha, P.K. and Hong, Q.H. (1990). IEEE Trans. **MI–9**, 345–346.

Skilling, J., ed. (1989a). *Maximum Entropy and Bayesian Methods* (Kluwer, Dordrecht, Boston MA and London). Proceedings of the 8th MaxEnt Workshop, Cambridge, 1988.

Skilling, J. (1989b). Classic maximum entropy. In *Maximum Entropy and Bayesian Methods* (Skilling, J., ed.) pp. 45–52 (Kluwer Academic, Dordrecht, Boston MA and London).

Skilling, J. (1990). Quantified maximum entropy. In *Maximum Entropy and Bayesian Methods* (Fougère, P.F., ed.) pp. 341–350 (Kluwer, Dordrecht, Boston MA and London).

Skilling, J. and Bryan, R.K. (1984). Mon. Not. Roy. Astron. Soc. **211**, 111–124.

Sklansky, J. (1978). IEEE Trans. **C–27**, 923–926.

Slepian, D. (1964). Bell System Tech. J. **43**, 3009–3057.

Slepian, D. (1965). J. Math. Phys. **44**, 99–140.

slepian, d. (1976). on bandwidth. Proc. IEEE **64**, 292–300.

Slepian, D. (1978). Bell System Tech. J. **57**, 1371–1430.

Slepian, D. and Pollak, H. O. (1961). Bell System Tech. J. **40**, 43–63.

Slepian, D. and Sonnenblick, E. (1965). Bell System Tech. J. **44**, 1745–1760.

Slump, C.H. and Ferwerda, H.A. (1982). Optik **62**, 93–104 and 143–168.

Slump, C.H. and Ferwerda, H.A. (1986). Statistical aspects of image handling in low-dose electron microscopy of biological materials. Adv. Electron. Electron Phys. **66**, 201–308.

Smart, P. and Tovey, N.K. (1982). *Electron Microscopy of Soils and Sediments: Techniques*, Chapter 12 (Clarendon, Oxford).

Smart, P. and Tovey, N.K. (1988). Scanning **10**, 115–121.

Smith, K.C.A. (1972). Scanning Electron Microsc. 1–8.

Smith, P.R. (1978). Ultramicroscopy **3**, 153–160.

Smith, P.R. (1980). The Hague, vol. 2, pp. 698–699.

Smith, K.C.A. (1981). EMAG, pp. 109–114.

Smith, K.C.A. (1982). J. Microscopy **127**, 3–16.

Smith, K.C.A. (1985). J. Microscopy **139**, 177–185.

Smith, C.R. and Erickson, G.J., eds (1987). *Maximum-Entropy and Bayesian Spectral Analysis and Estimation Problems* (Reidel, Dordrecht, Boston, Lancaster PA and Tokyo). Proceedings of the 3rd workshop on maximum entropy and Bayesian methods, 1983.

Smith, C.R. and Grandy, W.T., eds (1985). *Maximum-Entropy and Bayesian Methods in Inverse Problems* (Reidel, Dordrecht, Boston and Lancaster PA). Papers from the MaxEnt workshops of 1981 and 1982.

Smith, P.R., Peters, T.M. and Bates, R.H.T. (1973). J. Phys. A: Math. Nucl. Gen. **6**, 361–382.

Smith, K.C.A., Unitt, B.M., Holburn, D.M. and Tee, W.J. (1977a). Scanning Electron Microsc. (I), 49–56 and 40.

Smith, K.T., Solmon, D.C. and Wagner, S.L. (1977b). Bull. Am. Math. Soc. **83**, 1227–1270 ; addendum (1978) **84**, 691.

Smith, D.J., Saxton, W.O., O'Keefe, M.A., Wood, G.J. and Stobbs, W.M. (1983). Ultramicroscopy **11**, 263–281.

Smith, K.C.A., Smith, T. and Catto, C.J.D. (1986). Kyoto, vol. 1, pp. 483–484.

Smith, C.R., Erickson, G. J. and Neudorfer, P. O., eds (1992). *Maximum Entropy and Bayesian Methods* (Kluwer, Dordrecht, Boston and London). Proceedings of the 11th MaxEnt Workshop, 1991.

Song, J. and Delp, E.J. (1990). Comput. Vision Graph. Im. Proc. **50**, 308–328.

Spence, J.C.H. (1974). Opt. Acta **21**, 835–837.

Spivak, G.V., Sasov, A.Yu. and Rau, E.I. (1980). The Hague, vol. 1, pp. 508–509.

Stark, H., ed. (1987). *Image Recovery: Theory and Application* (Academic Press, Orlando and London).

Staunton, R.C. (1989a). Image Vision Comput. **7**, 162–166.

Staunton, R.C. (1989b). Proc. SPIE **1008**, 23–27.

Steeds, J.W. (1984). Electron crystallography. In *Quantitative Electron Microscopy* (Chapman, J.N. and Craven, A.J., eds) pp. 49–96 (Scottish Universities Summer School in Physics, Edinburgh).

Steeds, J.W. and Carlino, E. (1992). In *Electron Microscopy in Materials Science* (Merli, P.G. and Antisari, M.V., eds) pp. 279–313 (World Scientific, Singapore, River Edge NJ and London).

Stefanescu, I.S. (1985). J. Math. Phys. **26**, 2141–2160.

Steinkilberg, M. and Schramm H.J. (1980). Hoppe–Seyler's Z. Physiol. Chem. **361**, 1363–1369.

Stens, R.L. (1983). Signal Proc. **5**, 139–151.

Sternberg, S.R. (1980). In *Pattern Recognition in Practice* (Gelsema, E.S. and Kanal, L.N., eds) pp. 35–44 (North-Holland, Amsterdam, New York and Oxford).

Sternberg, S.R. (1982). Cellular computers and biomedical image processing. In *Biomedical Images and Computers* (Sklansky, J. and Bisconte, J.-C., eds) pp. 294–319 (Springer, Berlin and New York).

Sternberg, S.R. (1986). Comput. Vision Graph. Im. Proc. **35**, 333–355.

Steven, A.C. (1986). EMSA **44**, pp. 62–65.

Steven, A.C., Trus, B.L., Unser, M., McDowall, A.W., Dubochet, J. and Podolsky, R.J. (1986). EMSA **44**, pp. 108–109.

Stewart, M. (1988). J. Electron Microsc. Tech. **9**, 301–324 and 325–358.

Stewart, M. (1991). Cambridge, pp. 3–9.

Stott, W.R. and Chatfield, E.J. (1979). Scanning Electron Microsc. (II), 53–60.

Streibl, N. (1984). Opt. Commun. **49**, 6–10.

Strintzis, M.G. (1976). Proc. IEEE **64**, 1255–1257.

Strintzis, M.G. (1977). Proc. IEEE **65**, 979–980.

Strintzis, M.G. (1978). IEEE Trans. **AC–23**, 801–809.

Suzuki, T., Ichioka, Y., Kawata, S. and Kondo, K. (1977). HVEM Kyoto, pp. 175–178.

Svalbe, I. and Jones, R. (1992). Patt. Rec. Lett. **13**, 123–129.

Swift, J.A. and Brown, A.C.A. (1975). J. Microscopy **105**, 1–14.

Tan, D.G.H., Qu, J.X. and Bates, R.H.T. (1986). Optik **73**, 25–29.

Tan, K.-C., Lim, H. and Tan, B.T.G. (1991). CVGIP: Graph. Models Image Proc. **53**, 491–500.

Taniguchi, Y., Ikuta, T., Endoh, H. and Shimizu, R. (1990a) J. Electron Microsc. **39**, 137–144.

Taniguchi, Y., Ikuta, T., Endoh, H. and Shimizu, R. (1990b). Technol. Repts Osaka Univ. **40** (1988), 19–24.

Taniguchi, Y., Shimizu, R., Chaya, M. and Ikuta, T. (1990c). Seattle, vol. 1, pp. 458–459.

Taniguchi,Y., Takai, Y., Ikuta, T. and Shimizu, R. (1992a). J. Electron Microsc. **41**, 21–29.

Taniguchi, Y., Takai, Y., Shimizu, R., Ikuta, T., Isakozawa, S. and Hashimoto, T. (1992b). Ultramicroscopy **41**, 323–333.

Taylor, L.S. (1980). J. Opt. Soc. Am. **70**, 1554–1556.

Taylor, L.S. (1981). IEEE Trans. **AP–29**, 386–391.

Taylor, K.A. and Crowther, R.A. (1986). EMSA **44**, 30–33.

Taylor, K.A. and Crowther, R.A. (1991). Ultramicroscopy **38**, 85–103.

Taylor, K.A. and Crowther, R.A. (1992). Ultramicroscopy **41**, 153–167.

Taylor, P.A. and Gopinath, A. (1973). EMAG, pp. 146–151.

Tee, W.J., Smith, K.C.A. and Unitt, B.M. (1977). EMAG, pp. 95–98.

Thon, F. (1968). Rome, vol. 1, pp. 127–128.

Thon, F. (1971). Phase contrast electron microscopy. In *Electron Microscopy in Material Science* (Valdrè, U., ed.) pp. 570–625 (Academic Press, New York and London).

Thon, F. and Willasch, D. (1970). Grenoble, vol. 1, pp. 3–4.

Thust, A. and Urban, K. (1992). Ultramicroscopy **45**, 23–42.

Tichelaar, W. and Heel, M. van (1990). J. Struct. Biol. **103**, 180–184.

Tietz, H. R. (1990). Seattle, vol. 1, pp. 528–529.

Tietz, H.R. (1992). Granada, vol. 1, pp. 131–135.

Titchmarsh, E.C. (1939). *The Theory of Functions* (Oxford University Press, London).

Toll, J.S. (1956). Phys. Rev. **104**, 1760–1770.

Tom, V.T., Quatieri, T.F., Hayes, M.H. and McClellan, J.H. (1981). IEEE Trans. **ASSP–29**, 1052–1058.

Toms, N., Chapman, J.N. and Ferrier, R.P. (1972). Manchester, pp. 422–423.

Toorn, P. van and Ferwerda, H.A. (1976a). Opt. Acta **23**, 457–468.

Toorn, P. van and Ferwerda, H.A. (1976b). Opt. Acta **23**, 469–481.

Toorn, P. van and Ferwerda, H.A. (1977). Optik **47**, 123–134.

Toorn, P. van, Huiser, A.M.J. and Ferwerda, H.A. (1978). Optik **51**, 309–326.

Toorn, P. van, Greenaway, A.H. and Huiser, A.M.J. (1984). Opt. Acta **31**, 767–774.

Toraldo di Francia, G. (1951). Atti Fond. G. Ronchi **6**, 73–79.

Toraldo di Francia, G. (1955). J. Opt. Soc. Am. **45**, 497–501.

Toraldo di Francia, G. (1956a). In *Problems in Contemporary Optics* (Proc. Florence Meeting, 10–15 September 1954. Fiorentini, A., ed.) pp. 67–75 (Istituto Nazionale di Ottica, Florence).

Toraldo di Francia, G. (1956b). Trans. IRE **AP–4**, 473–478.

Toraldo di Francia, G. (1969). J. Opt. Soc. Am. **59**, 799–804.

Toriwaki, J.-i. and Yokoi, S. (1981). Prog. Pattern Recognition **1**, 187–264.

Tovey, N.K. (1971a). EMAG, pp. 244–247.

Tovey, N.K. (1971b). Scanning Electron Microsc. 49–56.

Tovey, N.K. (1980). J. Microscopy **120**, 303–315.

Tovey, N.K. and Smart, P. (1986). Scanning **8**, 75–90.

Tovey, N.K. and Wong, K.Y. (1978). Scanning Electron Microsc. (I), 381–392.

Tovey, N.K., Smart, P., Hounslow, M.W. and Leng, X.L. (1989). Scanning Microsc. **3**, 771–784.

Tovey, N.K., Smart, P., Hounslow, M.W. and Desty, J.P. (1991). Cambridge, pp. 315–330.

Tricart, J.-P., Brevart, O., Baranger, R. and Martinez, L. (1991). Bull. Centres Rech. Explor. Prod. Elf–Aquitaine **15**, 279–305.

Trus, B.L., Steven, A.C., Unser, M., McDowall, A.W., Dubochet, J. and Podolsky, R.J. (1986). Kyoto, vol. 4, pp. 3103–3104.

Trus, B.L., Unser, M. and Steven, A.C. (1988). York, vol. 3, pp. 367–368.

Trus, B.L., Unser, M., Pun, T. and Steven, A.C. (1991). Cambridge, pp. 441–451.

Trussell, H.J. (1980). IEEE Trans. **ASSP–28**, 114–117.

Trussell, H.J. (1984). *A priori* knowledge in algebraic reconstruction methods. Adv. Comput. Vision Image Proc. **1**, 265–316.

Trussell, H.J. and Hunt, B.R. (1979). IEEE Trans. **C–28**, 57–62.

*Turner, J.N., ed. (1981). *Three-dimensional Ultrastructure in Biology*. Methods in Cell Biology, vol. 22 (Academic Press, New York and London).

Tsai, R.Y. and Huang, T.S. (1984). Multiframe image restoration and registration. Adv. Comput. Vision Image Proc. **1**, 317–339.

Twomey, S. (1963). J. Assoc. Comput. Mach. **10**, 97–101.

Tyan, S. G. (1981). Median filtering: deterministic properties. In *Two-dimensional Digital Signal Processing, II, Transforms and Median Filters* (Huang, T. S., ed.) pp. 197–217 (Springer, Berlin and New York).

Typke, D. (1981). Nucl. Instrum. Meth. **187**, 217–226.

Typke, D. and Hoppe, W. (1972). Manchester, pp. 72–73.

Typke, D., Hoppe, W., Sessler, W. and Burger, M. (1976). Jerusalem, vol. 1, pp. 334–335.

Typke, D., Burger, M., Lemke, N. and Lefranc, G. (1980). The Hague, vol. 1, pp. 82–83.

Typke, D., Pfeifer, G., Hegerl, R. and Baumeister, W. (1990). Seattle, vol. 1, pp. 244–245.

Typke, D., Dierksen, K. and Baumeister, W. (1991). EMSA **49**, pp. 544–545.

Typke, D., Hegerl, R. and Kleinz, J. (1992a). Ultramicroscopy **46**, 157–173.

Typke, D., Hegerl, R. and Kleinz, J. (1992b). EMSA **50**, pp. 1000–1001.

Unitt, B.M. (1975). J. Phys. E: Sci. Instrum. **8**, 423–425.

Unitt, B.M. and Smith, K.C.A. (1976). Jerusalem, vol. 1, pp. 162–167.

Unser, M. (1984). Signal Proc. **7**, 231–249.

Unser, M. (1990). Signal Proc. **20**, 3–14.

Unser, M., Steven, A.C. and Trus, B.L. (1986). Ultramicroscopy **19**, 337–348.

Unser, M., Eden, M. and Trus, B.L. (1987). Signal Proc. 12, 83–91.

Unser, M., Trus, B.L., Steven, A.C. and Eden, M. (1988). York, vol. 1, pp. 211–212.

Unser, M., Trus, B.L. and Steven, A.C. (1989a). Ultramicroscopy 30, 299–310.

Unser, M., Trus, B.L., Frank, J. and Steven, A.C. (1989b). Ultramicroscopy 30, 429–433.

Unser, M., Trus, B.L. and Eden, M. (1989c). Signal Proc. 17, 191–200.

Unwin, P.N.T. (1975). J. Mol. Biol. 98, 235–242.

Unwin, P.N.T. and Henderson, R. (1975). J. Mol. Biol. 94, 425–440.

Uyeda, N., Kirkland, E., Fujiyoshi, Y. and Siegel, B. (1978). Toronto, vol. 1, pp. 220–221.

Vainshtein, B.K. (1970). Kristallografiya 15, 894–902; Sov. Phys. Cryst. 15, 781–787.

Vainshtein, B.K. (1971). Dokl. Akad. Nauk SSSR 196, 1072–1075 ; Sov. Phys. Dokl. 16, 66–69.

Vainshtein, B.K. (1973). Usp. Fiz. Nauk 109, 455–497; Sov. Phys. Usp. 16, 185–206.

Vainshtein, B.K. (1978). Electron microscopical analysis of the three-dimensional structure of biological macromolecules. Adv. Opt. Electron Microsc. 7, 281–377.

Venot, A., Lebruchec, J.F. and Roucayrol J.C. (1984). Comput. Vision Graph. Im. Proc. 28, 176–184.

Verschoor, A., Frank, J., Radermacher, M., Wagenknecht, T. and Boublik, M. (1983). EMSA 41, 758–759.

Verschoor, A., Frank, J., Radermacher, M., Wagenknecht, T. and Boublik, M. (1984). J. Mol. Biol. 178, 677–698.

Verschoor, A., Frank, J. and Boublik, M. (1985). J. Ultrastruct. Res. 92, 180–189.

Verschoor, A., Frank, J., Wagenknecht, T. and Boublik, M. (1986). J. Mol. Biol. 187, 581–590.

Verschoor, A., Zhang, N.-y., Wagenknecht, T., Obrig, T., Radermacher, M. and Frank, J. (1989). J. Mol. Biol. 209, 115–126.

Vicario, E. and Pitaval, M. (1973). EMAG, pp. 142–145.

Vicario, E., Escudie, B. and Hellion, A. (1979). J. Microsc. Spectrosc. Electron. 4, 341–350.

Vogel, R.H. and Provencher, S.W. (1988). Ultramicroscopy 25, 223–240.

Vogel, R.H., Provencher, S.W., Bonsdorff, C.-H. von, Adrian, M. and Dubochet, J. (1986). Nature 320, 533–535.

Wackerman, C.C. and Yagle, A.E. (1991). J. Opt. Soc. Am. A8, 1898–1904.

Wagenknecht, T., Grassucci, R. and Frank, J. (1988a). J. Mol. Biol. 199, 137–147.

Wagenknecht, T., Frank, J., Boublik, M., Nurse, K. and Ofengand, J. (1988b). J. Mol. Biol. 203, 753–760.

Walker J.G. (1981). Opt. Acta 28, 735–738.

Walther, A. (1963). Opt. Acta 10, 41–49.

Weiland, H. and Schwarzer, R. (1985). BEDO **18**, 55–60.

Weiland, H. and Schwarzer, R. (1986). Kyoto, vol. 1, p. 451.

Weisman, A.D., Dougherty, E.R., Mizes, H.A. and Miller, R.J.D. (1992). J.Appl. Phys. **71**, 1565–1578

Wendt, P.D. (1990a). IEEE Trans. **ASSP–38**, 2099–2107.

Wendt, P. (1990b). Proc. SPIE **1247**, 204–211.

Wendt, P.D., Coyle, E.J. and Gallagher, N.C. (1986a). IEEE Trans. **ASSP–34**, 898–911.

Wendt, P.D., Coyle, E.J. and Gallagher, N.C. (1986b). IEEE Trans. **CAS–33**, 276–286.

Werman, M., Peleg, S. and Rosenfeld, A. (1985). Comput. Vision Graph. Im. Proc. **32**, 328–336.

Wernecke, S.J. (1977). In *Image Science Mathematics* (Wilde, C.O. and Barrett, E., eds.) pp. 148–152 (Western Periodicals, North Hollywood).

Wernecke, S.J. and d'Addario, L.R. (1977). IEEE Trans. **C–26**, 351–364.

White, E.W., McKinstry, H.A. and Johnson, G.G. (1968). Scanning Electron Microsc. 95–103.

White, E.W., Görz, H., Johnson, G.G. and McMillan, R.E. (1970). Scanning Electron Microsc. 57–64.

White, E.W., Görz, H. and Johnson, G.G. (1971). BEDO **4**(2), 415–424.

White, E.W., Mayberry, K. and Johnson, G.G. (1972a). Pattern Recognition **4**, 173–193.

White, E.W., Görz, H., Johnson, G.G. and Lebiedzik, J. (1972b). BEDO **5**, 609–624.

Whittaker, E.T. and Watson, G.N. (1927). *A Course of Modern Analysis*, 4th ed. (University Press, Cambridge).

Williams, D.B. and Hunt, J.A. (1992). Granada, vol. 1, pp. 243–247.

Wilson, S.S. (1989). IEEE Trans. **SMC–19**, 1636–1644.

Wilson, S.S. (1990). Proc. SPIE **1350**, 44–55.

Wilson, S. S. (1991). Proc. SPIE **1451**, 242–253.

Wilson, S. S. (1992a). Theory of matrix morphology. IEEE Trans. **PAMI–14**, 636–652.

Wilson, S.S. (1992b). Proc. SPIE **1769**, 332–343.

Wilson, S.S. (1993). Proc. SPIE **2030**, 78–87.

Wingham, D.J. (1992). IEEE Trans. Signal Proc. **40**, 559–570.

Wisse, E. and Zanger, R.B. de (1988). York, vol. 3, pp. 363–364.

Wolf, E. (1962). Proc. Phys. Soc. (London) **B80**, 1269–1272.

Won, M.C., Mnyama, D. and Bates, R.H.T. (1985). Opt. Acta **32**, 377–396.

Wong, E. (1968). SIAM J. Appl. Math. **16**, 756–770.

Wong, E. (1978). IEEE Trans. **IT–24**, 50–59.

Wong, K., Xu, P., Loane, R., Kirkland, E. and Silcox, J. (1991). EMSA **49**, pp. 1006–1007.

Wong, K., Kirkland, E., Xu, P., Loane, R. and Silcox, J. (1992). Ultramicroscopy **40**, 139–150.

Wood, J.W., Fiddy, M.A. and Burge, R.E. (1981). Opt. Lett. **6**, 514–516.

Wood, J.W., Hall, T.J. and Fiddy, M.A. (1983). Opt. Acta **30**, 511–527.

Woods, J.W. (1978). IEEE Trans. **AC–23**, 846–850.

Woods, J.W. (1981). Two-dimensional Kalman filtering. In *Two-dimensional Digital Signal Processing, I, Linear Filtering* (Huang, T.S., ed.) pp. 155–205 (Springer, Berlin and New York.

Woods, J.W. (1984). Image detection and estimation. In *Digital Image Processing Techniques* (Ekstrom, M. P., ed.) pp. 77–110 (Academic Press, Orlando and London).

Woods, J.W. and Radewan, C.H. (1977). IEEE Trans. **IT–23**, 473–482; correction: Woods, J.W. (1979). IEEE Trans. **IT–25**, 628–629.

Woodward, P.M. (1953). *Probability and Information Theory, with Applications to Radar* (Pergamon, Oxford and New York) ; 2nd ed., 1964.

Wu, J.K. and Burge, R.E. (1982). Comput. Graph. Image Proc. **19**, 392–400.

Ximen, J. (1986). Optik **75**, 16–19.

Ximen, J. (1990). Seattle, vol. 1, pp. 456–457.

Ximen, J. and Kapp, O. H. (1985). Optik **70**, 101–108.

Ximen, J. and Kapp, O. H. (1986). Optik **72**, 87–94 and 143–148 and **73**, 10–12.

Ximen, J.-y. and Li, Y. (1986). Optik **74**, 27–29.

Ximen, J. and Shao, Z. (1985). Optik **71**, 143–148.

Yew, N.C. and Pease, D.E. (1974). Scanning Electron Microsc. 191–197.

Yin, L., Astola, J.T. and Neuvo, Y.A. (1993). IEEE Trans. Signal Proc. **41**, 162–184.

Yli-Harja, O. (1989). Median filters: extensions, analysis and design. Lappeenranta Univ. Technol. Res. Papers, **13**, 173 pp.

Yli-Harja, O., Astola, J. and Neuvo, Y. (1991). IEEE Trans. Signal Proc. **39**, 395–410.

Yoshihara, T., Yamada, M., Kobayashi, T. and Overton, P. (1988). York, vol. 1, pp. 149–150.

Youla, D.C. (1978). IEEE Trans. **CAS–25**, 694–702.

Youla, D.C. (1987). Mathematical theory of image restoration by the method of convex projections. In *Image Recovery: Theory and Application* (Stark, H., ed.) pp. 29–77 (Academic Press, Orlando and London).

Youla, D.C. and Webb, H. (1982). IEEE Trans. **MI–1**, 81–94.

Zapata, E.L., Benavides, I., Rivera, F.F., Bruguera, J.D., Pena, T.F. and Carazo, J. M. (1992). Signal Proc. **27**, 51–64.

Zeitler, E. (1974). Optik **39**, 396–415.

Zemlin, F. (1978). Ultramicroscopy **3**, 261–263.

Zemlin, F. (1979). Ultramicroscopy **4**, 241–245.

Zemlin, F. (1988). York, vol. 1, pp. 139–140.

Zemlin, F. (1989). J. Electron Microsc. Tech. **11**, 251–257.

Zemlin, F. (1991). In *Electron Crystallography of Organic Molecules* (Fryer, J.R. and Dorset, D.L., eds) pp. 305–308 (Kluwer, Dordrecht, Boston MA and London).

Zemlin, F., Weiss, K., Schiske, P., Kunath, W. and Herrmann, K.-H. (1978). Ultramicroscopy **3**, 49–60.

Zeng, B., Gabbouj, M. and Neuvo, Y. (1991). IEEE Trans. **CAS–38**, 1003–1020.

Zhang, N. (1991). EMSA **49**, pp. 538–539.

Zhang, N.-y., Wagenknecht, T., Radermacher, M., Obrig, T. and Frank, J. (1987). EMSA **45**, pp. 936–937.

Zhuang, X., Østevold, E. and Haralick, R.M. (1987). The principle of maximum entropy in image recovery. In *Image Recovery: Theory and Application* (Stark, H., ed.) pp. 157–193 (Academic Press, Orlando and London).

Zinzindohoué, P. (1991). Optik **87**, 45–47.

Zinzindohoué, P. (1992). Optik **90**, 97–100.

Zinzindohoué, P., Bonnet, N., Ploton, D., Beorchia, A. and Laberrigue, A. (1988). York, vol. 1, pp. 179–180.

Zwick, M. and Zeitler, E. (1973). Optik **38**, 550–565.

Part XVI, Chapters 78–79

References to measurements of the coherence of electron sources have not been included in this Part since the present trend is to measure the envelope functions that attenuate the phase contrast transfer functions. The relevant references are listed in Part XV, Chapter 77. The value of $|\gamma_{12}(0)|$ has also been estimated by examining the visibility of the interference fringes produced with the aid of a biprism, notably by Hibi and Takahashi (1969), briefly summarized by Hibi (1974) and by Burge *et al.* (1975, 1976). Among subsequent studies using this technique, we mention Speidel and Kurz (1977), Hanszen *et al.* (1985), Pozzi *et al.* (1986). A procedure for measuring coherence length with the aid of a Wien filter has been tested by Möllenstedt and Wohland (1980). The effect of the Wien filter on longitudinal coherence has been studied at length by Nicklaus and Hasselbach (1993).

In addition to the work discussed in Chapter 66, the following studies on partial coherence in electron microscopy are directly relevant to this Part: Ferwerda (1976a, determination of the complex degree of coherence by the methods of Section 74.3; 1976b, conical illumination); Ferwerda and van Heel (1977, coherence of thermionic sources); Fertig and Rose (1977, coherence and dark-field imaging in TEM and STEM); Huiser and Hoenders (1981), determination of spatial coherence, q.v. for earlier papers); Medina and Pozzi (1990, spatial coherence in the context of electron interferometry and holography).

A form of the transmission cross-coefficient that depends on specimen thickness has been introduced by Coene *et al.* (1986a,b).

Coherence, especially in connection with field-emission sources, has been reappraised by Passow (1993).

Ade, G. (1975). Optik **42**, 199–215.

Agarwal, G.S., Foley, J.T. and Wolf, E. (1987). Opt. Commun. **62**, 67–72.

Apresyan, L.A. and Kravtsov, Yu.A. (1984). Photometry and coherence: wave aspects of the theory of radiation transport. Usp. Akad. Nauk SSSR **142**, 689–711; Sov. Phys. Usp. **27** , 301–313.

Baltes, H.P., Geist, J. and Walther, A. (1978). Radiometry and coherence. In *Inverse Source Problems in Optics* (Baltes, H.P., ed.) pp. 119–154 (Springer, Berlin and New York).

Barnett, M.E. (1974). Optik **39**, 470–474.

Bartelt, H.O., Brenner, K.-H. and Lohmann, A.W. (1980). Opt. Commun. **32**, 32–38.

Bastiaans, M.J. (1977). Opt. Acta **24**, 261–274.

Bastiaans, M.J. (1978). Opt. Commun. **25**, 26–30.

Bastiaans, M.J. (1979a). Opt. Commun. **30**, 321–326.

Bastiaans, M.J. (1979b). J. Opt. Soc. Am. **69**, 1710–1716.

Bastiaans, M.J. (1979c). Opt. Acta **26**, 1265–1272 and 1333–1344.

Bastiaans, M.J. (1981a). In *Optics in Four Dimensions–1980* (Machado, M.A. and Narducci, L.M., eds) pp. 292–312 (American Institute of Physics, New York ; Conf. Proc. No. 65).

Bastiaans, M.J. (1981b). Opt. Acta **28, 1215–1224**.

Beorchia, A. and Bonhomme, P. (1974). Optik **39**, 437–442.

Bonhomme, P., Beorchia, A. and Bonnet, N. (1973). C.R. Acad. Sci. Paris **B277**, 83–86.

Bonnet, N. and Bonhomme, P. (1980). Optik **56, 353–362**.

Brenner, K.-H., Lohmann, A.W. and Ojeda-Castañeda, J. (1983). Opt. Commun. **44**, 323–326.

Brenner, K.-H. and Ojeda-Castañeda, J. (1984). Opt. Acta **31**, 213–223.

Burge, R.E. and Dainty, J.C. (1976). Optik **46**, 229–240.

Burge, R.E. and Scott, R.F. (1975). Optik **43**, 503–507.

Burge, R.E. and Scott, R.F. (1976). Optik **44**, 159–172.

Burge, R.E., Dainty, J.C. and Thom, J. (1975). EMAG, pp. 221–224.

Burge, R.E., Dainty, J.C. and Thom, J. (1976). Jerusalem, vol. 1, pp. 256–258.

Cairns, B. and Wolf, E. (1987). Opt. Commun. **62**, 215–218.

Carter, W.H. (1978). Opt. Commun. **26**, 1–4.

Carter, W.H. (1988). Radio Science **23**, 1085–1093.

Carter, W.H. (1990). J. Mod. Opt. **37**, 109–120.

Carter, W.H. (1992). J. Mod. Opt. **39**, 1461–1470.

Carter, W.H. and Wolf, E. (1975). J. Opt. Soc. Am. **65**, 1067–1071.

Carter, W.H. and Wolf, E. (1977). J. Opt. Soc. Am. **67**, 785–796.

Carter, W.H. and Wolf, E. (1978). Opt. Commun. **25**, 288–292.

Carter, W.H. and Wolf, E. (1981). Opt. Acta **28**, 227–244 and 245–259.

Carter, W.H. and Wolf, E. (1985). J. Opt. Soc. Am. **A2**, 1994–2000.

Carter, W.H. and Wolf, E. (1987). Phys. Rev. **A36**, 1258–1269.

Castaño, V.M. (1988). Optik **81**, 35–37.

Castaño, V.M. (1989). In *Computer Simulation of Electron Microscope Diffraction and Images* (Krakow, W. and O'Keefe, M.A., eds) pp. 33–41 (Minerals, Metals and Materials Society, Warrendale, PA).

Castaño, V.M., Vázquez-Polo, G. and Gutierrez-Castrejón, R. (1991). Cambridge, pp. 415–422.

Cittert, P.H. van (1934). Physica **1**, 201–210.

Coene, W., Dyck, D. van and Landuyt, J. van (1986a). Optik **73**, 13–18.

Coene, W., Dyck, D. van and Landuyt, J. van (1986b). Kyoto, vol. 1, pp. 779–780.

Collett, E. and Wolf, E. (1979). J. Opt. Soc. Am. **69**, 942–950.

Collett, E. and Wolf, E. (1980). Opt. Commun. **32**, 27–31.

Erickson, H.P. (1973). Adv. Opt. Electron Microsc. **5**, 163–199.

Erickson, H.P. and Klug, A. (1970a). Ber. Bunsen-Ges. Phys. Chem. **74**, 1129–1137.

Erickson, H.P. and Klug, A. (1970b). EMSA, pp. 248–249.

Erickson, H.P. and Klug, A. (1971). Phil. Trans. Roy. Soc. (London) **B261**, 105–118.

Fertig, J. and Rose, H. (1977). Ultramicroscopy **2**, 269–279.

Ferwerda, H.A. (1976a). Opt. Commun. **19**, 54–56.

Ferwerda, H.A. (1976b). Optik **45**, 411–426.

Ferwerda, H.A. (1980). In *Imaging Processes and Coherence in Optics* (Schlenker, M., Fink, M., Goedgebuer, J.P., Malgrange, C., Viénot, J.C. and Wade, R.H., eds). pp. 85–94 (Springer, Berlin and New York).

Ferwerda, H.A. and Heel, M.G. van (1977). Optik **47**, 357–362.

Ferwerda, H.A. and Heel, M.G. van (1978). In *Coherence and Quantum Optics IV* (Mandel, L. and Wolf, E., eds) pp. 443–448 (Plenum, New York and London).

Foley, J.T. (1990). Opt. Commun. **75**, 347–352.

Foley, J.T. (1991). J. Opt. Soc. Am. **A8**, 1099–1105.

Foley, J.T. and Nieto-Vesperinas, M. (1985). J. Opt. Soc. Am. **A2**, 1446–1447.

Foley, J.T. and Wolf, E. (1985). Opt. Commun. **55,** 236–241.

Foley, J.T. and Wolf, E. (1988). J. Opt. Am. **A5**, 1683–1687.

Foley, J.T. and Wolf, E. (1991). J. Mod. Opt. **38**, 2053–2068.

Foley, J.T., Carter, W.H. and Wolf, E. (1986). J. Opt. Soc. Am. **A3**, 1090–1096.

Foley, J.T., Kim, K. and Nussenzveig, H.M. (1988). J. Opt. Soc. Am. **A5**, 1694–1708.

Frank, J. (1973). Optik **38**, 519–536.

Frank, J. (1976a). Optik **44**, 379–391.

Frank, J. (1976b). Jerusalem, vol. 1, pp. 97–98.

Friberg, A.T. (1979). J. Opt. Soc. Am. **69**, 192–198.

Friberg, A.T. (1981). Opt. Acta **28**, 261–277.

Friberg, A.T. (1986a). Opt. Acta **33**, 1369–1376.

Friberg, A.T. (1986b). Appl. Opt. **25**, 4547–4556.

Friberg, A.T. and Wolf, E. (1983). Opt. Acta **30**, 1417–1435.

Gabor, D. (1956). Rev. Mod. Phys. **28**, 260–276.

Gamliel, A. and Wolf, E. (1988). Opt. Commun. **65**, 91–96.

Gori, F., Guattari, G., Palma, C. and Padovani, C. (1988a). Opt. Commun. **66**, 1–5.

Gori, F., Guattari, G., Palma, C. and Padovani, C. (1988b). Opt. Commun. **67**, 1–4.

Gori, F., Marcopoli, G.L. and Santarsiero, M. (1991). Opt. Commun. **81**, 123–130.

Guigay, J.P. (1978). Opt. Commun. **26**, 136–138.

Hahn, M. (1973). *Theoretische und experimentelle Untersuchungen zum Nachweis von Einzelatomen mit Durchstrahlungs-Elektronenmikroskopen.* Dissertation, Düsseldorf.

Hanszen, K.-J. and Ade, G. (1975). Optik **42**, 1–22.

Hanszen, K.-J. and Ade, G. (1976). Jerusalem, vol. 1, pp. 446–447.

Hanszen, K.-J. and Trepte, L. (1970). Grenoble, vol. 1, pp. 45–46.

Hanszen, K.-J. and Trepte, L. (1971a). Optik **32**, 519–538.

Hanszen, K.-J. and Trepte, L. (1971b). Optik **33**, 166–198.

Hanszen, K.-J., Lauer, R. and Ade, G. (1985). Optik **71**, 64–72.

Hauser, H. (1962). Opt. Acta **9**, 121–140 and 141–148.

Hawkes, P.W. (1977a). EMAG, pp. 123–126.

Hawkes, P.W. (1977b). Optik **49**, 149–161.

Hawkes, P.W. (1978). Coherence in electron optics. Adv. Opt. Electron Microsc. **7**, 101–184.

Hawkes, P.W. (1994). In *Courants, Amers, Ecueils en Microphysique* (Lochak, G. and Lochak, P., eds; Fondation Louis de Broglie, Paris).

Hibi, T. (1974). Canberra, vol. 1, pp. 208–209.

Hibi, T. and Takahashi, S. (1969). Z. angew. Phys. **27**, 132–138.

Hopkins, H.H. (1951). Proc. Roy. Soc. (London) **A208**, 263–277.

Huiser, A.M.J. and Hoenders, B.J. (1981). Opt. Acta **28**, 1273–1276.

Humphreys, C.J. and Spence, J.C.H. (1981). Optik **58**, 125–144.

Ishizuka, K. (1986). Kyoto, vol. 1, pp. 771–772.

Ishizuka, K. (1990). Seattle, vol. 1, pp. 60–61.

James, D.F.V. and Wolf, E. (1989). Opt. Commun. **72**, 1–6.

Kim, K. and Wolf, E. (1987). J. Opt. Soc. Am. **A4**, 1233–1236.

Krivanek, O.L. (1975). Optik **43**, 361–372.

Krivanek, O.L. (1976). Jerusalem, vol. 1, pp. 263–264.

Lenz, F (1957). *Theoretische Untersuchungen über die Ausbreitung von Elektronenstrahlbündeln in rotationssymmetrischen elektrischen und magnetischen Feldern.* Habilitationsschrift, Aachen.

Littlejohn, R.G. and Winston, R. (1993). J. Opt. Soc. Am. **A10**, 2024–2037.

Mandel, L. (1961). J. Opt. Soc. Am. **51,** 1342–1350.

Mandel, L. and Wolf, E. (1965). Rev. Mod. Phys. **37**, 231–287.

Mandel, L. and Wolf, E. (1976). J. Opt. Soc. Am. **66**, 529–535.

Mandel, L. and Wolf, E. (1981). Opt. Commun. **36**, 247–249.

Marchand, E.W. and Wolf, E. (1972a). Opt. Commun. **6**, 305–308.

Marchand, E.W. and Wolf, E. (1972b). J. Opt. Soc. Am. **62**, 379–385.

Marchand, E.W. and Wolf, E. (1974a). J. Opt. Soc. Am. **64**, 1219–1226.

Marchand, E.W. and Wolf, E. (1974b). J. Opt. Soc. Am. **64**, 1273–1274.

Martínez-Herrero, R. (1979). Nuovo Cim. **B54**, 205–210.

Martínez-Herrero, R. and Durán, A. (1981). Opt. Acta **28**, 65–76.

Martínez-Herrero, R. and Mejías, P.M. (1981). Opt. Commun. **37**, 234–238.

Martínez-Herrero, R. and Mejías, P.M. (1984). J. Opt. Soc. Am. **A1**, 556–558.

Medina, F.F. and Pozzi, G. (1990). J. Opt. Soc. Am. A7, 1027–1033.

Menzel, E. (1958). Optik 15, 460–470.

Menzel, E. (1960). In *Optics in Metrology* (Mollet, P., ed.) pp. 283–293 (Pergamon, Oxford).

Menzel, E., Mirandé, W. and Weingärtner, I. (1973). *Fourier-Optik und Holographie* (Springer, Vienna and New York).

Misell, D.L. (1971). J. Phys A: Gen. Phys. 4, 782–797 and 798–812.

Misell, D.L. (1973a). Adv. Electron. Electron Phys. 32, 63–191.

Misell, D.L. (1973b). J. Phys A: Math. Nucl. Gen. 6, 62–78 and 205–217.

Misell, D.L. and Atkins, A.J. (1973). J. Phys. A: Math. Nucl. Gen. 6, 218–235.

Möllenstedt, G. and Wohland, G. (1980). The Hague, vol. 1, pp. 28–29.

Moreau, P., Coene, W., Dyck, D. van and Landuyt, J. van (1987). Micron Microsc. Acta 18, 237–238.

Munch, J. (1975). Optik 43, 79–99.

Munch, J. and Zeitler, E. (1974). EMSA, pp. 386–387.

Nicklaus, M. and Hasselbach, F. (1993). Phys. Rev. A48, 152–160.

O'Connell, R.F. and Wigner, E.P. (1981a). Phys. Lett. 83A, 121–126.

O'Connell, R.F. and Wigner, E.P. (1981b). Phys. Lett. 83A, 145–148.

O'Keefe, M.A. (1979). EMSA 37, pp. 556–557.

Papoulis, A. (1974). J. Opt. Soc. Am. 64, 779–788.

Passow, C. (1993). Optik 93, 127–136.

Pedersen, H.M. (1982). Opt. Acta 29, 877–892.

Polo, G.V., Acosta, D.R. and Castaño, V.M. (1992). Optik 89, 181–183.

Pozzi, G., Matteucci, G. and Carpenter, R.W. (1986). Kyoto, vol. 1, pp. 275–276.

Pulvermacher, H. (1981). Optik 60, 45–60.

Saxton, W.O. (1978). *Computer Techniques for Image Processing in Electron Microscopy.* Adv. Electron. Electron Phys. Suppl. 10.

Scherzer, O. (1980). Optik 56, 333–352.

Schmid, H. (1984). Budapest, vol. 1, pp. 285–286.

Slansky, S. (1959). J. Phys. Radium 20, 13S–14S .

Slansky, S. (1960). Rev. Opt. 39, 555–577.

Slansky, S. (1962). Opt. Acta 9, 277–294.

Slansky, S. and Maréchal, A. (1960). C. R. Acad. Sci. Paris 250, 4132–4134.

Speidel, R. and Kurz, D. (1977). Optik 49, 173–185.

Spence, J.C.H. (1980, 1988). *Experimental High-resolution Electron Microscopy* (Oxford University Press, New York and Oxford); 2nd ed. (1988).

Spence, J.C.H. and Zuo, J.M. (1992). *Electron Microdiffraction* (Plenum, New York and London).

Sudarshan, E.C.G. (1969). J. Math. Phys. Sci. **3**, 121–175.

Sudol, R.J. and Friberg, A.T. (1984). In *Coherence and Quantum Optics, V* (Mandel, L. and Wolf, E., eds) pp. 423–430 (Plenum, New York and London).

Toorn, P. van and Robinson, D.W. (1980). Optik **56**, 323–331.

Troyon, M. (1977). Optik **49**, 247–251.

Walther, A. (1968). J. Opt. Soc. Am. **58**, 1256–1259.

Walther, A. (1973). J. Opt. Soc. Am. **63**, 1622–1623.

Walther, A. (1974). J. Opt. Soc. Am. **64**, 1275.

Walther, A. (1978). J. Opt. Soc. Am. **68**, 1606–1610.

Walther, A. (1978). Opt. Lett. **3**, 127–129.

Winston, R. and Ning, X. (1986). J. Opt. Soc. Am. **A3**, 1629–1631.

Winston, R. and Welford, W.T. (1990). Opt. Commun. **76**, 191–193.

Winston, R. and Welford, W.T. (1991). Opt. Commun. **81**, 155–156.

Winthrop, J.T. (1972). J. Opt. Soc. Am. **62**, 1234–1235.

Wolf, E. (1976). Phys. Rev. D **13**, 869–886.

Wolf, E. (1978a). J. Opt. Soc. Am. **68**, 6–17.

Wolf, E. (1978b). J. Opt. Soc. Am. **68**, 1597–1605.

Wolf, E. (1981a). Opt. Commun. **38**, 3–6.

Wolf, E. (1981b). In *Optics in Four Dimensions–1980* (Machado, M.A. and Narducci, L.M., eds) pp. 42–48 (American Institute of Physics, New York ; Conf. Proc. No. 65).

Wolf, E. (1982). J. Opt. Soc. Am. **72**, 343–351.

Wolf, E. (1984). Opt. Lett. **9**, 387–389.

Wolf, E. (1986a). J. Opt. Soc. Am. **A3**, 76–85.

Wolf, E. (1986b). Phys. Rev. Lett. **56**, 1370–1372.

Wolf, E. (1987a). Nature **326**, 363–365.

Wolf, E. (1987b). Opt. Commun. **62**, 12–16.

Wolf, E. (1987c). Phys. Rev. Lett. **58**, 2646–2648.

Wolf, E. (1992a). In *Huygens' Principle 1690–1990, Theory and Applications* (Blok, H., Ferwerda, H.A. and Kuiken, K.K., eds) pp. 113–127 (North-Holland, Amsterdam).

Wolf, E. (1992b). J. Mod. Opt. **39**, 9–20.

Wolf, E. and Carter, W.H. (1975). Opt. Commun. **13**, 205–209.

Wolf, E. and Carter, W.H. (1976). Opt. Commun. **16**, 297–302.

Wolf, E. and Carter, W.H. (1978). J. Opt. Soc. Am. **68**, 953–964.

Wolf, E. and Carter, W.H. (1984). Opt. Commun. **50**, 131–136.

Wolf, E. and Collett, E. (1978). Opt. Commun. **25**, 293–296.

Wolf, E. and Devaney, A.J. (1981). Opt. Lett. **6**, 168–170.

Wolf, E. and Nieto-Vesperinas, M. (1985). J. Opt. Soc. Am. **A2**, 886–890.

Wolf, E., Devaney, A.J. and Foley, J.T. (1981). In *Optics in Four Dimensions–1980* (Machado, M.A. and Narducci, L.M., eds). pp. 123–130 (American Institute of Physics, New York ; Conf. Proc. No. 65).

Wolf, E., Devaney, A.J. and Gori, F. (1983). Opt. Commun. **46**, 4–8.

Wolf, E., Jannson, J. and Jannson, T. (1990). Opt. Lett. **15**, 1032–1034.

Zernike, F. (1938). Physica **5**, 785–795.

Conference Proceedings

The following list gives full publishing details of the series of International and European conferences on Electron Microscopy and of the conferences organized by the Electron Microscopy and Analysis Group (EMAG) of the British Institute of Physics. For the reader's convenience, a few other meetings are included, in particular those on charged particle optics and the Asia–Pacific congresses on electron microscopy. The irregular series of meetings on high-voltage electron microscopy is identified by the acronym HVEM.

The list does not contain details of the annual meetings of the Electron Microscopy Society of America, now the Microscopy Society of America; these are identified in the reference lists by EMSA (or MSA) and the meeting number. The Proceedings of the 25th–40th meetings (1967–1982) were published by Claitor's, Baton Rouge (LA) and those of subsequent meetings by the San Francisco Press, San Francisco (CA). The editors are as follows:

25th–32nd meeting (1967–1974), C.J. Arceneaux, ed.
33rd–47th meeting (1975–1989), G.W. Bailey, ed
48th meeting (1990) coincided with Seattle, 1990.
49th meeting (1991), G.W. Bailey and E.L. Hall, eds
50th meeting (1992), G.W. Bailey, J. Bentley and J.A. Small, eds, 2 Vols.
51st meeting (1993), G.W. Bailey and C.L. Rieder, eds

Many other national electron microscopy societies publish proceedings of their major meetings but few contain much optics. A notable exception is the series of All-Union meetings held in the former Soviet Union, the proceedings of which are mainly published in *Izv Akad. Nauk SSSR (Ser. Fiz.)*, translated as *Bull Acad. Sci. USSR (Phys. Ser.)*, from which the acronyms SSSR and USSR have now been removed; a few papers appear in *Radiotekhnika i Elektronika (Radio Engineering and Electronic Physics and later Soviet Journal of Communications Technology and Electronics)*. Brief details of these are given at the end of the main list. The other noteworthy exceptions are Japan and Italy; abstracts of Japanese national meetings are published regularly and rapidly in the *Journal of Electron Microscopy*. Extended abstracts of the biennial Italian meetings are published as a bound supplement to *Microscopia Elettronica* and details are included here under the acronym SIME.

Delft, 1949: *Proceedings of the Conference on Electron Microscopy*, Delft, 4–8 July, 1949 (Houwink, A.L., Le Poole, J.B. and Le Rütte, W.A., eds; Hoogland, Delft, 1950).

Paris, 1950: *Comptes Rendus du Premier Congrès International de Microscopie Electronique*, Paris, 14–22 September, 1950 (Editions de la Revue d'Optique Théorique et Instrumentale, Paris, 1953) 2 Vols.

Washington, 1951: *Electron Physics. Proceedings of the NBS Semicentennial Symposium on Electron Physics*, Washington, 5–7 November, 1951. Issued as National Bureau of Standards Circular **527** (1954).

London, 1954: *The Proceedings of the Third International Conference on Electron Microscopy*, London, 1954 (Ross, R., ed.; Royal Microscopical Society, London, 1956).

Gent, 1954: *Rapport Europees Congrès Toegepaste Electronenmicroscopie*, Gent, 7–10 April, 1954, edited and published by G. Vandermeersche (Uccle-Bruxelles, 1954).

Toulouse, 1955: *Les Techniques Récentes en Microscopie Electronique et Corpusculaire*, Toulouse, 4–8 April, 1955 (C.N.R.S., Paris, 1956).

Stockholm, 1956: *Electron Microscopy. Proceedings of the Stockholm Conference*, September, 1956 (Sjöstrand, F.J. and Rhodin, J., eds; Almqvist and Wiksells, Stockholm, 1957).

Tokyo, 1956: *Electron Microscopy. Proceedings of the First Regional Conference in Asia and Oceania*, Tokyo, 1956 (Electrotechnical Laboratory, Tokyo, 1957).

Berlin, 1958: *Vierter Internationaler Kongress für Elektronenmikroskopie*, Berlin, 10–17 September, 1958, *Verhandlungen* (Bargmann, W., Möllenstedt, G., Niehrs, H., Peters, D., Ruska, E. and Wolpers, C., eds; Springer, Berlin, 1960) 2 Vols.

Delft, 1960: *The Proceedings of the European Regional Conference on Electron Microscopy*, Delft, 1960 (Houwink, A.L. and Spit, B.J., eds; Nederlandse Vereniging voor Elektronenmicroscopie, Delft n.d.) 2 Vols.

Philadelphia, 1962: *Electron Microscopy. Fifth International Congress for Electron Microscopy*, Philadelphia, Pennsylvania, 29 August to 5 September, 1962 (Breese, S.S., ed.; Academic Press, New York, 1962) 2 Vols.

Prague, 1964: *Electron Microscopy 1964. Proceedings of the Third European Regional Conference*, Prague (Titlbach, M., ed.; Publishing House of the Czechoslovak Academy of Sciences, Prague, 1964) 2 Vols.

Calcutta, 1965: *Proceedings of the Second Regional Conference on Electron Microscopy in Far East and Oceania*, Calcutta 2–6 February, 1965 (Electron Microscopy Society of India, Calcutta)

Kyoto, 1966: *Electron Microscopy 1966. Sixth International Congress for Electron Microscopy*, Kyoto (Uyeda, R., ed.; Maruzen, Tokyo, 1966) 2 Vols.

Rome, 1968: *Electron Microscopy 1968. Pre-Congress Abstracts of Papers Presented at the Fourth Regional Conference*, Rome (Bocciarelli, D.S., ed.; Tipografia Poliglotta Vaticana, Rome, 1968) 2 Vols.

HVEM Monroeville, 1969: *Current Developments in High Voltage Electron Microscopy (First National Conference)*, Monroeville, 17–19 June, 1969. Proceedings not published but *Micron* **1** (1969) 220–307 contains official reports of the meeting based on the session chairmen's notes.

Grenoble, 1970: *Microscopie Électronique 1970. Résumés des Communications Présentées au Septième Congrès International*, Grenoble (Favard, P., ed.; Société Française de Microscopie Electronique, Paris, 1970) 3 Vols.

HVEM Stockholm, 1971: *The Proceedings of the Second International Conference on High-Voltage Electron Microscopy*, Stockholm, 14–16 April, 1971; published as *Jernkontorets Annaler* **155** (1971) No. 8.

EMAG, 1971: *Electron Microscopy and Analysis. Proceedings of the 25th Anniversary Meeting of the Electron Microscopy and Analysis Group of the Institute of Physics*, Cambridge, 29 June - 1 July, 1971 (Nixon, W.C., ed.; Institute of Physics, London, 1971) Conference Series **10**.

Manchester, 1972: *Electron Microscopy 1972. Proceedings of the Fifth European Congress on Electron Microscopy*, Manchester (Institute of Physics, London, 1972).

HVEM Oxford, 1973: *High Voltage Electron Microscopy. Proceedings of the Third International Conference*, Oxford, August, 1973 (Swann, P.R., Humphreys, C.J. and Goringe, M.J., eds; Academic Press, London and New York, 1974).

EMAG, 1973: *Scanning Electron Microscopy: Systems and Applications*, Newcastle-upon-Tyne, 3–5 July, 1973 (Nixon, W.C., ed.; Institute of Physics, London, 1973) Conference Series **18**.

Canberra, 1974: *Electron Microscopy 1974. Abstracts of Papers Presented to the Eighth International Congress on Electron Microscopy*, Canberra (Sanders, J.V. and Goodchild, D.J., eds; Australian Academy of Science, Canberra, 1974) 2 Vols.

Sarajevo, 1974: *Electron Microscopy 1974. Pre-congress Abstracts of Papers presented at the First Balkan Congress on Electron Microscopy*, Sarajevo, 22–26 May, 1974 (Devidé, Z., Dobardžić, R., Jerković, L., Marinković, V., Pantić, V., Pejovski, S. and Pipan, N., eds).

HVEM Toulouse, 1975: *Microscopie Electronique à Haute Tension. Textes des Communications Présentées au 4e Congrès International*, Toulouse, 1–4 Septembre, 1975 (Jouffrey, B. and Favard, P., eds; SFME Paris, 1976).

EMAG, 1975: *Developments in Electron Microscopy and Analysis. Proceedings of EMAG 75*, Bristol, 8–11 September, 1975 (Venables, J.A., ed.; Academic Press, London and New York, 1976).

Jerusalem, 1976: *Electron Microscopy 1976. Proceedings of the Sixth European Congress on Electron Microscopy*, Jerusalem (Brandon, D.G. (Vol. I) and Ben-Shaul, Y. (Vol. II), eds; Tal International, Jerusalem, 1976) 2 Vols.

HVEM Kyoto, 1977: *High Voltage Electron Microscopy 1977. Proceedings of the Fifth International Conference on High Voltage Electron Microscopy*, Kyoto, 29 August to 1 September, 1977 (Imura, T. and Hashimoto, H., eds; Japanese Society of Electron Microscopy, Tokyo, 1977); published as a supplement to *Journal of Electron Microscopy* **26** (1977).

EMAG, 1977: *Developments in Electron Microscopy and Analysis. Proceedings of EMAG 77*, Glasgow, 12–14 September, 1977 (Misell, D.L., ed.; Institute of Physics, Bristol, 1977) Conference Series **36**.

Istanbul, 1977: *Abstracts of Communications, Second Balkan Congress on Electron Microscopy*, Istanbul, 25–30 September, 1977 (Erbengi, T., Chairman Sci. Prog. Comm.; Istanbul Faculty of Medicine and Turkish Society of Electron Microscopy, Istanbul).

Toronto, 1978: *Electron Microscopy 1978. Papers Presented at the Ninth International Congress on Electron Microscopy*, Toronto (Sturgess, J.M., ed.; Microscopical Society of Canada, Toronto, 1978) 3 Vols.

EMAG, 1979: *Electron Microscopy and Analysis, 1979. Proceedings of EMAG 79*, Brighton, 3–6 September, 1979 (Mulvey, T., ed.; Institute of Physics, Bristol, 1980) Conference Series **52**.

The Hague, 1980: *Electron Microscopy 1980. Proceedings of the Seventh European Congress on Electron Microscopy*, The Hague (Brederoo, P. and Boom, G. (Vol. I), Brederoo, P. and Priester, W. de (Vol. II), Brederoo, P. and Cosslett, V.E. (Vol. III), and Brederoo, P. and Landuyt, J. van (Vol. IV), eds). Vols. I and II contain the proceedings of the Seventh European Congress on Electron Microscopy, Vol. III those of the Ninth International Conference on X-Ray Optics and Microanalysis, and Vol. IV those of the Sixth International Conference on High Voltage Electron Microscopy (Seventh European Congress on Electron Microscopy Foundation, Leiden, 1980).

Giessen, 1980: *Charged Particle Optics. Proceedings of the First Conference on Charged Particle Optics*, Giessen, 8–11 September, 1980 (Wollnik, H., ed.) Nucl. Instrum. Meth. **187** (1981) 1–314.

EMAG, 1981: *Electron Microscopy and Analysis, 1981. Proceedings of EMAG 81*, Cambridge, 7–10 September, 1981 (Goringe, M.J., ed.; Institute of Physics, Bristol, 1982) Conference Series **61**.

Hamburg, 1982: *Electron Microscopy, 1982. Papers Presented at the Tenth International Congress on Electron Microscopy*, Hamburg (Deutsche Gesellschaft für Elektronenmikroskopie, Frankfurt, 1982) 3 Vols.

HVEM Berkeley, 1983: *Proceedings of the Seventh International Conference on High Voltage Electron Microscopy*, Berkeley, 16–19 August, 1983 (Fisher, R.M., Gronsky, R. and Westmacott, K.H., eds). Published as a Lawrence Berkeley Laboratory Report, LBL-16031, UC-25, CONF-830819.

EMAG, 1983: *Electron Microscopy and Analysis, 1983. Proceedings of EMAG 83*, Guildford, 30 August – 2 September, 1983 (Doig, P., ed.; Institute of Physics, Bristol, 1984) Conference Series **68**.

Budapest, 1984: *Electron Microscopy 1984. Proceedings of the Eighth European Congress on Electron Microscopy*, Budapest 13–18 August 1984 (Csanády, Á.,Röhlich, P. and Szabó, D., eds; Programme Committee of the Eighth European Congress on Electron Microscopy, Budapest, 1984) 3 Vols.

Singapore, 1984: *Conference Proceedings 3rd Asia Pacific Conference on Electron Microscopy*, Singapore, 29 August – 3 September, 1984 (Chung Mui Fatt, ed.; Applied Research Corporation, Singapore).

Ocean City, 1984: *Electron Optical Systems for Microscopy, Microanalysis and Microlithography. Proceedings of the 3rd Pfefferkorn Conference*, Ocean City (MD), 9–14 April, 1984 (Hren, J.J., Lenz, F.A., Munro, E. and Sewell, P.B., eds; Scanning Electron Microscopy, AMF O'Hare, IL).

EMAG, 1985: *Electron Microscopy and Analysis, 1985. Proceedings of EMAG 85*. Newcastle-upon-Tyne, 2–5 September, 1985 (Tatlock, G.J., ed.; Institute of Physics, Bristol, 1986) Conference Series **78**.

Albuquerque, 1986: *Charged Particle Optics. Proceedings of the Second International Conference on Charged Particle Optics*, Albuquerque, 19–23 May, 1986 (Schriber, S.O. and Taylor, L.S., eds) Nucl. Instrum. Meth. Phys. Res. **A258** (1987) 289–598.

Kyoto, 1986: *Electron Microscopy 1986. Proceedings of the XIth International Congress on Electron Microscopy*, Kyoto, 31 August – 7 September, 1986 (Imura, T., Maruse, S. and Suzuki, T., eds; Japanese Society of Electron Microscopy, Tokyo) 4 Vols; published as a supplement to *Journal of Electron Microscopy* **35** (1986).

Beijing, 1986: *Proceedings of the International Symposium on Electron Optics*, Beijing, 9–13 September, 1986 (Ximen, J.-y., ed.; Institute of Electronics, Academia Sinica, 1987).

EMAG, 1987: *Electron Microscopy and Analysis, 1987. Proceedings of EMAG 87*, Manchester, 8–9 September, 1987 (Brown, L.M., ed.; Institute of Physics, Bristol and Philadelphia, 1987) Conference Series **90**.

Niagara Falls, 1987: *Image and Signal Processing in Electron Microscopy. Proceedings of the 6th Pfefferkorn Conference*, Niagara Falls (Canada), 28 April – 2 May, 1987 (Hawkes, P.W., Ottensmeyer, F.P., Saxton, W.O. and Rosenfeld, A., eds; Scanning Microscopy Int., AMF O'Hare, Chicago IL, 1988) Scanning Microsc. Supplement 2.

SIME, 1987: *Atti del XVI Congresso di Microscopia Elettronica*, Bologna, 14–17 October, 1987; Supplement to **8**(2) of *Microscopia Elettronica*.

York, 1988: *Proceedings of the Ninth European Congress on Electron Microscopy*, York, 4–9 September, 1988 (Goodhew, P.J. and Dickinson, H.G., eds; Institute of Physics, Bristol and Philadelphia, 1988) Conference Series **93**, 3 Vols.

Bangkok, 1988: *Electron Microscopy 1988. Proceedings of the IVth Asia–Pacific Conference and Workshop on Electron Microscopy*, Bangkok, 26 July - 4 August, 1988 (Mangclaviraj, V., Banchorndhevakul, W. and Ingkaninun, P., eds; Electron Microscopy Society of Thailand, Bangkok).

EMAG, 1989: *EMAG–MICRO 89. Proceedings of the Institute of Physics Electron Microscopy and Analysis Group and Royal Microscopical Society Conference*, London, 13–15 September, 1989 (Goodhew, P.J. and Elder, H.Y., eds; Institute of Physics, Bristol and New York, 1990) Conference Series **98**, 2 Vols.

Athens, 1989: *Proceedings III Balkan Congress on Electron Microscopy*, Athens, 18–22 September, 1989 (Margaritis, L.H., ed.).

SIME, 1989: *Atti del XVII Congresso di Microscopia Elettronica*, Lecce, 4–7 October, 1989; Supplement to **10**(2) of *Microscopia Elettronica*.

Toulouse, 1990: *Charged Particle Optics. Proceedings of the Third International Conference on Charged Particle Optics*, Toulouse, 24–27 April, 1990 (Hawkes, P.W., ed.) Nucl. Instrum. Meth. Phys. Res. **A298** (1990) 1–508.

Seattle, 1990: *Electron Microscopy 1990. Proceedings of the XIIth International Congress for Electron Microscopy*, Seattle WA, 12–18 August, 1990 (Peachey, L.D. and Williams, D.B., eds; San Francisco Press, San Francisco) 4 Vols.

EMAG, 1991: *Electron Microscopy and Analysis 1991. Proceedings of EMAG 91*, Bristol, 10–13 September, 1991 (Humphreys, F.J., ed.; Institute of Physics, Bristol, Philadelphia and New York, 1991) Conference Series **119**.

Cambridge, 1991: *Signal and Image Processing in Microscopy and Microanalysis. Proceedings of the 10th Pfefferkorn Conference*, Cambridge, 16–19 September, 1991 (Hawkes, P.W., Saxton, W.O. and O'Keefe, M.A., eds; Scanning Microscopy Int., AMF O'Hare, Chicago IL, 1992) Scanning Microsc. Supplement 6 (published 1994).

SIME, 1991: *Atti del XVIII Congresso di Microscopia Elettronica*, Padova, 24–28 September 1991; Supplement to **12**(2) of *Microscopia Elettronica*.

Granada, 1992: *Electron Microscopy 92. Proceedings of the 10th European Congress on Electron Microscopy*, Granada, 7–11 September 1992 (Ríos, A., Arias, J.M., Megías-Megías, L. and López-Galindo, A. (Vol. I), López-Galindo, A. and Rodríguez-García, M.I. (Vol. II) and Megías–Megías, L., Rodríguez-García, M.I., Ríos, A. and Arias, J.M. (Vol. III), eds; Secretariado de Publicaciones de la Universidad de Granada, Granada) 3 Vols.

Beijing, 1992: *Electron Microscopy I and II. 5th Asia–Pacific Electron Microscopy Conference*, Beijing, 2–6 August, 1992 (Kuo, K.H. and Zhai, Z.H., eds; World Scientific, Singapore, River Edge NJ, London and Hong Kong) 2 Vols.

EMAG, 1993: *Electron Microscopy and Analysis 1993. Proceedings of EMAG 93*, Liverpool, 15–17 September, 1993 (Craven, A.J., ed.) Institute of Physics, Bristol, Philadelphia and New York, 1994. Conference Series, **138**.

SIME, 1993: *Proceedings Multinational Congress on Electron Microscopy*, Parma, 13–17 September 1993; Supplement to **14**(2) of *Microscopia Elettronica*.

The Proceedings of the Soviet All-Union conferences on electron microscopy are to be found in the volumes of *Izv. Akad. Nauk (Ser. Fiz.)* or *Bull. Acad. Sci. (Phys. Ser.)* indicated:

1. Moscow 15–19/12/1950; **15** (1951) Nos 3 and 4 (no English translation).
2. Moscow 9–13/5/1958; **23** (1959) Nos 4 and 6.
3. Leningrad 24–29/10/1960; **25** (1961) No. 6.
4. Sumy 12–14/3/1963; **27** (1963) No. 9.
5. Sumy 6–8/7/1965; **30** (1966) No. 5.
6. Novosibirsk 11–16/7/1967; **32** (1968) Nos 6 and 7.
7. Kiev 14–21/7/1969; **34** (1970) No. 7.
8. Moscow 15–20/11/1971; **36** (1972) Nos 6 and 9.
9. Tbilisi 28/10–2/11/1973; **38** (1974) No. 7.
10. Tashkent 5–8/10/1976; **41** (1977) Nos 5 and 11.
11. Tallin 10/1979; **44** (1980) Nos 6 and 10.
12. Sumy 1982; **48** (1984) No. 2.
13. Sumy 10/1987; **52** (1988) No. 7.
14. Suzdal 10 and 11/1990; **55** (1991) No. 8.

Additional Conference Proceedings

Paris, 1994: *Electron Microscopy 1994. Proceedings of the 13th International Congress on Electron Microscopy*, Paris, 17–22 July, 1994 (Jouffrey, B., Colliex, C., Chevalier, J.P., Glas., F., Hawkes, P.W., Hernandez-Verdun, D., Schrevel, J. and Thomas, D., eds; Editions de Physique, Les Ulis) 5 Vols.

Tsukuba, 1994: *Charged Particle Optics. Proceedings of the Fourth International Conference on Charged Particle Optics* (CPO4), Tsukuba, 3–6 October (Ura, K., Hibino, M., Komuro, M., Kurashige, M., Kurokawa, S., Matsuo, T., Okayama, S., Shimoyama, H. and Tsuno, K., eds) *Nucl. Instrum. Meth. Phys. Res.* **A363** (1995) 1–496.

MSA 1994: *Proceedings Fifty-second Annual Meeting Microscopy Society of America,* New Orleans LA, 31 July – 5 August 1994 (Bailey, G. W. and Garratt-Reed, A.J., eds; San Francisco Press, San Francisco 1994). Note that subsequent proceedings volumes are published by Jones and Begall, New York, in conjunction with the *Journal of the Microscopy Society of America.*

EMAG, 1995: *Electron Microscopy and Analysis 1995. Proceedings of EMag 95,* Birmingham, 12–15 September, 1995 (Cherns, D., ed.; Institute of Physics, Bristol, Philadelphia and New York, 1995) Conference Series **147**.

SIME, 1995: *Atti XX Congresso di Microscopia Elettronica*, Rimini, 11–14 September 1995; Supplement to **16**(2) of *Microscopia Elettronica.*

MSA 1995: *Proceedings Microscopy and Analysis 1995, Microscopy Society of America 53rd Meeting*, Kansas City MO, 13–17 August 1995 (Bailey, G.W., Ellisman, M.H., Hennigar, R.A. and Zaluzec, N.J., eds; Jones and Begell, New York 1995).

The 15th Russian conference on electron microscopy was held in Chernogolovka in May 1994 and the proceedings appear in *Izv. Akad. Nauk (Ser. Fiz).* or *Bull. Russ. Acad. Sci. (Phys.)* **59** (1995) No. 2.